THE IDENTIFICATION OF DARK MATTER

Proceedings of the Sixth International Workshop on

THE IDENTIFICATION OF DARK MATTER

Rhodes, Greece 11–16 September 2006

edited by

Minos Axenides and George Fanourakis

Institute of Nuclear Physics, Greece

John Vergados

University of Ioannina, Greece

World Scientific

NEW JERSEY • LONDON • SINGAPORE • BEIJING • SHANGHAI • HONG KONG • TAIPEI • CHENNAI

Published by

World Scientific Publishing Co. Pte. Ltd.

5 Toh Tuck Link, Singapore 596224

USA office: 27 Warren Street, Suite 401-402, Hackensack, NJ 07601

UK office: 57 Shelton Street, Covent Garden, London WC2H 9HE

British Library Cataloguing-in-Publication Data
A catalogue record for this book is available from the British Library.

THE IDENTIFICATION OF DARK MATTER
Proceedings of the Sixth International Workshop

ISBN-13 978-981-270-852-6
ISBN-10 981-270-852-9

Printed in Singapore by World Scientific Printers (S) Pte Ltd

PREFACE

The present volume contains in written form the scientific talks presented at the 6th International Workshop on the Identification of Dark Matter (IDM2006) which was held in the Island of Rhodes-Greece in the period of 6-11 September 2006. It is the first one to be held outside the United Kingdom, a continuation of a biannual series of Similar Workshops commenced in 1996 at Sheffield (IDM96).

Their aim is to bring together active researchers from across the globe, both experimentalists and theorists, working on the problem of fundamental importance, such as the Identity of Dark Matter. Questions such as what it takes to observe it, in terms as of state of the art technology, past-present and future observational techniques, as well as theoretical issues of the many possible Dark Matter constituents as elementary particle relics from the Early Universe, their role in Structure Formation (both large and small scale) and galactic Dynamics dominate the agenda of the Workshop. The recent developments of the inclusion of an unidentifiable Homogeneous Dark Energy component in the Dark Sector of our Universe enlarged the scope and content of the topics tackled.

Following the format of previous meetings (IDM96, IDM98, IDM2000, IDM2002, and IDM2004) reviews on all of the above general areas are included with short talks making up the core of the contributed papers. A special session is devoted to axion searches as well as one alike to neutrino telescopes. These two in addition to the ones devoted to standard topics such as direct and indirect searches for WIMPs, supersymmetric dark matter, accelerator searches for dark matter…

The social part of the Workshop consisted of two afternoon sightseeing tours, the first one to the Rhodes Magistrate Palace followed up by the conference banquet which was kindly offered by the City Mayor. The second one took place at the castle of Lindos at the closing of the Workshop.

We are thankful to the Astroparticle Physics group of Sheffield University and the Institute of Nuclear Physics of NCSR Demokritos who contributed to the success of the Workshop. For their generous financial support many thanks go to the Ministries of Tourism and of the Aegean and Island Policy, the Prefecture of South Aegean and Municipality of Rhodes and last but not least The Greek Atomic Energy Commission and NCSR Demokritos.

Minos Axenides
George Fanourakis
John Vergados

INTERNATIONAL SCIENTIFIC COMMITTEE

D. Akerib (CWRU)
E. Aprile (Columbia)
F. Avignone (S. Carolina)
R. Bernabei (Rome)
B. Carr (QMW)
G. Charpak (CERN)
D. Cline (UCLA)
J. Ellis (CERN)
E. Fiorini (Milano)
K. Freese (Michigan)
G. Gerbier (Saclay)
Y. Giomataris (Saclay)
G. Gounaris (AUTH)
S. Katsanevas (IN2P3)
H. Klapdor-Kleingrothaus (MPI)
H. Kraus (Oxford)
L. Krauss (Case Western)
V. Kudryavtsev (Sheffield)
A. Lahanas (Athens)
G. Lazarides (AUTH)
N. Mavromatos (London)
J. Morales (Zaragoza)
P. Nath (NEU)
M. Rowan-Robinson (ICL)
L. Roszkowski (Sheffield)
A. Rubbia (CERN)
B. Sadoulet (Berkeley)
N. Smith (RAL)
N. Spooner (Sheffield)
T. Summer (ICL)
K. Tamvakis (Ioannina)
K. Van Bibber (LLNL)
K. Zioutas (Patras)

ORGANIZING COMMITTEE

M. Axenides (Demokritos)
G. Fanourakis (Demokritos)
T. Geralis (Demokritos)
C. Kokorelis (Demokritos)
V. Kudryavtsev (Sheffield)
L. Perivolaropoulos (Ioannina)
M. Robinson (Sheffield)
M. Sakellariadou (Athens, London)
G. Savvidy (Demokritos)
N. Spooner (Sheffield)
N. Tetradis (Athens)
L. Thompson (Sheffield)
J.D. Vergados (Ioannina)

SESSION CHAIRMEN

J. Vergados
L. Perivolaropoulos
V. Kudryavtsev
M. Axenides
I. Giomataris
H. Kraus
G. Gounaris
K. Van Bibber
I. Irastorza
D. Cline
N. Spooner
K. Zioutas
S. Katsanevas
G. Cantatore
S. Tzamarias
E. Daw

CONTENTS

*Contribution not received.

Session A2: WIMPs

Session A3: Axions

Session A4: Neutrinos

Session A5: Background Studies

Session A6: Advanced Detectors

SESSION B. COSMOLOGY AND LARGE SCALE STRUCTURE

SESSION C. DARK ENERGY

SESSION D. UNDERGROUND LABS/ NEW FACILITIES/FUTURE ASTROPARTICLE FACILITIES

COSMOLOGY AND DARK MATTER AT THE LHC

RICHARD ARNOWITT,[1] ADAM AURISANO,[1] BHASKAR DUTTA,[1]
TERUKI KAMON,[1] NIKOLAY KOLEV,[2] PAUL SIMEON,[1]
DAVID TOBACK[1] and PETER WAGNER[1]

[1]*Department of Physics, Texas A&M University, College Station, TX 77843-4242, USA*

[2]*Department of Physics, University of Regina, Regina, SK S4S 0A2, Canada*

We examine the question of whether neutralinos produced at the LHC can be shown to be the particles making up the astronomically observed dark matter. If the WIMP alllowed region lies in the SUGRA coannihilation region, then a strong signal for this would be the unexpected near degeneracy of the stau and neutralino i.e., a mass difference $\Delta M \simeq (5-15)$ GeV. For the mSUGRA model we show such a small mass difference can be measured at the LHC using the signal 3τ+jet+$E_T^{\rm miss}$. Two observables, opposite sign minus like sign pairs and the peak of the $\tau\tau$ mass distribution allows the simultaneous determination of ΔM to 15% and the gluino mass $M_{\tilde{g}}$ to be 6% at the benchmark point of $M_{\tilde{g}}$=850 GeV, A_0=0, μ >0 with 30 fb^{-1}. With 10 fb^{-1}, ΔM can be determined to 22% and one can probe the parameter space up to $m_{1/2}$=700 GeV with 100 fb^{-1}.

1. Introduction

Supersymmetry (SUSY) offers the possibility of solving a number of theoretical problems of the Standard Model (SM). Thus the cancelations implied by the bose-fermi symmetry resolves the gauge hierarchy problem, allowing one to consider models at energies all the way up to the GUT or Planck scale. Further, using the SUSY SM particle spectrum with one pair of Higgs doublets. (pairs of Higgs doublets are needed on theoretical grounds to cancel anomalies and on phenomenological grounds to give rise to both u and d quark masses) the renormalization group equations (RGE) show that grand unification of the SM gauge coupling constants occurs at $M_G \simeq 10^{16}$ GeV opening up the possibility of SUSY GUT models. However, none of this can actually occur unless a natural way of spontaneously breaking SUSY occurs, and this very difficult to do with global supersymmetry. The problem was resolved by promoting supersymmetry to a gauge symmetry, supergravity (SUGRA),[1] where spontaneous breaking of supersymmetry can easily occur. One can then build SUGRA GUT models[2,3] with gravity playing a key role in the construction. A positive consequence of this promotion was that the RGE then show that the breaking of supersymmetry at the GUT scale naturally leads to the required $SU(2) \times U(1)$ breaking at the electroweak scale, thus incorporating all the successes of the SM, without any prior assumptions of negative $(\text{mass})^2$ terms.

In spite of the theoretical successes of SUGRA GUTs, there has been no experimental evidence for its validity except for the verification of grand unification (which has in fact withstood the test of time for over a decade). However, one expects SUSY particles to be copiously produced at the LHC. Further, models with R parity invariance predict the lightest neutralino $\tilde{\chi}_1^0$ to be a candidate for the astronomically observed dark matter (DM) and models exist[4] consistent with the amount of dark matter observed by WMAP,[5] and being searched for in the Milky Way by dark matter detectors. Thus it is possible to build models that both cover the entire energy range from the electroweak scale to the GUT scale and go back in time to $\sim 10^{-7}$ seconds after the Big Bang when the current relic dark matter was created.

The question then arises can we verify if the dark matter particles in the galaxy is the neutralino expected to be produced at the LHC? In principle this is doable. Thus assuming the DM detectors eventually detect the dark matter particle, they will measure the mass and cross sections, and these can be compared with those measured at the LHC. However, this may take a long time to achieve. More immediately, can we look for a signal at the LHC that is reasonably direct consequence of the assumption that the neutralino is the astronomical DM particle and in this way experimentally unify particle phenomena with early universe cosmology? To investigate this question it is necessary to choose a specific SUSY model, and for simplicity we consider here the minimal mSUGRA (though a similar analysis could be done for a wide range of other SUGRA models).

2. The mSUGRA model

The mSUGRA model depends on four soft breaking parameters and one sign. These are; $m_{1/2}$ (the universal gaugino soft breaking mass at M_G); m_0 (the universal scalar soft breaking mass at M_G); A_0 (the universal cubic soft breaking mass at M_G); $\tan\beta =< H_2 > / < H_1 >$ at the electroweak scale (where $< H_2 >$ gives rise to u quark masses and $< H_1 >$ to d quark masses); and the sign of μ parameter (where μ appears in the quadratic part of the superpotential $W^{(2)} = \mu H_1 H_2$).

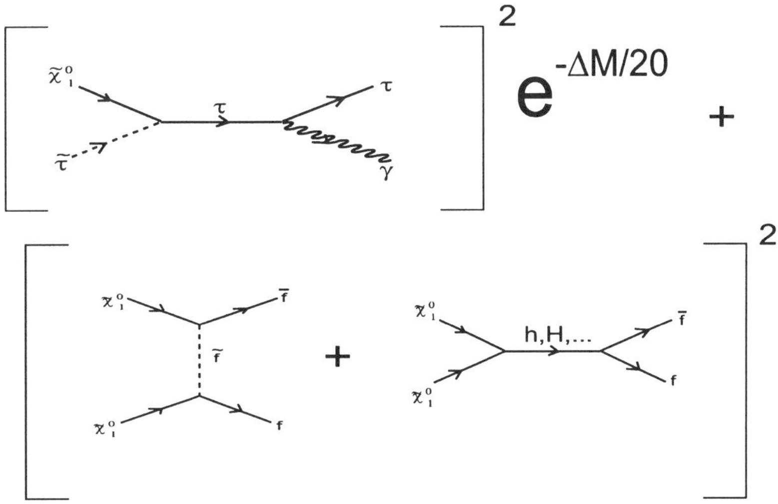

Fig. 1. The feynman diagrams for annihilation of neutralino dark matter in the early universe

Current experimental data significantly constrains these parameters. the main accelerator constraints are: The Higgs mass $m_H > 114$ GeV;[6] the lightest chargino mass $M_{\tilde{\chi}_1^\pm} > 104$ GeV; the $b \to s\gamma$ branching ratio 2.2 $\times 10^{-4} < Br(b \to s\gamma) < 4.5 \times 10^{-4}$;[7] and the muon g-2 anomaly[8] which now deviates from the SM prediction by 3.4 σ. The astronomical constraint is the WMAP determination of the amount of dark matter and we use here a 2 σ range:[5]

$$0.094 < \Omega_{\tilde{\chi}_1^0} h^2 < 0.129. \tag{1}$$

The WMAP constraint limits the parameter space to three main regions arising from the diagrams of Fig.1. (1) The stau-neutralino ($\tilde{\tau}_1 - \tilde{\chi}_1^0$) coannihilation region. Here m_0 is small and $m_{1/2} \leq 1.5$ TeV. (2)The focus region where the neutralino has a large Higgsino component. Here $m_{1/2}$ is small and $m_0 \geq 1$ TeV. (3) The funnel region where annihilation proceeds through heavy Higgs bosons which have become relatively light. Here both m_0 and $m_{1/2}$ are large.

(In addition there is a small bulk region) Note that a key element in the coannihilation region is the Boltzman factor from the annihilation in the early universe at $kT \sim 20$ GeV: $\exp[-\Delta M/20]$ where $\Delta M = M_{\tilde{\tau}_1} - M_{\tilde{\chi}_1^0}$. Thus significant coannihilation occurs provided $\Delta M \leq 20$ GeV.

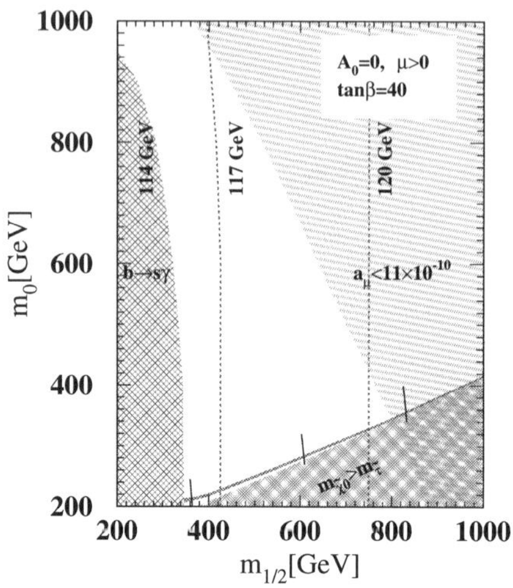

Fig. 2. Allowed parameter space for $\tan\beta = 40$ with $A_0 = 0$ and $\mu > 0$.

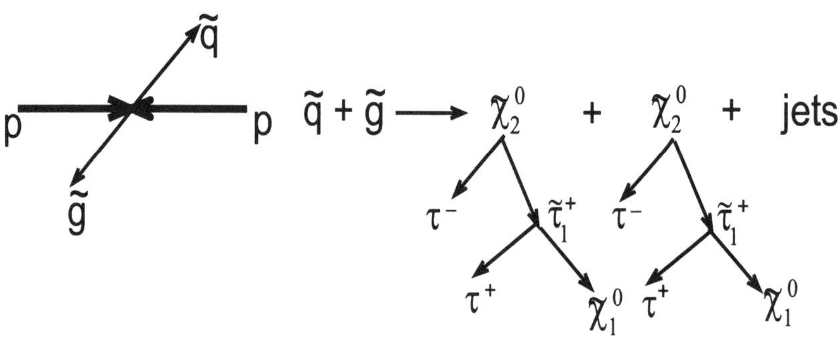

Fig. 3. SUSY production and decay channels

The accelerator constraints further restrict the parameter space and if the muon g-2 anomaly maintains, $\mu > 0$ is prefereed and there remains mainly the coannihilation region. This is illustrated in Fig.2 which shows the allowed narrow coannihilation band (for the case $\tan\beta = 40$, $A_0 = 0$, $\mu > 0$) where $\Delta M = (5 - 15)$ GeV and $m_{1/2} \leq 800$ GeV. (There is a small focus region for small $m_{1/2}$ and $m_0 > 1$ TeV since the $b \to s\gamma$ constraint ceases to opperate at $m_0 > 1$ TeV.)

The coannihilation band is narrow ($\Delta M = 5 - 15$ GeV) due to the Boltzman factor in Fig.1, the range in ΔM corresponding to the allowed WMAP range for $\Omega_{\tilde{\chi}_1^0} h^2$. The dashed verticle lines are possible Higgs masses.

One may ask two questions; (1) Can such a small stau-neutralino mass difference (5-15 GeV) arise in mSUGRA, i.e. one would naturally expect these SUSY particles to be hundreds of GeV apart and (2) Can such a small mass difference be measured at the LHC? If the answers to both these questions are affirmative, the observation of such a small mass difference would be a strong indication that the neutralino is the astronomical DM particle since it is the cosmological constraint on the amount of DM that forces the near mass degeneracy with the stau, and it is the accelerator constraints that suggests that the coannihilation region is the allowed region.

3. Can ΔM be Small in SUGRA models?

At the GUT scale $m_{1/2}$ governs the gaugino masses, while m_0 the slepton masses. Thus, at M_G one would not expect any degeneracies between the two classes of particles. However, at the electroweak scale the RGE can modify this result. To see analytically this possibility , consider the lightest selectron $\tilde{e}^c$ which at the electroweak scale has mass

$$m_{\tilde{e}^c}^2 = m_0^2 + 0.15 m_{1/2}^2 + (37\text{GeV})^2 \tag{2}$$

while the $\tilde{\chi}_1^0$ has mass

$$m_{\tilde{\chi}_1^0}^2 = 0.16 m_{1/2}^2 \tag{3}$$

The numerical accident that the coefficients of $m_{1/2}^2$ is nearly the same for both cases allows a near degeneracy. Thus for $m_0 = 0$, the $\tilde{e}^c$ and $\tilde{\chi}_1^0$ become degenerate at $m_{1/2}$=(370-400) GeV. For larger $m_{1/2}$, the near degeneracy is maintained by increasing m_0, so that one can get the narrow corridor in the m_0-$m_{1/2}$ plane seen in Fig.2. Actually the case of the stau $\tilde{\tau}_1$ is more complicated since the large t-quark mass causes left-right mixing in the stau mass matrix and results in the $\tilde{\tau}_1$ being the lightest slepton (not the selectron). However, a result similar to Eqs. (1,2) occurs, with a $\tilde{\tau}_1 - \tilde{\chi}_1^0$ coannihilation corridor resulting.

We note that the results of Eqs.(1,2) depend only on the U(1) gauge group and so coannihilation can occur even if there were non-universal scalar mass soft-breaking or non-universal gaugino mass soft breaking at M_G. Thus, coannihilation can occur in a wide class of SUGRA models, and is not just a feature of mSUGRA.

4. Coannihilation signal at the LHC

At the LHC, the major SUSY production processes are gluinos ($\tilde{g}$) and squarks ($\tilde{q}$) e.g., $p+p \to \tilde{g}+\tilde{q}$. These then decay into lighter SUSY particles and Fig.3 shows a major decay scheme. The final states involve two $\tilde{\chi}_1^0$ giving rise to missing transverse energy $E^T_{\rm miss}$) and four τ's, two from the $\tilde{g}$ and two from the $\tilde{q}$ decay chain for the example of Fig 3. In the coannihilation region, two of the taus are high energy ("hard" taus) coming from the $\tilde{\chi}_2^0 \to \tau\tilde{\tau}_1$ decay (since $M_{\tilde{\chi}_2^0} \simeq 2M_{\tilde{\tau}_1}$) while two are low energy ("soft" taus) coming from the $\tilde{\tau}_1 \to \tau + \tilde{\chi}_1^0$ decay since ΔM is small. The signal is thus $E_T^{\rm miss}$+ jets $+\tau$'s, which should be observable at the LHC detectors.

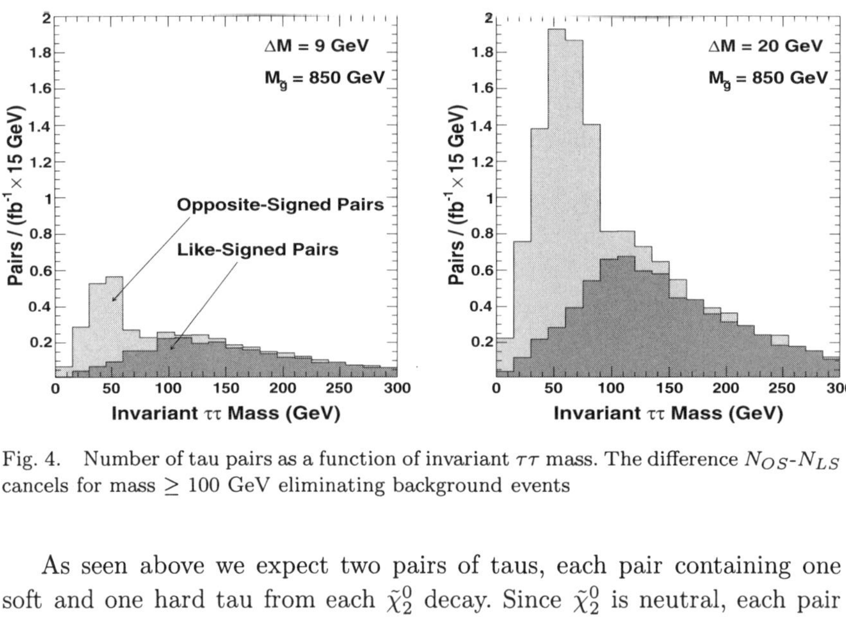

Fig. 4. Number of tau pairs as a function of invariant $\tau\tau$ mass. The difference N_{OS}-N_{LS} cancels for mass $\geq$ 100 GeV eliminating background events

As seen above we expect two pairs of taus, each pair containing one soft and one hard tau from each $\tilde{\chi}_2^0$ decay. Since $\tilde{\chi}_2^0$ is neutral, each pair should be of opposite sign(while SM and SUSY backgrounds, jets faking taus will have equal number of like sign as opposite sign events). Thus one can suppress backgrounds statistically by considering the number of oppsite sign events N_{OS} minus the like sign events N_{LS}. The four τ final state has the smallest background but the acceptance and efficiency for reconstructing all four taus is low. Thus to implement the above ideas we consider here the three τ final state of which two are hard and one is soft. (The two τ final state with higher acceptance but larger backgrounds was discussed in,[9] and an analysis of the coannhilation signal at the ILC was given in.[10]

We label three taus by their transverse energies with $E_1^T > E_2^T > E_3^T$ and form the pairs 13 and 23. For signal events one of the two pairs should be coming from a $\tilde{\chi}_2^0$ decay and have opposite sign(OS) while the other is not correlated. There are two measurables that can be formed. The number N and the mass of the pair M. To simulate the data we use ISAJET 7.64[11] and PGS detector simulator.[12] Events are chosen with $E_T^{\rm miss}$ and 1 jet and three taus with visible momenta $p_T^{\rm vis} > 40$ GeV, $p_T^{\rm vis} > 40$ GeV, $p_T^{\rm vis} > 20$ GeV. We assume here that it is possible to reconstruct taus with $p_T^{\rm vis}$ as low as 20 GeV. Standard Model background is reduced by requiring $E_T^{\rm jet1} > 100$ GeV, $E_T^{\rm miss} > 100$ GeV with tevatron results, $E_T^{\rm jet1} + E_T^{\rm miss} > 400$ GeV. We also assume rate of a jet faking a τ ($f_{j\to\tau}$) to be $f_{j\to\tau} = 1\%$ (with a 20% error in $f_{j\to\tau}$) consistent with Tevatron results.

Fig 4. shows the number of events as a function of the $\tau\tau$ mass for gluino mass $M_{\tilde{g}} = 850$ GeV and $\Delta M = 9$ GeV and 20 GeV. One sees that the difference $N_{OS} - N_{LS}$ cancels out as expected for $\tau\tau$ mass $\geq$100 GeV (consistent with the fact that the signal events are expected to lie below 100 GeV).

Fig. 5 shows the behavior of N_{OS-LS} as a function of ΔM and $M_{\tilde{g}}$. The central black line is for the assumed 1%. rate for jets faking a τ, the shaded region around it is for a 20% uncertainty in $f_{j\to\tau}$. One sees that provided this uncertainty is not large, it produces only a small effect.

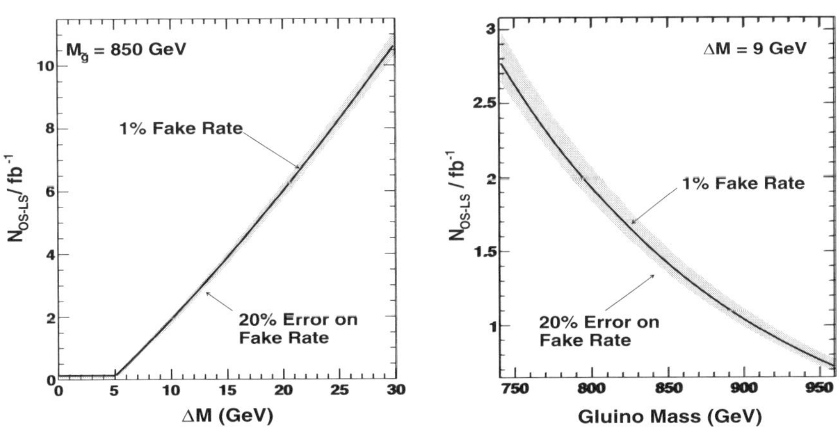

Fig. 5. N_{OS-LS} as function of ΔM (left graph) and as a function of $M_{\tilde{g}}$ (right graph). The central black line assumes a 1% fake rate, the shaded area representing the 20% error in the fake rate

Figs 4 and 5 show two important features. First, N_{OS-LS} increases with ΔM(since the τ acceptance increases) and N_{OS-LS} decreases with $M_{\tilde{g}}$(since the production cross section of gluinos and squarks decrease with $M_{\tilde{g}}$). Second, from Fig.4 one sees that N_{OS-LS} forms a peaked distribution.[9,13] The ditau peak position $M_{\tau\tau}^{\rm peak}$ increases with both ΔM and $M_{\tilde{g}}$. This allows us to use the two measurables N_{OS-LS} and $M_{\tau\tau}^{\rm peak}$ to determine both ΔM and $M_{\tilde{g}}$. Fig.6 shows this determination for the benchmark case of ΔM=9 GeV, $M_{\tilde{g}}$=850 GeV, A_0=0 and $\tan\beta$=40. Plotted there are constant values of N_{OS-LS} and constant values of $M_{\tau\tau}^{\rm peak}$ in the $\Delta M - M_{\tilde{g}}$ plane which exhibit the above dependance of these quantities on ΔM and $M_{\tilde{g}}$. With luminosity of 30 fb^{-1} one determines ΔM and $M_{\tilde{g}}$ with the following accuracy:

$$\delta\Delta M/\Delta M \simeq 15\%;\ \delta M_{\tilde{g}}/M_{\tilde{g}} = 6\% \tag{4}$$

Fig. 7 shows how the accuracy of the measurement changes with luminosity. One sees that even with 10 fb^{-1} (which should be available at the LHC after about two years running)one could determine ΔM to within 22%, which should be sufficient to know whether one is in the SUGRA coannihilation region.

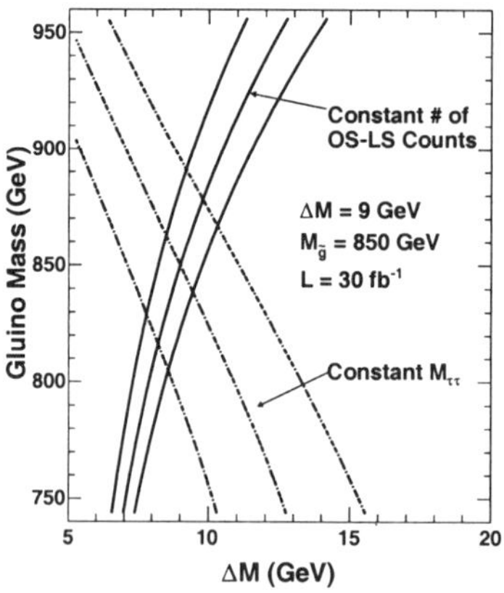

Fig. 6. Simultaneous determination of ΔM and $M_{\tilde{g}}$. The three lines plot constant N_{OS-LS} and $M_{\tau\tau}^{\rm peak}$ (central value and 1σ deviation) in the $M_{\tilde{g}}$-ΔM plane for the benchmark point of ΔM=9 GeV and $M_{\tilde{g}}$=850 GeV assuming 30 fb^{-1} luminosity

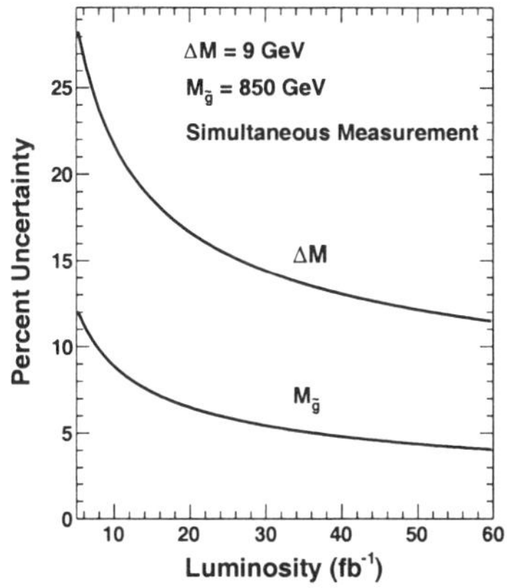

Fig. 7. Uncertainty in the determination of ΔM and $M_{\tilde{g}}$ as a function of luminosity.

5. Conclusions

We have examined here the question of how one might show that the $\tilde{\chi}_1^0$ particle produced at the LHC is the astronomically observed dark matter. If ΔM, the stau-neutralino mass difference lies in the coannihilation region of the SUGRA m_0-$m_{1/2}$ plane where $\Delta M = (5-15)$ GeV, this would be strong indication that the neutralino is the dark matter particle as otherwise the mass difference would not naturally be so small. We saw how it was possible to measure such a small mass difference at the LHC for the mSUGRA model using a signal of $E_T^{\rm miss}$+ 1 jet+3τ, and simultaneously determine the gluino mass $M_{\tilde{g}}$, provided it is possible at the LHC to reconstruct taus with $p_T^{\rm vis}$ as low as 20 GeV. With 30 fb^{-1} one could then determine ΔM with 15% accuracy and $M_{\tilde{g}}$ with 6%, at our benchmark point of ΔM=9 GeV, $M_{\tilde{g}}$=850 GeV, $\tan\beta = 40$, and A_0=0. Even with only 10fb^{-1} one would determine ΔM to within 25% accuracy, sufficient to learn whether the signal is in the coannihilation region.

While the analysis done here was within the framework of mSUGRA, similar analyses can be done for other SUGRA models provided the production of neutralinos is not suppressed. However, the determination of $M_{\tilde{g}}$ does depend on the mSUGRA universality of the gaugino masses at M_G to relate $M_{\tilde{\chi}_2^0}$ to $M_{\tilde{g}}$. Thus a model independent method of determining $M_{\tilde{g}}$ would allow one to to test the question of gaugino universality. However, it may not be easy to directly measure $M_{\tilde{g}}$ at the LHC for high $\tan\beta$ in the coannihilation region due to the large number of low energy taus, and the

ILC would require a very high energy option to see the gluino.

As mentioned above, one can also measure ΔM using the signal $E_T^{\rm miss}+$ 2 jets$+2\tau$.[9] This signal has higher acceptance but larger backgrounds. There, with 10 $\rm fb^{-1}$ one finds that one can measure ΔM with 18% error at the benchmark point assuming a separate measurement of $M_{\tilde{g}}$ with 5% error has been made. While we have fixed our benchmark point at $M_{\tilde{g}}$ = 850 GeV(i.e. $m_{1/2}$ =360 GeV), higher gluino mass would require more luminosity to see the signal. One finds that with 100 $\rm fb^{-1}$ one can probe $m_{1/2}$ at the LHC up to $\sim$ 700 GeV (i.e., $M_{\tilde{g}}$ up to $\simeq$ 1.6 TeV).

Finally, it is interesting to compare with possible measurements of ΔM at the ILC. If we implement a very forward calorimeter to reduce the two γ background, ΔM can be determined with 10% error at the benchmark point.[10] Thus in the coannihilation region, the determination of ΔM at the LHC is not significantly worse than at the ILC.

Acknowledgement

This work was supported in part by DOE grant DE-FG02-95ER40917, NSG grant DMS 0216275 and the Texas A&M Graduate Merit fellowship program.

References

1. D.Z. Freedman, P. Van Niewenhuisen, and S. Ferrara, Phys. Rev. **D13**, 1980 (1729); S. Deser and B. Zumino, Phys. Lett. **B65**, 369 (1976);Phys. Lett. **B65**, 1976 (369).
2. A.H. Chamseddine, R. Arnowitt, and P. Nath, Phys. Rev.Lett **49**, 1982 (970).
3. R. Barbieri, S. Ferrara, and C.A. Savoy, Phys. Lett. **B119**, 343 (1982); L. Hall, J. Lykken, and S. Weinberg,Phys. Rev. **D27**, 1983 (2359); P. Nath, R. Arnowitt, and A.H. Chamseddine, Nucl. Phys. **B227**, 121 (1983); For a review, see P. Nilles, Phys. Rept. **100**, 1984 (1).
4. J. Ellis, K. Olive, Y. Santoso, and V. Spanos, Phys. Lett. **B176**, 565 (2003); R. Arnowitt, B. Dutta, and B. Hu, hep-ph/0310103; H. Baer, C. Balazs, A. Belyaev, T. Krupovnickas, and X. Tata, JHEP **0306**, 2003 (054); B. Lahanas and D.V. Nanopoulos, Phys. Lett. **568**, 2003 (55); U. Chattopadhyay, A. Corsetti, and P. Nath, Phys. Rev. **D68**, ((2)003)035005; E. Baltz and P. Gondolo, JHEP **0410** (2004) 052; B. C. Allanach and C. G. Lester, Phys. Rev. **D73**, 2006 (015013); A. Djouadi, M. Drees, and J-L. Kneur, JHEP **0603**, 2006 (033).
5. WMAP Collaboration, D.N. Spergel *et al.*, Astrophys. J. Suppl. **148**, 2003 (175).
6. ALEPH, DELPHI, L3, OPAL Collaborations, G. Abbiendi *et al.* (The LEP Working Group for Higgs Boson Searches), Phys. Lett. B **565** (2003) 61; Particle Data Group, S. Eidelman *et al.*, Phys. Lett. **592**, 2004 (1).

7. M. Alam *et al.*, Phys. Rev. Lett **74**, 1995 (2885).
8. Muon $g-2$ Collaboration, G. Bennett *et al.*, Phys. Rev. Lett **92**, 2004 (161802); S. Eidelman, Talk at ICHEP 2006, Moscow, Russia.
9. R. Arnowitt, B. Dutta, T. Kamon, N. Kolev, and D. Toback, Phys. Lett. **B639**, 2006 (46).
10. V. Khotilovich, R. Arnowitt, B. Dutta, and T. Kamon, Phys. Lett. **B618**, 2005 (182).
11. F. Paige, S. Protopescu, H. Baer, and X. Tata, hep-ph/0312045. We use ISAJET version 7.64.
12. PGS is a parameterized detector simulator. We used the CDF detector information to obtain approximate jet finding. We used version 3.2 (see http://www.physics.ucdavis.edu/~conway/research/software/pgs/pgs4-general.htm).
13. I. Hinchliffe and F.E. Paige, Phys. Rev. **D61**, 2000 (095011).

EXPERIMENTAL SEARCHES FOR DARK MATTER

DAVID CLINE

Center for Advanced Accelerators,
Department of Physics & Astronomy,
University of California, Los Angeles, CA 90095 USA,
dcline@physics.ucla.edu

We review the various methods to search for dark matter particles based on the results from the recent symposium at Marina Del Rey [1]. At present, the most promising methods are: (a) Direct searches with earth-bound detectors made of Ge, Xe, Ar, etc. (b) Search for Gamma Rays from dark matter annihilation in the local galaxy. We note there is progress on (a) to date, the expectation for the current generation of detectors CDMS II, Edelweiss II, ZEPLIN II, WARP etc. and hopes for future one ton detectors. In the case of (b) the detection of Dark Matter Gamma Rays will depend on the clumping of Dark Matter in the Galaxy and Backgrounds. We will discuss this. We also discuss very briefly warm dark matter.

1. Introduction

The search for dark matter elementary particles is producing the most important scientific activity in the world (see Ref. 1). While the evidence for dark matter in galactic clusters goes back to 1933 and F. Zwicky [2], recently there has been a beautiful astrophysics proof of dark matter in the observation of colliding galactic clusters (Fig. 1). This observation that started in 2004 shows clearly that baryonic matter and dark matter are physically separated in destruction to the MOND theory. See Table 1 for a brief history. This is an important breakthrough in the understanding of the components of the universe.

There are suggested modifications to take into account the effects attributed to dark matter. The most noticeable is the MOND theory of Milgram (see Ref. 4 for references). Some say that this theory violates Einstein's Strong Equivalence principle---and therefore general relativity. Thus it is important to determine its correctness.

The MOND theory was invented to explain the star rotation curves for galaxies that are normally used to indicate the existence of cold dark matter. MOND assumes that gravity is modified beyond a gravitational acceleration of $a_0 \propto 10^{-8}$ cm/Sec^2 (see Ref. 4 for more information).

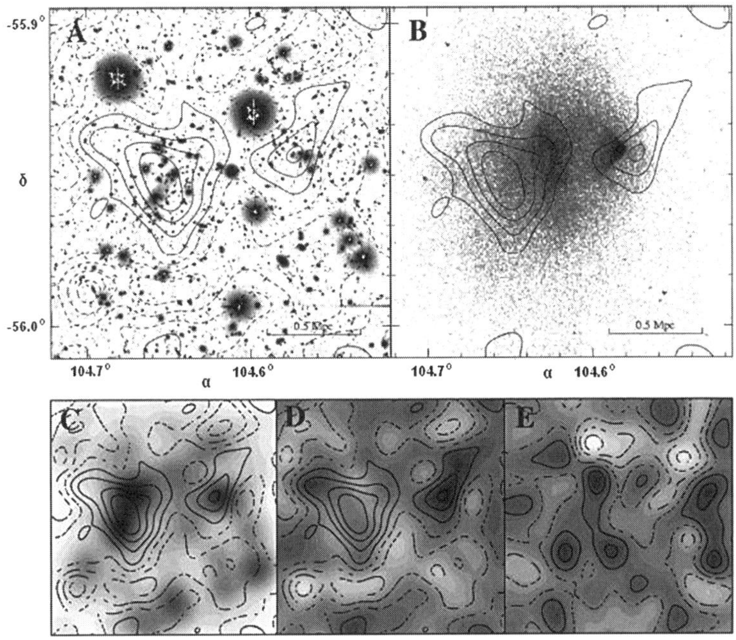

Fig. 1. Temperature profile from x-rays for a galactic cluster from Ref. 4.

Table 1. Brief history of the evidence for dark matter and dark energy.

Year	Event
~1933	F. Zwicky observes fast galaxies in comet clusters; suggests mixing mass in cluster is the cause.
~1960s	Astronomers realize that galaxies have fast moving stars in halo; suggest dark matter is the cause.
~1980	Suggestions for MOND to explain radiation curves by modifying Newtonian gravity [1][2]
~1998	Experiments on SN1A reports at dark matter meeting; indication of accelerating universe; dark energy is suggested cause
~2003	WMAP data strongly supports both dark matter and dark energy components of the universe
~2005	SDSS observes Baryon acoustic oscillations; provides additional proof for dark matter (several independent measurements of Ω_{DM} suggest a single origin: cold dark matter)
~2006	D. Clowe and collaborators show that a colliding cluster of galaxies demonstrate the existence of dark matter (Refs. 1,3)

For acceleration greater than this we get the normal law:

$$a_N = \frac{MG}{R^2} \tag{1}$$

for acceleration less than a_0 we get

$$a = \sqrt{a_0 a_N} = \sqrt{a_0 MG}\,\frac{1}{R} \tag{2}$$

The 1/R behavior then provides a good fit to the rotation curves of galaxies at large r.

There are three more (at least) ways to test MOND [4]:

a) Study of clusters in x-rays and weak lensing to determine whether dark matter is different from Baryonic matter (see Ref. 1)

b) Study of CMBR and Baryon oscillations that come from Z ~ 1000 – the surface of last scatter – two independent ways to measure Ω_{DM}

c) Direct detection of dark matter particles.

d) Collision of galactic clusters (see Fig. 1, [3]).

2. Types of Dark Matter: CDM or WDM

Dark matter can be generally classified into cold and warm dark matter, which refers to the velocity of the dark matter particle in the early universe. Cold dark matter has low velocity, warm (or hot) will have high velocity, even near the speed of light. While much of the work on dark matter has assumed CDM the jury is still out on warm dark matter. At the Dark Matter 2006 symposium there were even claims that there is new evidence for WDM from dwarf spheroidal galaxy dynamics. Fig. 2 and 3 show the conclusions of the talk of G. Gilmore at Dark Matter 2006.

2.1. *Dwarf spheroidal galaxies as pure dark matter systems*

At the Dark Matter 2006 meeting G. Gilmore gave an interesting talk about some implication of the study of 9d Sph systems and indicated some possibly remarkable conclusions. A figure of the mass to light ratio from his talk is shown in Figure (Fig. 2, 3) and Fig. 2 gives the conclusions [8].

If these conclusions are correct they imply that dark matter is very low mass and could be warm dark matter. Sterile neutrinos of a few KeV mass could be a condition [1].

summary

- A vast increase in precise stellar kinematic data allows more sophisticated derivation of mass profiles in the dSph.
- UMa – discovered 2005 – extends to M/L ~ 500
- All are consistent with:
- Central mass **cores**, not cusps
- Central mass density **≤20GeV/cc**
- Dispersion **~9km/s**
- Scale length **~few x100pc**
- DM minimum mass **~5x10^{7}M$_\odot$**
- **Somewhat preferring particle-mass<<GeV...**
- We have two new dSph under study (today), to extend the sample further, and see if these numbers are really meaningful

Fig. 2. From talk of G. Gilmore at DM06 (Ref. 1).

Systematic properties of DM –II

--minimum mass, length, dispersion?

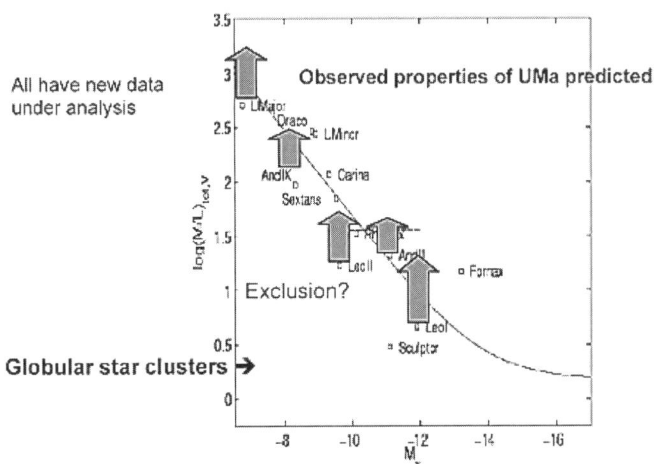

Fig. 3. Mass to light ratios for dwarf spheroidal galaxies (Ref. 1).

Under any circumstance the d Sphs systems consist of very little mass and are nearly pure dark matter systems. WIMPS could also be the dark matter in d Sphs and in this case the GLAST detection could see a signal of gamma rays from d Sphs.

3. Methods to Search for CDM; WIMPs and SUSY Particles

The bulk of the search for dark matter on earth involves searching for WIMPs by direct scattering in a matter detector.

The direct search for WIMP matter is one of the hardest set of experiments ever started in science. A very small energy release is searched for in a fairly massive detector (Kg to 100s of Kg). At the earth's surface such an experiment would be flooded with background. Therefore these experiments must be carried out deep underground. Even then the walls of the cavern and penetrating cosmic rays still endure a very large background. Finally there is intensive radioactive background in all materials on earth that must be dealt with.

Because of these facts most modern detectors have some form of powerful background rejection. There are two general classes of such detectors:

a) Cryogenic detectors that detect a signal of heat injection onto the system from the WIMP scatter;
b) Detectors at high temperatures that detect a signal of scintillation light from the WIMP scatter.

In both cases methods to discriminate between nuclear recoils and radioactive background have been devised and tested.

There are other very promising methods such as bubble detection or recoil track detection in a gas detector. However the detectors in (1) and (2) constitute the major number of detectors being deployed now to search for WIMPS (see Ref. 1).

Table 2 lists the methods whereby we may prove a dark matter discovery.

Table 2. How to prove that dark matter has been discovered.

Measurement	Comment
Observation of annual variation with discriminated signal (very hard to do)	Effect as few percent. Detection requires great stability over a long time
Observation of directional signal (very, very hard to do)	Requires a very large DRIFT-like detector. We don't know the size until σ_w measured ($\sigma_w \sim 10^{-10}$ pb) as being this building
A careful observation of the SI A^2 dependence (for Xenon/Argon A^2 ratio ~ 9)	Requires range of Target A detectors, all at the same time, if possible

4. The Experimental Search for SUSY-WIMPS

Are we in the $\sqrt{t}$ region yet? If there is significant background in a detector the background subtraction results in limited sensitivity.

Experiments are close to a cross section of 10^{-7} pb or will soon be there. If there is no background the sensitivity goes like

$$\sigma \propto \frac{1}{t} \tag{3}$$

For background limited samples

$$\sigma \propto \frac{1}{\sqrt{t}} \tag{4}$$

The CDMS group has reported one (possibly two) background events.

5. Indirect Search

The search for WIMPs by a signal from annihilation in the galaxy sun or galactic center is very promising. In some cases this search can be more sensitive than a direct search [1].

The signal strength will depend on the clumping of the dark matter in the galaxy. Recent studies have shown that tidal forces in the inner galaxy could destroy most of the clumps, making the detection of WIMPs much harder [1].

In the near future GLAST is very promising for the search as is demonstrated in the section on indirect search in the Proceedings of DM06 (see Chapter 4) [1].

The study of the positron energy spectrum or the rate of cosmic antiproteins is also very important. In the former case the recent launch of the PAMELA mission in Russia is extremely promising (PAMELA is mainly an Italian project).

6. One Ton Detectors

The direct search for dark matter particles is among the hardest experiments ever undertaken in science. Backgrounds exist for cosmic rays, natural radioactivity even at great depths underground. Early reviews can be found in Ref. 10. Therefore the next generation detector will almost certainly use a method to discriminate against background as well as an active veto shield to reduce the neutron flux from cosmic ray induced events even at great depths underground.

Table 3. One ton dark matter detector proposals.

Detector	Material	Method	Proposal	Current Prototype
GENIUS LNGS (a)	Ge	Ultrapure detector in LNGS	1997	10kg GENIUS test detector
ZEPLIN IV (Max) Boulby/ SNOLAB (b)	Xe	2 phase discriminating detector	~1999	ZII/III Detectors at Boulby
Super CDMS SNOLAB (c)	Ge/Si	Ionization and Phonons	~2001	CDMS II
XMass Japan (d)	Xe	2 phase (?)	~2000	prototype
Xenon LNGS (e)	Xe	2 phase Detector	~2001	prototype
WARP LNGS (f)	Ar	2 phase (possible larger than one ton)	~2003	prototype
LUX	Xe	2 phase	~2009	New collaboration from USA/UK

a. H.V. Klapdor-Kleingrothaus et al, Z. Physics A 359, 351 (1997).
b. D. Cline, H. Wang et al, UCLA DM 2000, published in Proceedings.
c. R. Gaitskell et al, 2001 Snowmass Proceedings.
d. Y. Suzuki, hep-ph/008296.
e. ZENON Collaboration, NSF Proposal 0201740, 2001.
f. C. Rubbia talk, UCLA DM 2004, to be published in the Proceedings.

The types of detectors can be generally classed as

1. Cryogenic
2. Liquid Xenon, Neon or Argon
3. Other methods such as bubble chamber or non-discriminating detector.

To get some sense of the number of detectors [2] and time scale we give a partial list of proposed one ton detectors in Table 3 [15].

The concept of discrimination of background is illustrated in Fig. 4 for a cryogenic detector and for a liquid Xenon detector (Fig. 5).

7. The Direct Search for Dark Matter Particles

7.1. *Cryogenic detectors*

For more than 15 years several groups around the world have been studying the possibility of constructing a low temperature detector to measure the recoil energy of the nucleus having been hit by a WIMP [3]. Since this energy is in the range of kiloelectron volts the detector must act as a bolometer to measure the "heat" produced by the recoil (see [6] and [7] for recent examples). Three groups have now made such detectors using this technique. These groups are:

- CDMS
- Edelweiss
- Cresst

All three groups have now reported limits in the in the search for dark matter particles. So far the nucleus of choice has been Ge or Si. However the Cresst group has worked with $Al0_2$ as well as CaW mixtures [1].

The best limits that have been set by these detectors came from the CDMS II detector operated at the Soudan underground laboratory [7]. These limits are well below the claimed signal by the DAMA group as shown in Fig. 7. All of these detectors are being upgraded to larger mass and 2005 will be a big year for these types of detectors.

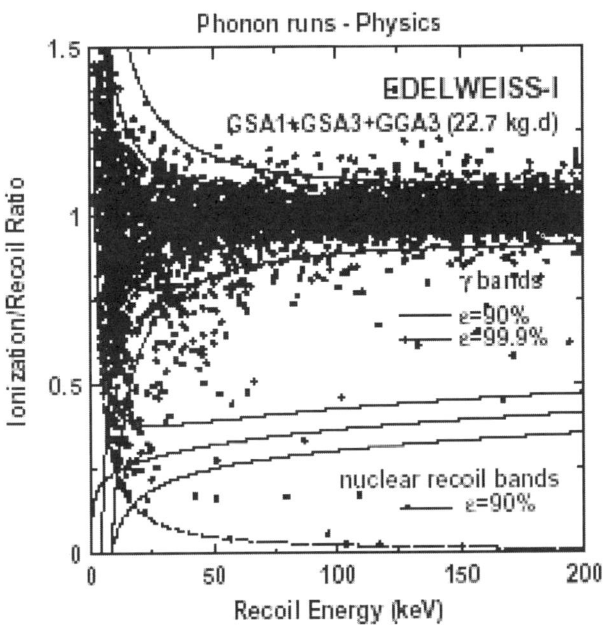

Fig. 4. Distribution of the ratio of the ionization energy to the recoil energy as a function of the recoil energy collected in the fiducial volume of three EDELWEISS detectors in the last run for 2003.

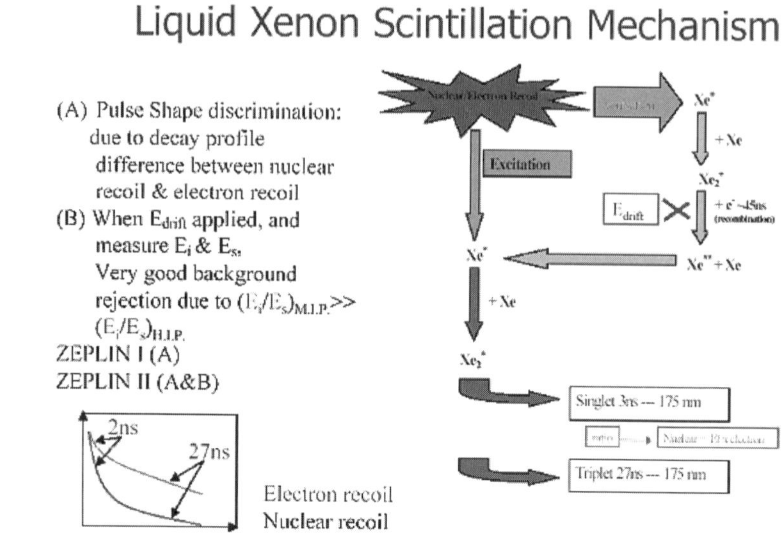

Fig. 5. Schematic of signals in a liquid Xenon detector [12].

7.2. *Liquid Noble Gas Detectors: Xenon, Argon and Neon*

Another promising method to detect dark matter is to use the scintillation light produced in Noble gas liquids (see [1] and [5] for other references). The process is very well known since excimer lasers use a similar concept. For example the very first excimer laser was made in Russia in 1970 using liquid Xenon. A key part of this method is to apply an electric field to the detector to drift out any electrons that are produced at the recoil vertex as a basis to discriminate against background (Fig. 5) [5].

7.3. *Dark matter in the Milky Way Galaxy: Halo Uncertainty and Streams*

In order to detect dark matter particles we must have an understanding of the flux of particles through any given detector on earth. Therefore we must understand the halo of dark matter for our galaxy [1]. In addition some models give clumps of dark matter, others give causes of dark matter. These effects can increase or decrease the rate of interaction in an earth-bound detector. At the recent Marina del Rey meeting we devoted an entire session to the knowledge of our halo. The expected halo velocity dispersion plots are shown in Fig. 6 [5, 9].

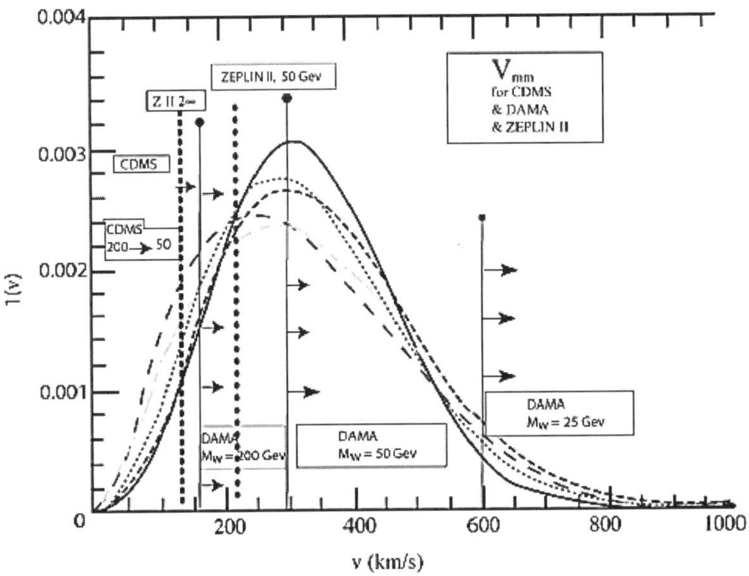

Fig. 6. Schematic of the halo velocity distribution with minimal velocities for CDMS, DAMA and ZEPLIN II; the figure is modified from A. Green, P.R.D. 68, 023004 (2003) (Reference 6).

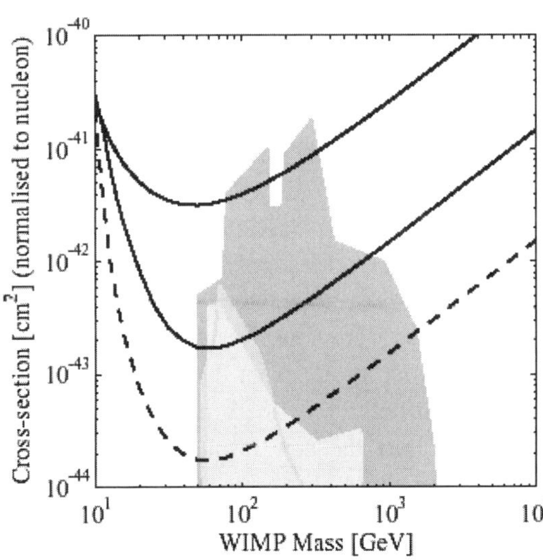

Fig. 7. CDMS results for a spin-independent WIMP nucleon cross section versus WIMP mass. Shown as the Si (upmost curve) and Ge (middle curve) data, as well as the prediction for the 5 towers run at Soudan (dashed curve).

The halo model is very important when attempting to compare different types of experiments, say, direct search and annual variation searches. There is no doubt that the ultimate test for the existence of dark matter will be the observation of an annual variation signal. However there is a strong debate among the experiments as to whether this annual variation signal should be carried out with discriminated events (reduced background) or with raw data (large background). We will show an example of the former from simulation of the ZEPLIN II detector later in this article.

There are also models of dark matter caustics by P. Sikivie and colleagues that can give the opposite sign of the annual variation to that expected in the standard isothermal sphere model. At the Marina del Rey meeting in 2004, two notable contributions were given by Anne Green and Larry Krauss at the last Marina del Rey meeting [9].

One concept is GENIUS, which will use one ton of 76Ge (also to be used for double β decay search) [see Table 3 for references]. While this detector has no discrimination it is to be produced of ultrapure material so that there is little or no radioactive background. The detector is submerged in a large bath of liquid Nitrogen to shield out neutrons from cosmic rays.

Another concept is to construct a "bubble chamber" to detect WIMPS by the formation of bubbles in the detector. More information on this novel scheme can be found in Ref. 1 (see papers in the 2004 Dark Matter meeting) [9].

8. Status of the Search for Dark Matter Particles

8.1. *Early history*

A serious search for dark matter particles started around 1995 with the use of NaI detectors at several locations. Fig. 8 traces the history of this early search.

By the end of 2003 there were considerable data on this search. In addition the CDMS I group carried out a joint fit of their data and the DAMA data, and claimed these data were inconsistent to 98 percent confidence level at Dark Matter 2002. One month later the new results from CDMS II at the Soudan underground laboratory were presented [9]. The ZEPLIN II detector is now taking data at Boulby. 1200 kg have been recorded.

8.2. *Current limits on search*

We will use as an example the recent results of the CDMS II detector [7].

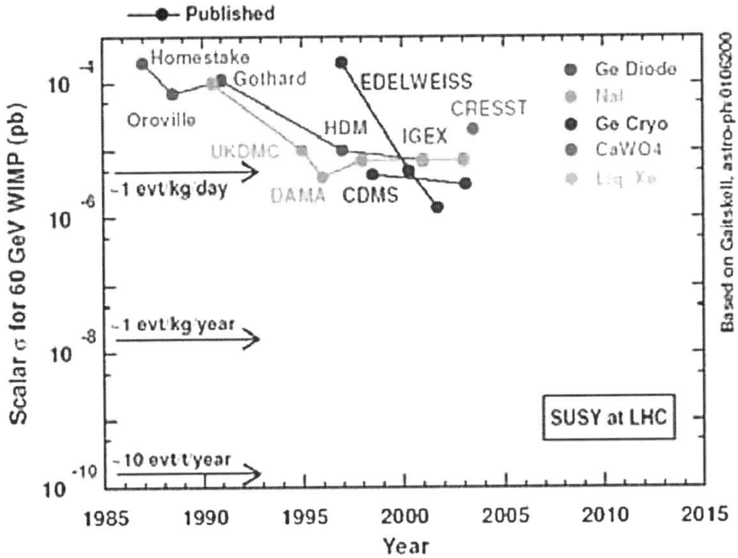

Fig. 8.

9. Future Sensitivity

There were many new estimates for the SUSY DM cross-section range given at the DM04 symposium [1, 9]. Note in these types of calculations that the most likely region of discovery is between 10^{-7} – 10^{-8} pb cross-section but that the signal could be as low as 10^{-9} – 10^{-10} pb. See also Ref. 1 for similar estimates.

While the next generation of detectors will likely reach 10^{-7} or even 10^{-8} pb (CDMS II, ZEPLIN II, Edelweiss II, etc.) there is no certainty that even 10^{-8} pb can be reached. For this case much larger detectors in the one ton range will be needed. Even if a tentative signal is observed at 10^{-8} a much larger detector will be needed to confirm this signal. Fig. 7 shows the current limit of CDMS II and projections for the future as an example [7].

A new, third-generation of detectors is being studied for this case. We consider the example of ZEPLIN IV/MAX here for such a detector [5]. In the case that a single one ton detector is to be constructed the detector will require some new concepts beyond that employed in the ZEPLIN II/III detectors. Of course the data from these detectors will be crucial to the understanding of how such a detector will work underground.

Acknowledgement

I wish to thank the organizers of the Dark Matter 2004 and Dark Matter 2006 meetings.

References

1. Most of the information for this report came from the results of the 7th Symposium for "Sources and Detection of Dark Matter and Dark Energy in the Universe", Nucl. Phys B (in press). See papers there for more information.
2. F. Zwicky, Helv. Phys Act 6 (1993), 110.
3. Clowe, D., Bradac, M., Gonzalez, A. H., Markevitch, M, Randall, S., Jones, C., & Zaritsky, D., "A Direct Empirical Proof of the Existence of Dark Matter" 2006, ApJ Letters, 648, 109.
4. See the references in D. Cline, "The Search for Dark Matter: Einstein's Cosmology and MOND", astro-ph/0510576.
5. D. Cline, "The Dark Universe: The Search for Dark Matter and the Nature of Dark Energy", in AIP Conference Proceedings Volume 809, Advanced Summer School in Physics 2005: Frontiers in Contemporary Physics EAV05, Oscar Rosas-Ortiz, Mauricio Carbajal, Omar Miranda, eds., Mexico City (Mexico), July 2005, pp. 11-22.
6. V. Sanglard, "Edelweiss II: Status and Future", in the Proceedings for the 7th Symposium on Sources and Detection of Dark Matter and Dark Energy in the Universe, in press.
7. D. Akerib et al, "CDMS, Supersymmetry and Extra Dimensions", in the Proceedings for the 7th Symposium on Sources and Detection of Dark Matter and Dark Energy in the Universe, in press.
8. G. Gilmore et al, "Observed Properties of Dark Matter: Dynamical Studies of dSph Galaxies", in the Proceedings for the 7th Symposium on Sources and Detection of Dark Matter and Dark Energy in the Universe, in press.
9. D.B. Cline, ed., *6th UCLA Symposium on Sources and Detection of Dark Matter and Dark Energy in the Universe, Marina del Rey, USA, 18-20 February 2005*, New Astron. Rev. 49, Number 2-6, (2005), 334 pp.

DARK MATTER IN SUGRA, STRINGS AND BRANES

PRAN NATH

Department of Physics, Northeastern University, Boston, MA 02115, USA
** E-mail: nath@neu.edu*

A brief review is given of dark matter in SUGRA, strings and branes. For SUGRA models the implications of Yukawa coupling unification on dark matter are discussed in the light of $g-2$ and $b \to s\gamma$ constraints. A brief discussion is given of the dark matter in orbifold string compactifications under constraints of modular invariance and radiative breaking of the electroweak symmetry. Finally a new candidate for dark matter - an extra-weakly interacting massive particle or an XWIMP- is discussed. Such dark matter can arise in a wide class of models, including the Stueckelberg extensions of MSSM, in U(1) extensions of MSSM with off diagonal kinetic energy, and possibly in a wider class of models which may have a string/D-brane origin. Satisfaction of the relic density of XWIMPs consistent with WMAP is also discussed.

Keywords: Dark matter, sugra unificaiion, strings, XWIMPS.

1. Introduction

In this talk we give a brief overview of the leading candidate for cold dark matter[1] in a broad class of models which includes the supergravity unified models (SUGRA), string models and brane models. In mSUGRA[2] the neutralino[3] arises as the least massive supersymmetric particle (LSP) over a broad region of the parameter space,[4] and with R parity it can be a candidate for cold dark matter. The nature of dark matter depends critcally on the type of soft breaking and it is this aspect that differentiates the nature of dark matter in SUGRA models, vs dark matter in strings and branes. For instance, in orbifold compactifications of the heteoric string, the constraints of modular invariance play an important role in the nature of soft breaking, and hence on the nature of neutralino dark matter. Aside from the neutralino, even in supersymmetry there exist other possible candidates such as the gravitino, and the sneutrino. Specifially, the gravitino possibility has resurfaced recently.[5] In addition other dark matter candidates abound such as the Kaluza-Klein states[6] in extra dimensional models, Q balls,[7] as well as a variety of other possibilities.[8–10] To this list we will add a new candidate - an extra weakly interacting massive particle or an XWIMP.[11] Such a particle can arise in a wide class of models including the Stueckelberg extensions of MSSM,[12–16] the $U(1)_X$ extensions with off diagonal kinetic energy terms,[11,17–19] and possibly in a broader class of models with string/D-brane origins.

The outline of the rest of the paper is as follows. In Sec. 2 we give an overview of dark matter in SUGRA models with focus on inclusion of Yukawa coupling unification. It is known that in this case the constraints of $b \to s + \gamma$, and the sign of the μ parameter play a central role in the analysis. Since the sign of μ is closely tied to the sign of the supersymmetric contribution to $g_\mu - 2$, the analysis in this case is highly constrained. In Sec. 3 we discuss dark matter in heterotic string models, and point out that here $\tan\beta$ in no longer a free parameter but is determined by the twin constraints of radiative breaking of the electroweak symmetry and by the constraints of modular invariance. In Sec. 4 we discuss the new candidate for dark matter- an extra-weakly interacting massive particle, and show that despite its extra weak interactions, it is possible to satisfy the WMAP relic density constraints.

2. Dark matter in SUGRA unifiication

Extensive investigations of the relic density in SUGRA models exist in the literature (for recent works see Refs. 20–22) and for the mSUGRA case this implies exploration of the parameter space spanned by the four conventional parameters: $m_0, m_{1/2}, A_0, \tan\beta$. Significant regions exist where WMAP constraints can be satisfied and these regions can be broadly labeled as the stau co-annilation region, where coannhiliation between the LSP and the stau produces relic densities consistent with WMAP, the resonance region where relic density constraint is satisfied due to the Higgs poles, and the hyperbolic branch (HB)[23] where the relic density is satisfied due to a relatively large higgsino component of the LSP. These analyses are very sensitive to the nature of soft breaking and thus the inclusion of non-universalites in the soft breaking produce significant effects in the analysis. Non-universalites can appear in a variety of ways but these must be consistent with the flavor changing neutral current constraints. Such constraints can be respected by inclusion of non-universalites in the Higgs sector and in the gaugino sector, and several analyses exist where the Higgs sector[24] and the gaugino sector[25–30] non-universalites have been included in dark matter analyses. In addition to the above analyses of dark matter are also sensitive to CP phases and we discuss this later.

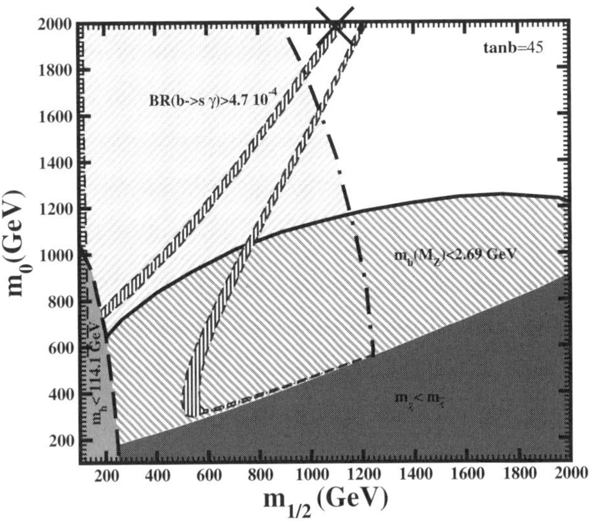

Fig. 1. Analysis of the neutralino relic density with the $b-\tau$ Yukawa unification constraints in the $m_0 - M_{1/2}$ plane when the soft terms are universal and real with $\mu < 0$ ($\theta_\mu = \pi$), $A_0 = 0$, $m_t = 178$ GeV, and $tan\beta = 45$. Areas contoured by the dashed line has a neutralino relic density which is inside the WMAP bounds.[31,32] The area above the solid line predicts $m_b(M_Z) > 3.10$ GeV, while the area inside the dashed (dot-dashed) line is excluded by the lower bound on m_h (the upper bound on $BR(b \to s+\gamma)$. On the lower dark area $m_\chi > m_{\tilde{\tau}}$ while on the upper side EWSB is not achieved. The thin dashed line indicates $m_{\chi^+} = 103$ GeV. Taken from Ref. 33

For the remainder of this section we will focus on the analysis of dark matter including the effects of Yukawa coupling unification. Thus in many unified models the b and τ Yukawa couplings are related at high scales, e.g., $h_b \simeq h_\tau$. These are evolved down to the electroweak scale and constrained by experiment $m_\tau = 1.7463$ GeV, and $2.69\text{GeV} < m_b(M_Z) < 3.10\text{GeV}$. In some models one extends the above to a full Yukawa unification $h_\tau \simeq h_b \simeq h_t$. We note in passing that while $b-t-\tau$ unification in $SO(10)$ models with 10 plet of Higgs for breaking of the electroweak symmetry requires a large $\tan\beta$, a large $\tan\beta$ is not a necessity when the symmetry breaking is achieved via alternative schemes (see, e.g., Ref. 34).

The sign of μ plays a central role in $b-\tau$ and $b-t-\tau$ unification. It is known that the supersymmetric contribution to $a_\mu = (g_\mu - 2)/2$ is directly correlated to the sign of μ[35] and further that one can infer this sign experimentally.[36] A positive μ is favored by the $b \to s\gamma$ since the parameter space of msugra and of other models is less stringently constrained by it.[37] On the other hand $b-\tau$ unfication seems to favor a negative μ.[38,39] This is so because $b - \tau$ unification requires a negative loop contribution to the b quark mass, and the major contribution to this loop comes from the gluino exchange and its sign depends on $\mu\tilde{m}_g$. Many analyses exist which have worked to resolve this problem.[40–43] One such possibility is to use non-universalities.[43] For example, in $SU(5)$ the gaugino masses transform like the symmetric product of (24×24) which can be expanded as $1+24+75+200$. Now for the singlet case one gets universality of gaugino masses at the GUT scale. However, if one considers the 24 plet case, then M_3, M_2, M_1 are in the ratio (-2, 3, 1), and one finds a relative minus sign for the gluino mass term. This gaugino mass pattern switches the sign of $(\mu.m_{\tilde{g}})$ from positive to negative, which allows one to achieve a $b - \tau$ unification with a positive μ. Experimentally, the most recent analyses appear to favor a positive μ.[44] Still we discuss both μ signs for Yukawas unification and dark matter.[a] For positive μ the analysis of dark matter is given in Ref. 43, while for negative μ it is given in Ref. 33, and an exhibition of one case is given in Fig. 1 using the WMAP relic density of Refs. 31, 32. The analysis of Fig. 1 shows that Yukawa unification constraint allows for a satisfaction of the relic density constraint consistent with WMAP.[31,32] CP phases also have a strong effect on dark matter[33,46] but here one needs to pay attention to the satisfaction of the edm constraints which, however, can be satisfied even for large phases via the cancellation mechanism.[47] In passing we draw attention to the recent improved analyses of $b \to s + \gamma$ which, as discussed above, has an important effect on dark matter. These improved analyses include the next to leading order (NLO) corrections enhanced by large $\tan\beta$.[48,51] The most recent analysis of Ref. 51 additionally includes the full arrary of CP violating effects and these results will be useful in future dark matter analyses. In unified models there is also a strong link between proton stability and dark matter[49,50] a topic which is beyond the scope of this talk.

3. Dark matter in heretotic string models

As in SUGRA models, dark matter in hererotic string models is largely governed by the soft breaking. In orbifold compactifications one typically has a large radius- small radius symmetry, so that $R \to \alpha'/R$, and more generally an $SL(2, Z)$ modular invariance symmetry. There are many analyses which have looked at soft breaking with modular invariance[52,53] and their implications (For a sample see, Refs. 54–57 and references therein). In

[a]For an analysis of dark matter with quasi-Yukawa unification see Ref. 45

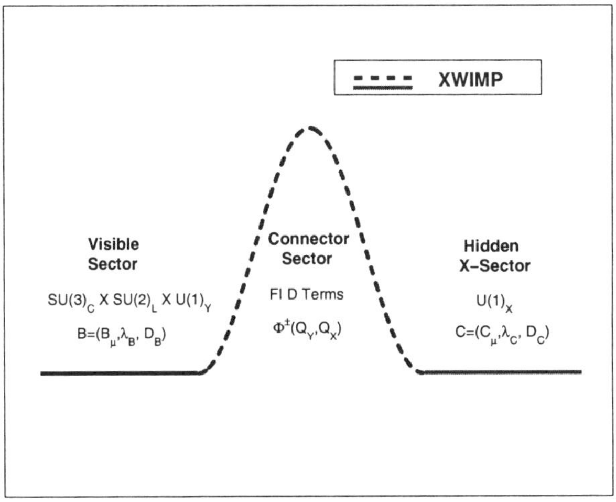

Fig. 2. An exhibition of the mechanism by which an XWIMP is generated. A neutral fermion from the connector sector merges with a gaugino of the hidden sector to form an XWIMP whose interactions with the MSSM particles are suppressed. Taken from Ref. 11

the analysis of Refs. 55, 56, the further constraint of radiative breaking of the electroweak symmetry was utilized. With the twin constraints of modular invariance and radiative breaking, $\tan\beta$ is no longer a free parameter but is determined in terms of α_{string} and the remaining soft parameters. Using this constraint an analysis of dark matter for $\mu > 0$ with WMAP constraint implies an upper limit on particle masses which lie within reach of the LHC, and further the neutralino-proton cross sections lie within reach of the dark matter detectors.[56]

4. Extra weakly interacting dark matter

Recently a new candidate for dark matter has been proposed whose couplings with matter are weaker than weak or extra weak.[11] The mechanism for its generation is exhibited in Fig. 2, and it depends on three sectors: a visible sector where the particles of MSSM reside, a hidden sector where fields do not have any direct interactions with the fields in the visible sector and a third sector[58] which connects both to the fields of the visible sector and of the hidden sector. We label this third sector, the connector sector. A spontaneous breaking in the connector sector produces mixings between the neutralino states in the visible sector and the neutralino states in the hidden sector. If the LSP of the hidden sector (XLSP) lies lower than the LSP of the visible sector, then the XLSP becomes the LSP of the entire system. This is the XWIMP. As an example we consider the $U(1)_X$ Stueckelberg extension[12–16] where one has the $U(1)_X$ gauge fields C_μ, λ_C, D_C. The connector sector is chosen to be a pair cf chiral fields $\phi^{\pm}$[59,60] which are charged under both $U(1)_X$ and $U(1)_Y$. We add to the mix a Fayet-Illiopoulos term[61]

$$\mathcal{L}_{\mathcal{FI}} = \xi_X D_C + \xi_Y D_B \tag{1}$$

Elimination of the D terms then leads to the potential

$$V = \frac{g_X^2}{2}\Big(Q_X|\phi^+|^2 - Q_X|\phi^-|^2 + \xi_X\Big)^2 + \frac{g_Y^2}{2}\Big(Y_\phi|\phi^+|^2 - Y_\phi|\phi^-|^2 + \xi_Y\Big)^2 \tag{2}$$

which on minimization gives $\langle\phi^+\rangle = 0$, $\langle\phi^-\rangle \neq 0$. After spontaneous breaking two new mass parameters emerge. $M_1 = \sqrt{2}g_X Q_X < \phi^- >$, $M_2 = \sqrt{2}g_Y Y_\phi < \phi^- >$, or alternately one can choose the new parameters to be M_1 and $\epsilon = M_2/M_1$. The hidden sector and the connector sector provide two additional neutralino fields, χ_S, λ_X which together with the four neutral fermionic states in the MSSM, $\lambda_Y, \lambda_3, \tilde{h}_1, \tilde{h}_2$ give a set of six Majorana spinors. In the basis $((\chi_S, \lambda_X); (\lambda_Y, \lambda_3, \tilde{h}_1, \tilde{h}_2))$ one finds a 6×6 Majorana mass matrix, whose eigen states we label $((\xi_1^0, \xi_2^0); (\chi_1^0, \chi_2^0, \chi_3^0, \chi_4^0))$, where χ_a^0 (a=1,2,3,4) are essentially the four neutralinos that appear in MSSM, and ξ_α^0, $(\alpha = 1, 2)$ are the new states.

Following the same procedure used to constrain extra dimensions[62] one can put also a constraint on ϵ, and one finds $\epsilon < 0.06$.[11] The LEP and the Tevatron data put further constraints on the model.[11] reducing further the size of ϵ. Because of the smallness of ϵ, the interactions of ξ_α^0 with the visible sector fields will be suppressed by an additional factor of ϵ and thus the interactions of the XWIMP with the visible sector will be extra weak. Further, if the mass of either ξ_1^0 or ξ_2^0 is smaller than the mass of all of the MSSM particles, then the XLSP will be the LSP of the entire system and hence with R parity conservation, a candidate for dark matter. It was shown in[11] that a similar situation arises for the $U(1)_X$ extension of MSSM with gauge kinetic energy mixing[17–19] involving a mixing of the $U(1)_X$ and the $U(1)_B$ gauge field strengths. Supersymmetric version of this model leads to a form of the neutralino mass matrix which, although different in form, also produces an XWIMP. Further, one may conceive of other models where considerations of the type outlines above lead to an XWIMP and a candidate for cold dark matter. Models of the type discussed here may have a string/D-brane origin.[64]

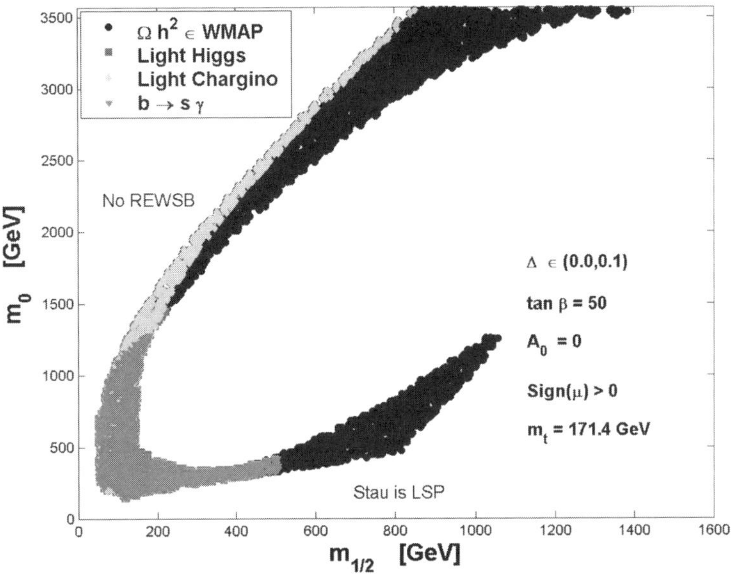

Fig. 3. The allowed parameter space in the $m_0 - m_{1/2}$ plane, under the 1σ WMAP3 constraint[63] in extended mSUGRA for the case $A_0 = 0$, $\tan\beta = (30, 50)$ (upper,lower), sign($\mu > 0$), $m_t = 171.4$ GeV, $m_{1/2} \in (0, 1.5)$TeV and $m_0 \in (0, 3.5)$TeV, and Δ in the range (0.0, 0.1). Regions eliminated by the light chargino mass constraint, by the light Higgs mass constraint, and by the $b \to s + \gamma$ constraint are also exhibited. Taken from Ref. 11

A priori it would appear that an XWIMP would not satisfy the relic density constraints as its extra weak interactions would not allow for an efficient annihilation of excess CDM in the early universe. However, if any of the MSSM particles lie close to the XWIMP mass, then the XWIMPS can co-annihilate quite efficiently and the relic density constraints can be satisfied. Thus, for example, if the NLSP turns out to be $\chi^0 = \chi_1^0$, and if $\Delta = (m_{\chi^0} - m_{\xi^0})/m_{\chi^0} > 0$, then the XWIMP $\xi^0 = \xi_1^0$ relic density is given by[11]

$$\sigma_{eff}(XWIMP) \simeq \sigma_{eff}^{\chi^0\chi^0}\left(\frac{Q}{1+Q}\right) \tag{3}$$

where $Q \sim (1+\Delta)^{3/2}e^{-x_f\Delta}$. If $\Delta << 1$ then $Q = O(1)$ and an efficient co-annihilation of XWIMPS can occur. A detailed analysis was carried out in Ref. 11 using the packages of Refs. 65, 66 and of Refs. 67, 68. In the analysis we have also examined the effects of the Z' pole on the relic density using the techniques of integration over the Z' pole in thermal averaging.[69,70]

The result of the analysis[11] is exhibited in Fig. 3 using the constraints of the three year WMAP data[63] and using the parameter space of mSUGRA accessible at the LHC.[71] Further, in the analysis of Ref. 11 the sensitivity to the top mass[72] was also investigated and the analysis found to have significant variations with a one σ variation in the top mass.

Acknowledgements

The work was supported in part by the U.S. National Science Foundation under the grant NSF-PHY-0546568.

References

1. For a recent work on the direct empirical evidence for the existence of dark matter in galaxies see, D. Clowe et.al., astro-ph/0608407.
2. A. H. Chamseddine, R. Arnowitt and P. Nath, Phys. Rev. Lett. **49** (1982) 970; R. Barbieri, S. Ferrara and C. A. Savoy, Phys. Lett. B **119** (1982) 343; L. J. Hall, J. D. Lykken and S. Weinberg, Phys. Rev. D **27** (1983) 2359; P. Nath, R. Arnowitt and A. H. Chamseddine, Nucl. Phys. B **227** (1983) 121. For a recent review see, P. Nath, arXiv:hep-ph/0307123.
3. H. Goldberg, Phys. Rev. Lett. **50**, 1419 (1983); J. R. Ellis, J. S. Hagelin, D. V. Nanopoulos, K. A. Olive and M. Srednicki, Nucl. Phys. B **238**, 453 (1984).
4. R. Arnowitt and P. Nath, Phys. Rev. Lett. **69**, 725 (1992).
5. J. R. Ellis, J. E. Kim and D. V. Nanopoulos, Phys. Lett. B **145**, 181 (1984); T. Moroi, H. Murayama and M. Yamaguchi, Phys. Lett. B **303**, 289 (1993); G. F. Giudice, A. Riotto and I. Tkachev, JHEP **9911**, 036 (1999); M. Fujii and T. Yanagida, Phys. Rev. D **66**, 123515 (2002); M. Bolz, A. Brandenburg and W. Buchmuller, Nucl. Phys. B **606**, 518 (2001) [arXiv:hep-ph/0012052]. J. L. Feng, A. Rajaraman and F. Takayama, Phys. Rev. Lett. **91**, 011302 (2003) J. R. Ellis, K. A. Olive, Y. Santoso and V. C. Spanos, Phys. Lett. B **588**, 7 (2004) K. Jedamzik, K. Y. Choi, L. Roszkowski and R. Ruiz de Austri, JCAP **0607**, 007 (2006) ; F. D. Steffen, JCAP **0609**, 001 (2006).
6. H. C. Cheng, J. L. Feng and K. T. Matchev, Phys. Rev. Lett. **89**, 211301 (2002); G. Servant and T. M. P. Tait, Nucl. Phys. B **650**, 391 (2003); K. Hsieh, R. N. Mohapatra and S. Nasri, Phys. Rev. D **74**, 066004 (2006).
7. A. Kusenko and P. J. Steinhardt, Phys. Rev. Lett. **87**, 141301 (2001).
8. L. Covi, L. Roszkowski, R. Ruiz de Austri and M. Small, JHEP **0406**, 003 (2004) ; A. Brandenburg and F. D. Steffen, JCAP **0408**, 008 (2004).
9. J. McDonald, Phys. Rev. D **50**, 3637 (1994) ; C. P. Burgess, M. Pospelov and T. ter Veldhuis, Nucl. Phys. B **619**, 709 (2001).
10. P. H. Frampton, arXiv:astro-ph/0607391.
11. D. Feldman, B. Kors and P. Nath, "Extra-weakly interacting dark matter," arXiv:hep-ph/0610133.
12. B. Körs and P. Nath, Phys. Lett. B **586**, 366 (2004).

13. B. Körs and P. Nath, JHEP **0412** (2004) 005 [hep-ph/0406167].
14. B. Körs and P. Nath, JHEP **0507**, 069 (2005); hep-ah/0411406.
15. D. Feldman, Z. Liu and P. Nath, Phys. Rev. Lett. **97**, 021801 (2006).
16. D. Feldman, Z. Liu and P. Nath, hep-ph/0606294. To appear in JHEP.
17. B. Holdom, Phys. Lett. B **166**, 196 (1986).
18. J. Kumar and J. D. Wells, hep-ph/0606183.
19. W. F. Chang, J. N. Ng and J. M. S. Wu, hep-ph/0608068.
20. J. R. Ellis, K. A. Olive, Y. Santoso and V. C. Spanos, Phys. Lett. B **565**, 176 (2003) U. Chattopadhyay, A. Corsetti and P. Nath, Phys. Rev. D **68** (2003) 035005 H. Baer and C. Balazs, JCAP **0305**, 006 (2003) H. Baer, C. Balazs, A. Belyaev, T. Krupovnickas and X. Tata, JHEP **0306**, 054 (2003) A. B. Lahanas and D. V. Nanopoulos, Phys. Lett. B **568** (2003) 55 C. Munoz, Int. J. Mod. Phys. A **19** (2004) 3093 M. Battaglia, A. De Roeck, J. R. Ellis, F. Gianotti, K. A. Olive and L. Pape, Eur. Phys. J. C **33**, 273 (2004) H. Baer, A. Mustafayev, S. Profumo, A. Belyaev and X. Tata, JHEP **0507** (2005) 065 H. Baer, A. Mustafayev, E. K. Park, S. Profumo and X. Tata, JHEP **0604**, 041 (2006) J. R. Ellis, K. A. Olive, Y. Santoso and V. C. Spanos, JHEP **0605** (2006) 063; V. Khotilovich, R. Arnowitt, B. Dutta and T. Kamon, Phys. Lett. B **618** (2005) 182 G. Belanger, F. Boudjema, S. Kraml, A. Pukhov and A. Semenov, Phys. Rev. D **73** (2006) 115007 G. B. Gelmini and P. Gondolo, Phys. Rev. D **74** (2006) 023510. ; T. Nihei, Phys. Rev. D **73** (2006) 035005; S. F. King and J. P. Roberts, hep-ph/0603095; D. A. Demir, M. Frank and I. Turan, Phys. Rev. D **73**, 115001 (2006).
21. V. Barger, C. Kao, P. Langacker and H. S. Lee, Phys. Lett. B **600**, 104 (2004) S. Nakamura and D. Suematsu, hep-ph/0609061; S. Profumo and A. Provenza, hep-ph/0609290; A. Provenza, M. Quiros and P. Ullio, hep-ph/0609059; B. de Carlos and J. R. Espinosa, Phys. Lett. B **407**, 12 (1997).
22. A. Djouadi, M. Drees and J. L. Kneur, Phys. Lett. B **624**, 60 (2005) A. Djouadi, M. Drees and J. L. Kneur, JHEP **0603**, 033 (2006).
23. K. L. Chan, U. Chattopadhyay and P. Nath, Phys. Rev. D **58**, 096004 (1998).
24. P. Nath and R. Arnowitt, Phys. Rev. D **56**, 2820 (1997) J. R. Ellis, K. A. Olive and Y. Santoso, Phys. Lett. B **539**, 107 (2002) J. R. Ellis, T. Falk, K. A. Olive and Y. Santoso, Nucl. Phys. B **652**, 259 (2003) H. Baer, A. Mustafayev, S. Profumo, A. Belyaev and X. Tata, JHEP **0507**, 065 (2005).
25. A. Corsetti and P. Nath, Phys. Rev. D **64**, 125010 (2001).
26. A. Birkedal-Hansen and B. D. Nelson, Phys. Rev. D **64**, 015008 (2001).
27. G. Belanger, F. Boudjema, A. Cottrant, A. Pukhov and A. Semenov, Nucl. Phys. B **706**, 411 (2005).
28. D. G. Cerdeno and C. Munoz, JHEP **0410**, 015 (2004).
29. U. Chattopadhyay and D. P. Roy, Phys. Rev. D **68**, 033010 (2003).
30. A. Bottino, F. Donato, N. Fornengo and S. Scopel, Phys. Rev. D **70**, 015005 (2004).
31. C. L. Bennett *et al.*, Astrophys. J. Suppl. **148**, 1 (2003).
32. D. N. Spergel *et al.* [WMAP Collaboration], Astrophys. J. Suppl. **148**, 175 (2003)

33. M. E. Gomez, T. Ibrahim, P. Nath and S. Skadhauge, Phys. Rev. D **72** (2005) 095008.
34. K. S. Babu, I. Gogoladze, P. Nath and R. M. Syed, Phys. Rev. D **72**, 095011 (2005). ; Phys. Rev. D **74**, 075004 (2006)
35. T. C. Yuan, R. Arnowitt, A. H. Chamseddine and P. Nath, Z. Phys. C **26**, 407 (1984) ; D. A. Kosower, L. M. Krauss and N. Sakai, Phys. Lett. B **133**, 305 (1983). ; J. L. Lopez, D. V. Nanopoulos and X. Wang, Phys. Rev. D **49**, 366 (1994) ; T. Moroi, Phys. Rev. D **53**, 6565 (1996) ; T. Ibrahim and P. Nath, Phys. Rev. D **61**, 095008 (2000); ; Phys. Rev. D **62**, 015004 (2000).
36. U. Chattopadhyay and P. Nath, Phys. Rev. Lett. **86**, 5854 (2001). See also L. L. Everett, G. L. Kane, S. Rigolin and L. T. Wang, Phys. Rev. Lett. **86**, 3484 (2001) ; J. L. Feng and K. T. Matchev, Phys. Rev. Lett. **86**, 3480 (2001) ; E. A. Baltz and P. Gondolo, Phys. Rev. Lett. **86**, 5004 (2001) ; S. Komine, T. Moroi and M. Yamaguchi, Phys. Lett. B **506**, 93 (2001) ; T. Ibrahim, U. Chattopadhyay and P. Nath, Phys. Rev. D **64**, 016010 (2001) ; J. R. Ellis, D. V. Nanopoulos and K. A. Olive, Phys. Lett. B **508**, 65 (2001) ; R. Arnowitt, B. Dutta, B. Hu and Y. Santoso, Phys. Lett. B **505**, 177 (2001) ; S. P. Martin and J. D. Wells, Phys. Rev. D **64**, 035003 (2001) ; H. Baer, C. Balazs, J. Ferrandis and X. Tata, Phys. Rev. D **64**, 035004 (2001).
37. P. Nath and R. Arnowitt, Phys. Rev. Lett. **74**, 4592 (1995) ; F. M. Borzumati, M. Drees and M. M. Nojiri, Phys. Rev. D **51**, 341 (1995) ; H. Baer, M. Brhlik, D. Castano and X. Tata, Phys. Rev. D **58**, 015007 (1998).
38. D. M. Pierce, J. A. Bagger, K. T. Matchev and R. j. Zhang, Nucl. Phys. B **491**, 3 (1997) ; H. Baer, M. A. Diaz, J. Ferrandis and X. Tata, Phys. Rev. D **61**, 111701 (2000).
39. W. de Boer, M. Huber, A. V. Gladyshev and D. I. Kazakov, Eur. Phys. J. C **20**, 689 (2001).
40. H. Baer and J. Ferrandis, Phys. Rev. Lett. **87**, 211803 (2001).
41. T. Blazek, R. Dermisek and S. Raby, Phys. Rev. Lett. **88**, 111804 (2002); Phys. Rev. D **65**, 115004 (2002); R. Dermisek, arXiv:hep-ph/0108249.
42. S. Komine and M. Yamaguchi, Phys. Rev. D **65**, 075013 (2002).
43. U. Chattopadhyay and P. Nath, Phys. Rev. D **65**, 075009 (2002); U. Chattopadhyay, A. Corsetti and P. Nath, Phys. Rev. D **66**, 035003 (2002).
44. For a recent review see, D. Stockinger, arXiv:hep-ph/0609168; J. Haestier, S. Heinemeyer, W. Hollik, D. Stockinger, A. M. Weber and G. Weiglein, arXiv:hep-ph/0610318.
45. C. Pallis and M. E. Gomez, arXiv:hep-ph/0303098; M. E. Gomez, G. Lazarides and C. Pallis, Phys. Lett. B **487**, 313 (2000).
46. U. Chattopadhyay, T. Ibrahim and P. Nath, Phys. Rev. D **60**, 063505 (1999) ; T. Falk, A. Ferstl and K. A. Olive, Astropart. Phys. **13**, 301 (2000) ; M. Argyrou, A. B. Lahanas, D. V. Nanopoulos and V. C. Spanos, Phys. Rev. D **70**, 095008 (2004) ; T. Nihei and M. Sasagawa, Phys. Rev. D **70**, 055011 (2004).
47. T. Ibrahim and P. Nath, Phys. Lett. B **418**, 98 (1998); ; Phys. Rev. D **58**, 111301 (1998) ; T. Falk, A. Ferstl and K. A. Olive, Phys. Rev. D **59**, 055009 (1999) ; M. Brhlik, G. J. Good and G. L. Kane, Phys. Rev. D **59**, 115004

(1999) ; A. Bartl, T. Gajdosik, W. Porod, P. Stockinger and H. Stremnitzer, Phys. Rev. D **60**, 073003 (1999) ; E. Accomando, R. Arnowitt and B. Dutta, Phys. Rev. D **61**, 115003 (2000). For a review see, T. Ibrahim and P. Nath, arXiv:hep-ph/0107325.

48. G. Degrassi, P. Gambino and G. F. Giudice, JHEP **0012**, 009 (2000); H. Baer, M. Brhlik, D. Castano and X. Tata, Phys. Rev. D **58**, 015007 (1998); M. Carena, D. Garcia, U. Nierste and C. E. M. Wagner, Phys. Lett. B **499**, 141 (2001); D. A. Demir and K. A. Olive, Phys. Rev. D **65**, 034007 (2002); A. Buras, P. Chankowski, J. Rosiek and L. Slawianowska, Nucl. Phys. **B659** (2003) 3-78; G. Degrassi, P. Gambino and P. Slavich, hep-ph/0601135.
49. R. Arnowitt and P. Nath, Phys. Lett. B **437**, 344 (1998).
50. P. Nath and P. F. Perez, arXiv:hep-ph/0601023.
51. M. E. Gomez, T. Ibrahim, P. Nath and S. Skadhauge, Phys. Rev. D **74**, 015015 (2006).
52. A. Brignole, L. E. Ibanez, C. Munoz and C. Scheich, Z. Phys. C **74** (1997) 157.
53. P. Binetruy, M. K. Gaillard and B. D. Nelson, Nucl. Phys. B **604** (2001) 32.
54. G. L. Kane, J. D. Lykken, S. Mrenna, B. D. Nelson, L. T. Wang and T. T. Wang, Phys. Rev. D **67**, 045008 (2003).
55. P. Nath and T. R. Taylor, Phys. Lett. B **548**, 77 (2002).
56. U. Chattopadhyay and P. Nath, Phys. Rev. D **70**, 096009 (2004). ; arXiv:hep-ph/0411364.
57. P. Binetruy, Y. Mambrini and E. Nezri, Astropart. Phys. **22**, 1 (2004).
58. P. Nath, Phys. Rev. Lett. **76**, 2218 (1996).
59. G. R. Dvali and A. Pomarol, Phys. Rev. Lett. **77**, 3728 (1996).
60. B. Körs and P. Nath, Nucl. Phys. B **711**, 112 (2005).
61. P. Fayet and J. Iliopoulos, Phys. Lett. B **51**, 461 (1974).
62. P. Nath and M. Yamaguchi, Phys. Rev. D **60**, 116004 (1999).
63. D. N. Spergel *et al.*, astro-ph/0603449.
64. I. Antoniadis, E. Kiritsis, J. Rizos and T. N. Tomaras, Nucl. Phys. B **660**, 81 (2003); P. Anastasopoulos, M. Bianchi, E. Dudas and E. Kiritsis, arXiv:hep-th/0605225; P. Anastasopoulos, T. P. T. Dijkstra, E. Kiritsis and A. N. Schellekens, arXiv:hep-th/0605226, and references therein.
65. G. Belanger, F. Boudjema, A. Pukhov and A. Semenov, [hep-ph/0607059]; Comput. Phys. Commun. **174**, 577 (2006); Comput. Phys. Commun. **149** (2002) 103.
66. A. Djouadi, J. L. Kneur and G. Moultaka, hep-ph/0211331.
67. J. Edsjo and P. Gondolo, Phys. Rev. D **56**, 1879 (1997); J. Edsjo, M. Schelke, P. Ullio and P. Gondolo, JCAP **0304**, 001 (2003); P. Gondolo, J. Edsjo, P. Ullio, L. Bergstrom, M. Schelke and E. A. Baltz, JCAP **0407**, 008 (2004).
68. F. E. Paige, S. D. Protopopescu, H. Baer and X. Tata, hep-ph/0312045.
69. K. Griest and D. Seckel, Phys. Rev. D **43**, 3191 (1991).
70. P. Nath and R. Arnowitt, Phys. Rev. Lett. **70**, 3696 (1993).
71. H. Baer, C. Balazs, A. Belyaev, T. Krupovnickas and X. Tata, JHEP **0306**, 054 (2003). ; H. Baer, A. Belyaev, T. Krupovnickas and J. O'Farrill, JCAP **0408**, 005 (2004) ; Y. Mambrini and E. Nezri, hep-ph/0507263; ; M. M. No-

jiri, G. Polesello and D. R. Tovey, JHEP **0603**, 063 (2006) ; E. A. Baltz, M. Battaglia, M. E. Peskin and T. Wizansky, hep-ph/0602187.

72. P. Nath, J. z. Wu and R. Arnowitt, Phys. Rev. D **52**, 4169 (1995) ; B. Allanach, S. Kraml and W. Porod, arXiv:hep-ph/0207314. ; J. R. Ellis, K. A. Olive, Y. Santoso and V. C. Spanos, Phys. Rev. D **69**, 095004 (2004) ; M. E. Gomez, T. Ibrahim, P. Nath and S. Skadhauge, Phys. Rev. D **70**, 035014 (2004); ; J. R. Ellis, S. Heinemeyer, K. A. Olive and G. Weiglein, JHEP **0502**, 013 (2005) ; B. C. Allanach and C. G. Lester, Phys. Rev. D **73**, 015013 (2006) ; A. Djouadi, M. Drees and J. L. Kneur, JHEP **0603**, 033 (2006).

STUDYING SUPERSYMMETRY WITH DARK MATTER EXPERIMENTS

DAN HOOPER

Theoretical Astrophysics,
Fermi National Accelerator Laboratory,
Batavia, IL 60510, USA
E-mail: dhooper@fnal.gov

A wide range of techniques have been developed to search for particle dark matter, including direct detection, indirect detection, and collider searches. The prospects for the detection of neutralino dark matter is quite promising for each of these three very different methods. Looking ahead to a time in which these techniques have successfully detected neutralino dark matter, we explore the ability of these observations to determine the parameters of supersymmetry. In particular, we focus on the ability of direct and indirect detection techniques to measure the parameters μ and m_A. We find that μ can be much more tightly constrained if astrophysical measurements are considered than by LHC data alone. In supersymmetric models within the A-funnel region of parameter space, we find that astrophysical measurements can determine m_A to roughly ± 100 GeV precision.

Keywords: Dark Matter, Supersymmetry

1. SUSY Parameters From Direct and Indirect Detection

If low scale supersymmetry exists in nature, it will likely be discovered in the next few years at the Large Hadron Collider (LHC), or perhaps even earlier at the Tevatron. In many supersymmetric scenarios, the LHC can identity the presence of the lightest neutralino in the form of missing energy in the cascade decays of squarks and/or gluinos. At approximately the same time, direct and indirect searches for dark matter will be reaching the level of sensitivity needed to discover neutralinos in a wide range of supersymmetric models.

Collider and astrophysical experiments tell us very different things about the nature of supersymmetry. The LHC is likely to reveal the approximate masses of a number of superpartners, including squarks, gluinos, the lightest neutralino, and in some cases sleptons and heavier neutralinos. Other properties of the supersymmetric model will remain unconstrained by the LHC, however. In particular, the composition of the lightest neutralino (the mixture of bino, wino, and higgsino components), and therefore its couplings, will be very difficult to deduce at a hadron collider. In contrast, the cross sections relevant to astrophysical dark matter experiments depend critically on the composition of the lightest neutralino.

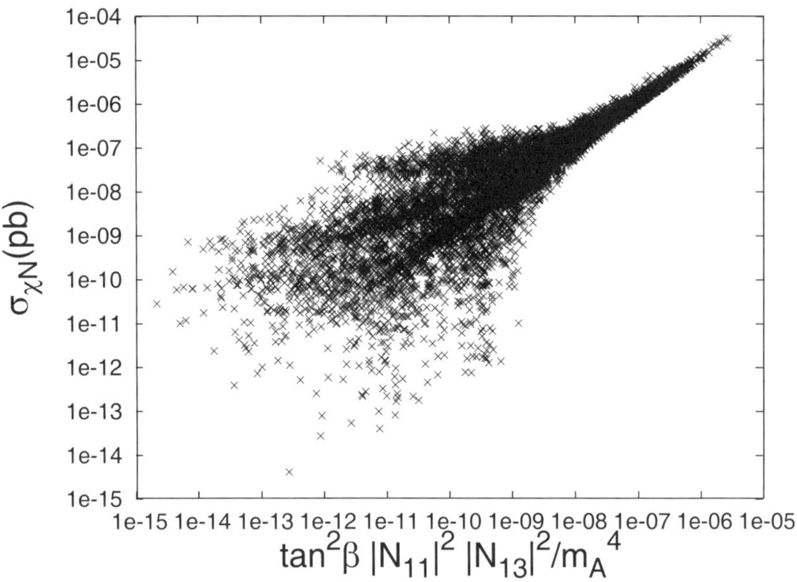

Fig. 1. The relationship between the quantity $|N_{11}|^2|N_{13}|^2 \tan^2\beta/m_A^4$ and the spin-independent neutralino-nucleon elastic scattering cross section.

In Fig. 1, we show how the rates in astrophysical dark matter experiments depend on the composition of the lightest neutralino. In this figure, we compare the scalar neutralino-nucleon elastic scattering cross section (which determines the rate in direct dark matter experiments) to the quantity: $\tan^2\beta|N_{11}|^2|N_{13}|^2/m_A^4$. Here, m_A is the mass of the CP-odd Higgs boson in the MSSM. The $|N|^2$'s are defined by the following: $\chi_1^0 = N_{11}\tilde{B} + N_{12}\tilde{W}^3 + N_{13}\tilde{H}_1 + N_{14}\tilde{H}_2$. The correlation shown is quite tight for models with large cross sections, which are produced through diagrams which exchange a heavy Higgs boson coupling to b and s quarks.[1]

Neutralinos scattering with the Sun can become gravitationally bound and accumulate in the Sun's core, eventually annihilating to produce high-energy neutrinos. The rate of those neutrinos being observed at a kilometer-scale neutrino telescope such as IceCube or KM3 is shown in Fig. 2 as a function of the quantity $|N_{13}|^2 - |N_{14}|^2$. This correlation comes from the χ^0-χ^0-Z coupling which largely determines the capture rate in the Sun in nearly all observable models.[2]

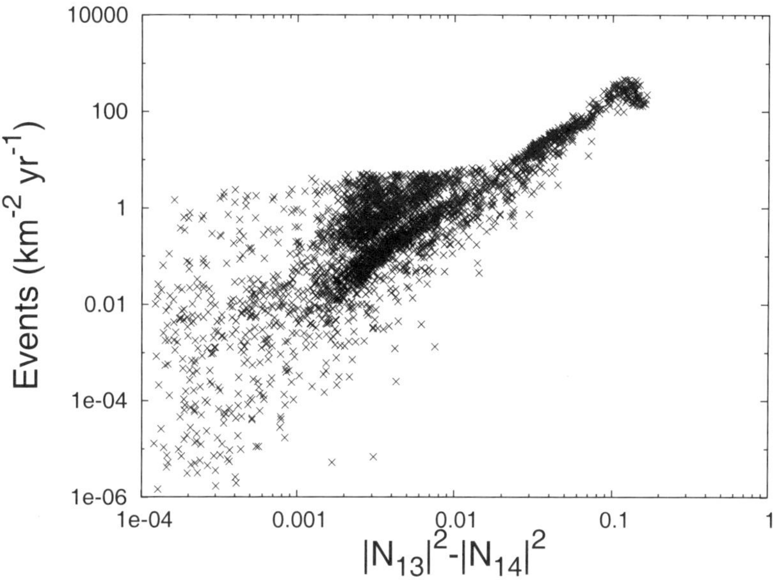

Fig. 2. The rate of neutrinos detected in a kilometer scale neutrino telescope, such as IceCube, from neutralino annihilations in the Sun as a function of the quantity $|N_{13}|^2 - |N_{14}|^2$. A constraint 100 times more stringer than the current CDMS bound has been applied in anticipation of increased sensitivity from direct detection experiments in the coming few years.

2. An Example of Constraining μ

We will next consider a benchmark model in order to study how direct and indirect detection information can aid in determining the composition of the lightest neutralino, and therefore the parameter μ. We will adopt the following benchmark: $M_2 = 472.9$ GeV, $\mu = 619.2$ GeV, $\tan\beta = 50.6$, $m_A = 396.5$ GeV, and a common sfermion mass scale of 2130 GeV. This model generates a thermal relic abundance of $\Omega_{\chi^0}h^2$ =0.098, which is consistent with the measurements of WMAP. The lightest neutralino in this model

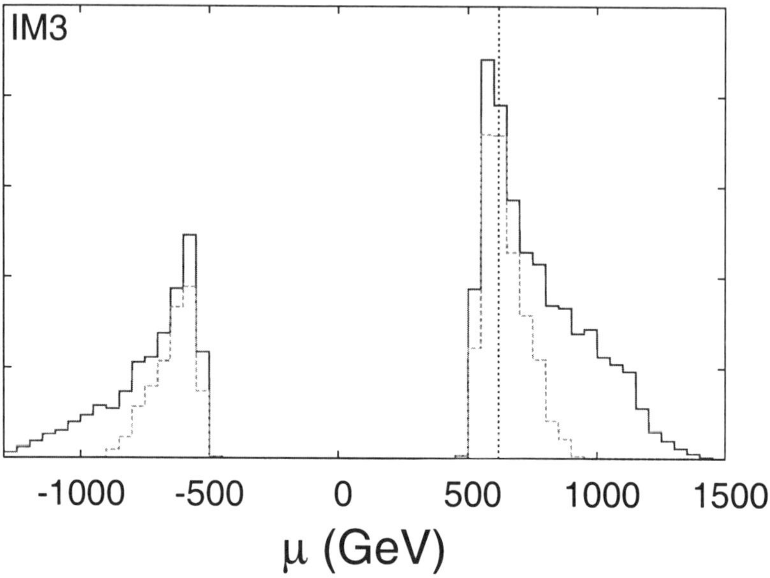

Fig. 3. The distribution of μ in the models found by our scan which match the features that would be observed at the LHC in our first benchmark model (upper curve). The low curve only contains those models which also match the features that would be observed by direct and indirect detection experiments.

is 99.1% bino, with a small higgsino admixture. Although IceCube/KM3 will not be sensitive to this model, the elastic scattering cross section of 9.6×10^{-9} pb will be within the reach of next generation direct detection experiments.

A number of supersymmetric particles will be within the reach of the LHC in this model. In addition to quarks, gluinos, and the lightest neutralino, $\tan\beta$ is large enough, and m_A small enough, to be determined though the channel $H/A \to \tau^+\tau^-$ at the LHC. In the upper curve of Fig. 3, we plot the distribution of SUSY models found with our scan that match the following constraints from LHC measurements: $m_{\tilde{q}} = 2130$ GeV $\pm 30\%$, $m_{\chi^0} = 236$ GeV $\pm 10\%$, $m_A = 397 \pm 3$ GeV, and $\tan\beta = 51 \pm 15\%$, in addition to the relic density constraint. The lower curve on this figure, in addition to the LHC measurements, is constrained by direct and indirect detection measurements. Whereas the LHC alone can only constrain $1500 \gtrsim |\mu| \gtrsim 500$ GeV, the inclusion of astrophysical measurements can be used to tighten the upper bound considerably: $800 \gtrsim |\mu| \gtrsim 500$ GeV.

3. An Example of Measuring m_A

In models with low to moderate $\tan\beta$ or somewhat heavy m_A, heavy Higgs bosons will be beyond the reach of the LHC. In models in which neutralinos annihilate primarily through A-exchange, however, known as the A-funnel region, the value of m_A can potentially be determined through dark matter observables.

Adopting a second benchmark model ($M_2 = 550.8$ GeV, $\mu = 1318$ GeV, $\tan\beta = 6.8$, $m_A = 580.2$ GeV, and a common sfermion mass scale of 2239 GeV), we plot in Fig. 4 the ability of astrophysical measurements to determine the value of m_A. In this case, we find a determination of $m_A \approx 620 \pm 100$ GeV. Since the sfermions are all heavy in this model, to avoid being overproduced in the early universe, either the neutralino must contain a substantial higgsino component, or be annihilating near the A-resonance. Direct detection experiments will find in this model $\sigma_{\chi N} \lesssim 10^{-10}$ pb, constraining $|N_{13}|^2 + |N_{14}|^2$ to be less than ~0.1. Through this constraint, $2m_{\chi^0} \approx m_A$ can be determined.

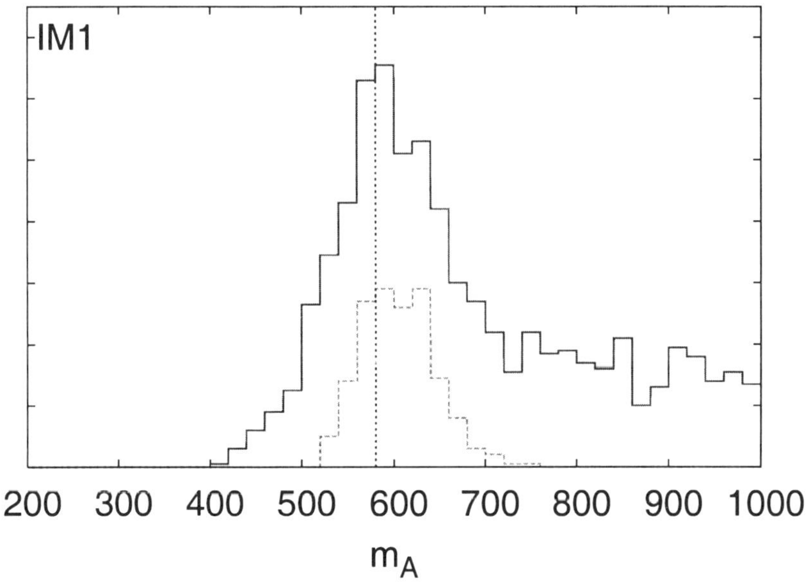

Fig. 4. The distribution of m_A in the models found by our scan which match the features that would be observed at the LHC in our second benchmark model (upper curve). The low curve only contains those models which also match the features that would be observed by direct and indirect detection experiments.

Acknowledgment

These proceedings have been based on work done with Andrew Taylor[3] and follow closely an earlier set of conference proceedings.[4] DH is supported by the Department of Energy and by NASA grant NAG5-10842.

References

1. M. Carena, D. Hooper and P. Skands, Phys. Rev. Lett. **97**, 051801 (2006) [arXiv:hep-ph/0603180].
2. F. Halzen and D. Hooper, Phys. Rev. D **73**, 123507 (2006) [arXiv:hep-ph/0510048].
3. D. Hooper and A. Taylor, hep-ph/0607086.
4. D. Hooper, Talk given at ULCA dark matter symposium, Marina Del Rey (2006), arXiv:hep-ph/0606289.

IS THE DARK MATTER INTERPRETATION OF THE EGRET GAMMA RAY EXCESS COMPATIBLE WITH ANTIPROTON MEASUREMENTS?

W. DE BOER, I. GEBAUER, C. SANDER, M. WEBER and V. ZHUKOV

Institut für Experimentelle Kernphysik
University of Karlsruhe, Physikhochhaus
Postfach 6980, 76128 Karlsruhe, Germany
E-mail: wim.de.boer@cern.ch

The diffuse galactic EGRET gamma ray data show a clear excess for energies above 1 GeV in comparison with the expectations from conventional galactic models. This excess shows all the features expected from Dark Matter WIMP Annihilation: a)it is present and has the same spectrum in all sky directions, not just in the galactic plane, as expected for WIMP annihilation b) it shows an interesting substructure in the form of a doughnut shaped ring at 14 kpc from the centre of the galaxy, where a ring of stars indicated the probable infall of a dwarf galaxy. From the spectral shape of the excess the WIMP mass is estimated to be between 50 and 100 GeV, while from the intensity the halo profile is reconstructed, which is shown to explain the peculiar change of slope in the rotation curve at about 11 kpc (due to the ring of DM at 14 kpc).

Recently it was claimed by Bergström et al. that the DM interpretation of the EGRET gamma ray excess is excluded by the antiproton fluxes, since in their propagation model with isotropic diffusion the flux of antiprotons would be far beyond the observed flux. However, the propagation can be largely anisotropic, because of the convection of particles perpendicular to the disc and inhomogeneities in the local environment. It is shown that anisotropic propagation can reduce the antiproton yield by an order of magnitude, while still being consistent with the B/C ratio. Therefore it is hard to use antiprotons to search for *light* DM particles, which yield a similar antiproton spectrum as the background, but the antiprotons are a perfect means to tune the many degenerate parameters in the propagation models.

1. Introduction

Cold Dark Matter (CDM) makes up 23% of the energy of the universe, as deduced from the WMAP measurements of the temperature anisotropies in the Cosmic Microwave Background, in combination with data on the Hubble expansion and the density fluctuations in the universe.[1] The nature of the CDM is unknown, but one of the most popular explanation for it is the neutralino, a stable neutral particle predicted by Supersymmetry.[2] The neutralinos are spin 1/2 Majorana particles, which can annihilate into pairs of Standard Model (SM) particles. The stable decay and fragmentation products are neutrinos, photons, protons, antiprotons, electrons and positrons. From these, the protons and electrons disappear in the sea of many matter particles in the universe, but the photons and antimatter particles may be detectable above the background, generated by particle interactions. Searches for the stable products of dark matter annihilation (DMA) (so-called indirect Dark Matter detection) have been actively pursued, see e.g the review by Bergström[3] or more recently by Bertone, Hooper and Silk.[4]

In previous papers we showed that the so-called EGRET excess of diffuse galactic gamma rays[5] exhibits all the features of DMA.[6–8] However, Bergström et al.[9] claimed that the DM interpretation of the EGRET gamma ray excess is incompatible with the antiproton fluxes, since in their propagation model with isotropic diffusion (based on DarkSusy) the flux of antiprotons would be far beyond the observed flux. In this contribution it is shown that more realistic propagation models could solve this problem.

After summarizing the DMA interpretation of the excess of gamma rays, the expected antiproton flux will be discussed based on the GALPROP propagation model[10] after implementing and retuning its parameters and taking into account the expected anisotropic propagation and the clumpiness of the gas distribution.

2. Gamma Rays from Dark Matter Annihilation

The thermally averaged annihilation cross section for any thermal relic is known from the inverse proportionality to the relic density.[2] This cross section comes out to be that of a Weakly Interacting Massive Particle (WIMP). The dark matter annihilation (DMA) is expected to yield predominantly mono-energetic quark pairs, since the kinetic energy is negligible for CDM. From the hadronization of the quarks one expects a large flux of gamma rays from the decay of the π_0 mesons, typically several tens of gamma rays per annihilation with energies of several GeV. The gamma ray spectrum from mono-energetic quarks has been studied in detail in the hadronization of quarks produced at electron-positron colliders. The DMA gamma ray spectrum is considerably harder than the background spectrum, which originates from inelastic scattering of cosmic rays (CR) on the interstellar gas. If the CR spectra are known and uniform in the Galaxy, the shape of the background gamma rays is known from accelerator experiments. The

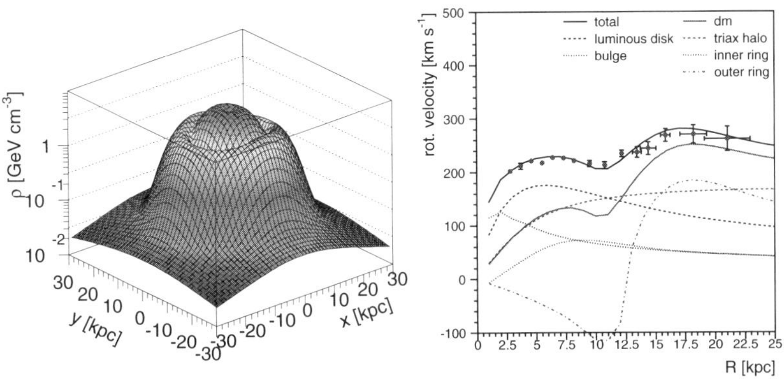

Fig. 1. CDM halo density in the galactic disk with the two ringlike substructures at 4 and 14 kpc (left) and the corresponding rotation curve (right). Adapted from Ref. 7.

absolute value of neither the background nor DMA is known, because of the large uncertainties in density of the interstellar medium, CR density and CDM density. Therefore the obvious way to search for DMA is to leave the absolute normalizations of the background and DMA contributions free and fit only the shapes of the background and DMA for a given sky direction.

Experimentally, the spectral shape of the diffuse Galactic gamma rays has been measured with the EGRET satellite; we use the EGRET data in the range 0.07 to 10 GeV in 8 energy bins. For the relative amount of electron- and nucleon-induced gamma rays the estimates from real data, as implemented in the publicly available "conventional" GALPROP model,[10] can be used, so one has only one normalization constant for the background instead of separate ones for the different background components.

Comparing the background with the EGRET data shows that above 1 GeV there is a large deficit of gamma rays, which reaches more than a factor of two towards the Galactic centre.[5] Fitting the background together with the DMA, yields a perfect fit in *all* sky directions for a CDM particle mass around 60 GeV.[6,7] The shape fit automatically finds from the free normalizations the relative amount of background and DMA. Furthermore, the results are consistent with Supersymmetry.[8] From the amount of excess in 180 independent sky directions one can reconstruct the CDM profile, which in turn can be used to calculate the rotation curve. The result explains the hitherto unexplained change of slope in the outer rotation curve,[7] as shown in Fig. 1. For the halo profile one is only interested in the relative contributions in the various sky directions, so here all experimental errors cancel,

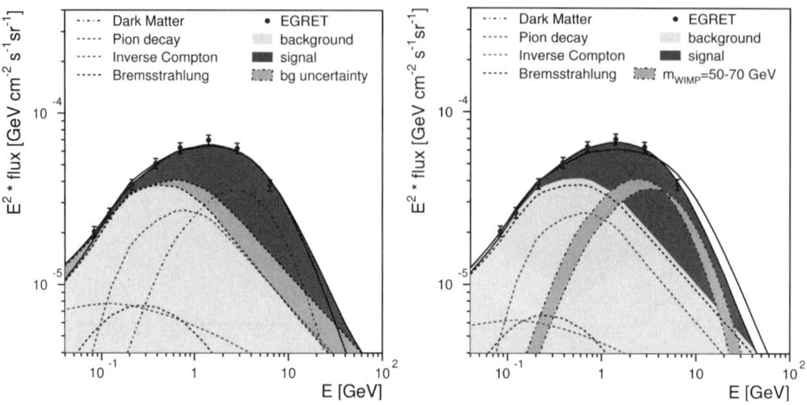

Fig. 2. The EGRET gamma ray spectrum fitted with CDM annihilation for a 60 GeV WIMP mass (left). The blue shaded area indicates the uncertainty from the shape of the CR spectra, which is dominated by the uncertainty in solar modulation (see text). On the right hand side the variation of the WIMP mass between 50 and 70 GeV is shown (blue shaded area), which is the range allowed by the EGRET data assuming that the locally observed CR spectra are representative for our Galaxy.

since the EGRET satellite does not care in which direction it measures. The EGRET errors, as discussed in Ref. 11 are not relevant, since we are not interested in predicting absolute gamma ray fluxes, but only fit the shapes with a free normalization. In this case only the point-to-point errors are relevant. Furthermore, since the systematic errors are dominating, every data point has approximately the same weight, so changing the total error does not change the solution for the minimum of the χ^2 distribution; larger errors only decrease its value. But in the fits of around 1400 data points the $\chi^2/d.o.f$ is already well below 1 with a 7% point-to-point error, suggesting that these errors for a shape fit are already overestimated.

Uncertainties from the background, which are dominated by the solar modulation uncertainty in the primary CR spectra, are shown in Fig. 2. Note that the solar modulation depletes the CR spectrum at low energies, but fitting the shape translates this into an uncertainty mainly at high energy. This is simply because at low energy the spectrum is almost purely background, so the expectations are effectively "normalized" to this low energy data by the fit, whatever the shape of the spectrum.

Clearly the uncertainties in the background shape cannot explain the excess, if one assumes that the locally observed CR spectra are representative for the spectra outside the heliosphere after correcting for solar modulation.

For nuclei the spectra are expected to be indeed similar everywhere because the diffusion is fast compared with energy loss times. So local variations of the spectra or intensities, as proposed in Ref. 12 to explain the excess, seem to us unlikely, especially since this needs in addition rather strong breaks in the CR injection spectra in order to keep the gamma rays below 1 GeV the same, but only increase the high energy gamma rays. Furthermore, these breaks are only applied to protons, not to other nuclei in order to maintain the B/C ratio. Also the "fresh" harder source component $\propto E^{-2}$ instead of $\propto E^{-2.7}$ is not expected to yield a significant effect, since this is only a small fraction of the total CR density. This is obvious for older Galaxies, where the amount of CR escaping to outer space (with an escape time of $\mathcal{O}(10^7 - 10^8)$ y) should be equal to the amount of generated CR (with a source life time of $\mathcal{O}(10^4 - 10^5)$ y), so the fresh component should be of $\mathcal{O}(10^{-2} - 10^{-3})$. That the shape of CR in the steady state is similar everywhere, is confirmed by a numerical solution of the diffusion equation, as used in GALPROP. For this "conventional" model of CRs having everywhere the spectrum of the locally observed one, the WIMP mass is rather well constrained (50-70 GeV), as shown on the right hand side of Fig. 2.

In summary, the gamma rays play a very special role for indirect CDM searches, since they point back to the source and are independent of propagation models. Therefore the gamma rays provide a perfect means to reconstruct the intensity (halo) profile of the CDM by observing the intensity of the gamma ray emissions in the various sky directions. This halo profile can in turn be used to check the shape of the rotation curve, thus providing a direct link between the excess of the gamma rays and the strongest evidence for CDM, the rotation curve.

3. Antiproton Fluxes

Contrary to gamma rays the charged particles change their direction by the interstellar magnetic fields, energy losses and scattering. Therefore one needs a detailed propagation model to calculate the amount of particles which will arrive from the source to the detector. Charged particles usually make a random walk process by changing their direction through interaction with the galactic magnetic field, which is thought to have a larger random (turbulent) component in the interstellar space. But galactic winds may lead to a strong convective transport of these magnetic turbulences perpendicular to either side of the galactic plane, which take the charged particles with them to outer space,[13] thus leading to strong anisotropic propagation. Furthermore, the transport and production of charged parti-

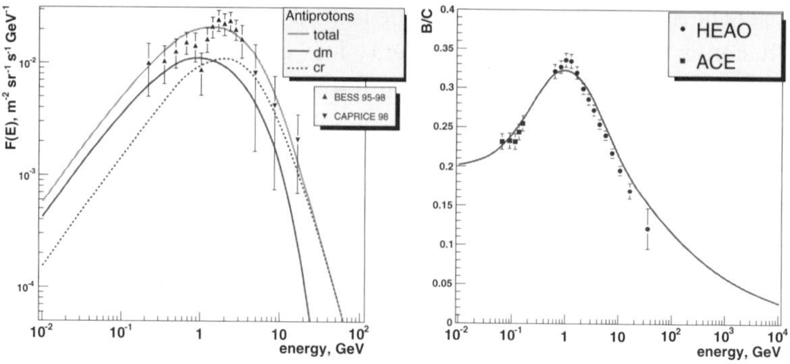

Fig. 3. The antiproton fluxes and the B/C ratio from the modified GALPROP code including DMA and anisotropic propagation. Note that roughly half of the antiprotons are coming from DMA, as for the gamma rays above 1 GeV, while DMA does not contribute to the B/C ratio.

cles can be strongly influenced by the neighbourhood of our solar system with its underdense local bubble and overdense clouds ("local fluff") and magnetic walls with as much as five orders of magnitude drop in the diffusion coefficient in the heliosphere.[14] Up to now all these highly uncertain details have not been studied. The common propagation models simply assume an isotropic diffusion in all directions in a large volume before the particles escape to outer space.

By increasing the convection perpendicular to the disk and implementing the local bubble, the local clouds and "magnetic walls" with slow diffusion in the solar neighbourhood one can change the antiproton flux by an order of magnitude, while still being consistent with the B/C ratio, as shown in Fig. 3. The DMA contribution explains the traditional EGRET "excess" of gamma rays without the need for assuming that the locally observed CR spectra are different from the CRs in the rest of the Galaxy. Here the GALPROP model was used after including DMA and retuning the diffusion and convection parameters. Traditionally these parameters have been determined by the B/C ratio and the cosmic clocks, like the $^{10}Be/^{9}Be$ ratio. The diffusion coefficient needed for the B/C ratio required a large halo with a distance of z=4 kpc to get a long enough trapping time for the cosmic clocks. This traps also the antiprotons from DMA, thus leading to the solution from Ref. 3 using DarkSusy. This results can be reproduced with GALPROP, if the isotropic diffusion dominates. In our case the antiprotons are blown to outer space by convection, which overtakes diffusion a few hundred parsec above the disk. In this case of large convection a one to two orders of magnitude smaller diffusion coefficient is needed, which is much closer to the values used in heliospheric propagation models.[14] The large B/C ratio is obtained by the local fluff with a size of 5 pc in the local bubble. Note that most of the molecular gas is concentrated in large molecular clouds, which occupy only a few % of the volume. These clouds act as localized sources of all secondary particles and are particularly strong if nearby, since the flux decreases as $1/r^2$. Naively one expects that if locally a large amount of secondary boron nuclei are produced (by fragmentation of CNO and heavier atoms on the gas), one expects a correspondingly large amount of secondary antiprotons. This is not true, since the latter require CR protons with an energy above 10 GeV (due to threshold effects), while for rigidities of CNO nuclei below 10 GeV the fragmentation cross sections just increase. Therefore changing the injection spectra of primary particles below 10 GeV by 10% or changing the energy dependence of the diffusion constants immediately changes the antiproton/B ratio for rigidities around 1 GeV by a factor of a few.

In summary, recent claims that the antiproton fluxes exclude the DMA interpretation of the EGRET excess should be considered in the light of the limitations of DarkSusy, which uses a simple analytical solution of the diffusion equation with unrealistic smooth gas distributions and isotropic diffusion coefficients. In order to allow for anisotropies in gas distributions, convection velocities and diffusion coefficients one has to resort to numerical solutions of the diffusion equation, as implemented in GALPROP after suitable modifications for DMA, non-equidistant grids and anisotropic diffusion and convection, i.e. D(r,z) and V(r,z). In the latter case a consistent flux of local CR spectra, antiproton fluxes, B/C ratio and gamma rays can be obtained.

4. Summary and Outlook

In summary, the excess of EGRET diffuse gamma rays has all the properties expected for DMA. Especially the excess has the shape expected for the annihilation of 60 GeV WIMPs and the distribution of the excess over the sky is in perfect agreement with the shape of the rotation curve of our Galaxy, which is the hallmark of a DMA signal.

Objections against the DMA interpretation of the EGRET excess[9] concerning a too high antiproton flux should be considered in the light of their simple diffusion model. Our preliminary investigations show that a more realistic propagation model can reduce the antiproton flux by more than an order of magnitude. Therefore it is hard to use antiprotons to search for *light* CDM particles, which yield a soft antiproton spectrum similar to the background. However, the antiprotons are perfect to tune the many parameters in more realistic propagation models, if the CDM halo is determined from the gamma rays.

Future data on high energy gamma rays (GLAST satellite) and high energy charged particles (space experiments PAMELA and AMS) will be of great interest in order to see if this picture of DMA is confirmed, while direct DM detection experiments and the new hadron collider LHC may be able to determine independently the WIMP mass. If they all find a WIMP mass in the range suggested by the EGRET excess, this would be great.

Acknowledgements

Very helpful discussions with V. Florinski, V.Ptuskin, R. Schlickeiser and H. Völk about propagation uncertainties are acknowledged. This work was supported by the DLR (Deutsches Zentrum für Luft- und Raumfahrt).

References

1. D.N. Spergel et al, 2003, ApJS, 148, 175;
   C.L. Bennett et al., 2003, ApJS, 148, 1;
2. G. Jungman, M. Kamionkowski and K. Griest, Phys. Rep. **267** (1996) 195.
3. L. Bergström, Rept. Prog. Phys. **63** (2000) 793 [arXiv:hep-ph/0002126].
4. G. Bertone, D. Hooper and J. Silk, Phys. Rept. **405**, 279 (2005) [arXiv:hep-ph/0404175].
5. Hunter, S. D. *et al.*, Astrophysical Journal **481**, 205 (1997)
6. W. de Boer, New Astron. Rev., **49** (2005) 213; arXiv:hep-ph/0408166.
7. W. de Boer et al., A&A **444** (2005) 51.
8. W. de Boer, et al., Phys. Lett. B **636** (2006) 13 [arXiv:hep-ph/0511154].
9. L. Bergström, J. Edsjo, M. Gustafsson and P. Salati, JCAP **0605** (2006) 006, [arXiv:astro-ph/0602632]
10. A. W. Strong and I. V. Moskalenko, Astrophys. J. **509**, 212 (1998); [arXiv:astro-ph/9807150].
    I. V. Moskalenko and A. W. Strong, Astrophys. Space Sci. **272** (2000) 247; [arXiv:astro-ph/9908032].
11. I. V. Moskalenko, S. W. Digel, T. A. Porter, O. Reimer and A. W. Strong, [arXiv:astro-ph/0609768].
12. A. W. Strong, I. V. Moskalenko and O. Reimer, Astrophys. J. **613**, 962 (2004); [arXiv:astro-ph/0406254].
13. V.S. Ptuskin, et al., A&A **321** (1997)434.
14. V. Florinski, G.P. Zank, N.V. Pogorelov, JORA**108** 12282003.

STATUS OF THE EDELWEISS II EXPERIMENT

F. SCHWAMM FOR THE EDELWEISS COLLABORATION
CEA Saclay, DAPNIA/SPP, 91191 Gif-sur-Yvette, France

EDELWEISS is an experiment searching for dark matter WIMPs (weakly interacting massive particles) by measuring the heat and ionisation signals of low temperature germanium detectors. Following the successful conclusion of the first phase of EDELWEISS in the summer of 2003, involving a total of 1 kg of detector mass, phase II of the experiment was set-up in the Modane underground laboratory (LSM) and is now undergoing commissioning runs. It will operate 9 kg of detectors in a first step and 40 kg ultimately. A short summary of the results and conclusions from EDELWEISS I data and an update on the status of EDELWEISS II will be given in this article.

1. Introduction

EDELWEISS is a dark matter search experiment situated in the Laboratoire Souterrain de Modane in the Frejus road tunnel in the French Alps. In this laboratory the remaining flux of cosmic muons is reduced to 4 μ m^{-2} d^{-1} due to 1700 m of rock (4800 m w.e.) above the laboratory. The remaining neutron flux is about 1500 n m^{-2} d^{-1} (E > 1 MeV) making it an ideal environment for rare event searches such as EDELWEISS [1,2].

EDELWEISS uses high purity germanium cryodetectors to look for dark matter WIMPs (weakly interacting massive particles). The detectors are double-Schottky diodes that are operated at a temperature of 17 mK and fitted with NTD-Germanium thermistors. Each particle interacting with the detectors hence produces two signals, an ionisation signal and a heat signal, which are both read out and recorded in the experiment.

Since electrons and photons interact with the electrons in the detector material while neutrons and WIMPs interact only with the nuclei, the ratio of the respective ionisation and heat signals are different. This ratio, the quenching factor, is about a factor of three higher for electron recoil events than for nuclear recoils (for a precise discussion see [3]). This makes it possible to identify background events with electronic interaction and remove them from the data, thus significantly reducing the overall experimental background. Hence, EDELWEISS is a powerful tool in the search for dark matter WIMPs.

2. Results of the first phase of the EDELWEISS experiment

During the first phase of EDELWEISS (EDW I) three 320 g high purity germanium detectors with a diameter of 70 mm and a thickness of 20 mm (Fig. 1) were operated during 65 days of lifetime thus giving a total exposure of 62 kg d. The detectors were set up in a specially designed shielding and cryostat. The shielding consisted of a 30 cm neutron shield made of paraffin, 15 cm of lead and 10 cm copper. An additional inner layer of 7 cm of roman lead shielded the detectors from the electronic components within the cryostat. The entire set-up was continuously flushed with nitrogen gas to avoid radon accumulation in the system.

The cryostat was a dilution refrigerator type with a base temperature of 15 mK and an experimental volume of one litre, which was built inside the Fréjus laboratory using low radioactivity materials (Fig. 1).

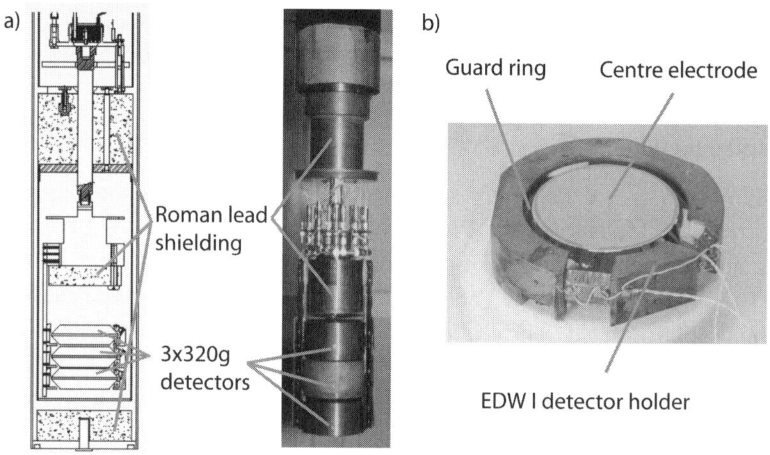

Fig. 1. a) Cryostat with mounted detectors of the EDW I experiment; b) 320 g EDELWEISS NTD/Ge detector in EDW I type holder.

The data taking took place from fall 2000 to Nov 2003 giving a total of 40 nuclear recoil events with energies above 15 keV, and only three events with recoil energies between 30 keV and 100 keV. The nuclear recoil detection efficiency for 100 GeV WIMPs was 50% at the threshold of 15 keV, 75% at 20 keV and reaching 90% at about 30 keV.

Comparisons of the measured nuclear recoil spectrum to simulated WIMP-spectra showed no agreement to any of these spectra, so at least a part of the spectrum is due to non-WIMP background (Fig. 2).

The analysis of this background can be found in [4]. It is shown to be mostly consistent with a surface pollution by radon and its descendents.

These surface events are an important source of background in the EDELWEISS experiment, since electron recoil events interacting close to the detector surfaces can have significantly reduced ionisation signals due to incomplete charge collection. This results in a reduced quenching, so that they can be mistaken for nuclear recoil events.

For the final analysis, the optimum interval method as described by S. Yellin in [5] was used to derive exclusion limits from the data (Fig. 3). A more detailed description of the results of EDW I can be found in [6].

Even though the results of EDW I already start exploring some of the more optimistic SUSY models, an increase of the experimental sensitivity by several orders of magnitude is necessary to explore a more significant part of the theoretical parameter space. Such a gain in sensitivity can only be reached by considerably increasing the detector mass and at the same time improving the background discrimination as much as possible, as is being done for the second phase of the EDELWEISS experiment, EDELWEISS II.

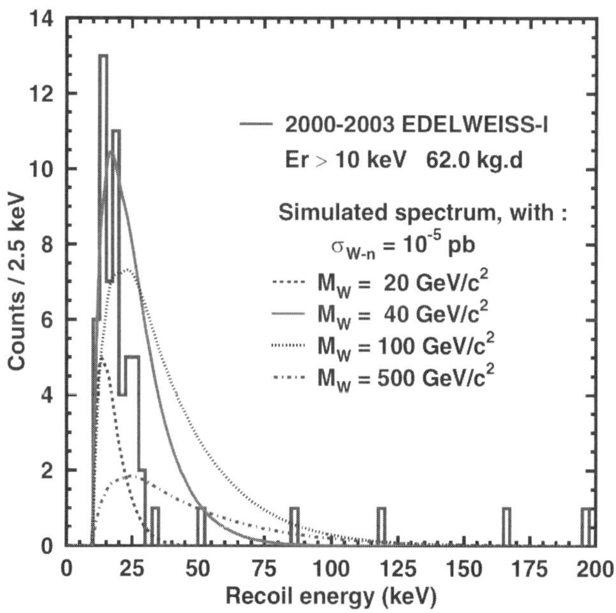

Fig. 2. Simulated WIMP spectra for different WIMP masses M_W and a scattering cross section of $\sigma_{W\text{-}N} = 10^{-5}$ pb. The measured spectrum shown is the nuclear recoil energy spectrum with E > 10 keV for the full data-set of 62 kg d of fiducial volume exposure.

3. Status of EDELWEISS II

The aim of EDELWEISS II (EDW II) is to increase the sensitivity of the experiment by two orders of magnitude to 2 10^{-8} pb. In order to be able to do this, at first 28 high purity germanium detectors with a mass of about 9 kg then up to 120 detectors with a total mass of 40 kg will be operated in a newly designed experimental system.

The dilution cryostat used in the experiment is of an inverted design, having the experimental chamber on the top of the structure (Fig. 4a). It is a nitrogen-free system, using pulse tubes for the first cooling stages. It also uses a Helium reliquefier, strongly reducing the Helium consumption of the system. The experimental volume is about 50 litres, giving enough space for the detectors to be arranged in a compact way, so that they can provide self-shielding for a further reduction of background events.

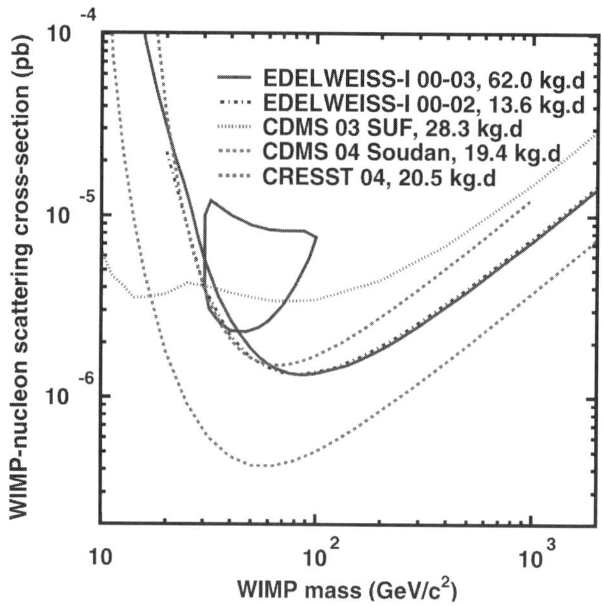

Fig. 3. EDELWEISS I spin independent exclusion limit (90% C.L.) for the full 62 kg d of fiducial exposure with $E_R > 15$keV (solid line). Other exclusion limits as given in the figure. The solid contour corresponds to the 3σ allowed region from the DAMA 1-4 annual modulation data [7].

The data acquisition system has been fully redesigned as well. In EDW II especially built 'bolometer boxes', modules that contain the complete 'warm' parts of the electronics for one detector, are used. They are situated outside the cryostat but within the shielding, connected to 'cold' FETs – the pre-amplifier stage – which are inside the cryostat and, via optical fibres, to the data acquisition computers outside the experimental area. Thanks to the modularity of the system, installation and removal of individual boxes, e.g. for maintenance, upgrades or repairs, can be easily done without interfering with the rest of the electronics system.

Most importantly several improvements of the shielding have taken place in comparison to EDW I. All materials used in the set-up of EDW II have been tested in dedicated low level counting experiments using HPGe detectors. The entire experiment, including the shielding, has been set-up in a class 10000 clean room, additionally, the experimental chamber of the cryostat can be put into a class 100 laminar flow screen for e.g. the mounting of the detectors. The clean room is supplied with deradonised air (Activity A< 0.1 Bq m^{-3}) delivered from a radon reduction facility [8].

The inner shielding of the cryostat consists now of 20 cm of lead, surrounded by 50 cm of polyethylene neutron shield and an active muon shield consisting of plastic scintillators with a total geometrical coverage of 99%.

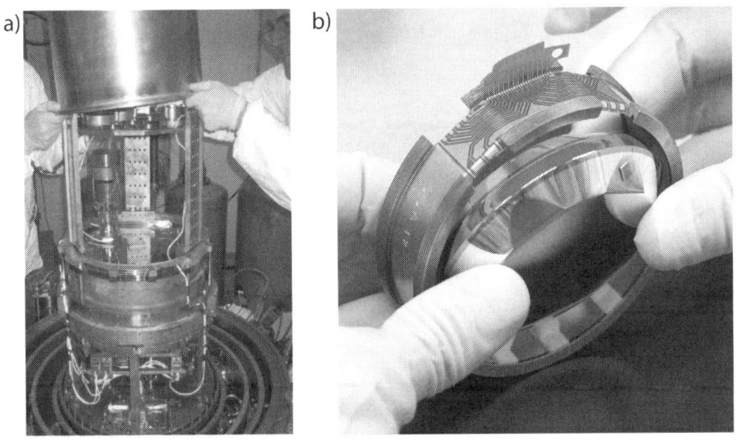

Fig. 4. a) Open experimental chamber of the reverse design cryostat of EDELWEISS II; b) EDELWEISS II detector in its new holder.

In the 28 detector stage, 21 detectors are of the same type and design as those already used in the first phase of the experiment. All detectors of this type are now having amorphous Ge or Si sublayers underneath the electrodes to improve the charge collection for surface events [9]. The used NTD thermistors are optimised in size and have homogenous working temperatures, providing a sub-keV energy resolution. The detector holders were fully redesigned, they now only consist of copper, the detectors are held in place by three teflon clamps (Fig. 4b).

In order to identify surface events the use of detectors that are sensitive to athermal phonons has been investigated. Prototype detectors using Nb/Si thin films to measure the athermal phonon signal have already been successfully tested in the EDW I experimental set-up [10]. They were able to reject 95% of surface events with a fiducial volume reduction of only 10% (Fig. 5).

To evaluate the long term performance of these detectors in comparison to the NTD-type detectors, seven of them will be operated in the 28 detector stage of EDW II.

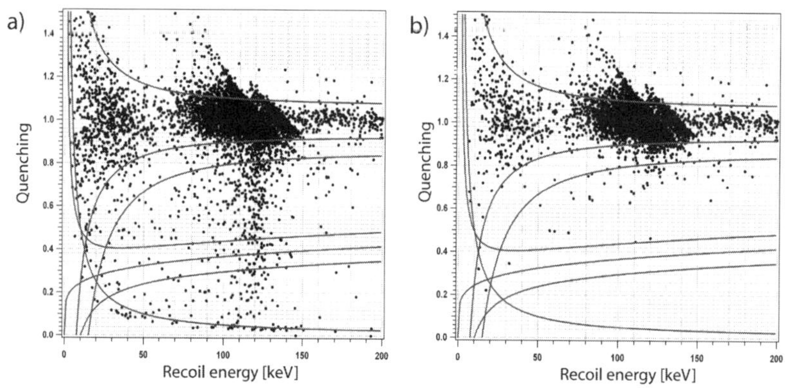

Fig. 5. Efficiency of Nb/Si thin film electrodes in identifying surface events: a) Sum spectrum of ^{57}Co and ^{137}Cs measurements including all surface events; b) The same spectrum after removal of all surface events identified by the Nb/Si electrodes.

At present EDW II is undergoing commissioning runs with eight germanium detectors, six of the 320 g Ge/NTD type in different variants of the new holders and two Nb/Si detectors with masses of 200 g and 400 g respectively. Aim of these runs is to evaluate and minimise the level of microphonics in the system, test the new holders and connectors of the detectors and the new electronics and data acquisition system.

4. Conclusion

The validation stage of EDW II is currently in progress. After the completion of initial data taking runs, the 28 detector phase will begin in early 2007, which will already provide considerably increased sensitivity in comparison to EDW I.

After a successful run of the 28 detector phase, the 120-detector stage is expected to be approved and funded in 2007. In this final phase, EDELWEISS II will reach its projected sensitivity of 2 10^{-8} pb.

The EDELWEISS experiment is therefore another step towards the EURECA ton-scale experiment [11].

Acknowledgements

This work has been partially supported by the EEC Applied Cryodetector Network under contract HPRN-CT-2002-00322 and the ILIAS integrating activity under contract RII3-CT-2003-506222.

References

1. V. Chazal et al., Astropart. Phys 9 (1998) 163
2. G. Chardin and G. Gerbier, Proc of the 4th Int. Workshop on Dark Matter, World Scientific, Singapore, 2003, p. 470
3. O. Martineau et al., NIM A 530 (2004) 426
4. S. Fiorucci et al., astro-ph/0610821
5. S. Yellin, Phys. Rev. D 66 (2002) 032005
6. V. Sanglard et al., Phys. Rev. D 71 (2005) 122002
7. R. Bernabei et al., Phys. Lett. B 480 (2000) 23
8. X. Sarazin et al., Nucl. Phys. B (Proc. Suppl.) 143 (2005) 221
9. X.F. Navick et al., Physica B 329-333 (2003) 1616
10. A. Juillard et al., NIM A 559 (2006) 393
11. Hans Kraus et al., these proceedings

XENON

MASAKI YAMASHITA
ON BEHALF OF THE XENON COLLABORATION
Physics Department and Astrophysics Lab, Columbia University,
New York, NY 10027, USA
** E-mail: yamashita@astro.columbia.edu*
http://xenon.astro.columbia.edu

The XENON experiment aims at the direct detection of dark matter in the form of Weakly Interacting Massive Particles (WIMPs) via their elastic scattering off Xe nuclei. The final detector will have a fiducial mass of 1000 kg, distributed in 10 independent liquid xenon time projection chambers (TPCs). Such an experiment will be able to probe the lowest interaction cross-section predicted by SUSY models. The TPCs are operated in dual (liquid/gas) phase, to allow a measurement of nuclear recoils down to < 10 keV energy, via simultaneous detection of the ionization, through secondary scintillation in the gas, and primary scintillation in the liquid. The distinct ratio of primary to secondary scintillation for nuclear recoils from WIMPs (or neutrons), and for electron recoils from background, is used for the event-by-event discrimination. As part of the R&D phase, we built a first XENON module (XENON10) with 15 kg fiducial mass and installed it underground, at the Laboratori Nazionali del Gran Sasso (LNGS), on March 2006. XENON10 has accumulated an exposure of more than 30 live days, operating with quite stable condition. A preliminary analysis of the background data is presented here.

Keywords: Dark matter, WIMP, liquid xenon

1. Introduction

Weakly Interacting Massive Particles (WIMPs) populating the halo of our galaxy can be detected directly, via their interactions with nuclei in terrestrial detectors, or indirectly, via their annihilation products in the galactic center, halo, or sun.[1] XENON is a new generation dark matter detector, with the aim of observing the small energy released after a WIMP scatters off a Xe nucleus.[2] Using a target mass of 1000 kg, a low energy threshold and excellent event-by-event discrimination, XENON will reach a sensitivity of 10^{-46}cm^2 for spin-independent cross-sections, which is several orders of magnitude below the current best limit of 1.6×10^{-43}cm^2 by CDMS experiment.[3] Cryogenic noble liquids, such as Argon and Xenon, offer the best prospects for detectors with large mass and excellent background rejection capabilities. In the XENON Dark Matter experiment, the simultaneous detection of ionization and scintillation in a liquid Xenon (LXe) 3-D position sensitive time projection chamber (TPC) is used to identify nuclear recoils, produced by WIMPs but also neutrons, with energy as low as 10 keV, from electron recoils produced by gamma and beta background. Additionally, the event localization with millimeter resolution and the self-shielding capability of a large, homogeneous and dense liquid volume, provide additional background discrimination. R&D for the XENON concept started in Sep 2002 and has culminated in a 10 kg fiducial mass prototype (XENON10), which is now taking dark matter data at the Gran Sasso Underground Laboratory (LNGS)[4] in Italy. Together with results from small detectors, the performance of this first generation TPC demonstrate the capabilities of the XENON approach for dark matter detection.[5,6] From August 2006, XENON10 started to take WIMP search data and we have accumulated more than 30 live days. The XENON10 sensitivity goal of $\sim 2\times10^{-44}$ cm^2 will be achieved in 2007.

2. XENON10 Detector

XENON10 is a dual phase XeTPC filled with 22 kg of ultra pure LXe. The active volume is defined by a Teflon cylinder with an inner diameter of 20 cm and a height of 15 cm for a total active mass of about 15 kg of LXe. The TPC is equipped with four wire meshes, two in the liquid and two in the gas. The bottom mesh serves as cathode and the next one, positioned just below the liquid level, together with a series of field shaping rings, form the 15 cm drift region. The two last meshes, together with the one below the liquid level, serve to define the gas proportional scintillation region.

The active volume is viewed by 89 Hamamatsu R8520-06-Al 1" square PMTs, 35 mm high. The bottom array of 41 PMTs is located below the cathode, fully immersed in LXe, to detect the direct scintillation light.

The choice of the Hamamatsu R8520-06-Al for XENON10 was motivated by the combination of compact size, square geometry (for more photo-coverage), improved quantum efficiency (>20%) and above all, lower radioactivity, compared to the previously tested 2" tubes (R9288-06-MOD).

The ionization electrons produced by a charged particle interaction in the liquid are drifted towards the liquid surface by the electric field and extracted into the gas phase where they are accelerated by an appropriated

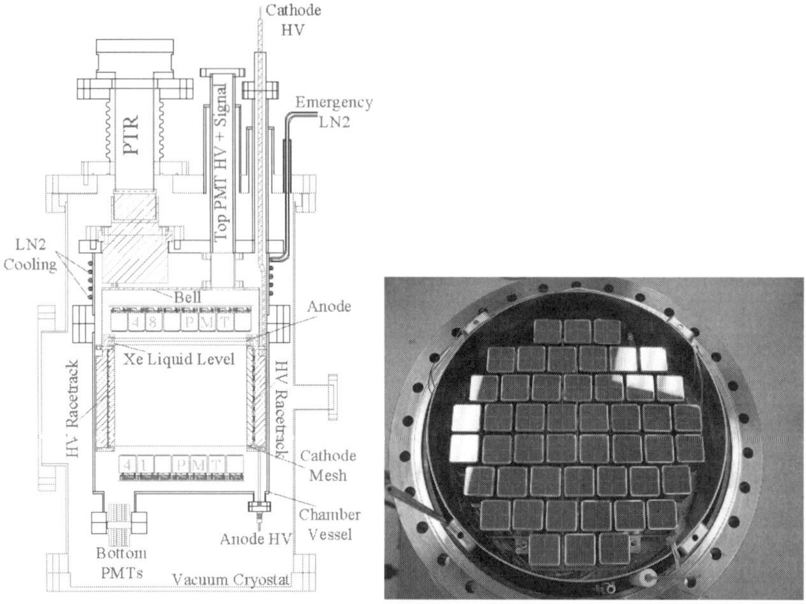

Fig. 1. (left) XENON10 technical design and (right) the top PMT array.

electric field and emit proportional scintillation light. The 48 PMTs of the top array, in the gas, are used to detect the proportional scintillation signal which is used for the event $x - y$ position reconstruction. A photo of the assembled top PMTs array is shown in Figure 1. The waveforms of a typical low energy gamma event are shown in figure 2 (center). The time separation between the two pulses of direct and proportional light, provides the event depth of interaction in the liquid (or z-coordinate).

Cooling for the XENON10 detector is provided by a Pulse Tube Refrigerator (PTR) mounted on the cryostat top flange, directly in contact with the Xe gas. The unit, especially designed for LXe temperature,[7] has been pivotal for the success of this detector. The 90 W maximum cooling power delivered to the detector is PID controlled. With this system we have been able to achieve the level of long term stability and reliability required for a dark matter experiment. Figure 2 (left) shows the remarkably stable pressure ($\Delta P < 0.006$ atm) and temperature ($\Delta T < 0.005°$C) values of XENON10 as monitored by the Slow Control System for over three months. Under such stable conditions the fluctuation of PMT gains is $< 2\%$.

High purity of the LXe is an essential requirement for a TPC like XENON10, with a drift gap of 15 cm. The purity must be preserved at all time during the detector operation in order to ensure stable performance.

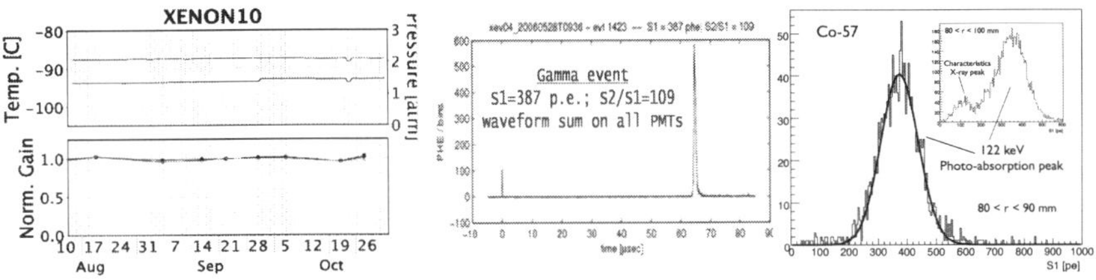

Fig. 2. (left) XENON10 detector's pressure, temperature and PMT gains during operation; (center) total waveform for a typical low energy gamma event; (right) S1 scintillation spectrum from ^{57}Co calibration.

We have fulfilled this requirement by continuous gas circulation through a high temperature getter. The electron lifetime inferred from recent data is (1.8 ± 0.4) ms which corresponds to $\ll 1$ ppb (O_2 equivalent) purity. The Xe gas used in XENON10 was procured from Spectra Gases,[8] with the required low level (< 10 ppb) of Kr contamination. A sample of 25 kg of Xe gas with lower Kr content (< 10 ppt) was also processed using a charcoal based Kr purification system. The ppb Kr level is however sufficient for the XENON10 sensitivity goal.

3. Calibration

To calibrate the detector with gamma and neutron sources, with minimum impact on WIMP search exposure time, the passive lead an polyethylene shield[9] was designed to allow introduction of external sources without exposing the detector cavity to outside air. The gamma calibration sources are ^{57}Co (122 keV) and ^{137}Cs (662 keV). The 662 keV photo-absorption peak of ^{137}Cs gives an average light yield of 1464 pe (2.2 pe/keV) and a sigma of 205 pe (14%). The light yield does not change substantially (less than 5%) for radius $r < 9$ cm. Figure 2 (right) shows a scintillation light spectrum from a ^{57}Co calibration, after drift time corrections. For r between 8 and 9 cm, the average light yield for 122 keV gamma rays is 374 pe (3.1 pe/keV), with a resolution of 17% (σ). The inset in figure 2 shows the clear detection of the characteristic X-ray peak from K-shell (30 keV).

An important background rejection feature of the XENON10 detector is the ability to localize events in 3D with a good spatial resolution. Since electron diffusion in LXe is small, the S2 pulse is produced in a small region with the same $x - y$ coordinates as the interaction site, allowing $x - y$ localization with an accuracy of a few mm. With the more precise z information, better than 1 mm, from the drift time measurement, the 3-D event localization provides powerful background discrimination via fiducial volume cuts. The determination of the z coordinate from the drift time not only has a good position resolution, but also a good multiple hit separation.

4. XENON10 Data and Discussion

The operation of XENON10 underground has allowed us: 1) to establish the low energy threshold; 2) to demonstrate the power of LXe self shielding; 3) to understand background sources; 4) and to demonstrate the electron recoil discrimination capability of the dual phase XeTPC. The total rate in XENON10, including events with only an S1 signal from interactions in the "passive" LXe outside the sensitive volume, is shown in figure 3 (black curve). By requiring an S2 signal and with only a few millimeter z-cut from above, the rate in the sensitive volume is reduced by three orders of magnitude at 10 keV(red curve). This demonstrates the efficient self shielding effect of LXe. The rate can be further reduced by a factor of 2 by using the x-y event localization and removing events in the outer 5 cm of LXe (green curve).

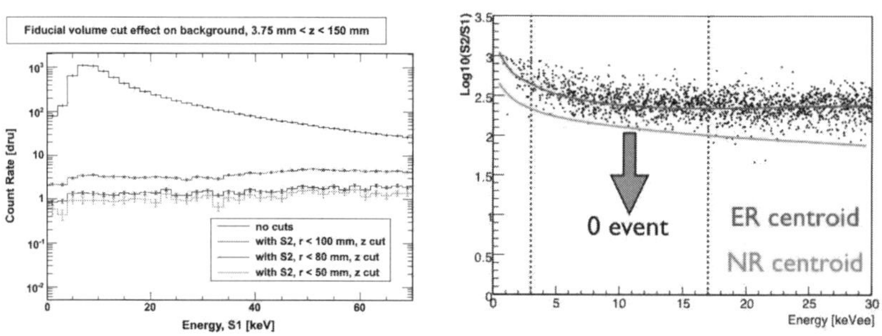

Fig. 3. (left) Background energy spectrum in XENON10, before and after selecting events with both light and charge signal and after fiducial volume cuts. (right) XENON10 ionization yield versus energy. The magenta and green lines correspond respectively to the centroid of the ER and the NR bands.

Monte Carlo simulations, using the GEANT4 software package, were carried out to understand the background sources in XENON10. As shown in figure 3 (right), the measured background rate can be explained by the radioactivity in the PMTs, the stainless steel used in the detector and cryostat, and the signal and high voltage ceramic feedthroughs (see also Ref. 9).

The statistics of the dark matter search data accumulated to-date amounts to more than 30 days of live time. We continue the data taking, given the very stable detector's performance. Figure 3 shows a preliminary plot of log10(S2/S1) as a function of energy for 22.2 live days, giving a total exposure of 38 kg d after cuts. The magenta line corresponds to the centroid of the electron recoil band while the green one corresponds to the centroid of the nuclear recoil band, from the measurements with the small prototypes.[10]

In the electron equivalent energy region 3-16 keVee (corresponding to recoils energy 10-32 keVr), zero events are observed in the 50% acceptance window of the nuclear recoil band. In the region above 16 keVee, we observe several event in the NR band. Our understanding is that these events are double scatter events, with one interaction in the passive LXe layer between the cathode and the bottom PMT array, and one in the fiducial volume. Since no S2 signal is observed for interactions outside the sensitive volume, the additional S1 light from the passive region results in a smaller log10(S2/S1) ratio, making the event look like a nuclear recoil. Among the planned XENON10 upgrades in 2007, we will "blind" the PMTs from light produced in the bottom LXe Layer, and will test this hypothesis.

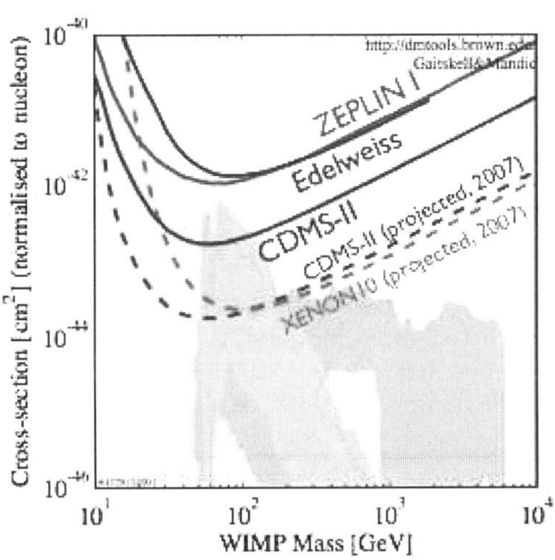

Fig. 4. WIMP-nucleon cross-section upper limit (90%C.L.) from direct dark matter search experiments CDMS,[3] Edelweiss[11] and ZEPLINI.[12] The projected upper limits from CDMS-II and XENON10 experiments in 2007 are shown as dashed lines. The shaded areas are allowed parameters from some supersymmetry models.[13,14]

Until recently, the radioactivity associated with the PMTs readout in XENON10 was assumed to make the largest contributionto the total gamma background rate. It turns out that the latest version of the R8520-06-AL PMT, specifically developed by Hamamatsu Photonics Company for the XENON project, has extremely low radioactivity. A sample of 7 PMTs was recently screened using the very sensitive HPGe detector available at the LNGS low background facility, yielding values of 0.17/0.24/10/0.56 mBq/PMT for (U/Th/K/Co), respectively. In comparison, the HV divider circuit mounted on a Cirlex board directly on the PMT, gives a larger contribution.

Figure 4 shows the WIMP-nucleon cross-section upper limit from direct dark matter search experiments. By replacing the feedthroughs and the stainless steel vessel and with the PMT activity as recently measured, the expected background is estimated as 0.15 events/keVee/kg/day, one order of magnitude lower than the current background. If this background can be achieved, we will reach a sensitivity of $2\times10^{-44}\mathrm{cm}^2$ by 2007.

Acknowledgments

This work was funded by NSF Grants No. PHY-03-02646 and No. PHY-04-00596, and DOE Grant No. DE-FG02-91ER40688.

References

1. K. Freese, astro-ph/0508279.
2. XENON: A Liquid Xenon Experiment for Dark Matter Proposal submitted to NSF, Particle and Nuclear Astrophysics, Proposal 0201740 (September 2001).
3. D.S. Akerib, *et al* (CDMS Collaboration) Phys. Rev. Lett. **96** (2006) 011302.
4. LNGS web page, http://www.lngs.infn.it
5. E. Aprile *et al.*, Phys. Rev. D **72**, 072006 (2005).
6. E. Aprile *et al.*, Phys. Rev. Lett. **97**, 081303(2006).
7. S. Mihara *et al.*, Cryogenics **44**, 223, (2004).
8. Spectra Gases, http://www.spectra-gases.com
9. P. Sorensen, *et al.* (XENON Collaboration), Proc. 6th IDM Workshop (2006) in the current proceedings.
10. T. Shutt *et al.*, arXiv:astro-ph/0608137 (2006)
11. V. Sanglard *et al.* Phys. Rev. D **71** 122002 (2005)
12. G. J. Alner *et al.* 2005 Astropart. Phys. **23** 444
13. E. A. Baltz and P. Gondolo, JHEP **0410** 052 (2004)
14. J. R. Ellis, Y. Santoso and V. C. Spanos, Phys. Rev. D **71** 095007 (2005)

MASS ESTIMATES OF DARK MATTER BODIES FROM THE TIMESCALE OF QUASAR VARIATION

M. R. S. HAWKINS*

Institute for Astronomy (IfA), University of Edinburgh, Royal Observatory, Blackford Hill, Edinburgh EH9 3HJ, Scotland, UK

**E-mail: mrsh@roe.ac.uk*

This paper sets out to measure the timescale of quasar variability with a view to new understanding of the size of accretion discs in active galactic nuclei. We combine data from two large scale monitoring programmes to obtain Fourier power spectra of light curves covering nearly three orders of magnitude in frequency in blue and red passbands. If the variations are interpreted as due to gravitational microlensing, then timescale measurements in the observer's frame imply a minimum mass for the microlensing bodies of around $0.4M_{\odot}$.

Keywords: quasars: general; galaxies: active

1. Introduction

Since the first observations of quasars, the timescale of variation in optical passbands has placed important constraints on their structure. Short timescale variations quickly led to a very small upper limit to the size of the emitting region, thus putting tight constraints on atempts to model the power source. A more difficult parameter to measure is the characteristic timescale of variation on longer timescales. If the variations are intrinsic to the quasar then this is a measure of the overall size of the accretion disc. However, in this paper we consider the possibility that the observed variations are dominated by microlensing [Hawkins 1996], and in this case the characteristic timescale of variation is a measure of the mass of the microlensing bodies.

The characteristic timescale of variation has proved difficult to measure in the past. This is due to a number of factors, includiing insufficiently large samples of quasar light curves, unevenly sampled data, an insufficiently long timespan of observations and inadequate analysis techniques. In this paper we report work which attempts to improve on these shortcomings to provide a measure of the characteristic timescale of variation, and hence an estimate of the mass of microlensing bodies assumed to make up the dark matter.

2. Observations

The data for the analysis of timescale came from two sources. Firstly, a set of some 800 light curves from a long term monitoring programme of ESO/SERC field 287 with the UK Schmidt telescope at Siding Spring, Australia ([Hawkins 1996]). This data includes yearly magnitude measures in blue and red passbands covering some 28 years, and for most epochs four plates were obtained. Quasars were selected on the basis of both variability and ultra-violet excess, resulting in a sample of 814 quasars with complete light curves. Data on short timescales was obtained from quasars observed by the MACHO project ([Geha et al. 2003]). By appropriate binning this gave data points regularly spaced at 50 days and 10 days. The MACHO quasars followed a similar distribution in redshift and luminosity as for the Field 287 quasars, although the sample was smaller, numbering around 60 quasars.

In order to measure timscales of variation, Fourier power spectra were obtained for sets of light curves with one year, 50 days and 10 days spacing from the two sets of data. The random noise for the one year and 50 day data was estimated by measuring the rms fluctuation of the light curves of stars, assumed to be constant, and also from the known photometric errors. This was then applied as a correction to the Fourier power spectra, in the form of a constant subtracted from all frequencies.

3. Measurement of Timescale

The combined Fourier power spectrum of the three sets of light curves with one year, 50 day and 10 day sampling intervals is shown in Fig. 1. No normalisation has been applied, and the fact that the power spectra for the three samples trace out the same curve can be seen as evidence that any features are real.

The combined Fourier power spectrum has an approximately linear central section. At high frequencies it was not possible to make a correction for the Gaussion noise with any confidence, and so the curve flattens out in this regime. At the low frequency end, there is some indication of a departure from the linear relationship, and if this is the case would indicate that a characteristic timescale for the variations could be measured.

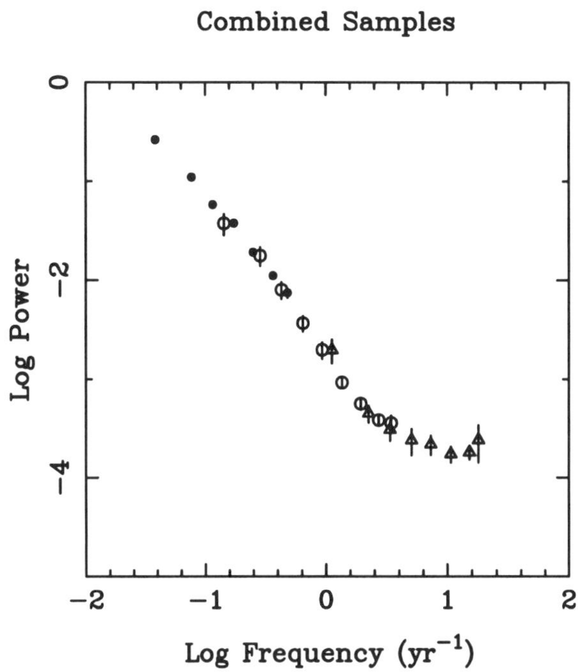

Fig. 1. Fourier power spectra of blue passband light curves in the observers frame for the sample of quasars in Field 287 sampled yearly (filled circles), the MACHO quasars with 50 day sampling (open circles) and 10 day sampling (open triangles).

In order to quantify any break in the power spectrum in Fig. 1, we fit the following curve P(f):

$$P(f) = \frac{C}{\left(1 + \frac{f}{f_c}\right)^a} + d \tag{1}$$

The central part of this function is a power law, which flattens off at high frequencies to match the increase in Gaussian noise in the data. At the low frequency end there is a break in the power law relation with a characteristic frequency f_c. This frequency corresponds to a characteristic timescale of variation, and in order to measure this timescale in the quasar light curves, the function $P(f)$ was iteratively fitted to the power spectrum in Fig. 1.

The result of this fitting procedure is shown in Fig. 2. The best fit for $P(f)$ is shown as a dashed line. Also shown, as a solid line, is the asymptotic power law relation for $P(f)$. The 'cut-off' frequency f_c, which is of central importance to this investigation, has a best fit value corresponding in the

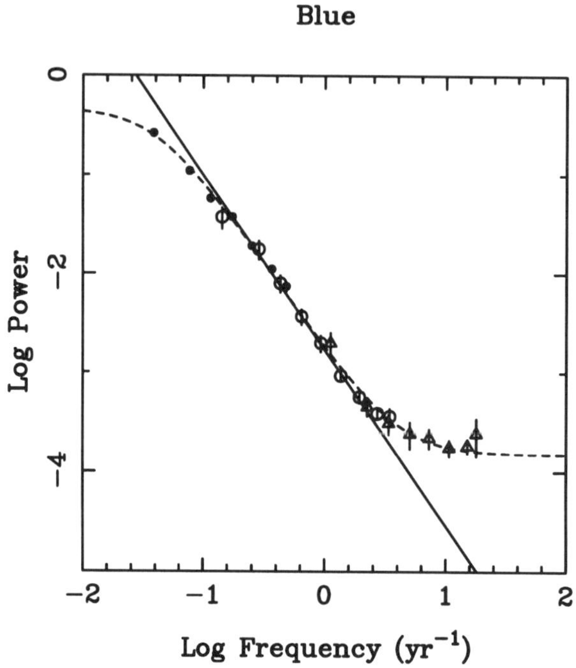

Fig. 2. Fourier power spectra of blue passband light curves in the observer's frame, for quasars in the Field 287 and MACHO samples. Symbols are as for Fig. 1. The dashed line is the best-fit curve of Eq. (1), and the solid line is the asymptotic power law relation for $P(f)$ as described in the text.

time domain to 24.0 ± 2.7 years. Given that the total run of data is only 25 years, this figure should be treated as a lower limit to the timescale of variation.

In addition to the blue passband light curves, a similar but smaller sample of red passband were also available as described above. Fourier power spectra of the combined Field 287 and MACHO samples were constructed for the red passband samples in the same way as for the blue, and are plotted in Fig. 3. The function $P(f)$ in Eq. (1) was iteratively fitted as for the blue data, and the best fit curves are also shown in Fig. 3. The best fit value for the cut-off frequency in the time domain was found to be 70.9 ± 33.4 years. Given that the data only cover 22 years, we shall take that as a lower limit for the timescale of variation.

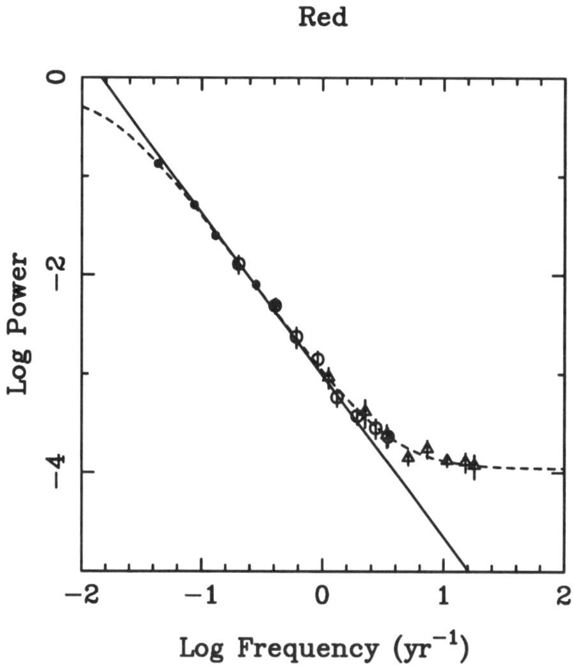

Fig. 3. Fourier power spectra of red passband light curves in the observer's frame for quasars in the Field 287 and MACHO samples. Symbols are as for Fig. 1. The dashed line is the best-fit curve of Eq. (1), and the solid line is the asymptotic power law relation for $P(f)$ as described in the text.

4. Microlens Masses

In the event that the quasar variations are dominated by the effects of microlensing, it is appropriate to use the light curves measured in the observer's frame, and the cut-off frequency f_c from Eq. (3) is a measure of the characteristic mass of the microlenses. In fact, it is easy to show that the microlens mass is proportional to the square root of the timescale of variation. If we adopt a characteristic value of 600 km s^{-1} for the transverse velocity of the lenses, and assume that most lenses lie at aredshift $z \sim 0.5$ ([Turner et al. 1984]) then by using the lower limit to the timescale of 24 years measured from the data in Figs. 2 and 3, we can obtain a value for the minimum microlens mass $m \sim 0.4 M_\odot$.

The lower limit of $m \sim 0.4 M_\odot$ for the mass of the microlensing bodies is considerably larger than earlier estimates using this approach ([Hawkins

1996]). Such objects would lie in the same mass range as those detected by the MACHO project ([Alcock et al. 1997]), although they would have to be more abundant to be responsible for the extent of variability observed. Estimates of lens mass from the study of microlensing events in multiply lensed quasar systems tend to give upper limits somewhat similar to the lower limit quoted above for the present work. For example, Kochanek (2004) finds an upper limit to the mean of the mass function of microlensing bodies in the system Q2237+0305 to be around $m \leq 0.2M_\odot$, and Refsdal et al. (2000) find an upper limit to the mass of microlensing bodies in the quasar system Q0957+561 of $m \leq 0.5M_\odot$.

References

Alcock et al. 1997. Alcock C. et al., 1997, ApJ, 486,
Geha et al. 2003. Geha M. et al., 2003, AJ, 125, 1
Hawkins 1996. Hawkins M.R.S., 1996, MNRAS, 278, 787
Kochanek 2004. Kochanek C.S., 2004, ApJ, 605, 58 697
Refsdal et al. 2000. Refsdal S., Stabell R., Pelt J., Schild R., 2000, A&A, 360, 10
Turner et al., 1984. Turner E.L., Ostriker J.P., Gott J.R., 1984, ApJ, 284, 1

NEW SIGNATURES IN DIRECT DARK MATTER SEARCHES

J. D. VERGADOS[(1)*], N. TETRADIS[(2)] and AMAND FAESSLER[(3)]

[(1)] *Theoretical Physics Division, University of Ioannina,*
Ioannina, Gr 451 10, Greece
[*] *E-mail: vergados@cc.uoi.gr*

[(2)] *Department of Physics, University of Athens,*
Zographou 157 84, Greece
[*] *E-mail: ntetrad@physics.uoa.gr*

[(3)] *Institute of Theoretical Physics, University of Tuebingen,*
Tuebingen Germany
[*] *E-mail: amand.faessler@Uni-tuebingen.de*

The event rates for direct cold dark matter detection, both directional and non directional, are presented using a velocity didtribution function obtained by considering a coupling between dark matter and dark energy. The magnitude of the effect depends on the strength κ of this new interaction relative to gravity. The resulting isothermal velocity distribution for dark matter in galaxy halos is still Maxwell-Boltzmann (M-B), but the characteristic velocity and the escape velocity are increased by $\sqrt{1+\kappa^2}$. We adopt a phenomenological approach and consider values of κ near unity. The phenomenological implications of such a coupling are discussed

1. Introduction

The combined MAXIMA-1,[1] BOOMERANG,[2] DASI[3] and COBE/DMR Cosmic Microwave Background (CMB)[4] observations imply that the Universe is flat[5] and most of its energy content is exotic.[6] These results have been confirmed and improved by the recent WMAP data.[7] The deduced cosmological expansion is consistent with the luminosity distance as a function of redshift of distant supernovae.[8,9] According to the scenario favored by the observations there are various contributions to the energy content of our Universe. The most accessible energy component is baryonic matter, which accounts for $\sim 5\%$ of the total energy density. A component that has not been directly observed is cold dark matter (CDM)): Its contribution to the total energy density is estimated at $\sim 25\%$. The component with the biggest contribution to the energy density has an equation of state similar to that of a cosmological constant and is characterized as dark energy. This component is responsible for $\sim 70\%$ of the total energy density and induces the observed acceleration of the Universe.[8,9]

Since a non-exotic component cannot exceed 40% of the CDM,[10] there is room for a component consisting of exotic weakly interacting massive particles (WIMPs). Supersymmetry naturally provides candidates for these dark matter constituents.[11,12] In the most favored scenario of supersymmetry, the lightest supersymmetric particle (LSP) can be described as a Majorana fermion, a linear combination of the neutral components of the gauginos and higgsinos.[11–13] Other extensions of the Standard Model yield viable cold dark matter candidates, like Kaluza Klein Particles (KKP) in theories wirh universal extra dimensions. All such candiadtes are called WIMPs (Weakly Interacting Massive Particles).

Even though there exists firm indirect evidence for a halo of dark matter in galaxies from the observed rotational curves, it is essential to detect such matter directly.[11–14] The possibility of their direct detection, however, depends on the nature of the dark matter constituents, especially their interactions, their mass, their density in our vicinity and velocity distributions. We do not know for sure what this distribution is, but it is not expected to depend on the nature of the WIMP. In other words all WIMPs are expected to have the same velocity distribution and matter density. The particle density, however, which enters the event rate, is expected to be inversely proportional to the WIMP mass. In the present work we will consider two variants of a Maxwellian distribution, differing in their characteristic velocities.

The WIMPs are expected to have energies in the keV region. So they are not likely to excite the nucleus. They can be directly detected mainly via the recoiling of a nucleus (A, Z) in elastic scattering. In the standard nuclear recoil experiments one has the problem that the reaction of interest does not have a characteristic feature to distinguish it from the background. So for the expected low counting rates the background is a formidable problem. Some special features of the LSP-nuclear interaction can be exploited to reduce the background problems. In this paper we will consider two such features, the time dependence of the event rate (modulation effect) and asymmetries in directional recoil experiments, i.e. measurements in which the direction of the recoiling nucleus is observed.[15–22]

We will concentrate here on the standard Maxwell-Boltzmann (M-B) distribution for the WIMPs of our galaxy and we will not be concerned with other non thermal distributions, even though they may yield stronger directional signals. Among those one should mention the late infall of dark matter into the galaxy, i.e caustic rings,[23–27] dark matter orbiting the Sun[19] and Sagittarius dark matter.[28]

In a previous paper we have found that the observed rate is correlated with the direction of the sun's motion.[21,22] The currently planned experiments, however, may not be able to distinguish the two possible senses along the line of nuclear recoil.[29] Such experiments cannot, e.g., measure the backward-forward asymmetry. In light of this possibility we will extend our previous directional calculations[21,22] and explore, which characteristics, if any, of the previous results persist, if the rates for both senses of the nuclear recoil are summed up.[30]

2. Rates

In the present work we will assume that the velocity distribution is Maxwell-Boltzmann (M-B) with a characteristic velocity υ_0 and an upper cut off equal to the escape velocity, $\upsilon_{esc} = 2.84\upsilon_0$. For comparison we will also consider M-B distributions with larger characteristic velocity $n\upsilon_0$ and cut off velocity $2.84n\upsilon_0$, with $n \geq 1$. This situation arises, if one considers the coupling of dark matter to dark energy via a scalar field. Then the gravitational interaction of matter is not affected. The interaction involving dark matter is increased. Via the virial theorem this results to an isothermal M-B distribution with higher temperature. We will not elaborate further on this point, but we refer the interested reader to the literature.[31,32]

We will distinguish two possibilities:

i)The direction of the recoiling nucleus is not observed.

Even though our main interest is in the directional rate for orientation purposes we will summarize the main points entering the standard (non directional) searches. The number of events in time t for the coherent WIMP-nucleus elastic scattering:[21,22,25,33] is given by:

$$R \simeq 1.60\ 10^{-3} \frac{t}{1\text{y}} \frac{\rho(0)}{0.3GeV cm^{-3}} \frac{m}{1\text{Kg}} \frac{\sqrt{\langle v^2 \rangle}}{280\text{k}ms^{-1}} \frac{\sigma^S_{p,\chi^0}}{10^{-6}\ \text{pb}} f_{coh}(A, \mu_r(A)) \quad (1)$$

with

$$f_{coh}(A, \mu_r(A)) = \frac{100\text{GeV}}{m_{\chi^0}} \left[\frac{\mu_r(A)}{\mu_r(p)}\right]^2 A\ t_{coh}\left(1 + h_{coh} cos\alpha\right) \quad (2)$$

In the above expression m is the target mass, A is the number of nucleons in the nucleus and $\langle v^2 \rangle$ is the average value of the square of the WIMP velocity (with $n = 1$) and α is the phase of the Earth (zero around June 3nd). The quantity of interest to us is $r = t_{coh}\,(1 + h_{coh} cos\alpha)$, which contains all the information regarding the WIMP velocity distribution and the structure of the nucleus. It also depends on the reduced mass of the system.

ii) The direction $\hat{e}$ of the recoiling nucleus is observed.
One can show that the event rate is still given by Eq. (1) except that now:

$$f_{coh}(A, \mu_r(A)) = \frac{100\text{GeV}}{m_{\chi^0}} \left[\frac{\mu_r(A)}{\mu_r(p)}\right]^2 A \frac{\kappa}{2\pi} t_{coh}\,(1 + h_m(coh) cos(\alpha + \alpha_m \pi)) \tag{3}$$

κ is a measure of the reduction in the event rate in directional experiments over and above the geometric factor of $1/(2\pi)$. We should mention that in the directional case we have two modulation amplitudes, h_c and h_s, proportional to $\cos\alpha$ and $\sin\alpha$ respectively. Thus:

$$h_m(coh) = \sqrt{h_c^2 + h_s^2},\ \alpha_m = -\frac{1}{\pi} \arctan \frac{h_s}{h_c} \tag{4}$$

The parameters κ, h_m and α_m depend, among other things, on polar angle Θ and azymouthal angle Φ $(0 \le \Theta \le \pi,\ 0 \le \Phi \le 2\pi)$ specifying the direction of observation $\hat{e}$.

iii) Partly directional experiments.
In this case one can specify the line the nucleus is recoiling but not the sense of direction on it. The results in this case are obtained from those of the previous section, by summing up the events in both directions specified by $(\Theta,\ \Phi)$ and $(\pi - \Theta,\ \Phi + \pi)$. A given line of observation is now specified by $\Theta,\ \Phi$ in the range:

$$0 \le \Theta \le \pi/2,\ 0 \le \Phi \le \pi$$

3. Applications

In this section we are going to apply the formalism of the previous section in the case of the light target ^{32}S appearing in CS_2 involved in the experiment DRIFT.[15]

The parameter t entering the non directional case is shown in Fig. 1.

We continue our analysis by plotting the parameter κ given above for a WIMP mass of 100 GeV (see Fig. 2).

From these figures we see that in the case of directional experiments the maximum value of κ is attained when the observation is made opposite

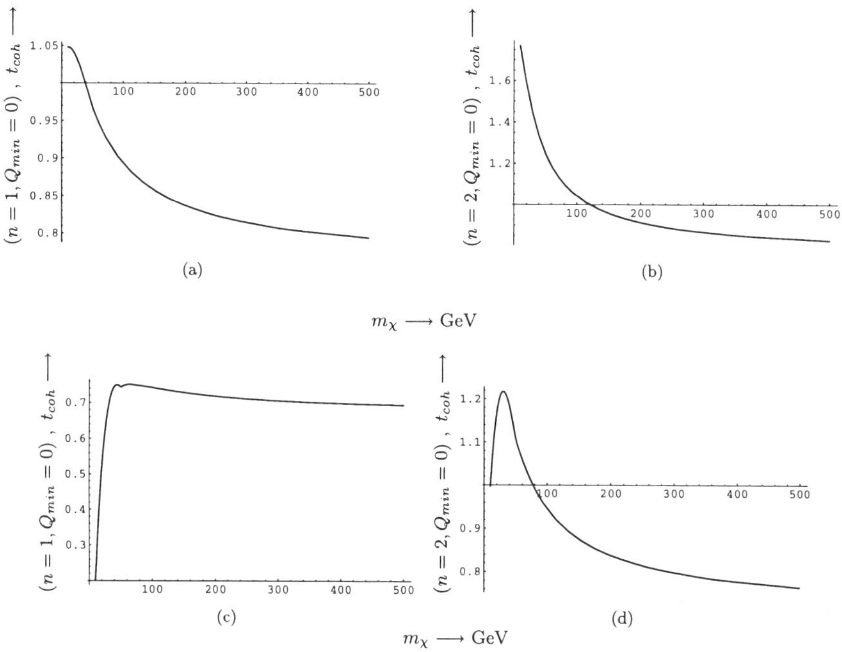

Fig. 1. The uantity t_{coh} in the case of the coherent mode is shown for zero energy cut off on the top and 10 keV at the bottom. The results correspond to $n = 1$ on the left and $n = 2$ on the rifgt.

to the sun's direction of motion. This is expected in the case of the M-B distribution.

We will next consider the modulation effect, i.e. the parameters h_{coh} and h_m, shown in Figs. 3 and 4, and α_m shown in Fig. 5. From Figs. 3 and 4 we see that the modulation for $n = 2$ is quite small both in the directional and non directional cases. The parameters κ, h_m and α_m shown in Figs. 2, 4 and 5 correspond to a WIMP mass of 100 GeV. For other WIMP masses the reader is referred to the literature.[30,32]

4. Summary and Conclusions

We have seen that, given a sufficient number of events, the directional experiments, in which one measures the direction of the nuclear recoil, provide an excellent signature to discriminate against background. Some of the requirements that should be met by such detectors have recently been discussed.[34,35] It is encouraging to know that the gaseous TPC detectors

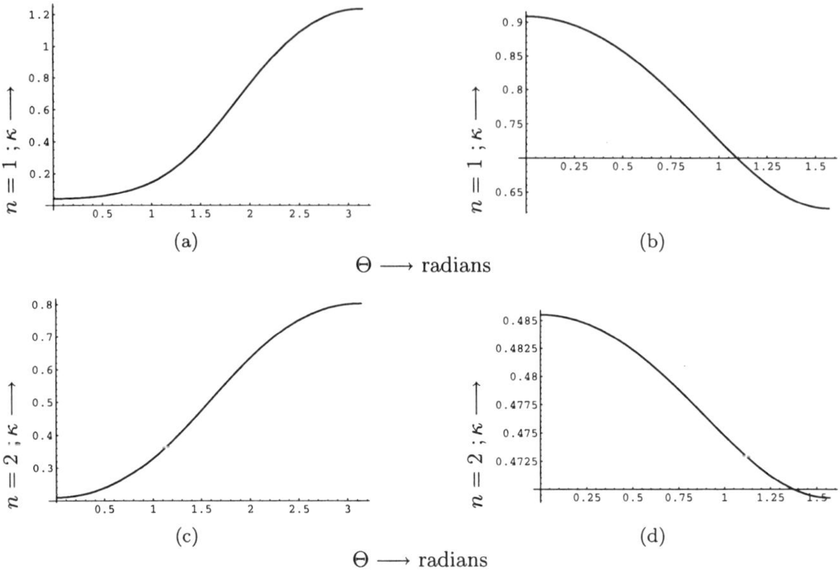

Fig. 2. The parameter κ defined in the text for a light target in the case of a WIMP mass of 100 GeV. (a) Directional in a given sense for $n = 1$. (b) The sum of both senses for $n = 1$. (c) The same as in (a) for $n = 2$. (d) The same as in (b) for $n = 2$.

will be able to register recoils in all directions and, once some interesting events appear, they can be further analyzed directionally. To fully exploit the advantages of such detectors one should be able to distinguish between recoils with momenta $\mathbf{p}$ and $-\mathbf{p}$ (sense of direction), which now appears to be feasible. Some of the predicted interesting features of directional event rates persist, even if turns out that the sense of motion of recoils along their line of motion cannot be measured.

The predictions regarding the directional event rates depend, of course, on the assumed WIMP velocity distribution. In the present work we selected to work with a M-B distribution: (i) the traditional one with characteristic velocity that of sun around the center of the galaxy and (ii) a variant obtained when dark matter is coupled to dark energy via a scalar field yielding an increase in the gravitational field for dark matter.[32] In the latter case the characteristic velocity in the WIMP distribution increases by a factor of $n \geq 1$.

In the most favored direction, opposite to the sun's direction of motion, the event rate is $\approx \frac{1}{2\pi}$ down from the standard non directional experiments. The modulation amplitude h_m in this direction $(\Theta = \pi)$ depends on the

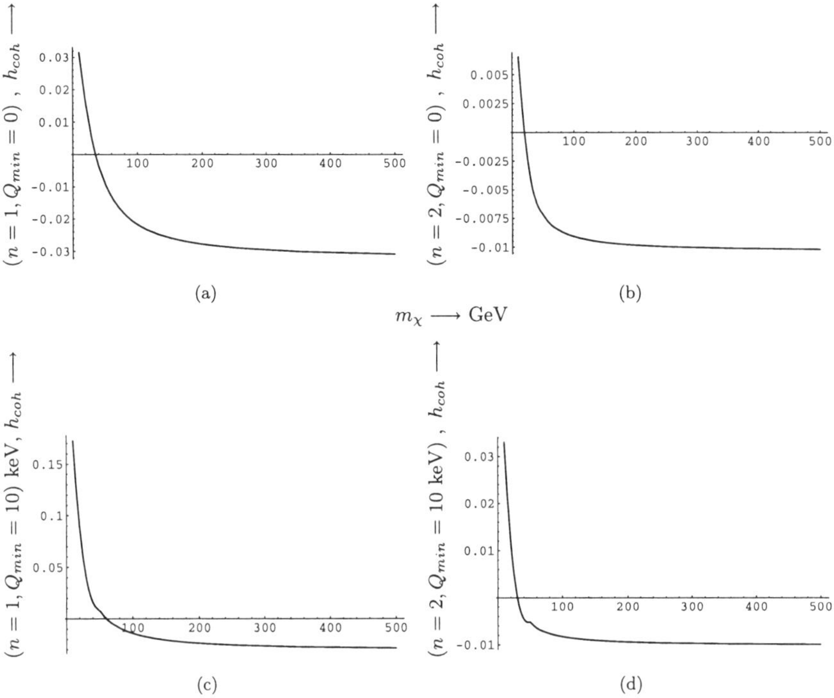

Fig. 3. The modulation amplitude h_{coh} in the case of coherent scattring for $Q_{min}=0$ on the top and $Q_{min}=10$ keV at the bottom. From left to right $n=1$ and 2.

WIMP mass. For a light target it ranges between 0.05 and and 0.1 depending on the WIMP mass. For a heavy target it is somewhat reduced, $h_m = 0.02 - 0.07$, but it remains higher than that expected in the non directional experiments. It is also characterized by a definite sign (maximum around Junc 3nd). Higher values $h_m - 0.25 - 0.35$, yielding $50 - 70\%$ difference between the maximum and the minimum rates, are expected by a judicious choice of the direction of observation ($\Theta \approx \pi/2$). Thus quite large asymmetries with seasonal dependence are predicted. The time the maximum occurs is also direction dependent, so it cannot be mimicked by seasonal variations of the background.

In partly directional experiments, i.e. experiments in which the sense of motion of the recoiling nucleus is not determined, one no longer can measure asymmetries. We still find, however, that the expected event rate is maximum along the line of motion. Equally large modulation amplitudes h_m

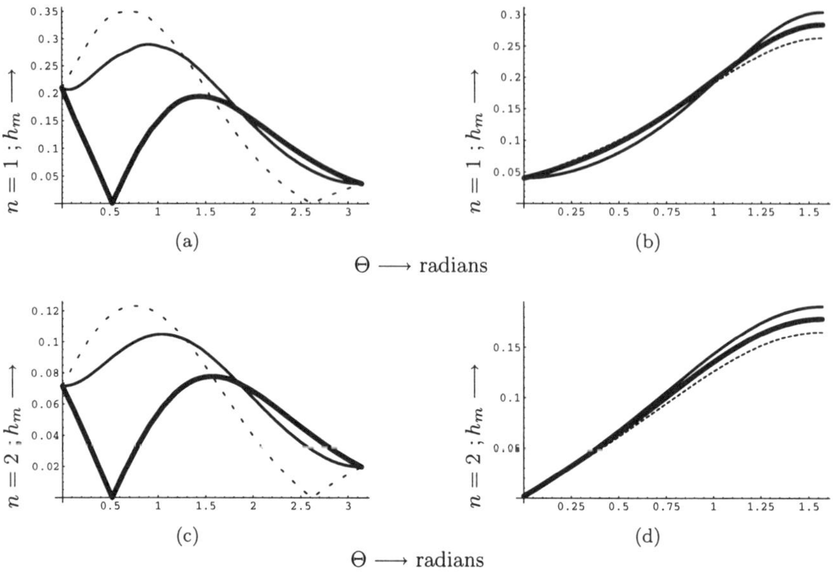

Fig. 4. The parameter h_m, defined in the text, as function of the polar angle Θ for a light target in the case of a WIMP mass of 100 GeV. Note that, when the sense is not distinguished, the maximum angle of the line of observation with the sun's direction of motion is $\pi/2$, while the maximum azimuthal angle is π (a) Directional in a given sense for $n = 1$. Various values of the azimuthal angle Φ have been considered. The fine solid, thick solid, dashed and long dashed curves correspond to $\Phi = 0, \pi/2, \pi$, and $3\pi/2$ respectively (the cases $\Phi = \pi/2$ and $3\pi/2$ cannot be distinguished). (b) The sum of both senses is shown for $n = 1$. Now the thick solid, fine solid, dashed and long dashed curve correspond to $\Phi = 0, \pi/4, \pi/2$ and $3\pi/4$ respectively (the cases $\Phi = \pi/4$ and $3\pi/4$ cannot be distiguished). (c) The same as in (a) for $n = 2$. (d) the same as in (b) for $n = 2$.

are predicted with a seasonal variation, which, again, cannot be mimicked by seasonal variations in the background.

The modulation amplitude is decreased, if dark matter is coupled with dark energy, in a fashion analogous to the non directional case. This is not unexpected, since the ratio of the velocity of the earth to the characteristic velocity becomes smaller. This suggests that one should test whether the above conclusions hold, by considering other velocity distributions. Thus one may be able to infer the velocity distribution, when the experimental data become available.

The directional experiments are quite hard. It is encouraging that the angular distribution is such that the average angle is given by $<\xi>=<\cos\theta>= 0.5-0.8$ depending on the angle Θ, the polar angle of the direction

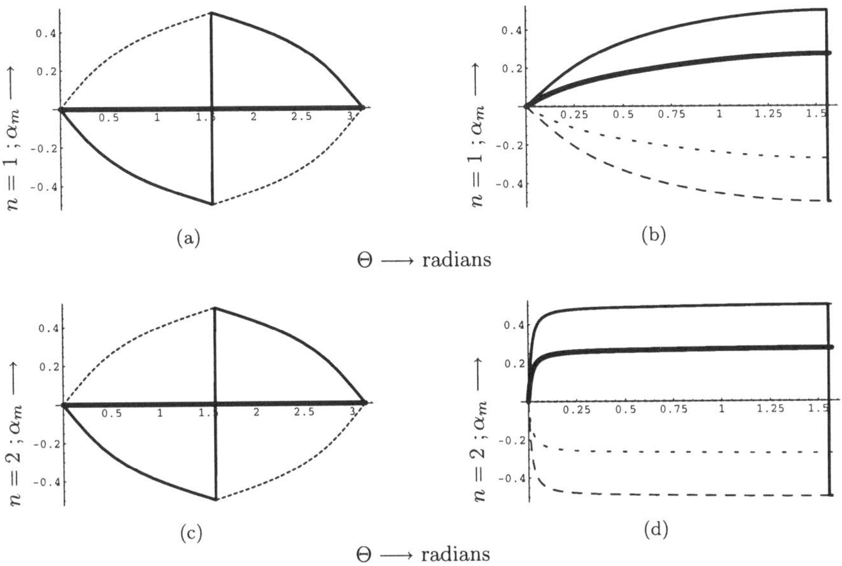

Fig. 5. The parameter α_m (in units of π), defined in the text, as function of the polar angle Θ for a light target. Otherwise the notation is the same as in Fig. 4.

of observation from the sun's direction of motion.[30] Anyway in experiments planned by DRIFT[15] one can register events recoiling in all directions. If some interesting events are found, one may further analyze them by the direction of recoil to discriminate against background.

References

1. S. Hanary *et al.*, *Astrophys. J.* **545**, L5 (2000);
   J.H.P Wu *et al.*, *Phys. Rev. Lett.* **87**, 251303 (2001);
   M.G. Santos *et al.*, *Phys. Rev. Lett.* **88**, 241302 (2002).
2. P. D. Mauskopf *et al.*, *Astrophys. J.* **536**, L59 (2002);
   S. Mosi *et al.*, *Prog. Nuc.Part. Phys.* **48**, 243 (2002);
   S. B. Ruhl *et al.*, astro-ph/0212229 and references therein.
3. N. W. Halverson *et al.*, Astrophys. J. **568**, 38 (2002)
   L. S. Sievers *et al.*, astro-ph/0205287 and references therein.
4. G. F. Smoot *et al.* (COBE Collaboration), Astrophys. J. **396**, L1 (1992).
5. A. H. Jaffe *et al.*, Phys. Rev. Lett. **86**, 3475 (2001).
6. D. N. Spergel *et al.*, Astrophys. J. Suppl. **148**, 175 (2003).
7. D.N. Spergel *et al.*, Three-Year WMAP Results: Implications for Cosmology, astro-ph/0603449;
   L. Page *et al.*, Three-Year WMAP Results: Polarization Analysis, astro-ph/0603450;

G. Hinsaw *et al.*, Three-Year WMAP Observations: Implications Temperature Analysis, astro-ph/0603451;
N Jarosik *et al.*, Three-Year WMAP Observations: Beam Profiles, Data Processing, Radiometer Characterization and Systematic Error Limits, astro-ph/0603452.

8. A. G. Riess *et al.*, [Supernova Search Team Collaboration], Astron. J. **116** (1998) 1009 [arXiv:astro-ph/9805201].
9. W. J. Percival *et al.*, [The 2dFGRS Collaboration], Mon. Not. Roy. Astron. Soc. **327** (2001) 1297 [arXiv:astro-ph/0105252].
10. D. P. Bennett *et al.*, Phys. Rev. Lett. **74**, 2867 (1995).
11. M. W. Goodman and E. Witten, Phys. Rev. D **31**, 3059 (1985).
12. J. Ellis and L. Roszkowski, Phys. Lett. B **283**, 252 (1992).
13. A. Bottino *et al.*, *Phys. Lett B* **402**, 113 (1997).
R. Arnowitt. and P. Nath, *Phys. Rev. Lett.* **74**, 4592 (1995); *Phys. Rev. D* **54**, 2374 (1996); hep-ph/9902237;
V. A. Bednyakov, H.V. Klapdor-Kleingrothaus and S.G. Kovalenko, *Phys. Lett. B* **329**, 5 (1994).
14. T. S. Kosmas and J. D. Vergados, Phys. Rev. D **55**, 1752 (1997).
15. B. Morgan, A. M. Green, and N. J. C. Spooner, Phys. Rev. D **71**, 103507 (2006); astro-ph/0408047.
16. Y. Shimizu, M. Minoa, and Y. Inoue, Nuc. Instr. Meth. A **496**, 347 (2003).
17. V.A. Kudryavtsev, Dark matter experiments at Boulby mine, astro-ph/0406126.
18. B. Morgan and A. M. Green, Phys. Rev. D **72**, 123501 (2005).
19. C. Copi, J. Heo, and L. Krauss, Phys. Lett. B **461**, 43 (1999).
20. C. Copi and L. Krauss, Phys. Rev. D **63**, 043507 (2001).
21. J. D. Vergados, Phys. Rev. D **57**, 103003 (2003); hep-ph/0303231.
22. J. D. Vergados, J. Phys. G **30**, 1127 (2004); 0406134.
23. P. Sikivie, Phys. Rev. D **60**, 063501 (1999).
24. P. Sikivie, Phys. Lett. B **432**, 139 (1998).
25. J. D. Vergados, Phys. Rev. D **63**, 06351 (2001).
26. A. M. Green, Phys. Rev. D **63**, 103003 (2001).
27. G. Gelmini and P. Gondolo, Phys. Rev. D **64**, 123504 (2001).
28. A. M. Green, Phys. Rev. D **66**, 083003 (2002).
29. N. Spooner, Private communication.
30. J. D. Vergados and Amand Faessler, Direct WIMP detection in directional experiments, hep-ph/0611230.
31. N. Brouzakis and N. Tetradis, JCAP **0601**, 004 (2006).
32. N. Tetradis and J.D. Vergados, Detection Rates of Dark Matter Coupled to Dark Energy, to be published; hep-ph/0609078.
33. J. D. Vergados, On The Direct Detection of Dark Matter- Exploring all the signatures of the neutralino-nucleus interaction, hep-ph/0601064.
34. C. J. Copi, L. M. Krauss, D. Simmons-Duffin and S. R. Stroiney, Assessing Alternatives for Directional Detection of a WIMP Halo, astro-ph/0508649.
35. A. M. Green and B. Morgan, Optimizing WIMP directional detectors, astro-ph/0609115.

PROSPECTS FOR DIRECT DARK MATTER SEARCHES IN THE CONSTRAINED MSSM

ROBERTO TROTTA*

Astrophysics Department, Oxford University
Denys Wilkinson Building, Keble Road, Oxford OX1 3RH, United Kingdom
** Corresponding author and speaker. E-mail: rxt@astro.ox.ac.uk*

ROBERTO RUIZ DE AUSTRI

Departamento de Física Teórica C-XI and Instituto de Física Teórica C-XVI, Universidad Autónoma de Madrid, Cantoblanco, 28049 Madrid, Spain

LESZEK ROSZKOWSKI

Department of Physics and Astronomy, University of Sheffield, Sheffield S3 7RH, England

We outline the WIMP dark matter parameter space in the Constrained MSSM by performing a comprehensive statistical analysis that compares with experimental data predicted superpartner masses and other collider observables as well as a cold dark matter abundance. We find that $10^{-10}\,\mathrm{pb} \lesssim \sigma_p^{SI} \lesssim 10^{-8}\,\mathrm{pb}$ for direct WIMP detection (with details slightly dependent on the assumptions made). We conclude that most of the 95% probability region for the cross section will be explored by future one–tonne detectors, that will therefore cover most of the currently favoured region of parameter space.

1. Introduction

Two of the most challenging questions facing particle physics today are the instability of the Higgs mass against radiative corrections (known as the "fine–tuning problem") and the nature of dark matter. Unlike the Standard Model (SM), weak scale softly broken supersymmetry (SUSY) provides solutions to both of them. Firstly, the fine–tuning problem is addressed via the cancellation of quadratic divergences in the radiative corrections to the Higgs mass. Secondly, assuming R–parity, the lightest supersymmetric particle (LSP) is a leading weakly interactive massive particle (WIMP) candidate for cold dark matter (CDM). Despite these and other attractive features, without a reference to grand unified theories (GUTs), low energy

SUSY models suffer from the lack of predictivity due to a large number of free parameters (*e.g.*, over 120 in the Minimal Supersymmetric Standard Model (MSSM)), most of which arise from the SUSY breaking sector.

The MSSM with one particularly popular choice of universal boundary conditions at the grand unification scale is called the Constrained Minimal Supersymmetric Standard Model (CMSSM).[1] The CMSSM is defined in terms of five free parameters: common scalar (m_0), gaugino ($m_{1/2}$) and tri–linear (A_0) mass parameters (all specified at the GUT scale) plus the ratio of Higgs vacuum expectation values $\tan\beta$ and sign(μ), where μ is the Higgs/higgsino mass parameter whose square is computed from the conditions of radiative electroweak symmetry breaking (EWSB). The economy of parameters in this scheme makes it a useful tool for exploring SUSY phenomenology.

Here we report results from a Bayesian exploration of the CMSSM parameter space, obtained through the use of Markov Chain Monte Carlo methods. In particular we focus on the prospects for direct neutralino dark matter detection with the next generation of dark matter searches. We refer the reader to Ref. 2 for full details. For other works applying a similar approach to the CMSSM, see Ref. 3 (with some relevant differences) and more recently Ref. 4. For a comprehensive discussion of implications for Higgs searches at the Tevatron, see Ref. 5.

2. Parameter space and data used

We restrict our analysis to the case sign$(\mu) = +1$, as motivated by the fact that the observed anomalous magnetic moment of the muon is positive, and since the sign of the SUSY contribution to it is the same as the sign of μ. We thus consider the 8 dimensional parameter space (θ, ψ), where θ is a vector of CMSSM parameters,

$$\theta = (m_0, m_{1/2}, A_0, \tan\beta) \tag{1}$$

while ψ is a vector of relevant SM parameters,

$$\psi = (M_t, m_b(m_b)^{\overline{MS}}, \alpha_{\rm em}(M_Z)^{\overline{MS}}, \alpha_s(M_Z)^{\overline{MS}}), \tag{2}$$

where M_t is the pole top quark mass, $m_b(m_b)^{\overline{MS}}$ is the bottom quark mass at m_b, while $\alpha_{\rm em}(M_Z)^{\overline{MS}}$ and $\alpha_s(M_Z)^{\overline{MS}}$ are the electromagnetic and the strong coupling constants at the Z pole mass M_Z, the last three evaluated in the $\overline{MS}$ scheme. We are not interested in constraining the value of the SM parameters, but rather in including the effect of the uncertainty in their experimental determination in our inferences. Therefore we marginalize over them in the end.

We employ a Bayesian approach coupled with a Markov Chain Monte Carlo method for exploring the CMSSM parameter space and derive posterior constraints on the neutralino mass and spin–independent scattering cross section, assuming that all of the cosmological dark matter abundance is made of neutralinos. From the CMSSM and SM parameters η we compute a series of derived observable quantities $f(\eta)$: the W gauge boson mass, the effective leptonic weak mixing angle $\sin^2\theta_{\text{eff}}$, the anomalous magnetic moment of the muon, $a_\mu \equiv (g-2)_\mu$, the branching ratios $BR(\bar{B} \to X_s\gamma)$ and $BR(B_s \to \mu^+\mu^-)$, the cosmological neutralino relic abundance $\Omega_{\text{CDM}}h^2$, the light Higgs mass and the superpartner masses. For all of those quantities, relevant measurements (summarized in the bottom section of Table 1) or experimental limits are included via the likelihood and used to constrain high posterior probability regions of the model. The likelihood also includes estimated theoretical uncertainties in the mapping from CMSSM and SM parameters to derived quantities, another major advantage of employing a Bayesian approach (see Ref. 2 for details). We then compute a spin–independent dark matter WIMP elastic scattering cross section on a free proton, σ_p^{SI}, including full supersymmetric contributions which have been

Table 1. Experimental mean μ and standard deviation for the nuisance parameters (top section) and derived observable quantities (bottom section, including a theoretical uncertainty describing the imprecise mapping of CMSSM and SM parameters onto observable quantities) used in the analysis.

Nuisance parameter	Mean value μ	Uncertainty experimental	Uncertainty theoretical
M_t	172.7 GeV	2.9 GeV	N/A
$m_b(m_b)^{\overline{MS}}$	4.24 GeV	0.11 GeV	N/A
$\alpha_s(M_Z)^{\overline{MS}}$	0.1186	0.002	N/A
$1/\alpha_{\text{em}}(M_Z)^{\overline{MS}}$	127.958	0.048	N/A
Derived observable			
M_W	80.425 GeV	34 MeV	13 MeV
$\sin^2\theta_{\text{eff}}$	0.23150	16×10^{-5}	25×10^{-5}
$\delta a_\mu^{\text{SUSY}}\times10^{10}$	25.2	9.2	1
$\text{BR}(\bar{\text{B}} \to \text{X}_s\gamma)\times10^4$	3.39	0.30	0.30
$\Omega_\chi h^2$	0.119	0.009	$0.1\,\Omega_\chi h^2$

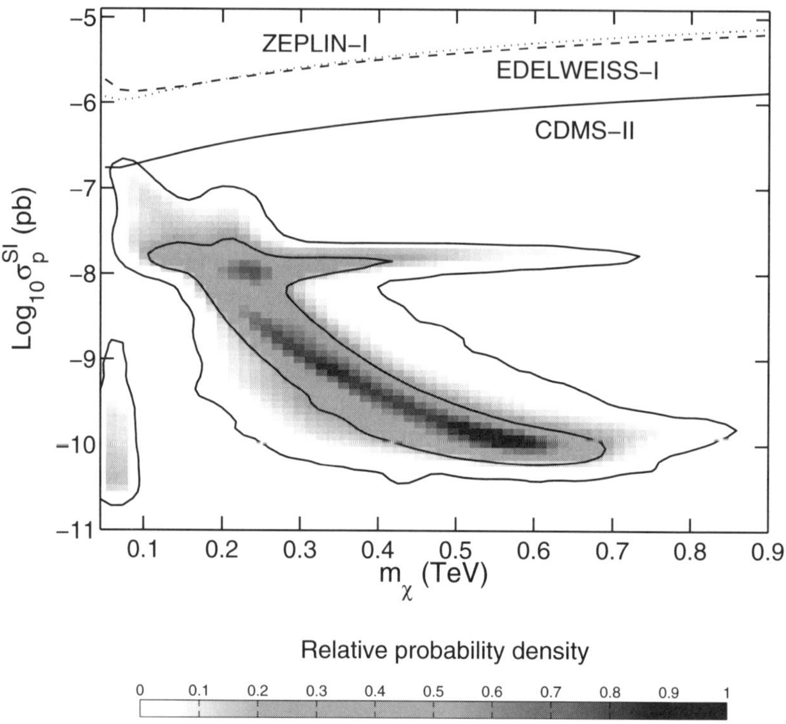

Fig. 1. The 2–dimensional probability density in the neutralino mass and spin–independent cross section plane in the CMSSM (all other parameters marginalized) with the contours containing 68% and 95% probability also marked. Current 90% experimental upper limits are also shown. A large fraction of the high–probability region lies just below current constraints and it will be probed by the next generation of dark matter searches, starting from the focus point region (horizontal region at $\sigma_p^{SI} \sim 10^{-8}$).

derived by several groups (see Ref. 6 and references therein), but we do not include current constraints in the likelihood, in view of the uncertainties in the structure of the Galactic halo (*e.g.*, existence of clumps of dark matter and therefore the value of the local halo mass density) as well as in the values of some hadronic matrix elements entering the computation of σ_p^{SI}.

3. Results for neutralino direct detection

Our Bayesian approach allows to easily compute the posterior pdf for the cross section or any other derived variable. In Fig. 1 we present the 2–dimensional posterior pdf for σ_p^{SI} and m_χ, with all other parameters marginalized over. For comparison, we also show current CDMS–II,[7]

Edelweiss–I[8] and UKDMC ZEPLIN–I[9] 90% CL upper limits, but we stress that this constraint has not been used in our analysis.

The biggest, banana–shaped region of high probability (68% regions delimited by the internal solid, blue curve) shows a well–defined anticorrelation between σ_p^{SI} and m_χ. It results from two allowed regions in the CMSSM parameter space: the bulk and stau coannihilation region and from the A–resonance at large $\tan\beta$. This region covers roughly the range $10^{-10} \lesssim \sigma_p^{SI} \lesssim 10^{-8}$ pb and $200 \lesssim m_\chi \lesssim 700$ GeV. In both cases the dominant contribution to σ_p^{SI} comes from a heavy Higgs exchange. At small $m_\chi \lesssim 100$ GeV we notice a small vertical band of fairly low probability density ($\lesssim 0.2$) at small σ_p^{SI}, a region where the light Higgs resonance contribute to reducing $\Omega_\chi h^2$ at small $m_{1/2}$ to meet the WMAP measurement. Finally, we can see a well pronounced region of high probability at fairly constant $\sigma_p^{SI} \sim 1.6 \times 10^{-8}$ pb for $m_\chi \lesssim 420$ GeV which at low m_χ partly overlaps with the previous region. At 95% this region extends up to $m_\chi \lesssim 720$ GeV for fairly constant σ_p^{SI}. This "high" σ_p^{SI} band results from the focus point (FP) region, basically independently of $\tan\beta$. This result has to be interpreted carefully, since there are large uncertainties associated with FP region, in particular with its location in the $(m_{1/2}, m_0)$ plane. Despite those outstanding questions, we believe that it is safe to expect that the FP will be the first to be probed by dark matter search experiments.

After marginalizing over all other parameters, we obtain the following 1–dimensional region encompassing 95% of the total probability:

$$0.5 \times 10^{-10}\,\text{pb} < \sigma_p^{SI} < 3.2 \times 10^{-8}\,\text{pb} \quad (95\% \text{ region}). \tag{3}$$

Currently running experiments (most notably CDMS–II but also Edelweiss–II and ZEPLIN–II) should be able to reach down to a few $\times 10^{-8}$ pb, on the edge of exploring this FP region. A future generation of "one–tonne" detectors is going to reach down to $\sigma_p^{SI} \gtrsim 10^{-10}$ pb, thus exploring much of the 95% interval.

We have checked that the above result is rather robust with respect to a change in the prior range used in the analysis or to the exclusion of the constrain coming from $g-2$.

4. Conclusions

We have presented a detailed investigation of the prospects for dark matter detection in the framework of the CMSSM parameter using state–of–the–art Bayesian methods. We have shown that the WIMP dark matter direct detection elastic scattering cross section σ_p^{SI} presents a wide spread of values (below today's limits) at around $10^{-9\pm1}$ pb and a strong anticorrelation with m_χ. In addition, a region at relatively large $\sigma_p^{SI} \simeq 1.6 \times 10^{-8}$ pb, and fairly independent of m_χ, appears to be a feature of the focus point region (despite large theoretical uncertainties) and will be the first to be tested in direct detection experiments.

We conclude that a large fraction of the high–probability parameter space of neutralino dark matter in the CMSSM will be within reach of currently running upgraded dark matter detector, and most of it will be explored by future one–tonne detectors. Our result is thus highly suggestive of a possible detection by the next generation of direct dark matter searches.

Acknowledgements

R.T. is supported by the Royal Astronomical Society through the Sir Norman Lockyer Fellowship. R.Rda is supported by the program "Juan de la Cierva" of the Ministerio de Educación y Ciencia of Spain. R.T. would like to thank the organizers of the workshop for a very enjoyable and highly interesting meeting.

References

1. G. L. Kane, C. F. Kolda, L. Roszkowski and J. D. Wells, *Study of constrained minimal supersymmetry*, *Phys. Rev.* **D49** (1994) 6173.
2. R. Ruiz de Austri, R. Trotta and L. Roszkowski, *A Markov Chain Monte Carlo analysis of the CMSSM* (2006) *JHEP* **05** (2006) 002.
3. E. A. Baltz and P. Gondolo, *Markov chain monte carlo exploration of minimal supergravity with implications for dark matter*, *JHEP* **0410** (2004) 052.
4. B. C. Allanach and C. G. Lester, *Multi-dimensional MSUGRA likelihood maps*, arXiv:hep-ph/0507283. B. C. Allanach, *Naturalness priors and fits to the constrained minimal supersymmetric standard model*, arXiv:hep-ph/0601089.
5. L. Roszkowski, R. R. de Austri and R. Trotta, arXiv:hep-ph/0611173.
6. Y. G. Kim, T. Nihei, L. Roszkowski and R. Ruiz de Austri, *Upper and lower limits on neutralino WIMP mass and spin–independent scattering cross section, and impact of new $(g-2)_\mu$ measurement*, *JHEP* **0212** (2002) 034.
7. D. S. Akerib, *et al.*, [CDMS Collaboration], *First results from the cryogenic dark matter search in the Soudan underground lab.*, *Phys. Rev. Lett.* **D93** (2004) 211301.
8. V. Sanglard *et al.* [EDELWEISS Collaboration], *Final results of the EDELWEISS–I dark matter search with cryogenic heat–and–ionization Ge detectors*, *Phys. Rev.* **D71** (2005) 122002.
9. G. J. Alner *et al.* [UK Dark Matter Collaboration], *First limits on nuclear recoil events from the ZEPLIN–I galactic dark matter detector*, *Astropart. Phys.* **23** (2005) 444.

MACCHIOs AS COLD DARK MATTER

B. J. CARR

Astronomy Unit, Queen Mary, University of London,
Mile End Road, London E1 4NS, England

This paper discusses the possibility that the cold dark matter may comprise dark baryons or primordial black holes as well as WIMPs. Some baryons are certainly dark and, although the most conventional repository for them would be hot or warm gas, they could also be contained in cold clouds or stellar remnants. Primordial black holes could provide dark matter if they are either too large to have evaporated or the stable relics of evaporating ones. All such objects would cluster inside halos, along with WIMPs, but if they were pre-galactic, there would also be an intergalactic distribution of them. The term MACCHIO is introduced as a useful generalization of MACHO. There is no conclusive evidence for MACCHIOs but they could be of astrophysical interest even if their cosmological density is low.

Keywords: WHIM, IMBH, MACHO, PBH

1. Introduction and Acronyms

It is now generally accepted that the Universe has almost exactly the critical density, with about 70% of this being in dark energy, 25% in Cold Dark Matter (CDM) and 5% in baryons. CDM is usually assumed to comprise some form of Weakly Interacting Massive Particle (WIMP) but various other forms of dark matter are dynamically "cold". For example, it is clear that an appreciable fraction of baryons are dark. Although the most conservative hypothesis is that these are contained in hot gas or a warm-hot intergalactic medium (WHIM), some of them could also be locked up in the remnants of a first generation of stars or in cold clouds of gas which never fragmented. It seems unlikely that standard stellar remnants (SSRs) could make an appreciable contribution to the halo density but it is likely that the first stars were non-standard. Another possibility is that some CDM is in the form of primordial black holes (PBHs) which formed in the early universe. These might be either the Planck-mass relics of evaporating black holes or PBHs bigger than 10^{15}g which would not have evaporated yet.

Since stellar remnants and PBHs are dynamically "cold", one would expect them to cluster along with WIMPs inside galactic halos, so they would both be classified as Massive Compact Halo Objects (MACHOs). One would therefore expect halo dark matter to be a mixture of WIMPs and MACHOs. For PBHs, the density of the two components could in principle be comparable. However, for stellar remnants the relative densities would depend on the formation epoch of the stars. If they were *pregalactic*, the mass ratio of MACHOs to WIMPs in the halo should just reflect the cosmological background ratio, which is about 0.1. If they were *protogalactic*, the ratio could be larger since the baryons could have dissipated and become more centrally concentrated than the WIMPs. In any case, microlensing results indicate that MACHOs cannot provide all the dark matter in halos, so one certainly needs some WIMPs.

It should be stressed that, except for the scenario in which one has protogalactic stellar remnants, there should also be an *intergalactic* distribution of the MACHOs. The constraints on these would generally be weaker and they could also have different observational consequences, so the distinction is important. We have noted that the dark baryons could also be in the form of cold clouds, although these would not usually be classed as MACHOs. The term MACCHIO (Massive Cloud/Compact Halo/Intergalactic Object) is therefore a useful generalization of the term MACHO.[a] The mass ranges of the various CDM candidates are indicated in Fig. 1, the other acronyms being defined later. Most of these ranges are fairly narrow, but it is noted that PBH masses span a much wider range and fill the gap between WIMPs and MACHOs.

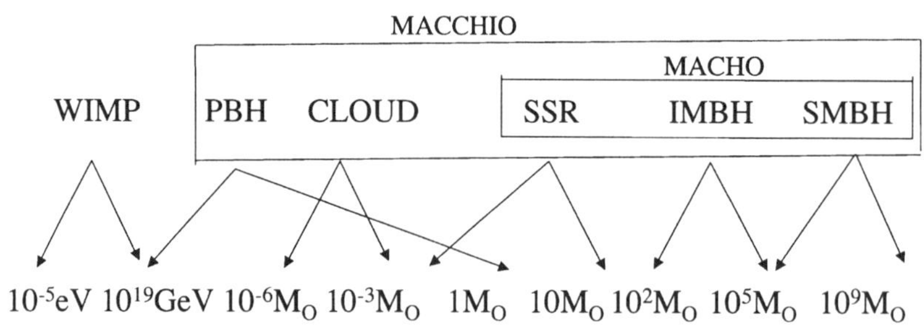

Fig. 1. Mass ranges of WIMPs, MACHOs and MACCHIOs

[a]This is also the brand name of a fashionable shirt, worn by the author for his talk!

The purpose of this talk is to review the various MACCHIO candidates. Since there is still no definite evidence for any of them, the emphasis will be largely historical.[b] Each of the candidates has waxed and waned in popularity over the years. Sometimes one of them has become the front-runnerf for a while, usually as a result of some unexpected observational claim, only to fall out of favour again when the observation has proved to be spurious. After reviewing the evidence for dark baryons in Section 2, I will consider the standard MACCHIO candidates (i.e. clouds, ordinary stellar remnants and SMBHs) in Section 3. Although these objects are likely to exist at some level, it has been known for some time that they are unlikely to make an appreciable contribution to the halo density.[1] Section 4 will focus in more detail on the Intermediate Mass Black Hole (IMBH) remnants of very massive stars. These are attracting a lot of interest at present since there is good evidence (both observational and theoretical) that the first stars were of this kind. Such objects were first proposed as halo candidates more than 20 years ago under the name Very Massive Object (VMO) and – although their halo contribution is now thought to be small – the cosmological density of IMBHs could still be comparable to that of the Supermassive Black Holes (SMBHs) thought to reside in galactic nuclei. Section 5 focusses on PBHs and we draw some conclusions in Section 6.

2. Evidence for Baryonic Dark Matter

There are three probes of the cosmological baryon density – the primordial light element abundances,[2] the microwave background anisotropies[3] and the baryon acoustic peak in the galaxy distribution[4,5] – and these all indicate a baryon density parameter $\Omega_B \approx 0.04$. This is also comparable to the baryon density associated with Lyman-α clouds at a redshift of 3. Indeed at least 80% of the baryons seem to be in gaseous form at this redshift,[6] which suggests that there was little room for any dark baryons before then. Therefore MACHOs may need to be protogalactic.

The baryon density is well below the CDM density $\Omega_C \approx 0.3$, so *most* of the dark matter cannot be baryonic. On the other hand, Ω_B is more than the density associated with visible baryons.[7,8] A review by Fukugita[9] gives the following contributions to the baryon density parameter: stars 0.0025; neutral gas 0.00062; molecular gas 0.00016; hot gas in clusters 0.002; warm and cool plasma 0.022. The dominant contribution is the last one

[b]The title of this paper was originally "**W**HIM, **I**MBHs, **M**ACHOs, **P**BHs or WIMPs?" but I judged the second-order acronym too obscure; see *keywords*.

and this includes any baryons in galactic coronae.[10] However, not all of this contrition needs to be in galaxies and this presumably includes the baryons in the warm/hot intergalactic medium (WHIM) advocated by Cen & Ostriker.[11] In the CDM scenario, they find that half the mass of baryons should be in warm ($10^5 - 10^7$K) intergalactic gas at the present epoch and there are now claims to have detected this through X-ray absorption lines in the spectra of distant quasars.[12,13] This may not seem to leave much room for other possible repositories of dark baryons, like low surface brightness galaxies,[14] dwarf galaxies[15] or MACCHIOS. But the error bars in the WHIM parameters are certainly large enough to allow some of these extra contributions.

A nice review of the history of MACHOs was given by Evans & Belokurov[16] at IDM 2004 and we briefly recap their account here. When halo dark matter was first discovered in the 1970s, MACHOs were the favoured explanation since WIMPs had not been thought of. They remained popular in the 1980s, especially when the evidence from cooling flows suggested that low mass stars could be made prolifically. Their topicality peaked in 1994 when the first detections of microlensing of stars in the LMC[17] by the MACHO collaboration indicated a halo fraction close to unity in 0.1 $M_\odot$ objects.[18] However, six years of further data changed this conclusion: with 13-17 microlensing events, it seemed that MACHOs could comprise at most 20% of the halo and the favoured mass became 0.5 $M_\odot$.[19] The results of the EROS survey were equally discouraging, with 8 years of monitoring producing only three candidates and implying even stronger constraints.[20] Evans & Belokurov also highlighted the results of microlensing surveys of stars in M31.[21,22] They again concluded that there is little evidence for MACHOs, even though the POINT-AGAPE collaboration (of which they are members) had detected six events and inferred that at least 20% of the halo must be in MACHOs.[24] So the situation is clearly ambiguous.

Evans & Belokurov entitled their paper "RIP: The MACHO era", which is reminiscent of the title, "The death of baryonic dark matter", of the paper by Freese et al. at IDM 2000.[25] In my view, both these titles are misleading or at least beg the question of when a proposal can be certified "dead". Although it is certainly implausible that *most* of the halo dark matter comprises MACHOs, they may still exist at some level and they could have important astrophysical consequences even if their cosmological density is small. For example, dynamical and lensing constraints preclude SMBHs from making up galactic halos but they are clearly important because they reside in galactic nuclei.

3. Assessing the MACCHIO Candidates

In the remaining sections we will discuss each of the MACCHIO candidates in turn, focussing particularly on their possible contribution to galactic halos. The emphasis will be primarily historical. We will indicate how the popularity of each candidate has evolved over time with a series of figures. The vertical axis in each figure is qualitative measure of the popularity of the candidate, although this measure is not defined precisely enough for the axis to be calibrated!

3.1. *Brown Dwarfs*

Brown dwarfs (BDs) are objects in the mass range $0.001 - 0.08\ M_\odot$ which never burn hydrogen. Their popularity history is indicated in Fig. 2, At first it was argued that such objects may form efficiently in pregalactic or protogalactic cooling flows.[26,27] Their direct detection is difficult but indirect evidence may come from extrapolating the initial mass function (IMF) of hydrogen-burning stars to lower masses. Although the IMF for Population I stars ($dN/dm \propto m^{-\alpha}$ with $\alpha < 1.8$) suggests that only 1% of them could be in BDs,[28] at one time there were observational claims that low-metallicity Population II stars had a much steeper IMF,[29] so the same might apply for Population III stars The prevalence of BDs also seemed to be indicated by the first MACHO microlensing results.[18] However, claims that low-metallicity objects have a steep IMF were later discredited, observations of Galactic and LMC globular clusters[30] and dwarf spheroidal field stars[31] indicating a universal IMF with $\alpha < 1.5$ at low masses. Also the 6-year MACHO results were found to exclude a large fraction of halo BDs.[19] The scenario was briefly revived by Hansen,[32] who suggested that there could be "Beige Dwarfs" in the mass range $0.1 - 0.3\ M_\odot$ which are larger than the traditional BD upper limit because they form by sufficiently slow accretion that they never ignite their nuclear fuel. Lynden-Bell & Tout[33] later argued that this proposal is implausible but Schild[34] has recently advocated objects in the BD range to order to explain microlensing in lensed quasars.

3.2. *Red Dwarfs*

The popularity history of red dwarfs (RDs) – low mass hydrogen-burning stars – is indicated in Fig. 3. It was known ten years ago that discrete source counts imply that the fraction of the halo mass in our own Galaxy contained in RDs must be less than 1%.[35,36] These limits would be weakened

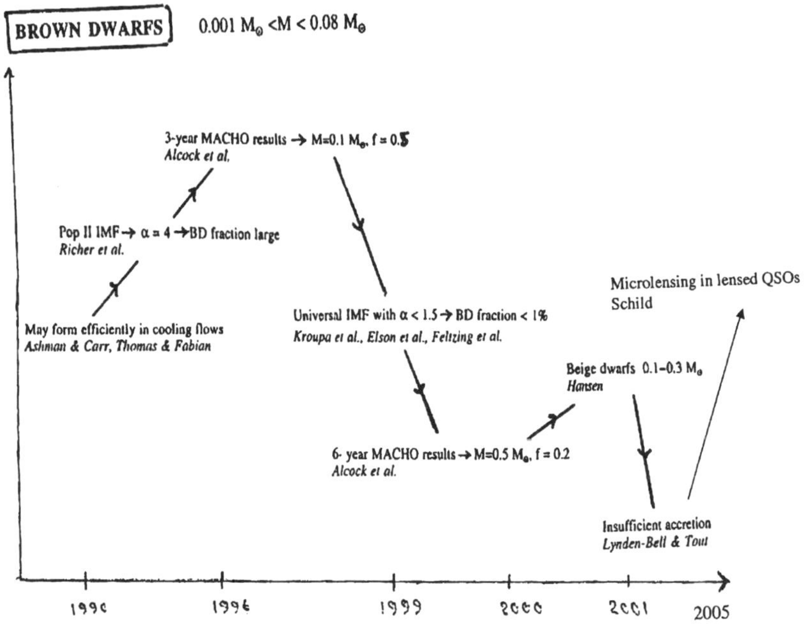

Fig. 2. History of popularity of brown dwarfs

if the stars were clustered[37] but not much. However, the discovery of a red halo around NGC 5907 by Sackett et al.,[38] apparently emanating from low mass stars with a density profile like that of the halo, was therefore a particularly interesting development. This detection was confirmed in the J and K bands by James & Casali.[39] However, the suggestion that the stars might be of primordial origin (with low metallicity) was contradicted by the results of Rudy et al.,[40] who found that the color indicated stars with solar metallicity. Furthermore, the red light around NGC 5907 has only been observed within a few kpc and no NIR emission is detected at 10-30 kpc.[41] Both these points go against the claim that the light is associated with MACHOs. Later it was suggested that the red light is more likely to derive from a disrupted dwarf galaxy, the stars of which would naturally follow the dark matter profile,[42] or to be a ring left over from a disrupted dwarf spheroidal galaxy.[43] In any case, NGC 5907 did not seem to be typical since ISO observations of four other edge-on bulgeless spiral galaxies gave no evidence for red halos.[44] Recently, however, Zackrisson et al.[45] have claimed to detect red halos in blue compact galaxies, while Zibetti et al.[46] have reported them in edge-on SDSS disc galaxies, so the jury is still out.

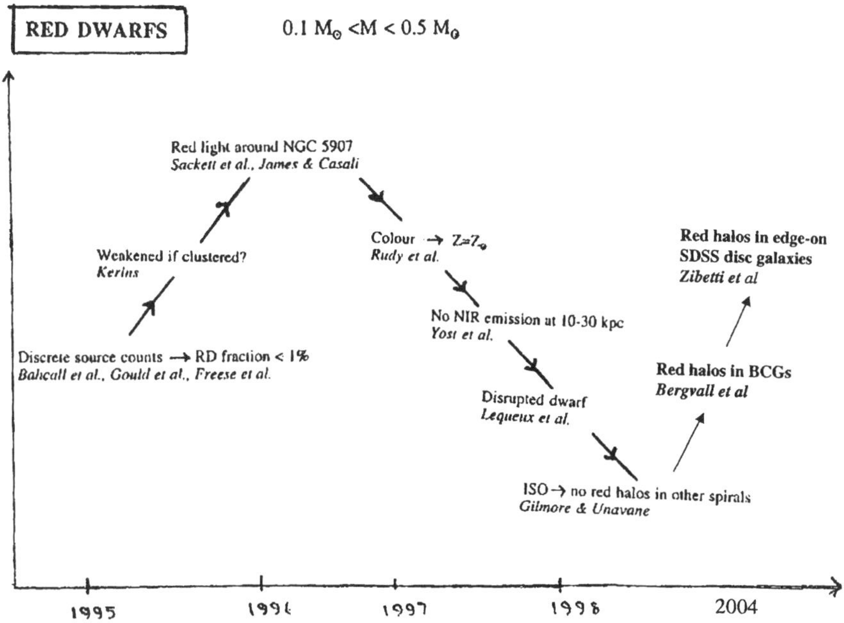

Fig. 3. History of popularity of red dwarfs

3.3. *White Dwarfs*

The popularity history of white dwarfs (WDs) is indicated in Fig. 4. Starting from 1990, a variety of constraints made WDs increasingly implausible dark matter candidates. One requires a very contrived IMF, lying between 2 $M_\odot$ and 8 $M_\odot$, in order to avoid excessive production of light or metals;[47] the fraction of WD precursors in binaries is expected to produce too many type 1A supernovae;[48] and the halo fraction is constrained to be less than 10% in order to avoid the luminous precursors contradicting the upper limits from galaxy counts.[49] Later calculations strengthened these limits. A study of CNO production[50] suggested that a halo comprised entirely of WDs would overproduce C and N compared to O by factors as large as 60 and a similar limit came from considering helium and deuterium production.[51] Also extragalactic background light limits required that the halo WD fraction be less than 6%.[52] However, the "many nails in the coffin" of the WD scenario were confounded by the 6-year microlensing results of the MACHO project, which suggested that the lens mass is probably in the WD range.[19] Theorists therefore tried to resuscitate the scenario. At least some of the afore-mentioned limits had to be reconsidered in view of claims by Hansen[53] that old WDs with pure-hydrogen envelopes could be much bluer and brighter than previously supposed. This suggestion was apparently supported by HST observations of Ibata et al.,[54] who claimed to detect five candidates of this kind. The objects were blue and isolated and showed high proper motion. They inferred that they were 0.5 $M_\odot$ hydrogen-atmosphere WDs with an age of 12 Gyr. However, Ibata et al. later withdrew this claim and the recent VIRMOS survey implies a WD fraction below 1%.[55]

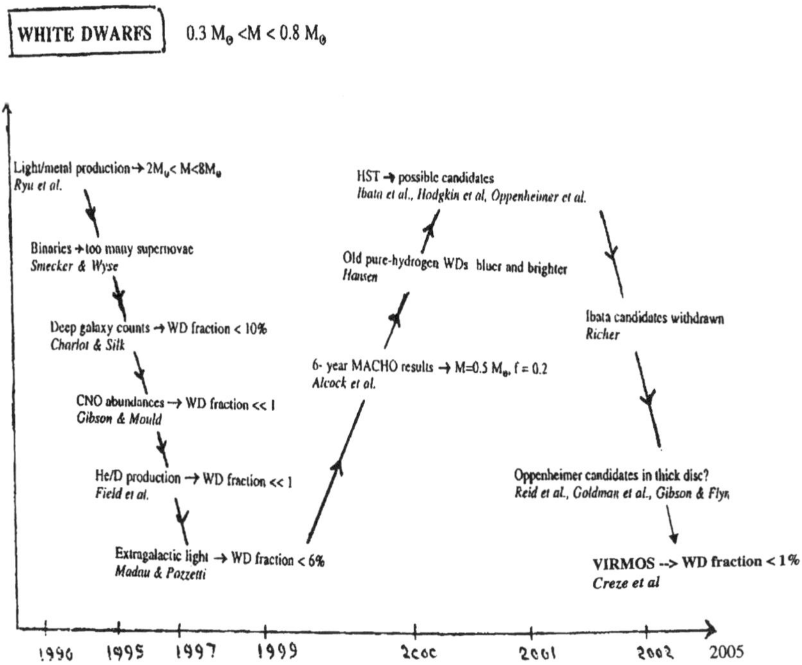

Fig. 4. History of popularity of white dwarfs

3.4. *Supermassive Black Holes and Clusters*

Stars larger than 10^5 $M_\odot$ are termed "Supermassive Objects" (SMOs) and would collapse directly to supermassive black holes (SMBHs) without any nucleosynthetic or background light production.[56,57] They would therefore be natural dark matter candidates if they could ever form. They are known to exist in galactic nuclei but the formation mechanisms for a halo population of SMBHs would have to be very different. Their popularity history is indicated in Fig. 5. In the 1980s halo objects of $10^6\ M_\odot$ were invoked to explain the observed heating of stars in the Galactic disc.[58] However, it was later found that heating by density waves or giant molecular clouds would be equally plausible.[59] Subsequently. theoretical arguments appeared to support the possibility that the first stars could be SMOs.[60,61] However, SMBHs would still have noticeable lensing and dynamical effects,[62] and these strongly constrain the fraction of halo dark matter in such objects. Dark clusters of lower-mass MACHOs would in some ways offer an alternative to SMBHs but these too would be strongly constrained if they were not to be disrupted by tidal effects.[63,64]

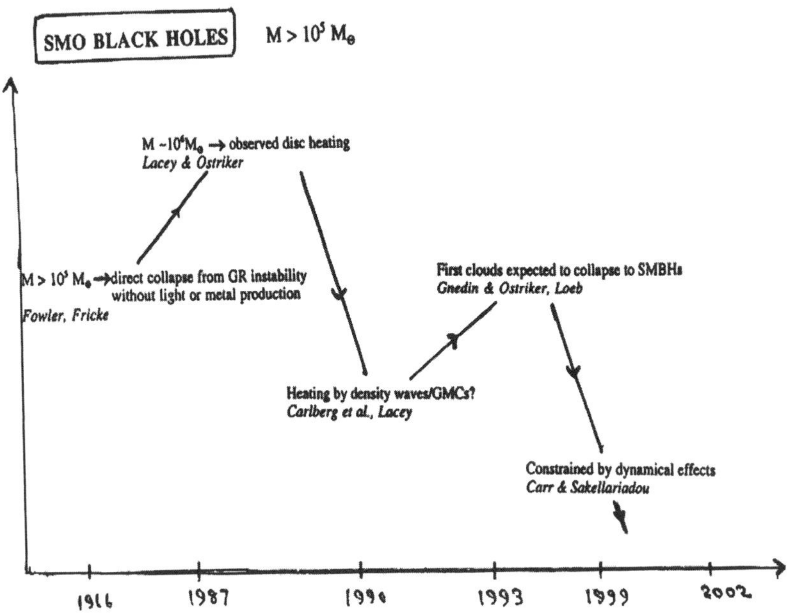

Fig. 5. History of popularity of SMO black holes (SMBHs)

3.5. *Cold Clouds*

The suggestion that dark matter in galaxies could be in cold clouds was first made by Pfenniger et al.[65] and the popularity history of these objects is indicated in Fig. 6. They envisaged the clouds having a mass of around $10^{-3}\ M_\odot$ and being distributed in a disc. This now seems dynamically rather implausible, but De Paolis et al.[66] then proposed a similar scenario with a *spheroidal* distribution of clouds. Walker[67] argued that such clouds could explain both the "Extreme Scattering Events" detected by radio observations in our own galaxy and the quasar microlening events claimed by Hawkins.[68,69] Indeed Walker & Wardle[70] advocated a model in which the halo entirely comprises such clouds, with visible stars being formed as a result of collisions between them. They claimed that this scenario naturally produces various observed features of galaxies. Although clouds of $10^{-3}\ M_\odot$ could not explain the LMC microlensing events, Draine[71] argued that such clouds might still produce microlensing through refraction effects. The interaction of cosmic rays with such clouds would also produce an interesting γ-ray signature.[72,73] Recently it has been pointed out that EGRET could detect clouds of $10^{-3}\ M_\odot$.[74] It is interesting that clouds are the only candidate whose popularity has risen monotonically!

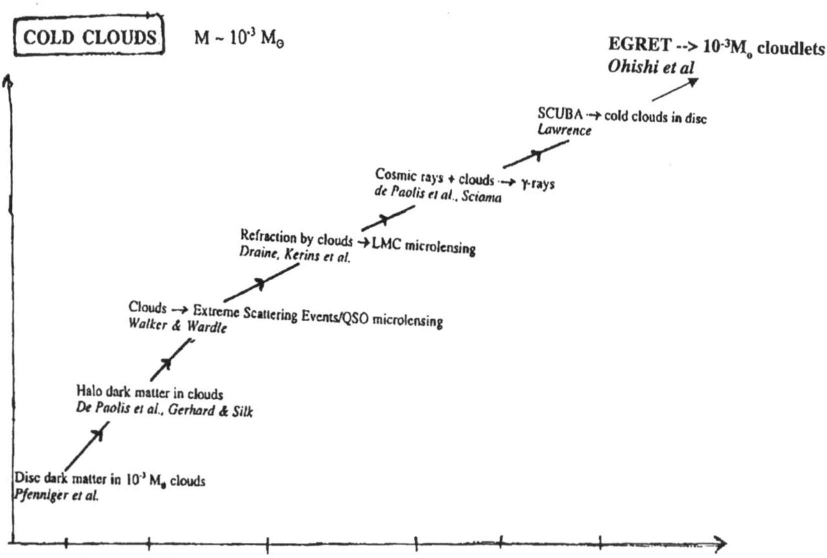

Fig. 6. History of popularity of cold molecular clouds

4. Intermediate Mass Black Hole Remnants

Stars in some range above 8 $M_\odot$ would leave neutron star (NS) remnants, while those in some range above about 20 $M_\odot$ would leave ordinary black hole remnants. However, neither of these would be plausible candidates for halo dark matter because their precursors would have unacceptable nucleosynthetic yields. Stars larger than 100 $M_\odot$ are sometimes termed "Very Massive Objects" (VMOs).[75] Although they are radiation-dominated and therefore unstable to pulsations and mass loss, they are expected to survive their hydrogen-burning stage (lasting a few million years) but to go unstable during their oxygen-burning stage due to the electron-positron pair-instability.[76] It has been known for 20 years from both theoretical and numerical calculations that this leads to complete disruption below some critical mass around 200 $M_\odot$ but complete collapse to (what are now termed) "Intermediate Black Holes" (IMBHs) above this.[75,77]

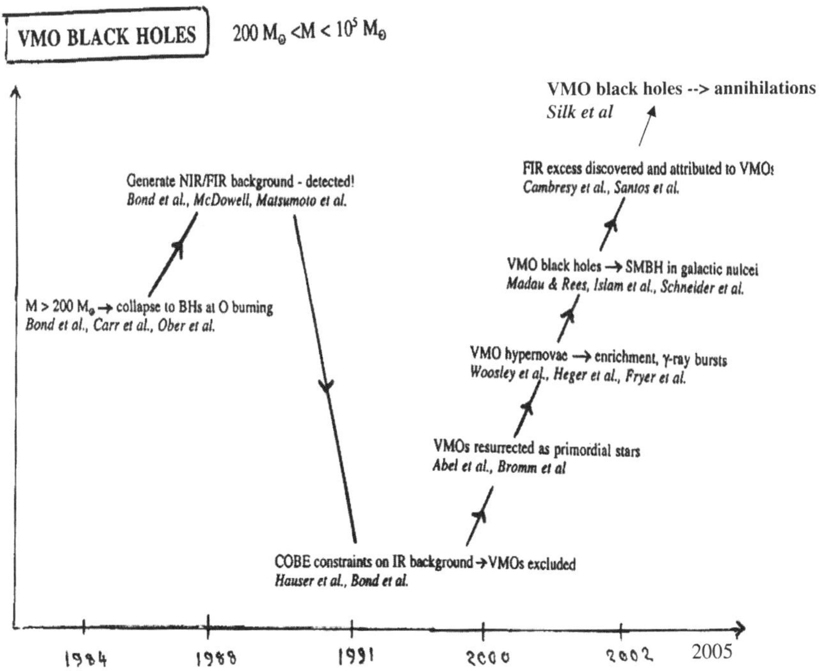

Fig. 7. History of popularity of VMO black holes (IMBHs)

The popularity history of this candidate is indicated in Fig. 7. IMBHs might be regarded as good dark matter candidates because their precursors would not have the nucleosynthetic consequences of ordinary stars.[78] However, during their main-sequence phase, VMOs would be expected to generate of a lot of background light. By today this should have been shifted into the infrared or submillimetre band, as a result of either redshift effects or dust reprocessing, so one would expect a sizeable infrared/submillimetre cosmic background.[79] In 1984 such a background was detected, which led to a short-lived popularity of the VMO picture.[80,81] However, the COBE measurements showed this was a false alarm and, since then, the IR background measurements has been mainly used to constrain the VMO scenario.[82] COBE does now seem to have detected a genuine infrared background[83] but this can probably be attributed to ordinary Population I and II stars. In any case, the current constraints on such a background strongly limit the density of any VMOs.

Despite the demise of VMOs as MACHO candidates, there has been a growing interest in the possibility that the first (zero-metallicity) stars to form in the universe may have been VMOs. Numerical calculations suggest that such stars could have an IMF which peaks in the VMO range[84] and evolutionary calculations confirm the existence of the critical mass of around 200 $M_\odot$ above which an IMBH forms.[85] The "hypernovae" generated by exploding VMOs might also be associated with gamma-ray bursts.[86,87] There have been several excellent reviews of this topic.[88–90] Here we will summarize some recent developments relating to the astrophysical consequences of halo IMBHs.

* Madau & Rees[91] proposed that IMBHs could form in minihalos at z=20 from 3σ fluctuations and have a cosmological density comparable to the SMBHs in AGN. They would be expected to cluster in bulges and dynamical friction would then drag drags some of them into galactic nuclei, where they would be captured by the SMBH and generate gravitational waves from coalescences.

* Schneider et al.[92] argued that the precursors of IMBHs would generate an initial enrichment ($Z \sim 10^{-4} Z_\odot$) sufficient to decrease the subsequent fragment mass to around 1 $M_\odot$. In drifting to galactic nuclei, the IMBHs would also generate off-centre X-ray sources. Islam et al.[93] considered how much the infall of IMBHs could contribute to the mass of the central SMBH. They also studied the distribution and dynamical and accretion effects of minihalos that do not reach the nucleus.

* Salvaterra & Ferrara[94] examined the redshifted light from a first generation of massive stars at z=9 and claimed that they could generate the observed NIR background if their mass fraction f_* is 10-50%. However, they would need to be VMOs to avoid overenrichment of the IGM.

* Zhao & Silk[95] pointed out that IMBHs in minihalos should be enshrouded by CDM minispikes. Gamma-rays resulting from the annhilation of WIMPs could then stand out above the smooth Galactic background and be detectable by GLAST or Air Cherenkov Telescopes. This proposal was studied in more detail for specific formation scenarios by Bertone et al.,[96] who emphasized that such observations probe both MACHOs and WIMPs.

* Mu & Totani[98] and Krolik[99] explored the possibility that Ulta-Luminous X-ray (ULX) sources could be the IMBH remnants of Population III stars passing through dense molecular clouds. They pointed out that they would be preferentially found at edges of large gas clouds in star-forming regions. Mapelli et al.[100] performed N-body simulations of halo IMBHs accreting as they pass through HI or H_2 regions and showed that constraints on ULX sources in the Milky Way imply $f_* < 0.001 - 0.01$.

* Inoue & Chiba[101] have recently proposed that IMBHs could be detected through lensing effects by observing their silhouettes in the images of multiply-imaged QSO. This would require an angular resolution of 0.03 mas, such as could be achievable by future VLBI space missions.

5. Primordial Black Holes

Black holes with a wide range of masses could have formed in the early Universe as a result of the great compression associated with the Big Bang.[102–104] A comparison of the cosmological density at a time t after the Big Bang with the density associated with a black hole of mass M shows that PBHs would have of order the particle horizon mass at their formation epoch: $M_H(t) \approx 10^{15}(t/10^{-23}\mathrm{s})\mathrm{g}$. PBHs could thus span an enormous mass range: those formed at the Planck time (10^{-43}s) would have the Planck mass (10^{-5}g), whereas those formed at 1 s would be as large as $10^5 M_\odot$, comparable to the mass of the holes thought to reside in galactic nuclei. By contrast, black holes forming at the present epoch could never be smaller than about $1 M_\odot$.

The high density of the early Universe is a necessary but not sufficient condition for PBH formation. One also needs density fluctuations, so that overdense regions can eventually stop expanding and recollapse. Indeed one reason for studying PBH formation is that it imposes important constraints on primordial inhomogeneities,[105] especially those generated by inflation.[106] PBHs may also form at various phase transitions expected to occur in the early Universe. In some of these one still requires pre-existing density fluctuations but it becomes easier for PBHs to form. For example, this applies if the equation of state becomes soft for a while.[107] In others the PBHs form spontaneously, even if the Universe starts off perfectly smooth. For example, PBHs could form from the collapse of cosmic strings[108,109] or from bubble collisions.[110,111]

The realization that PBHs might be small prompted Hawking to study their quantum properties. This led to his famous discovery[112] that black holes radiate thermally with a temperature $T \approx 10^{-7}(M/M_\odot)^{-1}$K, so they evaporate on a timescale $\tau(M) \approx 10^{64}(M/M_\odot)^3$y. Only black holes smaller than 10^{15}g would have evaporated by the present epoch, so this implies that this effect could be important only for black holes which formed before 10^{-23}s. Hawking's result was a tremendous conceptual advance, since it linked three previously disparate areas of physics - quantum theory, general relativity and thermodynamics. Even if PBHs never existed, it has therefore been useful to think about them! However, at first sight it was bad news for PBH enthusiasts. For since PBHs with a mass of 10^{15}g would be producing photons with energy of order 100 MeV at the present epoch, the observational limit on the γ-ray background intensity at 100 MeV immediately implied that their density could not exceed 10^{-8} times the critical density.[113] Not only did this render such PBHs unlikely dark matter candidates, it also implied that there was little chance of detecting black hole explosions at the present epoch.[114] Nevertheless, evaporating PBHs could still have many interesting cosmological consequences.[115] They might also contribute to cosmic rays,[116] in which case they could provide a unique probe of higher energy physics and extra dimensions.[117]

PBHs larger than 10^{15}g would not have evaporated yet and so the γ-ray limit does not preclude their having an appreciable cosmological density. Furthermore even evaporating black holes may leave stable Planck relics and these would only be indirectly constrained by the γ-ray limit.[118] In both cases, the PBHs would be dynamically "cold", so in the next two sections we focus on their role as CDM candidates. The popularity history of PBHs is indicated in Fig. 8.

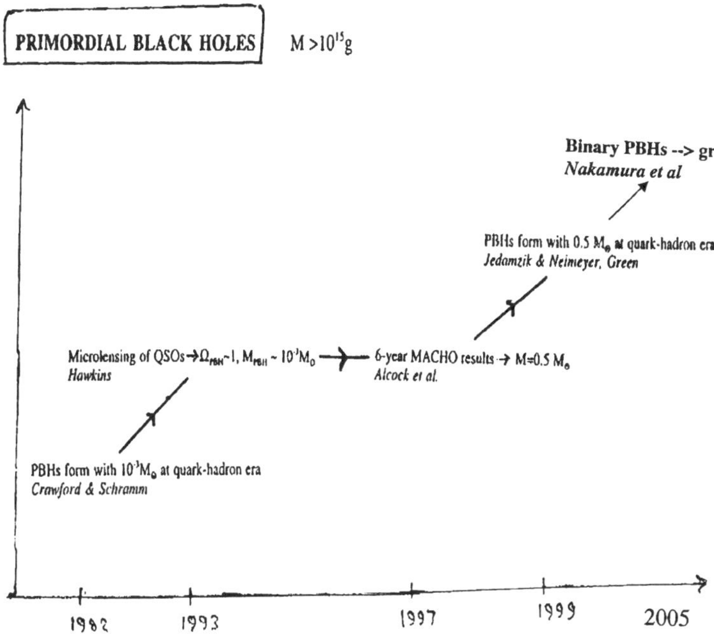

Fig. 8. History of popularity of Primordial Black Holes

5.1. *PBHS as Dark Matter*

It should first be stressed that the fraction β of the Universe's mass collapsing into PBHs at early times need only be tiny to contribute appreciably to the dark matter today. The current density parameter $\Omega_{\rm PBH}$ associated with PBHs which form at a redshift z or time t is related to β by $\Omega_{\rm PBH} = \beta\Omega_R(1+z) \approx 10^6\beta(t/s)^{-1/2} \approx 10^{18}\beta(M/10^{15}g)^{-1/2}$, where $\Omega_R \approx 10^{-4}$ is the density parameter of the microwave background.[105] The $(1+z)$ factor arises because the radiation density scales as $(1+z)^4$, whereas the PBH density scales as $(1+z)^3$. Any limit on $\Omega_{\rm PBH}$ therefore places a constraint on $\beta(M)$.[119,120]

One possibility is that PBHs with a mass of around $1M_\odot$ could have formed at the quark-hadron phase transition at 10^{-5}s because of a temporary softening of the equation of state then.[121] One only needs $\beta \approx 10^{-9}$ at this epoch to produce $\Omega_{PBH} \approx 1$. If the QCD phase transition is assumed to be 1st order, then hydrodynamical calculations show that the amplitude of the fluctuations required for PBH formation is reduced below the value which pertains in the radiation era.[122] This means that PBH formation will be strongly enhanced at the QCD epoch, with the mass distribution peaking at around the horizon mass then.[123–125] Such PBHs might be able to explain the MACHO microlensing results[19] and also the claimed microlensing of quasars.[68,69] Indeed this "lensing mass coincidence" is one reason why PBHs have become a prime CDM contender in the last decade.

The QCD transition is probably the latest phase transition that could generate PBHs like this but it is not the only one. For example, PBHs with a mass of around $10^{-7}M_\odot$ could form in TeV quantum gravity scenarios.[126] Whatever the PBH mass, one of the interesting implications of these scenarios is the possible existence of a halo population of *binary* black holes.[127] With a full halo of such objects, there could be a huge number of binaries inside 50 kpc and some of these could be coalescing due to gravitational radiation losses at the present epoch.[128] If the associated gravitational waves were detected, it would provide a unique probe of the halo distribution.[129] Gravity waves from binary PBHs would be detectable down to $10^{-5}M_\odot$ using VIRGO, $10^{-7}M_\odot$ using EURO and $10^{-11}M_\odot$ using LISA.[126]

There are some mass ranges in which PBHs are already excluded from providing dark matter in galactic halos. For example, femtolensing of gamma-ray bursts by PBHs precludes those in the mass range $10^{17}-10^{20}$g from having a critical density and microlensing of stars in the LMC exclude $10^{26}-10^{34}$g PBHs.[19] However, there are no constraints in the intermediate (sublunar) mass range $10^{20}-10^{26}$g.[130] Note that LISA might detect the gravitational impulse induced by any nearby passing PBH.[131,132] However, this method would not work below 10^{14}g (because the effect would be hidden by the Moon) or above 10^{20}g (because the encounters would be too rare).

PBHs might also play a role in the formation of large-scale structure.[133–136] For example, it was pointed out a long time ago that Poisson fluctuations in the number density of PBHs can generate appreciable density perturbations on large scales if the PBHs are big enough.[137–140] This idea has recently been revived to explain voids and Lyman-alpha clouds.[141] In such scenarios, it is important to know how much a PBH can grow through accretion and it has been argued that this could be significant.[142]

Several people have suggested that the $10^6-10^8M_\odot$ SMBHs thought to reside in galactic nuclei could be of primordial origin. For example, a 1st order phase transition could produce clusters of PBHs, around which a single SMBH might then condense.[143] Alternatively, it has been proposed that inflation could produce closed domain walls, which then collapse to form $10^8M_\odot$ PBHs.[135] Another scheme invokes more modest mass PBHs of $10^3 M_\odot$, resulting from a feature in the inflaton potential, which then grow to $10^8 M_\odot$ by accretion.[144] A related idea comes from Bean & Magueijo,[145] who have suggested that PBHs may accrete from the quintessence field which is invoked to explain the acceleration of the Universe. However, they use a Newtonian formula for the accretion rate[102] and this is known to be questionable.[104] Recent studies of the accretion of a scalar field by a PBH also indicate that this is unlikely.[146]

5.2. *Planck Mass Relics as Dark Matter*

Some people have speculated that black hole evaporation could cease once the hole gets close to the Planck mass.[147–149] For example, in the standard Kaluza-Klein picture, extra dimensions are assumed to be compactified on the scale of the Planck length. This means that the influence of these extra dimensions becomes important at the energy scale of 10^{19}GeV and such effects could conceivably result in evaporation ceasing at the Planck mass. Various non-quantum-gravitational effects (such as higher order corrections to the gravitational Lagrangian or string effects) could also lead to stable relics but the relic mass is usually close to the Planck scale. Another possibility, as argued by Chen & Adler,[150] is that stable relics could arise if one invokes a "generalized uncertainty principle". This replaces the usual uncertainty principle with one of the form $\Delta x > \hbar/\Delta p + l_P^2 \Delta p/\hbar$, where the second term is supposed to account for self-gravity effects and l_p is the Planck lengfth. This means that the black hole temperature reduces to the standard Hawking form for $M \gg M_P$ but remains finite instead of diverging at the Planck mass itself.

Whatever the cause of their stability, Planck mass relics would provide a possible cold dark matter candidate.[118,120,151] Producing a critical density of relics obviously requires fine-tuning of the collapse fraction β but no more so than for non-evaporating PBHs.[152] In particular, it has been argued that hybrid inflation could produce relics from 10^5g PBHs formed at 10^{-32}s.[153,154]

6. Conclusions

There are many forms of CDM besides WIMPs and these may be collectively described as MACCHIOs. Some MACCHIO candidates – such as cold gas clouds or stellar remnants – are baryonic. There are certainly *some* dark baryons and, although one cannot exclude these being in hot gas or a warm-hot intergalactic medium, there is certainly some room for baryonic MACCHIOs. However, the evidence for such objects is ambiguous: the popularity of each candidate has vacillated considerably over the years in response to various observational claims and we cannot be certain that any of them exist in appreciable numbers. Currently, Intermediate Mass Black Holes are attracting much interest and these could have important astrophysical effects even if their cosmological density is low. The other MACCHIO candidates are primordial black holes – either ones large enough not to have evaporated by now or the Planck mass relics of evaporating ones. Such objects would be regarded as non-baryonic since they form when the Universe is radiation-dominated. In all scenarios, the best probe of MACCHIOs is microlensing surveys. The LMC and M31 results suggest that there may be MACCHIOs of around 0.5 $M_\odot$ but the evidence is still ambiguous. In any case, it is clear that MACCHIOs do not provide *all* the mass in halos, so the important question is how small the fraction needs to be before the scenario is regarded as uninteresting. One cannot declare MACCHIOs "dead" without answering this question.

References

1. B.J. Carr, Ann.Rev.Astron.Astrophys. 32 (1994) 531.
2. D. Kirkmann et al., Ap.J. Supp. 149 (2003) 1.
3. D.N. Spergel et al., Ap.J. Supp. 148 (2003) 175.
4. D.J. Eisenstein et al., Ap.J. 633 (2005) 560.
5. S. Cole et al., MNRAS 362 (2005) 505.
6. D.H. Weinberg et al., Ap.J. 490 (1997) 564.
7. M. Persic & P. Salucci, MNRAS 258 (1992) 14P.
8. M. Fukugita, C.J. Hogan & P.J.E. Peebles, Ap.J. 503 (1998) 518.
9. M. Fukugita, IAU Symp. 220, eds. S. D. Ryder et al. (Astron. Soc. Pac. 2004), p.227.
10. M. Fukugita & P.J.E. Peebes, Ap.J. 616 (2004) 643.
11. R. Cen & J.P. Ostriker, Ap.J. 514 (1999) 1.
12. F. Nicastro et al., Nature 433 (2005) 495.
13. F. Nicastro, M. Elvis, F. Fiore & S. Mathur, Adv. Space Res. 36 (2005) 721.
14. W.J.G. De Blok & S. McGaugh, MNRAS 290 (1997) 533.
15. J. Loveday et al., Ap.J. 489 (1997) L29.
16. N.W. Evans & V. Belokurov, in The Identification of Dark Matter, ed. N.J. C. Spooner & V. Kudryavtsev World Scientific, 2005) p.141; astro-ph/0505167.
17. B. Paczynski, Ap.J. 304 (1986) 1; 308 (1986) L43.
18. C. Alcock et al., Ap.J. 486 (1997) 697.
19. C. Alcock et al., Ap.J. Lett. 550 (2001) L169.
20. T. Lasserre et al., Astron.Astrophys. 355 (2000) L39.
21. R. Ansari et al., Astron.Astrophys. 344 (1999) L49.
22. A.P.S. Crotts & A.B. Tomaney, Ap.J. 473 (1997) L87.
23. J.H. An et al., MNRAS 351 (2004) 1071.

24. S. Calchi Novati et al., Astron. Astrophys. 443 (2005) 911.
25. K. Freese, B.D. Fields & D.S. Graff, in The Identification of Dark Matter, ed. N. J. C. Spooner & V. Kudryavtsev (World Scientific, 2001) p. 213
26. K.A. Ashman & B.J. Carr, MNRAS 249 (1991) 13.
27. P. Thomas & A.C. Fabian, MNRAS 246 (1990) 156.
28. P. Kroupa, C. Tout & G. Gilmore, MNRAS 262 (1993) 545.
29. H.B. Richer & G.G. Fahlman, Nature 358 (1992) 383.
30. R.A. Elson et al., in Stellar Populations in the Magellanic Clouds (1999).
31. S. Feltzing, G. Gilmore & R.F.G. Wyse, Ap.J. 516 (1999) L17.
32. B.M.S. Hansen, Ap.J., 517 (1999) L39.
33. D. Lynden-Bell & C. Tout
34. R.E. Schild, in The Identification of Dark Matter, ed. N. J. C. Spooner & V. Kudryavtsev. (World Scientific Publishing Company, 2005) p.183.
35. J.N. Bahcall et al., Ap.J. 435 (1995) L31.
36. A. Gould C. Flynn & J. Bahcall, Ap.J. 503 (1998) 798.
37. E. Kerins, Astron.Astrophys. 328 (1997) 5.
38. P.D. Sackett et al., Nature 370 (1994) 441.
39. P. James & M.M. Casali, Spectrum 9 (1996) 14.
40. R.J. Rudy et al., Nature 387 (1997) 159.
41. S.A. Yost et al., Ap.J. 535 (1999) 644.
42. J. Lequeux et al., Astron.Astrophys. 334 (1998) L9.
43. Z. Zheng et al., Astron.J. 117 (1999) 2757.
44. G. Gilmore & M. Unavane, MNRAS 301 (1998) 813.
45. E. Zackrisson et al., Ap.J. 650 (2006) 812.
46. S. Zibetti, S.D.M. White & J. Brinkman, MNRAS 347 (2004) 556.
47. D. Ryu, K.A. Olive & J. Silk, Ap.J. 353 (1990) 81.
48. T.A. Smecker & R.F.G. Wyse, Ap.J. 372 (1991) 448.
49. S. Charlot & J. Silk, Ap.J. 445 (1995) 124.
50. B.K. Gibson & J.R. Mould, Ap.J. 482 (1997) 98.
51. B.D. Field, K. Freese & D.S. Graff, Ap.J. 534 (2000) 265.
52. P. Madau & L. Pozzetti, MNRAS 312 (2000) L9.
53. B.M.S. Hansen, Nature, 394 (1998) 860.
54. R.A. Ibata, H.B. Richer, R.L. Gilliland & D. Scott, Ap.J. 524 (1999) L95.
55. M. Crézé et al., Astron.Astrophys. 426 (2004) 65.
56. W. Fowler, Ap.J. 144 (1966) 180.
57. K.J. Fricke, Ap.J. 183 (1973) 941,
58. C.G. Lacey & J.P. Ostriker, Ap.J. 299 (1985) 633.
59. R.G. Carlberg et al., Ap.J. 294 (1985) 674.
60. N. Yu. Gnedin & J.P. Ostriker, Ap.J. 400 (1992) 1.
61. A. Loeb, Ap.J. 403 (1993) 542.
62. B.J. Carr & M. Sakellariadou, Ap.J. 516 (1998) 195.
63. B.J. Carr & C.G. Lacey, Ap.J. 316 (1987) 23.
64. R.B. Metcalf & J. Silk, Ap.J. 519 (1999) L1.
65. D. Pfenniger, F. Combes & L. Martinet, Astron.Astrophys. 285 (1994) 79.
66. F. De Paolis et al., Astron.Astrophys. 295 (1995) 567; 299 (1995) 647.
67. M. Walker, MNRAS 306 (1999) 504; 308 (1999) 551.
68. M.R.S. Hawkins, Nature 366 (1993) 242.

69. M.R.S. Hawkins, Mon.Not.Roy.Astr.Soc. 278 (1996) 787.
70. M. Walker & M. Wardle, Ap.J. 498 (1998) L125.
71. B.T. Draine, Ap.J. 509 (1998) L41.
72. F. De Paolis, G. Ingrosso, Ph. Jetzer & M. Roncadelli, Ap.J. 510 (1999) L103.
73. D.W. Sciama, MNRAS 312 (2000) 33.
74. M. Ohishi, M. Mori & M. Walker, Ap.J. 610 (2004) 868.
75. J.R. Bond, W.D. Arnett & B.J. Carr, Ap.J. 280 (1984) 825.
76. W. Fowler & F. Hoyle, Ap.J. Supp. 9 (1964) 201.
77. W.W. Ober, M.F. El Eid & K.L. Fricke, Astron. Astroiphys. 119 (1983) 61.
78. B.J. Carr, J.R. Bond & W.D. Arnett, Ap.J. 277 (1984) 445.
79. J.R. Bond, B.J. Carr & C.J. Hogan, Ap.J. 306 (1986) 428.
80. T. Matsumoto, M. Akiba & H. Murakami, Ap.J. 332 (1988)575.
81. J.C. McDowell, MNRAS223 (1986) 763.
82. J.R. Bond, B.J. Carr & C.J. Hogan, Ap.J. 367 (1991) 420.
83. D.J. Fixsen et al., Ap.J. 508 (1998) 123.
84. T. Abel, G. Bryan & M.L. Norman, Ap.J. 540 (2000) 39.
85. V. Bromm, A. Ferrara, P.S. Coppi & R.B. Larson, MNRAS 328 (2001) 969.
86. A. Heger & S.E. Woosley, Ap.J. 567 (2002) 532.
87. C.L. Fryer, S.E. Woosley & A. Heger, Ap.J. 550 (2001) 372.
88. M. Coleman-Miller & E.J.M. Colbert, Inter. J. Mod. Phys. D. 13 (2004) 1.
89. R.P. Van der Marrel, Bull. Amer. Astron. Soc. 35 (2003) 1370.
90. H. Baumgardt, J. Makino, P. Hut, S. McMillan & S. Portegies Zwart, Ap.J.Lett. 589 (2003) L25.
91. P. Madau & M.J. Rees, Ap.J. Lett. 551 (2001) L27.
92. R. Schnieder, A. Ferrara, P. Natarajan & K. Omukai, Ap.J. 571 (2002) 30.
93. R.R. Islam, J.E. Taylor & J. Silk, MNRAS 340 (2003) 647.
94. R. Salvaterra & A. Ferrera, MNRAS 339 (2003) 973.
95. H.S. Zhao & J. Silk, Phys. Rev. Lett. 95 (2005) 011301.
96. G. Bertone, A.R. Zentner & J. Silk, Phys. Rev. D. 72 (2005) 103517.
97. M. Volonteri & R. Perna, MNRAS 358 (2005) 913.
98. H. Mu & T. Totani, Ap.J. 628 (2005) 873.
99. J.H. Krolik, Ap.J. 615 (2004) 383.
100. M. Mapelli, A. Ferrar & N. Rea, MNRAS 368 (2006) 1340.
101. K.T. Inoue & M. Chiba, in Science and Technology (Kinki University Press, 2006), astro-ph/05126481.
102. Ya.B. Zeldovich and I.D. Novikov, Sov.Astron.A.J. 10 (1967) 602.
103. S.W. Hawking, MNRAS 152 (1971) 75.
104. B. J. Carr and S. W. Hawking, MNRAS 168 (1974) 399.
105. B.J. Carr, Ap.J. 201 (1975) 1.
106. B.J. Carr and J.E. Lidsey, Phys.Rev.D., 48 (1993) 543.
107. M.Yu. Khlopov and A.G. Polnarev, Phys.Lett.B., 97 (1980) 383.
108. S.W. Hawking, Phys. Lett. B. 231 (1989) 237.
109. A.G. Polnarev and R. Zemboricz, Phys.Rev.D. 43 (1988) 1106.
110. S.W. Hawking, I. Moss and J. Stewart, Phys.Rev.D., 26 (1982) 2681.
111. H. Kodama, K. Sato, M. Sasaki and K. Maeda, Prog.Theor.Phys. 66 (1981) 2052.

112. S.W. Hawking, Nature 248 (1974) 30.
113. D.N.Page and S.W. Hawking, Ap.J. 206 (1976) 1.
114. N.A. Porter and T.C. Weekes, Nature 277 (1979) 199.
115. B.J. Carr, Ap.J. 206 (1976) 8.
116. J.H. MacGibbon and B.J. Carr, Ap.J. 371 (1991) 447.
117. Y. Sendouda, K. Kohri, S. Nagataki and K. Sato, Phys.Rev.D. 70 (2004) 063512.
118. J.H. MacGibbon, Nature 329 (1987) 308.
119. I.D. Novikov, A.G. Polnarev, A.A. Starobinsky and Ya.B. Zeldovich, Astron. Astrophys. 80 (1979) 104.
120. B.J. Carr, J.H. Gilbert and J.E. Lidsey, Phys.Rev.D. 50 (1994) 4853.
121. M. Crawford & D.N. Schramm, Nature 298 (1982) 538.
122. K. Jedamzik and J. Niemeyer, Phys.Rev.D. 59 (1999) 124014.
123. K. Jedamzik, Phys.Rev.D. 55 (1997) R5871; Phys. Rep. 307 (1998) 155.
124. M. Patel and G.M. Fuller (2000), astro-ph/0003062.
125. J. Yokoyama, Astron. Astrophys. 318 (1997) 673.
126. K.T. Inoue and T. Tanaka, Phys.Rev.Lett. 91 (2003) 021101.
127. T. Nakamura, M. Sasaki, T. Tanaka and K. Thorne, Ap.J. 487 (1997), L139.
128. J.R. Bond and B.J. Carr, MNRAS 207 (1986) 585.
129. K. Ioka, T. Tanaka and T. Nakamura et al., Phys. Rev. D. 60 (1999) 083512.
130. D. Blais, C. Kiefer and D. Polarski, Phys.Lett.B. 535 (2002) 11.
131. A.W. Adams and J.S. Bloom (2004), astro-ph/0405266.
132. N. Seto and A. Cooray, Phys.Rev.D. 70 (2004) 063512.
133. A.D. Dolgov, P.D. Naselsky and I.D. Novikov (2000), astro-ph/0009407.
134. S. Kainer and W.K. Rose, AIP Conf. Proc. 666 (2003) 237.
135. M.Yu.Khlopov, S.G. Rubin and A.S. Sakharov (2004), astro-ph/0401532.
136. S.G. Rubin, A.S. Sakharov and M.Yu. Khlopov, JETP 91 (2001) 921.
137. B.J. Carr, Astron.Astr. 56 (1977) 377.
138. B. J. Carr and J. Silk, Ap.J. 268 (1983) 1.
139. K. Freese, R. Price and D. N. Schramm, Ap.J. 275 (1983) 405.
140. P. Meszaros, Astron.Astrophys. 38 (1975) 5.
141. N. Afshordi, P. McDonald and D.N. Spergel, Ap.J.Lett. 594 (2003) L71.
142. K.J. Mack and J.P. Ostriker, Amer.Astron.Soc. Meeting 205 (2004).
143. V.I. Dokuchaev, Yu.N. Eroshenko and S.G. Rubin (2004), astro-ph/0412418.
144. N. Duchting, Phys.Rev.D. 70 (2004) 064015.
145. R. Bean and J. Maguiejo, Phys.Rev.D. 66 (2002) 063505.
146. T. Harada and B.J. Carr, Phys.Rev.D. 71 (2005) 10410.
147. M.J. Bowick et al., Phys.Rev.Lett. 61 (1988) 2823.
148. S. Coleman, J. Preskill and F. Wilczek, Mod.Phys.Lett.A. 6 (1991) 1631.
149. K. Maeda, Class.Quant.Grav. 3 (1986) 233.
150. P.Chen and R.J. Adler, Nuc.Phys.B. 124 (2003) 103.
151. J.D. Barrow, E.J. Copeland and A.R. Liddle, Phys.Rev.D. 46 (1992) 645.
152. A. Barrau, D. Blais, G. Boudoul and D. Polarski, Ann.Phys. 13 (2004) 115.
153. P. Chen, New Astron. Rev. 49 (2005) 233.
154. K.A. Thompson, in *22nd Texas Symposium on Relativistic Astrophysics* (2005), astro-ph/0504606.

NEUTRALINO CLUMPS AND COSMIC RAYS

P. SALATI*

LAPTH, 9 Chemin de Bellevue, BP110,
F–74941 Annecy–le–Vieux Cedex, France
**E-mail: salati@lapp.in2p3.fr*
http://lappweb.in2p3.fr/ salati/

The halo of the Miky Way might contain numerous and dense substructures inside which the putative weakly interacting massive particles (suggested as the main constituent of the astronomical dark matter) would produce a stronger annihilation signal than in the smooth regions. The closer the nearest clump, the larger the positron and antiproton cosmic ray fluxes at the Earth. But the actual distribution of these substructures is not known. The predictions on the antimatter yields at the Earth are therefore affected by a kind of cosmic variance whose analysis is the subject of this contribution. The statistical tools to achieve that goal are presented and Monte Carlo simulations are compared to analytic results.

Keywords: Dark Matter Clumps; Neutralinos; Antimatter Cosmic Rays

1. Motivations

The universe contains a significant fraction of invisible and non–baryonic matter whose nature is still unresolved. This dark matter could be made of weakly interacting massive particles, such as the supersymmetric neutralinos, whose annihilations inside the galactic halo might produce a sizeable yield of positrons and antiprotons. These rare antimatter cosmic rays will be detected with improved accuracy by the forthcoming experiments. Numerical simulations of structure formation indicate that dark matter (DM) is clumpy on small scales. Because neutralinos are more tightly packed inside the DM substructures, their mutual annihilation is enhanced. The resulting increase of the associated antimatter signals at the Earth has been so far accounted for by an overall boost factor and by shifting upwards the cosmic ray fluxes obtained in the case of a smooth distribution of DM particles.

This approach is too simple. If a substructure were to lie in our immediate vicinity, we would observe a significantly distorted spectrum as

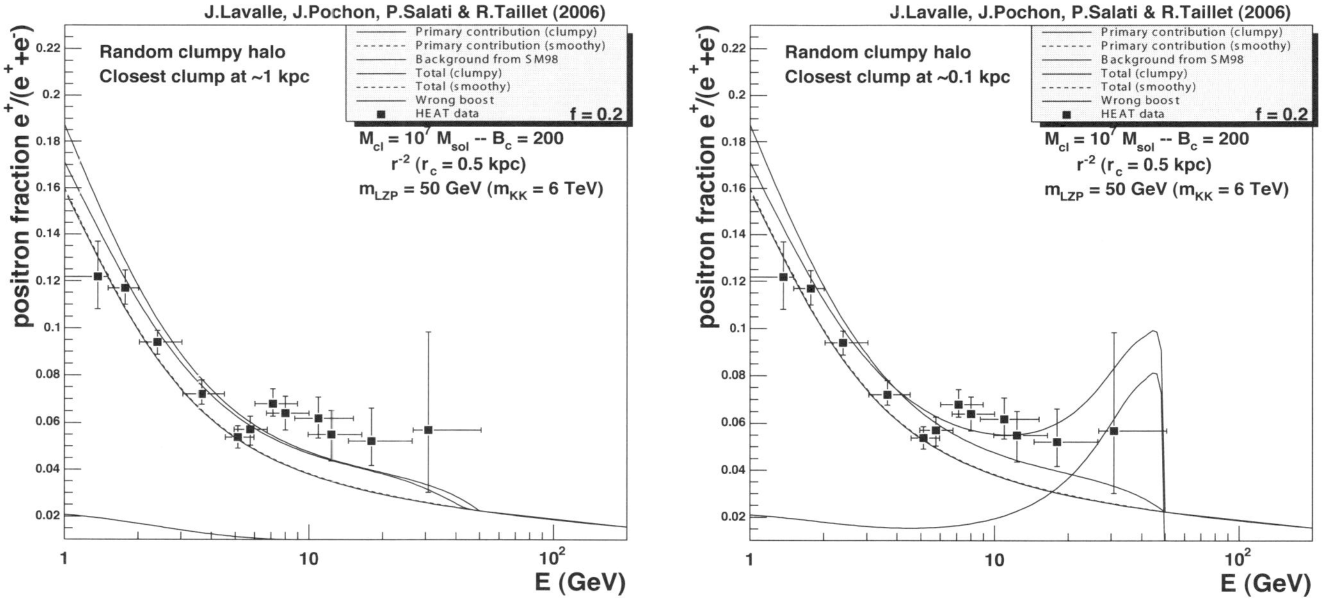

Fig. 1. The positron fraction is derived for a 50 GeV Kaluza–Klein inspired particle[1] and compared to the HEAT excess[2] which cannot be explained by the pure conventional background[3] in black. An isothermal profile has been assumed for the dark matter halo of the Milky Way, a fraction $f = 0.2$ of which lies in clumps with mass 10^7 $M_\odot$ and intrinsic boost $B_c = 200$. Two random realizations of that clumpy DM halo are featured in blue. The distance of the closest substructure has been set equal to 1 kpc in the left panel and decreased to 0.1 kpc in the right panel. The green curve corresponds to the traditional and wrong shift by a factor of $f \times B_c = 40$ of the positron spectrum which a completely smooth DM halo would yield.

illustrated in Fig. 1 for a 50 GeV LZP species. The question then naturally arises to know if such a possibility is probable or very rare. Should we know the exact location of each DM clump, the positron and antiproton spectra would be derived exactly. On the contrary, they are affected by a kind of cosmic variance because an infinite number of different halo realizations are possible. The theoretical predictions on the detectable fluxes at the Earth must take into account that lack of knowledge.

We have therefore built specific tools[4] to address this issue. We will concentrate here on the pedagogical example where a positron line is produced through the mutual annihilations of the DM particles χ into e^+e^- pairs. This scenario is plausible in the framework of Kaluza–Klein inspired theories. In the absence of any substructure, the DM distribution ρ_s is smooth and the resulting flux is given by

$$\phi_s(E) = \mathcal{S} \int_{\text{DM halo}} G(\mathbf{x}) \left\{ \frac{\rho_s(\mathbf{x})}{\rho_0} \right\}^2 d^3\mathbf{x} \ , \tag{1}$$

where the factor $\mathcal{S}$ is defined as

$$\mathcal{S} = \frac{1}{8\pi} v_{e^+}(E) \left\langle \sigma_{\text{ann}} \left(\chi\chi \to e^+ e^- \right) v \right\rangle \left(\frac{\rho_0}{m_\chi} \right)^2 \ . \tag{2}$$

The probability for a positron injected at $\mathbf{x}$ to propagate towards the Earth which it reaches with the degraded energy $E \leq E_S$ is described by the Green function $G(\mathbf{x})$. In the presence of clumps, the DM distribution is given by the superposition $\rho = \rho'_s + \delta\rho$ where a smooth component ρ'_s is still present while $\delta\rho$ accounts for the substructures. The propagator G has already been discussed in the literature. Positrons loose energy through synchrotron radiation and inverse Compton scattering on stellar light and on the CMB. The farther they originate, the smaller their energy E at the Earth for a fixed value of E_S. Positrons that are detected at the energy E have been produced within a typical range λ_{D} which decreases as E increases towards E_S but is nevertheless much larger than the size of the DM substructures. The positron flux becomes

$$\phi(E) = \phi'_s + \left(\phi_r = \sum_i \varphi_i \right) \ , \tag{3}$$

where the contribution from the ith clump is $\varphi_i = \mathcal{S} \times G(\mathbf{x}_i) \times \xi_i$. That minihalo produces as many positrons as if the entire volume

$$\xi_i \equiv \frac{B_i \, M_i}{\rho_0} = \int_{\text{ith clump}} \left\{ \frac{\delta\rho(\mathbf{x})}{\rho_0} \right\}^2 d^3\mathbf{x} \tag{4}$$

were filled with the density ρ_0. The volume ξ_i can also be expressed as a function of the substructure mass M_i and **intrinsic** boost B_i.

An infinite set of halo realizations must be taken into account, each of which produces a different flux ϕ_r. The boost factor $B = \phi/\phi_s$ **at the Earth** is not unique and must be treated as a random variable. We present in the next section a procedure to determine the statistical law according to which B is distributed and will show that its average value and variance depend on the energy E.

2. Computing the Odds of the Galactic Lottery

The statistical properties of the random flux ϕ_r and of the associated boost factor B have been thoroughly investigated in a recent analysis.[4] I will just point out its salient features and summarize the hypotheses on which it is based.

(i) To simplify the discussion and without loss of generality, we assume identical clumps with mass M_c and intrinsic boost B_c. The random flux ϕ_r simplifies into

$$\phi_r = \mathcal{S} \times \xi_c \times \sum_i G_i \;, \tag{5}$$

where the effective volume $\xi_c = B_c\, M_c/\rho_0$ is the same for each minihalo.

(ii) The actual distribution of DM substructures is one particular realization to be taken from the statistical ensemble made up by all the possible **random** distributions. The flux contributed by the clumps and the boost factor must be averaged on that infinite set to yield $\langle\phi_r\rangle$ and $B_{\rm eff} = \langle B = (\phi/\phi_s)\rangle$ whereas the variance is defined as

$$\sigma_r^2 = \langle\phi_r^2\rangle \;-\; \langle\phi_r\rangle^2 \quad \text{and} \quad \sigma_B = \sigma_r/\phi_s \;. \tag{6}$$

(iii) Clumps are distributed independently of each other. This is a strong assumption that generally holds because the substructure correlation length is smaller than the propagation range $\lambda_{\rm D}$. We just need then to determine how a single clump is distributed inside the Milky Way halo in order to derive the statistical properties of an entire population of N_H such substructures. If φ denotes the contribution of a single minihalo, we get

$$\langle\phi_r\rangle = N_H\,\langle\varphi\rangle \quad \text{and} \quad \sigma_r^2 = N_H\,\sigma^2 = N_H\left\{\langle\varphi^2\rangle \;-\; \langle\varphi\rangle^2\right\} \;. \tag{7}$$

(iv) The set of the random distributions of one single clump inside the Milky Way halo $\mathcal{D}_H$ makes up the statistical ensemble $\mathcal{T}$ which we eventually need to consider. An event from that ensemble consists in a clump

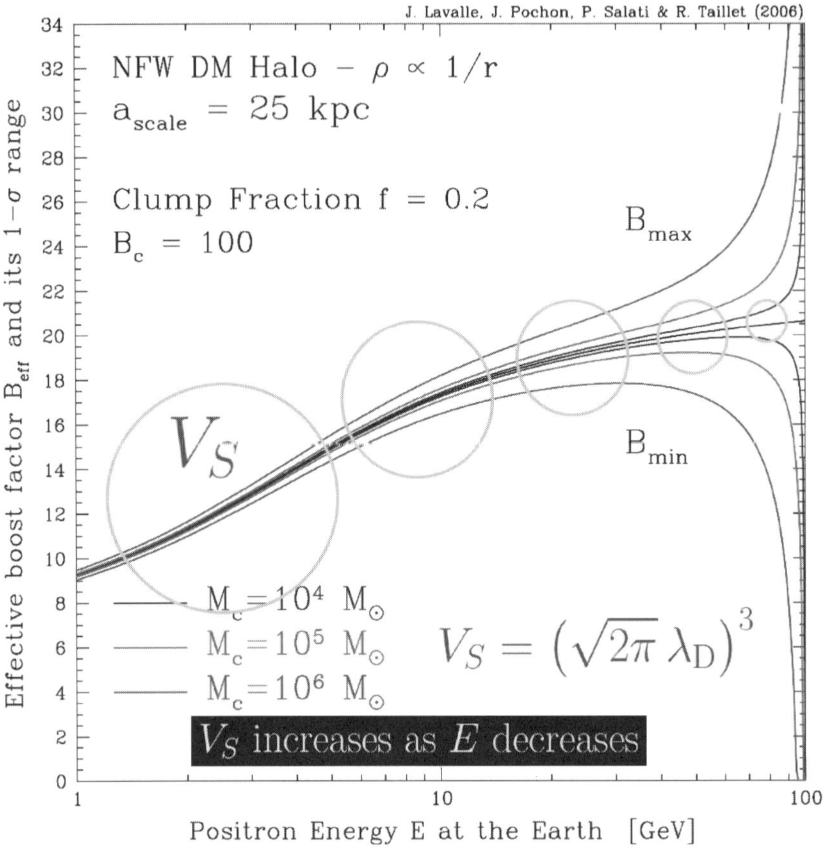

Fig. 2. Positrons are injected with the energy $E_S = 100$ GeV and detected at the energy E. The effective boost factor $B_{\rm eff}$ (black line) accounts for the average enhancement of the positron signal resulting from DM clumpiness. Although it has been so far considered to behave as a constant, the boost actually depends on the energy. Furthermore, as E approaches E_S, the boost variance σ_B increases significantly. The region from which the positrons detected at the Earth originate shrinks and the number N_S of clumps which it contains decreases.

located at position $\mathbf{x}$ within the elementary volume $d^3\mathbf{x}$ and is weighted by the probability $p(\mathbf{x})\,d^3\mathbf{x}$. The spatial distribution $p(\mathbf{x})$ and the flux density of probability $\mathcal{P}\,(\varphi)$ are related by

$$\mathcal{P}\,(\varphi)\;d\varphi \;=\; dP \;=\; \int_{\mathcal{D}_\varphi} p(\mathbf{x})\,d^3\mathbf{x} \;, \tag{8}$$

where $\mathcal{D}_\varphi$ is the region of the halo in which a clump must lie in order to contribute a flux φ at the Earth. We readily conclude that any function $\mathcal{F}$

of the flux φ (for instance φ itself or its square φ^2) is given on average by

$$\langle \mathcal{F} \rangle = \int \mathcal{F}(\varphi)\, \mathcal{P}(\varphi)\, d\varphi = \int_{\mathcal{D}_H} \mathcal{F}\{\varphi(\mathbf{x})\}\, p(\mathbf{x})\, d^3\mathbf{x} \; . \tag{9}$$

Once $p(\mathbf{x})$ is known, $B_{\rm eff}$ and σ_B can be derived.

3. Results and Perspectives

We have applied that procedure in the case where $\rho'_s \equiv (1-f) \times \rho_s$ and have finally assumed that the clumps trace the smooth DM distribution (which is not the case in general) by imposing that $p(\mathbf{x}) \propto \rho_s(\mathbf{x})$. The main results are summarized in Figs. 2 and 3. To commence, the average boost $B_{\rm eff}$ depends on the energy. As E decreases, the part of the halo which contributes effectively to the positron signal at the Earth grows. Regions located close to the galactic center come into play, with a much larger density ρ_s than in the solar neighborhood. At fixed intrinsic boost B_c, neutralino annihilation inside clumps is relatively less enhanced in these dense regions than in our vicinity, hence the behaviour exhibited in Fig. 2. We also find that the boost variance σ_B increases significantly as E approaches E_S. The above mentioned domain of the halo, inside which the positrons detected at the Earth with the energy E have been produced, has a typical volume $V_S = (\sqrt{2\pi}\,\lambda_{\rm D})^3$ which shrinks as E gets near to E_S. The number N_S of the substructures which it contains falls down and the boost variance increases because

$$\frac{\sigma_r}{\langle \phi_r \rangle} \simeq \frac{\sigma_B}{B_{\rm eff}} \simeq \frac{1}{\sqrt{N_S}} \; . \tag{10}$$

Finally, in the large N_S regime, the boost distribution follows a Maxwellian law as is featured in Fig. 3. This result may be derived with the help of the central limit theorem. In the opposite regime where $N_S \leq 1$, the boost distribution is driven by the product $N_H \times \mathcal{P}(\varphi)$ where the density of probability $\mathcal{P}(\varphi)$ for the flux generated by a single clump comes into play. The method outlined here can be applied in particular to the intermediate mass black hole scenario.[5] An effective boost $B_{\rm eff}$ of ~ 2 to 3×10^3 is found for both antiprotons and positrons with a large variance.

Acknowledgments

I would like to thank the organizers for the warm and friendly atmosphere of this exciting and inspiring conference as well as for their financial support. This work could not have been performed without the help of the french programme national de cosmologie PNC.

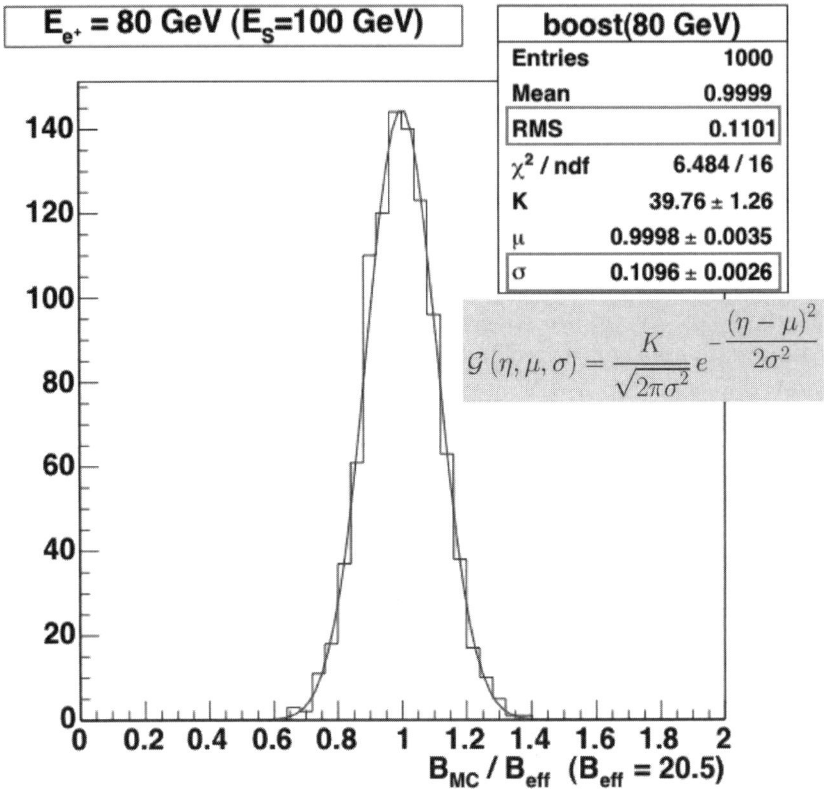

Fig. 3. In this Monte–Carlo simulation, clumps contribute a fraction $f = 0.2$ to the mass of the Milky Way DM halo (assumed to follow a NFW profile with a scale radius of 25 kpc). Each substructure has a mass of 10^5 $M_\odot$. A thousand different realizations of the DM halo have been generated here, each of them involving 271,488 clumps. The positron injection energy is $E_S = 100$ GeV. The number of realizations has been plotted as a function of the boost ratio $\eta = B/B_{\text{eff}}$. The rms value of 0.1101 yielded by the Monte–Carlo for η is in excellent agreement with the anaytic value $\sigma_\eta = \sigma_B/B_{\text{eff}} = 0.1097$.

References

1. K. Agashe and G. Servant, *Phys. Rev. Letters* **93**, 231805 (2004).
2. S. Coutu *et al.*, Positron Measurements with the Heat-Pbar Instrument, in *International Cosmic Ray Conference*, 2001.
3. I. V. Moskalenko and A. W. Strong, *Astrophys. J.* **493**, 694 (1998).
4. J. Lavalle, J. Pochon, P. Salati and R. Taillet, `astro-ph/0603796`, to be published in A&A (2007).
5. G. Bertone, A. R. Zentner and J. Silk, *Phys. Rev.* **D72**, p. 103517 (2005).

SEARCH FOR A DARK MATTER ANNIHILATION SIGNAL IN THE TEV RANGE WITH H.E.S.S.

E. MOULIN* ON BEHALF OF THE H.E.S.S. COLLABORATION†

Laboratoire de Physique Théorique et Astroparticules,
Montpellier Cedex 5, 34095, France
E-mail: moulin@in2p3.fr
† *www.mpi-hd.mpg.de/hfm/HESS/*

The H.E.S.S. (High Energy Stereoscopic System) array of imaging atmospheric Cherenkov telescopes is designed to detect very high energy (E_γ>100GeV) γ-rays in the TeV range. A strong signal from the high energy source HESS J1745-290 has been detected in the direction of the Galactic Center. Constraints have been derived on a possible component due to dark matter particles annihilation. Diffuse γ-ray emission has been highlighted along the Galactic plane. Prospects of detection from Sagittarius Dwarf galaxy scheduled for observations will be discussed.

Keywords: Dark Matter, Gamma ray astronomy

1. Introduction

H.E.S.S. (High Energy Stereoscopic System) is an array of four imaging atmospheric Cherenkov telescopes located in the Khomas Highlands of Namibia (1800 m a.s.l). Each telescope of 13 m diameter consists of an optical reflector of about 107 m^2 effective area.[1] Their cameras equipped with 960 fast photomultiplier tubes[2] collect the Cherenkov light emitted by showers of secondary particles induced by the interaction of the primary γ-ray with Earth's upper atmosphere. The large field of view of 5° of each camera combined with the stereoscopic technique allows for accurate reconstruction of the direction and the energy of the primary γ-rays as well as efficient rejection of the cosmic ray induced background. The energy threshold of H.E.S.S near zenith is about 100 GeV and the angular resolution of about 0.1° per γ-ray. The point source sensitivity of the H.E.S.S. instru-

*Present address : DAPNIA/SPP, CEA Saclay, Gif-sur-Yvette 91191, France ment is better than $2.0 \times 10^{-13} \mathrm{cm}^{-2}\mathrm{s}^{-1}$ above 1 TeV for a 5σ detection in 25 hours of observations.[3]

This paper reports on the detection of the Galactic Center source HESS J1745-290 and the diffuse emission along the Galactic plane. The strong emisssion from HESS J1745-290 is discussed in terms of a Dark Matter (DM) annihilation signal. Prospects of indirect detection in the nearby dark matter dominated Sagittarius dwarf galaxy are investigated in MSSM and Kaluza-Klein scenarios.

2. Dark matter interpretation of the VHE γ-ray emission from the source HESS J1745-290

The observations carried out by the H.E.S.S. collaboration in the Galactic Center (GC) region reveal a strong source of TeV γ-rays, HESS J1745-290.[4] The GC region has been re-observed in 2004 and the initial discovery has been confirmed with a significance of about 38σ. Observations from 2003 and 2004 shows no indication for variability. The source position derived from the 2004 dataset assuming a point-like source lies at an angular distance of 5" $\pm$ 10"$_{\mathrm{stat}}$ $\pm$ 20"$_{\mathrm{syst}}$ from the supermassive central black hole Sgr A*. A fainter source associated to the pulsar wind nebula G 0.9+0.1 is located near l=-1°. Even though the pointing accuracy of the instrument is as low as 20", the SNR Sgr A East and the recently discovered PSR G359.95-0.04 can not be ruled out as sources of the observed TeV γ-ray flux. The energy spectrum of the excess presented on Fig. 2 is well fitted with a power law in the energy range 160 GeV - 30 TeV with a photon index $\Gamma = 2.25 \pm 0.04_{\mathrm{stat}} \pm 0.10_{\mathrm{syst}}$.

Besides plausible astrophysical origins, an alternative explanation is the annihilation of DM in the central cusp of our Galaxy. The spectrum presented on Fig. 2 shows no indication for γ-ray lines. The observed γ-ray flux may result from secondaries of DM annihilation such as $b\bar{b}$, etc. When plotted $E^2 d\Phi/dE$ as a function of the γ-ray energy E, the neutralino annihilation is characterized by a plateau for γ-ray energies E_γ much lower than the neutralino mass m_χ. A rapid decrease is observed when $E_\gamma \sim m_\chi$. We observe that the spectrum extended up to masses about 10 TeV requires large neutralino masses (>10 TeV) which are unnatural in phenomenological MSSM scenarii. The Kaluza-Klein models provide harder spectrum but significatively deviate from the measured one. Non minimal version of the MSSM may yield flatter spectrum with mixed 70% $b\,\bar{b}$ and 30% $\tau^+\tau^-$ final states. Even this scenario does not fit to the measured spectrum. The hypothesis that the spectrum measured by H.E.S.S. results from DM particle annihilation only is highly disfavoured.

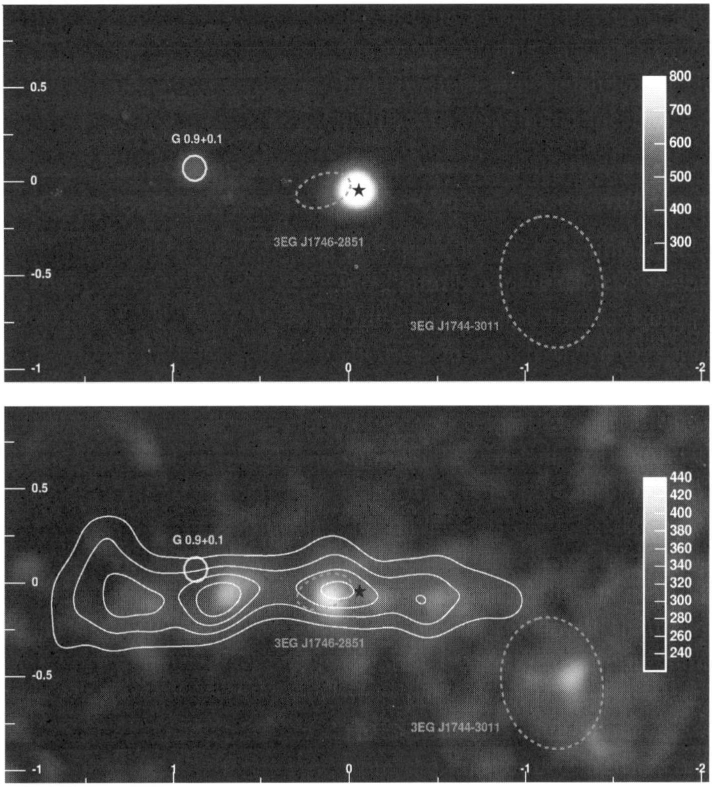

Fig. 1. Top: VHE γ-ray images of the Galactic Center region.[9] The smoothed count map for 2004 H.E.S.S. data shows a strong excess of VHE γ-rays from HESS J1745-290 close to the location of the Sgr A* marked with a black star. 95% confidence region for the position of the unidentified EGRET sources are shown as dashed green ellipses.[12] Bottom : H.E.S.S. count map after subtraction of point sources at the location of HESS J1745-290 and the supernova remnant G 0.9+0.1. Diffuse γ-ray emission is observed along the Galactic Plane. Contours from CS maps[10] trace molecular hydrogen and correlate with the high energy emission seen by H.E.S.S.

On the other hand, the observed signal may result from the superposition of a DM annihilation signal and a power-law astrophysical background spectrum. In this case, using a NFW profile, a constraint on the velocity-weighted annihilation cross-section $\langle \sigma v \rangle$ can be derived.[5] Fig. 3 shows the 99% C.L. limits on $\langle \sigma v \rangle$ from H.E.S.S. data in the case of pMSSM and KK DM models. Predictions from pMSSM models calculated with DarkSUSY4.1[6] and KK models from[8] allowed by particle physics are shown as

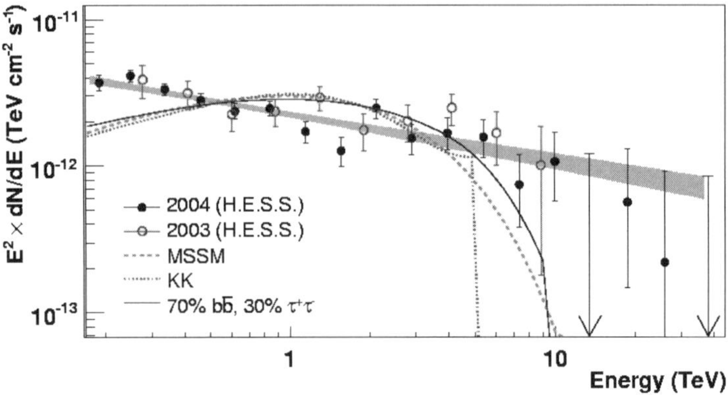

Fig. 2. Spectral energy density $E^2 \times dN/dE$ of γ rays from HESS J1745-290 for 2003 (red empty circles) and 2004 (black filled circles) datasets of the H.E.S.S. observation of the Galactic Centre. The shaded area shows the best power-law fit to the 2004 data points. The spectra expected from the annihilation of a MSSM-like 14 TeV neutralino (dashed green line), a 5 TeV KK DM particle (dotted pink line) and a 10 TeV DM particle annihilating into 70% b $\bar{b}$ and 30% $\tau^+\tau^-$ in final state (solid blue line) are presented.

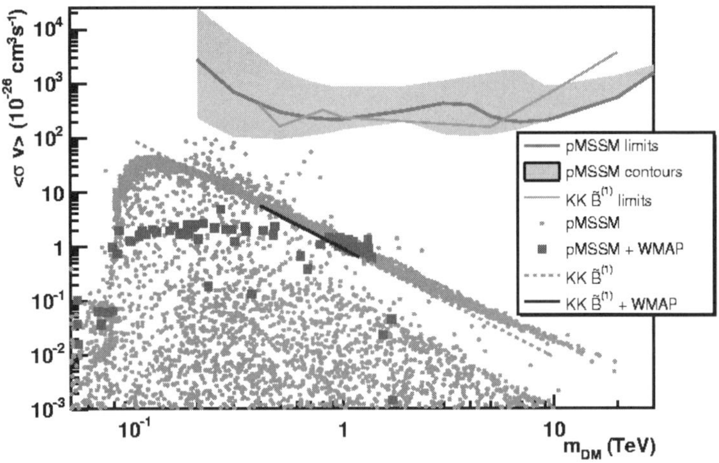

Fig. 3. Exclusion limits on $\langle \sigma v \rangle$ versus the DM particle mass for NFW DM halo. Limits derived from H.E.S.S. data for pMSSM (grey line and light grey contours) and KK DM models (green line) are shown. The grey points and dashed green lines correspond to pMSSM and KK models respectively, allowed by particle physics. Pink boxes and blue segment represent DM models allowed by the cosmological constraints of WMAP.

well as those satisfying cosmological constraints. The limits on $\langle\sigma v\rangle$ derived from H.E.S.S. data for both DM models are $\mathcal{O}(10^{-24})\mathrm{cm}^3\mathrm{s}^{-1}$. With a NFW profile, no $\langle\sigma v\rangle$ from SUSY or KK DM models can be ruled out. With assumptions on the annihilation cross section, limits on the density profile can be calculated. A Moore profile can be excluded for most models.[7]

3. Diffuse emission along the Galactic plane

An extended component of γ-ray emission is observed in the Sgr A region. The subtraction of the two points sources HESS J1745-290 and HESS J1747-281 associated to G 0.9+0.1, highlights a clear excess along the Galactic plane between -1° and 1° in longitude and a rms width of about 0.2° in latitude.[9] This excess correlates well with the dense molecular clouds traced by CS emission.[10] The spectrum of the diffuse component is well described by a power law of index $\Gamma = 2.29 \pm 0.07_{\mathrm{stat}} \pm 0.20_{\mathrm{syst}}$. The VHE emission can be explained by a large number of sources or by cosmic ray interactions in the central molecular zone (CMZ). In the first case, the emission is due to inverse Compton scattering of electrons. Nevertheless, the sources should appear point-like to H.E.S.S. due to the intense ambient magnetic field. In the second case, since the spectrum is harder than expected for γ-rays from galactic cosmic ray interaction, the cosmic rays should have been recently produced by an accelerator located in the central molecular zone. Examples are the SNR Sgr A East or the black hole Sgr A*.

4. Detection prospects from Dwarf Spheroidal galaxies

Recently discovered dwarf spheroidal galaxies from the Local Group such as Sagittarius, Canis Major or Draco are amongst the most extreme DM dominated environements. Dwarfs are ideal astrophysical systems to probe the nature of DM since they usually consist of stellar systems with no hot or warm gas, no cosmic ray population and little dust. Sagittarius Dwarf is located at a distance of about 25 kpc with a size of $3^\circ \times 8^\circ$. It presents a large M/L ratio and a cuspy luminous profile[11] and thus is a source of interest for indirect dark matter search.

Predicted γ-ray fluxes have been derived for Sagittarius dwarf galaxy from calculations in the framework of effective pMSSM and KK models. The expected number of γ-ray and well as the signal significance have been calculated. Results are presented in Tab. 1.

Table 1. H.E.S.S. sensitivity to neutralino and KK $\tilde{B}^{(1)}$ DM annihilation in the Sagittarius dwarf spheroidal galaxy. Values are derived assuming a NFW profile. Acceptance and rejection efficiency are taken from the GC analysis.

Models	Exposure (hrs)	$N\gamma$	$N\sigma$
Best pMSSM	50	650	13
KK $\tilde{B}^{(1)}$	50	39	0.8

5. Conclusion

The TeV γ-ray energy spectrum measured by H.E.S.S. in the GC region is unlikely to be interpreted in terms of neutralino or KK $\tilde{B}^{(1)}$ DM annihilations. Loose constraints on the velocity-weighted annihilation cross-section have been derived in the case of a background component with a power-law spectrum. Neither pMSSM nor KK models can be ruled out. Indirect DM searches with H.E.S.S. will be complemented with observations of the Sagittarius dwarf spheroidal galaxy.

Acknowledgements

The support of the Namibian authorities and of the University of Namibia in facilitating the construction ond operation of H.E.S.S. is gratefully acknowledged, as is the support by the German Ministry for Education and Research (BMBF), the Max Planck Society, the French Ministry for Research, the CNRS-IN2P3 and the Astroparticle Interdisciplinary Programme of the CNRS, the U.K. Particle Physics and Astronomy Research Council (PPARC), the IPNP of the Charles University, the South African Department of Science and Technology and National Research Foundation, and by the University of Namibia. We appreciate the excellent work of the technical support staff in Berlin, Durham, Hamburg, Heildelberg, Palaiseau, Paris, Saclay, and in Namibia in the construction and operation of the equipment.

References

1. K. Bernlöhr, *et al.*, Astropart. Phys. **B20** (2003) 111
2. P. Vincent *et al.*, Proc. of the 28th International Cosmic Ray Conference, T. Kajita *et al.*, (Eds. Universal Academy Press, Tokyo, 2003), 2887
3. F. A. Aharonian, *et al.* (H.E.S.S. collaboration), A&A **457** (2006) 899
4. F.A. Aharonian, *et al.* (H.E.S.S. Collaboration), A&A **425** (2004) L13
5. F. A. Aharonian *et al.* (H.E.S.S. collaboration), to be published in *Phys. Rev. Lett.* (2006)
6. P. Gondolo *et al.*, JCAP **0407** (2004) 008

7. J. Ripken *et al.*, Proc. of the 29th ICRC, 00 (2005) 101
8. G. Servant and T. Tait, Nucl. Phys. **B650** (2003) 391
9. F.A. Aharonian, *et al.* (H.E.S.S. Collaboration), Nature **439** (2006) 695
10. M. Tsuboi, H. Tsohihiro and N. Ukita, Astrophys. J. Supp. **120** (1) 1999
11. L. Monaco, *et al.*, MNRAS **356** (2005) 1396
12. J. Mattox, *et al.*, Astrophys. J. Supp. **135** (2001) 155

DARK MATTER CLUMPS IN THE GALAXY

V. S. BEREZINSKY

Laboratori Nazionali del Gran Sasso, INFN,
67010 Assergi (AQ), Italy
E-mail: berezinsky@lngs.infn.it

V. I. DOKUCHAEV and Yu. N. EROSHENKO

Institute for Nuclear Research of the Russian Academy of Sciences,
60th Anniversary of October Prospect 7a, Moscow 117312, Russia
E-mail: dokuchaev@inr.npd.ac.ru; erosh@inr.npd.ac.ru

It is described the distribution of small-scale dark matter (DM) clumps in the Galactic halo. These clumps are efficiently destructed by tidal forces at early stages of structure formation starting from time of clump detachment from the expansion of the Universe. Only a small fraction of these clumps survives the stage of hierarchical clustering. The survived clumps can be further destructed in the Galaxy by tidal interactions with stars in the Galactic bulge, disk and halo. It is shown that collective gravitational field of the Galactic disc provides the dominant contribution to the destruction of small-scale clumps. The resulting enhancement (boost factor) of annihilation signal due to the halo clumpiness strongly depends on the primordial perturbation spectrum and varies in the range 10-100.

Keywords: Dark mater; Galactic halo; Neutralino; Annihilation.

1. Introduction

Annihilation in the Galactic halo is a promising indirect manifestation of Dark Matter (DM) particles.[1] An intriguing indication on the possible DM annihilation in the Galactic halo was found in the EGRET gamma ray data, which shows a noticeable excess at energies above 1 GeV in comparison with expectations from the conventional galactic models in all sky directions.[2]

The inflation cosmological models predict the near power-law primordial fluctuation spectrum with an power index $n_p \simeq 1.0$ (Harrison-Zeldovich spectrum). Both analytical calculations[3–6] and numerical simulations[7–9] with the inflationary-produced adiabatic density fluctuations predict the existence of DM clumps in the Galactic halo. The small-scale clumps are formed earlier than the larger ones and captured by the larger clumps in the process of a hierarchical clustering in the expanding Universe. Eventually all clumps consist in part of the smaller ones and of the separate DM particles. An effective index of the density perturbation power spectrum $n \to -3$ at small-scales (when mass inside the perturbation $M \to 0$). This means that a gravitational clustering of small-scale structures proceeds very fast. As a result the formation of new clumps and their capturing by the larger ones are nearly simultaneous processes.

The resulting enhancement of the annihilation signal due to the presence of clumpy substructures in the Galactic halo strongly depends on the fraction of the most dense small-scale clumps.[5,10] The most essential characteristics of clumps for calculations of DM annihilation in the Galactic halo are the minimum mass and distribution function of clumps. A mass distribution of small-scale clumps survived in the hierarchical structuring was derived in Ref. 5:

$$\xi_{\rm int}\,\frac{dM}{M} \simeq 0.01(n+3)\,\frac{dM}{M}, \tag{1}$$

where M is a clump mass, n is a power-index of density perturbations at a mass-scale M. The distribution function $\xi_{\rm int}$ is a mass fraction of DM in the form of clumps in the logarithmic mass interval $d\log M$. The minimal mass of DM clumps $M_{\rm min}$ is determined by the leakage of DM particles from the growing density fluctuations (the diffuse leakage and free streaming) and depends on the modelled properties of DM particles[4,5,11–15] and varies e.g. in the range of the Moon and Earth masses, $M_{\rm min} \sim (10^{-8} - 10^{-6})M_{\odot}$ for some specific neutralino models.

2. Tidal destruction of small-scale DM clumps

The small-scale DM clumps are gravitationally loosely bounded objects. As a result the tidal destruction by stars is a crucial process for a survival of clumps in the Galaxy.[10,16,17] An additional source for tidal heating of clumps is the Galactic disk shocking when clumps intersect the Galactic plane.[18] The Galactic disk shocking turns out to be the most important process of tidal destruction of clumps in the Galaxy. All small-scale DM clumps are totally destroyed by stars inside the Galactic bulge at distances $r \leq 3$ kpc. As a result there must be a void in small-scale clump distribution within the central ~ 3 kpc.

See in the Fig. 1 the modelled local fraction of survived clumps in the Galactic halo $P(r)$ depending on their internal density.[18] For destruction by halo stars there is an additional weak dependence on the clump mass M.

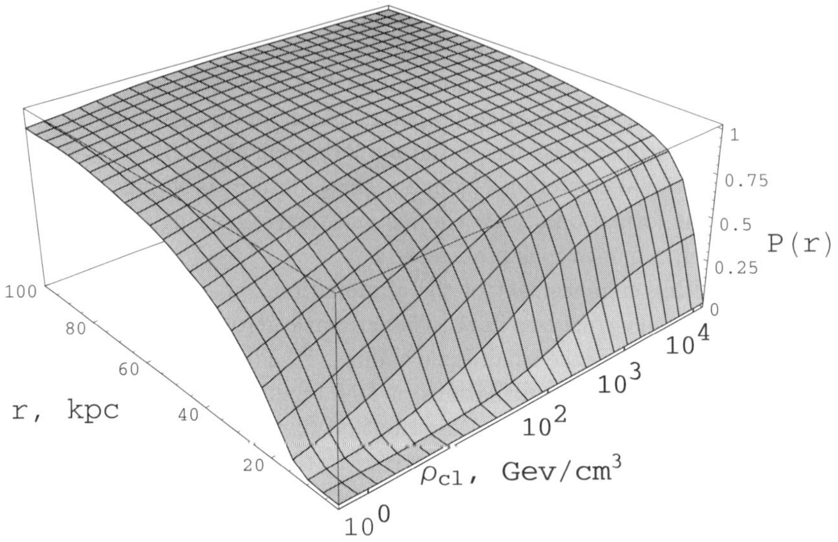

Fig. 1. A survived fraction of small-scale clumps $P(r)$ in the Galaxy halo inside the radial distance 100 kpc according to Ref. 18. The mean internal density of clump $\rho_{\rm cl}$ is in GeV cm^{-3}.

At the same time the destruction of clumps by halo stars is much weaker in comparison with the destruction by disk. With a good accuracy we may neglect the M dependence. In the Fig. 2 it is shown the corresponding model calculations of the local fraction (or survival probability) of clumps with mass $M = 2 \times 10^{-8} M_{\odot}$ in the Galactic halo.

3. Enhancement of annihilation signal by clumps

The relative *enhancement* (*idem* boost-factor or clumpiness-factor) of the annihilation signal due to the presence of DM clumps in the Galactic halo can than be written as $\eta(r) = (I_{\rm dif} + I_{\rm cl})/I_{\rm dif}$, where $I_{\rm dif}$ and $I_{\rm cl}$ are an annihilation signal from diffuse DM in the halo and clumps respectively. The corresponding integrated along the line of sight (observed) enhancement factor $\eta(\theta)$ see in the Fig. 3 for the model case of a simple isothermal spherical symmetric halo model.

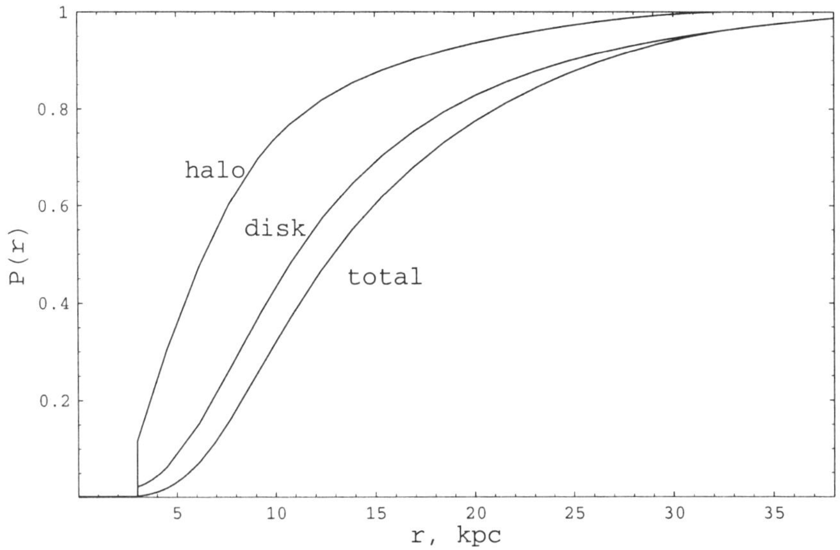

Fig. 2. The fraction of clumps with mass $M = 2 \times 10^{-8} M_{\odot}$ and peak-height $\nu = 2$ survived a tidal destruction in the Galactic disc P_d, in the Galactic halo $P_{\rm H}$ and the resulting total fraction $P(r) \equiv P_{\rm tot} = P_{\rm H} P_d$ as a function of distance from the Galactic center. The cutoff at $r \simeq 3$ kpc is due to a total destruction of clumps by stars inside the bulge.[18]

4. Conclusions

The number density distribution of small-scale DM clumps in the Galactic halo was calculated in dependance on a clump mass M, radius R (expressed through the fluctuation peak-height ν) and radial distance r to the Galactic center. These calculations were performed by taking into account the tidal destruction of clumps in the early hierarchical clustering and later in the Galaxy.[18] In the process of hierarchical clustering the small clumps are captured by the bigger ones, and so on. Thus the hierarchical structure is formed, when all clumps consist in part the smaller ones and the free DM particles. Some part of DM clumps are tidally disrupted in the gravitational field of the bigger host clumps. In this scenario we calculated the differential distribution of the survived clumps as a function of two independent parameters: e.g. a clump mass M and fluctuation peak-height ν (or a clump mass M and radius R). The corresponding integral mass function is given by (1), where $\xi_{\rm int}$ gives the mass fraction of clumps survived the tidal destruction in the hierarchical structuring. The predicted differential number

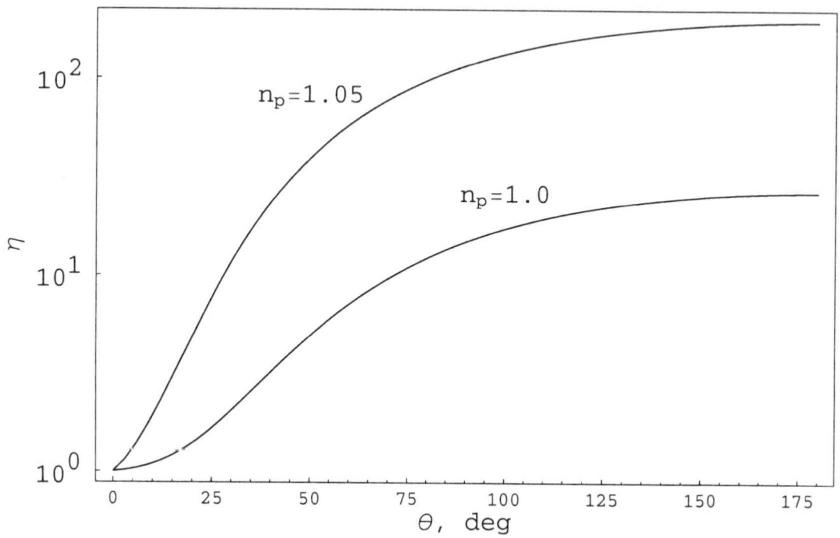

Fig. 3. An integrated along the line of sight (observed) enhancement factor $\eta(\theta)$ for clump internal density profile $\rho_{\rm cl}(r) \propto r^{\beta}$, $\beta = 1.8$, minimal clump mass $M_{\rm min} = 2 \cdot 10^{-8}$, $n_p = 1.0$ and $n_p = 1.05$ in the case of isothermal spherical symmetric halo model.

density of small clumps is very close to our previous calculations,[5] and both are in a good agreement with the recent results of numerical simulations.[10]

Our calculations are valid only for small-scale clumps with masses $M \leq 10^3 M_{\odot}$. The physics of larger mass clumps is rather different. For large-scale clumps the dynamical friction, tidal stripping and accretion of new clumps into the halo proceed in a different way. Nevertheless, the calculated mass function is in a good agreement with a mass function of the large clumps (obtained in the numerical simulations) in the intermediate mass-range.

The mutual tidal destruction of small-scale DM clumps is effective only at the early stage of hierarchical clustering. At later stages the DM clumps are additionally destructed by stars and by the collective gravitational field of the Galactic disc. In the Galaxy at radial distance $r \leq 3$ kpc all small-scale clumps are destructed by stars in the central bulge. At radial distances in the range $r = 3 - 40$ kpc the DM clumps are destructed by stars from the halo and by the tidal shocking in the Galactic disk. The latter provides the major contribution to the tidal destruction of clumps outside the bulge.

The tidal destruction of clumps by the Galactic disk and stars affects the annihilating signal mainly in the central region of the Galaxy where destructions are most effective. Therefore, a growing fraction of survived clumps $P(r)$ smooths the anisotropy of the awaited annihilation signal at the Sun position. A local annihilation rate is proportional to the clumps number density and, respectively, to $P(r)$. For example, at the position of the Sun the 17% of clumps survive, and so the local annihilation rate more then 5 times less in comparison with the $P = 1$ case.

Acknowledgments

This work was supported in part by the Russian Foundation for Basic Research grants 04-02-16757 and 06-02-16342, and the Russian Ministry of Science grant LSS 5573.2006.2.

References

1. J. Silk and H. Bloemen, *Astrophys. J. Lett.* **313**, 47 (1987).
2. W. de Boer, C. Sander, V. Zhukov, A. V. Gladyshev and D. I. Kazakov, *Astron. Astrophys.* **444**, 51 (2005); *Phys. Rev. Lett.* **95**, 209001 (2005); *Phys. Lett.* **B636**, 13 (2006).
3. J. Silk and A. Stebbins, *Astrophys. J.* **411**, 439 (1993).
4. A. V. Gurevich, K. P. Zybin and V. A. Sirota, *Sov. Phys. - Usp.* **167**, 913 (1997).
5. V. Berezinsky, V. Dokuchaev and Yu. Eroshenko, *Phys. Rev.* **D 68**, 103003 (2003).
6. P. Brax, C. van de Bruck, A. -C. Davis and A. M.Green, *Phys. Lett.* **B633**, 441 (2006).
7. B. Moore et al., *Astrophys. J.* **524**, L19 (1999).
8. S. Ghigna, B. Moore, F. Governato, G. Lake, T. Quinn and J. Stadel, *Astrophys. J.* **544**, 616 (2000).
9. A. Klypin, S. Gottlober, A. V. Kravtsov and A. M. Khokhlov, *Astrophys. J.* **516**, 530 (2002).
10. Juerg Diemand, Ben Moore and Joachim Stadel, *Nature* **433**, 389 (2005).
11. S. Hofmann, D. J. Schwarz and H. Stocker, *Phys. Rev.* **D 64**, 083507 (2001).
12. D. J. Schwarz, *Annalcn Phys.* **12**, 220 (2003).
13. A. M. Green, S. Hofmann and D. J. Schwarz, *Mon. Not. Roy. Astron. Soc.* **353**, L23 (2004).
14. A. Green, S. Hofmann and D. Schwarz, `arXiv:astro-ph/0508553`.
15. A. M. Green, S. Hofmann and D. J. Schwarz, *JCAP* **0508**, 003 (2005).
16. H. S. Zhao, J. Taylor, J. Silk and D. Hooper, `arXiv:astro-ph/0502049`.
17. H. S. Zhao, D. Hooper, G. W. Angus, J. E. Taylor and J. Silk, `arXiv:astro-ph/0508215`.
18. V. Berezinsky, V. Dokuchaev and Yu. Eroshenko, *Phys. Rev.* **D 73**, 063504 (2006).

THE BULLET CLUSTER 1E0657-56: IMPLICATIONS FOR DARK MATTER AND GALAXY EVOLUTION

ANTHONY H. GONZALEZ[*,†], D. CLOWE[‡], M. MARKEVITCH[§], S. RANDALL[§], M. BRADAČ[¶], S. CHUNG[*], C. JONES[§] and D. ZARITSKY[‖]

We present an overview of work on the merging cluster system 1E0657-56, commonly known as the "bullet" cluster. This system is one of only two known supersonic cluster mergers, and its unique properties enable fundamental constraints on the nature of dark matter. In a series of papers our group has used this system to demonstrate the existence of dark matter and place upper limits on the dark matter self-interaction cross section. In this proceedings we summarize these findings and discuss ongoing work to improve these constraints and quantify the impact of a supersonic merger on cluster galaxy evolution.

Keywords: Dark Matter, Galaxy Clusters, 1E0657-56

1. Introduction

Galaxy clusters provided the first astrophysical evidence for dark matter in the 1930's, when Zwicky observed that the virial mass of a galaxy cluster greatly exceeds the mass inferred from the stars alone.[1] In the intervening seven decades, the astrophysical importance of dark matter has grown dramatically. The presence of dark matter was used to explain the discovery that spiral galaxy rotation curves are flat rather than Keplerian at large radii,[2] and during the last twenty years cold, non-baryonic dark matter has become a central feature in standard cosmological models of structure formation.[3] Indeed, the ability of structure formation models that include cold dark matter and a cosmological constant (ΛCDM models) to reproduce the cosmic microwave background power spectrum and present day large scale structure is one of the most remarkable successes of modern cosmology.

*Department of Astronomy, University of Florida, Gainesville, FL 32611-2055, USA
†E-mail: anthony@astro.ufl.edu ; www.astro.ufl.edu/~anthony
‡Ohio University, Athens, OH
§Harvard-Smithsonian Center for Astrophysics, Cambridge, MA
¶Kavli Institute for Particle Astrophysics and Cosmology, Stanford, CA
‖Steward Observatory, University of Arizona, Tucson, AZ

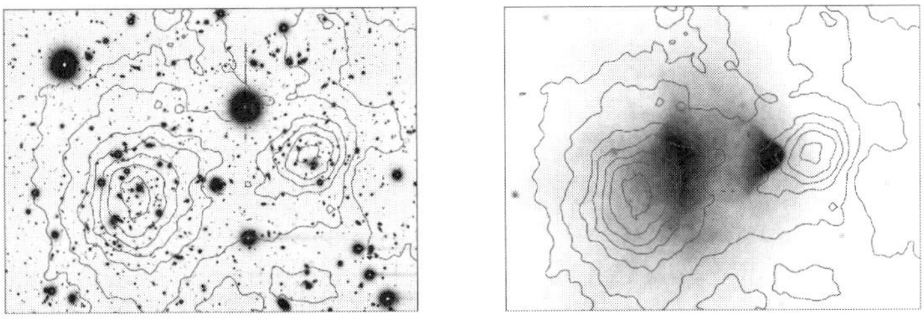

Fig. 1. The two panels show a Magellan R–band image (left) and a 522 ks *Chandra* X-ray (right) image of 1E0657-56. Overlaid are contours from the weak lensing convergence reconstruction. The offsets seen in the *Chandra* image between the intracluster gas and convergence peaks are significant at 8σ.[4]

Together these astrophysical arguments for the existence of dark matter provide a compelling motivation for terrestrial dark matter detection experiments. The caveat is that in each of these physical cases the baryons and dark matter are coincident, permitting one to explain the observations by modifying the gravitational force law rather than invoking the presence of dark matter.[5] This approach has successfully reproduced flat galaxy rotation curves,[6] and recently a Lorentz covariant formalism (TeVeS) has been developed that enables attempts to reproduce the fluctuation spectrum of the microwave background in the absence of dark matter.[7,8] The cleanest way to discriminate between these two alternatives – dark matter and modified gravity – is to observe an astrophysical system in which one expects the baryons and dark matter particles to be spatially segregated.

The merging galaxy cluster pair 1E0657-56, commonly known as the bullet cluster, is the one known astrophysical object that satisfies this criteria. *Chandra* observations reveal a supersonic merger of two massive clusters occurring nearly in the plane of the sky ($i < 10°$).[9] Leading the gas "bullet" associated with the lower mass cluster is a prominent bow shock which corresponds to a subcluster velocity of 4740^{+710}_{-550} km s^{-1}.[10] The lower mass cluster is currently exiting the core of the main cluster after an central passage $\sim$ 0.15 Gyr ago. The fortuitous geometry of the system ($i < 10°$) is crucial because it enables unambiguous interpretation of the gas, galaxy, and total mass distributions.

The intent with these proceedings is to provide a general overview of our work on the bullet cluster, which includes a demonstration of the existence of dark matter and derivation of upper limits on the dark matter self-interaction cross section. For more detailed discussion we encourage the reader to refer to the papers cited in the text.

2. Implications of 1E0657 for Dark Matter

2.1. *Direct Evidence for Dark Matter*

The two principle baryonic components in a galaxy cluster are the hot intracluster gas and cluster galaxies. The intracluster gas behaves as a collisional fluid that is subject to momemtum loss due to ram pressure during the merger. Cluster galaxies in contrast act as collisionless particles, with the result that during the merger the hot gas lags the cluster galaxies (Figure 1). For a massive cluster like 1E0657-56 the intracluster gas contains the bulk of the baryons ($\sim 90\%$). The galaxies can therefore be considered a tracer population that map out the expected distribution of collisionless particles during the merger while contributing negligibly to the total mass. Because dark matter particles are generally presumed to be collisionless (or nearly so),[11] we are left with the expectation that the peak of the cluster mass distribution should lie coincident with the galaxies if dark matter dominates the cluster mass budget. In contrast, if baryons dominate the mass budget the mass distribution should be coincident with the lagging hot gas. To discriminate between these two possibilities, we use the *Chandra* data and gravitational lensing to reconstruct the gas and total mass (or more precisely lensing convergence) distributions, respectively.

2.1.1. *Gas Distributions*

We use the 0.8-4 keV luminosity from *Chandra* to map the gas distribution.[10,12] Because the X-ray luminosity is proportional to the line of sight integral of the square of the gas density,[a] we can locate the gas peak directly from the *Chandra* image (Figure 1). We also derive a 3D model of the gas distribution, with independent components for each physically distinct region (e.g. the bullet and shock front). We determine optimal model parameters by computing the projected gas density and X-ray flux associated with the model and comparing with observations.[10]

2.1.2. *Lensing Convergence Maps and the Total Mass Distribution*

The deflection of light from background galaxies by the cluster potential – gravitational lensing – can be used to reconstruct the total gravitating mass distribution in the context of a given theory of gravity. Images of background galaxies are changed in both shape (shear, $\vec{\gamma}$) and size (convergence,

[a]The effect of gas temperature gradients across the cluster is a small second-order effect.

κ) by the cluster potential. Both quantities are second derivatives of the surface potential, and the latter, κ is linearly proportional to the surface mass density in general relativity. The association between κ and surface mass density is more complex and can be non-local for alternate gravity theories, but for any presumed gravity law the convergence map yields a surface mass distribution.

The terms strong and weak gravitational lensing refer to the two regimes in which the intrinsic ellipticities of the lensed galaxies are smaller and larger than the ellipticity change due to lensing, respectively. We use a statistical approach in the weak lensing regime to derive a convergence map.[4,13] We spatially average the ellipticities of the background galaxy population, for which the intrinsic orientations are uncorrelated and the mean intrinsic ellipticity is hence small. The observed ellipticities of the background galaxies, after correction for smearing by the point spread function, are a measurement of the reduced shear, $\vec{g} = \vec{\gamma}/(1-\kappa)$. The convergence can then be obtained by combining derivatives of $\vec{g}$. For this program we use a combination of high-resolution imaging from HST/ACS and wider field ground-based imaging. The HST/ACS imaging, by enabling shape measurements on > 70 galaxies arcmin^{-1} in the cluster core, enables a high resolution κ map at the location of the bullet, while the wide-field imaging is necessary to determine the normalization of the κ map and hence the total mass of the system. In the resulting convergence map we find that the gas peaks are offset from the convergence peaks for both the main cluster and the bullet at $\sim 8\sigma$ significance (Figure 1), with the errors determined via bootstrap resampling. There is no evidence of unassociated massive background structures contributing to the mass peak, and the probability of associated filaments producing this peak is negligible.

Complementary to the weak lensing analysis, we have also generated a convegence map that incorporates information from the strongly lensed gravitational arcs that are visible in the cores of both the main cluster and the bullet. This analysis yields consistent results for locations of the convergence peaks – 10σ and 6σ offsets from the X-ray gas for the main cluster and bullet, respectively.[14]

The most significant remaining unresolved issue in the lensing analysis is a discrepancy in the total lensing masses derived via our strong and weak gravitational lensing analyses. The locations of the peaks derived from the two analyses are well-determined and consistent, but the weak and strong lensing analyses currently yield enclosed masses within the central 500 kpc that are discrepant by roughly a factor of two. While such discrepancies are not uncommon in cluster lensing analyses, we are currently exploring the origin of this disagreement and working to refine the strong lensing map by obtaining redshifts for the strong arcs, which will remove the most significant remaining uncertainty in the strong lensing analysis.

2.1.3. *Conclusions and Alternative Explanations*

The presence of a dominant dark matter component in 1E0657-56 is the most simple explanation of our observed offset between the gas and κ peaks. One can attempt to avoid dark matter by constructing alternate gravity theories in which the correspondence between κ and surface mass density is complex and non-local, such as TeVeS. In this class of theories the convergence for the central mass peak in a multiple peak system can be supressed.[15] Recent work indicates however that even in the TeVeS framework it is only possible to produce a convergence map simliar to that of the bullet clusters if $> 70\%$ of the mass in the system is non-baryonic.[16] Moreover, the best model from this work still produces excess convergence at the location of the gas peaks despite using an initial bullet gas mass that is a factor of two lower than observed. A general problem for this class of theories is that peak suppression requires a symmetric arrangement of the mass peaks with the suppressed peak directly between the other peaks. In 1E0657-56 the required linear symmetry is broken, with the gas peaks offset from the line connecting the convergence peaks. For this and other reasons discussed elsewhere, it is difficult to see how our data can be explained in the absence of dark matter. Nonetheless, to facilitate detailed comparisons by other groups we have now released the convergence and gas mass maps.[b]

2.2. *The Dark Matter Self-Interaction Cross Section*

Terrestrial experiments place strong upper limits on the cross section for interactions between dark matter particles and baryons, but typically have little leverage in constraining the dark matter self-interaction cross section. This self-interaction cross section is normally presumed to be zero, but values in the range of $\sigma/m = 10^{-23} - 10^{-24}$ cm^2 GeV^{-1} have been proposed to explain several astrophysical phenomenon.[11,17] For the bullet cluster, a non-zero self-interaction cross section would have several effects. The most intuitive effect is that the dark matter mass peak associated with the substructure would be expected to lag the collisionless galaxies, as does the gas.

[b]Available at `http://flamingos.astro.ufl.edu/1e0657/`.

The observed peak of the mass distribution is coincident with the centroid of the galaxy distribution to within 1σ,[4,13] which allows us to place an upper limit on the cross section of $\sigma_m < 2 \times 10^{-23}$ cm^2 GeV^{-1}. A more stringent constraint can be obtained by considering the survival of the subcluster dark matter halo during the merger. The merger velocity is roughly a factor of four larger than the escape velocity from the subcluster. Consequently, depending on the cross section, dark matter particle collisions during the merger can remove a significant fraction of the subcluster mass – but the luminosity of the collisionless galaxies remains unchanged. We determined the mass-to-light ratio (M/L) of the subcluster to test for evidence of mass loss.[18] The consistency of the subcluster M/L with values derived for the main cluster and other more quiescent systems argues that the subcluster has lost $< 30\%$ of it's mass, and therefore $\sigma/m < 1.8 \times 10^{-24}$ cm^2 GeV^{-1}. Using numerical simulations (Randall et al., in prep), this bound has now been improved to $\sigma/m < 1.2 \times 10^{-24}$ cm^2 GeV^{-1}.

3. Astrophysics with 1E0657-56

While the focus of this meeting is dark matter, we would be remiss if we failed to mention the astrophysical questions that can be also be addressed with this unique system. As one example, we are currently using 1E0657-56 to quantify the impact of ram pressure from the shock upon the cold gas reservoirs – and hence star formation – in cluster galaxies. Using the IRAC instrument on the *Spitzer* Space Telescope, we have measured the ratio of polyaromatic hydrocarbon (PAH) emission, which is a tracer of the dust and cool gas required for star formation, to stellar emission in cluster galaxies as a function of distance from the shock. We find a significant decrement in the PAH emission for galaxies in the main cluster that have already passed through the shock front, when compared to those that remain in front of the shock. The transition is sharp at the shock boundary, suggesting that ram pressure is dissociating the dust and quenching star formation. This mechanism for quenching star formation in the cluster environment is relatively unexplored, but could be important for understanding the difference between cluster and field galaxy populations if supersonic shock fronts are present in most clusters at some point in their formation history.

4. Conclusions

The bullet cluster 1E0657-56 is an exceptionally rare supersonic cluster merger that is fortuitously oriented in the plane of the sky. The unique properties of this system have enabled us to demonstrate the existence of dark matter by detection of an $\sim 8\sigma$ spatial offset between the hot gas, which contains the bulk of the baryons, and the peak of the gravitational lensing convergence, which in general relativity is linearly related to the gravitational mass. We argue that alternative of modified gravity in the absence of dark matter is untenable, and note that the only present Lorentz-covariant modified gravity theory still required that the bulk of the matter be non-baryonic to explain our observations. We also find that the survival of the subcluster dark matter halo requires a dark matter self-interaction cross section $\sigma/m < 1.2 \times 10^{-24}$ cm^2 GeV^{-1}. Finally, we see evidence for the ram pressure actively quenching star formation in this cluster.

References

1. F. Zwicky, *Helvetica Physica Acta* **6**, 110 (1933).
2. V. C. Rubin, N. Thonnard and W. K. Ford, Jr., *ApJ* **238**, 471(June 1980).
3. G. R. Blumenthal, S. M. Faber, J. R. Primack and M. J. Rees, *Nature* **311**, 517(Octtober 1984).
4. D. Clowe, M. Bradač, A. H. Gonzalez, M. Markevitch, S. W. Randall, C. Jones and D. Zaritsky, *ApJ* **648**, L109(September 2006).
5. M. Milgrom, *ApJ* **270**, 365(July 1983).
6. K. G. Begeman, A. H. Broeils and R. H. Sanders, *MNRAS* **249**, 523(April 1991).
7. J. D. Bekenstein, *Phys. Rev. D* **70**, 083509(Octtober 2004).
8. C. Skordis, D. F. Mota, P. G. Ferreira and C. Bœhm, *Physical Review Letters* **96**, 011301(January 2006).
9. M. Markevitch, A. H. Gonzalez, L. David, A. Vikhlinin, S. Murray, W. Forman, C. Jones and W. Tucker, *ApJ* **567**, L27(March 2002).
10. M. Markevitch et al., *in prep* (2007).
11. D. N. Spergel and P. J. Steinhardt, *Physical Review Letters* **84**, 3760(April 2000).
12. D. Clowe, S. W. Randall and M. Markevitch, *ArXiv Astrophysics e-prints* (November 2006).
13. D. Clowe, A. Gonzalez and M. Markevitch, *ApJ* **604**, 596(April 2004).
14. M. Bradač, D. Clowe, A. H. Gonzalez, P. Marshall, W. Forman, C. Jones, M. Markevitch, S. Randall, T. Schrabback and D. Zaritsky, *ApJ* **652**, 937(December 2006).
15. G. W. Angus, B. Famaey and H. S. Zhao, *MNRAS* **371**, 138(September 2006).
16. G. W. Angus, H. Shan, H. Zhao and B. Famaey, *ArXiv Astrophysics e-prints* (September 2006).
17. K. Ahn and P. R. Shapiro, *Journal of Korean Astronomical Society* **36**, 89(September 2003).
18. M. Markevitch, A. H. Gonzalez, D. Clowe, A. Vikhlinin, W. Forman, C. Jones, S. Murray and W. Tucker, *ApJ* **606**, 819(May 2004).

OVERVIEW OF THE ZEPLIN TWO-PHASE PROGRAM

DAVID CLINE

Astrophysics Division,
Department of Physics & Astronomy
University of California, Los Angeles, CA 90095 USA
dcline@physics.ucla.edu

We present the history of the ZEPLIN two-phase concept and the early research and development for this detector. ZEPLIN II is now taking data at Boulby. ZEPLIN IV, a one-ton detector, is pre-approved for SNOLAB. We present the argument that a Liquid Xenon detector is crucial to discover and prove the existence of SUSY dark matter and the progress with ZEPLIN II and ZEPLIN IV.

1. Introduction and History of Two Phase Dark Matter Proposal

In the mid-1980s the ICARUS Liquid Argon project was born. It required a long drift of electrons to the collection system with an imposed electric field. Around 1990 some of us started to consider the possibility that LAr could also be used for the detection of dark matter due to the inherent very large dynamic range of the LAr detector. In about 1990 I gave a talk at an Oxford conference proposing a LAr dark matter detector. This is published in the proceedings.

In the early 1990s a UCLA/Torino group within the ICARUS collaboration started to study liquid Xenon for dark matter detection. At a 1993 meeting in Waseda University in Tokyo I presented some of these results along with another ICARUS talk. At this time the key players were P. Picche from Torino and Hanguo Wang [1]. Soon Hanguo enrolled in the PhD program at UCLA and we built a 2kg LXe detector to search for dark matter at CERN. This was taken to the Mont Blanc laboratory. Fortunately it was brought back to CERN just before a major fire in the Mont Blanc tunnel. At about this time in 1996 this small group considered a two-phase option for the LXe detector (Fig. 2)[1, 23].

The concept was to drift electrons from an interaction in the detector into gas and thereby amplify the signal. This was Hanguo Wang's thesis project at UCLA [2]. His thesis was actually published by the entire ICARUS team. One ton detectors like ZEPLIN IV have been proposed by the UCLA, UKDMC group [4].

Table 1. A brief history of the ZEPLIN II dark matter detector.

Year	Event
~1980-90	ICARUS team at CERN studies properties of liquid Argon.
~1990	DBC talk at Oxford meeting on LAr for dark matter detection.
~1992-93	Liquid Xenon properties studied by the UCLA/Torino group; LiXe dark matter detector invented.
~1992-97	Study 2Kg LiXe detector at CERN; take detector to Mont Blanc Lab.
1998	H. Wang thesis at UCLA.
1995-98	Form ZEPLIN collaboration: UCLA, Torino, UKDMC
2000	Publication of ZEPLIN concept by UCLA/Torino/CERN group astroparticle physicists [2]
2001-04	Construction of ZII at UCLA/TAMU/RAL
2004-05	Turn-on of detector and move to Boulby
2006	Start data-taking with ZII; goal 8000 kg day of data

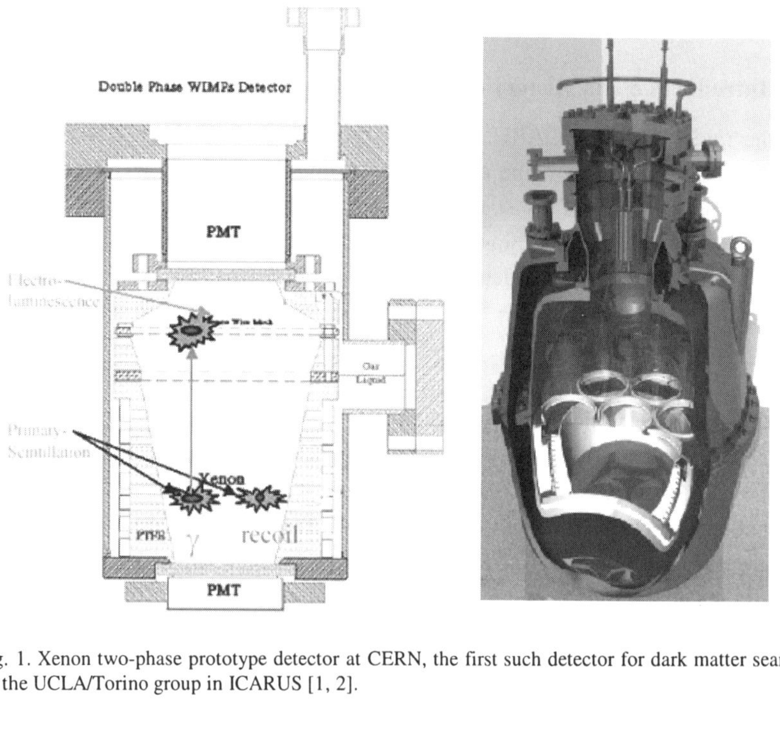

Fig. 1. Xenon two-phase prototype detector at CERN, the first such detector for dark matter search by the UCLA/Torino group in ICARUS [1, 2].

Since we did not have a place to search for dark matter due to the Mont Blanc fire we searched for a new underground location. Fortunately Peter Smith, director of the Boulby (UK) underground labs, was a visitor to UCLA. At this time the concept of the ZEPLIN two-phase detector was born—even the name, and a collaboration with the UK dark matter collaboration formed. Table 1 gives a history of the ZEPLIN two-phase program. The first two phase Xenon detector constructed at CERN is shown in Fig. 1.

2. The Construction of ZEPLIN II

The inner detector for ZEPLIN II was constructed at UCLA under the direction of Hanguo Wang (see Fig. 2). The UK team was in charge of the underground location and other key parts of the detector. Unfortunately the construction took longer than expected. In 2001 some believe that a person from Sheffield visited Columbia University and gave all of the ZEPLIN II information to a future competitor. This was a fundamental mistake.

3. Operation of ZEPLIN II at Boulby

In 2004 the ZEPLIN II detector was constructed. The central detector was built at UCLA. There were major delays in this project.

In 2004/5 the U.S. team helped turn the detector on at RAL with a gas fill. This was the first indication that the two-phase concept worked for ZEPLIN II. However delays at RAl slowed the installation at the Boulby mine. Finally in 2005 the detector was assembled underground and then later filled with liquid Xenon. At first the parity was very poor and again the U.S. team helped construct a system to improve the parity by orders of magnitude.

By 2006 real data taking started (the shielded detector is shown in Fig. 3). An example of a background event is shown in Fig. 4. The different intensifiers and the PMT allow an x,y position resolution of a few [can't read this word]. The resolution in Z (along the electron drift detector) is much better. Some examples of events in the detector are shown in Fig. 5a,b. Calibration sources of photons and neutrons were used underground and Fig. 5a,b shows that these particles are clearly separated by the two-phase discriminator method invented for dark matter by the UCLA/Torino teams in 1995.

In Fig. 6 a very early analysis of possible signal background is shown (this figure is from P. Smith [5].

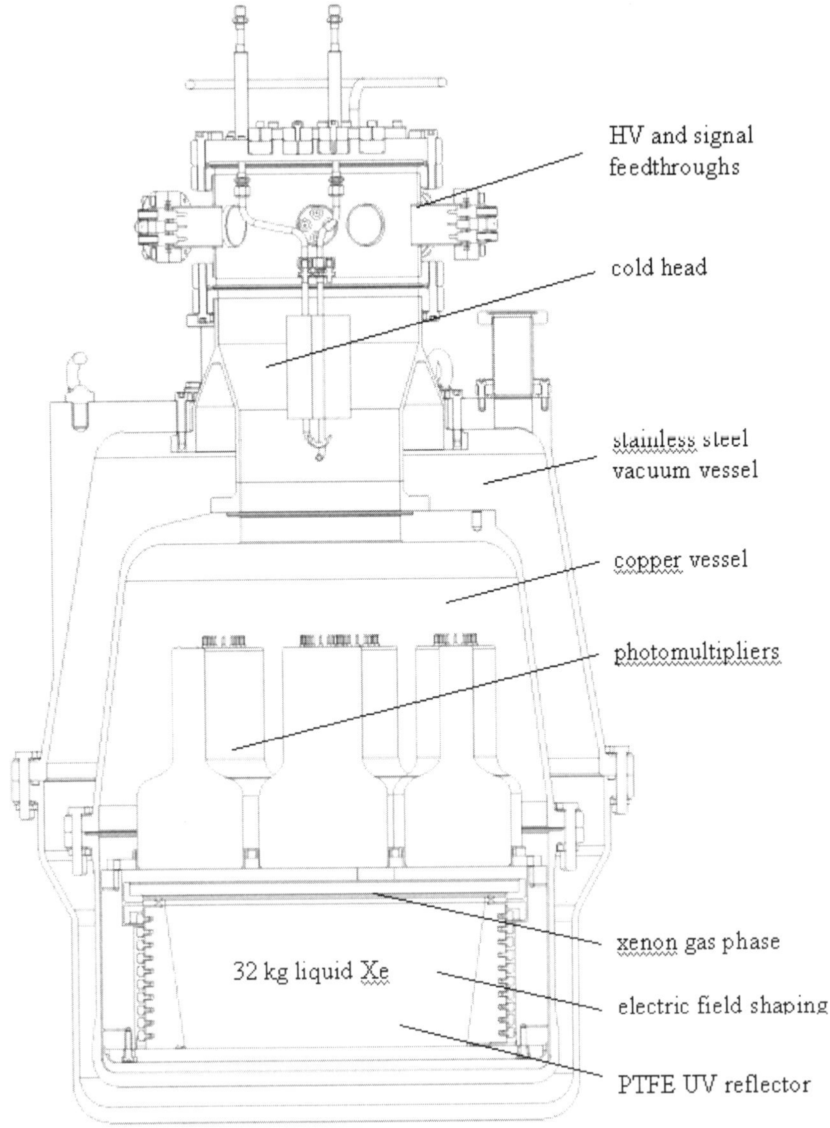

Fig. 2. Cross-section of ZEPLIN II detector. The inner detector was constructed at UCLA.

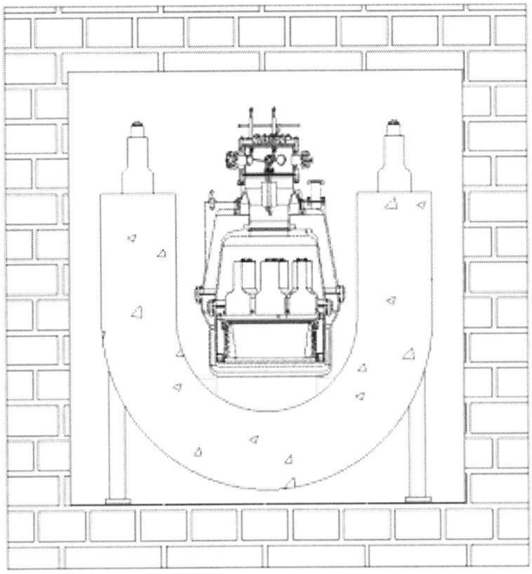

Fig. 3. Arrangement of ZEPLIN II within liquid scintillator veto and lead shielding (hydrocarbon shielding blocks above the detector not shown).

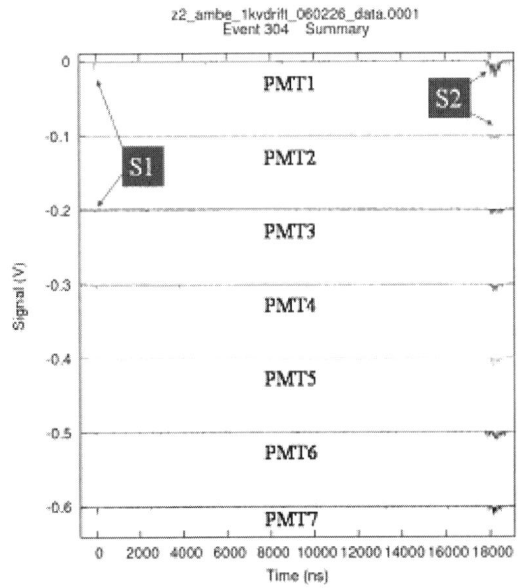

Fig. 4. Typical pulse registering in all seven photomultipliers.

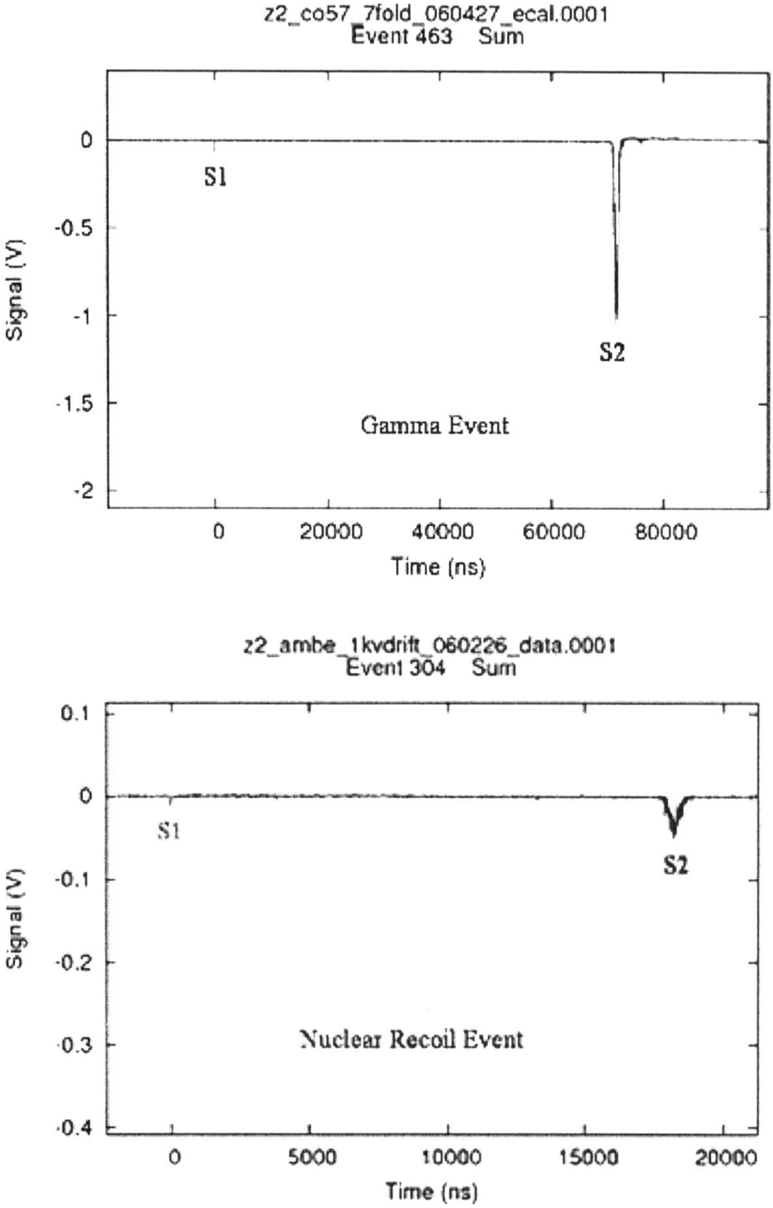

Fig. 5. Events summed from all PMTs. Upper: typical gamma event showing primary and secondary pulses separated by 70μs; lower: typical nuclear recoil event with smaller value of S2/S1.

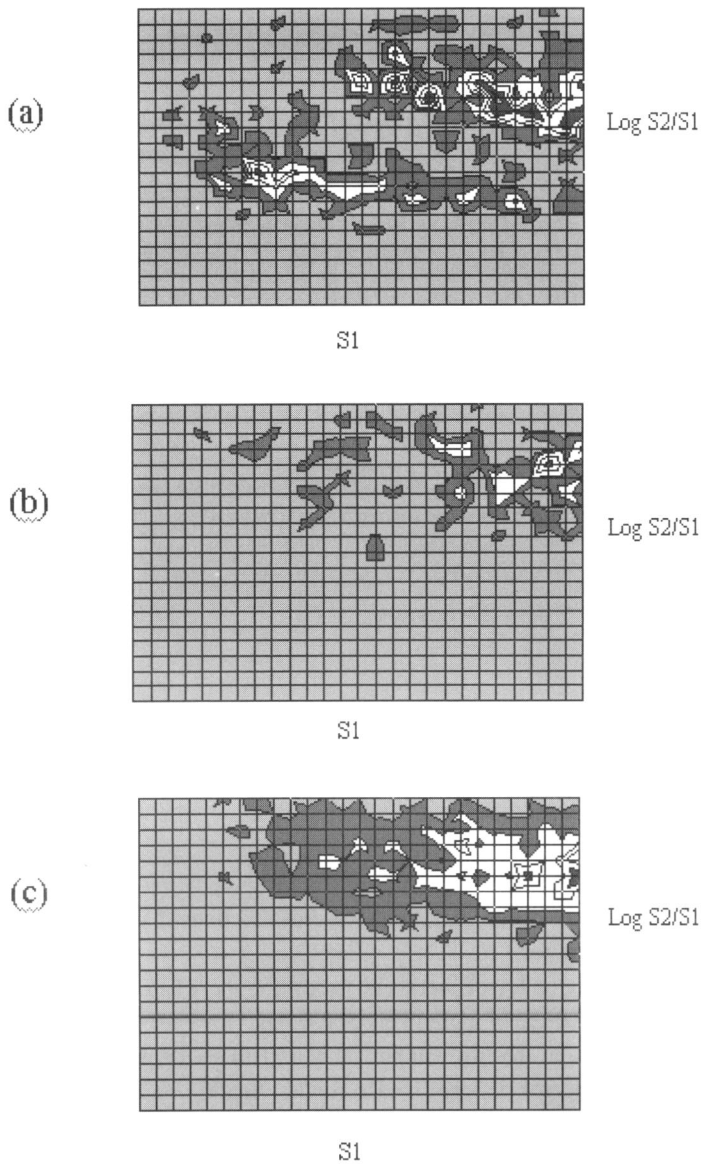

Fig. 6. Contour plots of data binned in 1 pe intervals and 0.1 intervals of log (S2/S1). Plot range is 4 to 30 pe (horizontal axis) and 1.0 to 3.0 in log (S2/S1) (vertical axis)[5]

a) AmBe source showing separated lower (neutron) and upper (gamma) populations

b) ^{60}Co source showing single population with little overlap with neutron region of (a)

c) Background runs, showing single population and little overlap with neutron region.

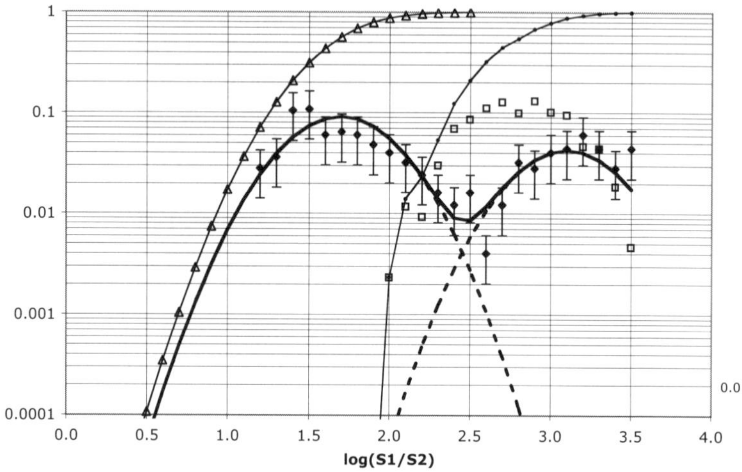

Fig. 7. Analysis of overlap of gamma and neutron populations for S1 range 3 to 16 photoelectrons.
diamonds: AmBe calibration with two-population fit (thick full line)
triangles: integral of neutron component
squares: background data
small circles and fine line: integral of background data [5].

More recently the ZEPLIN II team has discovered that the detector has a significant radon background that must be removed soon. Fig. 7 shows that the base discrimination works.

4. A Larger Liquid Xenon Detector: ZEPLIN IV

In order to cover the full SUSY range of cross-sections it will be necessary to build a ton-size detector.

There has been a lot of work and study for the ZEPLIN IV detector [4] aimed for either DUSEL of more likely SNOLAB. However, things are changing so rapidly in this field that we will not show a ZEPLIN IV design since the new concept LUX detector (Large Underground Xenon) is being proposed to NSF/DOE by the UCLA. TAMU, Rochester, Brown, Case, LLNL and LNBL team.

Acknowledgment

I wish to thank H. Wang for many discussions as well as P. Smith and the UKDMC team.

References

1. D.B. Cline et al, Astroparticle Physics 12 (2000) 373.
2. H. Wang (Thesis, UCLA 1999).
3. See papers in Proc. Int. Conferences on Identification of Dark Matter (World Scientific, 2003, 2005) and the Marina del Rey conferences (2002, 2004, 2006).
4. D. Cline et al, Proc 5th Int. Conf. Detection of Dark Matter, Nucl Phys B Proc. Suppl. 124 (2002).
5. P. Smith, private communication.

THE SECOND PHASE OF THE PICASSO EXPERIMENT

C. B. KRAUSS# FOR THE PICASSO* COLLABORATION

Department of Physics, Queen's University,
Kingston, Ontario K7L 2N6, Canada
#E-mail: ckrauss@owl.phy.queensu.ca
www.queensu.ca

PICASSO (Project in Canada to Search for Supersymmetric Objects) is entering its second experimental phase. A new setup has been installed underground at SNOLAB which will have a total of 2.5kg of active mass. For this a new detector with new purification techniques, new environmental control system and a new data acquisition have been developed. Data taking with the first new detectors has started at the time of the IDM 2006 conference.

Keywords: Dark Matter; Bubble Detectors; IDM 2006; SNOLAB

1. Superheated Droplet Detector

PICASSO (Project in Canada to Search for Supersymmetric Objects) uses the superheated droplet technique to detect nuclear recoils from interactions between WIMPs and Fluorine nuclei, the active mass. Since the Fluorine nucleus carries an unpaired nucleon, PICASSO is predominantly sensitive to WIMP-nucleus interactions that are spin dependent.

2. The PICASSO Detector

The PICASSO detector consists of a liquid, superheated medium (freon) in the form of superheated droplets which are suspended in a gel matrix. The gel can be loaded with up to 2% volume of active medium in droplets with a size of approximately 200 μm. The recoil of a Fluorine nucleus in these Freon droplets causes the liquid droplet to become gaseous. This transition

*PICASSO is a collaboration between the Université de Montréal, Queen's University, University of Indiana South Bend, Laurentian University, BTI, Chalk River and the Prague Technical University.

causes an acoustic shock-wave which is detected with a piezoelectric sensor. The gel matrix is a viscous, water based polymerized compound that serves three purposes:

- Keeping the droplets in suspension and avoid coagulation of the droplets
- Propagating the shock-wave from the phase transition to the sensors
- Keeping radioactive contaminants away from the droplets

The bubbles which expand during operation are later re-compressed with a hydraulic pressure system.

Fig. 1. Picture of the first four 4.5 L detectors in a temperature and pressure control system underground.

PICASSO has so far used two different types of detectors: a 1 L volume detector with two piezoelectric sensors and a 4.5 L detector with nine sensors. Figure 1 shows the first four 4.5 L detectors underground. The smaller type was used in 2003 for a proof of principle run which led to the most recent publication and exclusion curve from PICASSO.[1] The larger detector type was developed for phase II of the experiment. The new detectors use a larger droplet size that increases the active mass and reduces the sensitivity to internal contamination by alpha emitters. Superheated droplet detectors are threshold detectors with a threshold that can be varied by changing the detector temperature. Since nuclear recoils have a different energy deposition than alpha particles, the detector response differs between alphas and WIMP recoils. PICASSO uses this difference in temperature response in the search for dark matter WIMPs. Details about the detector principle can be found in Ref. 2.

The advantage of the superheated droplet detectors is that they are almost completely insensitive to a large fraction of radioactive backgrounds,

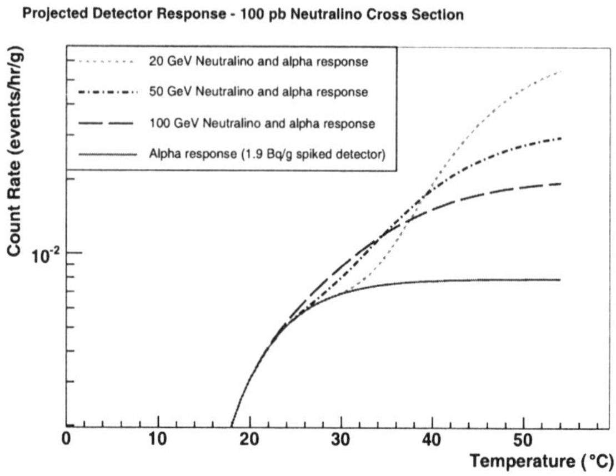

Fig. 2. Typical alpha and WIMP response in a PICASSO detector.

namely MIPs. The only backgrounds that need to be controlled are therefore the internal alpha emitting impurities and external neutrons. In the second phase of PICASSO a water shielding of one cubic foot water blocks is used. For the reduction of Uranium and Thorium chain elements in the gel matrix the ingredients of the gel need to be purified.

3. Purification

The internal purity of the detectors is improved by purifying all ingredients with the HTiO method developed for SNO[3] or with a new HZrO method which has a higher extraction efficiency. This method was developed for the purification of CsCl which is the most radioactively contaminated ingredient of the gel. With the HZrO method the heavy salt can be purified without first diluting it in an aqueous solution and later re-concentrating it.

All handling of detector materials is done in a clean room. In Montreal and Kingston clean room facilities have been in use for PICASSO since 2002. In a further development stage of the detector production all materials will be kept under a N2 cover gas to reduce the radon contamination.

4. Calibration

PICASSO detectors are calibrated with neutrons, both from an Ac/Be source and mono-energetic neutrons from an accelerator. These calibrations show a very good agreement between the sensitivity expected for the amount of active mass and the observed sensitivity. For in-situ calibration

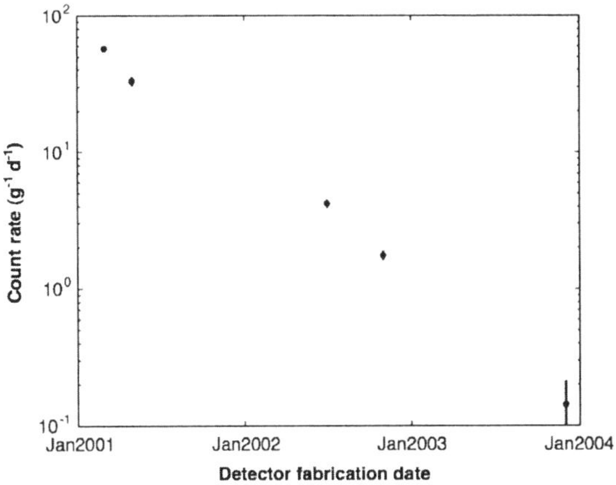

Fig. 3. Count rate of detectors produced between 2001 and 2004.

a switchable Am/Be source will be used. This source is pneumatically actuated to mechanically separate the Americium from the Beryllium. The source container is vacuum tight to prevent any radioactive material from escaping the container.

5. DAQ

For phase II of the experiment a new data acquisition system has been developed. It is a FPGA based system which allows the readout of 9 channels in each module with a trigger logic for each channel. The front-end electronics has a buffer which allows the storage of complete waveforms. The data is transferred from the front-end to a VME system by a digital LVDS link. This system is developed in cooperation with the groupe electronique of the Université de Montréal.

6. Temperature Control System

The first phase temperature control system was based on commercial units which offered only a limited temperature accuracy and homogeneity. A new system which houses four 4.5 L detectors was developed. This system was successfully operated for months at the participating institutions and the first units have been installed underground. The design temperature accuracy of 1/16°C was easily reached and the homogeneity also reached that design goal.

7. Data Analysis

The count rates for each detector are measured at several different temperatures. The temperature response is then used to fit the (known) alpha response in combination with the most likely neutralino response (see figure 2). In the past run (2003 data) no evidence for a neutralino signal could be established. The 95% C.L. limit from the three lowest contamination detectors is 1.3 pb/proton for a 29GeV neutralino.[1]

8. Installation at SNOLAB

The second phase setup of PICASSO is currently being installed at SNOLAB in the same location the 2003 PICASSO data were taken. The largest advantage of the underground location in SNOLAB is the low flux of cosmic muons, less than 0.27 muons/m^2/day. Furthermore, an excellent clean room environment is provided. The water shielding takes up an area of 220 cm x 220 cm. The final number of 32 detectors will be housed in 8 segments in two layers, each individually temperature controlled.

Fig. 4. Setup in SNOLAB with neutron water shield installed.

The production of detector containers, detectors, temperature and pressure control systems and the respective ancillary systems is well underway or fully completed.

Once the second phase system is fully installed we expect to reach a sensitivity two orders of magnitude below the previous one (see figure 5) - the final sensitivity depends on the level of internal contamination that the purification effort achieves. This will have to be determined underground since the background level determined on surface is limited by activity induced by cosmic radiation.

The PICASSO collaboration strives to build a 100 kg active mass experiment once this phase of data taking is successfully concluded. In order to make this experiment viable a further reduction in internal radioactive contaminants is necessary.

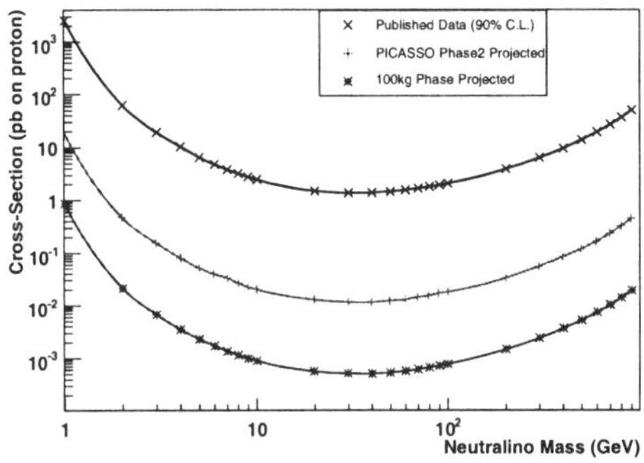

Fig. 5. Projected sensitivity for the next phases compared to published result.

References

1. Barnabe-Heider M *et al* 2005 *Phys. Lett.* B **624** 186-194
2. Barnabe-Heider M *et al* 2005 *Nucl. Instrum. Meth.* A **555** 184-204
3. Andersen T C *et al* 2003 *Nucl. Instrum. Meth.* A **501** 386-398

INVESTIGATIONS ON DARK MATTER AND ON SOME RARE PROCESSES BY DAMA: PAST, PRESENT AND FUTURE

R. BERNABEI, P. BELLI, F. MONTECCHIA and F. NOZZOLI

Dip. di Fisica, Università di Roma "Tor Vergata" and INFN, sez. Roma Tor Vergata, I - 00133 Rome - Italy

F. CAPPELLA, A. INCICCHITTI and D. PROSPERI

Dip. di Fisica, Università di Roma "La Sapienza" and INFN, sez. Roma, I - 00185 Rome - Italy

R. CERULLI

INFN - Laboratori Nazionali del Gran Sasso, I-67010 Assergi (Aq) - Italy

C. J. DAI, H. L. HE, H. H. KUANG, J. M. MA, X. D. SHENG and Z. P. YE

IHEP, Chinese Academy, P.O. Box 918/3, Beijing 100039 - PR China

The DAMA/NaI experiment ($\simeq$ 100 kg highly radiopure NaI(Tl)) at the Gran Sasso National Laboratory of the I.N.F.N. has pointed out - by a model independent approach - the presence of Dark Matter particles in the galactic halo at 6.3 σ C.L. with an exposure larger than 10^5 kg $\times$ day collected over seven annual cycles. Several corollary model dependent quests for the candidate particle have also been carried out; others (of the many possible) are in progress. At present the second generation DAMA/LIBRA set-up ($\simeq$ 250 kg highly radiopure NaI(Tl)) is in data taking deep underground. A 3-rd generation R&D towards a possible 1 ton set-up, we proposed in 1996, is also in progress. Many searches for other rare processes have been and are under investigation as well.

Keywords: Dark Matter; underground physics; rare events searches.

1. Introduction

A large number of possibilities exist for candidates as Dark Matter (DM) particles in the Universe either produced at rest or non relativistic at decoupling time. DM particles at galactic scale can be directly investigated through their interaction on suitable deep underground target-detectors.

The detector's signal can be induced either by recoiling nucleus or, in case of inelastic scattering, also by successive de-excitation gamma's; moreover, other possibilities, which directly involve practically only ionization/excitation phenomena in the detector, are open (see e.g. later) as well as the excitation of bound electrons in scatterings on nuclei giving rise contemporaneously at recoiling nuclei and electromagnetic radiation, etc.

The DAMA/NaI experiment was realized having the main aim to investigate in a model independent way the presence of DM particles in the galactic halo. For this purpose, we planned to exploit the effect of the Earth revolution around the Sun on the DM particles interactions in a suitable low background set-up placed deep underground. In fact, as a consequence of its annual revolution, the Earth should be crossed by a larger flux of DM particles around roughly June 2nd (when its rotational velocity is summed to the one of the solar system with the respect to the Galaxy) and by a smaller one around roughly December 2nd (when the two velocities are subtracted). This annual modulation signature – originally suggested in the middle of '80 in ref. [1] – is very distinctive since a seasonal effect induced by DM particles must simultaneously satisfy all the following requirements: (i) the rate must contain a component modulated according to a cosine function; (ii) with one year period; (iii) a phase roughly around $\sim$ June 2nd; (iv) this modulation must only be found in a well-defined low energy range, where DM particles can induce signals; (v) it must only apply to those events in which just one detector of many actually "fires", since the probability that DM particles would have multiple interactions is negligible; (vi) the modulation amplitude in the region of maximal sensitivity must be $\leq 7\%$ for usually adopted halo distributions, but it can be significantly larger in case of some possible scenarios such as e.g. those of refs. [2,3].

The DAMA/NaI experiment was located deep underground in the Gran Sasso National Laboratory of I.N.F.N.. It has been part of the DAMA project, which is also composed by several other low background set-ups, such as: i) DAMA/LXe [4], see also Cerulli's contribution in these Proceedings; ii) DAMA/R&D [5]; iii) the new second generation larger mass NaI(Tl) radiopure set-up DAMA/LIBRA; iv) DAMA/Ge detector for sample measurements. Detailed descriptions of the DAMA/NaI set-up and of its performances have been given e.g. in [6–9]. Thanks to its radiopurity and features, DAMA/NaI has also investigated other approaches for DM particles and several other rare processes [10].

2. DAMA/NaI model-independent result

The DAMA/NaI set-up has pointed out the presence of a modulation satisfying the many peculiarities of a DM particle induced effect, reaching an evidence at 6.3 σ C.L. over seven annual cycles [3,6–8,11–16]. In particular, in Fig. 1-left the time behaviour of the residual rate of the *single-hit* events in the cumulative (2-6) keV energy interval is reported. The data favour the presence of a modulated cosine-like behaviour at 6.3 σ C.L. and their fit for this cumulative energy interval offers modulation amplitude equal to (0.0200 ± 0.0032) cpd/kg/keV, a phase $t_0 = (140 \pm 22)$ days and a period $T = (1.00 \pm 0.01)$ year, all parameters kept free in the fit. The period and phase agree with those expected in the case of an effect induced by DM particles in the galactic halo ($T = 1$ year and t_0 roughly at $\simeq 152.5^{th}$ day of the year). The χ^2 test on the (2–6) keV residual rate disfavours the hypothesis of unmodulated behaviour giving a probability of $7 \cdot 10^{-4}$ ($\chi^2/d.o.f. = 71/37$).

The same data have also been investigated by a Fourier analysis, where a clear peak corresponding to a period of $\simeq 1$ year is present (Fig. 1-right). Modulation is not observed above 6 keV (we remind that the set-up has taken data up to the MeV region despite the optimization was done for the keV energy range).

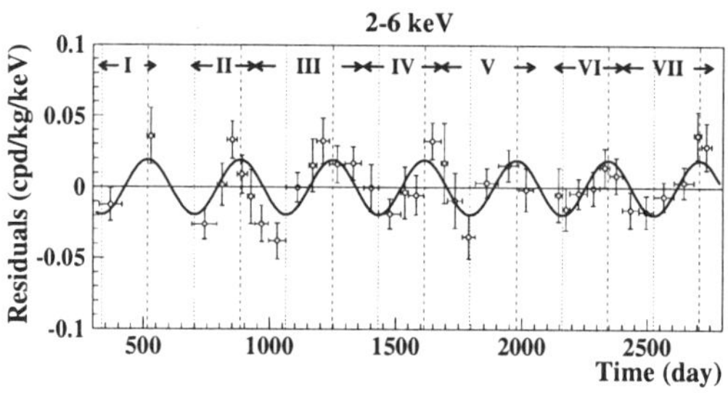

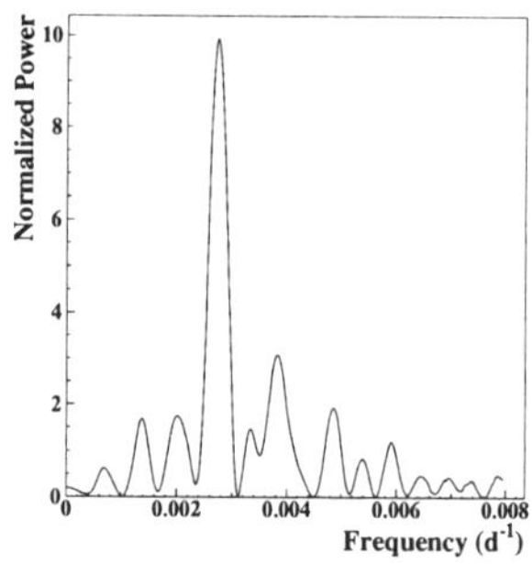

Fig. 1. *On the left*: experimental residual rate for *single-hit* events in the cumulative (2–6) keV energy interval as a function of the time over 7 annual cycles (total exposure 107731 kg $\times$ day); end of data taking July 2002. The superimposed curve represents the cosinusoidal function behaviour expected for a Dark Matter particle signal with a period equal to 1 year and phase exactly at 2nd June. *On the right*: power spectrum of the measured *single-hit* residuals for the cumulative (2–6) keV energy interval. The principal mode corresponds to a frequency of $2.737 \cdot 10^{-3}$ d^{-1}, that is to a period of $\simeq 1$ year. See ref. [6,7].

A careful quantitative investigation of all the known possible sources of systematics and side reactions has been regularly carried out and published at time of each data release. No systematic effect or side reaction able to account for the observed modulation amplitude and to satisfy all the requirements of the signature has been found; for a detailed quantitative discussion see e.g. ref. [6–8].

As a further relevant investigation, the *multiple-hits* events collected during the DAMA/NaI-6 and 7 running periods (when each detector was equipped with its own Transient Digitizer with a dedicated renewed electronics) have been studied and analysed by using the same identical hardware and the same identical software procedures as for the case of the *single-hit* events (see Fig. 2). The fitted modulation amplitudes are:

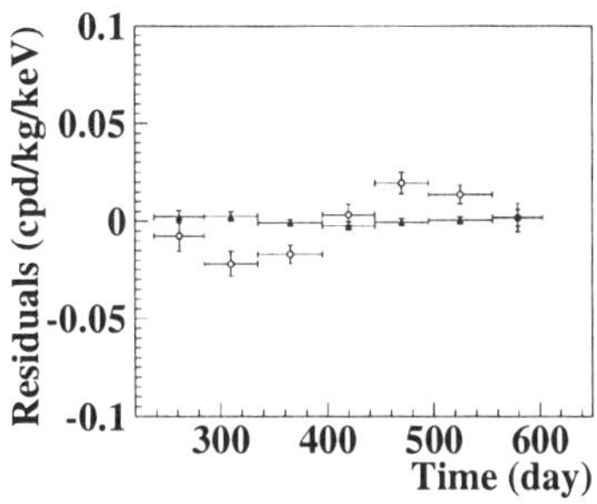

Fig. 2. Experimental residual rates over seven annual cycles for *single-hit* events (open circles) – class of events to which DM particle events belong – and over the last two annual cycles for *multiple-hits* events (filled triangles) – class of events to which DM particle events do not belong – in the (2–6) keV cumulative energy interval. They have been obtained by considering for each class of events the data as collected in a single annual cycle and using in both cases the same identical hardware and the same identical software procedures. The initial time is taken on August 7th. See text. [7]

$A = (0.0195 \pm 0.0031)$ cpd/kg/keV and $A = -(3.9 \pm 7.9) \cdot 10^{-4}$ cpd/kg/keV for *single-hit* and *multiple-hits* residual rates, respectively. Thus, evidence of annual modulation with proper features is present in the *single-hit* residuals (events class to which the DM particle-induced signals belong), while it is absent in the *multiple-hits* residual rate (event class to which only background events belong), offering an additional strong support for the presence of DM particles in the galactic halo further excluding any side effect either from hardware, from software procedures or from background.

3. Some corollary model-dependent quests

On the basis of the obtained 6.3 σ C.L. model-independent result, corollary investigations can also be pursued on the nature of the DM particle candidate. This latter investigation is instead model-dependent and - considering the large uncertainties which exist on the astrophysical, nuclear and particle physics assumptions and on the parameters needed in the calculations - has no general meaning (as it is also the case of exclusion plots and of DM particle parameters evaluated in indirect detection experiments) [3,6–8,11–16].

Low and high WIMP mass candidates interacting with ordinary matter via: i) mixed SI&SD coupling; ii) dominant SI coupling; iii) dominant SD coupling; iv) preferred SI inelastic scattering; have been considered for the WIMP class of DM particles [6,7].

Fig. 3 shows (filled areas) few examples of slices of the 4-dimensional allowed volume for the general case of a WIMP with mixed SI&SD coupling and the allowed region for the case of a WIMP with dominant SI interaction in given model frameworks [6,7]. Here, ξ is the fractional amount of local

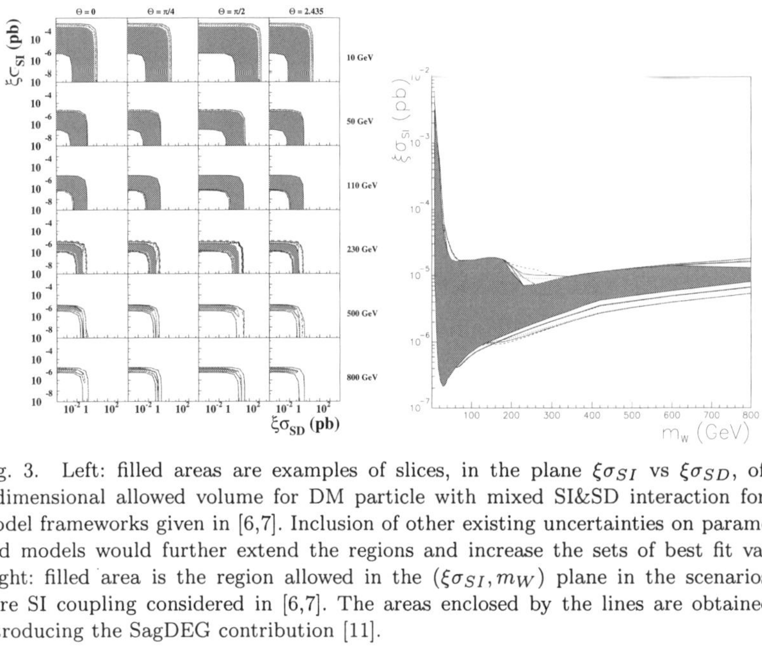

Fig. 3. Left: filled areas are examples of slices, in the plane $\xi\sigma_{SI}$ vs $\xi\sigma_{SD}$, of the 4-dimensional allowed volume for DM particle with mixed SI&SD interaction for the model frameworks given in [6,7]. Inclusion of other existing uncertainties on parameters and models would further extend the regions and increase the sets of best fit values. Right: filled area is the region allowed in the $(\xi\sigma_{SI}, m_W)$ plane in the scenarios for pure SI coupling considered in [6,7]. The areas enclosed by the lines are obtained by introducing the SagDEG contribution [11].

density of DM particles, σ_{SI} and σ_{SD} are the point-like SI and SD DM particle-nucleon cross sections and tgθ is the ratio between the effective SD coupling strengths of the DM particle with neutron and proton (θ is defined in the $[0,\pi)$ interval). The DAMA/NaI allowed regions of Fig. 3 are well compatible with theoretical expectations for neutralino in MSSM (see e.g. [17]). For completeness it is worth to note that no direct comparison is possible among results on purely SI coupled WIMPs achieved by using different nuclei, although apparently all the presentations generally refer to cross section on the nucleon.

In the case of WIMP candidates with dominant SD interaction a 3-dimensional allowed volume (σ_{SD},m$_W$, tgθ) is obtained (see refs. [6,7]). It is worth to note that experiments using either nuclei largely insensitive to SD coupling (like e.g. ^{nat}Ge, ^{nat}Si, ^{nat}Ar, ^{nat}Ca, ^{nat}W, ^{nat}O) or nuclei in principle all sensitive to such a coupling, but having a different unpaired nucleon (neutron in odd spin nuclei, such as ^{129}Xe, ^{131}Xe, ^{125}Te, ^{73}Ge, ^{29}Si, ^{183}W) with respect to the proton in ^{23}Na and ^{127}I, cannot explore most of the 4- and 3- dimensional volumes allowed by DAMA/NaI.

Also the inelastic DM particle scenario (where heavy nuclei are favoured with the respect to lighter ones) of ref. [2] has been analyzed obtaining an allowed volume in the 3-dimensional space ($\xi\sigma_p$, δ, m_W). For a discussion and some slices of such an allowed volume see ref. [6,7].

Furthermore, galaxy hierarchical formation theories, numerical simulations, the discovery of the Sagittarius Dwarf Elliptical Galaxy (SagDEG) in 1994 and more recent investigations suggest that the dark halo of the Milky Way can contain non thermalized substructures. Thus, recently the contribution of the SagDEG has been investigated by analyzing the consequences of its DM stream contribution to the galactic halo on the DAMA/NaI annual modulation data; see Fig. 3 and for details ref. [11].

In addition, the DAMA collaboration has also pursued a corollary quest for the candidate particle considering a light ($\simeq$ keV mass) bosonic candi-

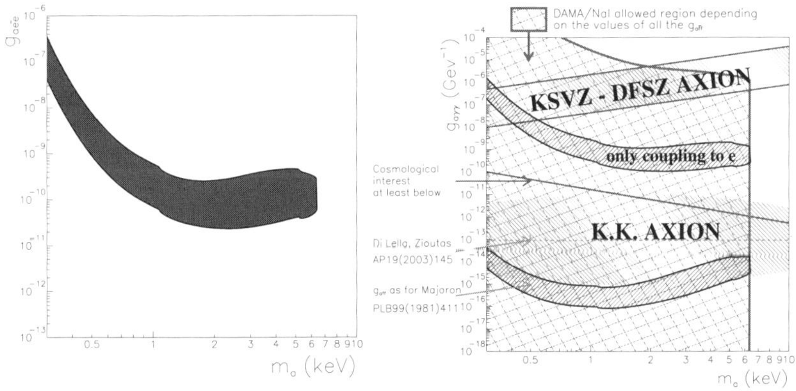

Fig. 4. Case of a pseudoscalar boson candidate (here named a). Left: region allowed at 3σ C.L. in the plane $g_{a\bar{e}e}$ vs m$_a$ by the DAMA/NaI annual modulation data in the considered model framework. Right: allowed region in the plane $g_{a\gamma\gamma}$ vs m$_a$ (crossed hatched region). All the configurations in this region can be allowed depending on the values of all the $g_{a\bar{f}f}$ (couplings to fermions). In the plot examples of the expectations for some models are also shown. The solid line corresponds to a particle with lifetime equal to the age of the Universe; at least all the $g_{a\gamma\gamma}$'s below this line are of cosmological interest. For more information and for other cases see ref. [3].

date, either with pseudoscalar or with scalar coupling, as DM component in the galactic halo [3] (see Fig. 4). For these candidates, the direct detection process is based on the total conversion in NaI(Tl) crystal of the mass of the absorbed bosonic particle into electromagnetic radiation. Thus, in these processes the target nuclei recoil is negligible and is not involved in the detection process; therefore, signals from these light bosonic DM candidates are lost in activities applying rejection procedures of the electromagnetic contribution to the counting rate (such as e.g. CDMS, Edelweiss, Cresst, Warp, etc.). For more information and for other cases see ref. [3].

Several other investigations are available in literature and other studies are in progress.

4. The new DAMA/LIBRA and beyond

In 1996 DAMA proposed to realize a ton set-up and a new R&D project for highly radiopure NaI(Tl) detectors was funded and carried out for several years in order to realize - as an intermediate step - the second generation highly radiopure DAMA/LIBRA experiment (successor of DAMA/NaI). Thus, as a consequence of the results of this second generation R&D, the new experimental set-up DAMA/LIBRA (Large sodium Iodide Bulk for RAre processes), $\simeq$ 250 kg highly radiopure NaI(Tl) scintillators (matrix of twenty-five $\simeq$ 9.70 kg NaI(Tl) crystals), was funded and realized. DAMA/LIBRA is taking data since March 2003 and the first data release will, most probably, occur at end 2008.

At present a 3-rd generation R&D towards the possible creation of a ton set-up is also in progress.

5. Conclusions

DAMA/NaI has experimentally pointed out at 6.3 σ C.L. the presence of DM particles in the galactic halo by investigating the model independent annual modulation signature over seven annual cycles. No systematic effect or side reaction able to account for the observed modulation amplitude and to satisfy all the many peculiarities of the signature has been found. Several corollary quests for the investigation on the nature of the DM candidates have also been pursued with the respect to the many possible at the present poor knowledge on the many astrophysical, nuclear and particle physics related aspects. Some other corollary studies are available in literature by other authors and further ones are in progress. Moreover, some positive hints have also been pointed out by space investigations and are not in conflict with the result of DAMA/NaI. The new second generation DAMA/LIBRA set-up is in data taking and a third generation R&D is in progress.

References

1. K.A. Drukier et al., Phys. Rev. D 33 (1986) 3495; K. Freese et al., Phys. Rev. D 37 (1988) 3388.
2. D. Smith and N. Weiner, Phys. Rev. D 64 (2001) 043502 and hep-ph/0402065.
3. R. Bernabei et al., Int. J. Mod. Phys. A 21 (2006) 1445.
4. P. Belli et al., Astropart. Phys. 5 (1996) 217; Il Nuovo Cim. 19 (1996) 537; Phys. Lett. B 387 (1996) 222; Phys. Lett. B 389 (1996) 783 (err.); Phys. Lett. B 465 (1999) 315; Phys. Rev. D 61 (2000) 117301; R. Bernabei et al., New Journal of Physics 2 (2000) 15.1; Phys. Lett. B 493 (2000) 12; Nucl. Instrum & Meth. A 482 (2002) 728; Eur. Phys. J. direct C 11 (2001) 1; Phys. Lett. B 527 (2002) 182; Phys. Lett. B 546 (2002) 23; Phys. Lett. B 436 (1998) 379.
5. R. Bernabei et al., Astropart. Phys. 7 (1997) 73; Il Nuovo Cim. A 110 (1997) 189; P. Belli et al., Astropart. Phys. 10 (1999) 115; Nucl. Phys. B 563 (1999) 97; R. Bernabei et al., Nucl. Phys. A 705 (2002) 29; P. Belli et al., Nucl. Instrum & Meth. A 498 (2003) 352; R. Cerulli et al., Nucl. Instrum & Meth. A 525 (2004) 535.
6. R. Bernabei el al., La Rivista del Nuovo Cimento 26 n.1 (2003) 1-73
7. R. Bernabei et al., Int. J. Mod. Phys. D 13 (2004) 2127.
8. R. Bernabei et al., Phys. Lett. B 480 (2000) 23; R. Bernabei et al., Eur. Phys. J. C 18 (2000) 283.
9. R. Bernabei et al., Il Nuovo Cim. A 112 (1999) 545.
10. R. Bernabei et al., Phys. Lett. B 389 (1996) 757; Phys. Lett. B 389 (1996) 757; Phys. Lett. B 408 (1997) 439; Il Nuovo Cim. A 112 (1999) 1541; P. Belli et al., Phys. Rev. C 60, 065501 (1999); Phys. Lett. B 460 236 (1999); R. Bernabei et al., Phys. Rev. Lett. 83 (1999) 4918; F. Cappella et al., Eur. Phys. J. direct C 14 1 (2002); R. Bernabei et al., Phys. Lett. B 515 (2001) 6; Eur. Phys. J. A 23 7 (2005); Eur. Phys. J. A 24 51 (2005).
11. R. Bernabei et al., Eur. Phys. J. C. 47 (2006) 263.
12. R. Bernabei et al., Phys. Lett. B 424 (1998) 195; R. Bernabei et al., Phys. Lett. B 450 (1999) 448.
13. P. Belli et al., Phys. Rev. D 61 (2000) 023512.
14. R. Bernabei et al., Phys. Lett. B 509 (2001) 197.
15. R. Bernabei et al., Eur. Phys. J. C 23 (2002) 61.
16. P. Belli et al., Phys. Rev. D 66 (2002) 043503.
17. A. Bottino et al., Phys. Rev. D 67 (2003) 063519; Phys. Rev. D 69 (2003) 043506; Phys. Rev. D 68 (2003) 043506; Phys. Rev. D 69 (2004) 037302; Phys. Rev. D 72 (2005) 083521; D. Hooper and T. Plehn, Phys. Lett. B 562 (2003) 18; G. Belanger et al., hep-ph/0212227.

THE FIRST HALOS

DOMINIK J. SCHWARZ
Fakultät für Physik, Universität Bielefeld, Postfach 100131, 33501 Bielefeld, Germany
E-mail: dschwarz at physik dot uni-bielefeld dot de

The size and time of formation of the first gravitationally bound objects in the Universe is set by the microphysical properties of the dark matter. It is argued that observations seem to favour cold and thermal candidates for the main contribution to the dark matter. For that type of dark matter, the size and time of formation of the first halos is determined by the elastic cross sections and mass of the CDM particles. Consequently, the astrophysics of CDM might allow us to measure some of the fundamental parameters of CDM particles. Essential for observations is the survival rate and spatial distribution of the very first objetcs, which are currently under debate.

Keywords: dark matter; structure formation

1. Motivation

There are at least four motivations to study the issue of the smallest structures and first objects in the Universe. Firstly, a scale invariant power spectrum, without cut-off at some small scale, would lead to an infinite amount of energy stored in acoustic waves (density fluctuations) in the Universe and would be inconsistent. Thus a cut-off at some scale is necessary and it is interesting to know where it is. Secondly, we lack a fundamental understanding of dark matter. All dark matter candidates give rise to the same structure at the largest scales, but differ in their predictions at smaller scales. E.g. light neutrinos and weakly interacting massive particles (WIMPs) lead to different structures at large subhorizon scales and can be discriminated by present large scale structure observations. Similarly, various WIMP candidates give rise to different smallest scale structure. Thus we hope that the astrophysics of dark matter might allow us one day to exclude some of the many dark matter candidates. A third motivation are direct and indirect search experiments. Although experimentalists and observers typically deliver limits on some cross section as a function of dark matter mass, they actually constrain the rate. However, the rates depend on some cross section times the local dark matter density. The local dark matter density might be significantly different from the matter density in the isothermal halo model. It is thus important to understand the small scale distribution of dark matter down to the scale of dark matter (also the velocity distribution is of concern, but not discussed in this contribution). Finally, a complete understanding of hierarchical structure formation must include the study of the very first objects in the hierarchy.

2. Primordial power spectrum

We can only make sensible statements on the power of modes that are generated by quantum fluctuations during cosmological inflation. Larger modes are certainly not observable today and we have no theory for that regime. The smallest primordial modes cross the horizon close to the end of inflation. The energy scale at the end of inflation lies between the Planck scale and nucleosythesis, which means that the smallest primordial scales are somewhere between 1 mm and 1 pc, measured in comoving length scales. If we argue that the matter-antimatter asymmetry in the Universe cannot be generated below the electroweak scale, the energy scale at the end of inflation should be at least 1 TeV. In that case the maximal mode number is $k_{\rm max} > 10^6/\rm pc \approx 1/(0.5au)$. For length scales smaller than the Hubble scale at the end of inflation, no squeezing of quantum fluctuations occurs and thus it seems that those modes will never become "classical" density inhomogeneities. For our discussion it is enough to realise, that the scenario of cosmological inflation, together with a successful mechanism of baryogenesis, allows us to speak about the primordial power spectrum down to at least $10^{12}/\rm Mpc$ or, expressed in cold matter mass, $10^{-24}M_\odot$.

3. Classification of dark matter candidates

There are basically two important criteria according to which dark matter candidates can be classified. The first one introduces the notion of HOT and COLD dark matter, but has nothing to do with temperature. Hot dark matter obeys a relativistic equation of state ($p \sim \epsilon$) at the time of matter-radiation equality, whereas cold dark matter is non-relativistic at that moment of time ($p \ll \epsilon$). The matter- radiation equality is the relevant moment of time, because growth of structure happens in the matter dominated epoch only (decoupled dark matter inhomogeneities grow only logarithmically in the radiation dominated epoch).

Table 1. Classification of dark matter candidates.

	HOT (relativistic)	COLD (non-relativistic)
THERMAL	light νs, ...	WIMPs (heavy ν, LSP, ...), ...
NON-THERMAL	string gas, ...	misalignment axions, primordial black holes, ...

Whether a dark matter candidate was ever in thermal equilibrium with the hot radiation fluid provides a second criterion. If so, it is called THERMAL, if not NON-THERMAL. Some examples of all four categories are given in Table 1.

4. What is the dominant dark matter?

We may ask, given that classification, is any one favoured by observation? A key here is that thermal candidates must have isentropic (often called adiabatic) initial conditions, whereas the non- thermal ones can have more general initial conditions (isocurvature/entropy perturbations). From WMAP and large scale structure data, it follows that the isentropic mode must be the dominant one, a contribution of order 10% of isocurvature modes is not excluded (see e.g. Ref. 1) . This suggests that thermal dark matter candidates are compatible with observations, but non-thermal candidates must provide a mechanism that surpresses isocurvature modes. An especially interesting case in this aspect is the axion, which seems on the first sight to be disfavoured by the data, as it allows isocurvature perturbations. However, as the onset of axion mass is triggered by the temperature of the coloured degrees of freedom in the hot plasma, the axion density perturbations are also largely isentropic, if inflation happens after Peccei-Quinn symmetry breaking. In the case of PQ symmetry breaking after inflation, the axion comes with isocurvature perturbations (see e.g. Ref. 3). To summarise, the data do not exclude non-thermal candidates in general, but they require that their isocurvature power is suppressed to the few % level.

What remains is the question hot versus cold dark matter. The current limits on light neutrinos can be taken representative for any hot and thermal component. They would lead to a different matter-radiation equality and to a damping of smaller structures due to neutrino free streeming. Fitting WMAP data and large scale structrure data indicates that not more than 10% of the dark matter can be hot (see e.g. Ref. 2). We thus conclude that observations are most naturally explained by candidates that are classified as cold and thermal dark matter. There are plenty of candidates in that category: the lightest supersymmetric particle (in many scenarios the neutralino), the lightest Kaluza-Klein particle, a heavy fourth generation neutrino (mass of order TeV), any other WIMP, etc.

5. Thermal and non-relativistic dark matter

In the folloing, we focus our attention on this class of candidates.

5.1. *Kinetic decoupling*

In contrast to relativistic thermal dark matter, chemical decoupling (freeze-out) and the kinetic decoupling happens at different moments of time.[4–6] For WIMPs with mass m, the freeze out happens at $T_{\rm cd} \sim m/25$, whereas kinetic decoupling happens much later; in order to keep thermal equilibrium only elastic scattering is necessary. (A detailed discussion of the relevant elastic rates for neutralinos is given in Refs. 5 and 7.) The decoupling time can be estimated with help of the relaxation time $\tau_{\rm relax}$. In each collision the typical change of momentum is

$$\frac{\Delta p}{p} \sim \sqrt{\frac{T}{m}}.$$

The CDM particle makes a random walk in momentum space, $(\Delta p/p)_N \sim \sqrt{N}\Delta p/p$, and thus has to scatter $N \sim m/T$ times to significantly change to the WIMP's momentum. When the relaxation time $\tau_{\rm relax} \sim N\tau_{\rm coll}$ reaches the Hubble time $t_{\rm H}$ the CDM decouples. For a WIMP with elastic cross sections

$$\sigma_{\rm el} \sim \frac{m^2}{m_Z^4}\left(\frac{T}{m}\right)^{1+l},$$

where $l = 0$ or 1 for domination of s- or p-wave scattering, we find that kinetic decoupling happens at

$$T_{\rm kd} \sim \left(\frac{m_Z^4 m^l}{m_{\rm Pl}}\right)^{\frac{1}{3+l}},$$

ranging from a few MeV to a few 10 MeV.[8,9] Profumo et al.[10] performed a more detailed calculation for specific models of supersymmetry and found that the decoupling temperture might be as large as a few GeV. They also invesitgated models with universal extra dimensions, which are in concordance with our naive estimates.

More recently Bertschinger[11] improved our estimate of the decoupling temperature by solving the appropriate Fokker-Planck equation. His results agree with our estimates at $\sim 20\%$.

5.2. *Damping and growth of inhomogeneities*

During and after kinetic decoupling the primordial inhomogeneities (from cosmological inflation) are washed out on the smallest scales. Before decoupling the CDM starts to deviate from the prefect fluid behaviour of the radiation fluid and this is very well described by an imperfect fluid initially. Hofmann et al.[5] calculated the viscosities and used the linearised Navier-Stokes equations to obtain a first estimate of this collisional damping epoch.

After decoupling the CDM particles continue to damp the smallest inhomogeneities due to free streaming.[5,6,12,13] The comoving free steaming length $l_{\rm fs}$ is a function of time, but becomes approximately constant as $z \ll z_{\rm eq}$. It defines the damping scale, which, expressed as the mass of CDM in a homogeneous sphere of radius $l_{\rm fs}/2$, is estimated as

$$M_{\rm fs} \sim 10^{-6} M_\odot \left(\frac{(1 + \ln(T_{\rm kd}/30\ {\rm MeV})/19.2)^2}{(m/(100\ {\rm GeV})(T_{\rm kd}/30\ {\rm MeV})} \right)^{3/2} .$$

Here we assumed a matter density of $\omega_{\rm m} = 0.14$. Collisional damping and free streaming have been incorporated by Green et al.[8,9] to estimate a typical cut-off scale in the CDM power spectrum. It turns out that for the generic WIMP the free streaming is the dominant contribution to damping and thus sets the scale for the smallest structures.

Loeb and Zaldarriaga[13] pointed out correctly, that the acoustic oscillations and the drag of the radiation fluid cannot be neglected at decoupling itself and they showed numerically that this has the effect to give rise to even more damping, however the order of magnitue is correctly estimated by free streaming only. Most recently Bertschinger[11] showed that the ad hoc treatment of the radiation drag in reference[13] can be improved with help of the appropriate the Fokker-Planck equation. These studies lead consistently to the conclusion that the order of magnitude of the cut-off in wave number is provided by the free streaming calculation. To fix the order of magnitude of the mass scale a more detailed calculation is needed, as a factor of 2 in wave number is a factor of 8 in mass! Close to the cut-off, the CDM power spectrum inherits the peaks and dips from the acoustic oscillations in the radiation fluid. The typical wavelength of these oscillations is given by the

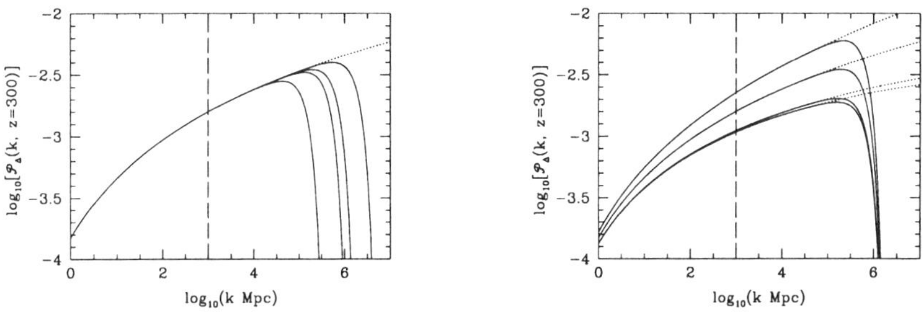

Fig. 1. CDM power specrum at $z = 300$ for four different WIMP models (left) and different scenarios of cosmological inflation (right) [from Green et al.[9]]. The dotted lines show the power spectra without damping due to CDM microphysics.

sound horizon at kinetic decoupling, in terms of CDM mass

$$M_{\rm osc} \sim 10^{-5} M_\odot \left(\frac{30\ {\rm MeV}}{T_{\rm kd}}\right)^3 .$$

The damping of inhomogeneities must be convoluted with the primordial spectrum to provide the cut-off scale and the scale of the maximal primordial power.[9] The CDM becomes fully decoupled only after kinetic decoupling and inhomogeneities in the CDM start to grow logarithmically. This is most easily seen from the evolution equation on subhorizon scales

$$a^2 \frac{{\rm d}^2\Delta_{\rm cdm}}{{\rm d}a^2} + \frac{3}{2}\left(1 - \frac{p}{\epsilon}\right) a \frac{{\rm d}\Delta_{\rm cdm}}{{\rm d}a} - \frac{3}{2}\frac{\epsilon_{\rm cdm}}{\epsilon}\Delta_{\rm cdm} = 0,$$

where $\Delta_{\rm cdm}$ is the CDM density contrast, a the scale factor, and ϵ and p are energy density and pressure, respectively. During the radiation dominated epoch $\epsilon_{\rm cdm} \ll \epsilon$ and $p \approx \epsilon/3$, which leads to $\Delta_{\rm cdm} = A + B\ln(a/a_{\rm kd})$, where A and B are constants to be fixed by the initial conditions. It is also possible to take the effects of baryons into account analytically. After CDM kinetic decoupling they are still tightly coupled to the radiation fluid and suppress the growth of structure until they decouple at around $z_{\rm b} \sim 150$.

Figure 1 shows the influence of the WIMP microphysics, whereas figure 2 shows how the maximum depends on the details of the primordial power spectrum.

5.3. *The first objects*

According to theory and consistent with present observations, primordial fluctuations are gaussian distributed. The growth of matter perturbations on small scales (compared to the Hubble horizon) is surpressed during the radiation dominated epoch — they grow logarithmically. After matter- radiation equality all subhorizon fluctuations grow proportional to the scale factor and collapse once the overdensity becomes non-linear. For approximately scale-invariant primordial perturbations, these ingredients lead to the hierarchical scenario of structure formation. The smallest scales are the first that start to grow and are thus the first to reach the nonlinear regime. The redshift when that happens is estimated for the scales $M > M_{\rm fs}$ to be $z_{\rm nl} = 40$ to 80, depending on the details of the WIMP microphysics and the details of the primordial power spectrum.

The very first objects however are formed much earlier, as they are not typical. For gaussian distributed inhomogeneities the $N\sigma$ fluctuations would collapse at $Nz_{\rm nl}$. Let us look at some numbers for $N = 3$ (see Ref. 9): consider the comoving volume the Milky Way with a mass collecting radius of about 1 Mpc. Let us fix the scale of the very first objects to be $10^{-6}M_\odot$ and let us assume that $z_{\rm nl} = 60$. The density contrast of these CDM clouds would be $\Delta \sim 10^7$ today, which is an order of magnitude above the mean density contrast of the galactic disc. There would be CDM cloud in a $\rm pc^3$ within the galaxy. However, the survival of those objects has to be understood before a conclusion for their observational relevance can be drawn. These studies must include numerical simulations and analytic estimates.[14,16–18]

In a numerical simulation by Diemand et al.,[14] it was shown that indeed structure on all scales down to the cut-off scale survive the merging of the first halos. The typical fluctuations of a given mass scale merge to give larger ones, however the rare ($N\sigma$, $N > 1$) fluctuations are denser and are less likely to be disrupted in the merger process. Rare fluctuations survive with a certain probability until the formation of the first stars. Once compact baryonic objects (stars and planets) have been formed, encounters of them with the smallest CDM clouds could lead to their disruption.[15] This process was studied in some detail in references,[17,18] and the conclusion is that the clouds would loose mass, but their core remains intact, i.e. the cores of the very first objects in the Universe might populate our and other galaxies today. Their distribution in mass, size and space is still a subject of research.

6. Non-thermal and non-relativistic dark matter

The best studied candidate for non-thermal CDM is probably the axion.[3] Let us just mention that in the case of axions there is a preferred mass scale on which miniclusters of $\sim 10^{-12}M_\odot$ are formed.[19]

7. Effects from the hot plasma

A sequence of transitions, like the QCD transition, took place in the radiation dominated epoch of the Universe.[21] Especially the QCD transition[20] and the $e^{\pm}$ annihilation[11] are of interest to the structure formation at the smallest scales. Extra features are imprinted in the CDM power spectra at scales of $10^{-8}M_{\odot}$ (most relevant in the for non-thermal CDM) and $50M_{\odot}$ (relevant for any CDM), respectively.

Acknowledgements

I would like to thank Anne Green and Stefan Hofmann for our enjoyable collaboration and Ed Bertschinger for a very useful e-mail conversation.

References

1. R. Trotta, astro-ph/0608116. R. Keskitalo, H. Kurki-Suonio, V. Muhonen & J. Valiviita, astro-ph/0611917.
2. M. Tegmark *et al.*, astro-ph/0608632; U. Seljak, A. Slosar & P. McDonald, JCAP **0610** (2006) 014.
3. P. Sikivie, astro-ph/0610440.
4. C. Schmid, D. J. Schwarz & P. Widerin, Phys. Rev. D **59** (1999) 043517.
5. S. Hofmann, D. J. Schwarz & H. Stöcker, Phys. Rev. D **64** (2001) 083507.
6. V. Berezinsky, V. Dokuchaev & Y. Eroshenko, Phys. Rev. D **68** (2003) 103003.
7. X. l. Chen, M. Kamionkowski & X. M. Zhang, Phys. Rev. D **64** (2001) 021302.
8. A. M. Green, S. Hofmann & D. J. Schwarz, Mon. Not. Roy. Astron. Soc. **353** (2004) L23.
9. A. M. Green, S. Hofmann & D. J. Schwarz, JCAP **0508** (2005) 003.
10. S. Profumo, K. Sigurdson & M. Kamionkowski, Phys. Rev. Lett. **97** (2006) 031301.
11. E. Bertschinger, Phys. Rev. D **74** (2006) 063509.
12. C. Boehm, P. Fayet & R. Schaeffer, Phys. Lett. B **518** (2001) 8.
13. A. Loeb & M. Zaldarriaga, Phys. Rev. D **71** (2005) 103520.
14. J. Diemand, B. Moore & J. Stadel, Nature **433** (2005) 389.
15. H. S. Zhao, J. Taylor, J. Silk & D. Hooper, arXiv:astro-ph/0502049.
16. V. Berezinsky, V. Dokuchaev & Y. Eroshenko, Phys. Rev. D **73** (2006) 063504.
17. A. M. Green & S. P. Goodwin, astro-ph/0604142.
18. T. Goerdt, O. Y. Gnedin, B. Moore, J. Diemand & J. Stadel, astro-ph/0608495.
19. C. J. Hogan & M. J. Rees, Phys. Lett. B **205** (1988) 228; E. W. Kolb & I. I. Tkachev, Phys. Rev. D **50** (1994) 769; Astrophys. J. **460** (1996) L25.
20. C. Schmid, D. J. Schwarz & P. Widerin, Phys. Rev. Lett. **78** (1997) 791.
21. D. J. Schwarz, Annalen Phys. **12** (2003) 220; D. Boyanovsky, H. J. de Vega & D. J. Schwarz, arXiv:hep-ph/0602002.

GRAVITINO DARK MATTER AND THE IMPLICATIONS*

KI-YOUNG CHOI

Department of Physics and Astronomy, University of Sheffield, Sheffield, S3 7RH, England
**E-mail: K.Choi@sheffield.ac.uk*

In this talk, I present the possibility of gravitino dark matter in the framework of the Constrained MSSM. Additionally we find the prospects of the gravitino for the implication on the vacuum structure of our universe and the solutions to the cosmic lithium problems.

Keywords: Supersymmetric Effective Theories, Cosmology of Theories beyond the SM, Dark Matter, Supersymmetric Standard Model.

1. Introduction

Gravitino should be present in the models where gravity is incorporated into supersymmetry (SUSY) via local SUSY, or supergravity (SUGRA). Since gravitino interactions with ordinary matter are strongly suppressed, it was realized early on that the particle was facing various cosmological problems [1,2]. If the gravitino is not the lightest superpartner (LSP), it decays late ($\sim 10^{2-10}$s) into the LSP (*e.g.*, the neutralino) and associated electromagnetic (EM) or hadronic (HAD) radiation. These high energetic particles may destroy the standard Big Bang nucleosynthesis (BBN). Since the number density of gravitinos is directly proportional to the reheating temperature $T_{\rm R}$, this leads to an upper bound of $T_{\rm R} < 10^{6-8}\,{\rm GeV}$ [3–6]). On the other hand, when the gravitino is the LSP and stable (the case considered in this work), ordinary sparticles can decay into it and associated EM and/or HAD radiation. A combination of this and the overclosure argument ($\Omega_{\widetilde{G}}h^2 < 1$) has lead in this case to a rough upper bound $T_{\rm R} \lesssim 10^9\,{\rm GeV}$ [5,7].

*Talk presented at IDM2006, Rhodes, Greece, 11 - 16 September 2006. Based on the work collaborated with D.G. Cerdeño, K. Jedamzik, L. Roszkowski, and R. Ruiz de Austri.

We study this gravitino as the lightest superpartner and a candidate for cold dark matter in the Universe in the framework of the Constrained MSSM. We require the gravitino to have the right abundance of gravitino for dark matter and satisfy the current accelerator and cosmological constraints. Assuming the gravitino mass $m_{\widetilde{G}}$ in the GeV to TeV range, and for natural ranges of other supersymmetric parameters, the neutralino region is excluded, while for smaller values of $m_{\widetilde{G}}$ it becomes allowed again. The gravitino relic abundance is consistent with observational constraints on cold dark matter from BBN and CMB in some well defined domains of the stau region but, in most cases, only due to a dominant contribution of the thermal population [8,9].

In the later sections, we address the implications from gravitino dark matter with regards to the vacuum structures and the cosmic lithium problems.

2. Gravitino in the MSSM

There are some generic ways through which gravitino LSP can be produced in the early Universe. One obvious mechanism is that the next–to–lightest superpartner (NLSP), at much later times, decays into the gravitino. Such a process does not depend on the previous thermal history of the Universe (so long as the freeze–out temperature is significantly lower than the reheating temperature $T_{\rm R}$ after inflation). We will call it a non–thermal production (NTP) mechanism.

To evaluate the abundances of light elements produced during BBN in the presence of EM/HAD showers, we need to know the energy ϵ_i^X transferred to each decay channel i and their respective branching fractions to EM/HAD showers B_i^X as well as the NLSP lifetime τ_X [14,15]. All the above quantities depend on the NLSP and (with the exception of the yield) on its decay modes and the gravitino mass. For the cases of interest (χ and $\tilde{\tau}_1$) these have been recently evaluated in detail in [16,17] and summarized in [8,9].

For the neutralino NLSP the dominant decay mode is $\chi \to \widetilde{G}\gamma$ and this decay produces mostly EM energy. Above kinematic thresholds, the neutralino can also decay via $\chi \to \widetilde{G}Z, \widetilde{G}h, \widetilde{G}H, \widetilde{G}A$ for which the decay rates are given in [17,18]. These processes contribute to HAD fluxes because of large hadronic branching ratios of the Z and the Higgs bosons ($B^Z_{had} \simeq$ 0.7, $B^h_{had} \simeq 0.9$). In this case we can evaluate the transferred energy ϵ_i^X and branching fraction B_i^X [8,9].

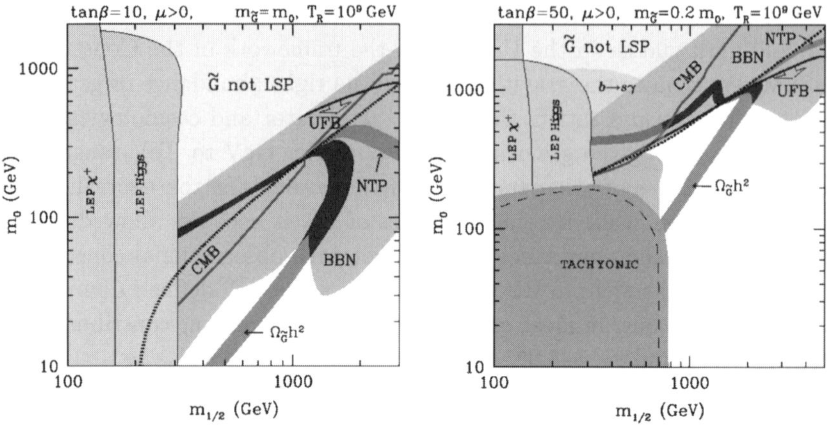

Fig. 1. The plane $(m_{1/2}, m_0)$ for $\tan\beta = 10$ (left window) and $\tan\beta = 50$ (right window) and for $A_0 = 0$ and $\mu > 0$. The light brown regions labelled "LEP χ^+" and "LEP Higgs" are excluded by unsuccessful chargino and Higgs searches at LEP, respectively. In the right window the darker brown regions labelled "$b \to s\gamma$" is excluded assuming minimal flavor violation and the dark grey region below the dashed line is labelled "TACHYONIC" because of some sfermion masses becoming tachyonic and is also excluded. In the rest of the grey region (above the dashed line) the stau mass bound $m_{\tilde{\tau}_1} > 87\,\text{GeV}$ is violated.) In both windows, the dark green bands labelled "$\Omega_{\widetilde{G}}h^2$" denote the total relic abundance of the gravitino from both thermal and non–thermal production is in the favored range, while in the light green regions (marked "NTP") the same is the case for the relic abundance from NTP processes alone. The UFB constraints are denoted by blue lines.

The dominant decay mode of the stau $\tilde{\tau}_1$ is $\tilde{\tau}_1 \to \widetilde{G}\tau$ which contributes basically only to EM showers. The leading contribution to HAD showers come from 3–body decays $\tilde{\tau}_1 \to \widetilde{G}\tau Z, \widetilde{G}\nu_\tau W$, or from 4–body decays $\tilde{\tau}_1 \to \widetilde{G}\tau\gamma^*/Z^* \to \widetilde{G}\tau q\bar{q}$. One typically finds $B_{had}^{\tilde{\tau}_1} \sim 10^{-5} - 10^{-2}$ when 3–body decays are allowed and $\sim 10^{-6}$ from 4–body decays [17].

The results are shown in Fig. 1 in the usual $(m_{1/2}, m_0)$ plane. We note that a robust constraint from D/H and Y_p excludes basically the whole neutralino NLSP region. This remains true so long as $m_{\widetilde{G}} > \mathcal{O}(1\,\text{GeV})$ as we will discuss later. Next, the important role played by the constraints from $^3He/D$ and $^6Li/^7Li$ in excluding additional regions of the $(m_{1/2}, m_0)$.

It is thus clear that, so long as $T_\text{R} \lesssim 10^9\,\text{GeV}$, one finds sizable regions of rather large $m_{1/2}$ and much smaller m_0 consistent with the preferred range of CDM abundance, where stau is NLSP and satisfy all the accelerator constraints.

3. False Vacua

However in SUSY theories the presence of several scalar fields carrying color and electric charge allows for a possible existence of dangerous charge and color breaking (CCB) minima [10,11] which would render the physical (Fermi) vacuum unstable.

The UFB–3 direction involves the scalar fields $\{H_u, \nu_{L_i}, e_{L_j}, e_{R_j}\}$ with $i \neq j$ and thus leads also to electric charge breaking. Since it yields the strongest bound among all the UFB and CCB constraints, and for future convenience, let us briefly give the explicit form of this constraint.

In Fig. 1 we mark the regions corresponding to a one–loop corrected UFB–3 direction becoming the global CCB minimum. These are encircled by a solid blue line and marked UFB. We can see that in both cases the whole previously allowed (white) and cosmologically favored (green) regions correspond to a false vacuum.

One cannot exclude the possibility that the color and electric charge neutral (Fermi) vacuum that the Universe ended up in after inflation is not a global one but merely a long–lived local minimum. For this reason, while the regions presented in Fig. 1 which correspond to the Fermi vacuum being a false one cannot be firmly excluded, should SUSY searches at the LHC find sparticles with masses indicating such a false vacuum, valuable information will be gained about the state of the Universe and about early Universe cosmology.

4. Cosmic Lithium Problems

Finally we consider the cosmic lithium problems. An intriguing further consequence of the gravitino LSP is that the decay of stau NLSP at around 10^3sec can reduce ^{7}Li abundance and produce ^{6}Li to give the agreement with observations. That gives a possible explanations for cosmic lithium problems [13].

Results of our lithium abundances with decaying NLSPs in the CMSSM are shown, for the particular case of $\tan\beta = 10$, $\mu > 0$, and $A_0 = 0$ in Fig. 2 for different choices of $m_{\tilde{G}}$. (We assume low enough $T_{\rm R}$ so that the TP contribution is negligible.) The regions imply: (a) green : to solve the origin of the high ^{6}Li abundance. (b) red : ^{7}Li/H ratio is considerably reduced by relic particle decay with the ^{6}Li produced due to other pre–galactic sources. (c) blue : both lithium problems are solved by relic particle decay. (d) pink : solve ^{7}Li problem with ^{6}Li excess which is reduced by some stellar depletion.

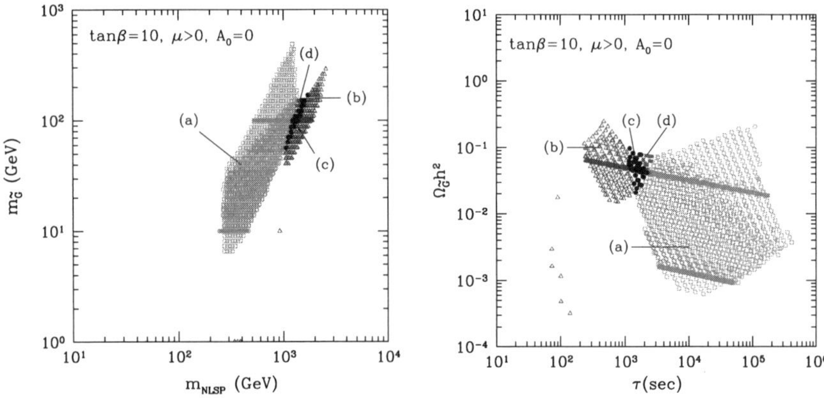

Fig. 2. Left: Gravitino mass $m_{\tilde{G}}$ as a function of NLSP mass $m_{\rm NLSP}$. Right: Present day gravitino abundance $\Omega_{\tilde{G}}h^2$ as a function of NLSP decay time.

It is encouraging that the CMSSM coupled to gravity provides solutions to both lithium problems. Essentially all of these solutions are obtained in the parameter space where the stau is the NLSP.

Decay times and final present–day gravitino abundance are shown in Fig. 2. Here only the NTP component of gravitinos produced during NLSP decay is shown. It is seen that the cosmologically most appealing points occur indeed for NLSP decay times around 10^3sec. Furthermore, it is intriguing to note that those points (blue and pink) at the same time may provide a significant fraction, or all, of the by WMAP required dark matter density $\Omega_{\rm DM}h^2 \approx 0.1126$ in form of the created gravitinos during the decay. Fig. 2 shows the stau and gravitino masses. The favorite region (blue and pink) obtains for stau masses around 1 TeV and gravitino masses between 30 and 200 GeV. Points which only explain the origin of ^{6}Li are obtained for lighter staus, whereas points which only solve the ^{7}Li discrepancy are found for heavier staus.

5. Conclusion

Gravitino is also well motivated candidate for CDM. In the context of CMSSM the neutralino as NLSP is excluded by the constraint from BBN. We have shown that in most of the stau NLSP region consistent with BBN and CMB constraints the usual Fermi vacuum is not the global minimum of the model. Instead, the true vacuum, while located far away from the Fermi vacuum is color and charge breaking. Further there exists a supersymmetric parameter space where the origin of ^{6}Li and the reduction of ^{7}Li may be explained by stau decay during or after BBN.

Acknowledgments

K.-Y. Choi would like to thank to the IDM2006 Local Committee for their kind invitation and for organizing a stimulating meeting. K.-Y. Choi is funded by PPARC.

References

1. H. Pagels and J.R. Primack, Phys. Rev. Lett. **48**, 1982 (223).
2. S. Weinberg, Phys. Rev. Lett. **48**, 1982 (1303).
3. M.Y. Khlopov and A. Linde, Phys. Lett. **B 138**, 1984 (265).
4. J.R. Ellis, J.E. Kim and D.V. Nanopoulos, Phys. Lett. **B 145**, 1984 (181).
5. J.R. Ellis, D.V. Nanopoulos and S. Sarkar, Nucl. Phys. **B 259**, 1985 (175).
6. D.V. Nanopoulos, K.A. Olive and M. Srednicki, Phys. Lett. **B 127**, 1983 (30).
7. T. Moroi, H. Murayama and M. Yamaguchi, Phys. Lett. **B 303**, 1993 (289).
8. L. Roszkowski, R. Ruiz de Austri and K.-Y. Choi, JHEP **0508**, 2005 (080) [hep-ph/0408227].
9. D.G. Cerdeño, K.-Y. Choi, K. Jedamzik, L. Roszkowski and R. Ruiz de Austri, JCAP **0606**, 2006 (005) [hep-ph/0509275].
10. J.A. Casas, A. Lleyda and C. Muñoz, Nucl. Phys. **B 471**, 1996 (3) [hep-ph/9507294].
11. J.A. Casas, A. Lleyda and C. Muñoz, Phys. Lett. **B 380**, 1996 (59) [hep-ph/9601357] and Phys. Lett. **B 389**, 1996 (305) [hep-ph/9606212].
12. for a discussion of the ^{7}Li isotope refer to: D. L. Lambert, astro-ph/0410418.
13. K. Jedamzik, K.-Y. Choi, L. Roszkowski and R. Ruiz de Austri, JCAP **0607**, 2006 (007) [hep-ph/0512044].
14. M. Kawasaki, K. Kohri and T. Moroi, Phys. Lett. **B 625**, 2005 (7) [astro-ph/0402490]; Phys. Rev. **D 71**, 2005 (083502) [astro-ph/0408426].
15. K. Jedamzik, hep-ph/0604251.
16. J.L. Feng, A. Rajaraman and F. Takayama, Phys. Rev. **D 68**, 2003 (063504) [hep-ph/0306024].
17. J.L. Feng, S. Su and F. Takayama, Phys. Rev. **D 70**, 2004 (063514) [hep-ph/0404198] and Phys. Rev. **D 70**, 2004 (075019) [hep-ph/0404231].
18. J. Ellis, K.A. Olive, Y. Santoso and V. Spanos, Phys. Lett. **B 588**, 2004 (7) [hep-ph/0312262].
19. A. Cuoco, F. Iocco, G. Mangano, G. Miele, O. Pisanti and P. D. Serpico, Int. J. Mod. Phys. A **19**, 4431 (2004); A. Coc, E. Vangioni–Flam, P. Descouvemont, A. Adahchour and C. Angulo, Astrophys. J. **600**, 544 (2004); R. H. Cyburt, B. D. Fields and K. A. Olive, Phys. Lett. B **567**, 227 (2003).

WHAT AN ASTROPHYSICIST CAN TELL ABOUT THE NATURE OF DARK MATTER?

S. COLAFRANCESCO*

INAF - Osservatorio Astronomico di Roma,
Monteporzio, 00040, Italy
**E-mail: Sergio.Colafrancesco@mporzio.astro.it*
www.mporzio.astro.it/~ cola/index.html

The nature of Dark Matter is still elusive to the ongoing experimental search. We discuss some astrophysical techniques to search for the nature of DM through a multi-frequency analysis of cosmic structures on large scales, from dwarf galaxies to galaxy clusters.

Keywords: Cosmology; Dark Matter; Cosmic structures.

1. Signals from the dark universe

There is overwhelming evidence that we live in a flat ($\Omega_0 \approx 1$), dark universe dominated by a dark form of matter (Dark Matter, DM) and an obscure form of energy (Dark Energy, DE). Dark Matter provides a fraction $\Omega_m \approx 0.23 \pm 0.04$ of the overall matter-energy content, the rest being provided by Dark Energy with $\Omega_{DE} \approx 0.73 \pm 0.04$ with the baryonic contribution limited to $\Omega_b \approx 0.044 \pm 0.004$ [15]. Given the basic properties that DM is observed to have it has been not difficult to think of possible candidates [2], but despite the large experimental efforts for their studies, the nature of the DM basic constituents is still unknown. Direct detection is the cleanest and most decisive discriminant. However, it would be interesting if astronomical techniques were to reveal some of the fundamental properties of DM particles. The dark side of the universe sends us, in fact, signals of the presence and of the nature of DM that can be recorded using different astrophysical probes. These probes are of *inference* and *physical* character.

Inference probes (i.e., the CMB anisotropy spectrum [15], the dynamics of galaxies [16], the hydrodynamics of the hot intra-cluster gas, (see e.g. [1]) and the gravitational lensing distortion of background galaxies by the intervening potential wells of galaxy clusters (see [3]) tell us about the presence, the total amount and the spatial distribution of DM in the large scale structures but cannot provide detailed information on the nature of DM. *Physical probes* tell us about the nature and the physical properties of the DM particles and can be obtained by studying the astrophysical signals of their interaction/annihilation in the atmospheres of DM-dominated structures (like galaxy cluster and galaxies). These probes can be recorded over a wide range of frequencies from radio to γ-rays and prelude to a full multi-frequency search for the nature of DM in cosmic structures.

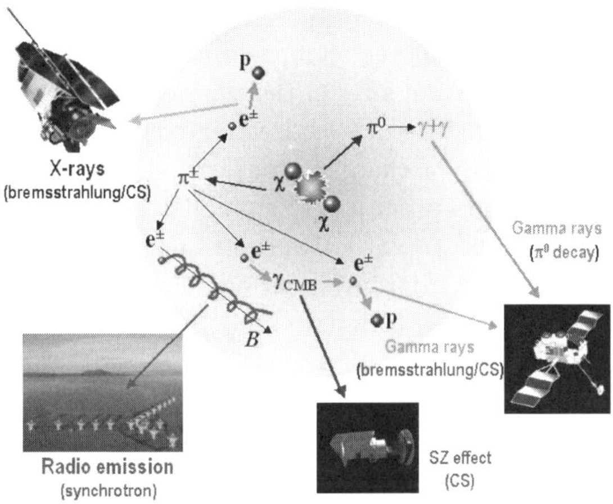

Fig. 1. The basic astrophysical mechanisms underlying the multi-frequency search for the nature of (χ) DM particles in large-scale structures (e.g., galaxy clusters and galaxies). These mechanisms are, among others: γ-ray emission from $\pi^0 \to \gamma + \gamma$, relativistic bremsstrahlung of secondary $e^{\pm}$ and ICS of CMB photons by secondary $e^{\pm}$; X-ray/UV emission due to bremsstrahlung and ICS of background photons by secondary $e^{\pm}$; synchrotron emission by secondary $e^{\pm}$ diffusing in the ambient magnetic field; SZ_{DM} (ICS of CMB photons by secondary $e^{\pm}$) effect.

2. A test case: neutralino DM

Among the viable competitors for having a cosmologically relevant DM species, the leading candidate is the lightest supersymmetric (SUSY) particle, plausibly the neutralino χ, with a mass M_χ in the range between a few GeV to a few hundreds of GeV [2].

The χ annihilation rate $R = n_\chi(r)\langle\sigma v\rangle$, depends on the χ number density $n_\chi(r) = n_{\chi,*}g(r)$, with a spatial profile given by the general formula $g(r) = (r/r_s)^{-\eta}(1 + r/r_s)^{\eta-\xi}$ with $\eta = 1$ and $\xi = 3$ reproducing the

NFW [14] profile, and from the $\chi\chi$ annihilation cross section $\langle\sigma v\rangle$ averaged over a thermal velocity distribution at freeze-out temperature [13]. The range of M_χ and $\langle\sigma v\rangle$ in the most general supersymmetric DM setup is extremely wide [8,9]. In the following discussion, we will consider two representative SUSY models: a soft $b\bar{b}$ model with $M_\chi = 40$ GeV and a hard W^+W^- model with $M_\chi = 81$ GeV, with their appropriate annihilation cross-sections (see [8] for details).

Neutralinos which annihilate inside a DM halo produce quarks, leptons, vector bosons and Higgs bosons, depending on their mass and physical composition. Electrons are then produced from the decay of the final heavy fermions and bosons. The different composition of the $\chi\chi$ annihilation final state will in general affect the form of the electron spectrum [7] [8].

Secondary $e^\pm$ are produced through various prompt generation mechanisms and by the decay of charged pions $\pi^\pm \to \mu^\pm \nu_\mu(\bar{\nu}_\mu)$, with $\mu^\pm \to e^\pm + \bar{\nu}_\mu(\nu_\mu) + \nu_e(\bar{\nu}_e)$ and are subject to spatial diffusion and energy losses. Both spatial diffusion and energy losses contribute to determine the evolution of the $e^\pm$ source spectrum into their equilibrium spectrum. The time evolution of the $e^\pm$ spectrum is described by the transport equation:

$$\frac{\partial n_e}{\partial t} = \nabla\left[D\nabla n_e\right] + \frac{\partial}{\partial E}\left[b_e(E)n_e\right] + Q_e(E,r)\,, \tag{1}$$

where $Q_e(E,r)$ is the $e^\pm$ source spectrum, $n_e(E,r)$ is the equilibrium spectrum and b_e (given here in units of GeV/s) is the $e^\pm$ energy loss per unit time, $b_e = b_{ICS} + b_{synch} + b_{brem} + b_{Coul}$, with $b_{ICS} \approx 2.5 \cdot 10^{-17}(E/GeV)^2$, $b_{synch} \approx 2.54 \cdot 10^{-18}B_\mu^2(E/GeV)^2$, $b_{brem} \approx 1.51 \cdot 10^{-16}(n_{th}/cm^{-3})\left(\log(\Gamma/n_{th}) + 0.36\right)$, $b_{Coul} \approx 7 \cdot 10^{-16}(n_{th}/cm^{-3})\left(1 + \log(\Gamma/n_{th})/75\right)$. Here n_{th} is the ambient gas density and $\Gamma \equiv E/m_e c^2$. The DM source spectrum, $Q_e(E,r)$, is constant over time and the $e^\pm$ population can be described by a quasi-stationary ($\partial n_e/\partial t \approx 0$) transport equation from which $n_e(E,r)$ reaches its equilibrium configuration mainly due to synchrotron and IC losses at $E \gtrsim 150$ MeV and to Coulomb losses at smaller energies. Spatial diffusion can be neglected in galaxy clusters while it is relevant on galactic and sub-galactic scales [8,9].

2.1. *Spectral Energy Distribution from DM annihilation*

The astrophysical signals of DM annihilation cover the entire e.m. spectrum, from radio to γ-ray frequencies (see Fig. 2).

Gamma rays. Gamma-ray emission is predominantly due to the hadronization of the decay products of $\chi\chi$ annihilation with a continuum

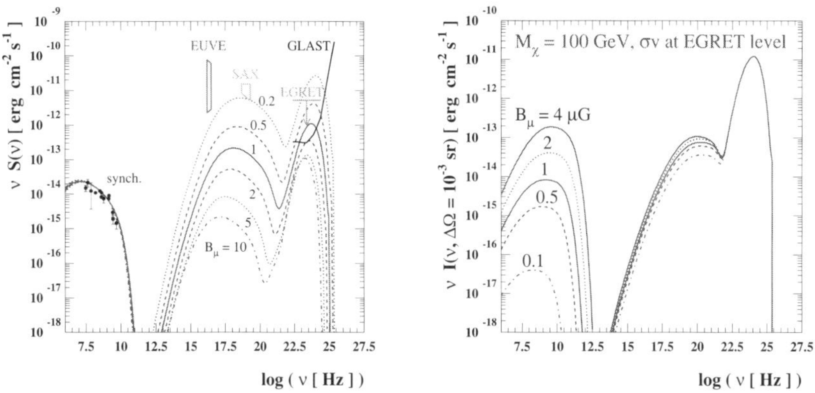

Fig. 2. Left. Multi-frequency spectrum of the best fit models of the radio halo flux of Coma: $M_\chi = 40$ GeV ($b\bar{b}$). The halo profile is the best fit NFW [14] profile with $M_{vir} = 0.9\,10^{15}h^{-1}$ and $c_{vir} = 10$, with subhalo setup as given in [8]. The scaling of the multi-frequency spectrum with the value for the mean magnetic field B_μ in Coma is shown. Right. The multi-frequency spectrum of Draco dwarf galaxy for $M_\chi = 100$ GeV ($b\bar{b}$), and the effect of varying the magnetic field strength. The $\chi\chi$ annihilation rate has been tuned to give a γ-ray signal at the level of the EGRET upper limit.

spectrum due to the decay $\pi^0 \to \gamma + \gamma$ [7] [8], even though the direct $\chi\chi$ annihilation results in a line emission feature at an energy $\sim M_\chi$. Gamma-ray emission is also expected from secondary $e^\pm$ through bremsstrahlung and ICS of CMB photons up to high energies (see Fig. 2). Gamma-ray emission from DM annihilation could be revealed provided that i) sufficient spectral and spatial resolution can be achieved by the upcoming experiments and ii) a clear understanding of other competing emission mechanisms expected to work in cosmic structures [5] will be obtained.

For the Coma cluster, the γ-ray flux produced by the χ model here considered is dominated by the continuum $\pi^0 \to \gamma\gamma$ component and it is a factor ~ 5 lower than the EGRET upper limit (see Fig. 2, left panel). For Draco, the dominant γ-ray emission is still given by the continuum $\pi^0 \to \gamma\gamma$ component while the dominant IC emission (i.e. that of the IC on CMB photons) peaks at much lower frequencies and is a factor $\sim 10^2$ less intense.

Radio emission. Secondary $e^\pm$ produced by $\chi\chi$ annihilation can produce synchrotron emission in the magnetized atmosphere of galaxy clusters (as well as galaxies) and can be observed as a diffuse radio emission centered on the DM halo. Observations of cluster radio-halos are, in principle, very effective in constraining the neutralino mass and composition [7–9] under

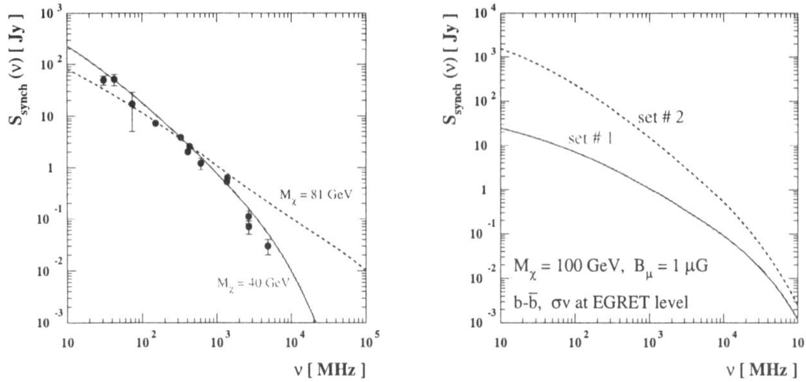

Fig. 3. Left. The radio flux density spectrum for Coma for a soft spectrum due to a $b\bar{b}$ annihilation final state (solid line) with $M_\chi = 40$ GeV, and a hard spectrum due to a W^+W^- channel (dashed line) with $M_\chi = 81$ GeV. Right. Radio flux density spectrum of Draco for a χ $b\bar{b}$ model with $M_\chi = 100$ GeV with $\langle\sigma v\rangle$ tuned to give 2 events in EGRET. Results are given for two choices of propagation parameters: a conservative choice ("set # 1") or a more extreme choice ("set # 2") (see [9] for details).

the hypothesis that DM annihilation provides a major contribution to the radio-halo flux. Under this hypothesis, a soft DM model ($b\bar{b}$ with $M_\chi = 40$ GeV) is able to reproduce both the overall radio-halo spectrum of Coma and the spatial distribution of its surface brightness [8] (see Fig. 3).

For the case of Draco, radio emission is strongly affected by propagation effects. Kolmogorov-type diffusion, $D(E) = D_0/B_\mu^{1/3}\,(E/1\,\mathrm{GeV})^{1/3}$, with $D_0 = 3\cdot 10^{28}$ cm^2 s^{-1} (set up #1) induce a depletion of the $e^\pm$ populations with a significant fraction leaving the diffusion region, while for $D(E) = D_0\,(E/1\,\mathrm{GeV})^{-0.6}$ with $D_0 = 3\cdot 10^{26}$ cm^2 s^{-1} (set up #2) they are more efficiently confined within the diffusion region, but still significantly misplaced with respect to the emission region. Diffusion effects produce also a steeper spectral slope when the $e^\pm$ are more efficiently confined within the diffusion region (set up #2) w.r.t. the case (set #1) where there is a larger depletion of the $e^\pm$ populations [9].

ICS of CMB: from infrared to γ-rays. Secondary $e^\pm$ up-scatter CMB (and other background) photons that redistribute on a wide frequency range up to γ-ray frequencies (see Fig. 2). The soft χ model with $M_\chi = 40$ GeV and $\langle\sigma v\rangle = 4.7\cdot 10^{-25} cm^3 s^{-1}$, with $B_\mu = 1.2$ that fits the Coma radio halo data [8] yields UV and hard X-ray fluxes which are much fainter than the data in order to be consistent with the EGRET upper limit (see Fig. 2). It is, therefore, impossible to fit all the non-thermal emission features of

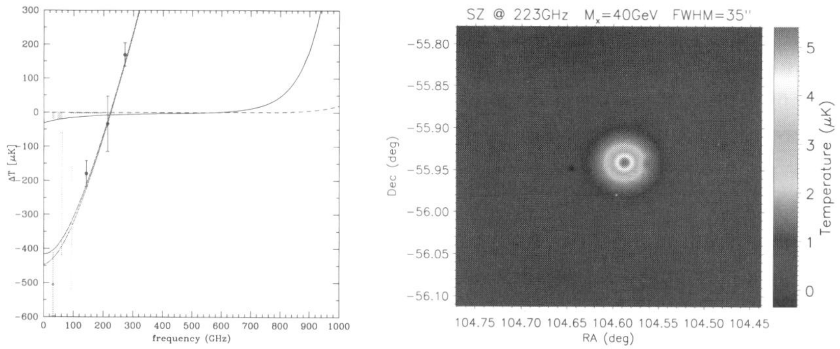

Fig. 4. **Left**. The SZ_{DM} effect in Coma produced by the $b\bar{b}$ model with $M_\chi = 40$ GeV (black solid curve) and by the W^+W^- model with $M_\chi = 81$ GeV (black dashed curve) are shown in comparison with the thermal SZ effect (blue curve). The red curves represent the overall SZ effect. The sensitivity of PLANCK (18 months, 1 σ) is shown for the LFI detector at 31.5 and 53 GHz channels (cyan shaded regions) and for the HFI detector 143 and 217 GHZ channels (green and yellow shaded areas). **Right**. The simulated SZ maps of the cluster 1ES0657-556 as observable with the SPT telescope at $\nu = 223$ GHz for a $b\bar{b}$ χ model with $M_\chi = 40$ GeV (see [10] for details). The SZ_{DM} signals are spatially resolved as negative spots centered on the DM clumps.

Coma for a consistent choice of the DM model and of the cluster magnetic field [8]. For Draco the dominant IC on CMB component produces an X-ray flux of $\sim 10^{-15} - 10^{-14}$ erg cm^{-2} s^{-1} when the γ-ray flux is normalized to the EGRET upper limit. The constraints obtainable by the coming GLAST observation will set much more realistic expectations for the diffuse X-ray emission produced from DM annihilation in Draco which could eventually be tested with high sensitivity X-ray experiments.

SZ effect from DM annihilation. Secondary $e^\pm$ produced by DM annihilation interact with the CMB photons and up-scatter them to higher frequencies producing a SZ effect with specific spectral and spatial features [6]. The χ model with $M_\chi = 40$ GeV produces a temperature decrement in Coma of $\sim$ 40 to 15 μK in the range $\sim$ 30 to 150 GHz (see Fig. 4). The presence of a substantial SZ_{DM} effect is likely to dominate the overall SZ signal at $\nu \sim 220 - 250$ GHz providing a negative total SZ effect. These signals are still within the uncertainties of the available SZ data for Coma but they could be detectable with higher sensitivity experiments.

The SZ_{DM} effect in Draco, even though could be a definite probe of the DM annihilation in such DM-dominated systems [6,11], is quite low due to the effects of secondary $e^\pm$ spatial diffusion.

2.2. *Optimal astrophysical laboratories for DM search*

The analysis of the spatial and spectral intensity of the astrophysical signals coming from $\chi\chi$ annihilation could be a powerful tool to unveil the elusive nature of Dark Matter. However, such DM-induced signals are expected to be confused or even overcome by other astrophysical signals originating from the ambient plasmas (thermal and/or non-thermal), especially when all these components are co-spatially distributed with the DM component. An ideal system to detect DM annihilation signals would be, therefore, a system which is either devoid of diffuse emitting material (this is the case of dark galaxies) or a system with a clear spatial separation between the various matter components (this is the case of the cluster 1ES0657-556).

Dwarf (dark) galaxies are among the best sites for the astrophysical search for DM but their multi-frequency SED is usually quite dim. The combination of radio and γ-ray observations of dwarf galaxies with the coming high-sensitivity instruments (SKA, LOFAR, EVLA, GLAST) could, nevertheless, set strong constraints on the nature of the DM particles [9].

The cluster 1ES0657-556 (where the spatial distribution of DM is clearly offset w.r.t. that of the IC gas) is an optimal laboratory for astrophysical DM search because one expects that the $\chi\chi$ annihilation signals are clearly spatially separated from the other astrophysical signals originating in the atmosphere of this cluster [10]. However, the expected γ-ray emission associated to the DM clumps of this cluster is too low ($\lesssim 1$ count vs. ~ 10 background counts at $E > 1$ GeV) and cannot be resolved by GLAST from other possible sources of γ-ray emission, both from the cluster and from AGNs in the field. Radio telescopes have, in principle, excellent resolution and sensitivity to probe the different spectra and brightness distribution of the DM-induced synchrotron emission (the DM induced radio emission from the largest DM clump is $\sim 3 - 10$ mJy at $\nu = 100$ MHz (for a smooth or smooth plus 50% mass clumpiness NFW DM profile, soft $B\bar{b}$ model with $M_\chi = 40$, with a $B = 1$ μG), still marginally detectable by LOFAR [10]), but the uncertainties associated to the radio emission of the DM clumps of 1ES0657-556 render the prediction of these signals quite uncertain.

In such a context, the possible observation of the SZ_{DM} effect for this system [10] (see Fig. 4 for a map of the cluster at 223 GHz) will provide an important complementary, and maybe unique, probe of the nature of DM.

3. Conclusions

Viable DM models which are consistent with WMAP and with the viable structure formation scenario are able to produce substantial astrophysical signals especially detectable at radio, microwave and gamma-ray frequencies. The constraints that the multi-frequency astrophysical observations can set on the $\langle \sigma v \rangle$-M_χ plane, are able to efficiently restrict the available nautralino models [8,9]. Additional restrictions of this plane may be obtained by comparing the astrophysical constraints to those coming from both accelerator physics and from other astrophysical probes (e.g., the study of IMBHs, Bertone et al. 2005; the study of the galactic center region, see Fornengo 2006 for recent review). Direct DM detection experiments have already explored large regions of the most optimistic SUSY models, and the planned next-generation experiments will probably be able to explore also the core of the SUSY models. In this context, the astrophysical study of DM annihilation proves to be complementary, but hardly competitive, especially when a full multi-frequency approach is chosen. When combined with future accelerator results, such multi-frequency astrophysical search might greatly help us to unveil the elusive nature of Dark Matter.

References

1. Arnaud, M. 2005, astro-ph/0508159
2. Baltz, E. 2004, astro-ph/0412170
3. Bartelmann, M. & Schneider, P. 1999, Phys.Rep., 340, 291
4. Bertone, G., Zentner, A: and Silk, J. 2005, PRD, 72, 103517
5. Colafrancesco, S. 2004a, in Frontiers of Cosmology, A. Blanchard & M. Signore Eds., p.75 and p.85
6. Colafrancesco, S. 2004b, A&A, 422, L23
7. Colafrancesco, S. & Mele, B. 2001, ApJ, 562, 24
8. Colafrancesco, S., Profumo, S. & Ullio, P. 2006, A&A, 455, 21
9. Colafrancesco, S., Profumo, S. & Ullio, P. 2007, PRD, 75, 3513
10. Colafrancesco, S. et al. 2007, A&A, in press
11. Culverhouse, T., Ewans, W. and Colafrancesco, S. 2006, MNRAS, 368, 659
12. Fornengo, N. 2006, astro-ph/0612786
13. Jungman, G., Kamionkowski, M. & Griest, K. 1996, PhR, 267, 195
14. Navarro, J., Frenk, C. & White, S.D.M. 1997, ApJ, 490, 493; Navarro, J.F. et al. 2004, MNRAS, 349, 1039
15. Spergel, D. et al. 2003, ApJS, 148, 175
16. Zwicky, F. 1933, Helv.Phys.Acta, 6, 110

MIMAC-He3: MICRO-TPC MATRIX OF CHAMBERS OF ^{3}He

D. SANTOS, O. GUILLAUDIN, TH. LAMY, F. MAYET and E. MOULIN

Laboratoire de Physique Subatomique et de Cosmologie,
CNRS/IN2P3 et Université Joseph Fourier (Grenoble-1),
53, avenue des Martyrs, 38026 Grenoble cedex, France

The project of a micro-TPC matrix of chambers of ^{3}He for direct detection of non-baryonic dark matter is outlined. The privileged properties of ^{3}He are highlighted. The double detection (ionization - projection of tracks) will assure the electron-recoil discrimination. The complementarity of MIMAC-He3 for supersymmetric dark matter search with respect to other experiments is illustrated.The modular character of the detector allows to have different gases to get A-dependence. The pressure degreee of freedom gives the possibility to work at high and low pressure. The low pressure regime gives the possibility to get the directionality of the tracks. The first measurements of ionization at very few keVs for ^{3}He in ^{4}He gas are described.

Keywords: Dark Matter, TPC, Helium-3, Spin-dependent interaction.

1. Introduction

Strong evidence in favor of the existence of non-baryonic dark matter arises from different cosmological observations. The cosmic microwave backgroung (CMB) data [1,2] in combination with high redshift supernovæ analysis [3] and large scale structure surveys [4] seem to converge on an unified cosmological model [5]. The non-baryonic cold dark matter (CDM) would consist of still not detected particles, among those being the generally referred to as WIMPs (Weakly Interacting Massive Particles) the privileged ones. Among the different possible WIMPs, the lightest supersymmetric particle, in most scenarii, the lightest neutralino $\tilde{\chi}$ predicted by SUSY theories with R-parity conservation, stands as a well motivated candidate.

In the last decades, huge experimental efforts on a host of techniques in the field of direct search of non-baryonic dark matter have been performed [8–10,12]. Several detectors reached sufficient sensitivity to begin to test regions of the SUSY parameter space. However, Wimp events have not yet been reported. Besides the fact that the cross section could be very weak, the energy threshold effect combined with the use of a heavy target nucleus leads to significant sensitivity loss for relatively light WIMPs (6 GeV $\leq$ $M_{\tilde{\chi}} \leq$ 40 GeV).

As reported elsewhere [13,15,16], the use of ^{3}He as a target nucleus is motivated by its privileged features for dark matter search compared with other target nuclei. First, ^{3}He being a spin 1/2 nucleus, a detector made of such a material will be sensitive to the spin-dependent interaction, leading to a natural complementarity to most existing or planned Dark Matter detectors (ν telescopes, scalar direct detection as well as proton based spin-dependant detectors). In particular, it has been shown [17,18] that an ^{3}He based detector will present a good sensitivity to low mass $\tilde{\chi}$, within the framework of effective MSSM models without gaugino mass unification at the GUT scale [6,7].

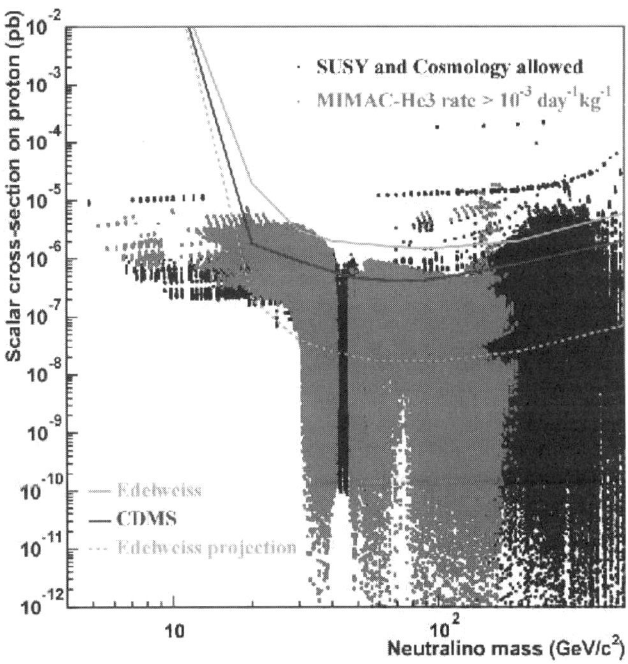

Fig. 1. SUSY non minimal models, calculated with DarkSusy code [7]. In grey the models giving an axial cross section ($\tilde{\chi}$ - ^{3}He) higher than the exclusion plot of MIMAC-^{3}He with 10kg [18]. These models are compared with exclusion plots of scalar experiments and their projections. There are models, in grey, that will be very difficult to get with only a scalar approach.

The ^{3}He presents in addition the following advantages with respect to other sensitive materials for WIMPs detection :
- a very low Compton cross-section to gamma rays, two orders of magnitude weaker than in Ge : 9×10^{-1} barns for 10 keV γ-rays
- the neutron signature made possible by the capture process :

$$n + {}^3\text{He} \rightarrow p + {}^3\text{H} + 764\,\text{keV}$$

Indeed it allows for an easy discrimination with $\tilde{\chi}^0$signal (E $\lesssim$ 6 keV). This property is a key point for Dark Matter search as neutrons in underground laboratories are considered as the ultimate background.

Any dark matter detector should be able to separate a $\tilde{\chi}$ event from the neutron background. Using energy measurement and electron-recoil discrimination, MIMAC-He3 presents a high rejection for neutrons due to capture and multi-scattering of neutrons [20]. The MIMAC project propose a modular detector in which different gases (^{3}He , CF_4) can be used to have a dependence on the mass of the target. The ^{19}F is other good target nucleus choice to have the axial interaction open, but proton based, increasing the attractiveness of the detector.

The MIMAC detector has two different regimes of work: i) high pressure (1,2 or 3 bar) and ii) low pressure (100 - 200 mbar). These two regimes allow us to have Wimp events at high pressure and search for correlation with the galactic halo apparent movement at low pressure. This last possibility should be validated with a special read out electronics as an important step of the project.

2. Micro-TPC and ionization-track projection detection

The micro time projection chambers with an avalanche amplification using a pixelized anode presents the required features to discriminate electron - recoil events with the double detection of the ionization energy and the track projection onto the anode. In order to get the electron-recoil discrimination, the pressure of the TPC should be such that the electron tracks with an energy less than 6 keV could be well resolved from the recoil ones at the same energy convoluted by the quenching factor. The electrons produced by the primary interactions will drift to the amplification region (mesh) in a diffusion process following the well known distribution characterized by a radius of $D \simeq \lambda\sqrt{(L[cm])}$ where λ is tipically 200 μm for ^{3}He at 1 bar and L is the total drift in the chamber up to the mesh. This process has been simulated with Garfield and the drift velocities estimated as a function of the pressure and the electric field. A typical value of 26μm/ns is obtained for

1 kV/cm in pure ^{3}He at a pressure of 1 bar. To prevent confusion between electron track projection and recoil ones the total drift length should be limited to L$\simeq$15 cm. It defines the elementary cell of the detector matrix and the simulations performed on the ranges of electrons and recoils suggest that with an anode of 350μm the electron-recoil discrimination required can be obtained. The quenching factor is an important point that should be addressed to quantify the amount of the total recoil energy recovered in the ionization channel. No measurements of the quenching factor (QF) in ^{3}He have been reported. However, an estimation can be obtained applying the Lindhard calculations [11]. The estimated quenching factor given by Lindhard's theory for ^{3}He shows up to 70 % of the recoil energy going to the ionization channel for 5 keV ^{3}He recoil.

3. Source MIMAC

In order to measure the QF for ^{3}He and ^{4}He we have developed at the LPSC a dedicated facility producing very light ions at a few keV energies. This facility, called source MIMAC, incorporates an ECR ion source coupled to a Wien filter, selecting q/m, and a high voltage extraction going up to 50 kV.

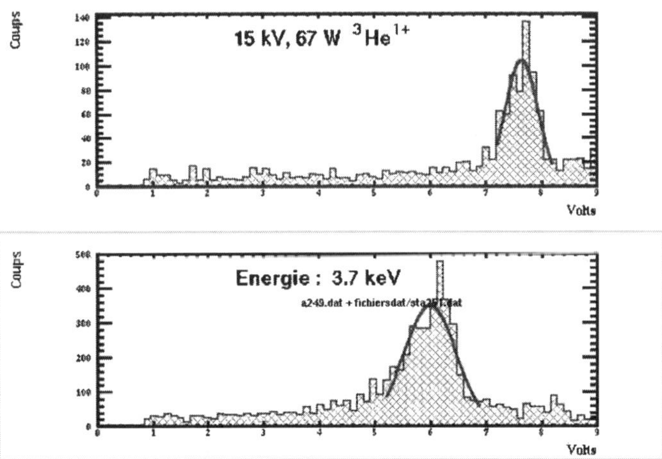

Fig. 2. Time of flight measurements performed with the MIMAC source. The figure shows the spectra at the two different positions (close to and far from the interface (source-chamber)) used to measure the ^{3}He$^+$ ions output energy when they have been accelerated at 15 kV.

The characterization of the output energies is made by a separate time of flight measurements as we can see on Fig. 2 for the case of ^{3}He ions accelerated at 15 kV having a mean output energy of 3.7 keV. Using this facility we can explore the ionization at very low energies for ^{3}He ions. We have measured by TOF, five output energies going from 13.7 keV up to 3.7 keV corresponding to five values from 30 to 15 kV of accelerating voltage extraction. Ionization measurements have been performed, with a standard micromegas grid in a gas chamber (95% of ^{4}He and 5% of isobutane at 1 bar). A linear calibration fits very well the points measured and extrapolating to even lower voltage extraction, we can estimate the maximum output energy corresponding to 10.5 kV to 800 eV. On Fig. 3 the spectrum of the ionization left in the chamber by ^{3}He at 800 eV is shown. On the same spectrum we show an internal conversion electron spectrum of ^{57}Co during the two minutes the beam of ^{3}He was on. This ^{57}Co source will allow us to get an idea of the equivalent electron energies. We can differentiate on the spectrum the peak of ionization well separated from the electronic noise.

Fig. 3. A two minutes spectrum showing ionization peak corresponding to a beam, produced by the MIMAC source, of ^{3}He at an energy estimated to 800 eV. An internal conversion source of ^{57}Co spectrum is shown on the same spectrum. This source will help us to get the electron energy calibration for the QF measurement.

This spectrum shows clearly that we can expect to get the ionization left by ^{3}He recoils in a chamber up to energies lower than 1 keV using the micromegas detector technology of our collaborators at Saclay [14].

The next steps will be to perform the QF measurements with the setup described above and to show the track of ^{3}He recoils with a dedicated read-out electronics that is building up at our laboratory.

References

1. A. Benoît *et al.*, Astron. & Astrophys. **399** (2003) L25, M. Tristram *et al.*, Astron. & Astrophys. **436** (2005) 785
2. D. Spergel *et al.*, Astrophys. J. Suppl. **148** (2003) 175
3. S. Perlmutter *et al.*, Phys. Rev. Lett. **83** (1999) 670
4. M. Tegmark *et al.*, Astrophys. J. **606** (2004) 702
5. U. Seljak *et al.* astro-ph/0604335.
6. G. Bélanger *et al.* hep-ph/0502079.
7. P. Gondolo *et al.* 2004 JCAP 0407 008.
8. A. Benoît *et al.*, Phys. Lett. **B545** (2002) 43
9. D. Akerib *et al.*, Phys. Rev. Lett. **93** (2004) 211301
10. W. Seidel *et al.*, Proc.of the 4th Intern. Conf. on Dark Matter in Astro and Particle Physics (DARK 2002), Feb. 2002, Cape Town (South Africa), Eds. H.-V. Klapdor-Kleingrothaus *et al.*, Springer, pp. 517
11. J. Lindhard *et al.*, Mat. Fys. Medd. K. Dan. Vidensk. Selsk. 33(1963)1-42.
12. E. Moulin *et al.*, Astron. & Astrophys. **453** (2006) 761
13. D. Santos *et al.*, Proc. of the 4th Intern. Symposium on Sources and Detection of Dark Matter and Dark Energy in the Universe (DARK 2000), Feb. 2000, Marina Del Rey (CA, USA), Ed. D.B. Cline, Springer, pp. 469,
14. Y. Giommataris *et al.* Nucl. Instr. and Meth. **A376** (1996) 29
15. F. Mayet *et al.*, Nucl. Instr. and Meth. **A455** (2000) 554
16. D. Santos *et al.*, J. Phys. Conf. Ser. **39** (2006) 154
17. F. Mayet *et al.*, Phys. Lett. **B538** (2002) 257
18. E. Moulin *et al.*, Phys. Lett. **B614** (2005) 143
19. D. Santos, F. Mayet *et al.*, *in preparation*
20. E. Moulin, PhD Thesis, 2005, Université J. Fourier, Grenoble, France

THE TRANSITION RADIATION DETECTOR OF AMS-02 TO SEARCH FOR DARK MATTER IN SPACE

F.R. SPADA

Università La Sapienza,
Rome, I-00185, Italy
E-mail: francesca.spada@roma1.infn.it
Url: http://www.cern.ch/spada

The AMS-02 (Alpha Magnetic Spectrometer) detector will operate on the International Space Station performing spectroscopy of cosmic rays searching for dark matter, trying to identify the products of the annihilation of WIMP candidates such as neutralinos.

The annihilation of a neutralino of a few hundred GeV mass could produce an excess in the 10–300 GeV region of the cosmic rays positron spectrum. With the joint use of a Transition Radiation Detector (TRD) and of an electromagnetic calorimeter, providing an overall 10^6 rejection factor for protons, AMS can detect a positron with a 90% efficiency.

The techniques used to achieve the physical and technological goals, together with the current status of the detector, will be discussed.

Keywords: Cosmic rays; Transition radiation; Dark matter.

1. Introduction

The AMS-02 (Alpha Magnetic Spectrometer) will operate on the International Space Station performing spectroscopy of cosmic rays, directly searching for antimatter (i.e., antihelium and higher Z nuclei) and indirectly searching for dark matter, trying to identify the products of the annihilation of candidates such as neutralinos.

The annihilation of a neutralino of a few hundred GeV mass could produce an excess in the 10–300 GeV region of the cosmic rays positron spectrum. Here, protons are about 10^4 times more abundant than positrons, as shown in figure 1. With the joint use of a Transition Radiation Detector (TRD) and of an electromagnetic calorimeter, providing an overall 10^6 rejection factor for protons, AMS can detect a positron with a 90% efficiency. Deviations from the ordinary cosmic ray positron spectrum in that

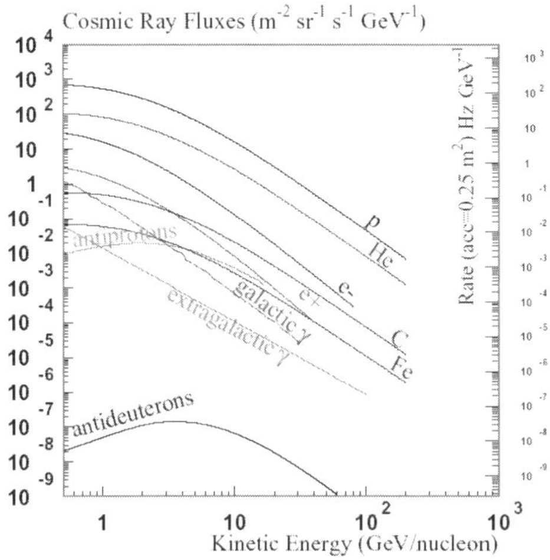

Fig. 1. Absolute fluxes of cosmic ray particles.

energy range have already been observed by several experiments: AMS-02 will explore it with unprecedented sensitivity (see figure 2).

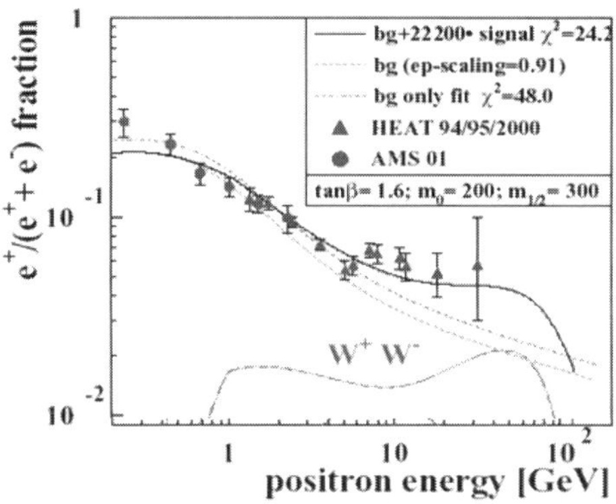

Fig. 2. Existing measurements of positron fraction vs. particle energy in cosmic rays.

2. The Transition Radiation Detector of AMS-02

The AMS-02 TRD construction is a joint effort of Rheinisch-Westflischen Technischen Hochschule (RWTH), Aachen, Germany; Kyungpook National

University (KNU), Daegu, Korea; Institut fr Experimentelle Kernphysik (IEKP), Karlsruhe, Germany; Massachussets Institut of Technology (MIT), Boston, USA; Università "La Sapienza" and Istituto Nazionale di Fisica Nucleare (INFN), Rome, Italy.

2.1. *Transition radiation emission by charged particles*

A relativistic ($\gamma \sim 1000$ or greater) charged particle crossing the interface between two media with different dielectric constants has a probability of the order of 1% to emit a photon in the soft X-ray region.[1] The energy of the emitted photon depends from the particle's mass and momentum, thus allowing for particle discrimination at very high energies where e.g. Čerenkov detectors are no longer useful.

The probability of transition radiation emission increases with the number of boundaries that the particle has to cross, but a compromise with the correspondent photon absorption increase must be found.

2.2. *The TRD layout*

In the TRD of AMS-02 the radiator is a fleece of 10 μm thick propylene fibres (LRP 375 BK), with a density of 0.06 g/cm^3.[2] The photon emission probability in the whole detector thus increases to 50% for electrons, and the X-rays can be collected, together with the ionization photons, in the proportional straw tubes of 6 mm diameter filled with a Xe:CO_2 [80:20] gas mixture. The tubes provide large nuclear measurements, fast drift of ionization electrons with no collection losses, high gain in the amplification region around the sense wire.

The overall detector consists of 328 modules of 16 tubes each interleaved with the radiator. To minimize the absorption, the carbon stiffeners and the support structure are as thin as possible compatibly with the mechanical stability of the module, that has to withstand the launch vibrations.[3] The detector has also to match weight requirements: this is why even though the proton-positron separation capability increases with the number of radiator–detector layers, this has been limited to 20, arranged into an octagonal aluminum-carbon structure as shown in figure 3. The modules have different sizes, the shortest one being 0.8 m and the longest one 2 m. The upper and lower four layers lay parallel to the magnetic field, while the twelve central ones are orthogonal to it, to increase the tracking capability in the bending plane.

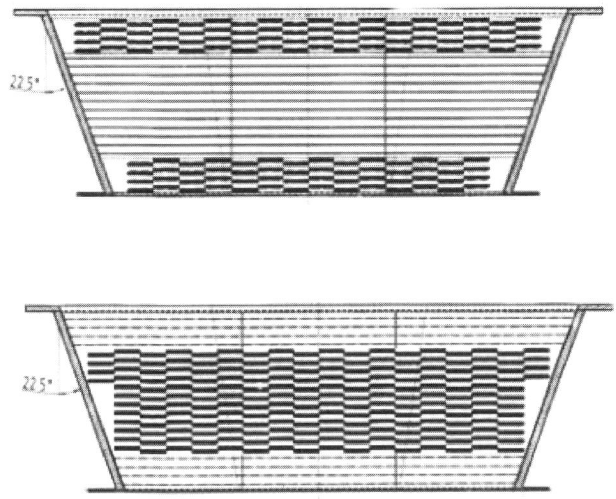

Fig. 3. The layout of the TRD (two projections are shown).

The safety factor – i.e. the ratio between the gas system operating period and the forseen duration of the mission – averaged over all the modules selected for flight, has been measured to be 7.9 with and RMS of 1.6.

2.3. *Gas distribution system and electronics*

The Xenon and CO_2 used to form the gas mixture are kept in two tanks of 50 cm and 40 cm of diameter respectively, containg an initial amount of 46 liters of Xenon and 4 liters of carbon dioxide at pressures of 107 bar and 65 bar respectively. This guarantees three to five years of operation.

The gas mixture filling the detector is refreshed every day, after controlling the composition with cross checks by a spirometer and a multi-channel analyzer, to ensure the stability of the gas gain.

The pumps circulate the gas in the TRD modules at a pressure of approximately of 1.4 bar. At all points of the gas system valves, pumps and pressure sensors have a two-fold redundancy, as in the scheme shown in figure 4.

The gas circulation system includes a Monitoring and Control Computer (JMDC), and a Power Distribution Box (PDB), which distributes the power supply from the 120 VDC provided by the Space Station. The electronics boards are connected to the Monitor and Control Computer and main data

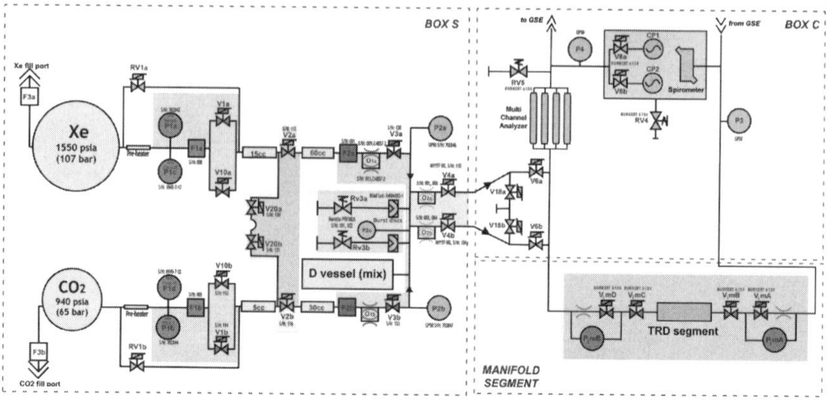

Fig. 4. The TRD gas distribution system.

acquisition via CAN-BUS. A monitor software tests the status of the gas system against pre-conditions and executes operational commands.

In case of power and communications failure, the gas system will be shut down safely.

3. The TRD Performances

A full 20-layers prototype was tested to verify the proton rejection power at the CERN T9, X7 and H6 beam lines, and about three million tracks were recorded. The particle energies ranged between 15 and 250 GeV.

The absolute energy scale was calibrated with a ^{55}Fe source, and the correlation between the density of the Xe-CO_2 mixture and the gas gain was measured to be $\Delta G/G = -5.5\Delta\rho/\rho$ at 1450 V. The spectra of protons and electrons in single track events are shown in figure 5.

In the data analysis the proton-electron separation was based on a likelihood algorythm. The Monte Carlo simulation was based on Geant3[4] with additional code to include transition radiation photon generation and absorption[5,6] and to improve the simulation of dE/dX Landau-fluctuations in thin gas layers.[7] The proton rejection factor was determined as the inverse proton selection efficiency with a likelihood cut set for an electron efficiency of 90%. The results are shown in figure 6.

The rejection factor obtained is well above the value required even at very high energies, thus ensuring the AMS-02 capability to reach the desired sensitivity on the measured positron spectrum on the whole interesting energy range.

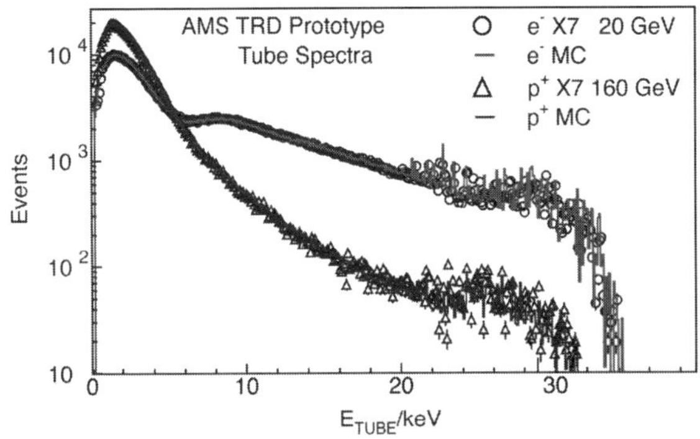

Fig. 5. Single track energy spectra. The proton spectrum (triangles) shows the expected Landau fluctuations. In the electron spectrum (circles) an enhancement due to the transition radiation component is clearly visible above 6 keV.

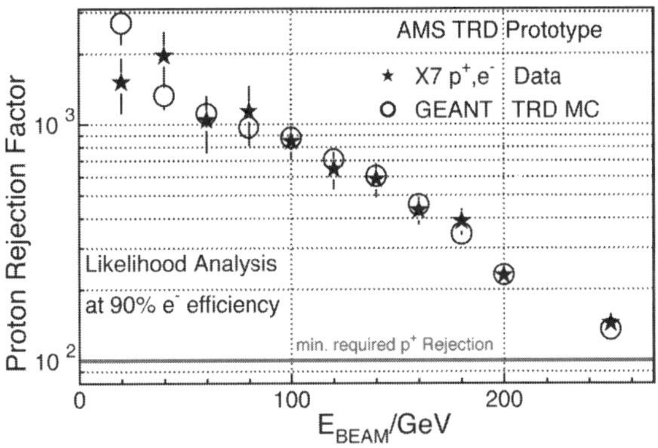

Fig. 6. The TRD proton rejection factor as a function of the particle's energy.

References

1. R. Battiston, *Nucl. Phys. Proc. Suppl.* **44**, 274 (1995); J. Alcaraz *et al.*, *Nuovo Cim.* **112 A**, 1325 (1999); W.J. Burger, *Nucl. Phys. Proc. Suppl.* **113**, 139 (2002).
2. T. Siedenburg *et al.*, *Nucl. Phys. Proc. Suppl.* **113**, 154–158 (2002).
3. O. Toker *et al.*, *Nucl. Instr. Meth.* **A 340**, 572 (1994); B. Alpat *et al.*, *Nucl. Instr. Meth.* **A 446**, 552 (2000).
4. R. Brun *et al.*, GEANT, CERN DD/EE/84-1 (Geant 3.21 Revised 1987).
5. V. Egorytchev, V. Saveliev, S.J. Aplin, *Nucl. Instr. Meth.* **A 453**, 364 (2000).
6. M. Cherry, *Phys. Rev.* **D 10**, 3594 (1974).
7. V. Ermilova *et al.*, *Nucl. Instr. Meth.* **A 145**, 555 (1977).

THE CRESST DARK MATTER SEARCH

I. BAVYKINA, A. BENTO, D. HAUFF, P. HUFF, R. LANG, B. MAJOROVITS, E. PANTIC, F. PETRICCA, F. PRÖBST, W. SEIDEL and L. STODOLSKY

Max Planck Institut für Physik, München, Germany
email: bela@mppmu.mpg.de

S. HENRY, J. IMBER, S. INGLEBY, H. KRAUS, M. MALEK, R. MCGOWAN, V. MIKHAILIK and B. TOLHURST

Dept. of Physics, University of Oxford, Keble Road, OX1 3RH, England

C. CIEMNIAK, C. COPPI, F. VON FEILITZSCH, C. ISAILA, J. LANFRANCHI, S. PFISTER, W. POTZEL and W. WESTPHAL

Technische Universität München, Germany

M. BAUER, J. JOCHUM, M. KIMMERLE, K. ROTTLER and S. SCHOLL

Universität Tübingen, Germany

C. BUCCI

Laboratori Nazionali del Gran Sasso, Italy

CRESST II is a WIMP search experiment using cryogenic detectors to detect heat and scintillation signals in $CaWO_4$ crystals simultaneously. Competitive limits on the coherent WIMP-nucleon scattering cross section and other results from runs from the upgrade to CRESST II are shown. New features and upgrades installed on the CRESST II setup are described and the status of the experiment is presented.

1. Introduction

Despite persuasive indirect evidence for the existence of dark matter in the universe and in galaxies, the direct detection of dark matter remains one of the outstanding experimental challenges of present-day physics and cosmology. Weakly Interacting Massive Particles (WIMPs) are a well motivated candidate for cold dark matter in the form of the lightest supersymmetric particle. Laboratory experiments can be used to detect WIMP-

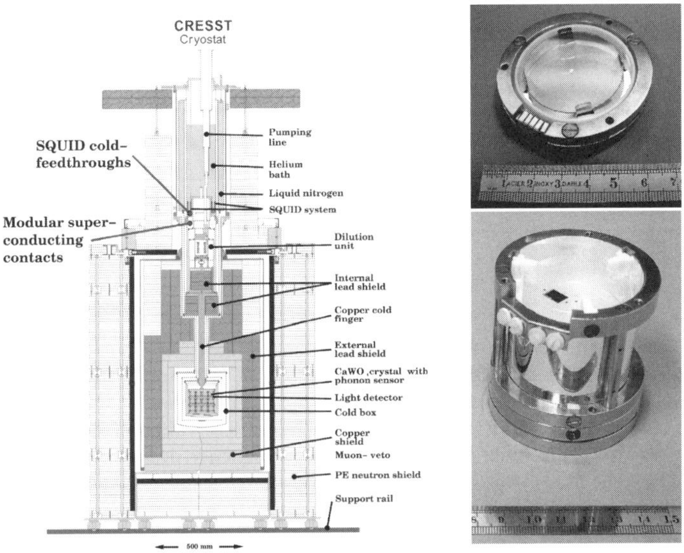

Fig. 1. Left: Schematic view of the CRESST II setup. Right: CRESST II detector module consisting of a $CaWO_4$ crystal (bottom) and a silicon on sapphire wafer as light detector (top).

nucleon interaction, and among these, those using low-temperature cryogenic calorimeters are particularly well suited due to their ability of detecting the very small energy deposit expected [1].

2. The CRESST Setup

CRESST uses cryogenic methods to detect WIMP scattering by target nuclei via the nuclear recoil they impart on these. The target crystals are cooled by a dilution refrigerator to approximately 10 mK. The energy of the recoiling nuclei is converted into an electrical signal by superconducting phase transition thermometers. The detectors are thermally coupled to the mixing chamber by a cold finger that separates the low background space containing the detectors from the dilution refrigerator unit that contains materials that are problematic in regard to radiopurity. Fig. 1 shows the CRESST setup, installed at LNGS, with the detector volume contained by shields of copper and lead.

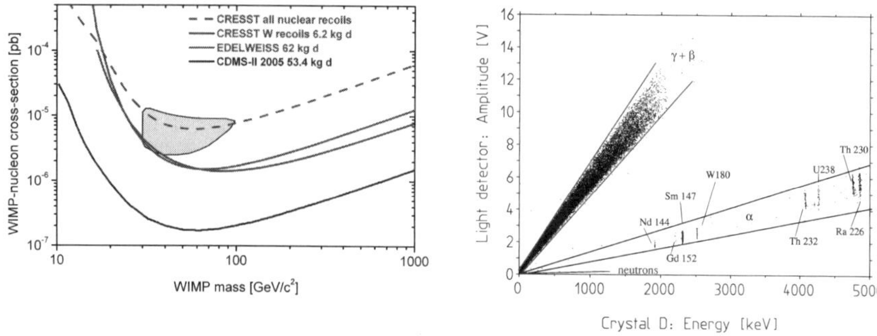

Fig. 2. Left: Limits for the coherent WIMP-nucleon scattering cross-section derived from a run to test the performance of the CRESST II detectors. Also shown are the limits derived from data of other cryogenic experiments (Edelweiss [7], CDMS [8]) and the DAMA evidence region [9]. Right: First unambiguous observation of α decay of ^{180}W. The light detector amplitude is plotted against the energy deposition in the phonon channel. The α–band shows several peaks from rare earth elements. At 2516 keV a peak appears that is corresponding to the α–decay of ^{180}W.

3. Earlier Results

Results from the first phase of CRESST using sapphire detectors have been reported earlier [2]. CRESST II is measuring simultaneously the phonon signal (heat) from energy deposition in $CaWO_4$ absorbers and its scintillation. This allows identification of nuclear recoils due to the scintillation yield being a function of interaction type [3] and the target nucleus [4]. The main background can be identified as for α–,β–and γ–particles energy deposition predominantly results from electromagnetic interactions with the electron system. Above 12 keV neutrons are scattered predominantely by oxygen nuclei; in contrast to WIMPs, for which coherent scattering is expected to be dominated off tungsten nuclei. Fig. 1 (right) shows one detector module of the CRESST II setup. It consists of the $CaWO_4$ crystal (top) and a sapphire wafer coated with silicon, acting as light detector (bottom). The light detectos is attached to the $CaWO_4$ holder to face the $CaWO_4$ crystal. Tungsten SPTs on both detectors convert the heat and light signals into electrical signals.

4. First Results

In a run with a four channel readout system two detector modules were operated for three months. The results have been published earlier [5]. Competitive limits could be set for the coherent WIMP-nucleus scattering cross-section as a function of the WIMP mass. They are shown in Fig. 2. This run also lead to the first unambiguous observation of the α–decay of ^{180}W [6]. The right panel of Fig. 2 shows the light detector response as a function of the energy in the phonon channel. Events resulting from α–decay can clearly be distinguished from interactions of γ–and β–particles. Several α–peaks from U and Th contaminations as well as rare earth elements that have been detected independently by ICP-MS measurement can be observed. Additionally a peak appears at 2516 keV. This corresponds to the Q-value of the α–decay of ^{180}W. A half-life of 1.8×10^{18} y was deduced.

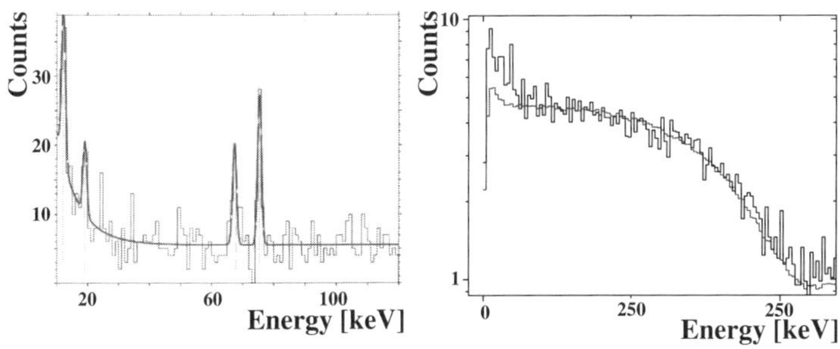

Fig. 3. Left: Low energy background spectrum of run 21. Peaks can be identified that are due to cosmogenic activation. Right: Background spectrum of run 28. The shape of the spectrum can be described by the β–decay of ^{90}Sr. Monte Carlo calculations of this decay are indicated by the shaded histogram.

5. Identified Background Contributions

The earlier runs that were dedicated to test the performance of the new CRESST II detector modules have been analyzed for their background composition. Fig. 3 shows low energy spectra in the phonon channel of run 21 (left) and run 28 (right). In run 21 peaks are seen at (67.4±0.4) keV and (75.5±0.2) keV. These can be attributed to the decays of ^{179}Ta and ^{181}W, respectively. These cosmogenic nuclei are produced through the channels ^{183}W(p,t)^{181}W and ^{182}W(p,α)^{179}Ta. Comparing the low energy spectrum shown with that from run 28, (see Fig. 3 in [5]) no peak is observed at 46.5 keV. The additional peak in the run 28 data is attributed to a ^{210}Pb contamination of the CuBe clamps of the light-detector. CuBe material in the direct vicinity of the detector is thus to be avoided.

The energy spectrum in the phonon channel up to 500 keV exhibits a continuous shape that is fit well to the energy spectrum of the β–decay energies of anthropogenically produced ^{90}Sr. In the histogram the spectrum from a Monte Carlo simulation is also shown. Sr is homologue to Ca and thus difficult to remove from the Ca used for crystal growth. As a direct consequence for future crystals it should be avoided to use Ca from ores above ground.

Fig. 4. The CRESST II detector holding structure can house 33 detector modules. Each module can be removed individually. The holding structure is resting on six CuSn6 springs (obscured by the detectors in this picture).

6. CRESST II Modifications

To allow data taking with 33 detector modules a new SQUID readout system was installed [10]. A detector holding structure that allows for a modular handling of the individual crystals was designed and built. It is made out

Table 1. Results from GDMS measurement of Cu, Sn and CuSn6 used in the CRESST II setup.

	NOSV copper	Sn	CuSn6
^{238}U Conc. [ppm]	$\leq$0.0001	$\leq$0.001	$\leq$0.0001
$A_{^{238}U}$ $[mBq/kg]$	$\leq$0.4	$\leq$4	$\leq$0.4
^{232}Th Conc. [ppm]	$\leq$0.0001	$\leq$0.001	$\leq$0.0001
$A_{^{232}Th}$ $[mBq/kg]$	$\leq$1.2	$\leq$12	$\leq$1.
^{40}K Conc. [ppm]	$\leq$ 0.01	$\leq$ 0.01	$\leq$ 0.01
$A_{^{40}K}$ $[mBq/kg]$	$\leq$ 0.31	$\leq$ 0.31	$\leq$ 0.31

of low activity electroformed copper. The holding structure is suspended to the cryostat cold finger on six CuSn6 springs that have been produced out of high purity tin and low activity copper (see Fig. 4). The CuSn6 was also used to replace the CuBe springs holding the silicon on sapphire wafer of the light detector. The used materials were analyzed by GDMS measurements. The results are shown in Tab. 1.

To shield against the neutron flux a 50 cm PE-shield was built around the setup. Additionally a muon veto system was installed. These two new features are shown in Fig. 5.

7. Outlook

At present CRESST II is taking data with the new detector holding structure, neutron shield and muon veto installed. First results are expected soon.

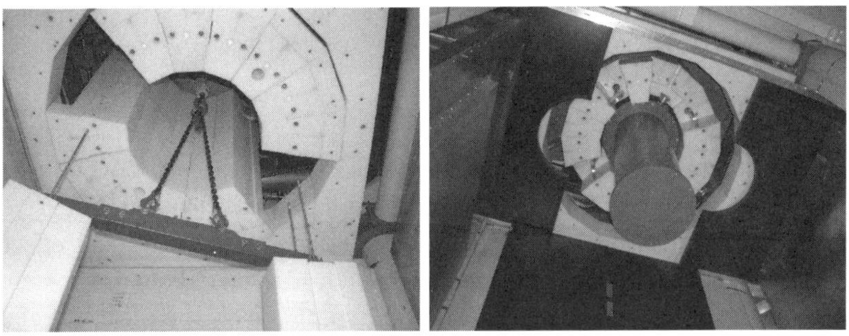

Fig. 5. Left: Mounting of the PE-neutron shield. Right: The newly installed muon veto surrounding the cryostat (black panels).

References

1. M.W. Goodman and E. Witten, *Phys. Rev. D* **31**, 3059 (1985).
2. G. Angloher et al., *Astropart. Phys.* **18**, 43 (2002).
3. P. Meunier et al. *Appl. Phys. Lett.* **75**, 1335 (1999).
4. J. Ninkovic et al., *Nucl. Instr. Meth. A* **564**, 567 (2006).
5. G. Angloher et al., *Astropart. Phys.* **23**, 325 (2005), astro-ph/0408006.
6. C. Cozzini et al., *Phys. Rev. C* **70**, 064606(2004).
7. A. Benoit et. al., *Phys. Lett. B* **545**, 43 (2002).
8. D.S. Akerib et al., *Phys. Rev. D* **73**, 011102 (2006).
9. R. Bernabei et al., *Phys. Lett. B* **480**, 23 (2000).
10. S. Henry et al., *Nucl. Instr. Meth. A* **520**, 588 (2004).

THE ArDM EXPERIMENT

A. BUENO ON BEHALF OF THE ArDM COLLABORATION
Dpto Física Teórica y del Cosmos & C.A.F.P.E., University of Granada,
Granada 18071, Spain
E-mail: a.bueno@ugr.es

We describe the Argon Dark Matter (ArDM) experiment. It consists on a one-ton liquid argon detector able to read independently ionization charge and scintillation light. This device has been optimized to perform a direct search for Weakly Interacting Massive Particles (WIMPs).

Keywords: Dark Matter; Liquid Argon; Direct Search.

1. Introduction

Nowadays there is strong evidence on the fact that one quarter of the mass-density of the Universe is made of non-baryonic dark matter.[1] This new form of matter has been observed only through its gravitational effects, therefore its ultimate nature remains unknown to us. From the particle physics point of view, the most favoured hypothesis on the nature of dark matter is that it is composed of WIMPs. A WIMP particle can be detected through elastic scattering from a target nucleus. A complete review for theoretical motivations and experimental WIMP searches can be found elsewhere.[2] So far only one experiment has claimed a positive WIMP signal.[3] This claim has not been confirmed yet by other experiments, being the best limit the one from the CDMS collaboration.[4]

More sensitive searches demand more massive detectors with improved background rejection capabilities. Among the new proposed methods, much attention has been devoted to detectors based on noble liquids as sensitive medium. The XENON,[5] ZEPLIN[6] and WARP[7] collaborations use liquid argon and xenon targets to look for WIMPs. The ArDM experiment aims at building a one-ton detector to measure the WIMP elastic scattering off argon nuclei by the observation of free electrons from ionization and photons from scintillation, which are produced by the recoiling nucleus interacting with neighbouring atoms. The ionization and scintillation signals can be measured with dedicated readout techniques, which constitute a fundamental part of the detector. In order to get a high enough target mass the noble gas argon is used in liquid phase as target material. Since the boiling point of argon is at 87 K at normal pressure, the operation of the detector requires a cryogenic system. For a recoil energy threshold of 30 KeVr, this target mass corresponds to an event rate of approximately 100 events per day at a cross section of 10^{-6} pb or 0.01 events per day at 10^{-10} pb. The most important source of backgrounds to this search are: a) neutrons, produced by detector components or in the rock of the underground cavern where we plan to operate our detector; b) the ^{39}Ar isotope which is a beta emitter and can be found in natural argon as a consequence of the interactions between cosmic rays with argon nuclei present in the atmosphere.

In what follows, we describe in detail the construction status of the ArDM experiment, its physics potential and future prospects.

2. Detector

We believe that the best R&D towards future dark matter detectors is to build realistically-sized prototypes and operate them under realistic conditions. This is what moved us[a] to propose a one-ton liquid argon detector as a tool to perform a direct search for dark matter.[8] Table 1 summarizes the main parameters of our detector. It is a double phase (gas-liquid) detector (see figure 1). Ionization electrons are drifted, by means of an intense electric field, towards the liquid surface where they are extracted (the maximum drift length is about 120 cm). The HV system must be able to produce a huge electric field (a total of 500 kV for a field of 4 kV/cm). We have adopted as a solution to use a Greinacher/Cockroft-Walton circuit mounted on the field shaping rings. The circuit will be composed of 200 stages of 82 nF polypropylene capacitors that will provide 2.5 kV per stage. So far a circuit of 80 stages has been successfully tested in liquid nitrogen and observed to work in stable conditions at 120 kV.

The extracted charge is amplified and readout with a Large Electron Multiplier (LEM) device.[9] Our tests in the laboratory have shown that it is feasible to obtain amplification factors of 10^4 at liquid argon temperatures by using a two stage LEM. This opens up the possibility to detect the ionization signals even in the presence of highly quenched nuclear recoils, like

---

[a]ETH Zürich, Granada University, Zürich University, Soltan Institute Warszawa, CIEMAT

in the case of a WIMP interaction. LEMs are segmented and will provide X and Y coordinates of the event, the drift time will provide the third spatial coordinate. Thanks to this segmentation, an image of the event will be obtained, retaining the salient features of the ICARUS imaging technology,[10] although with a much lower energy threshold.

The scintillation light produced by de-excitation of argon dimers is readout by PMTs located at the bottom of the detector, behind a transparent cathode.[11] This light has a wavelength of 128 nm and therefore must be shifted to the visible in order to be detected by our PMTs. To this purpose the inner detector is covered by a reflective material (Al-MgF_2 coated mylar foil with $> 95\%$ reflectivity) plus a wavelength shifter (Tetra-Phenil-Butadiene, TPB) deposited on it. TPB will be deposited on PMT windows as well in order to collect the non-negligible fraction of direct light.

Table 1. Main parameters of the ArDM detector.

Target mass	850 kg
Drift length	120 cm
Drift field	1-5 kV/cm
HV system	Greinacher/Cockroft-Walton circuit
Charge readout	LEM
Light readout	14 8" PMTs

3. Physics Performance

The combination of charge and light readout should provide a powerful background rejection. Since nuclear recoils due to WIMP or neutron interactions have a much higher characteristic light over charge ratio than electron and gamma interactions, these types of events are in principle well distinguishable if the light over charge ratio can be measured precisely (see figure 2). This is a mandatory condition for the rejection of predominant background signals like the internal ^{39}Ar signal. The ^{39}Ar isotope is present in natural argon liquefied from the atmosphere.[12] It produces a background rate due to beta decay of approximately 1 kHz in one ton of liquid argon. The alternative possibility of using ^{39}Ar-depleted argon extracted from underground natural gas is currently also studied. It should be noted that ^{39}Ar decays are uniformly distributed in the detector and thus provide a precise calibration and monitoring of the detector response as a function of time and position.

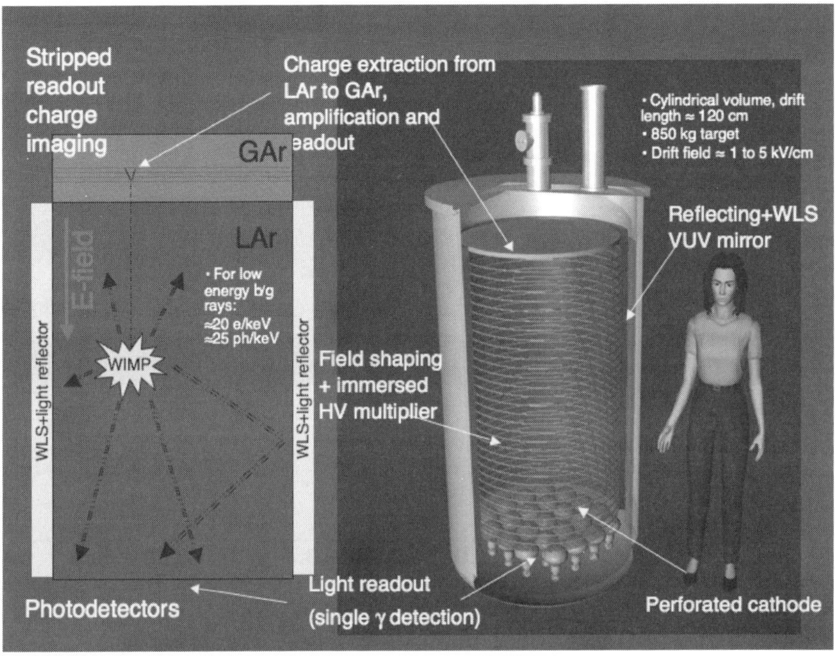

Fig. 1. The ArDM detection principle.

The comparison of fast and slow light components can also be used to further discriminate between nuclear recoils and background events.[13] Due to different ionization densities, the population of argon excimer states is different for nuclear recoil and electron/gamma events, leading to characteristic contributions to the fast and slow scintillation light components.

Neutron background coming from surrounding rock and muon-induced interactions can be efficiently shielded with an external veto and therefore is less important than the one due to neutrons produced by the detector components. The use of low background material for the LEM and PMTs is key to reduce the expected neutron contamination by an order of magnitude. An advantage of a large detector, like the one proposed here, is that neutrons have a large probability to scatter twice inside the fiducial volume. In our simulations we observed that 55% of the neutrons suffer more than one elastic scattering inside the imaging region and therefore do not constitute a background.[14] We estimate that less than 100 neutron events per year will look like a WIMP signal in the energy range of 30 to 100 KeVr.

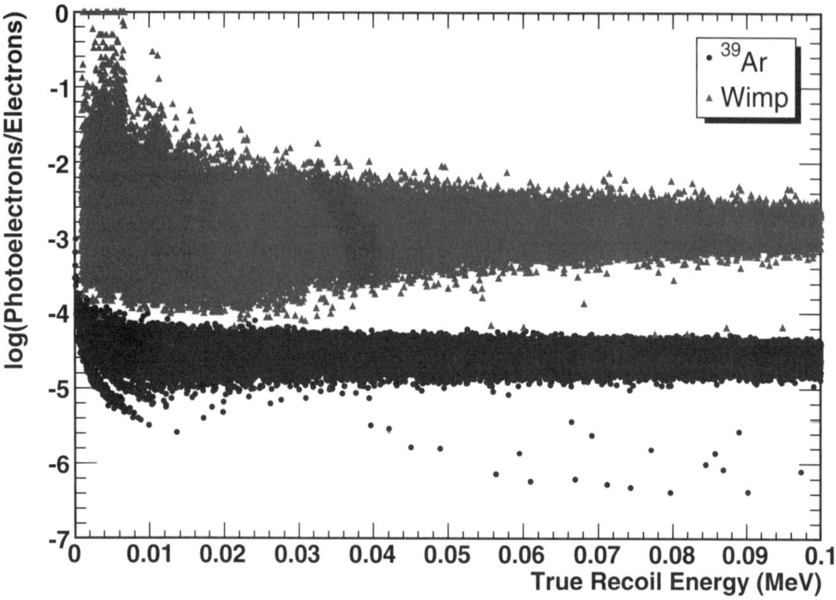

Fig. 2. Light over charge ratio for ^{39}Ar decays and WIMP interactions.

This should be compared with the expected signal of about 3500 WIMP events for a cross section of 10^{-43} cm^2.

4. Future Prospects

The ArDM detector is currently assembled at CERN. As described above, separate tests have been carried out for light readout, charge readout and the high voltage supply devices. The mounting of the detector above ground is planned to take place by the end of 2006 and beginning of 2007. In the phase of running the detector above ground, our first milestone is a proof of principle. The readout devices and the high voltage supply will be tested and tuned in order to assess the detector performance characteristics as its energy threshold, its resolution and its background rejection power. In a second phase we will address the questions of neutron shielding and the needed requirements on detector radio-purity. By the end of 2007 or beginning of 2008, the detector is planned to be moved underground. An Expression of Interest has been submitted to the Canfranc Underground Laboratory in Spain. During the run underground, it will hopefully be possible either to detect the WIMP or to confirm/place limits on the WIMP-nucleon cross section. Beyond the one-ton version, the detector size can be increased without fundamentally changing its technology. A ten ton liquid argon detector is a thinkable and realistic expansion possibility for ArDM, due to the scalable technologies used and the low cost of argon. Current experiments for Dark Matter detection at a mass scale of 1 kg to 100 kg with negative results demonstrate the necessity of larger mass experiments, if one wants to know what dark matter really is. We believe this technology will eventually allow us to reach sensitivities on the spin-independent cross section below 10^{-44} cm^2.

References

1. D. N. Spergel *et al.*, astro-ph/0603449.
2. R. J. Gaitskell, *Ann. Rev. Nucl. Part. Sci.* **54**, 315 (2004).
3. R. Bernabei *et al.*, *Riv. Nuovo. Cim.* **26**, 1 (2003).
4. D. S. Akerib *et al.*, *Phys. Rev. Lett.* **96**, 011302 (2006).
5. E. Aprile *et al.*, *New Astronomy Reviews* **49**, 289 (2005).
6. H. M. Araujo *et al.*, *Astropart.Phys.* **26**, 140 (2006). D. B. Cline, astro-ph/0111098.
7. R. Brunetti *et al.*, *New Astronomy Reviews* **49**, 265 (2005).
8. A. Rubbia, *J. Phys. Conf. Ser.* **39**, 129 (2006).
9. R. Chechick *et al.*, *Nucl. Instrum. Meth. A* **535**, 303 (2004) and references therein.
10. S. Amerio *et al.*, *Nucl. Instrum. Meth. A* **527**, 329 (2004).
11. C. Regenfus, these proceedings.
12. H. Loosli, *Earth and Planetary Science Letters* **63**, 51 (1983) 51: *Nachweis von* ^{39}*Ar in atmosphärischem Argon*, PhD thesis, University Bern, 1968.
13. M. G. Boulay and A. Hime, *Astropart. Phys.* **25**, 179 (2006).
14. L. Kaufman, these proceedings.

DAMA/LXe AT LNGS: RESULTS AND PERSPECTIVES

R. BERNABEI, P. BELLI, F. MONTECCHIA and F. NOZZOLI

Dipartimento di Fisica, Università di Roma "Tor Vergata" and INFN, Sezione di Roma "Tor Vergata", I-00133, Roma, Italy

F. CAPPELLA, A. INCICCHITTI and D. PROSPERI

Dipartimento di Fisica, Università di Roma "La Sapienza" and INFN, Sezione di Roma, I-00185, Roma, Italy

R. CERULLI

INFN - Laboratori Nazionali del Gran Sasso, I-67010 Assergi (Aq), Italy
**E-mail: riccardo.cerulli@lngs.infn.it*

C.J. DAI

IHEP, Chinese Academy, P.O. Box 918/3, Beijing 100039, China

V.Yu. DENISOV and V.I. TRETYAK

Institute for Nuclear Research, MSP 03680 Kiev, Ukraine

The liquid Xenon set-up of the DAMA project (DAMA/LXe) is a low background experiment located deep underground at LNGS. It is in operation since time and has been uograded several times. It can be filled with Kr-free Xenon enriched either in ^{129}Xe at 99.5% or in ^{136}Xe at 68.8%. This set-up has already given several competitive results investigating various kinds of rare processes. In this paper we shortly remind the main recent results on Dark Matter, double beta decay, charge non-conserving processes, nucleon, di-nucleon and tri-nucleon decay.

Keywords: Liquid Xenon scintillator; dark matter; double beta decay; underground physics; rare events searches

1. The DAMA/LXe experiment

Since the end of 80's we realized several liquid xenon prototype detectors. Then, preliminarily we put in measurement the set-up used in the data taking of ref. [1,2]; it was firstly upgraded at fall 1995 (see ref. [3–8]). DAMA/LXe has an inner vessel filled by $\simeq 6.5$ kg (i.e. $\simeq 2$ l) Kr-free liquid Xenon enriched in ^{129}Xe at 99.5% [8] or – since 2000 (when a deep modification was performed) – in ^{136}Xe at 68.8% [9,10]. The inner vessel is enclosed in low-background passive shield made of 5-10 cm of low radioactive copper inside a vacuum insulation vessel (to shield, at the most, mainly possible background contribution from PMTs), 2 cm of steel (wall of the insulation vacuum vessel), 5-10 cm of low radioactive copper, 5 cm of Polish lead and 10 cm of Boliden lead, $\simeq$1 mm cadmium and $\simeq$10 cm of polyethylene and paraffin. Finally, an external plexiglass box, sealed and continuously fluxed with HP nitrogen closes the external shield.

A vacuum/filling/purification/recovery system with two different lines allows the separate store, handling and use of the two enriched Xenon gases [8].

The main features of the set-up, details on the electronic chain, on the data acquisition, on the cryogenic and vacuum systems and on the running parameters control can be found in [6–8,10,11]. In particular, the energy scale in the high energy region is determined with the help of external standard gamma sources, having identified the possibility to perform high energy calibrations with external sources near a flange located on the top of the detector opening a limited upper part of the external shield. The energy resolution for ^{208}Tl line ($E_\gamma = 2614$ keV) is $\sigma \simeq 220$ keV, while for ^{137}Cs line ($E_\gamma = 662$ keV), which is the source usually exploited for the routine calibrations, σ yields $\simeq 70$ keV [7,10,11]. The low energy region is calibrated by using a ^{109}Cd source (22 keV and 88 keV photons) directly inserted in the liquid Xenon bulk by means of a suitable handler.

2. Main recent results

We pointed out the interest in using liquid Xenon as target-detector for particle Dark Matter (DM) investigation deep underground since ref. [12]. After preliminary measurements both on elastic and inelastic DM particles-^{129}Xe scattering [1,3], new results have been obtained with the renewed set-up [4,6] by exploiting – in the case of elastic interaction – the pulse shape discrimination between recoils and electromagnetic background (see Fig. 1). In fact, the recoil/electron light ratio and pulse shape discrimination capability in a similar pure LXe scintillator have carefully been measured by using: Am-B neutron source and 14 MeV ENEA-Frascati neutron generator [6], and later also the 2.5 MeV ENEA-Frascati neutron generator (see ref. [5] for details and comparisons).

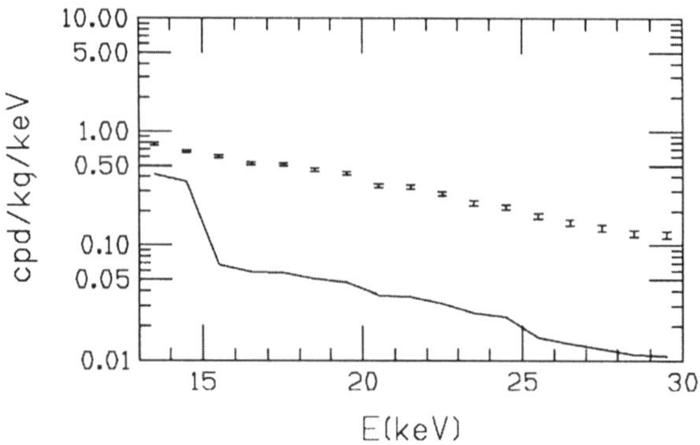

Fig. 1. Low energy distribution collected by DAMA/LXe corresponding to an exposure of 1763.2 kg × day [6]. The continuous line represents the upper limits at 90% C.L. on the recoil fraction. For details see ref. [6].

It has also been investigated double beta decay modes in ^{134}Xe and ^{136}Xe by measuring Kr-free Xenon enriched in ^{136}Xe at 68.8% that also contains 17.1% of ^{134}Xe. In particular, after the preliminary results of ref. [9,13] a joint analysis of the $0\nu\beta\beta$ decay mode in ^{134}Xe and in ^{136}Xe (as suggested in ref. [14]) has been carried out [10]. New lower limits on various $\beta\beta$ decay modes have been obtained: for the $0\nu\beta\beta(0^+ \rightarrow 0^+)$ decay mode in ^{134}Xe and in ^{136}Xe the limits at 90% C.L are $T_{1/2} = 5.8 \cdot 10^{22}$ yr and $T_{1/2} = 1.2 \cdot 10^{24}$ yr, respectively (see Fig. 2); they correspond to a limit value on *effective light Majorana neutrino mass* ranging from 1.1 eV to 2.9 eV (90% C.L.), depending on the adopted theoretical model. For the neutrinoless double beta decay with Majoron (M) in the ^{136}Xe isotope the limit is $T_{1/2} = 5.0 \cdot 10^{23}$ yr (90% C.L.); finally, for the $2\nu\beta\beta(0^+ \rightarrow 0^+)$ and the $2\nu\beta\beta(0^+ \rightarrow 2^+)$ decay modes in ^{136}Xe the limits at 90% C.L. are $1.0 \cdot 10^{22}$ yr and $9.4 \cdot 10^{21}$ yr, respectively. It is worth to note that the experimental limit on the $2\nu\beta\beta(0^+ \rightarrow 0^+)$ decay mode is in the range of the theoretical estimate by [15] ($2.11 \cdot 10^{22}$ yr) and about a factor 5 higher than that of ref. [16][a]. The overcoming of some previously considered theoretical estimates has also been recently confirmed by high-pressure proportional counters at Baksan laboratory [17].

[a] On the other hand, similar theoretical estimates suffer of the large uncertainties typically associated to the calculations of the nuclear matrix elements.

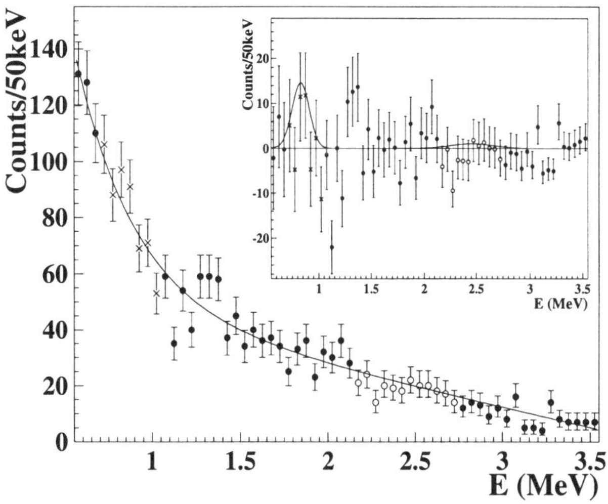

Fig. 2. Experimental energy distribution considered for the investigation of the $0\nu\beta\beta$ decay mode in ^{134}Xe and ^{136}Xe; crosses, filled and open circles are the experimental data while the superimposed continuous line represents the background contribution estimated by excluding the crosses and the open circles data points. In the inset, the residuals as a function of the energy is reported; the Gaussian peaks represent the expected signals for the $0\nu\beta\beta$ decay mode when $T_{1/2}(^{134}Xe) = 5.8 \times 10^{22}$ y and $T_{1/2}(^{136}Xe) = 1.2 \times 10^{24}$ y. See ref. [10].

Further profiting of the low-background characteristics of DAMA/LXe, several other rare processes have been searched for obtaining competitive results.

In particular, different charge non-conserving processes (CNC) have been studied. In this context, limits on the lifetime of the electron decay in both the disappearance and the $\nu_e + \gamma$ channels were set in ref. [2]. The latter has been subsequently improved up to: $2.0(3.4) \cdot 10^{26}$ yr at 90% (68%) C.L. [18] by investigating the presence of a gamma quanta with energy about 255 keV which could follow the possible decay of any electron in the LXe scintillator and in its surroundings.

Furthermore, new lifetime limits on the charge non-conserving electron capture with excitation of ^{129}Xe nuclear levels have also been established. In this process an electron disappears from the atomic shell and the nucleus is left in an excited state. This decay is equivalent to an electron capture without any nucleus charge changing. By searching for gammas emitted in the subsequent de-excitation of ^{129}Xe, limits in the range $(1-4) \cdot 10^{24}$ yr at 90% C.L. for the different excited levels of the isotope have been obtained

[19]. In addition, stringent restrictions on the relative strengths of charge non-conserving processes have been consequently derived: $\epsilon^2_W < 2.2 \cdot 10^{-26}$ and $\epsilon^2_\gamma < 1.3 \cdot 10^{-42}$ at 90% C.L. [19].

It has also been exploited – for the first time – the CNC process for the ^{136}Xe isotope by considering its possible charge non-conserving decay into ^{136}Cs [7], by using data collected during 8823.54 h. The used approach has been the investigation of the CNC processes by the search for the possible CNC decay firstly considered in [20]: if in a β decay $(A, Z) \rightarrow (A, Z+1) + e^- + \overline{\nu}_e$ some massless uncharged particle would be emitted instead of the electron (e.g., ν_e or γ or Majoron), an additional 511 keV energy release would occur. Thus, usually forbidden decays to the ground state or to the excited levels of the daughter nuclei would become energetically possible. The presence of the $(A, Z+1)$ isotope or of its daughter products in a sample, initially free from them, would indicate the existence of the CNC decay searched for. Large advantages arise when the so-called "active-source" technique (source = detector) is considered as in the case described here. In particular, after the possible ^{136}Xe CNC decay, the daughter nucleus ^{136}Cs will be created. It is β unstable ($T_{1/2} = 13.16$ d) with quite high energy release ($Q_\beta = 2.548$ MeV) [7]. Comparing the experimental energy distribution with the expected response function, no evidence for the effect searched for has been found. Thus, the life-time limit is: $\tau_{CNC}(^{136}\text{Xe}\rightarrow^{136}\text{Cs}) > 1.3 \cdot 10^{23}$ yr at 90% C.L. This limit is one of the highest available limit for similar processes [7]; however, the bound on the charge non-conserving admixture in the weak interactions, which can be derived according to ref. [21], is modest mainly due to the high degree of forbiddeness of the considered CNC transition.

Nucleon and di-nucleon decay into invisible channels in the ^{129}Xe isotope have also been investigated by exploiting a new approach [22]. It consisted in a search – in a real-time experiment such as DAMA/LXe – for radioactive decay of unstable daughter nuclei created as result of the possible N or NN disappearance in parent nucleus. If half-life of the daughter nucleus is of order of 1 s or greater, such a decay will be time-resolved from prompt products if they were emitted and observed in a detector. This approach has the advantage of a branching ratio close to 1 and – if the parent and daughter nuclei are located in the detector itself – also of an efficiency close to 1. The obtained limits at 90% C.L. are: $\tau(p \rightarrow$ invisible channel$) > 1.9 \cdot 10^{24}$ yr; $\tau(pp \rightarrow$ invisible channel$) > 5.5 \cdot 10^{23}$ yr and $\tau(nn \rightarrow$ invisible channel$) > 1.2 \cdot 10^{25}$ yr. These limits are similar or better than those previously available; moreover, they are valid for every possible disappearance channel [22] and the limits for the di-nucleon decay in $\nu_\tau \bar{\nu}_\tau$ are set there for the first time.

The same approach has been also followed by investigating N, NN and NNN instabilities in the ^{136}Xe isotope [23]. The presence of daughter nuclei created in the process has been searched for in the data collected during 8823.54 h. In particular, comparing the expected response functions of the LXe detector for the considered decays with the measured energy distribution, limits on the probability of the processes investigated have been extracted (see Fig. 3). The obtained results are summarized in Table 1 [23].

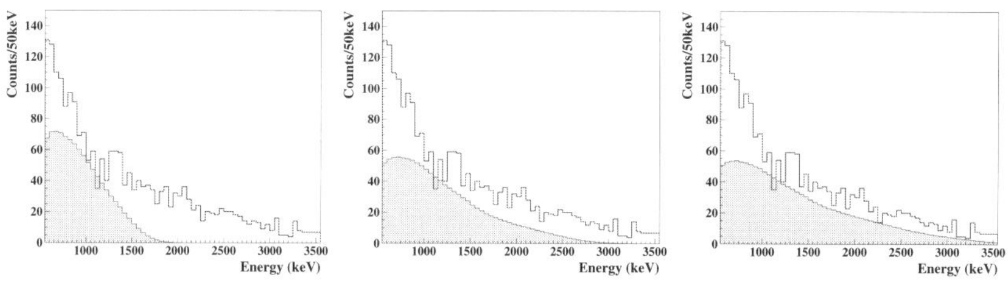

Fig. 3. Comparison between the experimental spectrum measured during 8823.54 h (thick histogram) and the expected signal (shaded histogram) for: *left*) the nnp disappearance with $\tau_{nnp} = 1.4 \cdot 10^{22}$ yr excluded at 90% C.L.; *center*) the npp disappearance with $\tau_{npp} = 2.7 \cdot 10^{22}$ yr excluded at 90% C.L.; *right*) the ppp disappearance with $\tau_{ppp} = 3.6 \cdot 10^{22}$ yr excluded at 90% C.L. [23].

Table 1. Experimental limits (90% C.L.) obtained by DAMA/LXe on the lifetime values of the possible N, NN and NNN decays into invisible channels in the ^{136}Xe isotope [23].

Decay	τ_{limit} years (90% C.L.)
n	$3.3 \cdot 10^{23}$
p	$4.5 \cdot 10^{23}$
nn	–
np	$3.2 \cdot 10^{23}$
pp	$1.9 \cdot 10^{24}$
nnn	–
nnp	$1.4 \cdot 10^{22}$
npp	$2.7 \cdot 10^{22}$
ppp	$3.6 \cdot 10^{22}$

Note that all the limits achieved hold for every invisible decay channel, including disappearance in extra-dimensions or decay in particles which weakly interact with the matter.

3. Conclusion

DAMA/LXe experiment is operative since long time in various configurations. Its low-background characteristics have already allowed to obtain competitive results studying several topics. Further data taking is foreseen and new investigations are in preparation.

References

1. P. Belli et al., Il Nuovo Cim. C 19 (1996) 537
2. P. Belli et al., Astrop. Phys. 5 (1996) 217
3. P. Belli et al., Phys. Lett. B 387 (1996) 222 and Phys. Lett. B 389 (1996) 783 (erratum)
4. R. Bernabei et al., New J. Phys. 2 (2000) 15.1
5. R. Bernabei et al., Eur. Phys. J. direct C 11 (2001) 1
6. R. Bernabei et al., Phys. Lett. B 436 (1998) 379
7. R. Bernabei et al., in the volume "Beyond the Desert 2003", Springer (2003) 365
8. R. Bernabei et al., Nucl. Instr. & Meth. A482 (2002) 728
9. R. Bernabei et al., Phys. Lett. B 527 (2002) 182
10. R. Bernabei et al., Phys. Lett. B 546 (2002) 23
11. F. Cappella, PhD Thesis, Università di Roma "Tor Vergata", 2005.
12. P. Belli et al., Il Nuovo Cim. 103A (1990) 767
13. R. Bernabei et al., in the volume "Technique and application of Xenon detectors", World. Sci. pub. (2002) 50
14. F. Simkovic, P. Domin, A. Faessler, hep-ph/0204278
15. A. Staudt, K. Muto and H. V. Klapdor, Europhys. Lett. 13 (1990) 31
16. E. Caurier et al., Nucl. Phys. A654 (1999) 973
17. Ju. M. Gavriljuk et al., nucl-ex/0510071
18. P. Belli et al., Phys. Rev. D 61 (2000) 117301
19. P. Belli et al., Phys. Lett. B 465 (1999) 315
20. G. Feinberg, M. Goldhaber, Proc. Nat. Acad. Sci. U.S.A. 45 (1959) 1301
21. J.N. Bahcall, Rev. Mod. Phys. 50 (1978) 881; J.N. Bahcall, *Neutrino Astrophysics* (Cambridge University Press, 1989) 359
22. R. Bernabei et al., Phys. Lett. B 493 (2000) 12
23. R. Bernabei et al., Eur. Phys. J. A 27 s01 (2006) 35

A POSSIBLE LMC HALO CONTRIBUTION TO MICROLENSING

PHILIPPE JETZER

Institute of Theoretical Physics University of Zürich, Winterthurerstrasse 190, CH-8057 Zürich, Switzerland
E-mail: jetzer@physik.unizh.ch

We discuss the possibility that some of the microlensing events found by the MACHO collaboration towards the LMC might be due to lenses located in the LMC halo, rather then in the halo of the Milky Way. Indeed, the dark matter fraction in form of MACHOs of both Milky Way and LMC halos need not to be the same. For a MACHO mass in the range 0.1-0.3 $M_{\odot}$, the LMC halo fraction can be larger than the Milky Way one.

1. Introduction

The microlensing surveys towards the LMC and the SMC probed the existence of compact halo objects along these lines of sight; however, the assessment of these results with respect to the actual fraction of the Milky Way (hereafter "MW") halo in the form of MACHOs is still highly debated. The MACHO collaboration reported the detection of 13-17 microlensing candidates towards the LMC [1], arguing in favour of a MACHO halo fraction of $\sim$ 20% of objects of $\sim 0.5\,\mathrm{M}_{\odot}$ and estimating a microlensing optical depth towards the LMC of $\tau = 1.2^{+0.4}_{-0.3} \times 10^{-7}$. The reported microlensing rate towards the LMC significantly exceeds the expected one from known visible components of our Galaxy. Further analysis mainly confirmed these conclusions [2]. On the other hand, the EROS collaboration, from observations towards both the LMC and the SMC, puts an upper limit on the MACHO contribution to galactic halos, which is hardly compatible with the MACHO results [3].

These questions have also been addressed by microlensing surveys towards M31, and a first evidence of a MACHO contribution along this line of sight have been reported [4], although challenged in [5]. Overall, the picture remains unclear, in particular the problem of the nature and location of the observed events.

We have already addressed some of these questions in [6] and [7]. In particular we came to the conclusion that the observed events do not necessarily all belong to the same lens population, and that self lensing alone cannot explain all the events. Furthermore we extended the analysis by fully considering the LMC and MW dark matter halo MACHO lens populations [8]. The main question we addressed is whether and to what extent events due to the lens population residing in the LMC halo can contribute to the observed rates.

2. The microlensing rate

We consider the LMC as composed of a luminous part, the bar, and the disc, plus a stellar and a dark matter halos. We also include the Galactic dark matter halo. Following [9], who present observational evidence based on the rotation curves, we include a significant LMC dark matter halo component. We assume a total LMC mass of 8.7×10^9 $M_\odot$ within 8.9 kpc with a truncation radius of 15 kpc [9], which we consider to also include the contribution from the stellar halo. We assume a spherical isothermal model with a core radius of 2 kpc. For the Galactic halo dark matter we consider the "standard" isothermal profile with a core radius of 5 kpc, local density $7.9 \times 10^6\, M_\odot\, \mathrm{kpc}^{-3}$, and a distance from the Galactic centre of 8.5 kpc.

Table 1. Expected duration, median values with 68% CL errors, and expected number of events for a full dark matter halo, respectively averaged and summed over the MACHO fields, as a function of the MACHO mass.

lens mass	MW		LMC	
$M_\odot$	T_E (days)	N_{exp}	T_E (days)	N_{exp}
10^{-2}	$8.0^{+6.5}_{-3.1}$	115	$13^{+8.7}_{-4.4}$	17.3
0.1	$20^{+15}_{-8.0}$	97	36^{+21}_{-13}	9.9
0.2	26^{+20}_{-10}	82	48^{+27}_{-17}	7.6
0.5	41^{+29}_{-16}	59	75^{+46}_{-27}	5.0
1	55^{+38}_{-21}	44	103^{+57}_{-37}	3.4

The main tool of investigation we use is $\frac{d\Gamma}{d T_E}$, the differential rate of microlensing events with respect to the Einstein time T_E (see Fig. 1). This allows us to make predictions on the timescale, the number, and the spatial distribution of the expected events, which we can compare with the corresponding observed quantities. The differential microlensing rate is evaluated, for each different lens population we consider, along the lines of sight towards the MACHO fields. In particular for each lens population, we calculate the expected number of events per field as

$$N_{\text{field}} = E_{\text{field}} \int_0^\infty \frac{\mathrm{d}\Gamma}{\mathrm{d}T_{\mathrm{E}}}\, \epsilon(T_{\mathrm{E}})\, \mathrm{d}\, T_{\mathrm{E}} \ , \tag{1}$$

where the "field exposure" E_{field} is defined in [1], as the product of the number of distinct light curves per field and the relevant time span.

In Table 1 we report, for both MW and LMC MACHO lens populations, the total number of the expected events in all the MACHO fields, in the case of a full MACHO halo, together with the expected timescales. Notice that very short timescales are strongly suppressed by the detection efficiency function $\epsilon(T_{\mathrm{E}})$. The predicted durations turn out to be almost independent of the position for both the halo populations we consider.

3. The LMC MACHO contribution to microlensing events

A straightforward conclusion which follows from the above expected number of events due to the dark matter MW and LMC halos is that, with the implicit hypothesis that the halo fractions in both the MW and the LMC halos are the same, most of the lenses should indeed belong to the Galactic halo. The expected timescale distributions for microlensing events due to lenses either in the LMC or in the MW halo are almost independent of the position. This property allows us to carry on an analysis in which we compare the observed timescales to the predicted one for each population.

We perform the Kolmogorov-Smirnov (KS) test, which allows to test the null hypothesis that the events are drawn from a given population. The resulting KS coefficient gives the significance level of the test. We introduce a parameter α, defined as the ratio of the MW events over the total (MW plus LMC events), in order to explore the possibility that an intermediate solution, with the two populations mixed, has to be preferred. To this purpose we look for the value of α that minimises the corresponding KS coefficient.

When we consider the MW and the LMC halos separately, the solutions with the highest level of confidence are found for a MACHO mass of ≈ 0.5 $\mathrm{M}_\odot$ and ≈ 0.15 $\mathrm{M}_\odot$, respectively. For the case of the combined populations, we find that the coefficient ks_α presents two minima near the two values of mass found in the previous case. The absolute minimum, which we note is lower than the values obtained in the case of the separate test analysis, is found at $m = 0.15\,\mathrm{M}_\odot$ with $\alpha \sim 0.1$, and it gives us the parameters with the highest confidence level. Moreover we observe that the α parameter grows

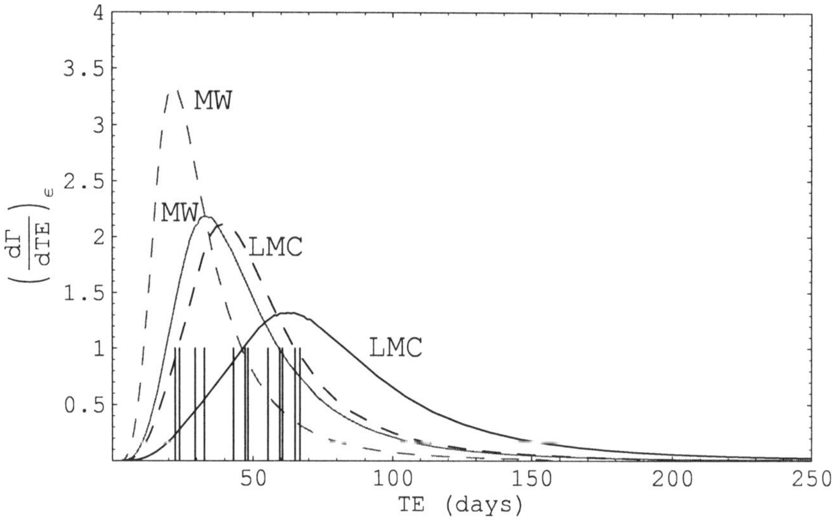

Fig. 1. Normalised differential rate $(d\Gamma/dT_{\rm E})_\varepsilon$ for both MW and LMC halos for 0.2 and 0.5 $M_\odot$, dashed and solid lines, respectively. Superimposed is the value of the observed durations. The y-axis values are in 10^{-2} units.

monotonically as a function of the mass from 0 up to 1 (Fig. 2). Thus the statistical analysis made on the duration of the events gives a first suggestion that a significant fraction of the observed events could belong to the LMC MACHO halo population.

We considered also the spatial distribution of the observed events. Indeed, the optical depth profiles clearly show that the LMC halo events are characterised, with respect to LMC self–lensing and to MW halo ones, by a strong asymmetry with respect to the line of nodes. A detailed analysis [8] of the spatial distribution of the observed events, shows that, independent of the value of the MACHO mass, the observed distribution better matches that expected for a LMC halo population than for a MW halo population.

As a next step we drop the hypothesis of equal halo MACHO fractions in the Galactic and the LMC halo. To that purpose we use a likelihood function with $f_{\rm MW}$ and $f_{\rm LMC}$ being the halo fractions for the MW and the LMC, respectively.

The main outcome of this analysis is that in a significant range of the MACHO mass, the LMC and MW dark matter halo fractions are *not* expected to be equal. In particular, from 0.1 up to 0.3 $M_\odot$, both a high value for $f_{\rm LMC}$ and a low one for $f_{\rm MW}$ are expected.

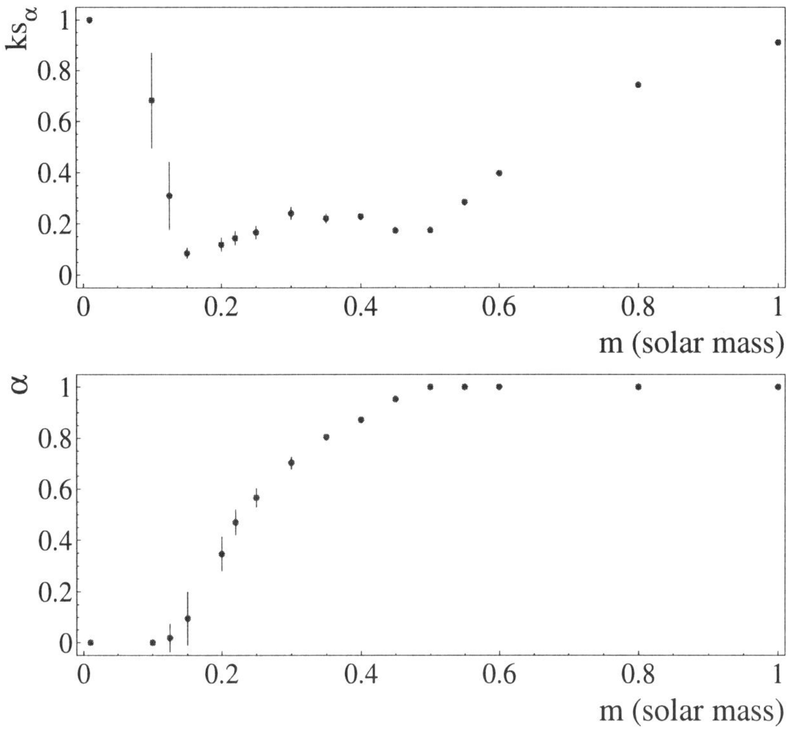

Fig. 2. Kolmogorov-Smirnov and α coefficients as a function of the MACHO mass, with α the ratio of the MW over the total number of dark matter halo events.

This behaviour is mainly due to the observed timescales. It is therefore not surprising that this result is coherent with the one obtained with the KS test, where we found a preferred value of the mass of about $0.2\,M_\odot$, with a significant expected contribution from LMC MACHO halo lenses. The likelihood analysis gives, for $m = 0.2\,M_\odot$, $f_{MW} \sim 4\%$ and $f_{LMC} \sim 80\%$. At face value, given the number of expected events, we get to the overall prediction of about 6-7 events to be attributed to the LMC dark matter halo lens population, 2-3 to the MW halo one, to be looked for among those with the shorter durations, still allowing for a contribution of about 2-3 events, distributed among self–lensing and LMC stellar halo lenses.

For higher values of the mass, the LMC MACHO halo fraction turns out to be almost degenerate though compatible with zero, and, for $0.5\,M_\odot$, we recover the result of a Galactic MACHO halo fraction of about 20% [1]. This is again coherent with the issue of the KS test.

4. Conclusions

Clearly this latter analysis has to be looked at together with the outcomes of the previous analysis on the timescale and the spatial distributions of the observed events. Overall, they indicate that *i*) a sizeable fraction of the observed events show characteristics in accord with those expected for a MACHO LMC halo population; *ii*) such a contribution may be expected by dropping the hypothesis that the halo fractions in the form of MACHOs in the Galactic and the LMC dark matter halo are equal.

A possible explanation of the origin of different halo fractions could come from the different formation histories of the two galaxies, or more simply, could be related to the fact that one observes all the LMC halo but, practically, only a line of sight through the Galactic halo.

Clearly our conclusions are based on the overall implicit assumption of the validity of the MACHO results, whereas they are actually challenged by the EROS collaboration. Furthermore, an intrinsic limit for all the analyses carried out up to now is the result of the lack of available statistics. Our analysis shows, however, that it is in principle possible to characterise and distinguish the two halo lens populations and, moreover, challenge the usual implicit assumption of equal halo fractions in the form of MACHO for both the Galactic and the LMC dark matter halos.

References

1. Alcock, C. et al. 2000, ApJ, 542, 281
2. Bennett, D. P., Becker, A. C., & Tomaney, A. 2005, ApJ, 631, 301
3. Tisserand, P. et al. 2006, astro-ph/0607207
4. Calchi Novati, S. et al. 2005, A&A, 443, 911
5. de Jong, J. T. A. et al. 2006, A&A, 446, 855
6. Jetzer, P., Mancini, L., & Scarpetta, G. 2002, A&A, 393, 129
7. Mancini, L., Calchi Novati, S., Jetzer, P., & Scarpetta, G. 2004, A&A, 427, 61
8. Calchi Novati, S., De Luca, F., Jetzer, P., & Scarpetta,G. 2006, A&A, 459, 407
9. van der Marel, R. P., Alves, D. R., Hardy, E., & Suntzeff, N. B. 2002, AJ, 124, 2639

LIMITS ON THE MACHOS FROM EROS-2

J. RICH FOR THE EROS-2 COLLABORATION

DAPNIA-SPP, CEA-Saclay,
91191 Gif-sur-Yvette, France
**E-mail: rich@hep.saclay.cea.fr*

We present the results of the EROS-2 search for microlensing of stars in the Magellanic Clouds. Such events could be caused by macho dark matter in the Milky Way halo. Unlike all previous studies of Magellanic microlensing, we use only a subsample of 7×10^6 bright stars. This strategy minimizes backgrounds due to variable stars and ensures accurate determination of lensing parameters by minimizing source confusion (blending). Using this sample of bright stars, only one candidate event was found, whereas ~ 42 events would have been expected if the Halo were entirely populated by objects of mass $M \sim 0.4 M_{\odot}$. Combined with the results of EROS-1, this implies that the optical depth toward the Large Magellanic Cloud (LMC) due to such lenses is $\tau < 0.36 \times 10^{-7}$ (95%CL), corresponding to a fraction of the halo mass of less than 7%. This limit on the optical depth is at variance with the measurement of the MACHO collaboration near the center of the LMC, implying that a significant fraction of their events are due to lenses in the LMC itself.

Keywords: Galaxy:halo; Cosmology: dark matter; Gravitational Lensing.

Since Paczyński's proposal [1] that dark matter in the form of faint compact objects (machos) could be found through gravitational microlensing, the EROS, MACHO, OGLE, MOA and SuperMACHO collaborations have monitored millions of stars in the Magellanic Clouds to search for microlensing events. Such events would be due to a lensing object passing near the line of sight toward a background Magellanic star, causing a transient magnification of the star's primary image as well as creating a secondary image. Neither the image separation nor the image size are normally resolvable, so the only easily observable effect during an event is an apparent transient amplification of the star's flux. The optical depth for microlensing, i.e. the probability that at a given time a given star is amplified by more than a factor 1.34, is proportional to the fraction, f, of the halo mass comprised of machos. Numerically, one finds $\tau_{\rm lmc} \sim 5.1 f \times 10^{-7}$

for a spherically symmetric isothermal halo. The characteristic duration of the events is the Einstein radius crossing time, $t_{\rm E}$. It is proportional to the square root of the macho mass: $\langle t_{\rm E}\rangle \sim 70(M/M_\odot)^{1/2}$ days.

Much excitement was generated by the claim by the MACHO collaboration that the rate of microlensing events they observe toward the LMC suggests that a significant amount of the Milky Way's Halo is comprised of machos. Their latest analysis [2] used 17 observed events[a] to measure an optical depth of $\tau_{\rm lmc}/10^{-7} = 1.2^{+0.4}_{-0.3}$ ($stat$) with an additional 20% to 30% systematic error. This would correspond to a Halo fraction $0.08 < f < 0.50$ (95% c.l.). The mean $t_{\rm E}$ of their events was 40 day corresponding to machos in the mass range $0.15M_\odot < M < 0.9M_\odot$. On the other hand the EROS collaboration [3] had placed only an upper limit on the halo fraction, $f < 0.2$ (95% c.l.) for objects in this mass range, ruling out a large part of the range of f favored by the MACHO collaboration.

In this work, we extend our previous analysis to find $\tau_{\rm lmc} < 0.36 \times 10^{-7}$ (95% c.l.) for $M \sim 0.4M_\odot$, corresponding to $f < 0.07$. More details of the analysis can be found in Ref. 4. Unlike the previous EROS limit, this is nominally in clear conflict with the MACHO result. Our new analysis is conservative in that, unlike all previous analyses, we use only a bright, well-measured subsample of the Magellanic stars, about 20% of the total. We believe that the use of this bright subsample gives a much more reliable measurement of the optical depth than measurements using faint stars. There are two reasons for this. First, bright stars have well reconstructed light curves that permit discrimination of intrinsically variable stars. Second, the use of bright stars limits so-called blending effects where reconstructed fluxes can receive contributions from more than one star, complicating the interpretation of events.

EROS-2 used the Marly 1 meter telescope at ESO, La Silla. The telescope was equipped with two 0.95 deg^2 CCD mosaics to monitor 93 deg^2 in the Magellanic Clouds, 63 deg^2 in the Galactic Bulge, and 28 deg^2 in the spiral arms of the Milky Way. The observations were performed between July 1996 and February 2003 (JD between 2,450,300 and 2,452,700).

Many hundreds of events were found in the direction of the Galactic Bulge. These events were expected because of the many normal low-mass stars in the Milky Way disk and bulge. EROS [5] MACHO and OGLE have found optical depths in agreement with the predictions of models of

[a] 13 of these events satisfied their so-called A criteria intended to identify high signal-to-noise events. The other 4 events, so-called B events, have lower signal-to-noise.

the Milky Way bulge and disk. On the other hand, EROS-2 found only one candidate event towards the Magellanic Clouds among the sample of bright stars. Its light curve is shown in fig. 1. It is identical to the MACHO collaboration candidate MACHO-97-SMC-1. No candidates were found by EROS-2 in the LMC.

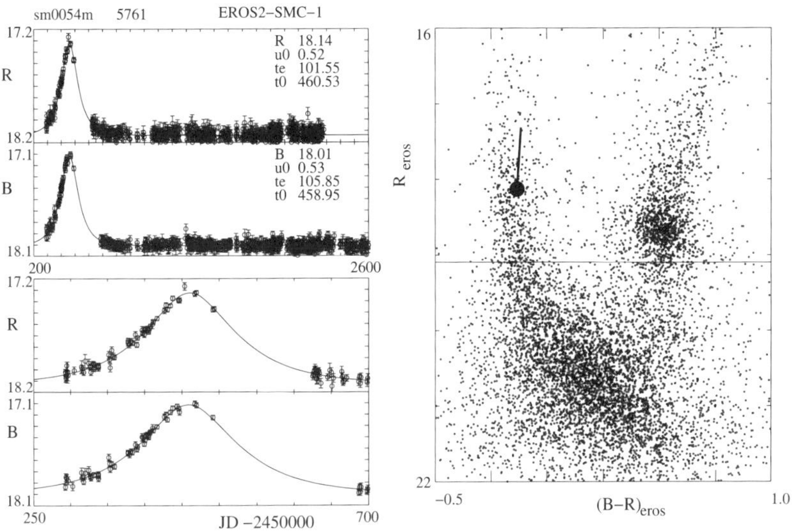

Fig. 1. The light curves of EROS-2 microlensing candidate EROS2-SMC-1 (star sm0054m-5761). Also shown is the color-magnitude diagram of the star's CCD-quadrant and the excursion of the event.

The expected number of LMC events as a function of the optical depth τ is $N_{\rm ex} = (2/\pi) T_{\rm obs} N_{\rm star} \tau \langle\epsilon\rangle / \langle t_{\rm E}\rangle$ where $T_{obs} = 2500$ day is the total EROS-2 observing time and $N_{star} = 5.5 \times 10^6$ is the number of bright LMC stars. For $M = 0.4 M_\odot$, we have $\langle t_{\rm E}\rangle = 44$ day and a mean efficiency of $\langle\epsilon\rangle = 0.35$ giving 35 LMC events for EROS-2 for $\tau = 5.1 \times 10^{-7}$. We add 7 LMC events for EROS-1 to give a total of 42 expected events.

For no observed events ($N < 3.0$, 95% CL), the 95% CL upper limit on the optical depth is $\tau_{lmc} < 0.36 \times 10^{-7}$ for 42 expected events. The limit on $\tau_{\rm lmc}$ as a function of M is shown in Figure 2. This result is in conflict with the value of the MACHO collaboration, $\tau_{\rm lmc} = 1.2^{+0.4}_{-0.3} \times 10^{-7}$ (to which an additional 30% systematic error may be added). The MACHO result predicts that EROS would see ~ 10 LMC events whereas none are seen. However there are considerable differences between the EROS and MACHO data sets that may help to resolve the conflict. Generally speaking, MACHO

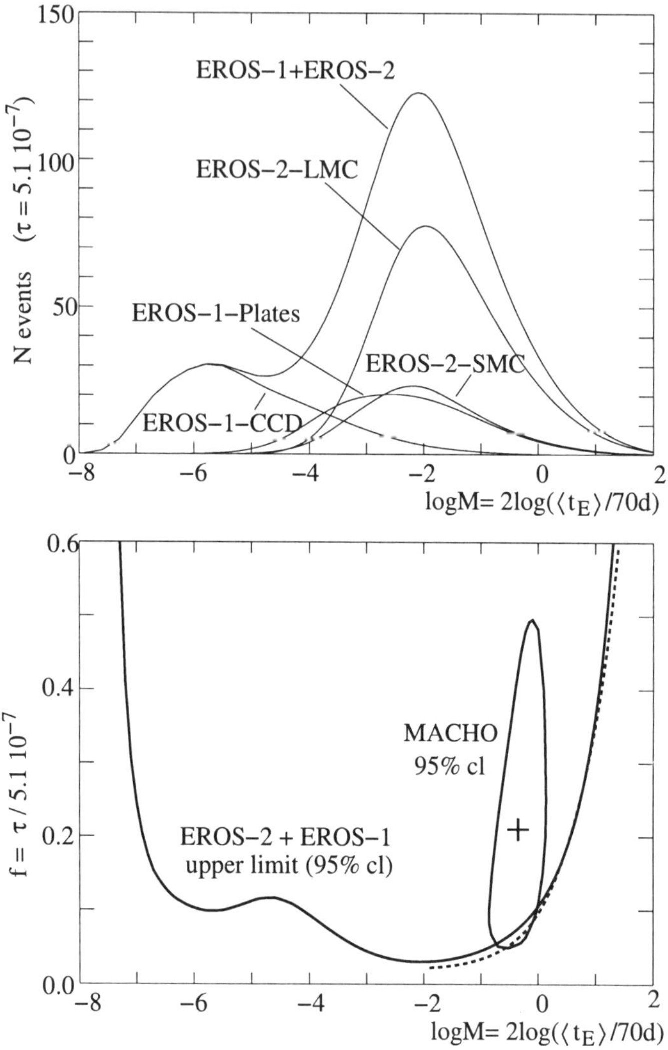

Fig. 2. The top panel shows the numbers of expected events as a function of macho mass M for the S model of Ref. 2. The expectations for EROS-2-LMC, SMC (this work) are shown along with those of EROS-1 [6] with contributions from the photographic plate program [7] and CCD program [8]. The number of events for EROS-2-SMC supposes $\tau_{smc} = 1.4\tau_{lmc}$. In the lower panel the solid line shows the EROS 95% CL upper limit on $f = \tau_{\rm lmc}/5.1 \times 10^{-7}$ based on no observed events in the EROS-2 LMC data and the EROS-1 data. The dashed line shows the EROS upper limit on $\tau_{\rm lmc}$ based on one observed SMC event in all EROS-2 and EROS-1 data assuming $\tau_{\rm smc-halo} = 1.4\tau_{\rm lmc}$. The MACHO 95% CL. curve is taken from Figure 12 (A, no lmc halo) of Ref. 2.

uses faint stars in dense fields (1.1×10^7 stars over 13.4 $\deg^2$) while EROS-2 uses bright stars in sparse fields (0.7×10^7 stars over 84 $\deg^2$). Of these bright EROS-2 stars, only 0.2×10^7 are in MACHO fields. The higher optical depth measured by MACHO in their dense fields suggests that their optical depth may be due, in part, to self-lensing (lensing of LMC stars by other LMC stars) in the inner parts of the LMC.

The use of faint stars by MACHO could cause two problems. The first would be contamination by variable stars that, in our Bright-Star Sample, are either not present or identified as such because of superior photometric precision. The low photometric precision for the faint LMC stars makes most of the events less convincing than the events on bright stars in the Galactic Bulge or the one EROS event in the SMC. Indeed, one of the MACHO A events, MACHO-LMC-23, has been identified as a variable star by a second variation in the EROS-2 data [4]. The second problem with the use of faint source stars is the large blending effects that must be understood. The MACHO collaborations [2] suggest a 30% systematic error to reflect this uncertainty.

Since several of the MACHO events are very convincing microlensing candidates, it seems unlikely that there exists a single explanation for the EROS-MACHO conflict. Hopefully, further data from the OGLE-3 and SuperMacho collaborations will clarify the issue.

References

1. B. Paczyński, *Ap. J.* **304**, 1 (1986).
2. C. Alcock, et al. (MACHO coll.), *Ap. J.* **542**, 281 (2000).
3. C. Afonso, et al. (EROS-2 coll.), *A. & A.* **400**, 951 (2003).
4. P. Tisserand, et al. (EROS-2 coll.) astro-ph/0607207, submitted to *A. & A.*.
5. C. Hamadache, et al (EROS-2 coll.), *A. & A.* **454**, 185 (2006).
6. C. Renault, et al. (EROS-1 coll.), *A. & A.* **324**, L69 (1997).
7. R. Ansari, R., et al. (EROS-1 coll.), *A. & A.* **314**, 94 (1996).
8. C. Renault, et al. (EROS-1 coll.), *A. & A.* **329**, 522 (1998).

STATUS OF THE GENIUS-TEST-FACILITY WITH NAKED GERMANIUM DETECTORS IN LIQUID NITROGEN

H. V. KLAPDOR-KLEINGROTHAUS and I. V. KRIVOSHEINA*

Max-Planck-Institut für Kernphysik,
P.O. Box 10 39 80, D-69029 Heidelberg, Germany

GENIUS-TF-II is a setup of *six* naked high purity Ge detectors (15 kg) in liquid nitrogen in Gran Sasso. It has been installed in October, 2004 - after the first four naked Ge detectors had been installed on May 5, 2003 (GENIUS-TF-I). The GENIUS-Test-Facility (GENIUS-TF) is **the first** and up to now **only** setup ever testing the novel technique aiming at extreme background reduction in search for rare decays in particular underground. The goal of GENIUS-TF was to test some key operational parameters of the full GENIUS project proposal in 1997.[1,2,3,4,5,6] Simultaneous physical goal is to search for the annual modulation of the Dark Matter signal.[10,23]

After operation of GENIUS-TF over three years with finally six naked Ge detectors (15 kg) in liquid nitrogen in Gran Sasso we realize the following problems: 1. Background from ^{222}Rn diffusing into the setup, on a level far beyond the expectation. 2. Limited long-term stability of naked detectors in liquid nitrogen as result of increasing leakage current. None of the six detectors is running after three years with the nominal leakage current. Three of the six detectors do not work any more at all. These findings require further investigation on the way to a full-size GENIUS-like project.

1. Introduction

Some years ago, the status of cold dark matter search, of investigation of neutrinoless double beta decay and of low-energy solar neutrinos all required new techniques of *drastic* reduction of background in the experiments. For this purpose we proposed the GENIUS (GErmanium in liquid NItrogen Underground Setup) project in 1997.[1,2,3,4,5,6] The idea of GENIUS (and GENIUS-TF) is to operate 'naked' Ge detectors in liquid nitrogen (as applied routinely already for more than 20 years by the CANBERRA Company for short term technical functions tests,[8] and as later 'rediscovered'[22]), and thus, by removing all materials from the immediate vicinity of the Ge

*On leave from Radiophysical-Research Institute, Nishnii-Novgorod, Russia.

crystals, to reduce the background considerably with respect to conventionally operated detectors. The liquid nitrogen acts both as a cooling medium and as a shield against external radioactivity.

After the success of the HEIDELBERG-MOSCOW experiment in $0\nu\beta\beta$ decay[17,18,19,20] GENIUS is no more needed for $\beta\beta$ decay experiments with ^{76}Ge, but may still be required for dark matter and solar neutrino experiments using Ge as target.[15] Therefore, we continued the research on the GENIUS Test Facility. *Monte Carlo simulations* for the GENIUS project, (and for GENIUS-TF) and investigation of the *new physics potential* of the project have been performed in great detail, and have been published elsewhere.[1,2,6,7,10,14] We were **the first** to show (in our HEIDELBERG low-level facility already **in 1997**) that such device can be used for *spectroscopy*[1,3,5] at least over *short* measuring times.

The small scale version of GENIUS, the GENIUS-Test-Facility has as main goal to test the long-term stability of the detectors under liquid nitrogen conditions, and also other operational parameters. A detailed description of the GENIUS-TF project is given in Refs. 10, 13.

The GENIUS-Test-Facility has been approved by the Gran Sasso Scientific Committee in March 2001.

Figure 1. Location of GENIUS-TF-I was the building on the right (car in front), opposite to the HEIDELBERG-MOSCOW experiment building (left side). Location of GENIUS-TF-II is the building of the HEIDELBERG-MOSCOW experiment (left).

Additionally to the investigation of some key operational parameters of GENIUS, the GENIUS-TF was planned to extend our work on WIMP-nucleon cross sections with the HEIDELBERG-MOSCOW and HDMS experiments,[28,29] and to test the claimed evidence for WIMP dark matter from the DAMA experiment.[30,31] The relatively large mass of Ge in the full scale GENIUS-TF compared to existing experiments would permit to search directly for a WIMP signature in form of the predicted seasonal modulation of the event rate.[23]

GENIUS-TF which is now operating over three years in Gran Sasso is the only up to now existing test facility underground for a project operating naked Ge detectors in liquid nitrogen such as GENIUS[2] and its copies (Cameo, Gerda).

The first four naked detectors (in total 10 kg) were installed on May 5, 2003 (GENIUS-TF-I). This has been reported in CERN Courier[12] and in Ref. 13. In Ref. 13 first energy calibration spectra are shown which demonstrate the (initial) good energy resolution obtained.

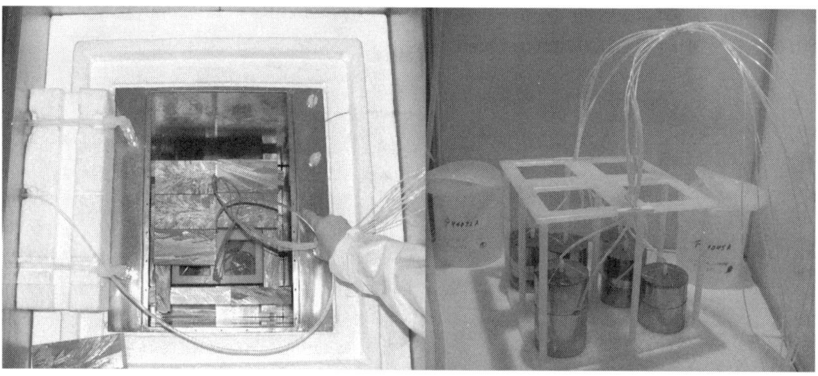

Figure 2. Left - View from the top on the GENIUS-TF-II setup during installation in October 2004. Right - The six contacted naked Ge detectors.

2. The GENIUS-TF-II and III Setups

In October 2004 we have installed a new setup GENIUS-TF-II (see Figs. 1,2,3,4,5), containing now **six naked** Ge detectors (in total 15 kg). Fig. 3 gives a schematic view of a detector with its *extremely* low amount of material used for the mechanical holder system. Fig. 4 shows the construction of the signal contact, which is a gold foil pressed to the bottom of the bore of the crystal by a spring made from a special steel wire.

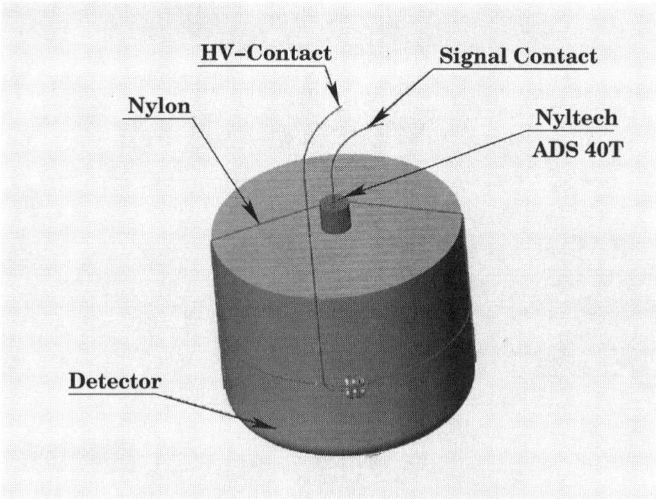

Figure 3. Schematic view of a naked Ge detector for the GENIUS-TF experiment and its holder construction.

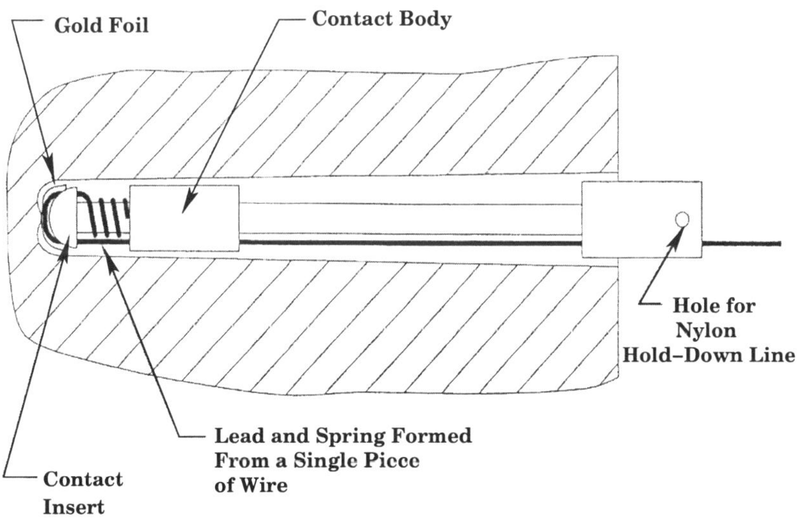

Figure 4. The construction of the signal contact of a naked Ge detector for the GENIUS-TF experiment (courtesy Ortec company).

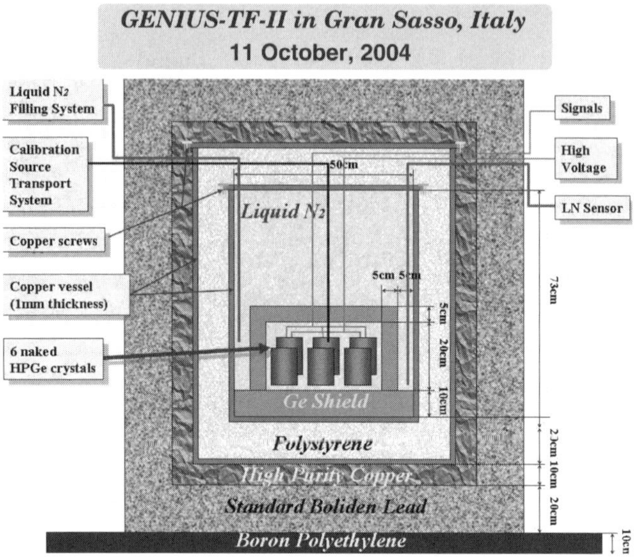

Figure 5. Schematic view of the GENIUS-TF-II setup.

As technical improvement in GENIUS-TF-II **a second copper vessel**, for further shielding of the Radon, has been additionally installed. That ^{222}Rn diffusing into the setup has been a problem for GENIUS-TF-I, has been described by analysis of the measured spectra in detail in Ref. 9. The inner shielding by bricks of (5-10) cm super-clean polycristalline Germanium (~300 kg) was used also in this setup forming the inner highly efficient shield of the Ge detectors (see Fig. 5).

The thin wall (1 mm) inner copper box containing the liquid nitrogen is made of high-purity electrolytic copper and is thermally shielded by 20 cm of special low-level styropor, the outer copper box (also made of electrolytic copper) is followed by a shield of 10 cm of electrolytic copper (15 tons) and 20 cm of low-level (Boliden) lead (>35 tons).

The high-purity liquid nitrogen used, is produced by the BOREXINO nitrogen plant, which has been extended for increase of the production capacity to be able to provide enough nitrogen also for GENIUS-TF (see Refs. 9, 10, 13).

GENIUS-TF-III started operation in beginning of March 2005 after the inner shield of polycrystalline Ge bricks had to be removed according to contract with Kurchatov institute.

3. Development of Operational Parameters

3.1. *Background from ^{222}Rn*

The unexpected (according to our Monte Carlo simulations[10]) high background from ^{222}Rn in GENIUS-TF-I has been reduced in GENIUS-TF-II by about a factor of 5.[11] It has been shown in Ref. 9, that the Rn background originates from diffusion of ^{222}Rn into the liquid nitrogen. The observed count rate corresponded to a ^{222}Rn activity of 1.2 mBq/m^3 [9] for GENIUS-TF-I (this concentration refers to the nitrogen in the gaseous phase). With the improvement of a factor of five in GENIUS-TF-II, the count rate in the energy interval 0 - 50 keV originating from ^{222}Rn is about 2 counts/kg yr keV. This value is about a factor of 400 higher than expected in the GENIUS-TF proposal[10] from Monte Carlo calculations and the measured ^{222}Rn contents in the liquid nitrogen after purification by the BOREXINO-GENIUS-TF Low Temperature Adsorber (LTA).[9,13] A background of this order of magnitude would be still compatible with the goal of GENIUS-TF to search for dark matter,[23] but will be a *serious problem for any full* GENUS-like experiments, because the ^{222}Rn leads to the 'famous' background from ^{214}Bi lines near the Q-value for double beta decay of ^{76}Ge.[16,17,21]

Some reduction of the ^{222}Rn background might be obtained by an improved transport system for the liquid nitrogen from the LTA to the GTF setup (see Ref. 9). But the above result shows that it is probably unavoidable to install permanent cleaning procedures of the liquid nitrogen from ^{222}Rn in full GENIUS-like experiments. We suggested already in Ref. 7 to install a liquid nitrogen recycling device (through condensation) inside the tank.

3.2. *Long-Term Stability*

The most dramatic result is obtained for the long-term stability of the detector operation in liquid nitrogen. It is shown in Table 1. As a result of increasing leakage currents, finally from initially six detectors after three years only three were still working and *not one of them* with the nominal high voltage (see Table 1). The reasons for a deterioration of the surface purity with time which could lead to the increasing leakage currents, need further investigation. There might, however, be also other reasons. However, for some detector experts a surface purity problem does not come very surprising.[24,25,26] From their experience, partly in detector producing companies, they expect this for a naked detector, having no shield of its

Table 1. The high voltages applied to the detectors after installation in GENIUS-TF I, II and III as function of time, and the nominal voltages.

Det-s	D1	D2	D3	D4	D5	D6
GENIUS-TF-I, from 10.12.2003, till 25.09.2004						
10.12.03	2404	2603	2879	2301	n. inst.	n. inst.
06.04.04	2600	2220	2879	2301	-	-
04.05.04	2600	2220	3200	2500		
GENIUS-TF-II, from 18.11.2004, till 28.02.2005						
08.10.04	250	1296	261	954	1253	502
18.11.04	364	2200	347	2298	3501	1015
20.01.05	"	"	"	"	"	1000
GENIUS-TF-III, from 15.03.2005						
15.03.05	80	1802	20	2153	3501	980
03.05.05	0	1700	0	1500	3501	980
09.03.06	-	1700	-	-	3500	911
19.05.06	-	1600	-	-	2100	850
Nom.	**3000**	**2600**	**3200**	**2500**	**3500**	**2000**

Table 2. Resolution (in keV) for all detectors of GENIUS-TF for the lines 356 keV of ^{133}Ba, 344 keV from ^{152}Eu and 1173 keV from ^{60}Co respectively.

Detectors	D1	D2	D3	D4	D5	D6
GENIUS-TF-I, data taking from 10.12.2003, till 25.09.2004						
for line 356 keV ^{133}Ba						
07.03.04	2.98	1.87	1.86	1.86	not inst.	not inst.
for line 1173 keV ^{60}Co						
15.04.04	4.2	4.17	5.82	3.06	not inst.	not inst.
GENIUS-TF-II, data taking from 18.11.2004, till 28.02.2005						
for line 356 keV ^{133}Ba						
08.10.04	11	8.2	-	8	11	9
10.10.04	13	10.4	-	9.2	8.7	8.31
26.10.04	-	5.4	-	4.96	5.68	5.27
30.10.04	-	3.94	-	2.86	3.06	3.08
22.11.04	-	3.46	-	3.39	3.55	3.59
06.12.04	-	3.37	-	3.19	3.53	3.85
31.01.05	-	3.31	-	3.1	3.4	3.9
07.02.05	-	3.3	-	2.56	3.18	3.76
GENIUS-TF-III, data taking from 15.03.2005						
15.03.05	-	5.34	-	2.9	7.37	6.14
for line 344 keV ^{152}Eu						
15.03.05	-	5.08	-	2.83	7.24	4.26
15.12.05	-	5.85	-	-	7.65	5.75
27.05.06	-	6.08	-	-	9.98	4.3

surface against an open surrounding (the liquid nitrogen). In fact it would not be surprising that the high voltage of several thousand volts lying at the detector surface would attract ions from the surrounding liquid nitrogen. Perhaps this suspicion might be checked by mass spectroscopic investigation of the detector surface material. The energy resolution on long terms also decreases systematically[27] (see also Table 2).

4. Conclusions

GENIUS-TF is the *only setup with naked Ge detectors world-wide running underground and over a longer time period.* It has lead to important insight into the conditions of technical operation of naked Ge detectors in liquid nitrogen. The relatively large background from ^{222}Rn diffusion is a problem unsolved up to now. The main problem realized, is, however, the increase of leakage current after long running of the detectors, which could be caused among others, by increasing surface impurity of the crystals. This led to serious restrictions of the high voltages applicable and finally to destruction of the detectors. The problems observed from GENIUS-TF after three years of operation require further investigation on the way to a full-size GENIUS-like project.

5. Acknowledgement

The authors would like to thank their colleagues from MPI Heidelberg: Herrn H. Strecker, T. Apfel, M. Reissfelder, M. Saueressig for their help during installation of GENIUS-TF-II. The authors would like to thank the technical staff of the Max-Planck Institut für Kernphysik and of the Gran Sasso Underground Laboratory. We acknowledge the invaluable support from BMBF and DFG, and LNGS of this project.

References

1. H.V. Klapdor-Kleingrothaus, Int. J. Mod. Phys. A13 (1998) 3953.
2. H.V. Klapdor-Kleingrothaus, J. Hellmig and M. Hirsch, *GENIUS-Proposal*, 20 Nov. 1997.
3. J. Hellmig and H.V. Klapdor-Kleingrothaus, Z. Phys. A 359 (1997) 351-359
4. H.V. Klapdor-Kleingrothaus, M. Hirsch, Z. Phys. A 359 (1997) 361-372.
5. H.V. Klapdor-Kleingrothaus, J. Hellmig and M. Hirsch, J. Phys. G24 (1998) 483-516.
6. H.V. Klapdor-Kleingrothaus, CERN Courier, Nov. 1997, 16-18.
7. H.V. Klapdor-Kleingrothaus et al. **MPI-Report MPI-H-V26-1999**, and Preprint: hep-ph/9910205, and in Proceedings of the 2nd Int. Conf. on Par-

ticle Physics Beyond the Standard Model BEYOND'99, Castle Ringberg, Germany, 6-12 June 1999, edited by H.V. Klapdor-Kleingrothaus and I.V. Krivosheina, *IOP Bristol* (2000) 915 - 1014.
8. J. Verplancke, *CANBERRA Company*, priv. commun. 5.03.2004.
9. H.V. Klapdor-Kleingrothaus et al., Nucl. Instr. Meth. A 530 (2004) 410-418.
10. H.V. Klapdor-Kleingrothaus et al., Nucl. Instrum. Meth. A 481 (2002) 149-159.
11. H.V. Klapdor-Kleingrothaus et al., to be publ.
12. H.V. Klapdor-Kleingrothaus, CERN Courier 43 Nr.6 (2003) 9.
13. H.V. Klapdor-Kleingrothaus et al., Nucl. Instr. Meth. A 511 (2003) 341;
H.V. Klapdor-Kleingrothaus, CERN Courier 43 Nr.6 (2003) 9;
H.V. Klapdor-Kleingrothaus et al., Proc. of the 3nd Intern. Conf. on Particle Physics Beyond the Standard Model, BEYOND02, Castle Ringberg, Germany, 2002, IOP 2003, ed. H.V. Klapdor-Kleingrothaus, 499.
14. H.V. Klapdor-Kleingrothaus, "60 Years of Double Beta Decay - From Nuclear Physics to Beyond the Standard Model", WS (2001) 1281 p.
15. H.V. Klapdor-Kleingrothaus, in Proc. of 2th Int. Workshop on Low Energy Solar Neutrino Detection, Tokyo, Japan, 4-5 Dec. 2000 (World Scientific, Singapore 2001) p.116.
16. H.V. Klapdor-Kleingrothaus, et al., Phys. Lett. B 578 (2004) 54-62.
17. H.V. Klapdor-Kleingrothaus, I.V. Krivosheina, A. Dietz et al., Phys. Lett. B 586 (2004) 198-212.
18. H.V. Klapdor-Kleingrothaus, I.V. Krivosheina, A. Dietz et al., Nucl. Instrum. Meth. A 522 (2004) 371-406.
19. H.V. Klapdor-Kleingrothaus et al. hep-ph/0201231 and Mod. Phys. Lett. A 16 (2001) 2409-2420.
20. H.V. Klapdor-Kleingrothaus, I.V. Krivosheina, Phys. Scripta T127 (2006) 52.
21. K.Ya. Gromov et al., J.Part. Nucl. Lett. 3 (2006) 30-41.
22. G. Heusser, Ann. Rev. Nucl. Part. Sci. 45 (1995) 543-590.
23. C. Tomei, A. Dietz, I. Krivosheina, H.V. Klapdor-Kleingrothaus, Nucl. Instrum. Meth. A 508 (2003) 343-352.
24. G.F. Knoll, "*Radiation Detection and Measurement*", second ed. 1989, John Wiley & Sons.
25. E. Flyckt (*Philips Photonics Comp.*, Brave, France) private comm. May 2006.
26. A.B. McDonald (SNO collaboration), Queen Univ., Canada, private comm. May 2006.
27. H.V. Klapdor-Kleingrothaus, I.V. Krivosheina et al., Gran Sasso Ann. Report 2005, p.125-133.
28. HEIDELBERG-MOSCOW collab., Phys. Rev. D 59 (1998) 022001.
29. H.V. Klapdor-Kleingrothaus, I.V. Krivosheina and C. Tomei, Phys. Lett. B 609 (2005) 226-231.
30. R. Bernabei et al., Phys. Lett. B 424 (1998) 195.
31. R. Bernabei et al., Riv. Nuovo Cim. 26(2003) 1-73; Phys. Lett. B 424 (1998) 195; 450 (1999) 448; 480 (2000) 23.

INDIRECT DARK MATTER SEARCHES WITH THE MAGIC TELESCOPE

M. DORO[a] FOR THE MAGIC COLLABORATION[b]

[a] *Department of Physics G.Galilei, University of Padua and Sezione INFN, 35131 Padua, Italy*
E-mail: michele.doro@pd.infn.it
[b] *An updated list of the authors can be found at*
http://wwwmagic.mppmu.mpg.de/collaboration/members/index.html

In many Dark Matter (DM) frameworks, the annihilation of DM particles can produce gamma rays that extend up to very high energies. Since the 17m 'Major Atmospheric Gamma-ray Imaging Cherenkov (MAGIC) Telescope' has the lowest energy threshold of all existing Cherenkov telescopes, it is best suited to search for these signals. For the ongoing cycle-II observation period, part of the observation time has been allocated to indirect DM searches.

This review will include a short introduction to MAGIC and the characteristics of the observations, as well as the strategy for indirect DM searches using ground based Cherenkov telescopes.

Keywords: MAGIC; Dark Matter.

1. The scientific case

Since the discovery some decades ago that an invisible ("dark") amount of matter was contributing meaningfully to the total amount of mass in the Universe, a plethora of candidate particles has been proposed to fulfill the lack. Barionic candidates were soon discarded, while only recently also neutrinos have been rejected as principal constituent of the dark matter.

Nowadays most of the theories agree that the DM candidate must be a weakly interacting, very stable, neutral, massive particle, also called a WIMP (Weakly Interacting Massive Particle). Among the WIMPs, in a Cold Dark Matter (CDM) Universe scenario, particularly interesting is the lightest particle in the SUSY models.[1] The neutralino is a mix of the *binos* ($\tilde{\mathrm{B}}$), *winos* ($\tilde{\mathrm{W}}_3$) and *higgsinos* ($\tilde{\mathrm{H}}_1^0$ and $\tilde{\mathrm{H}}_2^0$). These states mix into four Majorana eigenstates, each called *neutralinos*: $\tilde{\chi}_1^0$, $\tilde{\chi}_2^0$, $\tilde{\chi}_3^0$ and $\tilde{\chi}_4^0$ in order of increasing mass. The lightest is also called *the* neutralino and $\chi \equiv \tilde{\chi}_1^0$.

Currently a big effort is made by several experiments in search of the neutralino. In the past, experiments and models put coarse limits on the mass of this particle and now it is commonly believed that the neutralino mass should stay in the range 0.1–3 TeV/c^2, even if those values can be different in other models.

The DM, independently of its nature, has low cross sections and interact with gravitation and weak interaction. The DM usually form spherical structure and around galaxies, DM concentrate in structure called "galactic halos" which extend several times the extension of the barionic galaxy, up to 120 kpc. The distribution of the DM within the halo is not very precisely know even if many models are proposed since many years.[2–4] The basic idea is that the profile is a power law, with increasing slope going to the outer regions. More under debate is what happens very nearly the core. In some models, a flat plateau is present, while in others a softening of the density spectral index is predicted. Adiabatic compression of the DM in this region can also enhance the DM density.[5]

2. The indirect DM observation with IACTs

The neutralino is a Majorana particle and interact with itself in a self–annihilation process. As a result, fermion and anti–fermion pairs are produced. Quark induced jets and τ decays produce π_0 that decay into high energetic gamma. Additionally, a direct annihilation of the neutralinos into two gammas is possible even if it is strongly limited in term of branching ratio.

$$\chi\chi \rightarrow \mathrm{h}\gamma \quad \chi\chi \rightarrow \mathrm{Z}\gamma \quad \chi\chi \rightarrow \gamma\gamma \tag{1}$$

The gammas emitted in the former process have a broad–band spectrum, while the direct annihilation provide gammas with line emission. Unfortunately, these lines emitting processes are loop-suppressed and therefore any signal is very weak. The precise branching ratios depend on the model applied.

After the gammas are emitted, they travel through the astrophysical medium and eventually reach the Earth. In case the emitter at the order of galactic distances, the gamma rays flux does not suffer relevant absorption and the flux is still enough large to be observable. Thus at the Earth, the flux of the gamma rays from annihilation of DM is:

$$\Phi(E, D) = \Phi_0 \frac{dN}{dE}\left(\frac{<\sigma v>}{10^{-26}\text{cm}^3/s}\right)\left(\frac{m_\chi}{100\text{GeV}}\right)^2 \\ \cdot\left(\frac{D}{\text{kpc}}\right)^{-2}\left(\frac{\rho(r_{sp})}{100\text{GeVcm}^{-3}}\right)\left(\frac{\rho_{sp}}{\text{pc}}\right)^{14/3}\left(\frac{\rho_{cut}}{10^{-3}\text{pc}}\right)^{-5/3} \quad (2)$$

where E is the energy, D is the distance of the source, σv is the product of cross section and velocity, r_{sp} is the radius of the spike and r_{cut} is the radius at which the DM profile flattens.

The first line of Eq. (2) contains terms related to the particle characteristics, while the second line deals with astronomical properties of the DM region and it also called the *cosmological factor*. Due to the composition of this factors, a clear estimation of the flux is difficult, and for a source can span over many orders of magnitude. This makes very difficult to realize predictions on the signal from a source.

The spectrum from annihilation should be a continuum up to the DM mass, where an exponential cut-off take place. The latter feature is very interesting because many different objects containing DM should present the same cut-off in gamma spectra.

2.1. *The IACT technique*

Gammas produced through annihilation travel through the space and eventually reach the Earth atmosphere. On the top atmosphere, they decay into a positron-electron pairs, which then emits energetic gammas via bremsstrahlung. The secondary gammas in turn emit positron-electron pairs and the so-called electromagnetic cascade is initiated. The e^- and e^+ in this cascade are highly relativistic and therefore do emit Cherenkov light. This light travel through the lower atmosphere and at the ground level illuminates an area which has an diameter of the order of 100 m (also depending on energy and inclination of the incoming gamma ray). This light is collected by the reflective surface of the reflector, and focussed onto a multipixel camera which record the shape of the image produced by the shower which has an elliptic shape pointing to the center of the camera.

The typical duration of a Cherenkov flash is few nanoseconds, and it ranges in the near UV – visible domain, being peaked towards the blue. Thus the first characteristic required by this detector is fast response to incoming light in the visible range, which is usually now obtained with high quantum efficiency PMT.

3. The MAGIC telescope

Within the framework of the IACTs of the second generation, the MAGIC telescope[6,7] is the largest existing, with a reflecting surface of 17 m diameter (236 m^2 area). The huge collection area allows to lower the energy threshold of the detector. Pushing the yield below 100 GeV is a fundamental task because a gap from 1 GeV to 300 GeV has been left open by the past gamma astronomy (ground–based and satellite telescopes). MAGIC is located at the Canarian island of La Palma (Spain) in the northern hemisphere. It started full operations in fall 2004, and now is entering the second year (Cycle II) of full scientific observations.

Having an energy threshold of 70–100 GeV at zenith, it is able to observe signals of the type described in the Eq. (2) (with best results the higher the DM mass). The flux sensitivity is for 10^{-11} ph·cm^{-22}·s^{-1} corresponding to 2.5% of Crab Nebula in 50 hours (the Crab Nebula is a SNR Type Ia and is the standard candle for gamma astronomy due to its steady flux) with an energy resolution of 20–30% depending on energy.

4. Past observations: the Galactic Center

It is known that DM concentrates in the center of the galaxies. For this reason and because of its vicinity, in the past *our* Galactic Center (GC) has been subject of studies and observation.[8–10] Many Cherenkov telescopes observed the GC during the years 2004–2006 (Veritas, Cangaroo, HESS, MAGIC). Even if the results of the experiments are different, there is a good agreement between MAGIC and HESS flux and spectrum. In a recent article,[11] the GC observation is presented in details.

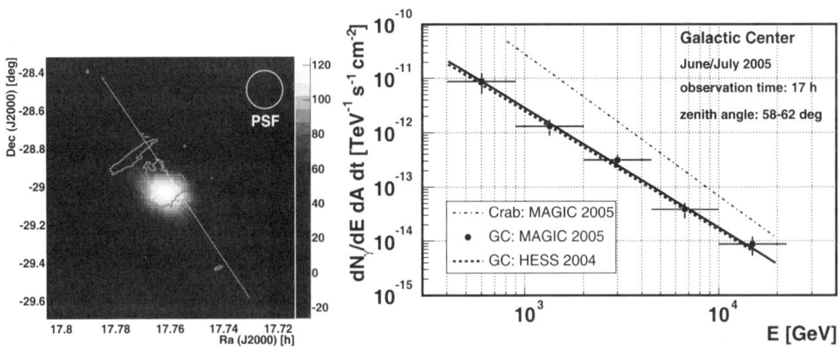

Fig. 1. The Galactic Center observed by MAGIC.

In the case of MAGIC detection, due to the position of the telescope in the northern hemisphere, the GC was observed only at high zenith angles ($> 50°$) resulting in a high energy threshold ($\sim 700\ GeV$). The source is a VHE emitter, spatially consistent with the massive black hole SgrA* as well as the SNR SgrA East. The emission is also steady. The signal is an untruncated power law with spectral index -2.2 up to several TeV, which is a strong mark for an astronomical emitters. This evidence does not rule out the presence of dark matter signal from annihilation, but shows that — even if the signal exists— it is overcome by at least one order of magnitude by the presence of the stronger astrophysical signal.

This is a first indication that the search of Dark Matter is best suited for places in the sky where other astrophysical emitters are not present and that the background of other emitters can be challenging to overcome. In the next chapter we will discuss two examples of such places.

5. Present fields of interest

5.1. *Dwarf Spheroidal Galaxies (dSph)*

Dwarf spheroidal galaxies are a subclass which show a large mass to luminosity ratio taking as reference the values for the Sun. This could be related to a larger presence of DM than in other objects. These objects are therefore considered good target for DM searches. Due to the distance, many gamma signals of astrophysical origin cannot reach us and thus an annihilation signal coming from them should be clean from other astrophysical emitters.

Two candidates are actually of big interest: the Draco Galaxy, which has a M/L 300, is quite close (around 80 kpc) and is well studied.[12] A second candidate can also be found in the Milky Way satellite galaxies,[13] and in particular the Ursa Mayor, 100 kpc distant, with M/L that could be larger than 500.

5.2. *Unidentified EGRET sources*

The model described in Ref. 14 shows that studying the evolution of super massive black holes, one faces the fact that a number of intermediate mass black holes (IMBH) do not suffer major merging and interaction with barions along the evolution of the Universe. Those IMBHs wander in the galactic halos and evolve undisturbed. DM accrete on IMBH in adiabatic growth in a way that the final radial profile is spiky. According to the evolution model, a galactic halo should host 10^2–10^5 of such IMBHs. In the spike, the DM density is increased and the annihilation rate would be enhanced, so that the IMBHs could be bright gamma emitters. Object like this would have the same spectrum and cut–offs, providing a smoking gun of the existence of dark matter. Additionally, according to the model, the interplay between the spike density distributions which depends on the DM mass and cross section, produce a fainter dependence of the flux from the these two parameters.

The best way at first approach to look for such objects is within the unidentified EGRET sources, which were observed in the HE domain. Those signals should be correlated up to the VHE range in case of DM origin of the signal.

6. Conclusions

DM particle in a SUSY framework can annihilate and produce gammas in the final states. The flux coming from cosmic object is largely unpredictable due to uncertainties in the DM mass and cross-section. Nevertheless, the actual IACTs have good sensitivity for low fluxes. MAGIC in particular has the lower energy threshold and is therefore suitable for this observation. DM searches concentrated in the past on the GC, where the signal demonstrated to be overcome by other astrophysical emitters. Actual objects of interest are dwarf spheroidal galaxies, where a huge amount of DM is expected and spike DM distribution around IMBHs which can be several in the galactic halo.

References

1. J. Wess and J. Bagger, *Supersymmetry and Supergravity* (University of Princeton, Princeton (USA), 1992).
2. J. F. Navarro *et al.*, *Mon. Not. Roy. Astron. Soc.* **349**, p. 1039 (2004).
3. D. Reed *et al.*, *Mon. Not. Roy. Astron. Soc.* **357**, 82 (2005).
4. T. Fukushige, A. Kawai and J. Makino, *Astrophys. J.* **606**, 625 (2004).
5. F. Prada, A. Klypin, J. Flix, M. Martinez and E. Simonneau (2004).
6. C. Baixeras *et al.*, *NIM A* **518**, p. 188 (2004).
7. M. Mariotti Prepared for 7th Workshop on Towards a Network of Atmospheric Cherenkov Detectors 2005, Palaiseau, France, 27-29 Apr 2005.
8. F. Aharonian and A. Neronov, *ApJ* **619**, p. 306 (2005).
9. A. Atoyan and C. Dermer, *ApJ* **617**, p. L123 (2004).
10. D. Horns *et al.*, *Phys.Lett* **B607**, p. 225 (2005).
11. J. Albert *et al.*, *ApJ Letters 638* **638**, p. L101 (2006).
12. M. Odenkirchen *et al.*, *AJ* **122**, p. 2538 (2001).
13. B. Willman *et al.*, *Astrophys.J* **626** (2005).
14. G. Bertone, A. R. Zentner and J. Silk, *Phys. Rev.* **D72**, p. 103517 (2005).

DARK MATTER NaI SEARCHES AT CANFRANC: STATUS OF ANAIS

J. AMARÉ, S. BORJABAD, A. CALLEJA, S. CEBRIÁN, E. GARCÍA, H. GÓMEZ, M. MARTÍNEZ, J. MORALES, A. ORTIZ DE SOLÓRZANO, C. POBES, J. PUIMEDÓN, A. RODRÍGUEZ, J. RUZ, M.L. SARSA, J.A. VILLAR

Laboratorio de Física Nuclear y Astropartículas.
Universidad de Zaragoza, 50009 Zaragoza, Spain
E-mail: Maria.Martinez@unizar.es

A 107 *kg* NaI(Tl) experiment is projected at the Canfranc Underground Laboratory to look for dark matter particles by the annual modulation in their interaction rate. It will consist of 10 crystals selected from a set of 14 that remain underground since 1986. A first stage aiming at background and threshold reduction was performed with a 10.7 *kg* NaI(Tl) prototype, resulting in a new encapsulation design for the crystals. Low energy gamma sources were used for energy calibration and threshold determination. Background near threshold and natural chains internal contamination of some of the available crystals were studied. Results of the prototype and prospects of the future experiment are presented.

1. Introduction

The ANAIS experiment, intended to look for the anual modulation effect in the expected interaction rate of WIMPs in NaI(Tl) detectors, will be installed at the Canfranc Underground Laboratory (LSC). It is strongly motivated by the modulation evidence reported by the DAMA/NaI experiment after 7 years of data-taking at the LNGS,[1–3] up to now the unique positive result in direct detection, whose comparison with other experiments using different target materials is highly model dependent. ANAIS will operate around 100 kg of NaI(Tl) (10 crystals selected among 14 hexagonal prism-shaped ones that have been used in previous experiments at Modane[4] and the LSC[5,6] since 1986) inside an adequate shielding. The first steps of the experiment can be traced back to NaI32,[7] one of the first searches of the annual modulation effect. After that, a first prototype of the ANAIS experiment was mounted at Canfranc,[8] consisting of one crystal in its original housing (a stainless steel vessel with a quartz window coupled to one PMT). By pulse shape discrimination (PSD) techniques the energy threshold was reduced down to 4 keV, but the analysis evidenced that further reduction required two PMT working in coincidence. An improved version of this prototype I, replacing the original PMT by an ultra-low background one (EMI 9302 LTD) coupled through a 10 *cm* long light guide, obtained a background reduction of around 40% in the $200-600$ keV region and more than 50% in the 1461 keV peak coming from ^{40}K contamination, but it also suggested a non-negligible internal contribution to the measured background.

The next step was to extract one of the crystals from the stainless steel vessel and to couple it to two PMTs in an adequate housing. To deal with this task a second prototype was developed, conceived as a flexible set-up to try different configurations (light guides of several lengths, different thickness of Teflon as diffuser, effect of a polymeric reflecting foil inside the Teflon) to reduce both background and threshold.[9,10] The crystal and the light collecting system are inside a copper box in dry and clean nitrogen atmosphere, that prevents the detector from humidity damage.

Light collection tests allowed to determine the loss in collected light when longer guides are used (an average reduction of about 10% every 5 *cm* of light guide length) and the slight increase with larger thickness of Teflon (around 5% more light every cm of Teflon). Finally the selected configuration consisted of a cylinder of Teflon of 1.5 *cm* width, a polymeric reflecting foil covering the inner surface and 10 *cm* length conical guides made of methacrylate to couple to the 3 inches PMT (see Fig. 1 left).

2. Underground measurements with the prototype II

The second prototype was installed at Canfranc inside a shielding made of 10 *cm* of archeological lead, 10 *cm* of low activity lead and an anti-radon enclosure. Three main actions were performed with this prototype at Canfranc: a gamma calibration in the low energy region with several radioactive sources, a measurement of the background level near the threshold and a determination of the internal contamination of some of the crystals coming from natural chains by means of PSD of alpha events.

2.1. *Low energy calibration*

Four radioactive sources with gamma lines below 150 keV were used in the calibration: ^{55}Fe, ^{57}Co, ^{133}Ba and ^{137}Cs. The results showed that the linear approximation is a very good hypothesis down to 14 *keV* (see Fig. 1 right). Nevertheless, the 6 *keV* peak of the ^{55}Fe source was systematically found at around $60-80\%$ of its extrapolated value according to the linear calibration. This effect, previously observed by other groups,[11] could arise from crystal surface effects (the mean free path for 6 *keV* photons in NaI is 5 μm). A dedicated measurement is in preparation to clarify this point. In the linear hypothesis, a (2.0 ± 1.2) *keVee* (electron equivalent) energy threshold was achieved.

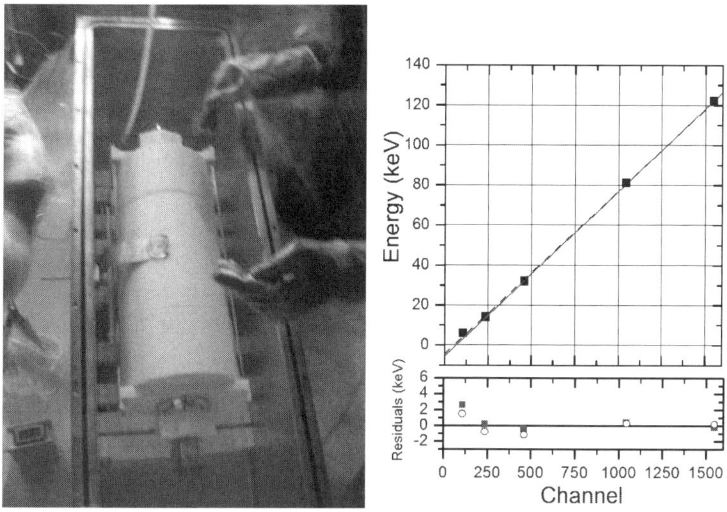

Fig. 1. Left: view of one of the configurations tested with the second prototype at Canfranc. Right: energy calibration below 150 *keV* with four radioactive sources (^{55}Fe, ^{57}Co, ^{133}Ba and ^{137}Cs) in the linear approximation, including the 6 *keV* peak (dashed line, open circles in the bottom panel) and without this peak (solid line, filled squares in the bottom panel).

2.2. *Background measurement*

After identifying the presence of a ^{232}Th contamination in the upper weldings of the copper box, an internal copper shielding of 2 *cm* thickness was installed over the Teflon bank inside the box. Fig. 2 shows the background measured after 1 month of data-taking at the LSC, with the two PMTs working in coincidence and the trigger at the level of one photoelectron in every PMT. The stability was controlled by periodic calibrations with ^{57}Co and ^{133}Ba sources and also monitoring the position of the

$\sim 50\ keV$ peak, that was identified as ^{210}Pb internal contamination (see next section).

The PMT pulses were recorded with an ultra-fast CompuScope 82G digitizer, with a sampling rate of 1 Gs/s. Standard PSD techniques were applied to reject the noise events near the threshold, coming mainly from the PMTs. The discrimination parameter selected for the analysis was:

$$P1 = \frac{area(100 - 600\ ns)}{area(0 - 600\ ns)} \tag{1}$$

The efficiency for the acceptance of scintillation events was estimated from the distribution of the discrimination parameter for a population of scintillation events collected from periodic calibrations with ^{57}Co source, and it was used to correct the spectrum shown in Fig. 2. It is worthwhile to note that despite the internal Cu shielding, the ^{232}Th contamination at the box has a non-negligible contribution to the measured low energy background.

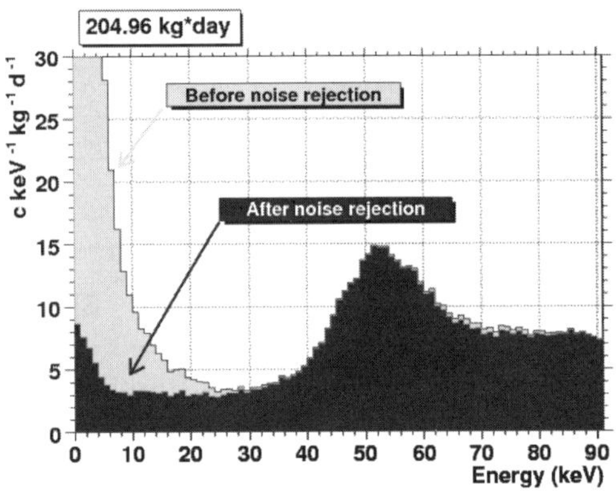

Fig. 2. Background measured with the second prototype at Canfranc with an exposure of 205 $kg \times day$ before and after noise rejection by standard PSD techniques and efficiency correction.

2.3. *Identification of internal contaminants*

To determine the amount of internal contaminants coming from natural chains a PSD analysis was done for the high energy pulses to discriminate alpha events.[12] Two of the crystals were studied, obtaining a rate of 0.027 α/s for first crystal (prototype II) and 0.017 α/s for second crystal, and a quite different shape for the alpha spectrum, so the internal contamination that can be inferred is also different. The chains were not in secular equilibrium, with a larger concentration of radium isotopes (^{226}Ra: 0.14 and 0.22 mBq/kg, ^{228}Ra: 0.07 and 0.02 mBq/kg) than that of the parents ^{238}U (0.02 and < 0.002 mBq/kg) and ^{232}Th (0.03 and < 0.002 mBq/kg). Also at ^{210}Pb equilibrium was broken and the estimated bulk concentration was clearly larger, specially in the first crystal (1.2 and 0.5 mBq/kg) and it is consistent with the peak at ~ 50 keV visible in the background spectrum. The background below 20 keV originated from the ^{238}U and ^{232}Th chains internal concentration was estimated by MC simulations, resulting in around $0.5 - 0.8$ $c\ keV^{-1}kg^{-1}d^{-1}$. As concerns internal ^{40}K, the measurements with the improved prototype I allowed to set an upper limit of 9 mBq/kg. This isotope is specially important in the low energy region because it produces a peak at 3 keV that can be one of the most important contributions to the background in the region of interest for WIMP direct searches, so work is in progress to determine its concentration in the crystals of the LSC.

3. Expected sensitivity

An analysis based on the statistical significance of the modulated signal has been performed considering different scenarios for ANAIS (107 kg). The results for some of them are shown at Fig. 3. Based on this analysis and supposing that the relevant parameters for the whole experiment will be similar to those derived from the second prototype, the coincidences between crystals could allow to reach the sensitivity to observe the DAMA/NaI modulation result with exposures of around five years.

4. Summary and future plans

Accepting the linear calibration from 122 to 14 $keVee$, a (2.0 ± 1.2) $keVee$ threshold has been achieved. The internal contamination coming from natural chains of two of the available crystals has been determined, and ^{40}K measurement is in preparation. The developed techniques will allow to prepare a protocol for crystal selection. Based on the results of the prototype, a new encapsulation for the crystals has been built, made of OFHC copper perfectly sealed, with a mylar window for low energy gamma calibration and two quartz windows for the photomultipliers. A crystal with this encapsulation and an improved shielding will start taking data at Canfranc in the next months. The photomultipliers are also being tested and new

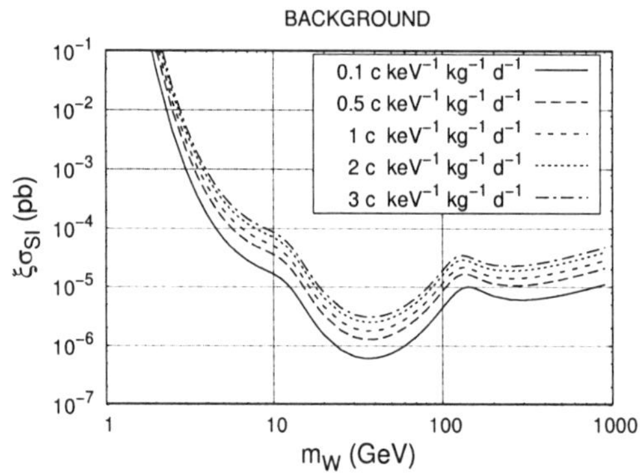

Fig. 3. Expected sensitivity of the ANAIS experiment (107 *kg*) for a neutralino with pure spin independent coupling and standard halo model, calculated for $\langle\delta^2\rangle = 5.6$.[13] It has been supposed a flat background, a threshold of 2 *keVee* and a period of exposure of 5 years.

voltage dividers are being designed. The next step will be to proceed with the re-encapsulation of the crystals.

Acknowledgments

This research was funded by the Spanish Commission of Science and Technology (MCyT and MEC) under contracts No. FPA2001-1767 and FPA2004-0974.

References

1. R. Bernabei *et al.*, *Phys. Lett. B* **480**, 23 (2000).
2. P. Belli *et al.*, *Phys. Rev. C* **66**, p. 043503 (2002).
3. R. Bernabei *et al.*, *Riv. Nuov. Cim.* **26**, 1 (2003).
4. A. Morales *et al.*, *Riv. Nuov. Cim. A* **100**, p. 525 (1988).
5. A. Morales *et al.*, *Riv. Nuov. Cim. A* **104**, p. 1581 (1991).
6. C. Sáenz *et al.*, *Phys. Rev. C* **50**, 1170 (1994).
7. M. L. Sarsa *et al.*, *Phys. Rev. D* **56**, 1856 (1997).
8. S. Cebrián *et al.*, *Nucl. Phys. B (Proc. Suppl.)* **114**, 111 (2003).
9. M. Martínez *et al.*, in *Proceedings of the International Workshop on the Identification of Dark Matter (IDM 04). Edimbourgh, September 2004*, (N. J. C. Spooner and V. Kudryavtsev. Published by World Scientific, 2004).
10. J. Amaré *et al.*, *J. Phys.: Conf. Ser.* **39**, 123 (2006).
11. G. Gerbier *et al.*, *Astrop. Phys.* **11**, 287 (1999).
12. J. Amaré *et al.*, *J. Phys.: Conf. Ser.* **39**, 201 (2006).
13. S. Cebrián *et al.*, *Astrop. Phys.* **14**, 339 (2001).

XMASS EXPERIMENT

ATSUSHI TAKEDA* FOR THE XMASS COLLABORATION

Kamioka Observatory, Institute for Cosmic Ray Research, University of Tokyo, Japan

** E-mail: takeda@suketto.icrr.u-tokyo.ac.jp*

The XMASS experiment aims to detect pp and ^{7}Be solar neutrinos, and neutrinoless double beta decay using ultra pure liquid xenon. It requires low background and a low threshold which will also enable us to search for dark matter in the galactic halo.

The first stage of XMASS project is concentrated on dark matter searches utilizes 800 kg liquid xenon detector. The detector design and expected sensitivity for dark matter based on Monte Carlo simulations will be presented.

Keywords: Dark matter searches; liquid xenon; low background

1. Introduction

XMASS is a multi purpose low-background detector with pure liquid xenon in Kamioka mine (2,700 m.w.e.), Japan. The physics goals of XMASS project are real-time measurement of pp and ^{7}Be neutrinos, detecting neutrinoless double beta decays, and direct dark matter searches.

The first stage of XMASS project is concentrated on direct dark matter searches utilizes an 800 kg liquid xenon detector. It requires an ultralow background in the fiducial volume of the detector. The key idea to reduce background is to utilize the self-shielding effect of xenon. Since xenon has a large atomic number (Z=54), several tens of centimeters of outer layer of liquid xenon can absorb and shield low energy external gamma rays. In addition, it is easy to purify even after the experiment started because of liquid.

This self-shielding effect has been confirmed by using a prototype detector. Several experiments with the prototype detector which consists of a 31 cm cubic oxygen-free-highpurity-copper (OFHC) chamber, 54 photomultipliers (PMTs), and heavy shield have performed good position and energy reconstruction and low background environment as well as self-shielding effect [1].

Based on those encouraging experimental results, we are now designing the 800 kg detector. In this paper, the 800 kg detector design will be reported.

2. The XMASS 800 kg detector

The 800 kg detector consists of a spherical OFHC chamber with about 90 cm diameter filled with pure liquid xenon, and about 800 PMTs immersed in liquid xenon. Fig. 1 shows the schematic view of the detector and PMT arrangement defined in our Monte Carlo (MC) simulations.

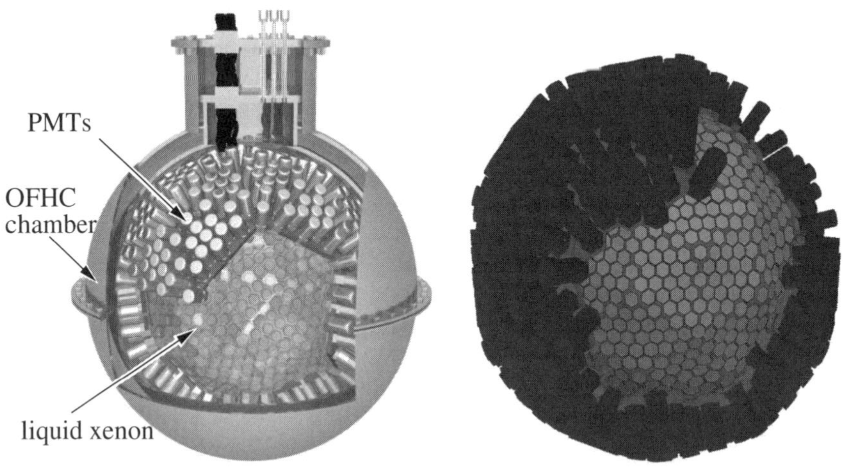

Fig. 1. The schematic view of the XMASS 800 kg detector consisting a spherical OFHC chamber with 90 cm diameter filled with pure liquid xenon, and about 800 PMTs immersed in liquid xenon (left), and PMT arrangement defined in MC simulations (right).

Physics performances of 800 kg detector have been studied by using MC simulations based on the geant4 simulation package. In order to set PMTs inside the detector as possible as dense, a shape of pentakisdodecahedron whose 60 triangle shape planes is chosen. A new PMT being developed for this project, Hamamatsu R8778 mod, whose hexagonal quarts window is set on a triangle shape plane. Since 10 PMTs per one plane are set, total 600 PMTs are set. In addition, other 212 PMTs are set on the gap between each triangle shape plane, therefore, total 812 PMTs are set. This arrangement achieves 67.0 % photo-coverage according to MC simulations. Averaged distance from the center of detector to photo-cathodes of PMTs is 44 cm. A volume of a sphere with 25 cm radius is used for fiducial.

The procedure of position and energy reconstruction is based on that of our prototype detector [1]. The position and energy are reconstructed from PMT charge patterns which are function of position and energy in the detector by using a likelihood method. Obtained position resolutions are 3.0 cm for 10 keV, and 4.8 cm for 5 keV at the boundary of fiducial volume. In general, the position resolution goes well far from the center because PMT is close to the position and number of detected photons increases.

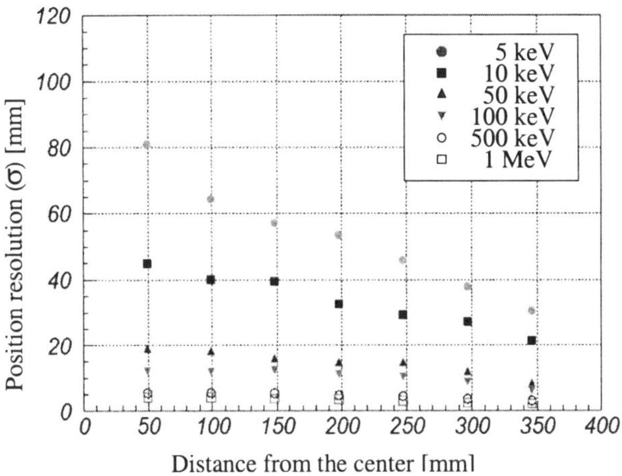

Fig. 2. Calculated position resolution. Gray closed circles, black closed squares, black closed triangles, gray closed triangles, open circles, and open squares are for 5 keV, 10 keV, 100 keV, 500 keV, and 1 MeV, respectively.

2.1. *Expected background level and sensitivity*

The main source of background achieved with the prototype detector is thought to be caused from the radio activities in R8778 PMTs. The measured radio activities of a R8778 PMT are $1.8 \pm 0.2 \times 10^{-2}$ Bq (U), $6.9 + 1.3 \times 10^{-3}$ Bq (Th), and $1.4 \pm 0.2 \times 10^{-1}$ Bq (^{40}K), and the expected background level from these activities is about 10^{-2} counts/day/kg/keV around 100 keV which explains the current background level of prototype detector. In the 800 kg detector, owing to much larger volume for self shielding further reduction is possible. In addition, we have a plan to develop one order of magnitude lower background PMTs.

We calculated gamma-ray background from the U chain radio activity in PMTs assuming that the activity is one order of magnitude lower (1.6×10^{-3} Bq) than that of current R8778 PMTs. Fig. 3 shows obtained energy spectra caused from U chain radio activity in PMTs. After fiducial volume cut of 20 cm from the wall, only one event was found below 100 keV with the statistic of 3.2 days which corresponds to $< 10^{-5}$ counts/day/kg/keV.

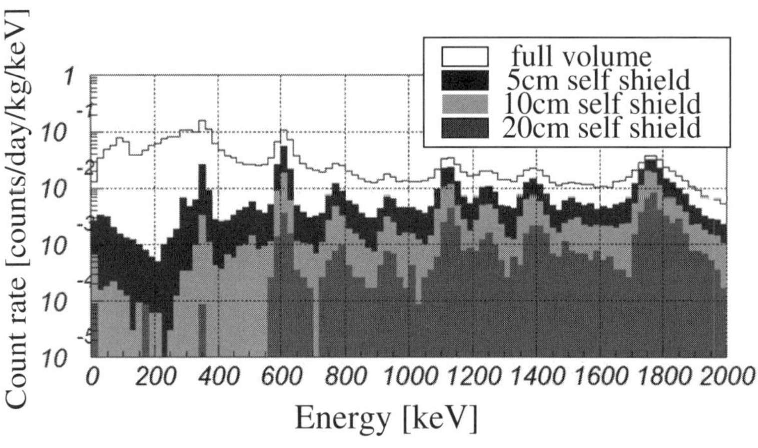

Fig. 3. Calculated background spectrum caused from U chain radio activity in PMTs. The open histogram shows the result obtained from the full volume. The black colored, light-gray colored, and gray colored histograms show the results after fiducial volume cuts of 5 cm, 10 cm, and 20 cm from the wall, respectively.

The background caused by external gamma rays and fast neutrons from the rock should be reduced below the PMT background level. Instead of a conventional lead and polyethylene shield, we selected a simple water shield against both gamma rays and fast neutrons. We estimated the required thickness of water shield surrounding the detector by using MC simulation. For external gamma rays, the initial energy spectrum from the rock calculated from the energy spectra measured by Ge detector without shield was used. As a result, more than 200 cm thickness is needed to reduce the background below the PMT background level. For fast neutrons generated by the (n, α) reaction from the radio activities in the rock, very conservative assumption was used, which is energies of all the fast neutrons are 10 MeV with measured flux of $1.15 \pm 0.12 \times 10^{-5}\,\mathrm{cm}^{-2}\mathrm{sec}^{-1}$ in Kamioka mine. As a result, more than 200 cm thickness is found to be sufficient.

Fig. 4 show the expected sensitivities for the spin-independent (SI) and

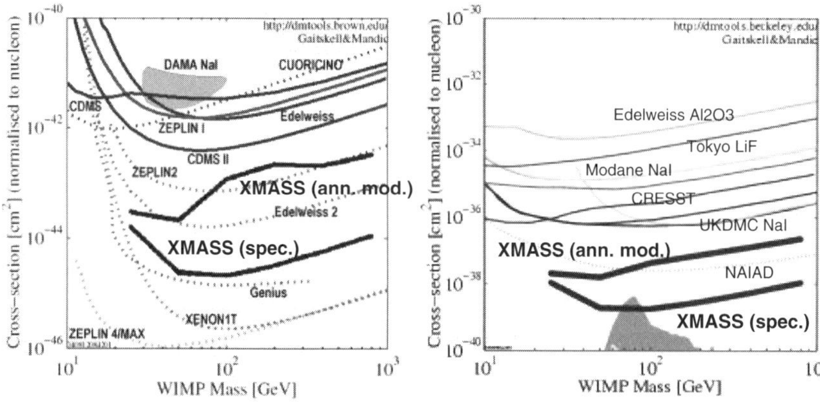

Fig. 4. Sensitivities for the spin-independent (left) and spin-dependent (right) cross sections of dark matter. Upper thick curves of each figure correspond to analyses to look for the annual modulation, and lower thick curves of each figure correspond to analyses to find an increase due to dark matter signal in the raw spectrum.

spin-dependent (SD) cross section of WIMPs, respectively, assuming that energy threshold of 5 keV ee and statistics of 0.5 ton · year and without any pulse shape discrimination.

3. Summary

XMASS experiment is a multi purpose low-background experiment with large mass liquid xenon. The first stage of XMASS project is constructing of 800 kg detector. We are now developing an 800 kg detector to detect dark matter and are going to start the dark matter search in the near future.

References

1. S. Moriyama, *In the proceedings of the International Workshop on Technique and Application of Xenon Detectors, Xenon01*, World Scientific, pp 123-135, University of Tokyo, Japan, 3-4 December, 2001; *In the proceedings of The Forth International Workshop on the Identification of Dark Matter, IDM2002*, World Scientific, pp 390-395, York, UK, 2-6 September, 2002; *In the proceedings of The Forth International Workshop on the Neutrino Oscillations and Their Origin, NOON2003*, World Scientific, pp 364-371, Kanazawa, Japan, 10-14 February, 2003; *In the proceedings of BEYOND 2003*, Springer Verlag, pp 385-396, Castle Ringberg, Tegernsee, Germany, 9-14 June, 2003; *In the proceedings of The Fifth International Workshop on the Identification of Dark Matter, IDM2004*, World Scientific, pp 248-253, Edinburgh, UK, 6-10 September, 2004.

MEASURING THE DAEMON-FLUX SEASONAL MAXIMA AT THE BAKSAN NEUTRINO OBSERVATORY

E. M. DROBYSHEVSKI and M. E. DROBYSHEVSKI

Ioffe Physico-Technical Institute, Russian Academy of Sciences St.Petersburg, 194021 Russia; e-mail: emdrob@ioffe.mail.ru

Detection of negative daemons (Dark Electric Matter Objects – presumably the multiply electrically charged Planckian objects) in low–background conditions in September 2005 and March 2006 has provided supportive evidence for the expected to occur at that times maxima in the flux of daemons with $V \approx 10$–15 km/s. These objects hit the Earth from the near–Earth, almost circular heliocentric orbits (NEACHOs). The ability of some FEU-167-1 PM tubes with a thicker inner Al coating to detect directly (without a scintillator) daemon passage through them has also been employed, an effect increasing ~100-fold the detector efficiency. As a result, the daemon flux recorded at the maxima was increased from $\sim 10^{-9}$ to 10^{-7} cm^{-2}s^{-1}. At the maxima, two phases in the observed flux can be discriminated. The first of them is associated with objects moving in outer NEACHOs, which catch up with the Earth and cross it. The intensity and direction of the flux during this phase which lasts about two weeks depend on the time of day and latitude of observations (therefore, synchronous measurements in the northern and southern Earth's hemispheres are desirable). In the second phase, where the flux consists primarily of few objects transferred into geocentric, Earth-surface-crossing orbits during the first phase, the day-time and latitude dependence becomes less pronounced. All the experimental results thus obtained either support the conclusions following from the daemon paradigm or find a simple interpretation within it.

1. Introduction

At the IDM-2004 we presented the experimental results [1] that supported our working hypothesis suggesting that the DM of the galactic disk contains daemons, which are, presumably, the relict elementary Planckian black holes carrying a negative electric charge ($m_\delta \approx 3\times10^{-5}$ g, $r_g \approx 2\times10^{-33}$ cm, $Ze \approx 10$e). They are detectable since: (*i*) when passing through the matter the daemons capture the atomic nuclei with releasing the bond energy (up to ~100 MeV for Fe or Zn); this leads to the ejection of atomic (and numerous refilling in metals and molecular complexes) electrons and nuclear nucleons, which excite the intensive scintillation; (*ii*) when entering a nucleus with $Z_n \geq 24/Z$, the daemon finds itself inside the proton and induces its rapid decay; as a final result, the daemon "digests" an old nucleus, and becomes capable to capture another nucleus;

(*iii*) owing to a combined action of the Sun and the Earth, the daemons of the galactic disk (whose flux is ~10^{-12} $cm^{-2}s^{-1}$ only) become transferred and accumulated in the near-Earth orbits; therefore, according to rough estimates [2], their flux could exceed the primary one for 4 to 5 orders of the magnitude.

On the grounds of this ideology, a detector was created whose modules contained two detached ~10 μm thick layers of horizontal and parallel to each other scintillators made of ZnS(Ag) (see [1,3,4] for details). Its exposition under ground-based conditions during 2000-2004 revealed that [1,4]: (*i*) there really exists a low-velocity (~5 to 30 km/s) flux of high-penetrating particles whose most expressive portion has the velocity of ~10-15 km/s (time of passing through the detector is Δt ~ 30 μs) in agreement with falling out of particles from the near-Earth, almost-circular, heliocentric orbits (NEACHOs), (*ii*) their measured flux was $f_\oplus$ ~ 10^{-9} $cm^{-2}s^{-1}$, and (*iii*) the flux changes with the period of half-year and with maxima in March and September.

The periodicity could be understood if one takes into account that the capture in the near-Earth orbits is most probable in zones where the projection of the velocity of the captured by the Sun object upon the Earth's orbit reaches its maximum. This condition is satisfied, by accident, in the vicinity of the vernal /autumnal equinox points (in March and in September) for the objects of the galactic disk population, relatively which the Sun moves with the velocity of about 20 km/s. Being sufficiently braked as a result of passing through the Sun, the daemons at first fall in the strongly elongated unclosed orbits whose perihelia lie inside the Sun. If such the object passes through the Earth's sphere of action, it would deviate as much as it would never fall on the Sun. After subsequent re-approaches to the Earth, the daemon could finally occurred in the NEACHO traversing the Earth's orbit. The points of the orbits' intersection are concentrated in the zones indicated above [1].

Despite fully prospecting character of the work, the reliability of the above presented conclusions was 99.9%, and the inferences themselves are well agreed with the initial daemon paradigm (with, of course, correcting some details). The obtained results also lead one to the conclusions that (*i*) a daemon-containing proton decays within ~10^{-6} s; (*ii*) there would exist daemon kernels in the Earth, Sun and galaxies, what suggests an explanation for a great number of previously incomprehensible phenomena (see [4] and refs. therein); and (*iii*) the existence of daemons is supported in some other ground-based experiments.

In particular, the attempts to measure the mass of the anti-neutrino basing on the peculiarities of the "tail" of the tritium β-spectrum are of interest. A corresponded half-year modulation of its position (so called "Troitsk anomaly") was revealed [5]. Surprisingly, but practically all its features are explainable in the framework of the daemon paradigm if the daemons capture the Nb-containing atomic clusters in super-conducting magnetic winds of the installation and carry them through the channel of tritium source; Auger electrons of observed energy are emitted there in the process of "digesting" the cluster's nuclei. It has been shown (or, if you wish, predicted) that the flux of daemons required for the explanation of the Troitsk anomaly should reach $\sim 10^{-7}$-10^{-6} $cm^{-2}s^{-1}$ [6].

2. On High Efficiency of Vacuum Detector

Our detector is of low efficiency. The point is that, that the digestion of the nucleus captured by a daemon lengthens up to tens μs, and during a significant part of this time the negative charge of the daemon remains compensated by the charge of the captured nucleus. Such a "poisoned" daemon is passive, and it can not capture a new nucleus with emitting particles that excite the intensive scintillation. Being in the poisoned state, the $V \sim 10$ km/s daemon passes ~10-50 cm (in dependence on the mass of nucleus), while having the negative unit charge, the daemon passes before the nucleus capture about 1 to 3 mm in air and only ~1 μm in the condensed matter [7]. So, one can see that the efficiency of our double -screen detector is quite low ($\sim 10^{-3}$). It is also clear that the efficiency would increase dramatically if one evacuates the detector [8]. In this case, as having even partially digested previously captured nucleus (up to the restoration of the negative charge), the daemon would not be poisoned again by capturing the nucleus from air and, therefore, when penetrating into the 10 μm layer of ZnS(Ag) it would capture the nucleus with the probability of 100% and with exciting the strong scintillation. Moreover, in the time interval of the flight in vacuum, an additional digest of a few protons more in the residual of the captured nucleus would take place, so that the negative charge of the complex "daemon + carried nucleus" could reach several units.

Despite the apparent simplicity, our detector has many "latent" parameters. In order to reveal them and obtain deliberate and reliable results, as well as to learn particular features of the daemon-matter interaction, we often varied the parameters of our installation and over-particularly studied the influence of every part of the system on the final result.

An important observation was made in March 2005 [8]. In addition to the fact that these measurements in combination with observations of 2000-2003-2004 raised up to 99.99% the reliability of the +30 μs maximum existence, we found that the photo-multiplier tubes (PMTs) of the same trademark but issued from different lots or different manufacturers respond differently on the daemon passage despite their identical photometric parameters. We used PMTs of FEU-167 type with a photo-cathode section of ∅120 mm and 45 mm in length. This section is conjugated by a conic insertion of 25-30 mm long with a cylinder leg of ∅50 mm, which contains dynodes. The Al mirror is deposited on the inside of the cylinder and rear conic parts of the photo-cathode section. Its thickness recommended by the designer should be ~0.1 μm. However, the manufacturer sometimes disregards this recommendation. We revealed that just those PMTs that had the Al coating with the thickness of 0.5-1 μm response on the passage of daemons through them [8]. It is quite reasonable, since according to our calculations, the complex "residual nucleus/daemon" with $Z = -1$ and $V \sim 10$ km/s passes before the capture of the nucleus in metal just ~1 μm [7]. If the thickness of the Al coating is equal to ~0.1 μm, the daemon, as a rule, goes through the metal without capturing. Then the nucleus capture occurs in glass (dielectric) with emitting an insignificant amount of electrons from the nearest environment. If the event of capture takes place in Al (in metal), one could think that many refilling electrons would also be emitted. Thus, the PMT gives a stronger signal.

3. Experiment in the Baksan Neutrino Observatory (BNO)

In a natural attempt to perform the detection of daemons under low-background conditions, we seized a kind invitation of Dr. V. Kuzminov to carry out the experiment in the BNO in latitude 42° N at a depth of 400 MWE.

Only a single detecting module (viz., the #3rd one [8]) was brought to the BNO from St.Petersburg. Only a top scintillating plate with the 10 μm ZnS(Ag) layer faced upward was used. To reduce the irradiation of the scintillator by the radon-emitted α-particles, two sheets of transparent 50 μm Lavsan film were put on the ZnS(Ag) layer.

The lower FEU-167 of the #3 module played a role of the bottom detector of the daemon passage now. It was, as usual, positioned at the distance 29 cм from the top scintillating plate, but this time its photo-cathode was screened by two sheets of the 6 μm Lavsan film with deposited Al layers 0.05 μm thick. The module was exposed during September 2005 and March 2006. In September the unit was blown by the liquid nitrogen vapors; in March it was not blown.

The main results are depicted in histograms Fig. 1a, b. The 30 μs maximum is predominating as before thus indicating the correctness of our decision made in 2000 when the distance of 29 cm between the upper scintillator and the bottom cover of the tinned iron cubic casing of the module was selected as a basic distance. An unexpected feature was a mirror position (at -30 μs) of the maximum in the September histogram.

Initially we supposed that the mirror maximum appeared as a result of positioning the layer of ZnS(Ag) on the upper surface of its substrate made of polystyrene sheet as distinct from all preceding experiments; the polystyrene could

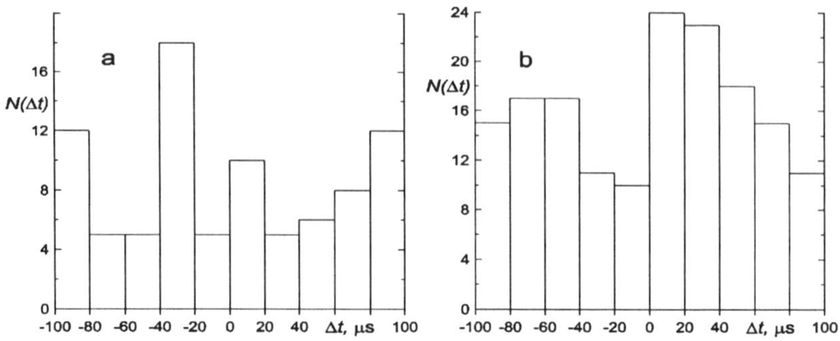

Figure 1. The $N(\Delta t)$ distribution of pair events in the time shift Δt between their beginnings. The top signal shape is typical for the heavy particle scintillations [3]. a – from 3 to 11 September 2005; b – from 3 to 18 March, 2006.

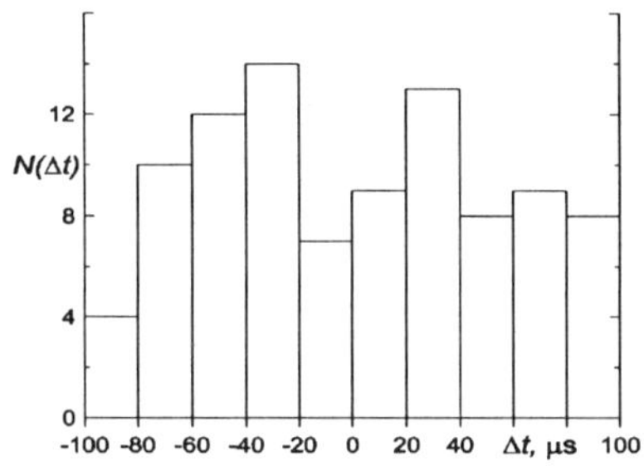

Figure 2. The same as Fig. 1, but for the time span from 18 to 24 March, 2006.

exhibit, to some extent, properties similar to vacuum in view of the low probability of the capture of carbon nucleus (the excitation potential of its first level reaches 4.4 MeV). However, the disappearance of the -30 μs maximum and the appearance of the habitual +30 μs in March 2006 rejected this supposition.

These facts could be understood if using the PMT as a sensitive element abolished the strong asymmetry of the detector sensitivity relative the upward /downward directions, and in the first half of September we really detect the daemon flux passing through the Earth bottom-up. [Besides, this favours our old interpretation of the signal excitation in the lower scintillator by the electrons emitted when a daemon falls on the lower cover of the tinned iron casing of the unit (when using a routine arrangement of the detector with two scintillating screens with the layers ZnS(Ag) faced downward [1,3,4,7]); when the daemon enters the iron cover from the bottom, the emitted electrons are captured in it.]

As far as in March the prevailing flux comes from above, and in September from below, and taking into account that the BNO is situated in the northern hemisphere, one should conclude that the daemons overtake the Earth during its motion in orbit (one should remember that in March the northern hemisphere "looks" back in regard to the direction of the Earth orbital motion, with an opposite situation (looks "forward") in September). This could take place if the principal NEACHO population of daemons is concentrated in the external NEACHOs, which only touch the Earth's orbit in March and in September. The accumulation of daemons in the external NEACHOs could be understood qualitatively if one assumes that they face difficulties when going out of the vicinity of the Earth's orbit to the zone of external planets. Indeed, to go there from the NEACHO, a daemon must obtain the velocity increment $\Delta V \approx V_{\oplus orb}(2^{1/2} - 1) = 12.3$ km/s, while a single passage through the Earth's sphere of action changes the value of its velocity for less than 11.2 km/s. Naturally, the problem needs to be considered in more details as well as the whole problem of the dependence of the near-surface flux of daemons on the season and latitude.

These underground experiments supported also our previous conclusion [7] that in addition to the rather well oriented flux of daemons falling onto the Earth from the NEACHOs and passing through its body, there exists a population captured in geocentric Earth-surface crossing orbits (GESCOs). The latter appears as a result of braking with resulting capture of a part ($\sim 10^{-3}$) of the NEACHO objects passing through the Earth. Inasmuch as the period of motion in GESCOs is measured in days and even in hours, the flux of this population loses rapidly its primary orientation, so that objects go both from the top and from the bottom. (See the data of the second half of March 2006 in Fig. 2. There are no similar data for September 2005, since that time our attention was focused on the reasons for the excess of events in $0 < \Delta t < 20$ μs bin. We came to the conclusion that this excess is not of the daemon-related nature but probably caused by the decay of certain short-term nuclides in the ZnS(Ag) layer mainly [9]).

4. Some Conclusions

When organizing the experiment in the BNO, we made bold to take risks as accepted two modifications of the system: (*i*) instead of the bottom scintillator, a "blind" PMT with thickened internal Al coating was used, and (*ii*) the top scintillating plate was turned over (layer ZnS(Ag) upward) and covered by the Lavsan sheets.

Our risk occurred to be justified. This evidences the adequate understanding of processes of interaction between the daemons and matter in the detector.

The underground experiments in the BNO brought the following results:

1. A possibility to use as vacuum sensing elements the PMTs, whose the photo-cathode section inner surface is coated by sufficiently thick metal layer (~1 μm), for detecting the daemons was shown.

2. As a result of using such the PMT, the sensitivity of the detector increased by a factor of $\sim 10^2$, so that the detected in September and March flux of the NEACHO daemons increased from $\sim 10^{-9}$ to $\sim 10^{-7}$ $cm^{-2}s^{-1}$ (the calculations were performed for the photo-cathode area of 110 cm^2 using the data of Fig. 1a). The value $f_\oplus \approx 10^{-7}$ $cm^{-2}s^{-1}$ is in agreement with the rough analytical estimate [2], as well as with the values predicted in February 2005 on the base of the analysis of the Troitsk anomaly [5] in the tritium β-spectrum [6].

3. It was found that in September the primary flux of daemons goes from underground, while in March it falls from above. Hence, the detected daemons moved in the external NEACHOs, pursuing the Earth in its orbital motion. In this light, the synchronous measurements in northern and southern hemispheres are highly desirable. The appearance itself of the maxima in September and March argues for capturing the NEACHO daemons from the population of the galactic disk by the Solar system moving relatively it.

4. A part of daemons of the primary NEACHO flux are captured in the GESCOs; they lose their primary orientation, shrink, and recede into the Earth. In the Earth's centre these daemons form a kernel whose existence explains numerous geophysical phenomena.

Although not all the results obtained now have a high confidence level, they fit unconstrainedly the daemon paradigm and point out directions for planning further experimental and theoretical studies.

Acknowledgments

The authors are deeply indebted to Dr. V.V. Kuzminov, Director of the BNO, for the help in organizing the experiments in the BNO. EMD thanks the Workshop Organizers and the Russian Foundation for Basic Research (Grant 06-02-272443z) for the financial support of his attending the IDM2006.

References

1. E.M.Drobyshevski, in: *Proc. of the 5th Intnl. Workshop on Identification of Dark Matter*, N.J.C.Spooner & V.Kudryavtsev (eds.), World Scientific, pp.408-413 (2005).

2. E.M.Drobyshevski, in: *Dark Matter in Astro- and Particle Physics*, H.V. Klapdor-Kleingrothaus & Y.Ramachers (eds.), World Scientific, pp.417-424 (1997).
3. E.M.Drobyshevski, *A&A Trans.*, **21**, 65-73 (2002).
4. E.M.Drobyshevski, *astro-ph*/0402367 (2004).
5. V.Lobashev, *Yadernaya Fizika (Physics of Atomic Nuclei)*, **63**(6), 1037-1043 (2000).
6. E.M.Drobyshevski, *hep-ph*/0502056 (2005).
7. E.M.Drobyshevski *et al.*, *A&A Trans.*, **22**, 19-32 (2003).
8. E.M.Drobyshevski, *A&A Trans.*, **25** (in press); *astro-ph*/0605314 (2006).
9. E.M.Drobyshevski & M.E.Drobyshevski, *A&A Trans.*, **25** (in press); *astro-ph*/0607046 (2006).

THRESHOLD IMPROVEMENTS IN SCINTILLATION EXPERIMENTS BY PHOTON INDIVIDUATION

C. POBES*, J. AMARÉ, S. BORJABAD, A. CALLEJA, E. GARCÍA, M. MARTÍNEZ, J. MORALES, A. ORTIZ DE SOLÓRZANO, J. PUIMEDÓN, M. L. SARSA and J. A. VILLAR

Laboratorio de Física Nuclear y Astropartículas.
University of Zaragoza, 50009 Zaragoza, Spain
**E-mail: cpobes@unizar.es*

Dark matter experiments face two main challenges: low rate and low energy detection. Background reduction techniques are widely in development in order to improve sensitivity. Energy threshold is more technique-dependent and so it requires specific efforts for each detection approach. In this work we present some improvements in this line that open the possibility of further reduction in the energy threshold of scintillation experiments. Some preliminary results within the projected NaI(Tl) experiment ANAIS are reported.

Keywords: Dark matter; scintillator; photon detection; energy threshold.

1. Introduction

Direct detection of Dark Matter is a cornerstone in today Physics. Several experiments using different techniques are dedicated to this goal: bolometers, scintillators, semiconductors or combination of some of them.[1] In any case, all have to face very low expected rates and very low energy depositions. To be sensitive to low rates, it is necessary to reduce background to very low levels. This is achieved by going underground, shielding properly the detectors, using clean materials and applying discrimination techniques when possible. All this approaches are somehow common to all experiments of rare events search. On the other hand, the detection of very low energy depositions is directly related to the energy conversion mechanism in the detector and so energy threshold is technique dependent.

In this presentation we focus on the optimization of the threshold in scintillation experiments. We show that photon counting could be used to improve analysis near threshold. The ideas presented have started to be used independently in different scintillator materials[2,3] and we discuss here the main common issues as well as the difficulties in applying the method to NaI(Tl). Some preliminary results in the framework of the ANAIS[4] experiment are presented.

2. Threshold in scintillation experiments

The minimum energy that can be detected with a scintillator (threshold) is directly related to the light quanta emitted per unit energy deposited (scintillation yield). Another important characteristic of scintillation is the decay time constant with which the photons are emitted. Both parameters for some scintillator materials are summarized in Table 1.

Table 1. Properties of some inorganic scintillators.

Scintillator	γ/keV	$\tau_s(ns)$
NaI(Tl)	43	230
CsI(Tl)	51.8	1000
$CaWO_4$	6	6000

Note: Extracted from ref.[5]

Light is converted into a readable signal usually with a photomultiplier (PMT) that produces an electrical pulse. A certain amount of light is lost in the propagation and around 75% is also lost in the photocathode. This imposes a restrictive limit to the energy threshold that can be achieved. A standard measure of the efficiency of the collection process is the so called photoelectrons per keV (γe/keV).

For NaI(Tl), typical values of 5-6 γe/keV are obtained for set-ups with two PMTs working in coincidence, although higher values may be reached. In this case, at 2 keV which is the lowest threshold usually considered in literature, the pulses consist of about 10 γe. Nonetheless, the high gain of photomultipliers makes single photoelectron detection possible. Several reasons make the analysis threshold to be higher than this hardware threshold. For instance, the difficulty to interpret the spectra with few photoelectrons, the presence of noise, or the sensitivity of the electronics (QDC, etc). Photoelectron individuation could improve the situation in some of these points.

3. Time structure at low energy

Ideally the pulse shape of a scintillation event is an exponential decay with the time constant of the scintillation emission, but at low energy very few photons are collected. Taking into account the width of the time response of typical photomultipliers ($\sim 10\ ns$) the individual arrival of photoelectrons is observed (see Fig. 1). Exploiting this peculiar shape, by integrating only the area under the peaks (A_{be}), (see Fig. 1 right) or ideally counting the number of photoelectrons, could improve both energy threshold and resolution at low energies. In the case of CsI(Tl)[2] or $CaWO_4$[3] for instance, the time constant is large enough to prevent photoelectron pulses from overlapping. In the case of NaI(Tl), peaks can overlap even with a low number of them (see Fig. 2). In this case, the area under the peaks becomes a good estimator for the energy, but for photon counting, a way to extract the underlying number of photoelectrons in the digitized forms must be implemented.

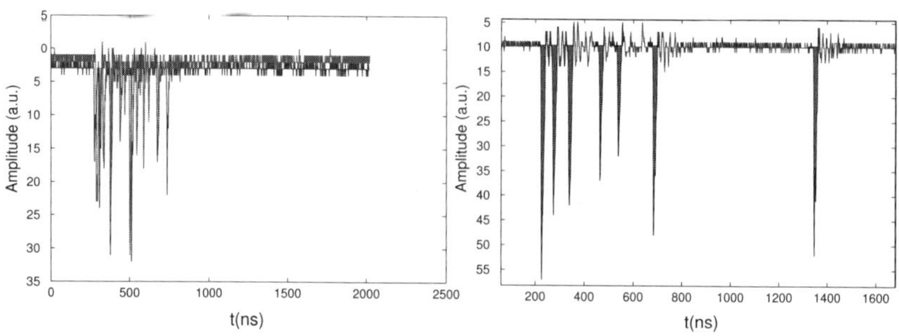

Fig. 1. ^{55}Fe pulse with the two channels superimposed (left) and integration of peaks (filled area) in a very low energy pulse (right).[6]

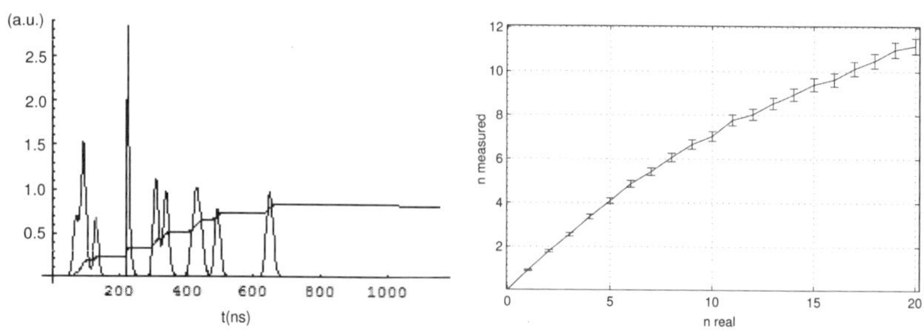

Fig. 2. Simulation of a pulse with 10 photoelectrons in NaI(Tl) (left) and simulation of the number of detected peaks as a function of the actual number for 10 ns time response width PMTs (right).[6]

4. Preliminary measurements

Within the ANAIS[4] experiment we performed several tests with a prototype. Gamma calibrations of a NaI(Tl) crystal with two photomultipliers were carried out at low energy and also measurements with the light of a pulsed LED adjusted at a low mean number of photons. The pulses were digitized and also integrated with a QDC for each of the two PMTs working in coincidence. The results (see Fig. 3) show the improvement in resolution at very low energy when using the area under the peaks. A relative variance $v = 0.200$ is obtained for the peak at channel $\sim$1000 for the sum of the QDCs while $v = 0.172$ is obtained for the peak around -2000 with the other method. Besides, down to very few photoelectrons can be identified in the spectra (see Fig. 4).

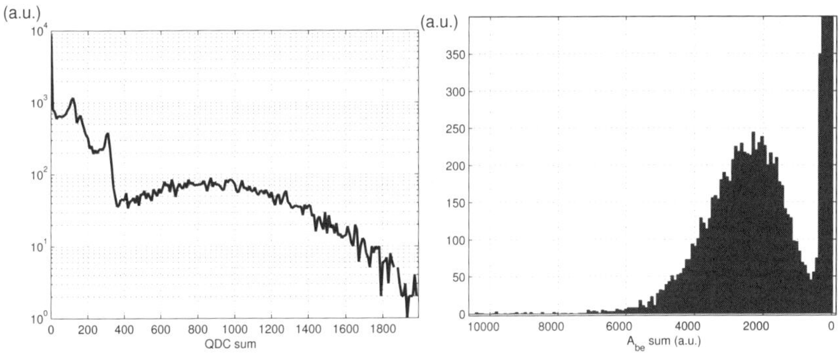

Fig. 3. Comparison of the spectra obtained with the QDCs (left) and integrating the peaks (negative area) in the digitized pulses (right) for a ^{55}Fe source.[6]

4.1. *Prospects for sensitivity improvements*

The number of photoelectrons at low energy is not gaussian distributed and the spectra can have structure. Besides, electronic noise or noise coming from the PMTs will be present. Applying Poisson statistics, and using the shape of pulses at low energy, the analysis could be extended down to a very few number of photoelectrons improving the energy threshold.[6,7] With this method it could be possible to construct the spectra in terms of photons down to very low light levels and translate it into energy with the photoelectrons per keV measured. It is possible to convert the expected rate per energy bin into a rate per number of photoelectron detected intro-

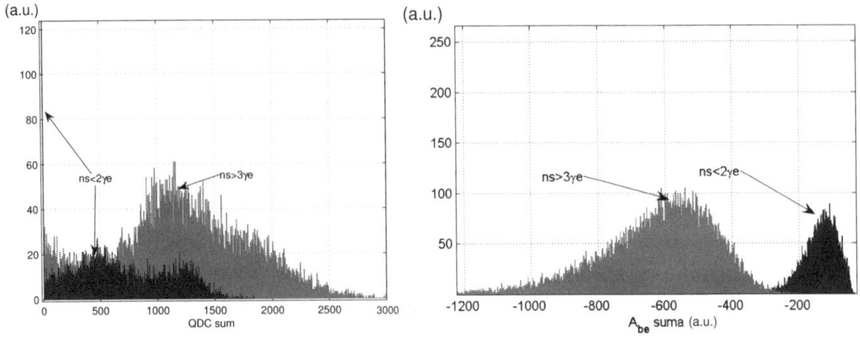

Fig. 4. Comparison of spectra obtained with the sum of the QDCs (left) or integrating the peaks (negative area) in the digitized pulses (right) using a pulsed LED. Populations with less and more than 3 peaks are highlighted.[6]

ducing Poisson statistics. In this way it could be feasible to reduce energy thresholds with the corresponding gain in sensitivity (see Fig. 5). Of course, a detailed control of the threshold, and evaluation of the efficiency of cuts at that energy should be guaranteed.

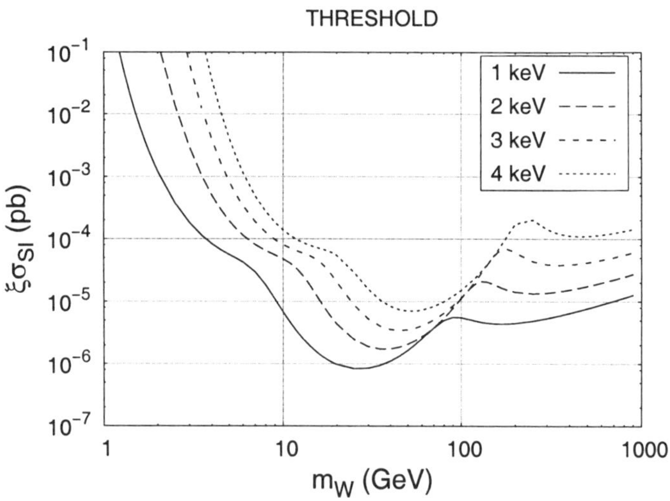

Fig. 5. Sensitivity plot for different energy thresholds achievable with the ANAIS experiment for pure spin independent coupling WIMPs. A background level of 1 $counts\ keV^{-1}kg^{-1}day^{-1}$ and a exposure of 5 years with 10.7 Kg of NaI have been assumed.[8]

5. Conclusions

In this short talk we propose some ideas to improve the energy threshold and extend the analysis of the spectra down in energy in direct dark matter scintillation experiments. Photon counting has been applied in various experiments and we show the difficulties in applying the method to NaI(Tl). Nonetheless, preliminary results within the ANAIS experiment show the interest of this technique for threshold reduction and so sensitivity improvement.

Acknowledgements

This research was founded by the Spanish Commission of Science and Technology (MCyT) under contracts No. FPA2001-1767 and FPA2004-0974.

References

1. S. Fiorucci, Direct non-baryonic dark matter search: An experimental review (2004), eprint: astro-ph/0406285.
2. H. Lee *et al.*, *Phys. Lett. B* **633**, 201 (2006).
3. J. Ninković *et al.*, *Nucl. Inst. and Met. A* **564**, 567 (2006).
4. J. Amaré *et al.*, *Journal of Physics: Conference Series* **39**, 123 (2006).
5. P. Lecoq *et al.*, *Inorganic Scintillators for Detector Systems* (Springer, Berlin, 2006).
6. C. Pobes, Optimización de la respuesta de un sistema multidetector de cristales de centelleo para un experimento de materia oscura (ANAIS), PhD thesis, University of Zaragoza, (Zaragoza, Spain, 2006).
7. J. D. Lewin and P. F. Smith, *Astrop. Phys.* **6**, 87 (1996).
8. M. Martínez, Diseño de un prototipo para un experimento de detección directa de materia oscura mediante modulación anual con centelleadores de ioduro de sodio, PhD thesis, University of Zaragoza, (Zaragoza, Spain, 2006).

MEASUREMENT OF THE SCINTILLATION EFFICIENCY OF Na RECOILS IN NaI(Tl) DOWN TO 10 keV NUCLEAR RECOIL ENERGY RELEVANT TO DARK MATTER SEARCHES

H. CHAGANI*, P. MAJEWSKI**, E. J. DAW, V. A. KUDRYAVTSEV and N. J. C. SPOONER

Department of Physics and Astronomy, University of Sheffield, Hicks Building, Hounsfield Road, Sheffield S3 7RH, United Kingdom
**E-mail: h.chagani@sheffield.ac.uk*
***E-mail: p.majewski@sheffield.ac.uk*

We present preliminary results of measurements of the quenching factor for Na recoils in NaI(Tl) at room temperature, made at a dedicated neutron facility at the University of Sheffield. Measurements have been performed with a 2.45 MeV mono-energetic neutron generator in the energy range from 10 keV to 100 keV nuclear recoil energy. A BC501A liquid scintillator detector was used to tag neutrons. Cuts on pulse-shape discrimination from the BC501A liquid scintillator detector and neutron time-of-flight were performed on pulses recorded by a digitizer with a 2 ns sampling time. Measured quenching factors range from 19% to 26%, in agreement with other experiments. From pulse-shape analysis, a mean time of pulses from electron and nuclear recoils are compared down to 2 keV electron equivalent energy.

1. Introduction

Thallium activated Sodium iodide (NaI(Tl)) crystals are a popular choice as a target material for dark matter experiments. This is because of their relatively high light yield and pulse shape differences between nuclear and electron recoils. Currently three experiments utilise these crystals: ANAIS [1]; DAMA/NaI [2]; and ELEGANT-V [3]. Although better discrimination can be reached with other target media, NaI(Tl) crystal-based detectors remain one of the best at determining spin-dependent WIMP-nucleon limits. For instance, for direct detection techniques, the NaIAD experiment [4] still holds the best spin-dependent limit on WIMP-proton interactions [5]. Hence, NaI(Tl) remains an important detector material in non-baryonic dark matter searches.

2. Measuring Scintillation Efficiencies

An important measurement to determine a scintillating target's sensitivity to dark matter particles is the ratio of light induced by a nuclear recoil (produced by WIMP/neutron collisions with nuclei) to that by an electron recoil of the same energy. This is known as the quenching factor.

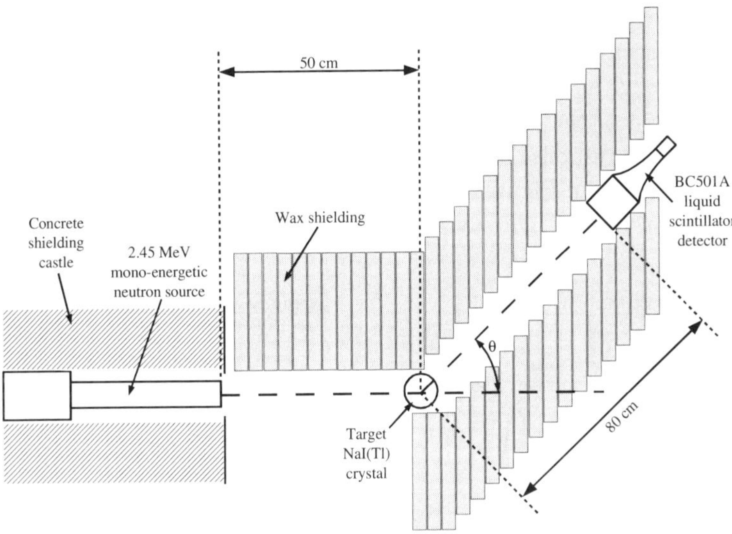

Fig. 1. Schematic view of the detector arrangement used to measure the scintillation efficiency of nuclear recoils in NaI(Tl).

The 2-inch diameter, encapsulated, NaI(Tl) crystal is placed in the path of a 2.45 MeV mono-energetic deuterium-deuterium neutron beam as shown in Figure 1. A 3-inch ETL 9265KB photomultiplier tube is optically coupled to the crystal. Neutrons that interact with the target nuclei scatter off at an angle that is dependent on the deposited neutron energy:

$$E_R \approx \frac{2m_A E_n m_n}{(m_A + m_n)^2} \cdot (1 - \cos\theta) \tag{1}$$

where E_R is the recoil energy, m_A is the mass of the target nucleus, E_n is the energy of the neutron beam, θ is the scattering angle and m_n is the mass of the neutron.

A BC501A liquid scintillator detector is placed at various angles around the target to detect scattered neutrons at nuclear recoil energies given by Eq. (1). Pulses from the target and liquid scintillator that are coincident within a 100 ns time window are sent to a 2-channel, 8-bit, Acqiris digitizer with a 500 MHz sampling rate, and written to disk.

3. NaI(Tl) Response to Electron Recoils

It is known that the response of NaI(Tl) to gamma-rays is non-linear at low energies [6,7]. The crystal is exposed to gamma-rays from a variety of low energy sources, between 29 keV (^{129}I Xe K_α X-rays) and 122 keV (^{57}Co γ-rays). A decrease in detector response is observed at the iodine K-shell absorption edge at 33.2 keV, consistent with other studies [6,7]. Therefore, determination of the energy scale must be performed in an energy region where a linear response is observed. The crystal is calibrated with a ^{57}Co source, as shown in Figure 2(a). The iodine escape peak at 90 keV is also seen. A light yield of 5.5 photoelectrons/keV is found.

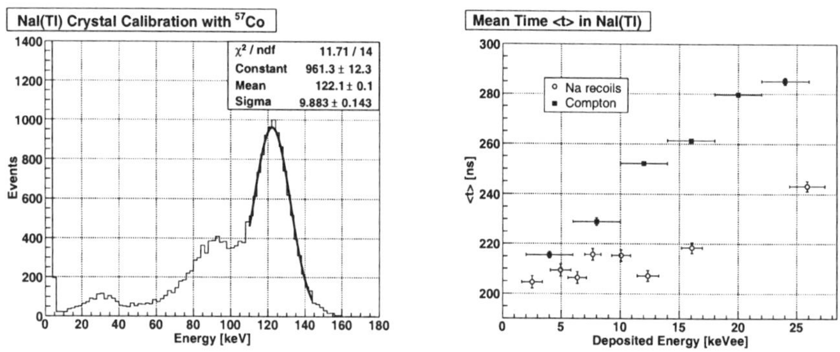

(a) (b)

Fig. 2. (a) Results from calibration of detector with 122 keV line from ^{57}Co source. (b) Mean time of the pulses as a function of deposited energy for sodium (Na) recoils and Compton electrons. Measurements of Na recoils are performed with the neutron beam. Compton electrons are induced by a ^{22}Na source.

4. Quenching Factor Measurements

The nuclear recoil energy spectrum measured at each scattering angle is converted to an electron equivalent energy scale (keVee) and compared to the expected nuclear recoil energy (keVnr) at that angle. The ratio between the measured nuclear recoil energy, and that calculated from Eq. (1), is the quenching factor.

In order to eliminate the background from gamma-ray interactions, nuclear recoils are discerned by discrimination of pulse shapes from the

BC501A detector and time-of-flight measurements.

Events that arise from gamma-ray interactions in the BC501A detector have shorter decay times relative to those from neutron collisions. Adequate discrimination between nuclear and electron recoils is achieved by integrating over the tail of the pulse. The ratio between this partial area and the total pulse area is then plotted as a function of the total area. An example is illustrated in figure 3(a).

Unlike gammas, neutrons that scatter off the target in this experiment are non-relativistic. Therefore, after interacting with a sodium nucleus in the crystal, a neutron takes approximately 40 ns to travel the 80 cm distance to the BC501A detector. By taking the time difference between coincident events in the target and BC501A detector, as shown in Figure 3(b), neutron and gamma events are separated.

Better discrimination at energy scales relevant to dark matter searches has been demonstrated in NaI(Tl) crystals using the mean time $\langle t \rangle$ [6,8] rather than the traditional double charge method [9]. The mean time is defined as: $\langle t \rangle = \frac{\sum_i A_i t_i}{\sum_i A_i}$, where A_i is the amplitude of the digitized pulse at the time bin t_i. After gamma events have been rejected by performing the cuts outlined above, mean time distributions for each nuclear recoil energy are investigated. An example is shown in Figure 3(c). This is compared with the mean time of Compton recoils induced by a ^{22}Na source in Figure 2(b). In agreement with previous work [10], it is clear that this difference is less prominent at energies less than 10 keVee, limiting the discrimination power of NaI(Tl) detectors. This cut serves to reduce the background from noise and gamma pulses, leading to a nuclear recoil peak as shown in Figure 3(d).

5. Results

The quenching factor varies between 19% to 26% in the range 10 keV to 100 keV nuclear recoil energy, which agrees with previous experimental results [6,10–12], as shown in Figure 4. From Figure 3(d), a scintillation efficiency of $25.2 \pm 6.4\%$ has been determined for 10 keVnr Na recoils.

Figure 4 shows that the quenching factor rises at low energies, agreeing with the trend outlined by the theoretical upper limit [13].

6. Summary

Scintillation efficiency measurements have been performed for Na recoils in a NaI(Tl) crystal. The results show an average value of 22.1% at energies less than 50 keVnr, and are in agreement with other measurements. Future

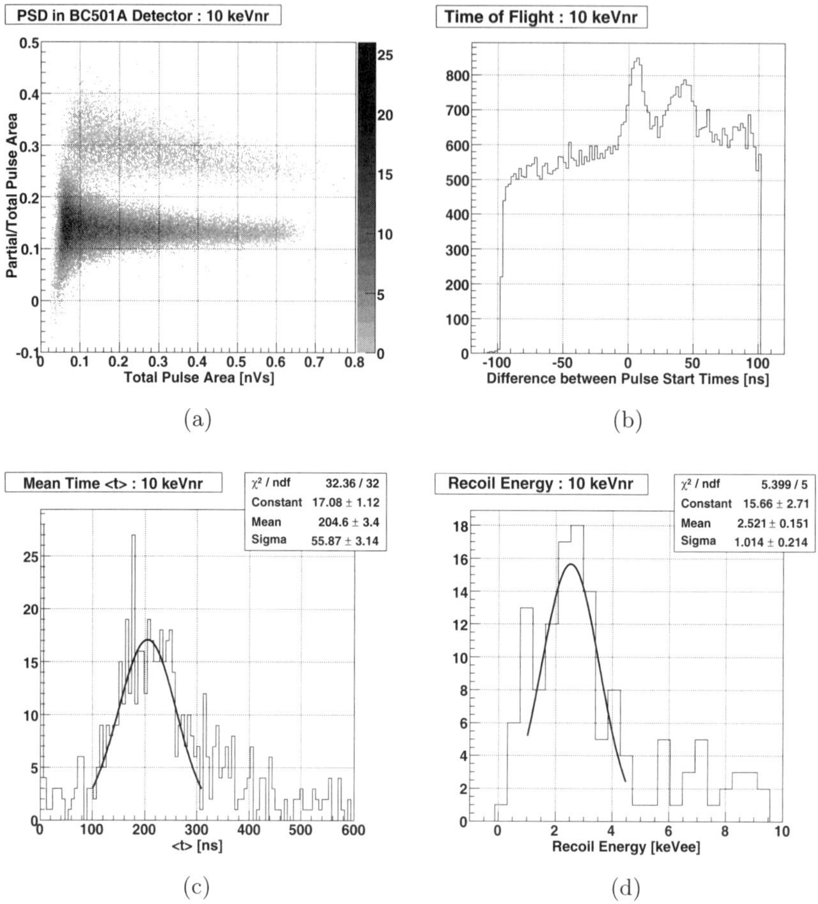

Fig. 3. (a) Pulse shape discrimination in the BC501A liquid scintillator detector for 10 keVnr Na recoils. The upper neutron event band is clearly distinguishable from gamma interactions. (b) Time-of-flight between the BC501A detector and NaI(Tl) crystal. The peaks at roughly 0 and 40 ns represent gamma and neutron events respectively. (c) Mean time of pulses from 10 keVnr Na recoils in the NaI(Tl) crystal. (d) Recoil energy in electron equivalent scale.

plans include simulating nuclear recoils in NaI(Tl) in an effort to decrease the errors in quenching factors.

Acknowledgements

The authors would like to thank Akira Hitachi for valuable discussions of the results, and the provision of an upper limit for the quenching factor. HC would also like to thank PPARC for a PhD studentship.

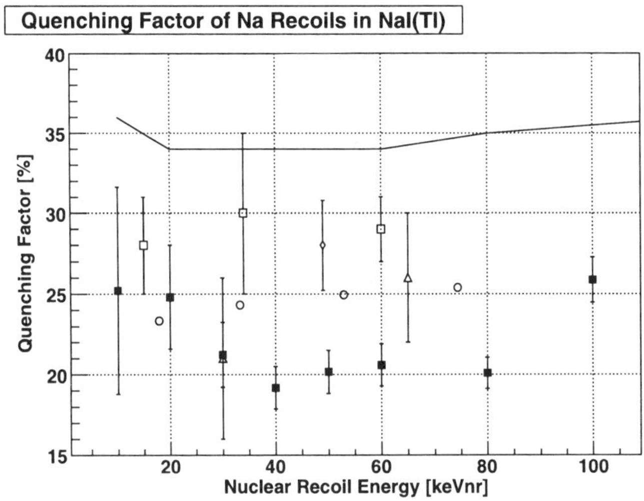

Fig. 4. Quenching factor measurements for Na recoils in NaI(Tl). Results from this work (closed black squares), Simon et al. [11] (open diamond), Gerbier et al. [6] (open circles), Tovey et al. [10] (open triangles), Spooner et al. [12] (open squares) and the theoretical prediction of an upper limit from Hitachi [13] (solid black line) are shown.

References

1. M. Martínez *et al.*, in these proceedings
2. R. Bernabei *et al.*, Phys. Lett. B **424** (1998) 195
3. K. Fushimi *et al.*, Astropart. Phys. **12** (1999) 185
4. B. Ahmed *et al.*, Astropart. Phys. **19** (2003) 691
5. G. J. Alner *et al.*, UKDMC, Phys. Lett. B **616** (2005) 17
6. G. Gerbier *et al.*, Astropart. Phys. **11** (1999) 287
7. L. R. Wayne *et al.*, Nucl. Instrum. Methods Phys. Res. A **411** (1998) 351; J. D. Valentine *et al.*, IEEE Trans. Nucl. Sci. **45** (1998) 1750; L. F. R. Ferreira *et al.*, Nucl. Instrum. Methods Phys. Res. A **516** (2004) 486
8. V. A. Kudryavtsev *et al.*, Phys. Lett. B **452** (1999) 167
9. C. L. Morris *et al.*, Nucl. Instrum. Methods **137** (1976) 397; M. S. Zucker & N. Tsoupas, Nucl. Instrum. Methods Phys. Res. A **299** (1990) 281
10. D. R. Tovey *et al.*, Phys. Lett. B **433** (1998) 150
11. E. Simon *et al.*, Nucl. Instrum. Methods Phys. Res. A **507** (2003) 643
12. N. J. C. Spooner *et al.*, Phys. Lett. B **321** (1994) 156
13. A. Hitachi, in these proceedings

CONSTRAINING KALUZA-KLEIN DARK MATTER THROUGH MULTIWAVELENGTH OBSERVATIONS OF SGR A*

L. PIERI*

INAF - Astronomical Observatory of Padova,
Vicolo dell'Osservatorio 5, 35122 Padua, Italy
and INFN- sezione di Padova
**E-mail: lidia.pieri@oapd.inaf.it*

In this paper we show the possibility of constraining Dark Matter properties through multiwavelenght observations of the Sgr A* region. We consider Kaluza-Klein Dark Matter annihilation and study the resulting X-ray synchrotron and gamma-ray emission. We show how the combination of these observations puts severe constraints on the shape of a Kaluza-Klein Dark Matter halo.

Keywords: Kaluza-Klein Dark Matter; X-rays; Gamma-rays; Galactic Center.

1. Introduction

Kaluza-Klein particles (KKP) in theories where there is a TeV^{-1} size universal extra dimension into which all standard model fields propagate are exciting candidates for the Dark Matter (DM).

Though the structure of the DM halo is unknown, N-body simulations suggest that the Galactic Centre (GC) may contain a DM cusp.

On the other hand it is clear from the observations that our Galaxy hosts a several times 10^6 $M_\odot$ supermassive black hole (SMBH) at the position of Sgr A*. The products of any DM annihilations will therefore be injected into the plasma falling into the central black hole.

In this work we calculate the signal expected from the synchrotron radiation due to the annihilation of KKP into electrons near the GC. All the references and details can be found in Ref. 1.

2. Multiwavelenght Observations of Sgr A*

Multi-wavelength observations of the GC (Fig. 1) do not lead to a unique interpretation. It is thought that the sub-Eddington accretion flow onto the black hole is fueled by stellar mass loss from the cluster of large mass stars which exists in that region. The radio, mm and infrared radiation is thought to come from the inner regions of this flow close to the black hole, whereas the X-ray emission observed by Chandra is thought to originate further from the black hole, close to the Bondi radius at the interface between the spherical inflow region and the stellar winds. The Bondi radius is thought to be at around 0.04 pc from the central black hole, rather close to the resolution of the Chandra telescope at this distance. HESS observations which have recently been confirmed by the MAGIC experiment show that there is significant TeV gamma-ray emission from the central 30 pc around the black hole. This emission might be due to the annihilation of DM or might have a more mundane origin, being created in the interactions of very high energy protons with the ambient photons and magnetic fields, or in proton-proton interactions or in bremmstrahlung and Inverse Compton processes involving high energy electrons, etc.

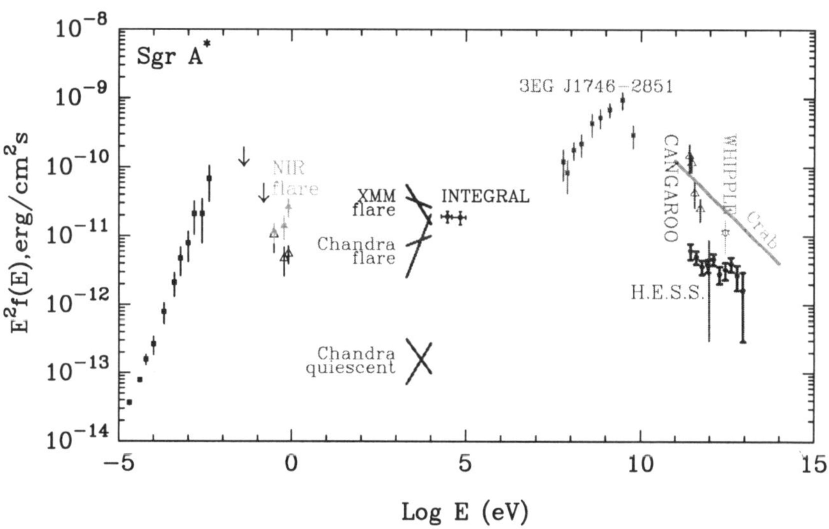

Fig. 1. Multiwavelength luminosity of Sgr A* in the quiescent state. Figure taken from Ref. 2, to which we refer for any reference.

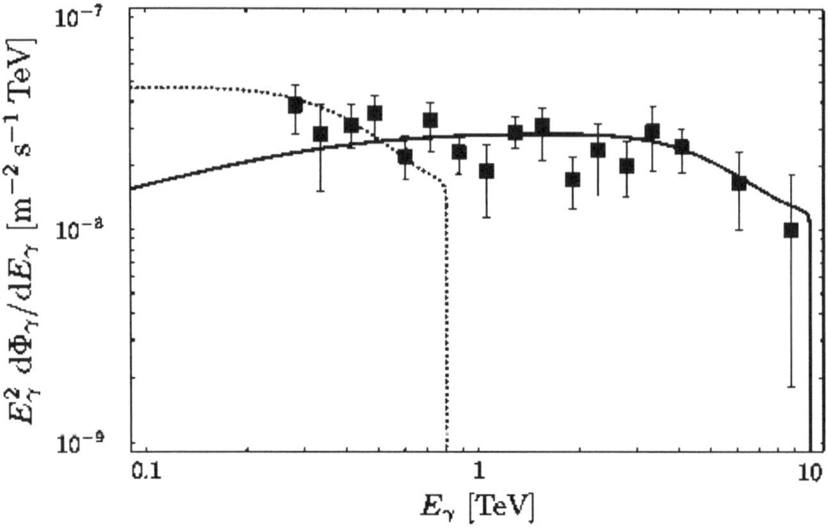

Fig. 2. HESS data compared with the gamma-ray flux expected from the annihilation of a 800 GeV KKP (dotted line), for an inner DM density slope of about 1.5, and from a hypothetical 10 TeV WIMP (solid line) which would require a much higher slope. Figure taken from Ref. 3.

3. X-Ray Synchrotron Radiation from KKP Annihilation

3.1. *The DM Density Profile*

We consider three density profiles for the DM distribution, all of which can be parametrised rather simply by the following expression

$$\begin{aligned} \rho(r) &= \rho(100\text{pc}) \left(\tfrac{100\text{pc}}{\text{r}}\right)^{\gamma_1} && r > r_{out} \\ \rho(r) &= \rho(r_{out}) \left(\tfrac{r_{out}}{r}\right)^{\gamma_2} && r_{out} > r > r_{in} \\ \rho(r) &= \rho(r_{in}) && r_{in} > r \end{aligned} \tag{1}$$

The three models we will consider are: A) the standard NFW $\gamma = 1$ profile with no adiabatic contraction or central spike: $\rho(100\text{pc}) = 25$ GeV/cm^3, B) The same $\gamma_1 = 1$ profile, but now with a central spike which has diffused away over time, considerably reducing its density: $\gamma_2 = 1.85$, $r_{out} = 7 \times 10^4 R_{BH}$, $r_{in} = 10 R_{BH}$, ($R_{BH} \sim 7 \times 10^{11}$ *cm* is the Schwarzschild radius for the black hole) C) A profile which has undergone adiabatic contraction on galactic scales due to the presence of baryons, and also has a central spike, also allowed to diffuse away over time: $\gamma_1 = 1.5$, $\gamma_2 = 1.82$, $\rho(100\text{pc}) = 360$ GeV/cm^3, r_{out} and r_{in} as for profile B.

3.2. *The Magnetic Field*

We consider two models for the Magnetic field at the GC: an equipartition magnetic field of strength

$$B_{eq}(r) = 3.9 \times 10^{-2} \left(\frac{0.01\text{pc}}{r}\right)^{\frac{5}{4}} \text{Gauss} \tag{2}$$

and a modified magnetic field of the form

$$\begin{aligned} B_m(r) &= B_{eq}(r) \quad r > 10^3 R_{BH} \\ B_m(r) &= B_{eq}(10^3 R_{BH}) \quad 3R_{BH} < r < 10^3 R_{BH} \\ B_m(r) &= B_{eq}(3R_{BH}) \left(\frac{r}{3R_{BH}}\right)^{-3} \quad R_{BH} < r < 3R_{BH} \end{aligned} \tag{3}$$

This comes from the consideration that the equipartition picture may not be correct at very small distances from the GC, where magnetic field line reconnection in the turbulent plasma may reduce the magnetic field.

3.3. *The Diffusion-Loss Equation*

When considering direct DM annihilation into electrons we will be interested in a delta function of electrons with energy m_{dm}. The solution of the diffusion-loss equation has then the following form

$$\frac{dn}{dE}(E,r) = \frac{1}{2}\left(\frac{\rho(r)}{m_{dm}}\right)^2 \langle\sigma_{tot}v\rangle N_{ee} b_{ee} \frac{1}{\dot{E}} \text{cm}^{-3}\text{GeV}^{-1} \tag{4}$$

which is valid over a range of energies, $E < m_{dm}$. $N_{ee} = 2$ is the overall number of electrons and positrons produced in each annihilation and $b_{ee} = 0.19$ is the branching ratio of annihilation in the electron-positron line.

The simplicity of the equation has been justified in Ref. 1, where it has been shown that synchrotron energy loss is dominant over Synchrotron Self-Compton and Inverse Compton Scattering off CMB and star photon, and that the synchrotron lenght scale is much bigger than the diffusion lenght scale, so that the diffusion terms can be ignored.

4. Results and Conclusions

Figure 3 shows the synchrotron spectra from a 1 TeV KKP annihilation at the GC, for the different density profiles described in Eq. 1 and for the different magnetic field assumptions of Eqs. 2 and 3, and the comparison to the CHANDRA data. We find that the steepest profile which is compatible with the x-ray data is $r^{-1.35}$. The gamma rays produced by such a profile

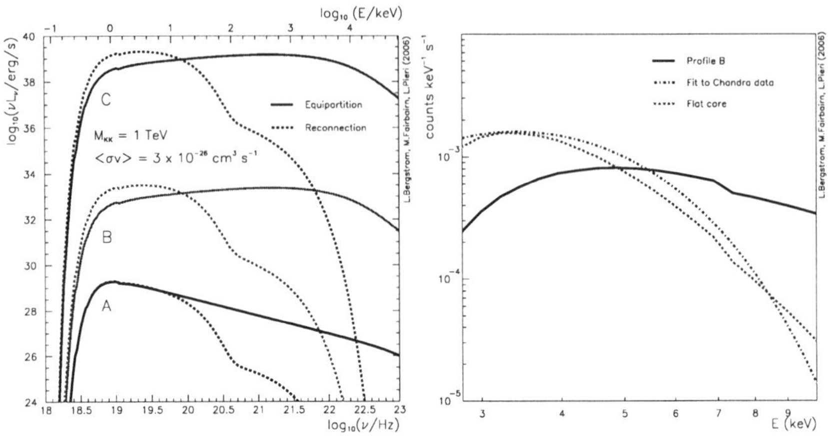

Fig. 3. Left: Synchrotron spectra from 1 TeV KK DM annihilation in the central 0.01 pc of the galaxy assuming the three density profiles (A,B,C) described in Eq. 1. The solid lines correspond to the equipartition magnetic field of Eq. 2 and the dotted lines are the spectra with the flux-reconnection magnetic field of Eq. 3. Right: Profile B compared to CHANDRA data. Figures taken from Ref. 1.

within the angular resolution of the HESS telescope array are much less than what is observed. Consequently we find that the X-ray observations from CHANDRA can be much more restrictive than the data from gamma ray telescopes with much larger angular uncertainties.

References

1. L. Bergström, M. Fairbairn and L. Pieri, accepted for publication in Phys. Rev. D (2006).
2. F. Aharonian and A. Neronov, Astrophys. J. **619**, 306 (2005).
3. L. Bergström, T. Bringmann, M. Eriksson and M. Gustafsson, Phys. Rev. Lett. **94** (2005) 131301.

PROSPECT OF AMS EXPERIMENT FOR DARK MATTER

MAURICE BOURQUIN ON BEHALF OF THE AMS COLLABORATION

Physics Department, University of Geneva, Quai Ernest-Ansermet 24
CH-1211 Geneva 4, Switzerland

AMS is a general purpose particle spectrometer constructed to measure cosmic rays and gamma rays on the International Space Station for a long duration flight. Selected performance simulations are reported for the search of dark matter signatures through the identification of positrons, antiprotons and gamma rays from WIMP annihilation in the Galaxy.

1. Introduction

In searches for Weakly Interacting Massive Particles (WIMP's), one expects a strong synergy between experiments at accelerators and experiments observing particles from the cosmos. If supersymmetric particles are discovered at the Large Hadron Collider (LHC), it is indeed not simple to verify that these particles represent the cold dark matter needed by astrophysics and cosmology. The direct or indirect detection of astrophysical WIMP's is thought to be necessary. Furthermore, part of the supersymmetric parameter space is not available at the LHC. For some benchmark models, supersymmetric particles which are not visible at the LHC are accessible to non-accelerator experiments.

Detectors above the Earth atmosphere can perform indirect dark matter searches through the identification of charged particles and gamma rays from WIMP annihilation in the Galaxy. In particular antiparticles like positrons, antiprotons and antideuterons produced in the annihilation of neutralinos would be trapped in the galactic magnetic fields and detected as an excess over the background generated by other well understood processes.

The Alpha Magnetic Spectrometer (AMS), when it will be operating in space on the International Space Station (ISS) for a period of at least 3 years, will have a high sensitivity to such dark matter signatures.

2. The AMS Experiment

AMS is a high energy particle physics detector in space. It allows the measurement of charged particles and antiparticles, from protons to iron nuclei, and gamma rays. It has a large acceptance (0.45 m^2 sr), a good energy resolution in the range of 1 GeV to a few TeV, and a good particle identification power.

The AMS-02 spectrometer draws experience from AMS-01, a detector flown successfully in the cargo bay of the NASA Space Shuttle Discovery in June 1998 (flight STS-91), but has much improved capabilities.

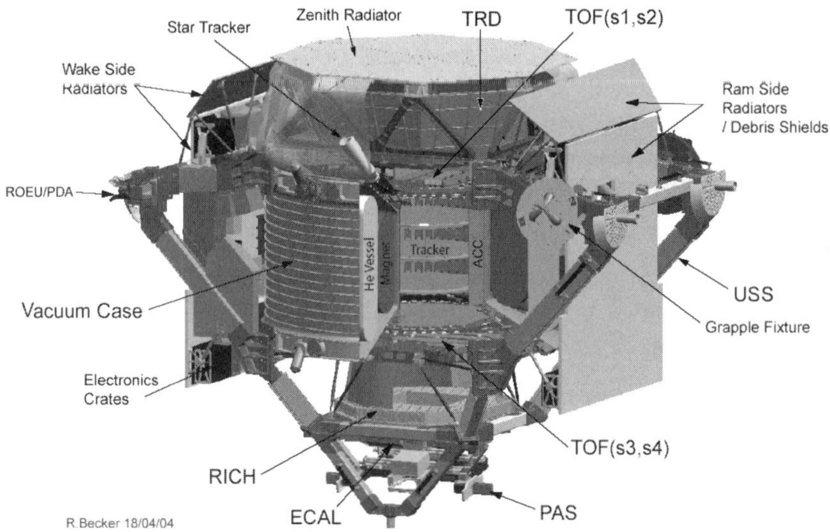

Figure 1. Schematic view of the AMS-02 detector configuration to be installed on the International Space Station.

The main detector components of AMS-02 are indicated on Figure 1. They include

- A Transition Radiation Detector (TRD), which provides electron and proton separation;
- A Time of Flight (TOF) scintillation counter hodoscope, which provides time of flight and dE/dx measurements, distinguishes downward from upward moving particles, and gives the primary trigger;
- A superconducting magnet, cooled by superfluid helium, which provides a bending power $BL^2 = 0.86$ Tm^2;

- A microstrip Silicon Tracker, which provides rigidity (momentum/charge) and dE/dx measurements, and determines the sign of the charge up to about 10 TeV;
- A system of Anti-Coincidence Counters (ACC) inside the magnet;
- A Ring Imaging Cerenkov Counter (RICH), which measures the velocity and the charge of particles;
- A sampling Electromagnetic Calorimeter (ECAL), which allows electron/ hadron separation and measures the energy of gamma rays and electrons.

With respect to the precursor flight STS-91, the higher field strength of the superconducting magnet will improve the rigidity resolution by a factor of about eight and the particle identification capability will be made more redundant.

3. AMS Prospects for Dark Matter

3.1. *Positron Measurements*

Figure 2 shows a simulation of the statistics accumulated by AMS-02 over one year for positrons. Data have been normalized to the HEAT 1994-1995 balloon experiment measurements [1]. The positron flux is expected to be measured up to about 450 GeV. Figure 3 shows a simulation for the ratio positron / (positron + electron), as compared to selected present measurements [2]. On Figure 4 a

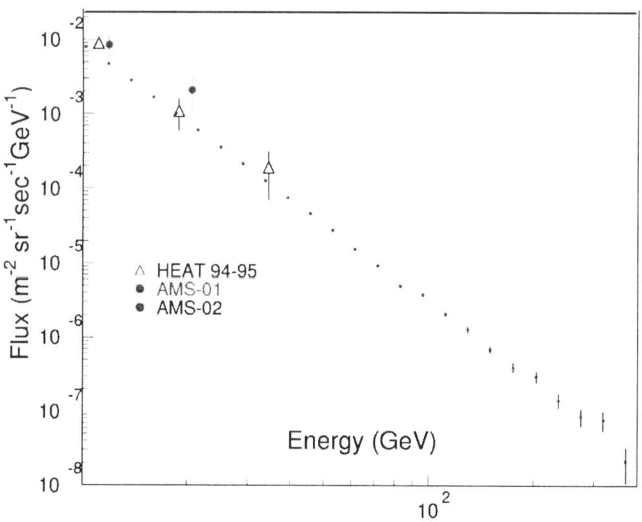

Figure 2. Simulation of the positron statistics to be measured by AMS-02 compared with data collected by the balloon experiments HEAT 1994-1995 and by AMS-01.

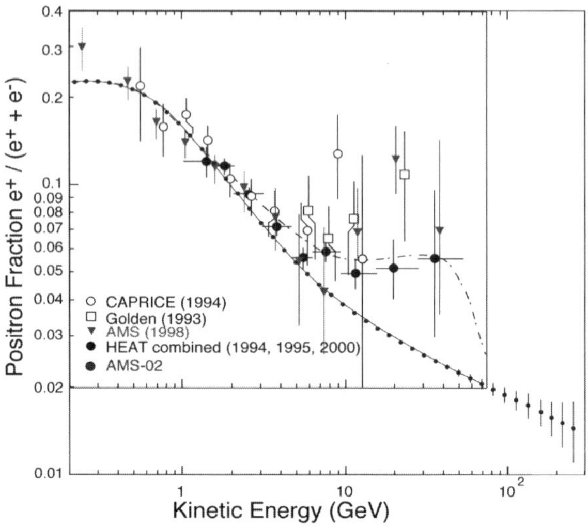

Figure 3. Simulation of the positron fraction statistics to be measured by AMS-02 as compared to present measurements collected in reference [2]. The dashed-dotted line corresponds to a generic SUSY model with a neutralino mass of 91 GeV.

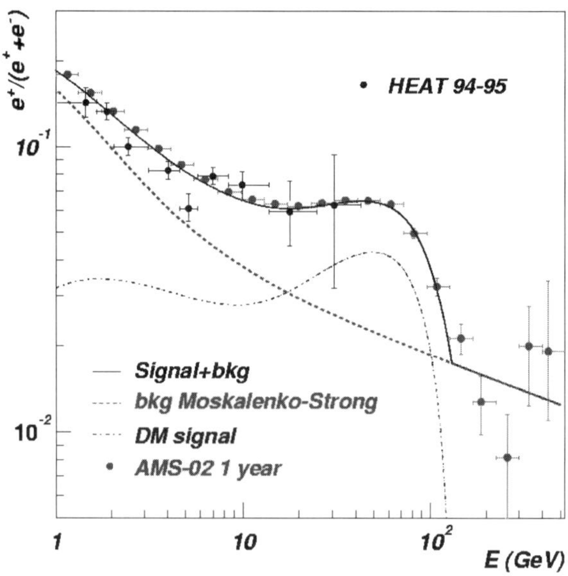

Figure 4. Simulation of the statistics accumulated by AMS-02 over one year for the ratio positron/(positron + electron). A dark matter signal produced by supersymmetric neutralino annihilation of 130.3 GeV mass has been superimposed over the background.

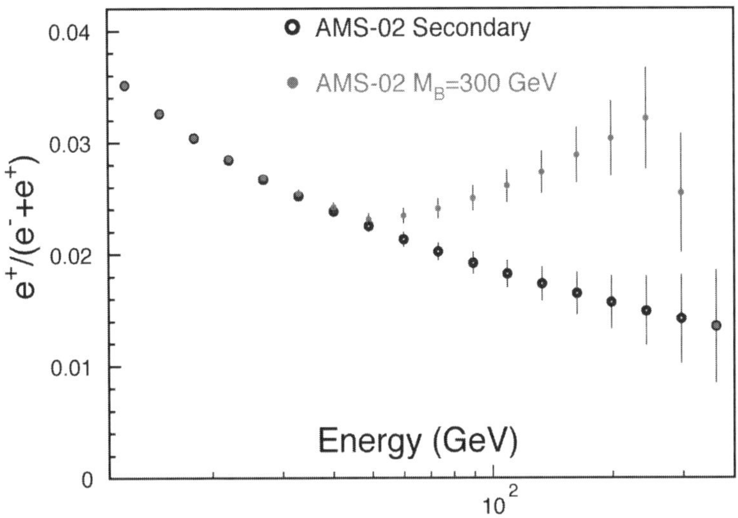

Figure 5. Simulation of the ratio positron / (positron + electron). A dark matter signal produced by Kaluza-Klein boson annihilation of 300 GeV mass in the galactic halo has been superimposed over secondary production.

dark matter signal produced by supersymmetric neutralino annihilation in the galactic halo, for a particular choice of SUSY parameters [3], has been superimposed over the background calculated by Moskalenko and Strong [4]. The signal would be observable at high energies.

The response of AMS-02 to Kaluza-Klein bosons as dark matter candidates with typical masses of 100 GeV to 1 TeV has been simulated. Following reference [5], AMS-02 expects a peak in the positron spectrum (Figure 5) and structures in the antiproton and antideuteron spectra (not shown).

3.2. *Antiproton Measurements*

Figure 6 shows a simulation of the statistics accumulated by AMS-02 for antiprotons as compared to selected present measurements. The flux is expected to be measured up to about 450 GeV. Figure 7 shows the expected antiproton to proton ratio. Detection of antiprotons produced in the annihilation of neutralinos would modify the shape of the distribution, as illustrated here for a neutralino mass of 206 GeV [6].

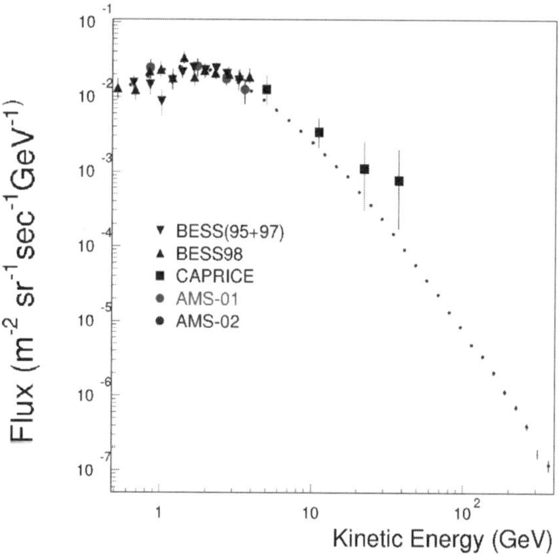

Figure 6. Simulation of the antiproton statistics for AMS-02, compared to selected present measurements.

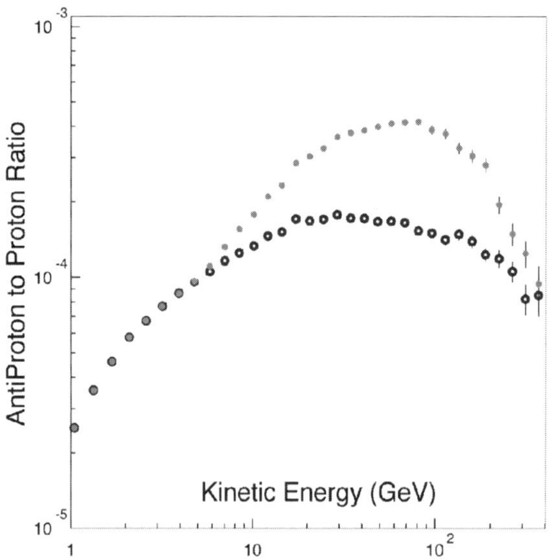

Figure 7. Projection of the AMS-02 antiproton to proton ratio. A dark matter signal (closed dots) has been superimposed over secondary antiproton production (open dots). The SUSY parameters include a neutralino mass of 206 GeV [6].

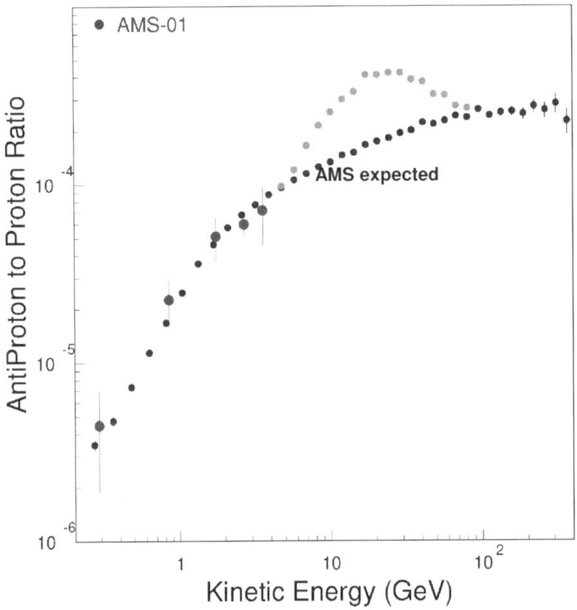

Figure 8. Projection of the AMS-02 antiproton to proton ratio. A dark matter signal has been superimposed over secondary antiproton production. The SUSY parameters include a neutralino mass of 840 GeV [7].

It should be noted that there exists regions of phase space where supersymmetric particles would not be visible at the LHC, but would be accessible to the AMS experiment. Such an example is the so-called benchmark "M" from reference [7], illustrated on Figure 8.

3.3. *Gamma Ray Measurements*

Two complementary methods can be applied to measure gamma rays in AMS [8]. The "conversion mode" consists in the identification and reconstruction of positron-electron pairs from photons converting in the material upstream of the Silicon Tracker. The "single photon mode" is based on the detection of photons in the electromagnetic calorimeter. These methods give AMS-02 good sensitivity to the diffuse gamma ray spectrum from the Galactic plane. Figure 9 illustrates such a study in the energy range from 100 GeV to 1 TeV.

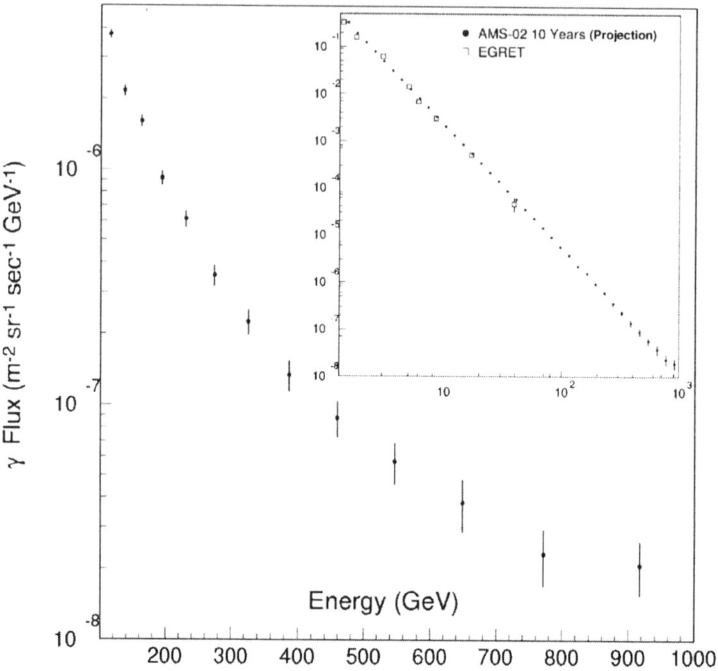

Figure 9. Simulation of the statistics of the high energy gamma ray flux to be obtained by AMS-02.

4. Conclusions

In the search for dark matter, the AMS experiment has the advantage that it can identify concurrently several different particles: antiprotons, positrons, antideuterons and gamma rays, Simulations have demonstrated the excellent performance of AMS to study dark matter signatures and provide a complementary view of possible discoveries in this domain at the LHC. By the end of the year 2008, the AMS-02 experiment will be ready at NASA Kennedy Space Center for launch to the International Space Station. Let us hope the best for the continuation of the Space Shuttle flights!

Acknowledgments

The contributions from colleagues of the AMS collaboration are gratefully recognized. The support of AMS funding agencies is acknowledged, including the essential support of NASA.

References

1. S.W. Barwick et al. (HEAT Collaboration), Ap.J. 498 (1998) 779.
2. J.J. Beatty et al., *Phys.Rev.Lett.* **93** (2004) 241102.
3. P. Maestro, Ph.D. Thesis, University of Sienna (2003).
4. I. Moskalenko and A.W. Strong, Ap.J. 493 (1998) 694.
5. H.C. Cheng, J.L. Feng and K.T. Matchev, *Phys.Rev.Lett* **V89**, N21 (2002) 211301-1.
6. P. Brun et al., private communication, 2006.
7. M. Battaglia et al., hep-ex/0106207.
8. M. Pohl, Int. J. of Mod. Phys. A17 (2002) 1809.
   G. Lamanna, Modern Physics Letters A18 (2003) 1951.

SEARCH FOR DARK MATTER WITH SPACE EXPERIMENTS

A. MORSELLI

INFN and University of Roma "Tor Vergata", Via della Ricerca Scientifica 00133 Rome, Italy

E-mail: aldo.morselli@roma2.infn.it

The detection of gamma-rays, antiprotons and positrons due to pair annihilation of dark matter particles in the Milky Way halo is a viable tecniques to search for supersymmetric dark matter candidates if there is the possibility to separate the signal from the backgroung generated by standard production mechanisms. Here we discuss the status of this indirect search and the prospective for the experiments GLAST and PAMELA.

Keywords: antiproton, gamma-rays, supersymmetry

1. Dark Matter Search and the PAMELA experiment

To uncover the dark matter, to connect what is astrophysically observed to what will be seen as new particles produced in the LHC, we need new measurements. The LHC will have either discovered putative dark matter related candidates or it will have set important new limits soon after the start of operations. If the former, the community will be intensely searching for any additional information about this new particle sector as well as clues to the relevance of the new particle(s) to the dark matter. Both can be obtained by so-called "indirect detection" of dark matter, i.e., the detection of signatures of dark matter particle annihilations (see Refs. [1,2] for recent reviews).

The PAMELA (a Payload for Antimatter Matter Exploration and Light-nuclei Astrophysics) experiment is a satellite-borne apparatus designed to study charged particles in the cosmic radiation with a particular focus on antiparticles (antiprotons and positrons).

Figure 1 shows a picture of the apparatus that consists of: a time of flight system, a magnetic spectrometer, an anticoincidence system, an electromagnetic imaging calorimeter, a shower tail catcher scintillator and a neutron detector. The orbit is elliptical and semi-polar, with an inclination of 70^0 and an altitude varying between 350 km and 600 km; the mission will last at least three years. The geometric acceptance in the standard trigger configuration is $\simeq 20.5$ cm^{2}sr. Moreover PAMELA is equipped with another trigger based only on the calorimeter signals: it permits to increase the acceptance of the $e^+ + e^-$ measurement at high energy up to about 600 cm^{2}sr [3].

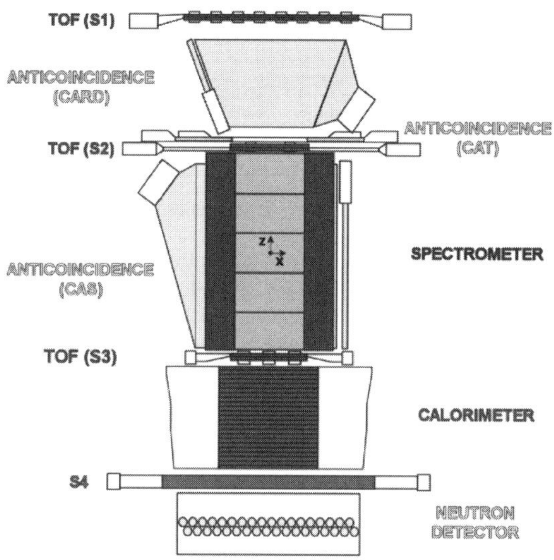

Fig. 1. A schematic overview of the PAMELA apparatus. The detector is approximately 1.3 m tall. The magnetic field lines in the spectrometer are oriented along the y direction.

The main PAMELA scientific primary goals and objectives are:

- Search for evidence of exotic matter as heavy antinuclei and nonbaryonic particles outside the Standard Model
- Understanding formation and evolution of our Galaxy and the Universe
- Exploring the cycles of matter and energy in the Universe.

Figure 2 shows the PAMELA expectations for the antiproton flux for the best standard production and propagation model [4] obtained with the use of its geometrical factor and detector characteristics and an example of a primary SUSY contribution to the $\bar{p}$ flux computed using the public code DarkSUSY for a neutralino mass of 1 TeV (obtained from a particular

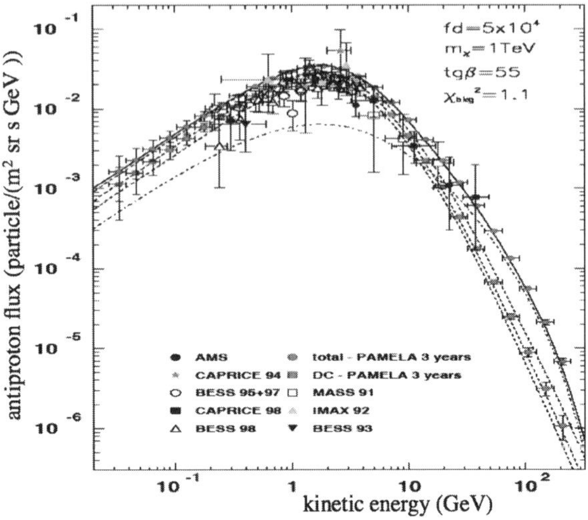

Fig. 2. Antiproton absolute flux: theoretical predictions for total uncertainty and best B/C fit for DC model (dashed lines). Experimental data are from [3]. The PAMELA expectations points (red squares) for DC background are for three years of data taking. The dash-dotted line is a neutralino induced contribution for a neutralino mass of 1 TeV (see text) and a clumpiness factor fd of 5 10^4 while the solid line is the total contribution calculated with the addition of the DC background and the red circles are the corresponding PAMELA point.

choice of mSUGRA parameters) and a clumpiness factor fd of 5 10^4 [4]. Higher neutralino masses improve high energy data fit but only with the increase of the clumpiness factor because of the dependence from the inverse neutralino mass squared m_χ in the $\bar{p}$ flux. For different values of the mSUGRA parameters one can found the minimal values of the clumpiness factors fd needed to disentangle a neutralino induced component in the antiproton flux with PAMELA. This factor can be computed as a function of the mSUGRA parameters. Fixing the less sensitive parameters A_0, $\tan\beta$ and sign(μ) $= +1$ the clumpiness factor becomes a function of m_0 and $m_{1/2}$. As the clumpiness factor is a function of m_0 and $m_{1/2}$ parameters the contour plots can be made calculating equi-clumpiness factors lines. The results is shown in figure 3 for $\tan\beta = 55$ and $\tan\beta = 60$ (bottom). Stripped regions are excluded either by accelerator bounds or because electroweak symmetry breaking is not achieved or because the neutralino is not the lightest supersymmetric particle. Red (dark shaded) are domains with the density of Dark Matter in the Universe compatible at 95 % C.L. with WMAP [5]. The upper green curve is the sensitivity in the case of

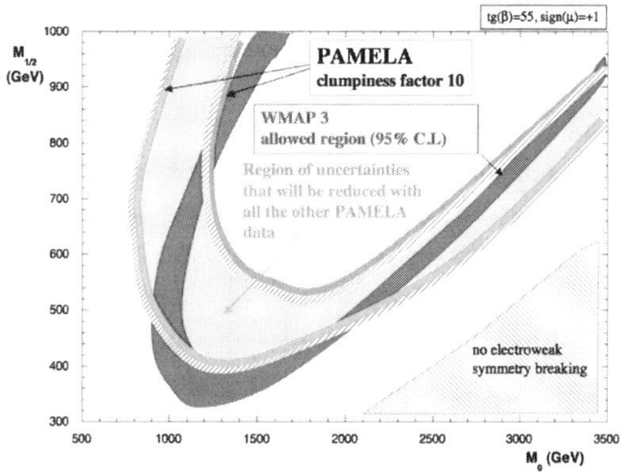

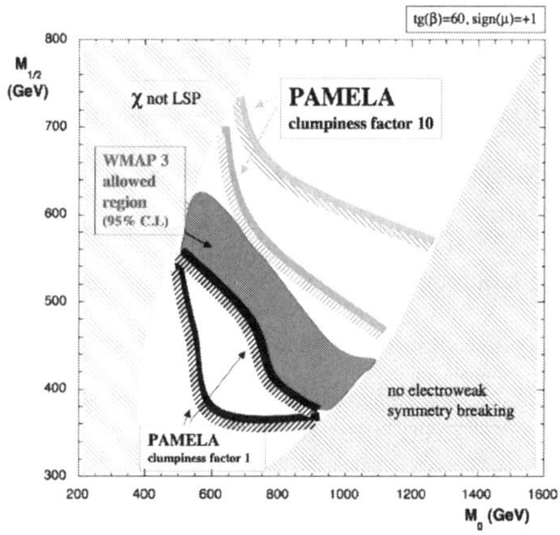

Fig. 3. Sensitivity plot for observation of mSUGRA with PAMELA for $\tan\beta = 55$ (top) and $\tan\beta = 60$ (bottom).

the lowest background flux showed in figure 2. Under this assumpion all the region below that curve can be explored by PAMELA. The lower green curve is the sensitivity in the case of the heightest background flux showed in figure 2. So the light green region between the two curve is an uncertanty region that depend on the uncertanties of the knoledge of the background. It is interesting to notice that this region will be reduced using all the other data that PAMELA will collect but the amount of this reduction is still under study.

It can be see that PAMELA will be able to disentangle a neutralino induced component for halo models that has fd as low as ~ 10 for $\tan\beta = 55$ and the situation is even more favorable with increasing value of $\tan\beta$ as

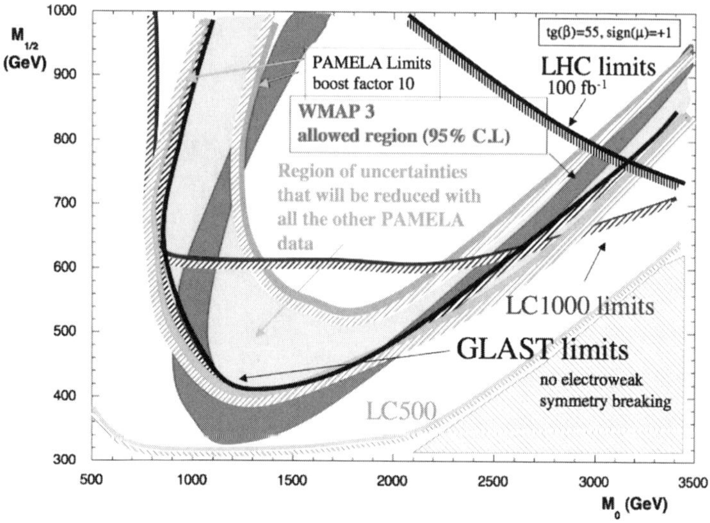

Fig. 4. GLAST sensitivity to a dark matter signal via the observation of WIMP annihilation photons (continuum spectrum) in the $M_{1/2}$ and M_0 mSUGRA parameter plane for $\tan\beta = 55$. GLAST 3σ sensitivity is shown at the blue line and below. Sensitivity plot for PAMELA and a number of accelerator experiments are indicated on the figure.

can be seen in figure 3 for $\tan\beta = 60$ where with a fd=10 all the WMAP region is completely covered and for fd greather than 1 some part of the region will be explored. A similar analysis using the PAMELA positron signal is presented in [6]

2. Dark Matter Search and the GLAST experiment

One example of a crucial observation of Dark Matter evidence is the measurement of a gamma ray continuum dark matter annihilation signal by GLAST. With such a measurement GLAST could establish a rough mass estimate for the WIMP to compare with the LHC particle mass estimates. In addition, the GLAST measurement would establish a lifetime scale for the particle consistent with expectations for WIMP dark matter. (LHC experiments can not place limits on particle lifetime beyond the microsecond regime.) If GLAST manages to observe line(s), WIMP annihilation into $\gamma\gamma$ and or γZ, the masses will be determined with higher precision, and we may glean even more information about the underlying Lagrangian from the relative strength of the observed lines or limits on one or the other. If the lines are not sufficiently bright to be observable, GLAST may still play a crucial role in identifying dark matter satellites in our galactic halo which can be followed up with atmospheric Cerenkov telescopes (ACTs) observing from the ground. These are much more sensitive at high energy, but they do not have the field of view necessary to do surveys at the sensitivity of GLAST. Thus, gamma ray lines might be observed from the ground in GLAST-discovered sources. That GLAST measurements can contribute with LHC measurements to help establish which WIMP theory is correct is particularly important until the ILC comes online sometime in the next decade. Figure 4 shows the PAMELA and GLAST limits together with the accelerator limits from [7]. The GLAST sensitivity is for a dark matter signal via the observation of WIMP annihilation photons (continuum spectrum) in the $M_{1/2}$ and M_0 mSUGRA parameter plane for $\tan\beta = 55$ [8]. The values of the neutralino mass is also shown. For the region in red, the cosmologically allowed WIMP region, the signal in GLAST above the blue line ($M_{WIMP} \sim 200GeV$) is not observable due the higher WIMP mass as one moves to higher $M_{1/2}$. The dark matter halo used for the GLAST indirect search sensitivity estimate is a truncated Navarro Frank and White (NFW) halo profile. For steeper halo profiles (like the Moore profile) the GLAST limits move up, covering a wider WMAP allowed region, while for less steep profile (like the isothermal profile) the GLAST limits move down, covering less WMAP allowed region.

As for the antiprotons, the situation is even more favorable with increasing value of $\tan\beta$.

References

1. L. Bergström, Rept. Prog. Phys. **63**, 793 (2000) [hep-ph/0002126];
2. G. Bertone, D. Hooper and J. Silk, Phys. Rept. **405**, 279 (2005) [hep-ph/0404175].
3. P. Picozza, A. Morselli, (2003) J. Phys. G: Nucl. Part. Phys., **29**, 903-911 [astro-ph/0211286]
   P. Picozza, A. Morselli, World Scientific Publishing Co. [astro-ph/0604207]
   P. Picozza et al. (2006), submitted to Astroparticle Physics [astro-ph/0608697]
4. A. Lionetto, A. Morselli, V. Zdravkovic (2005), J. Cosmol. Astropart. Phys. JCAP09, 010 [astro-ph/0502406]
5. D. N. Spergel, et al., the WMAP Collaboration, astro-ph/0603449
6. Y. Mambrini, C. Muñoz, E. Nezri, astro-ph/0607266
7. H. Baer, et al., JCAP 0408 (2004) 005
8. A. Cesarini, F. Fucito, A. Lionetto, A. Morselli and P. Ullio, Astropart. Phys. **21** (2004) 267 [astro-ph/0305075].

THE ZEPLIN III DETECTOR: RESULTS FROM SURFACE CALIBRATIONS

T. J. SUMNER*, H. M. ARAÚJO, A. BEWICK, D. DAVIDGE, J. DAWSON, B. EDWARDS, A. S. HOWARD, W. G. JONES, M. JOSHI, V. N. LEBEDENKO, I. LIUBARSKY, J. J. QUENBY, C. THORNE and R. J. WALKER

Blackett Laboratory, Imperial College London,
London, UK
**E-mail: t.sumner@imperial.ac.uk*
www.imperial.ac.uk

G. J. ALNER, C. BUNGAU, T. DURKIN, J. D. LEWIN, R. LÜSCHER, R. PREECE, N. J. T. SMITH and P. F. SMITH

Particle Physics Department, Rutherford Appleton Laboratory
Chilton, UK
www.cclrc.ac.uk

C. GHAG, E. KOROLKOVA, A. S. MURPHY and S. J. S. PLANK

School of Physics, University of Edinburgh,
Edinburgh, UK
www.ed.ac.uk

D. YU. AKIMOV, A. A. BURENKOV, A. KOVALENKO, I. S. KUZNETSOV and V. STEKHANOV

Institute for Theoretical and Experimental Physics,
Moscow, Russia
www.itep.ru

V. CHEPEL, A. LINDOTE, M. I. LOPES, F. NEVES, J. PINTO da CUNHA, C. SILVA and V. N. SOLOVOV

LIP–Coimbra, Department of Physics, University of Coimbra
Coimbra, Portugal
www.coimbra.lip.pt

M. J. CARSON, H. CHAGANI, E. DAW, T. GAMBLE, R. J. HOLLINGWORTH, K. MAVROKORIDIS, V. A. KUDRYAVTSEV, T. LAWSON, P. LIGHTFOOT, J. E. McMILLAN, P. MAJEWSKI, B. MORGAN, D. MUNA, G. G. NICKLIN,

S. M. PALING, M. ROBINSON, N. J. C. SPOONER, D. R. TOVEY and E. TZIAFERI
Physics and Astronomy Department, University of Sheffield
Sheffield, UK
www.shef.ac.uk

D. CLINE and H. WANG
Department of Physics & Astronomy, University of California
Los Angeles, USA
www.ucla.edu

J. WHITE
Department of Physics, Texas A&M University
Texas, USA
www.tamu.edu

F. WOLFS
Department of Physics & Astronomy, University of Rochester
New York, USA
www.rochester.edu

ZEPLIN-III is a direct dark-matter search instrument using liquid xenon as a target. Both scintillation light and ionisation charge are measured. The instrument has been built and tested in a surface laboratory, and we present some of the results here, including design thermal control system verification, photomultiplier calibration and two-phase operation showing simultaneous measurement of scintillation and charge and position reconstruction.

Keywords: Dark Matter; Xenon Detector.

1. The ZEPLIN III Instrument

ZEPLIN-III is a two-phase (liquid/gas) xenon instrument aiming to detect Galactic dark matter in the form of Weakly Interacting Massive Particles, or WIMPs. The instrument[1,2] has been tested at Imperial College, and will shortly move into our underground laboratory at the Boulby mine (North Yorkshire, UK).

Two-phase noble gas emission detectors[5] rely on different particle species generating different relative amounts of vacuum ultra-violet scintillation light and ionisation charge in liquid xenon. WIMPs scatter elastically off Xe atoms, much like neutrons, and a recoiling nucleus produces a different signature to the electron recoil from γ-rays. ZEPLIN III is a high-field xenon system.[6–9] Monte Carlo simulations[10,11] helped design the instrument and to assess its performance.[2]

Four requirements for a dark matter detector are: low energy threshold, particle discrimination, 3-D position reconstruction and low background. Figures 1 & 2 show ZEPLIN-III at concept and engineering levels.

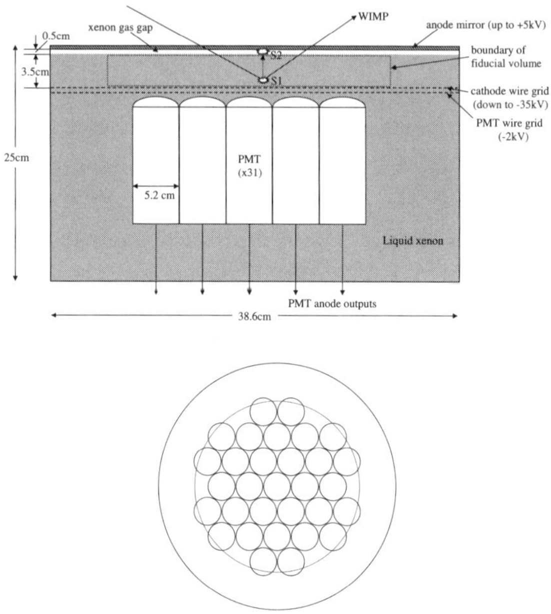

Fig. 1. Cross-sections of the target showing the concept. The top panel shows a side view. The bottom panel provides a top view of the PMT arrangement. The fiducial boundary is shown on both.

For low scintillation threshold our photomultipliers (PMTs) are in the liquid, removing two interfaces with refractive index mismatches, and we use a flat planar geometry giving the highest direct geometrical light collection.

Discrimination comes from simultaneously measuring scintillation and ionisation. The scintillation output, S1, is a prompt, fast signal (τ few tens of ns). Escaping ionisation is drifted, by an applied electric field, to the surface and extracted into the gas phase where it causes proportional electroluminescence, S2. S1 and S2 are separated in time (up to 17μs for the 3.5 cm depth) and read out by the same PMTs. S2/S1 depends on particle species[12] and 3-D position reconstruction identifies multiple scattering.

The S2 signal pattern in the 31 PMTs provides sub-cm $r - r\theta$ resolution even for single electrons extracted from the liquid.[16] The z co-ordinate, to $\sim 50\,\mu$m, comes from the S1 to S2 time delay. 3-D reconstruction defines the

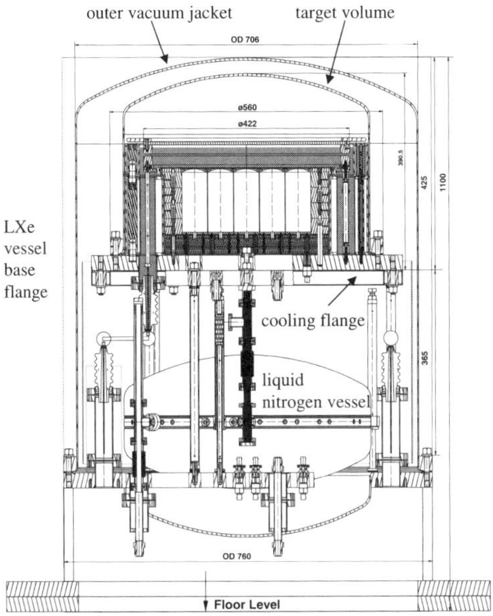

Fig. 2. Cross-section of the complete instrument showing the key subsystems. Dimensions are in mm and ZEPLIN-III is 1100 mm high with a diameter of 760 mm.

fiducial volume without reliance on physical surfaces. The fiducial diameter of 31.2 cm contains 8 kg of xenon.[2]

2. Commissioning Tests

The first cool-down verified the thermal control system and the PMT array. For this the electrodes were replaced by a copper plate 8 mm above the PMT array. 31 ^{241}Am sources were vacuum-sealed into it with a copper foil to stop the α-particles, leaving only low-energy (mainly 59.6 keV) photons. For subsequent tests the electrodes were installed.

2.1. *Cooling system, PMT array and scintillation output*

The initial cool-down, at $\sim$5°C/hour, uses 200 litres of liquid nitrogen (LN_2). The temperature on the cooling flange at the bottom of the target is used as a control temperature to automatically operate two valves: one exhausts straight from the gas volume of the LN_2 reservoir, and one exhausts through the cooling ring. In normal operations the temperature is stable to better than 0.2°C and the LN_2 usage drops to $\sim$ 20 litres/day.

Pulse height spectra, pulse waveforms and single photoelectron spectra (spes) were collected during the first cold-run both with the experiment DAQ electronics and with a pulse height analyser (PHA) set-up using a multichannel analyser (MCA). Scintillation pulses from low-energy ^{241}Am γ-rays showed the characteristic decay time of ~ 40 ns. The spes show well resolved peaks and were used to set the amplifiers to normalise all channels.

Figure 3 shows MCA spectra from one PMT. For the bottom spectrum the whole arrangement was covered with liquid xenon. The two spectral features are the 59.6 keV line and a blend of the 26.3 keV ^{241}Am γ-ray with the 30 keV escape line. Using the MCA single photoelectron spectrum gives a signal of $\sim$12 phe/keV at zero field. The top spectrum, with the liquid level below the source, has improved light collection ($\sim$ 17 phe/keV) due to total internal reflection at the liquid gas interface, and resolution.

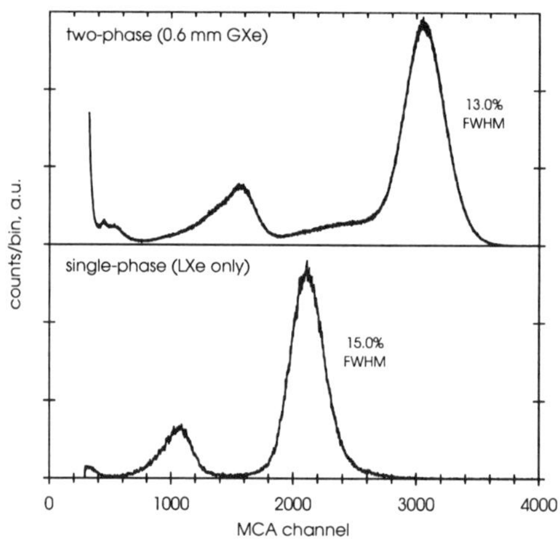

Fig. 3. Typical MCA scintillation spectra with ^{241}Am sources a few mm above the PMTs. The two panels have different liquid xenon levels.

2.2. *Two-phase operation*

In two-phase operation zero-field ^{57}Co γ-ray spectra were recorded with an uncollimated source above the detector. A photoelectron yield of 5 phe/keV was obtained from the fiducial volume, with a FWHM$\simeq$25%. This yield exceeds that predicted by Monte Carlo simulations (3.4–4.0 phe/keV),[2] which may indicate a higher LXe scintillation yield than previously thought.

Figure 5 shows γ-ray interactions with a field of 3.0 kV/cm in the liquid. Two fast S1 scintillations are seen with several broader S2 signals. For each event the time delay between S1 and S2 depends on the interaction depth and drift velocity at our field (~2.5 mm/μs). The S2 width depends on gas gap and electric field. Rise and fall times are a combination of extraction dynamics, diffusion and gas scintillation time-constant. S1 and S2

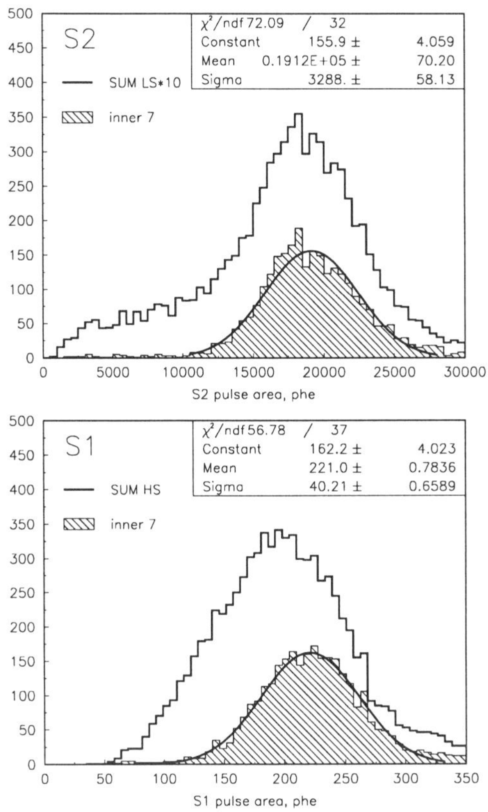

Fig. 4. S2 and S1 spectra from an uncollimated ^{57}Co source, at 3.0 kV/cm.

^{57}Co spectra at are shown in figure 4, They use all PMTs, with individual gains and QEs equalised. The shaded spectra only include events with peak signals in the inner 7 PMTs. The 'collimated' S1 spectrum has a yield of 1.8 phe/keV; ~35% of the zero-field value. A single-Gaussian fit to the 122.1 keV and 136.5 keV blend gives a $\simeq$40% FWHM.

A 3-D position recovery algorithm was developed providing sub-cm resolution in the horizontal plane.[16] Even before this is applied, the spatial sensitivity is demonstrated in figure 5, showing an event with two overlapping interactions. Without position sensitivity these cannot be unambiguously separated. However, using individual PMT traces (left-hand in figure 5) it is obvious that we have two double-Compton scatters.

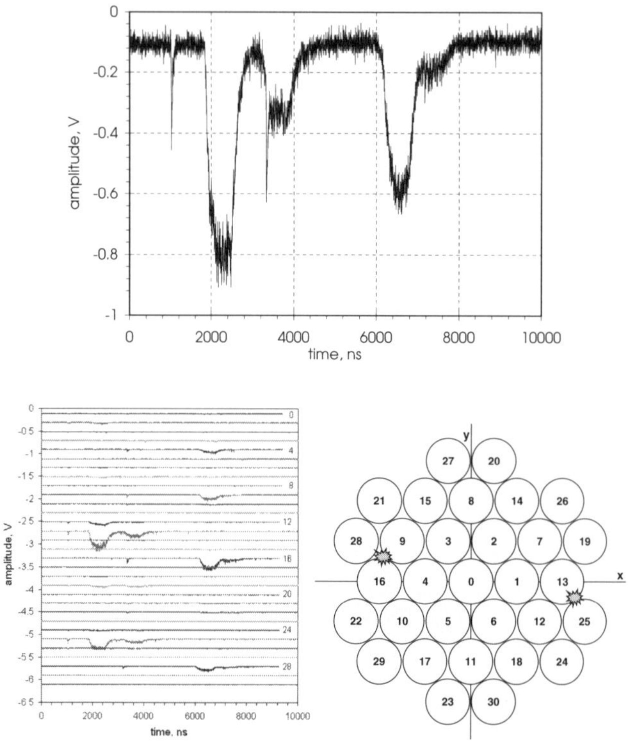

Fig. 5. Top: Summed waveforms from two overlapping γ-ray events. Lower: Individual waveforms from the 31 PMTs and reconstructed positions.

Figure 6, shows the S2/S1 ratio when a 10 mCi Am-Be (α,n) source is placed above the detector. The field within the liquid was 3 kV/cm. Discrimination between elastic recoils (population A) and γ-ray interactions (C) is clearly demonstrated. Population B is from neutron inelastic scattering off ^{129}Xe nuclei; some (quenched) nuclear recoil energy is augmented by a 40 keV γ-ray from nuclear de-excitation.

3. Acknowledgements

This work has been funded by the UK Particle Physics And Astronomy Research Council (PPARC).

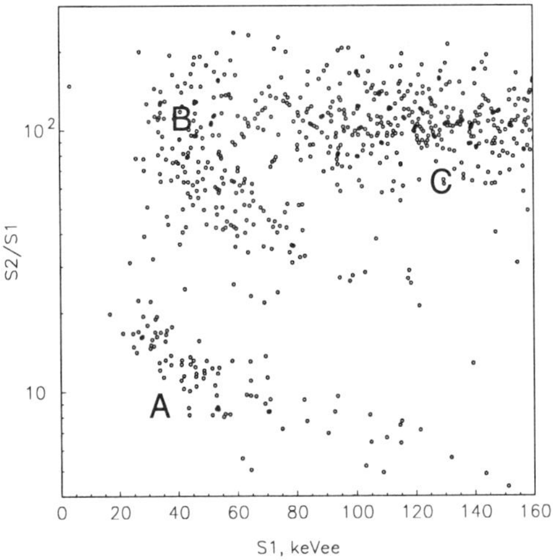

Fig. 6. S2/S1 ratio vs S1 for single-scatter events from an Am-Be source. For elastic nuclear recoil events (population A) the energy scale must be multiplied by 1.95, due to a combination of the inverse quenching factor, ($1/0.19^{13-15}$) and the S1 suppression factors at our operating field (~0.37). Population C are γ-rays.

References

1. D. Yu Akimov et al., Astropart. Phys., in press (2006)
2. H. M. Araújo et al., Astropart. Phys. 26 (2006) 140.
3. G. J. Alner et al., New Astronomy Reviews 49 (2005) 259.
4. G. J. Alner et al., Nucl. Instrum. Meth. A 555 (2005) 173.
5. B. A. Dolgoshein, V. N. Lebedenko & B. U. Rodionov, JETP Lett. 11 (1970) 513.
6. T. J. Sumner et al., Proc. 26th Int. Cosmic Ray Conf., D. Kieda, M. Salamon & B. Dingus (Eds), 2 (1999) 516.
7. A. S. Howard et al., Proc. 3rd Int. Workshop on the Identification of Dark Matter, N. J. C. Spooner & V. Kudryavtsev (Eds), Singapore: World Scientic (2001) 457.
8. D. Yu Akimov et al., Proc. 4th Int. Workshop on the Identification of Dark Matter, N. J. C. Spooner & V. Kudryavtsev (Eds), Singapore: World Scientific, (2003) 371.
9. T. J. Sumner, New Astronomy Reviews 49 (2005) 277.
10. D. Davidge, PhD Thesis, University of London (2003).
11. J. Dawson PhD Thesis, University of London (2003).
12. E. Aprile et al., (2006) - (ArXiv: astro-ph/0601552)
13. D. Akimov et al., Phys. Lett. B 524 (2002) 245.
14. E. Aprile et al., Phys. Rev. D 72 (2005) 072006.
15. V. Chepel et al., ArXiv:physics/0512136, (2005)
16. A. Lindote et al., Nucl. Instrum. Meth. A in press (2006)

WIMPs SEARCH BY EXCLUSIVE MEASUREMENTS WITH THIN MULTILAYER NaI(Tl) SCINTILLATORS (PICO-LON)

K. FUSHIMI*, H. KAWASUSO, K. YASUDA, Y. KAMEDA, N. KOORI and S. NAKAYAMA

Faculty of Integrated Arts and Sciences, The University of Tokushima
Tokushima city, Tokushima 770-8502, JAPAN
** E-mail: kfushimi@ias.tokushima-u.ac.jp*

K. ICHIHARA, M. NOMACHI and S. UMEHARA

Department of Physics, Osaka University
Toyonaka city, Osaka 560-0047, JAPAN

R. HAZAMA

Graduate School and Faculty of Engineering, Hiroshima University
Higashi-Hiroshima city, Hiroshima 739-8527, JAPAN

S. YOSHIDA

Department of Physics, Tohoku University
Sendai city, Miyagi 980-8678, JAPAN

H. EJIRI

Department of Physics, International Christian University
Mitaka, Tokyo 181-8585, JAPAN

K. IMAGAWA and H. ITO

Horiba Co. Ltd.
Kyoto city, Kyoto 601-8510, JAPAN

The WIMPs search project PICO-LON has been started with multilayer thin NaI(Tl) crystals. The thin (0.05cm) and wide area (5cm×5cm) NaI(Tl) crystals was successfully developed. The performances of thin NaI(Tl) scintillator was measured and they showed good energy resolution (20% at 60keV) and good position resolution (20% in 5cm×5cm wider area).

Keywords: WIMPs search, Thin NaI(Tl)

1. Introduction

Dark matter search are one of the most important subjects in nuclear- and particle-physics. The particle candidates for cosmic dark matter is a key subject not only astrophysics but also particle physics since the particle candidates for dark matter is proposed by various models of the beyond standard model.

The components of the universe has been clearly understood by many cosmological observations. [1–3]. Since, the most of the matter in the universe should be the cold dark matter, searching for WIMPs (Weakly Interacting Massive Particles) is quite important. Recently, a direct empirical evidence for dark matter in a galaxy has been reported [4,5]. The dark matter in the galaxy has become quite ensuring. One of the most promising candidate for WIMPs is SUSY neutralino, which interacts with the matter via only weak interaction.

The processes for WIMPs-nucleus interaction are schematically illustrated in Fig. 1. In Fig. 1(a), the scattering amplitude is summed coherently

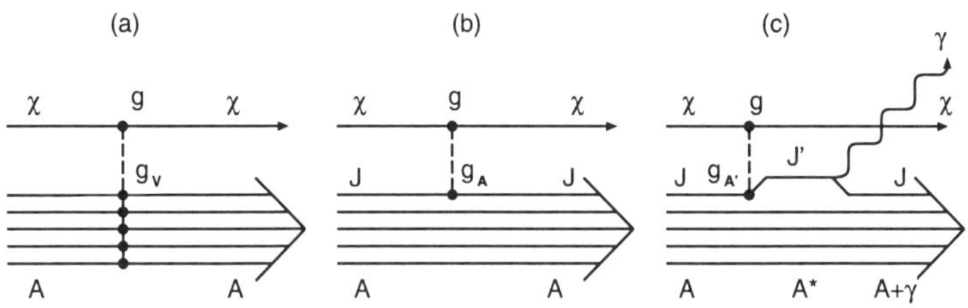

Fig. 1. Feinman diagrams for WIMPs-nucleus interaction. "A" and "χ" means the nucleus and the WIMPs. (a) Spin independent elastic scattering (SI). (b) Spin dependent elastic scattering (SD). (c) Spin dependent inelastic scattering (EX).

over all the nucleon. Thus the scattering cross section is coherently enhanced by the factor A^2, where A is the mass number of the target nucleus. In Fig. 1(b), only one nucleon which carries the nuclear spin contributes the cross section. Thus the cross section is a few orders of magnitude smaller than that of the previous case. The cross section depends on the nuclear spin-matrix element $\lambda^2 J(J+1)$. and has a large ambiguity because this matrix element has a large model dependence for the heavy nuclei [6]. In Fig. 1(c), the target nucleus will be excited to the low lying excited state, which is followed by the gamma ray emission. This process is arisen only if the WIMPs particle has enough kinetic energy. In this case, the matrix element is measured precisely by the nuclear de-excitation [6–8]. Thus the model dependence of the cross section is much smaller than the cross section of the SD case.

The segmentation of the detector is shown to be the best way to enhance the position sensitivity [9]. Coincidence measurements of nuclear recoils and γ rays for the inelastic excitation of ^{127}I enhanced the sensitivity using the highly segmented NaI(Tl) detector. Recently, ionized atomic electrons and hard X rays following WIMPs-nuclear interactions have been shown to be useful for the exclusive measurement of nuclear recoils from the elastic scatterings of WIMPs off nuclei [10,11]. On the other hand, the background events have their own characteristics of timing and spatial profiles. Because the event rate due to the background is reduced by segmentation, a probability of the accidental coincidence of individual background events is vastly reduced.

It has been shown that piling up many thin scintillators enhances the sensitivity for WIMPs search [9]. The highly sensitive detector system PICO-LON (Planar Inorganic Crystals Observatory for LOw-background Neutr(al)ino) has been developed. It consists of many thin NaI(Tl) crystals whose thickness is 0.05cm. Recently, PICO-LON-I which was made up three crystals of thin (0.05cm in thickness) and wide area (6.6cm×6.6cm) NaI(Tl) crystals has been developed. PICO-LON-II which was made up sixteen crystals of thin NaI(Tl) has been also developed. In this report, we describe the excellent performance of a single plate of thin NaI(Tl) which is the foundation of PICO-LON system.

2. The performance of thin NaI(Tl) scintillator

The performance of the thin NaI(Tl) scintillator was measured by irradiating a thin NaI(Tl) crystal whose dimension was 5cm×5cm×0.05cm with low-energy γ rays and X rays. The scintillation photons were collected at the four edges of the NaI(Tl) crystal using four photomultiplier tubes (PMTs), which were provided by Hamamatsu Photonics (R329-P).

The gamma rays and X rays were irradiated isotropically on the wider surface of the NaI(Tl) plate. Each PMT output signal was individually input into four discriminators. The threshold of the discriminators was set above that of the single-photoelectron signal; the corresponding hardware energy threshold was 0.8keV. The four PMT outputs were individually converted to digital data using a charge integrating analog-to-digital converter (RPC-022). The total charge outputs of the PMTs were summed event-by-event using an off-line analyzer.

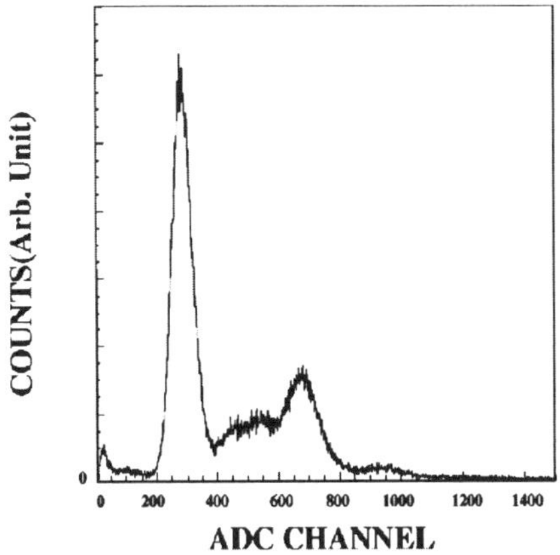

Fig. 2. Pulse height spectrum of ^{133}Ba.

The resulting photon outputs are shown in Fig. 2. In the pulse height spectrum of ^{133}Ba, high-energy gamma rays of energies above 200keV were not clearly observed because the detector was too thin to absorb the gamma rays. A photoelectric peak of 81keV and the corresponding X ray escape peak were observed at approximately 670ch and 500ch, respectively. The prominent peak at approximately 300ch was due to the K-X rays of cesium. Note that the small peak due to the low-energy L-X rays of cesium is observed approximately 30ch. It is important in the search for WIMPs to be able to observe low energy, and the present results correspond to the energy threshold being 2keV. The results showed that the thin NaI(Tl) scintillator displays great promise in the search for WIMPs.

The energy resolutions at FWHM (Full Width at Half Maximum) were calculated from the peaks and are shown in Table 1.

Table 1. Energy resolutions for low energy photons at FWHM. The calculated number of photoelectrons (P.E.) is listed in the fourth column. [12]

Source	Energy (keV)	$\Delta E/E$ (FWHM)	# of P.E.
^{133}Ba	81	0.17	197
^{241}Am	60	0.20	143
^{133}Ba	31	0.28	71
^{57}Co	14.4	0.40	35

The scintillation output has good linearity up to 120keV. From the pulse height spectra, the low energy threshold was found to be 2~3keV. The energy equivalent to a single photoelectron was also calculated using the photoelectron number N. The energy threshold of approximately 2$\sim$ 3keV corresponds to 4$\sim$5 photoelectrons. The results showed an excellent performance that is in accordance with the required performance for the advanced stage of the experiment.

The position resolution for the thinner directions is as good as 0.05cm because of the segmentation of the detector. Moreover, the position resolution for other directions was tested. Because the largest area has dimensions of 5cm$\times$5cm, good position information in the wider area enhances detector sensitivity. Position information was obtained by analyzing the ratio of the number of photons collected on the opposite sides of the detector. Precise position information on the largest area of the thin NaI(Tl) scintillator is important to ascertain the property of the events. By piling up the thin NaI(Tl) scintillator, the tracking of radiation such as cosmic rays and the multiple Compton scattering of photons is reconstructed precisely.

A collimator for low-energy γ rays was made of 1cm-thick lead brick with nine holes with a diameter of 2mm was used for position measurement. An ^{241}Am source was placed at the top of each hole. Position determination analysis was performed using 60keV gamma rays from ^{241}Am. The position resolution was calculated to be approximately 1cm in FWHM [12].

3. Estimated sensitivity for WIMPs

The case of inelastic scattering the nuclear spin-matrix element is experimentally deduced from the nuclear transition probability, consequently, the precise exclusion plot with small model dependence is obtained. The estimated sensitivity for SD type WIMPs is shown in Fig. 3. It is shown that the high sensitivity is expected by the small amount of NaI(Tl) crystal.

4. Future prospects

Test experiment with the three-layer NaI(Tl) detector (PICO-LON-I) has been performed at surface laboratory at Tokushima. The PICO-LON detector will be installed into Oto Cosmo Observatory in the south of Nara prefecture where is 150km east from Tokushima and 100km south from Osaka.

Oto Cosmo Observatory is covered with thick rock whose thickness is about 1200m.w.e. [13]. Thus the flux of cosmic ray is reduced by a factor of 10^{-5} [14]. The most serious origin of low energy background is ^{222}Rn in the air. Since the tunnel is opened at both ends and strong wind is running always, the concentration of Rn in the tunnel is the same as the one out

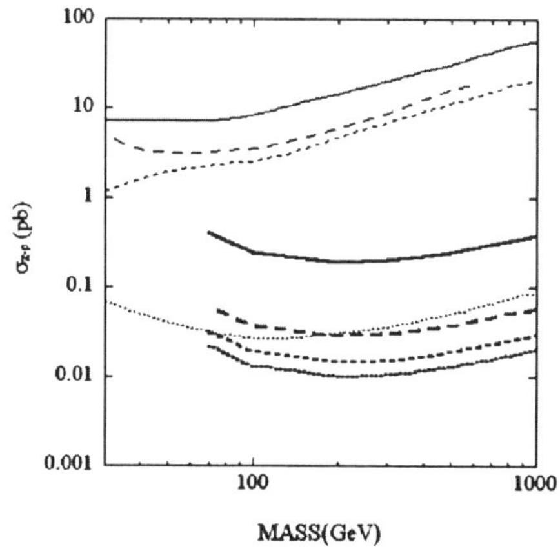

Fig. 3. Thick lines are the expected sensitivity for SD type WIMPs. The solid, long-dashed, sort-dashed and dotted lines are the expected sensitivity by 16, 256, 1024 and 2176 modules of NaI(Tl) array. Thin lines are results of CRESST (solid line), DAMA LXe (long-dashed line), ELEGANT V at OTO (short-dashed line) and the expected sensitivity by NAIAD (dotted line). [9]

of the tunnel. The measured Rn concentration in the air in the laboratory was measured by highly sensitive Rn monitor which can measure above 5mBq/m^3.

PICO-LON-I and PICO-LON-II will be installed into the shield with the 10cm thick OFHC(Oxygen Free High Conductive Copper) and 15cm thick old lead, which was used for the shield of ELEGANT V.

References

1. C.L.Bennett *et al.*, *ApJ Suppl.* **148**, p. 1 (2003).
2. W.J.Percival *et al.*, *MNRAS* **337**, p. 1068 (2002).
3. J.A.Peacock *et al.*, *Nature* **410**, p. 169 (2001).
4. D.Clowe *et al.*, *ApJ. Lett.* **648**, p. L109 (2006).
5. H.Gonzalez *et al.*, *Paper in this proceedings* (2006).
6. J. Ellis *et al.*, *Phys. Lett.* **B212**, p. 375 (1988).
7. H.Ejiri, K.Fushimi and H.Ohsumi, *Phys. Lett.* **B317**, p. 14 (1993).
8. K.Fushimi *et al.*, *Nucl. Phys. (Proc. Suppl.)* **B48**, p. 70 (1996).
9. K.Fushimi *et al.*, *Journal of Phys. Sci. of Japan* **74**, p. 3117 (2005).
10. Ch.C.Moustakidis, J.D.Vergados and H.Ejiri, *Nucl. Phys.* **B727**, p. 406 (2005).
11. H.Ejiri, Ch.C.Moustakidis and J.D.Vergados, arxiv:hep-ph/0510042.
12. K.Fushimi *et al.*, *Journal of Phys. Sci. of Japan* **75**, p. 064201 (2006).
13. S.Yoshida *et al.*, *Nucl. Phys. (Proc. Suppl.)* **B87**, p. 58 (2000).
14. N.Kudomi *et al.*, *Nucl. Phys.* **A629**, p. 527 (1998).

HIGH PERFORMANCE SCINTILLATING BOLOMETERS IN THE SEARCH FOR WIMPS: ROSEBUD EXPERIMENT

E. GARCÍA, J. PUIMEDÓN, M.L. SARSA* and L. TORRES

Laboratorio de Física Nuclear y Astropartículas, Universidad de Zaragoza, 50009 Zaragoza, Spain
** E-mail: mlsarsa@unizar.es*

N. CORON, P. DE MARCILLAC, J. GIRONNET, J. LEBLANC and T. REDON

Institute d'Astrophysique Spatiale, Bât 121, 91405 Orsay, Paris, France

The particle discrimination capability of various scintillating bolometers has been tested in the framework of the ROSEBUD (Rare Objects SEarch with Bolometers UnDerground) Dark Matter Search Collaboration. These detectors are well suited for dark matter searches. In particular, undoped sapphire has shown high light yield at very low temperature and low particle discrimination energy threshold. We will report on an estimate of the quenching factor in the heat signal for a scintillating bolometer made of this material, and we will present as well the prospects of the ROSEBUD experiment.

Keywords: dark matter search; bolometers; scintillation; quenching factor.

1. Introduction

The nature of the dark matter which is required to understand our Universe is still unknown, and is one of the most important challenges for the new century cosmology, astroparticle and particle physics. Direct dark matter search is focusing on the preparation of large mass experiments (ton scale) and background rejection strategies [1]. ROSEBUD is an experiment aiming at the search for galactic halo WIMPs at the Canfranc Underground Laboratory with bolometric detectors. It is a collaborative effort between the Institute d'Astrophysique Spatiale (IAS, Orsay, France) and the University of Zaragoza. The bolometers consist of an absorber with a small Neutron Transmutation Doped (NTD) Ge thermistor glued onto it in order to measure the temperature increase produced in the former by a particle interaction. They are suspended by Kevlar or Dacron wires inside copper frames thermally coupled to the mixing chamber of a small dilution refrigerator working at a temperature just below 20 mK. This dilution refrigerator is operated either at Canfranc for background measurements, where it is installed in an ultra-low radioactivity shielding, or at IAS-Orsay for bolometers testing.

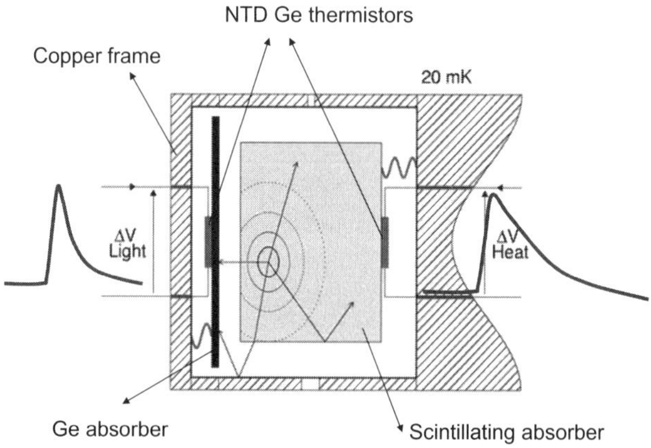

Fig. 1. Scheme of the double bolometer configuration used in the simultaneous measurement of light and heat. It consists of a scintillating bolometric target and an optical bolometer.

The first phase of the experiment [2] consisted of a series of runs performed with three small sapphire (Al_2O_3) bolometers and a total mass of 100 g. It was basically devoted to the understanding of the background and its progressive reduction by a careful selection of the components of the bolometers and cryostat. In the second phase ROSEBUD approached a new strategy based on the search for high-performing targets fitting the requirements imposed by the direct detection of the hypothetical dark matter: low energy threshold, β/γ background discrimination and wide absorber choice to match signal dependencies on nuclear spin and/or nuclear mass. In this second phase, three different absorbers were used: 50 g Al_2O_3, 67 g Ge and a prototype 54 g scintillating $CaWO_4$ [3]. It was the first time that three different target materials were operated simultaneously in a common experimental set-up. The $CaWO_4$ absorber was mounted in a double bolometer configuration: facing a high performance Ge optical bolometer (in the same copper frame) to measure the scintillation light escaping from the $CaWO_4$ after a particle interaction (see Fig. 1). The simultaneous measurement of scintillation light (as heat in the Ge bolometer) and heat (in the $CaWO_4$

target) allows to improve the sensitivity for dark matter searches thanks to the clear discrimination of the feebly scintillating nuclear recoils (like those produced by interacting WIMPs) against the electron recoils (produced by the usually much larger β/γ background) as shown in Fig. 2. ROSEBUD was the first experiment to implement this technique in a deep underground location [3].

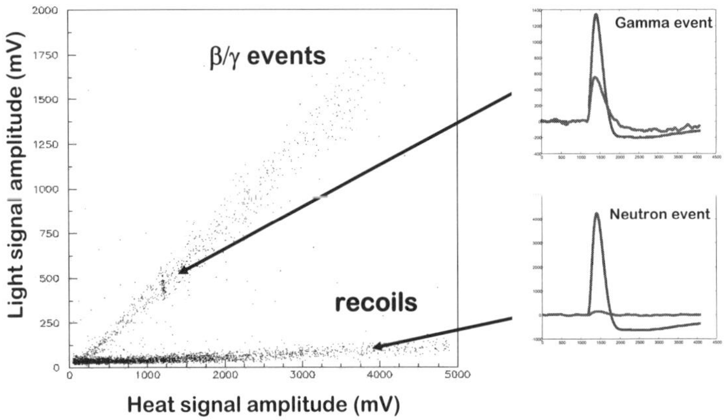

Fig. 2. Examples of light and heat pulses corresponding to gamma and neutron events in a Kyropoulos-grown undoped sapphire scintillating bolometer. The difference in relative scintillation yield is represented in a light vs. heat plot, where the anticorrelation mentioned in the text (Section 2.1) is observed at the 60 keV ^{241}Am gamma line (around 1.2 V heat signal amplitude).

2. Scintillating bolometers testing

Different materials have been tested at Orsay (BGO, Al_2O_3, TeO_2, CaF_2, SrF_2 and SiO_2, amongst others), showing all of them scintillation at temperatures around 20 mK and many of them excellent performances for particle discrimination [4]. In particular, undoped sapphire and BGO have shown a high light yield at low temperature and very good energy threshold for background discrimination and are going to be used in the next ROSEBUD run at the Canfranc Underground Laboratory.

2.1. *Undoped Sapphire Absorber*

Very interesting is the observation of a light-heat anticorrelation for γ monoenergetic events with scintillating bolometers made of undoped sapphire.

This anticorrelation has allowed to estimate the absolute light yield at very low temperature [5]. The scintillation mechanism responsible of the high light yield observed is not fully understood and further complementary tests are in progress. In particular, two different manufactured undoped sapphire bolometers have been tested, giving very different results concerning relative light yield (see Fig. 3) and fluorescence timing properties. The Kyropoulos-grown undoped sapphire has shown a higher beta-gamma vs. alpha light yield (10:1) with respect to the Verneuil-grown crystal (2:1), having both very similar light yield for alpha events. The differences in timing behavior of the scintillation, prompt for the former and delayed for the latter, hints at different scintillation mechanisms, probably related to very small amounts of impurities or defects.

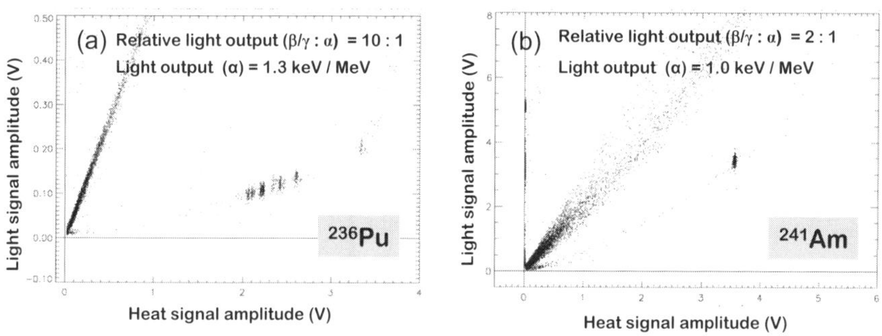

Fig. 3. Different behavior in the relative light yield obtained with two scintillating bolometers made of undoped sapphire crystals with powders from different origin and grown by different techniques: a) Kyropoulus and b) Verneuil.

An estimate of the heat signal relative efficiency factor (REF) for low-energy nuclear recoils against electron recoils, specially relevant for the dark matter search, is in progress in the Kyropoulos-grown undoped sapphire. Preliminary results will be presented here below. According to the conservation of the energy and assuming no energy loss into traps, those events showing smaller scintillation should be associated to larger heat depositions in the scintillating target because scintillation photons do effectively escape from the target to be absorbed in the optical bolometer. In the case of particles with associated high light yield, this should allow the nuclear recoils REF in the heat signal to be larger than one, making them very favorable detectors to be applied in the search for dark matter. The absolute light yield obtained in Ref. 5 for undoped sapphire assuming no energy losses and no self-absorption is 13%, implying that a REF of about 1.1 could be obtained for the heat signal coming from nuclear recoil events.

In order to investigate such a possibility, a calibration run with a ^{210}Po internal source has been carried out at IAS. The very thin source emits the α-particle but also the recoiling nucleus. The peak corresponding to the ^{206}Pb recoils is clearly seen in Fig. 4. From a calibration with gamma sources (^{57}Co and ^{109}Cd) and after fitting only the right side of the nuclear recoil peak to a gaussian, we associated to it an energy of $106.9 \pm 0.13_{stat}$ keV. This result implies a REF of 1.03 and it should be quite a conservative value provided that degradation of the nucleus energy in the source is possible. We did observe possible superficial effects in the heat signal of electrons from a calibrated ^{109}Cd source: electron events were observed at energies lower than nominal ones (see Fig. 4). Further investigation is required in order to understand the origin of such an effect and to analyze its possible influence in the previous recoil REF estimate.

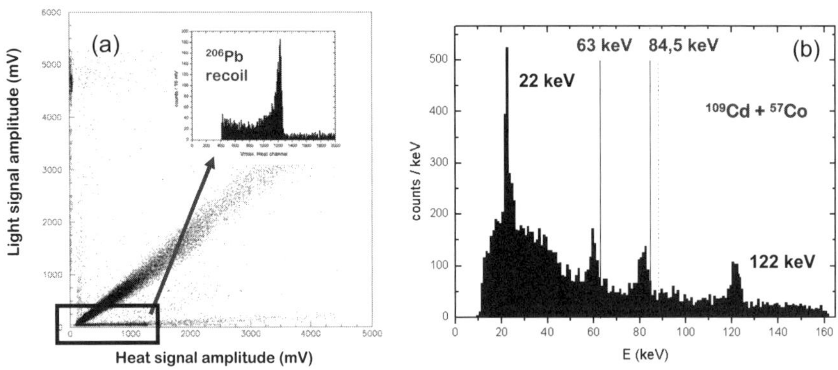

Fig. 4. (a) Results from a ^{210}Po calibration with an undoped sapphire scintillating bolometer: the light vs. heat plot shows the excellent discrimination between nuclear recoils from the source and the β/γ background. In the inset it is shown the discriminated recoil spectrum in the heat signal. (b) Results from a measurement with ^{109}Cd and ^{57}Co sources with undoped sapphire in the heat signal. The lines coming from electrons are shifted from the expected energy (solid vertical lines) if gamma calibration is used.

2.2. *BGO Absorber*

Results obtained with BGO as absorber are also very interesting [6]. A light output of 7.5 keV/MeV for β/γ events has been obtained and relative light output factors are 15:2.6:1 for β/γ, α and neutrons. A discrimination of β/γ

background larger than 90% C.L. above 23 keV is guaranteed. The very good discrimination capability did allow to observe the α-decay of ^{209}Bi [7] and although internal contaminants are important, the high discrimination expected could make BGO a very interesting target in the search for dark matter.

3. Prospects

In the frame of the Canfranc dark matter search program we plan to install two scintillating bolometers: 50 g Al_2O_3 and 92 g BGO already tested at IAS, Orsay. The goal is to obtain nuclear recoil discrimination thresholds of 15 keV and 10 keV for the BGO and Al_2O_3, respectively. Work on further background improvements, particularly that coming from materials in contact to or in the bolometers themselves is also in progress. A neutron shielding will be included in the experimental set-up. The final objective is to collect enough data with discrimination at low energies and conveniently chosen targets to explore WIMPs with low and intermediate masses (10-100 GeV) and with any of both, coherent or spin-dependent interactions.

Acknowledgments

This work has been partially supported by the French CNRS/INSU (MANOLIA and BOLERO projects), by the Spanish Commission for Science and Technology (MEC, Grant No. FPA2004-00974), the Gobierno de Aragón (Group in Nuclear and Astroparticle Physics), the EU Network Contract No. HPRN-CT-2002-00322 and the EU Project ILIAS Contract No. RII3-CT-2004-506222.

References

1. Proceedings of the 9th International Conference on Topics in Astroparticle and Underground Physics, TAUP2005, Zaragoza, Spain *Journ. of Physics (Conf. Series)* **39** (2006).
2. S. Cebrián *et al.*, *Astrop. Phys.* **10**, 361–368 (1999).
   S. Cebrián *et al.*, *Astrop. Phys.* **15**, 79–85 (2001).
3. S. Cebrián *et al.*, *Phys. Lett. B* **563**, 48–52 (2003).
   S. Cebrián *et al.*, *Astrop. Phys.* **21**, 523–533 (2004).
4. N. Coron *et al.*, *Nucl. Instr. Met. in Phys. Res. A* **520**, 159–162 (2004).
5. J. Amaré *et al.*, *Appl. Phys. Lett.* **87**, 264102–264104 (2005).
6. J. Amaré *et al.*, *Journ. of Physics (Conf. Series)* **39**, 133–135 (2006).
7. P. de Marcillac *et al.*, *Nature* **224**, 876–878 (2003).

DETECTOR PERFORMANCE OF THE NEWAGE EXPERIMENT

KENTARO MIUCHI[a], KAORI HATTORI[a], SHIGETO KABUKI[a], HIDETOSHI KUBO[a], SHUNSUKE KUROSAWA[a], HIRONOBU NISHIMURA[a], YOKO OKADA[a], ATSUSHI TAKADA[a], TORU TANIMORI[a], KEN'ICHI TSUCHIYA[a], KAZUKI UENO[a], HIROYUKI SEKIYA[b] and ATSUSHI TAKEDA[b]

Cosmic-Ray Group, Department of Physics, Graduate School of Science, Kyoto University Kitashirakawa-oiwakecho, Sakyo-ku, Kyoto, 606-8502, Japan[a] *E-mail: miuchi@cr.scphys.kyoto-u.ac.jp*

Kamioka Observatory, ICRR, The Univ. of Tokyo Higashi-Mozumi, Kamioka cho, Hida 506-1205 Japan [b]

NEWAGE(NEw generation WIMP search with an Advanced Gaseous tracking device Experiment) project is a direction-sensitive dark matter search experiment with a gaseous micro time-projection-chamber(μ-TPC). We report on the performance of the μ-TPC with a detection volume of $23 \times 28 \times 30$ cm^3 operated with a carbon-tetrafluoride (CF_4) of 0.2 bar.

Keywords: Time projection chamber; Micro-pattern detector; dark matter

1. Introduction

Weakly interacting massive particle(WIMPs) are thought to be one of the most plausible candidates of the dark matter. Most of the dark matter search experiments are designed to measure only the energy deposition on the nucleus by a WIMP-nucleus scatterings. Because the amplitude of an annual modulation signal is only a few % in the rate, positive signatures of the WIMPs are very difficult for detection with only the energy information. Owing to the motion of the solar system with respect to the galactic halo, the direction-distribution of the WIMP velocity observed at the earth is expected to show an asymmetry like a wind of WIMPs. Attempts to detect a positive signature of WIMPs by measuring the recoil angles have been carried out[1–5] ever since it was indicated to be an alternative and a reliable method.[6] Gaseous detectors are one of the most appropriate devices for detecting this WIMP-wind. DRIFT project has performed underground runs for more than two years with a $1m^3$ time projection chamber (TPC) filled with a low pressure CS_2 gas.[4] We proposed to use a carbon-tetrafluoride (CF_4) as a chamber gas of our time projection chambers with a micro-pixel chamber readout (μ-TPC) aiming to detect WIMPs via spin-dependent (SD) interactions.[7] In this paper, the performance of the μ-TPC is described.

2. Measurements

2.1. *Micro-TPC*

A μ-TPC is a time projection chamber with a micro pixel chamber (μ-PIC[8]) readout, developed for the detection of tracks of charged particles with fine spatial resolutions.[9] A μ-PIC is a gaseous two-dimensional position-sensitive detector manufactured by the printed circuit board (PCB) technology. With the PCB technology, large-area detectors can, in principle, be mass-produced, which is an inevitable feature for a dark matter detector. The pixel-pitch of the μ-PIC is 400 μm and the detection area is 31×31 cm^2. We had studied the performance of a small size ($10 \times 10 \times 10$ cm^3) μ-TPC with a 0.2 bar CF_4 gas.[9] We then developed a large-volume μ-TPC with a detection volume of $23 \times 28 \times 30$ cm^3, and studied its fundamental properties with an Ar-C_2H_6 gas mixture at a normal pressure.[10] The data acquisition system is described in Ref. 11. We will describe the performance of the μ-TPC with a 0.2 bar CF_4 gas in the following subsections.

2.2. *Energy calibration*

We calibrated the energy of the μ-TPC with α particles generated by the $^{10}B(n,\alpha)^7Li$(Q=2.7MeV) reaction. We set a glass plate with a size of $27 \times 70 \times 1mm^3$ coated with a thin 0.6 μm ^{10}B layer in the μ-TPC. The picture of the boron coated glass set in the μ-TPC is shown in the left panel of Fig. 1. Fast neutrons from ^{252}Cf were moderated and the thermalized neutrons were captured by the ^{10}B layer. Alpha particles are emitted from the layer and the integrated two-dimensional image is shown in the right panel of Fig. 1. It is seen that the α particles from the $^{10}Blayer$ are detected and the corresponding position has a high counting rate.

The measured and simulated spectra are shown in Fig. 2. The edge which corresponds to the full energy deposition of the α particle and the α+Li are seen in the both spectra. The elastic scatterings of the fast neutrons from the ^{252}Cf source gives an increase of the lower energy part of the measured energy spectrum.

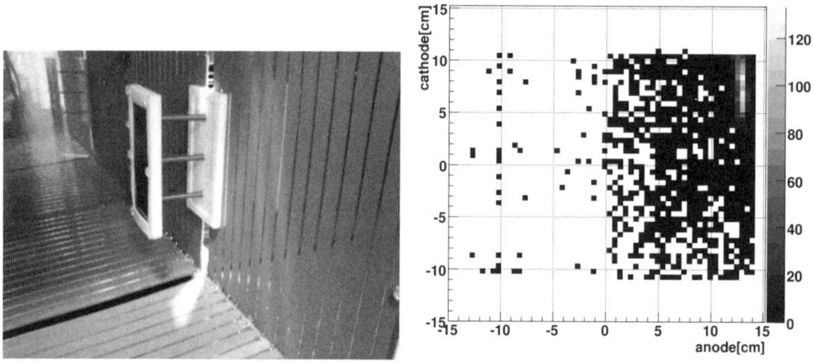

Fig. 1. Picture of the glass plate coated with a boron layer(right) and an integrated event display of a calibration measurement.

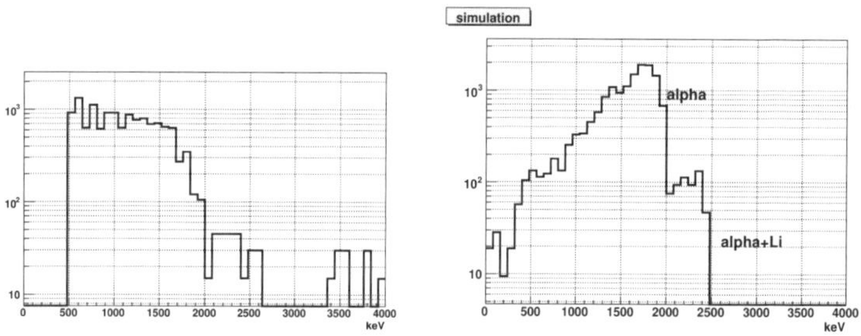

Fig. 2. Measured(left) and simulated spectra of the calibration.

Alpha peaks(5.6 MeV, 6.1 MeV, 7.2 MeV) form the decays of the radon daughters are also used for the energy calibration. The energy resolution was also measured with these alpha peaks and the energy resolution was 50%(FWHM) in the high energy(3-8 MeV) region.

2.3. *Absolute detection efficiency of nuclear recoils*

The detection and the event-selection efficiency was measured by irradiating the fast neutrons from ^{252}Cf. Here the events which satisfied the following three conditions were selected as a nuclear recoil events.

- The track is in the fiducial volume ($21.5 \times 22 \times 31$ cm^3).
- The track has more than three digital hits.
- The track is shorter than 1cm.

The first requirement is the fiducial cut to reject the protons from the drift wall, the second is for the direction determination, and the third one is for the gamma-ray rejection. The detection and the selection efficiency was about 40% at 100 keV and the efficiency in the energy region of 100-400keV (DM energy region) ϵ was fitted by $\epsilon = 1.0 \cdot erf((E-45.8)/165.2)$, where $erf(x) = \frac{2}{\sqrt{\pi}} \int_0^x exp(-x^2)dx$ is the error function and E is the detected energy. The measured efficiency and the best fit function are shown in the left panel of Fig. 3.

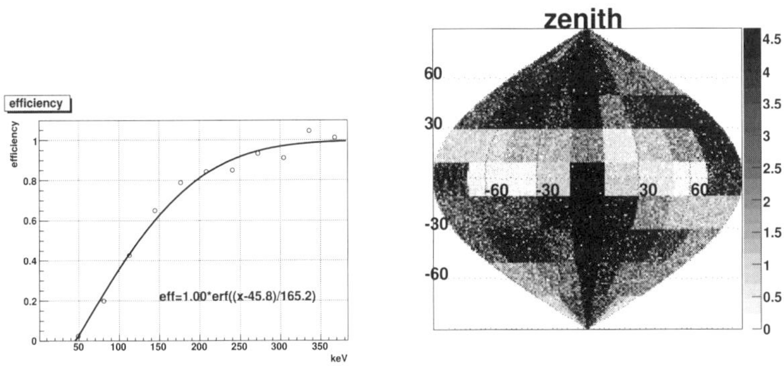

Fig. 3. Absolute detection efficiency of the nuclear recoils (left) and the relative direction-dependent response.

2.4. *Direction-dependent detector response*

The direction-dependent detector response is one of the most important properties for the direction-sensitive dark matter search. We irradiated the fast neutrons from various position to generate "a uniform recoils". We determined the direction of the selected events by fittings the digital hits and the direction-map is shown in the right panel of Fig. 3. Because we don't detect the head-and-tail of the tracks, this mas is restricted to the half sky, *i. e.* -90° < azimuth <90° and -90° < zenith <90° The response is not very flat nor symmetry because of the inhomogeneity of the detector. A much more flat response is expected for the next detector with an improved manufacture technologies.

2.5. *Gamma-ray rejection*

It is one of the outstanding advantage of the gaseous TPC that the track-length and the energy-deposition correlation provides a strong gamma-ray rejection.[9] We measured the gamma-ray rejection factor by irradiating the gamma-rays from a ^{137}Cs source. Measured spectra with and without the gamma-ray source and the background-subtracted spectrum are show in Fig. 4 We define the gamma-ray miss-identification factor (or the electron-detection efficiency) f_e as

$$f_e \equiv \frac{R_\gamma - R_{\mathrm{BG}}}{R_{\mathrm{sim}}}, \tag{1}$$

where R_γ and R_{BG} are the the measured counting rate DM energy region with and without the gamma-ray source and R_{sim}, is the expected rate of the electrons which has a initial energy in the DM energy region. The result was $f_e < 2.0 \times 10^{-4}$. This "rejection factor" was restricted by the background neutrons in the surface laboratory.

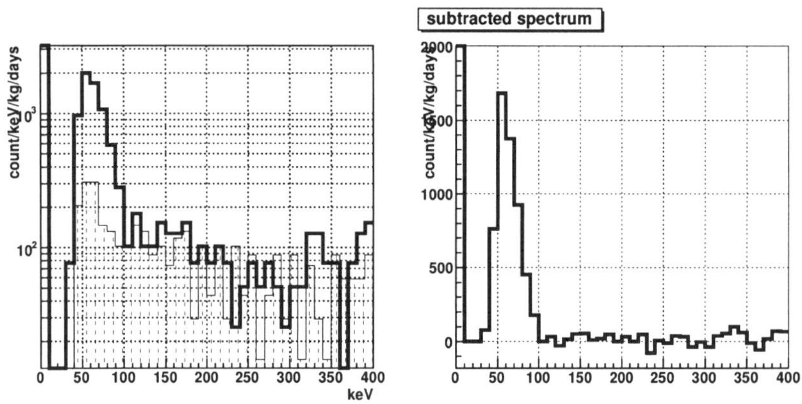

Fig. 4. The energy spectra measured with(solid) and without(hatched) the gamma-ray source(left). The background-subtracted(=solid-hatched) spectrum(right).

3. Summary

We measured the performance of the μ-TPC and obtained the following results.

- The detector is calibrated with α-particles.

- The absolute efficiency is about 0.4% at 100keV.
- The electron miss-identification probability is less than 2×10^{-4} in the 100-400 keV energy range.
- The direction dependent response is measured. After a pilot run in the surface laboratory, we will start an underground measurement in Kamioka Observatory in January, 2007.

Acknowledgments

This work was supported by a Grant-in-Aid in Scientific Research of the Japan Ministry of Education, Culture, Science, Sports, Technology; research-aid program of Toray Science Foundation; and Grant-in-Aid for the 21st Century COE "Center for Diversity and Universality in Physics".

References

1. P. Belli, R. Bernabei, A. Incicchitti, D. Prosperi, Nuovo Cimento C **15** (1992) 475; R. Bernabei, P. Belli, F. Nozzoli, A. Incicchitti, Eur. Phys. J. C 28 (2003)203.
2. K. N. Buckland, M. J. Lehner, G. E. Masek, and M. Mojaver, Phys. Rev. Lett. **73** (1994) 1067.
3. M. J. Lehner, K. N. Buckland, and G. E. Masek, Astroparticle Physics **8** (1997) 43.
4. G.J. Alner *et al.*, Nucl. Instrm. Methods A **555** (2005) 173; N Spooner, in these proceedings.
5. H. Sekiya *et al., Proceedings of IDM 2004", Edinburgh, UK, 2004*, edited by N. J.C. Spooner and V. Kudryavtsev, (World Scientific, 2005), p. 378.
6. D. N. Spergel, Phys. Rev. D **37** (1988) 1353.
7. T. Tanimori *et al.*, Phys. Lett. B **578** (2004) 241.
8. A. Takada *et al.*, proceedings of the conference of the 7th International Conference on Position Sensitive Detectors, Liverpool, England. To appear in Nucl. Instrm. Methods,doi:10.1016/j.nima.2006.10.283
9. H. Sekiya *et. al.*, proceedings of the conference of the 7th International, doi:10.1016/j.nima.2006.10.322
   Conference on Position Sensitive Detectors, Liverpool, England.
10. k. Miuchi *et al., Proceedings of the 8th International Workshop on Radiation Imaging Detectors (IWORID8)*,physics/0701085 Pisa, Italy, 2-6 July 2006.
11. H. Kubo *et. al.*, IEEE Nuclear Science Symposium Conference Record N14-30 (2005).

THE DRIFT-II DARK MATTER DETECTOR

D. N. MUNA and N. J. C. SPOONER
ON BEHALF OF THE DRIFT COLLABORATION
Department of Physics and Astronomy, University of Sheffield
Sheffield S3 7RH, United Kingdom
Email: d.muna@sheffield.ac.uk, n.spooner@sheffield.ac.uk

The DRIFT dark matter experiment is located in the Boulby mine located on the east coast of northern England. It is a directionally sensitive detector capable of reconstructing nuclear recoils in three dimensions. Motivation and status of the experiment are discussed.

Keywords: Dark Matter; Directional Detector; TPC; DRIFT.

1. Detecting a Directional Dark Matter Signal

Dark matter, thought to comprise 25–30% of the mass of the Universe, has not yet been positively and unambiguously identified, although there is a substantial list of observations that support its existence. The most promising candidate is a weakly interacting massive particle (WIMP), thought to form a galactic halo extending several times beyond visible matter. As our solar system orbits the galactic centre it passes through this halo — quite literally, given that the estimated cross section between WIMPs and baryonic matter is extremely small. Using the simplest form of the halo, a $1/r^2$ spacial distribution, the direction of WIMP-induced recoils will be preferentially strongest in the direction opposite to our motion through the halo. This effective WIMP "wind" presents an isotropy in the WIMP flux.

The Earth's ecliptic plane is inclined at an angle of 60° relative to the galactic equatorial plane, as illustrated in Figure 1. This geometry results in a difference of the Earth's velocity in the direction of solar motion. The Sun, and with it the solar system, travels around the galactic centre at 220 km/s. Earth's specific velocity around the galactic centre (and thus through the halo) varies according to the orbit geometry, being greatest in June and at a minimum in December. Correspondingly, the flux peaks and dips at these time, presenting an *annual modulation.* Beyond this, however,

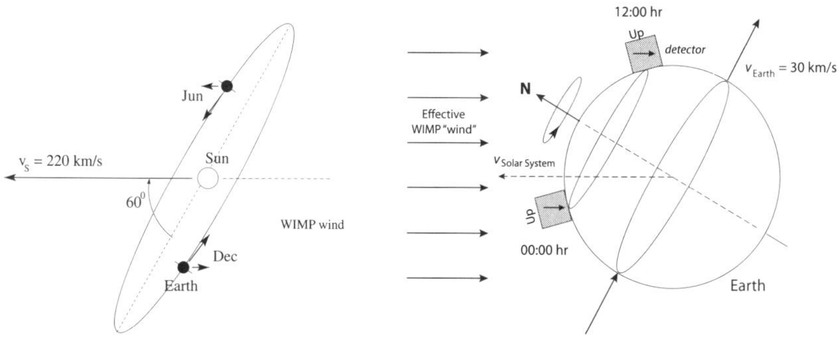

Fig. 1. Wimp wind.

there is another directional variation due to the Earth's daily rotation. The second diagram in Figure 1 illustrates a detector on Earth. At noon, the WIMP wind is directed towards one wall of the detector, while twelve hours later the wind is directed toward the bottom. (This particular geometry is specific to the latitude upon which the DRIFT detector sits.) This *daily modulation* over a sidereal day is a second directional signal of a WIMP halo. This is illustrated in Figure 2.

The DRIFT-II detector was designed to measure nuclear recoil tracks created by elastic collisions between WIMPs and nuclei in a target mass. The incident angle of the incoming WIMP cannot be measured, but the direction of the recoil can be. (More detail on how this works in the detector will follow below.) Assuming a zero background, the number of events required to positively identify a WIMP signal is on the order of tens assuming the sense (direction of the track) can be determined, although greater statistics can overcome that deficiency. Moving forward beyond a positive detection, the same directional information can then begin to distinguish different halo models, opening the door to "WIMP astronomy." The reader is directed to the paper by A. Green in this volume as well as Morgan, et al. 2005[1] and Morgan, Green, 2005[2] for more detailed statistical analyses.

2. The DRIFT Concept

The DRIFT experiment[3] is based on a negative ion time projection chamber (NITPC),[4,5] using TPC technology similar to that found in high energy particle physics. Figure 3 shows a simplified geometry. A steel box with a fiducial volume of 1 m^3 is filled with gaseous CS_2 and kept at a pressure of 40 Torr so that recoil tracks have measurable lengths on the order of mil-

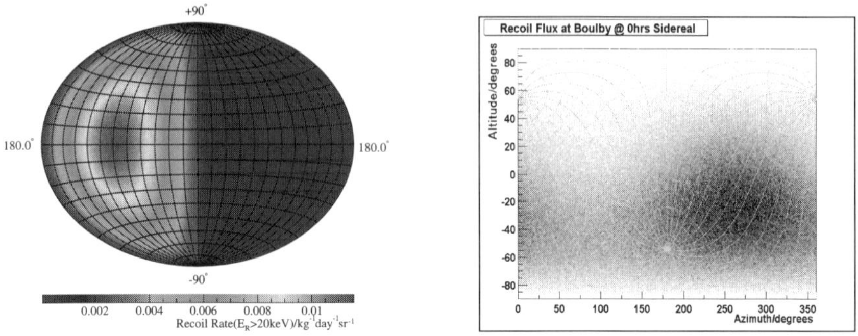

Fig. 2. The plot on the left is the result of a Monte Carlo calculating the recoil flux given the WIMP flux from the standard halo model. The anisotropy on this "all sky" map is very clear. The plot on the right demonstrates the directional variation of recoil flux due to the daily (sidereal) motion.

limetres. This gives a total target mass of 167 g. A central cathode, made from 20 μm wires with a 2 mm pitch, are kept at a voltage of 40 kV providing a 624 V/cm mean electric field in the chamber. This effectively splits it into two detectors. An incoming WIMP will pass through the detector until it elastically scatters off either a C or S nucleus, producing a track of ionisation as negative ion pairs (NIPs). CS_2 is an electronegative gas and will easily capture the free electrons, and now negatively charged ions will drift toward the multiwire proportional chamber (MWPC) endcap. The MWPC is comprised of two planes of 512 wires with a 2 mm pitch oriented perpendicularly. Charge is deposited onto anode wires via electron avalanche and induced on the grid (cathode) plane "above" it. The number of anode wires hit provides dx, the spread of the charge on the cathodes is dy, and the drift time is converted to dz via the drift velocity. Veto wires are positioned around the detector for discrimination — any charged particle (such as an alpha) entering the chamber will trigger the vetos and be rejected.

DRIFT-II is housed at the Boulby Laboratory, located in a working potash and salt mine 1100 m underground (~3000 metre water equivalent) on the east coast of England. It was designed, built, and made operational in one year. It has a potential sensitivity of 10^{-7} pb from 3 m^3 of CS_2 at 40 Torr over 3.4 years, run at a 20 keV S recoil energy threshold assuming a zero background. Rather than having to scale up a single detector, multiple modules can be run in parallel reducing cost and downtime due to any component failure. It is worth noting that if the pressure were to be increased to an achievable 160 Torr, a one tonne target would occupy

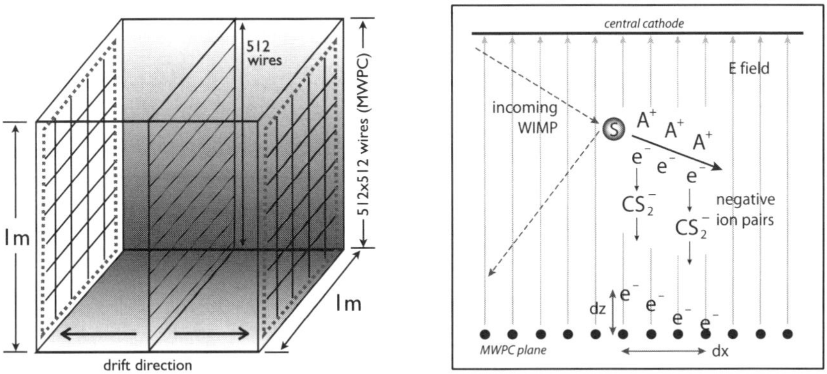

Fig. 3. A simple geometry of the DRIFT-II detector, described in the text.

about 25% of the volume of the MINOS neutrino experiment at the Soudan mine. Finally, if head-tail track discrimination can be achieved, the required target volume decreases by a factor up to ten. Thus, a scaling up of the DRIFT experiment is entirely possible.

Range is measured as a projection on the $x - -y$ plane as $R_2 = \sqrt{\Delta x^2 + \Delta y^2}$ and in three dimensions as $R_3 = \sqrt{\Delta x^2 + \Delta y^2 + \Delta z^2}$. It is convenient to use the energy unit of NIPs for the total integrated ionisation energy, and this value is proportional to the integrated voltage. The conversion rate is measured by performing a calibration of ^{55}Fe, where the peak of the energy spectrum is 300 NIPs. The R_2 and R_3 range values provide a powerful method of event discrimination; neutron tracks typically cross only a few wires while alphas can easily cross 20–200+ wires.[6] Similarly, the parameter dE/dx in conjunction with a low threshold discriminates gammas very well.

3. DRIFT-II Data Analysis

Energy calibration is performed by using an automated ^{55}Fe exposure every six hours. An example waveform is shown in Figure 4. The first plot shows the raw waveform, and it is difficult to pick out the 5.9 keV events from the noise. However, after removing some of the noise through a Fourier analysis and applying a Savitsky-Golay polynomial convolution filter,[7] the signal-to-noise ratio increases and the low energy events are easy to pick out. Calibration is performed independently for each MWPC and gives a left/right and anode/grid relative gain to 5%, and the smoothing allows for a relative gain calibration to better than 1%.

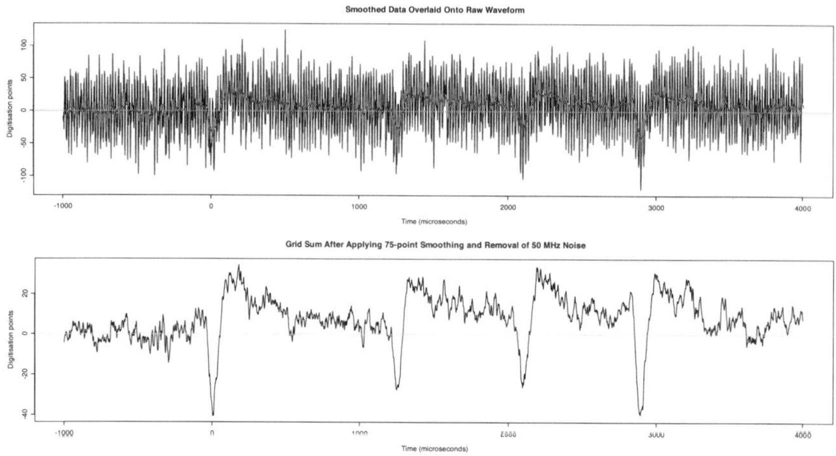

Fig. 4. Calibration waveform showing individual ^{55}Fe (5.9 keV) events. Above, raw data; below, post-processing by Fourier analysis and a polynomial convolution filter.

Figure 5 shows two typical events. On the left is an alpha track. Although there are 512 anode and grid (cathode) wires, they are grouped down to eight readouts. For this reason high energy events that produce long tracks, such as an alpha, will appear to "wrap around" as seen in this figure. It is not difficult to reconstruct these events to calculate, for example, how many wires were actually crossed. Most of the events that we are concerned with (such as WIMP events and neutron events for calibration) will not go beyond eight wires and so are fully contained by the available readouts. In the figures, the grid (cathode) readouts are labelled as G1–G8, the anodes are A1–A8, AS and GS indicate the sum of the anodes and grids respectively, GV and AV are the grid and anode veto readouts, and finally G–A is the anode veto readout subtracted from the grid veto readout (used for a coincidence cut).

Track reconstruction in three dimensions is possible by combining the anode (x), grid (y), and time (z) information. A detailed analysis of how this is done is presented in the paper by D. Muna elsewhere in this volume. The two events in Figure 5 are reconstructed in that paper.

The primary challenge of the data reduction process is to implement an automated way to identify events due to the background or electronic effects. A few of these cuts are implemented in real time by the DAQ which reduces the amount of data collected. Beyond this, other cuts applied identify

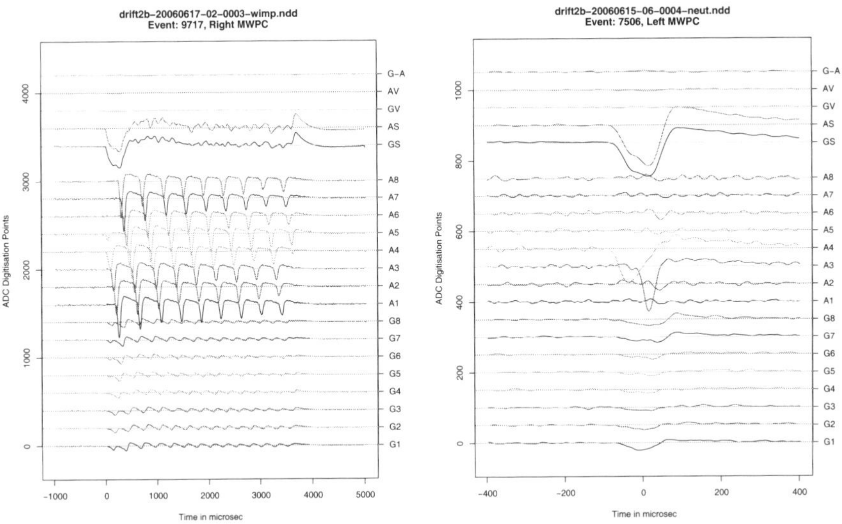

Fig. 5. Examples of event records: a typical alpha at left, and a neutron-induced S-recoil on the right. The MWPC is a grid of 512 wires that are grouped down to eight readouts. For this reason, the track appears to "wrap around." Track reconstruction unravels this into three dimensions (detailed in the paper by D. Muna elsewhere in this volume). Note the difference in both vertical and horizontal scales.

alphas, spurious signals in the pretrigger, vetos, large overshoots, sparks, narrow full-width half maximum signals, tracks that fall on non-consecutive wires, and others. These in total reduce the raw background by a factor of twenty. GEANT 4 simulations are performed to allow us to determine our sensitivity to ^{252}Cf and surrounding rock neutrons. The results for various types of runs (directed neutron runs, shielded runs, etc.) show our data after minimal cuts agree with simulations from 1–9%. Figure 6 shows the efficiency of simple cuts compared to more aggressive cuts designed to reduce the background.

DRIFT-II is very effective at rejecting gamma events. Five 0.52 mCi ^{60}Co sources were placed on top of the unshielded vacuum vessel and 0.575 days of livetime data were recorded. This is equivalent to $\sim$25 days of exposure to environmental gammas from U and Th. Zero events were detected, placing an upper limit of 2.44 at 90% confidence level. Table 7 shows the difference efficiencies at different energy levels (measured in NIPs).

The detector has performed as designed and has proven to work reliably over time. There are three main methods used to calibrate the de-

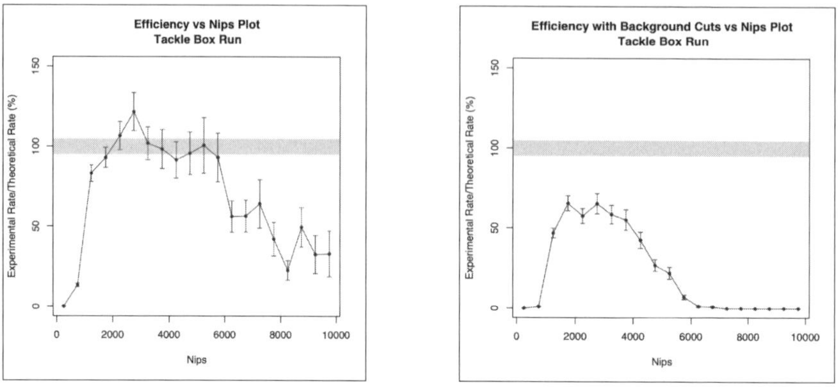

Fig. 6. Plots demonstrating neutron efficiency. The plot on the left is with simple event cuts, while the plot on the right employs more aggressive cuts to minimise the background.

NIPs Window	^{60}Co rate minus background rate (Hz)	Monte Carlo rate (Hz)	Rejection Factor Limits (90% C.L.)
1000–2000	$(-2\pm5)\cdot10^{-4}$	77 ± 1	$< 8\cdot10^{-6}$
2000–3000	$(-1\pm1)\cdot10^{-4}$	24.8 ± 0.5	$< 5\cdot10^{-6}$
3000–10000	$(2\pm6)\cdot10^{-5}$	27.2 ± 0.3	$< 3\cdot10^{-6}$
1000–10000	$(1\pm3)\cdot10^{-4}$	125.2 ± 0.7	$< 3\cdot10^{-6}$
1000–6000	$(1\pm3)\cdot10^{-4}$	128.9 ± 0.7	$< 3\cdot10^{-6}$

Fig. 7. Gamma rejection efficiency, ^{60}Co exposure.

tector: ^{55}Fe, ^{252}Cf, and ^{60}Co gammas, and the californium source itself has been well calibrated. DRIFT-IIa has had a stable response to neutrons over several months, and records every recoil event in the energy range of 1000-6000 NIPs. The theory for the response of the detector is well tested, including a GEANT simulation and a conversion of C and S energies to NIPs. The simulation assumed an active volume of 1 m^3 which is in complete agreement with the experimental result, showing the full volume is in use. Finally, we have been able to demonstrate that the livetime calculations of the DAQ are accurate.

4. Radon Progeny Recoils

A residual population of recoil-like background events were discovered in data from DRIFT-IIa. Coincident with this was a population of alpha events that were observed to both start and stop within the fiducial volume, having their origin within the gas.

Neutron simulations of the detector running in a number of different configurations could not account for the high rate of the recoil-like events. However, it has been found that the mean pulse-width of recoil-like events from a typical shielded background run is higher with a narrower distribution, compared to the same parameter extracted from shielded neutron runs with similar statistics. This fact is interpreted as follows; charge clusters deposited in the gas undergo a small diffusion as they drift to the readout. The longitudinal diffusion (i.e. in the drift direction) may be gauged from the width of the pulse - larger widths result from more diffusion and hence greater drift-distance, with the widest pulses being attributed to charge that has drifted from the vicinity of the central cathode. Thus, a cut on pulse-width allows a crude fiducialisation of events.

It was hypothesised that radon gas could be directly responsible for the alpha events and indirectly responsible for the anomalous recoil events. Because of the presence of trace-levels of uranium and thorium, detector components are expected to emanate radon into the detector gas volume. ^{238}U is the dominant contaminant, leading to production of ^{222}Rn, having a half-life $\tau_{\frac{1}{2}} = 3.8$ days and decay alpha energy $E_{alpha} = 5.49$ MeV. Decay of ^{222}Rn gives ^{218}Po which is produced in the form of a positive ion.[8] The central drift-cathode of DRIFT-II makes a very effective electrostatic precipitator, concentrating the charged progeny on the surface of the cathode wires, a process that has been observed elsewhere.[9]

Once at the cathode the progeny undergo further decays in which a high energy alpha particle and low energy recoiling nucleus are produced. There are three subsequent alpha decays between ^{218}Po and stable ^{206}Pb, however the rate for the final alpha-decay of ^{210}Po is dominated by the 22 year half-life of ^{210}Pb. Therefore, only the decays between ^{222}Rn and ^{210}Pb are in equilibrium. Approximately half the time the recoiling nucleus will be ejected into the drift volume (the alpha particle stoping within the wire). These alpha decay-generated nuclear recoils have energies typically of order 100 keV and ranges of order 600 μm — closely mimicking the expected signal events.

High energy alpha events have a very distinctive and easily-selectable signature. A subset of the alphas observed in the DRIFT-IIa dataset are refered to as 'cathode-crossers' (see Figure 5). These satisfy the following criteria: there are no veto signals associated with the event, the track is completly contained within the 4 ms time window of the event-record, and the track deposits ionization in both the left and right detectors.

It was assumed that all cathode-crosser alpha events could be attributed to the decay of ^{222}Rn, since the tracks, by virtue of the above criteria, start and stop in the gas, away from any solid surfaces; subsequent progeny in the chain would be expected to plate-out and hence not start in the gas and to appear only on one side. The rate of cathode-crossers was therefore used as a proportionate indicator of the rate of radon-decays in the one m^3 fiducial volume.

The possibility of low-energy events from Rn progeny first came to light when considering the variation of alpha event rate correlated with operational changes in the running of the detector.

The overall background alpha rate appeared to be stable at ~0.5 Hz, indicating the presence of a low-level alpha source. Initially it was feared that components used to support the ^{55}Fe calibration sources may be the contaminant. However, 200 μm thick vinyl windows were placed over the MWPC calibration apertures during an electronics upgrade part-way through 2005 operations. As expected, the ^{55}Fe rate was reduced by a factor ~10 but there was no observed reduction in alpha rate after this event.

An obvious rate-spike around day 194–195 (13–14 July 2005) was observed at around 1.5 Hz. The gas-flow through the detector was turned off for two days immediately prior to this spike (the only time this occured whilst data-taking continued). The high-rate data were acquired immediately after normal gas-flow had resumed. The assumption here is that radon emanation was allowed to build up within the vessel, where under normal operating conditions of continuous gas-flow it would be flushed out and kept at a lower equilibrium level.

In the period 17–19 August 2005 (day 229–232) the alpha rate started low (~0.3 Hz) but then rose and flattened off at the usual background rate of 0.5 Hz. Immediately prior to this the vessel had been opened, sealed and flushed-out several times with fresh CS_2. This behaviour is explained by the removal of Rn during the initial flushing operation. Once normal running resumed, the Rn level increased to its normal equilibrium level.

On the 9 June 2005 the detector was evacuated, flushed with fresh gas and refilled. Over the following 12 days the rate of cathode-crosser events was observed to rise towards an asymptotic value of ~650 per day. The rise in rate was fit to Equation 1 that depends on the lifetime of ^{222}Rn, $\tau_{\frac{1}{e}} = 5.52$ days, the mean lifetime of a particle within the detector as a result of the gas flow, $\rho_{\frac{1}{e}} = 3.15$ days, the start-time for the emanation, $t_0 = 159.5$ days (measured from 1 January 2005) and the inferred emanation rate, K. As K is the only free parameter, this was varied within the expression to give the best fit: $K = 2194 \pm 327$ per day.

$$R = \frac{K(1 - e^{(t_0 - t)/\tau} \cdot e^{(t_0 - t)/\rho})}{(1 + \frac{\tau}{\rho})} \tag{1}$$

K must then be multiplied by a geometric factor g to determine the total number of ^{222}Rn decays occurring within the 1 m^3 fiducial volume. The value of g was found using a simple Monte Carlo model in which track starting-points were randomly generated within the vessel volume and the ratio of tracks having at least their start point in the vessel volume to tracks meeting the cathode-crosser criteria determined. In each case, the tracks were required to have a range of 334 mm — this figure being the mean range determined from range-tables generated by SRIM 2003,[10] for alpha particles in CS_2 at 40 Torr. The geometric factor was thus determined to be $g = 117.0$, giving a total radon rate of 105,408±7,776 per day (= 1.22±0.08 Hz).

Measurements of radon emanation from a variety of materials used in DRIFT-IIa have revealed rates approximately 78% of the above figure. There are still other DRIFT-IIa materials that require testing, offering the possibility that the entire radon budget may be accounted for. Tests on alternative materials have been carried out and the results suggest that, of the materials tested, the radon-emanation rate may be reduced by a factor of <~25.

Acknowledgments

The DRIFT collaboration acknowledges the Integrated Large Infrastructure for Astroparticle Science (ILIAS) and the Particle Physics & Astronomy Research Council (PPARC).

References

1. B. Morgan, A. M. Green and N. J. C. Spooner, *Physical Review D* **71**, p. 103507(May 2005).
2. B. Morgan and A. M. Green, *Physical Review D* **72**, 123501(December 2005).
3. G. J. Alner and et al., *Nuclear Instruments & Methods A* **555**, 173 (2005).
4. C. J. Martoff, D. Snowden-Ifft, T. Ohnuki and M. J. Lehner, The drift concept-sensitive wimp dark matter search with a gas detector, in *The Identification of Dark Matter, Proceedings of the Second International Workshop*, eds. N. J. C. Spooner and V. Kudryavtsev1999.
5. D. P. Snowden-Ifft, C. J. Martoff and J. M. Burwell, *Physical Review D* **61**, 101301(May 2000).

6. D. P. Snowden-Ifft, T. Ohnuki, E. S. Rykoff and C. J. Martoff, *Nuclear Instruments & Methods A* **498**, 155(February 2003).
7. A. Savitsky and M. J. E. Golay, *Analytical Chemistry* **36**, 1627(July 1964).
8. W. W. Nazarof and A. Nero, *Radon and Its Decay Products in Indoor Air* (Wiley, Canada, 1988).
9. J. McLaughlin and G. Gath, *Radiation Protection Dosimetry* **82**, 257 (1999).
10. http://www.srim.org.

DETECTION OF VUV SCINTILLATION LIGHT IN ONE TON OF LIQUID ARGON

C. REGENFUS*

Physik-Institut, University of Zürich,
CH–8057, Zürich, Switzerland
** E-mail: regenfus@cern.ch*

In the framework of the ArDM project, a next-generation one ton liquid argon WIMP detector, we built a 5 litre prototype chamber. This setup is primarily used to develop the appropriate read out for argon VUV scintillation light (128 nm) using wave length shifting dyes. The final aim is a energy threshold of 30 keV corresponding to yield of roughly 2% for detected photo-electrons. Beside light yield studies we investigate purity effects as well as background suppression power from the light pulse shape.

Keywords: WIMPs; Liquid argon; VUV light; Triplet state quenching.

1. Introduction

The ArDM project was initiated in 2004 [1,2] with the aim to build a combined liquid argon TPC and calorimeter for the detection of WIMP recoils on a target mass around one ton. Free charge as well as the VUV scintillation light should be recorded with high efficiency to reach an order of 10^9 for the suppression of electron (background) versus nuclear recoil events. This should be achieved by the combined measurement of the charge to light ratio, the time structure of the light, the interaction position and the topologic information on a event-by-event base. β decays of ^{39}Ar, present in ^{40}Ar at a concentration around 10^{-15} [3], are supposed to be the dominant background and the main trigger source[a]. The fast and possible selective event trigger will be derived from the light readout.

The working principle of the bi-phase detector is shown in figure 1 (left). A WIMP interaction with liquid argon at 30 keV produces about 400 VUV photons together with a few free electrons. The strong vertical electric field

[a]Roughly 1 kHz in one ton of liquid argon

sweeps the charges to the top of the liquid and extracts them into the gas phase. After multiplication ($\approx 10^5$) in a two stage LEM[b] the charges are collected by a segmented anode. The VUV light is emitted isotropically from the interaction point and converted to blue light by wavelength shifter (WLS) on the side walls. At the bottom of the cryostat the shifted and reflected light is collected by 14 8" phototubes immersed in the liquid. The 3D CAD sketch (fig. 1 right) shows the mechanical design of the prototype detector and the cryostat. Not shown is the foil inside the field shapers with the applied wavelength shifter.

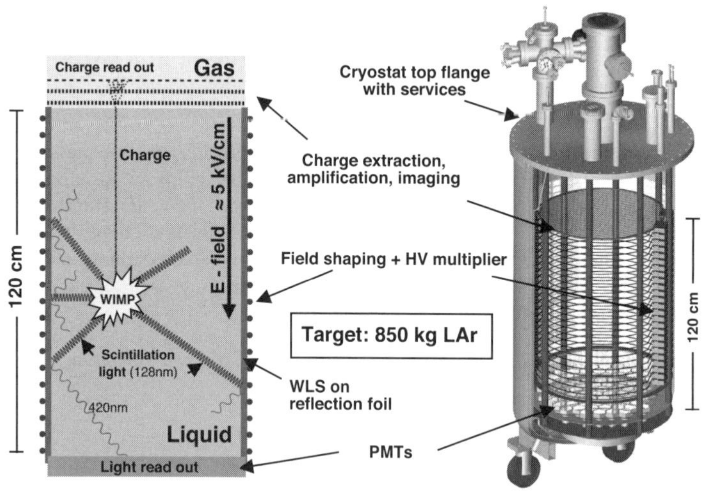

Fig. 1. *Detection principle and 3D drawing of the ArDM detector*

2. Test setup and measurements

A $5\,\ell$ cylindrical chamber (diam. 150 mm, no internal electrical field) was constructed with a variety of service connections (fig. 2 left). Without baking a typical vacuum of 10^{-5} mbar is reached (over 24 h) with a $60\,\ell/\mathrm{s}$ turbo pump. The residual gas is mainly composed of water. When the chamber is filled with liquid argon it is placed inside an open dewar with LAr for cooling. A 40 Bq internal ^{210}Pb source (5.3 MeV α, 1.2 MeV β) is used for the excitation of argon atoms. In argon gas at NTP α particles are fully absorbed, β particles only partially. Scintillation light and free charges are produced in the complex interplay of ionisation, excitation, recombination

[b]Large electron multiplier

and quenching [4–8] depending on the ionisation density. This leads to singlet and triplet argon eximers which decay radiatively (128 ± 5 nm) with two distinct decay times (τ_1,τ_2). The produced VUV photons are not absorbed by atomic argon and are inherently difficult to detect. Therefore we use optically active materials like TPB[c] for shifting the light into the blue (420 nm), which is well matched to the response of bialkali photocathodes. The wavelength shifter (WLS) material can either be dissolved and sprayed or evaporated on foils on the vessel walls. The optimum application technique and substrate material is the subject of developments, including different low temperature PMTs, which are fabricated with a thin Pt underlay. Presently we investigate two 3" prototypes, a square 8-dynode tube from Hamamatsu (R6237-MOD) and a round 12-dynode tube from ETL (D750UKFLB). The phototubes are placed in the upper part of the vessel being immersed in the liquid. To improve the overall light yield the

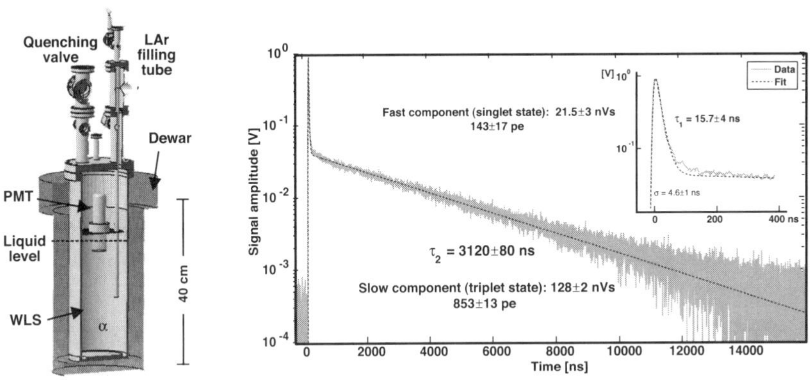

Fig. 2. *5 ℓ test setup and averaged signal of α particles in clean argon gas*

glass window of the phototubes can also be coated with WLS. However this operation is of minor importance since only a small fraction of VUV photons will hit the PMTs directly (Monte Carlo studies).

Liquid argon used for the measurements is delivered at an unspecified level of quality from Carbagas, Switzerland. Argon gas is taken from 50 ℓ cylinders of class Ar 60 (impurities ≤ 1.3 ppm [9]). A re-circulating gas cleaning system is under construction, while oxisorb (CuO) and hydrosorb (molecular sieve) cartridges can be used during filling[d].

[c]Tetraphenyl-butadiene, also P-terphenyl (PTP) is commonly used
[d]Hydrosorb is only used for filling with gaseous argon

The analogue signal from the PMT is sampled by a 1 GS/s digital oscilloscope (20 kS), stored to hard disk and analysed offline. A baseline correction is done with the fraction of the signal before trigger time. The noise is determined by the input amplifier (50 Ω) and shows a $1/f$ spectrum. The number of detected photons is calculated by dividing the integrated signal (charge) by the average single photon charge, which we find from distributions of dark counts and LED light pulses. Figure 2 (right) shows a semilogarithmic representation of 500 averaged α events taken in clean argon gas, the inset is a zoom of the first 400 ns. The dashed line represents a fit to the data by the sum of two exponential decays convoluted with a gaussian. A 6th free parameter was used to fit the event time t_0. The exponential decays model the light from singlet (fast) and triplet (slow) eximer states. From several measurements we find an average value for triplet to singlet yield $R = 6 \pm 1.8$, valid for 5.3 MeV α in argon gas at NTP. The error was estimated from fluctuations among the data sets. The largest measured value for the triplet decay time $\tau_2 = 3120$ ns was found for purest argon gas. This is in good agreement with literature (3200 ± 300 ns) [10].

3. Quenching of the triplet state luminescence

Impurities in argon gas were found to reduce the scintillation light by shortening the decay time τ_2. This effect was studied in detail by mixing argon

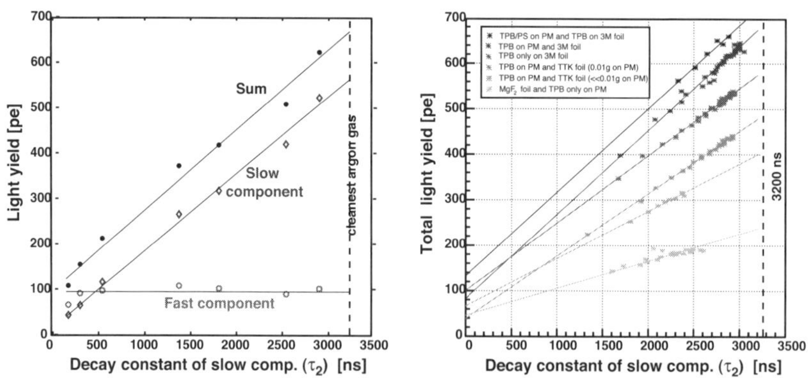

Fig. 3. *Number of photoelectrons (pe) versus decay time* τ_2

gas with air at partial pressures from 1 down to 10^{-6} mbar [11]. For the different air admixtures the averaged α signals were fitted and analysed as described above. Fig. 3 (left) shows the yields obtained for fast, slow and total light, plotted against the fit parameter τ_2. The data points of the slow component exhibit a linear increase with τ_2, while the fast component remains constant. We explain this behaviour with a non radiative destruction of the triplet eximer states by the gas impurities. This is presumably due to residual water, since cooling improves on the light yield (and τ_2). The linear relationship between the light yield and τ_2 is the nature of a PDF for an exponential decay. No evidence is found for a change of R nor for significant absorption of VUV light by the gas impurities.

These results establish our light yield calibration and allow for substantially faster measurements by simply flushing the vessel with argon gas without prior evacuation of the air. As an example, fig. 3 (right) shows measurements done with different coatings and reflectors. Knowing the yield of VUV photons (78 k per α particle [12]), we find the yields of the different designs by extrapolating to $\tau_2 = 3200$ ns.

A similar correlation of light yield and τ_2 seems to be present in liquid argon and is subject to current investigations. However stronger VUV absorption is expected due to the higher densities.

4. Measurements in liquid argon

Due to the high stopping power of liquid argon both, α and β particles from the ^{210}Pb source can be seen in the small setup. They cause different ionisation densities, hence show different values for R and light yield[e].

This effect was first observed in a measurement with TPB coated Teflon mesh (Tetratex) on the side walls of the test chamber. Since a low temperature PMT was not yet available we used a standard bialkali PMT (heated) with an acrylic light guide plunged into the liquid argon. Events from the internal ^{210}Pb and external γ source were recorded for several hours and analysed as described above. The signals were again found to be described by the sum of two exponential decays. A value of about 800 ns was measured for τ_2. Two populations of event types were identified, differing in the ratio R of triplet to singlet states. We identify these groups with signals from α ($R = 0.2$) and β particles ($R = 1.7$). The data are shown in fig. 4. The upper histograms are pulse height spectra (total light) of the ^{210}Pb source alone (left) and including an external γ source (right). The left peak is attributed to β and the right to α events. From the distribution means we find a quenching factor of about 0.4 for α particles. The lower plots add the information from the pulse shape to the ordinate, calculated from the

[e]Quenching: α particles lead to similar light yields as nuclear recoils

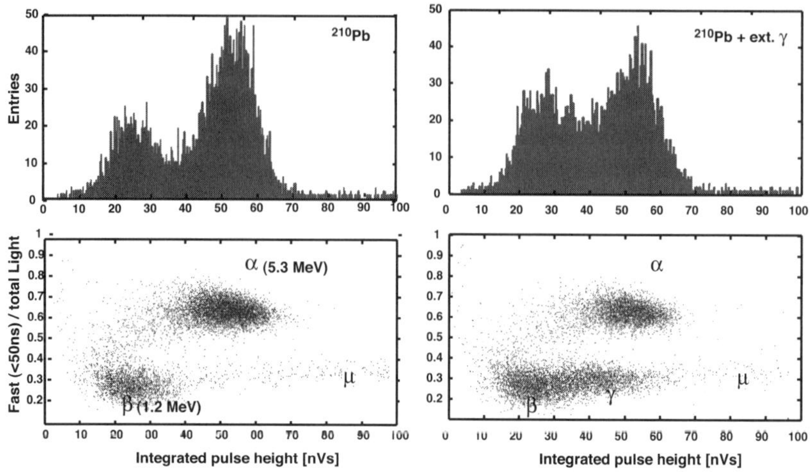

Fig. 4. *Signals in liquid argon, without (left) and with external γ source (right). The upper plots show the integrated pulse height spectra, the lower plots show the ratio of fast (<50 ns) to total light versus the integrated pulse height.*

fraction of fast (<50 ns) to total light. The two islands for α and β events are already well separated despite a modest light yield (0.4%).

5. Conclusions

Our R&D results for the VUV light yield are rather encouraging for a large detector volume. The goal is to reach 2%, while 1% has already been achieved. The acquired experience allows us to start the construction of the one ton prototype in 2007.

References

1. A. Rubbia, http://arxiv.org/abs/hep-ph/0510320 (2005)
2. The ArDM project, A. Bueno, these same proceedings
3. H. Loosli, Earth and Plan. Sci. Lett., **63**, 51 (1983)
4. J. Lindhard, Mat. Fys. Medd. Dan. Vid. Selsk. **33**, 1 (1963)
5. S. Kubota et al., J. Phys. C **11**, 2645 (1978)
6. M. Suzuki et al., NIM **192**, 565 (1982)
7. A. Hitachi et al., Phys. Rev. **B27**, 9, 5279 (1983)
8. T. Doke et al., NIMA **291**, 617 (1990)
9. Alphagaz product table: http://www.carbagas.ch
10. J. Keto et al., Phys. Rev. Letters 33, 1365 (1974)
11. A. Büchler, Bachelor thesis University Zürich 2006
12. R. Chandrasekharan et al., NIMA **546**, 426 (2005)

OPTIMIZING WIMP DIRECTIONAL DETECTORS

A. M. GREEN*

School of Physics and Astronomy, University of Nottingham,
University Park, Nottingham, NG7 2RD, UK.
** E-mail: anne.green@nottingham.ac.uk*

B. MORGAN

Department of Physics, University of Warwick,
Coventry, CV4 7AL, UK.
E-mail: ben.morgan@warwick.ac.uk

We study the dependence of the number of events required to directly detect a WIMP directional recoil signal on the capabilities of a directional detector. We consider variations in the nuclear recoil energy threshold, the background rate, whether the detector measures the recoil momentum vector in 2 or 3 dimensions and whether or not the sense of the momentum vector can be determined. The property with the biggest effect on the required exposure is the measurement of the momentum vector sense. If the detector cannot determine the recoil sense, the exposure required is increased by an order of magnitude for 3-d read-out and two orders of magnitude for 2-d read-out. For 2-d read-out the required exposure, in particular if the senses can not be measured, can be significantly reduced by analyzing the reduced angles with the, time dependent, projected direction of solar motion subtracted. The background rate effectively places a lower limit on the WIMP cross-section to which the detector is sensitive; it will be very difficult to detect WIMPs with a signal rate more than an order of magnitude below the background rate. Lowering the energy threshold also reduces the required exposure, but only for thresholds above 20 keV.

Keywords: dark matter; directional detection

1. Introduction

Weakly Interacting Massive Particle (WIMP) direct detection experiments aim to directly detect non-baryonic dark matter via the elastic scattering of WIMPs on detector nuclei, and are presently reaching the sensitivity required to detect the lightest neutralino (which in most supersymmetry models is the lightest supersymmetric particle and an excellent WIMP candidate). Since the expected event rates are very small, distinguishing a putative WIMP signal from backgrounds is crucial. The Earth's motion about the Sun provides two potential WIMP smoking guns: an annual modulation and a strong direction dependence of the event rate.

The direction dependence of the WIMP scattering rate has several advantages over the annual modulation. Firstly, the amplitude of the annual modulation is typically of order a few per-cent while the event rate in the direction of solar motion is roughly an order of magnitude larger than that in the opposite direction. Secondly, it is difficult for the directional signal to be mimicked by backgrounds. In most cases (a point source in the lab is a possible exception) a background which is anisotropic in the laboratory frame will be isotropic in the Galactic rest frame as the time dependent conversion between the lab and Galactic co-ordinate frames will wash out any lab specific features. Low pressure gas time projection chambers (TPCs), such as DRIFT (**D**irectional **R**ecoil **I**dentification **F**rom **Tr**acks) [1] and NEWAGE [2], seem to offer the best prospects for a detector capable of measuring the directions of sub-100 keV nuclear recoils.

We study the dependence of the number of events required to detect a WIMP signal on the detector properties: the energy threshold, background event rate, whether the read-out measures the recoil momentum in 2 or 3 dimensions (and if 2-d in which plane) and whether the sense (i.e. the absolute sign $+\vec{p}$ or $-\vec{p}$) of the recoil momentum vector can be measured (see also Refs. [3–7]).

2. Calculating the Nuclear Recoil Spectrum

2.1. *Modeling the Milky Way WIMP Halo*

We consider the simplest possible model of the Milky Way (MW) halo; an isotropic sphere with smooth density distribution $\rho \propto r^{-2}$ and a Maxwellian velocity distribution. This model is quite possibly not a good approximation to the real Milky Way, however in Refs. [4,5] we found that the number of events required to reject isotropy and detect a WIMP signal changed only weakly for observationally and theoretically motivated smooth halo models. It is possible (see e.g. Refs. [8]) that the WIMP distribution on the sub-milli-pc scales which are probed by direct detection is composed of streams. In this case the recoil momentum spectrum, and hence the detectability of WIMPs, would depend strongly on the number and directions of streams passing through the solar neighbourhood.

2.2. *Modeling the Detector Response*

In a TPC directional detector, the accuracy with which nuclear recoil directions can be reconstructed is limited by multiple scattering of the nucleus and diffusion of the ionisation generated along the recoil track. To estimate this angular resolution we to simulate recoils in a TPC filled with 0.05bar CS_2 gas, a 200μm pitch micropixel readout plane, a 10cm drift length over which a uniform electric field of 1kV cm^{-1} is applied, with recoil directions reconstructed via a moment analysis of the charge distributions. At present no simulations of the angular resolution function of a 2-d detector are available; projection effects combined with multiple scattering and diffusion will make the angular resolution a function of both the energy and primary recoil direction. We therefore assume perfect resolution for 2-d read-out.

2.2.1. *Energy threshold*

For primary recoil energies below 20 keV the short track length (3-4 pixels) and multiple scattering make it impossible to reconstruct the track direction in our simulated detector. We therefore define the energy threshold as the energy above which recoil directions can be reconstructed. It is important to note that this energy is higher than the energy threshold for simply *detecting* recoils. In addition, it may be lower than the energy above which recoils can be *discriminated* from electron backgrounds.

2.2.2. *Measurability of sense*

For sub-100keV energies, the dE/dx of nuclear recoils, and hence the ionisation density, is predicted to slowly decrease with decreasing recoil energy. Thus the sense of a recoil track is, in principle, measurable by determining the direction in which the ionisation density decreases along the track. However, the ionisation density distributions reconstructed from our simulated recoil tracks are close to uniform due to fluctuations in the production of ionisation and diffusion during the drift of this ionisation to the readout plane. It is therefore not possible to determine the absolute signs of the simulated reconstructed recoil vectors (i.e $+\vec{p}$ or $-\vec{p}$). Whether or not the sense of the recoils can be measured in reality is a question that needs to be addressed with experimental measurements. In our Monte Carlo simulations we therefore consider both possibilities. For clarity we refer to read-outs where the sense is measurable as vectorial, and those where it is not as axial.

2.3. *2-d Read-Out*

A 2-d read-out measures the projection of the recoil momentum vector into a plane fixed on the Earth. The recoil can be described via the angle ϕ between the projected recoil and some fiducial direction. It is also useful to measure the reduced angle ϕ_{red} between the projected recoil direction and the (time dependent) direction of solar motion projected into the plane. The degree of anisotropy in the distributions of ϕ and ϕ_{red}, and hence the detectability of a WIMP signal, is dependent on the orientation of the read-out plane. The anisotropy is maximised if the normal to the read-out plane is at 90° to the spin axis of the Earth. We focus on this optimal plane in our analysis.

2.4. *Background*

Zero background is the goal of the next generation of experiments made from low activity materials with efficient gamma rejection and shielding, located deep underground. We none–the–less investigate the effect of non-zero isotropic background.

3. Statistical Tests of Isotropy in 2-d and 3-d

Recoil directions in 3-d and 2-d constitute points on the unit sphere and circle respectively. For 3-d data the most powerful test for rejecting isotropy uses the fact that, for smooth halo models, the WIMP recoil distribution is expected to be peaked about the direction of solar motion $(l_{\odot}, b_{\odot})$. It involves calculating the average of the cosine of the angle between the direction of solar motion and the recoil direction. For 2-d data we found in Ref. [5] that the most powerful test is the Rayleigh test which uses the mean resultant length of the projected recoil vectors which, modulo fluctuations, should be zero for data drawn from an isotropic distribution. With axial data (i.e. unsigned lines) the standard procedure is to double the axial angles, reduce them modulo 360° and analyze the resulting vector data. We then use Monte Carlo simulations to find the number of event required to detect anisotropy at 90 (95)% confidence in 90 (95)% of experiments.

4. Results

As our baseline optimistic, but realistic, detector configuration we consider a detector with 3-d vector (i.e. with recoil sense measurable) read-out, energy threshold $E_{\rm TH} = 20$ keV, no background ($S/N = \infty$) and with the uncertainty in reconstructing the recoil directions taken into account. One or two of these specifications are varied in each of the alternative configurations. We consider two improved (and unrealistic) configurations, one with $E_{\rm TH} = 0$ keV and one with perfect recoil reconstruction (since the recoil reconstruction can not be taken into account for 2-d read-out, the number of events required for 2-d read-out should be compared with this later 3-d configuration). We also consider a number of 'degraded' configurations: 2-d read-out, axial read-out, a larger energy threshold and non-zero background. For the 2-d read-out we consider both the raw angles and the reduced angles (with the projected direction of solar motion subtracted). Throughout we assume, for illustrative purposes, a WIMP mass of $m_\chi = 100\,{\rm GeV}$.

Table 1 gives the details of the configurations considered and the number of events *above the energy threshold* required to reject isotropy (and hence detect a WIMP signal) at 90% (95%) confidence in 90% (95%) of experiments. The exposure required to accumulate the required number of events can be calculated, as a function of of the WIMP-nucleon elastic scattering cross-section using the. For the varying threshold configurations the number of events required decreases with increasing threshold as the anisotropy of the recoils increases with increasing energy. The recoil rate, however, decreases with increasing energy. The net effect is that the required exposure is fairly constant for $E_{\rm TH} < 20\,{\rm keV}$ and then increases with increasing threshold energy.

5. Summary

The property with the biggest effect on the detector performance is whether or not the sense of the recoils can be measured. If the detector cannot determine the recoil sense, the number of events required is increased by an order of magnitude for 3-d read-out and two orders of magnitude for 2-d read out. For 2-d read-out the performance, in particular if the senses can not be measured, can be significantly improved by recording and analyzing the reduced angles (with the projected direction of solar motion subtracted). The background rate effectively places a lower limit on the WIMP cross-section to which the detector is sensitive; it will be very difficult to detect a WIMP directional signal with a recoil rate more than an order of magnitude below the background rate. If the energy threshold is larger than about 20 keV significant improvements in the exposure required can be made by reducing the energy threshold (i.e. a factor of 2 if $E_{\rm th}$ is reduced from 50 keV to 20 keV and a factor of 5 if it is reduced from 100 keV to 50 keV).

difference from baseline configuration	N_{90}	N_{95}
none	7	11
$E_{\rm T} = 0$ keV	13	21
no recoil reconstruction uncertainty	5	9
$E_{\rm T} = 50$ keV	5	7
$E_{\rm T} = 100$ keV	3	5
$S/N = 10$	8	14
$S/N = 1$	17	27
$S/N = 0.1$	99	170
3-d axial read-out	81	130
2-d vector read-out in optimal plane, raw angles	18	26
2-d axial read-out in optimal plane, raw angles	1100	1600
2-d vector read-out in optimal plane, reduced angles	12	18
2-d axial read-out in optimal plane, reduced angles	190	270

Note: The dependence of the number of events *above the energy threshold* required to reject isotropy at 90(95)% confidence in 90(95)% of experiments, N_{90} (N_{95}), on the detector configuration. The baseline configuration has 3-d vector read-out, energy threshold $E_{\rm TH} = 20$ keV, no background ($S/N = \infty$) and the uncertainty in reconstructing the recoil directions taken into account. The second and third lines are (unrealistic) improved configurations. The subsequent lines are degraded configurations. For the non-zero background configurations the numbers given are the number of *signal* events required.

References

1. D. P. Snowden-Ifft, C. J. Martoff, and J. M. Burwell, Phys. Rev. D 61, 1 (2000) [astro-ph/9904064]; G. J. Alner *et al.*, Nucl. Inst. and Meth. A. 535, 644 (2004).
2. T. Tanimori *et al.*, Phys. Lett. B 578, 241 (2004) [astro-ph/0310638].
3. C. J. Copi, J. Heo and L. M. Krauss, Phys. Lett. B 461, 43 (1999) [astro-ph/990449]; C. J. Copi and L. M. Krauss, Phys. Rev. D 63, 043507 (2001) [astro-ph/0009467].
4. B. Morgan, A. M. Green and N. J. C. Spooner, Phys. Rev. D 71 103507 (2005) [astro-ph/0408047].
5. B. Morgan and A. M. Green, Phys. Rev. D 72, 123501 (2005) [astro-ph/0508134].
6. A. M. Green and B. Morgan [astro-ph/0609115].
7. C. J. Copi, L. M. Krauss, D. Simmons-Duffin and S. R. Stroiney, [astro-ph/0508649].
8. D. Stiff and L. M. Widrow, Phys. Rev. Lett. 90, 211301 (2003) [astro-ph/0301301]; B. Moore *et al.*, Phys. Rev. D 64, 063508 (2002) [astro-ph/010627].

THE ZEPLIN-II TWO-PHASE XENON DARK MATTER DETECTOR: STATUS REPORT

R. LÜSCHER*, ON BEHALF OF THE ZEPLIN-II COLLABORATION

(CCLRC/Rutherfore Appleton Laboratory, Imperial College, Edinburgh, Sheffield, Coimbra, ITEP, Texas A&M, UCLA, Rochester)

CCLRC/Rutherford Appleton Laboratory, Chilton, Didcot, Oxon, OX11 0QX, UK

** E-mail: r.luscher@rl.ac.uk*

Zeplin-II is a two-phase (liquid, gas) Xenon detector optimised for WIMP Dark Matter search. Both scintillation and ionisation is measured, in order to discriminate the radioactive background. It is currently running in the underground laboratory of Boulby Mine (UK). 1 ton x day exposure has been collected. The preliminary study of 10% of the data is presented.

Keywords: WIMP Dark Matter, two-phase detector, Xenon

1. Introduction

The quest for Dark Matter is among the most fundamental astroparticle physics questions. From the WMAP results on the cosmic microwave background[1] we expect that 23% of all matter of the universe is due to Cold Dark Matter. Further evidence of Dark Matter has been reported recently, following the observation of a galactic cluster merger.[2] The SUSY-WIMP model provides an excellent candidate, the neutralino. Extrapolated from collider data, the theoretical predictions for the rates of interaction between WIMPs and targets are at the level of 10^{-2} events/kg/day. To detect WIMPs hence requires very sensitive and massive targets.

The ZEPLIN project (Zoned Proportional scintillation in LIquid Noble gases), takes advantage of the particularly appropriate properties of Xe: in liquid form, it has a reasonably large density of 2.9 kg/lt. It has a heavy nuclei for a large spin-independent coupling and appreciable abundance of isotopes with spin for a large spin-dependent coupling. Low background Xe is available. LXe is known as a good scintillator: emitting in the UV region (175 nm), it enables a low energy threshold. Moreover, the interaction process in LXe owns characteristics which translate into a potentially high background discrimination.

2. Liquid Xenon as Dark Matter target

Any recoil in LXe gives rise to both ionisation and excitation of Xe atoms. The excitation result in the emission of a 175 nm photon from either a singlet (with decay time $\sim$ 3 ns) or a triplet state ($\sim$ 27 ns). The ratio single/triplet is 10 times bigger for nuclear recoil compared to electron recoils. In the absence of electric field, the ionisation recombines to produce further excited Xe atoms. The recombination time depends on the ionisation density: for nuclear recoil, it is very high and recombination is very fast; while for electron recoil, the lower ionisation density leads to longer times. This effect has been the design basis for Zeplin-I, a single-phase liquid Xenon detector.[3]

A two-phase design measures separately both the scintillation and the ionisation produced by the interacting particle. Any interacting particle produces excitation and ionisation within the liquid xenon target. An applied electric field partially suppresses recombination and drifts the electrons upwards towards the gaseous phase. There, mesh planes define a high field region in which avalanche occurs and electroluminescence is produced. The latter can then be recorded by the same PMTs which have already read the primary scintillation. The proportion of ionisation and excitation released depends on the dE/dx of the particle interaction. More ionisation is detected from an electron recoil than from a nuclear recoils, for a given scintillation pulse. This technology has shown an interesting discrimination potential.[4,5] Two detectors have been developed based on these properties: Zeplin-II, a relatively low drift field design (1 kV/cm), optimised for a larger target mass; reported in this contribution; Zeplin-III, a detector with an increased field in the target volume, optimised for a lower energy threshold. The latter is discussed in Ref. 10.

3. The Zeplin-II detector

The detector contains a target mass of about 30 kg confined in a PTFE cone (Fig. 1). Around it, 10 OFHC Cu field shaping rings homogenise the drift field across the active volume. The unavoidable field distortions around each field shaping ring are contained in the PTFE cone. The whole design is based on minimising dead regions, as the background discrimination power

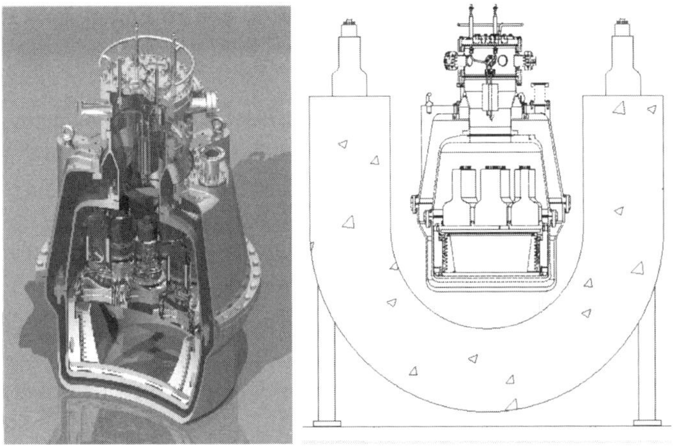

Fig. 1. The Zeplin-II detector: a two-phase designs.

relies on the efficiency to collect the ionisation charge. A fine mesh cathode is placed at the bottom of the active volume; it has a mesh transparency of 90%, so that the PTFE plate just below acts as a reflector. Two fine meshes are placed just below and above the liquid surface: the former defines the boundary between the active and the extraction regions, while the latter acts as an electroluminescence plate. The gap between these two mesh planes is 9 mm.

The level of liquid Xenon in this gap is chosen to optimise the electric field settings. A field of 4.2kV/cm in the liquid part of the gap ensures an efficiency of 90% to extract electrons into the gas phase.[8] To obtain a gain of 150 photon/electron in the electroluminescence process, a field of 12kV/cm is required.[6,7] This gain is ensuring that optical feedback of the electroluminescence light into the active volume is kept negligible. A liquid level of ~7 mm into the gap and a voltage of 5.8kV between the grids are matching these criteria.

Seven PMT — hemispherical 130 mm D742QKFLB tubes from Electron Tubes Ltd, with a platinum underlay to adapt for low temperatures — are placed in the gas region and are viewing the active target from above.

The target vessel has been built from OFHC Cu, using electron beam welding. The Zeplin-II is surrounded by the same PXE liquid scintillator Compton veto as Zeplin-I. Its performance is discussed in Ref. 11. An additional neutron shielding has been installed to close the open area on the top of the veto: Polyethylene slabs interleaved with Gd-loaded resin, with a thickness of $\geq$20 cm. The whole veto is then surrounded by a castle of 22.5 cm of Pb.

A Polycold PFC-330 unit from Megatech circulates refrigerant in a closed loop through the cold-head. The target is then cooled by convection of gas Xe created by this cold point. The heat load of the full target has been estimated to $\sim$25 W, by measuring the warm-up curves.

3.1. *Gas system*

The requirements of Xenon purity are stringent. To enable the ionisation electrons drifting to the gas phase and creating the electroluminescence, an electron lifetime of $>$100 μs is required — as the total drift length is $\sim$70 μs. This translates into concentration in electronegative contaminants (H_2O, O_2) at $\sim$ppb level. A purification system has been designed, which allows cleaning the Xenon prior to liquefaction in the target, by cryo-pumping it through a noble gas purification SAES getter. A constant recirculation is also in place during operation, by extracting vaporised Xenon, flushing it through the getter before re-liquefying it into the target.

An exposure of 1 ton $\times$ day has been recorded with optimised field settings and electron lifetime in excess of 100 μs.

3.2. *Data acquisition and electronics*

The 7 signals from the PMTs are recorded with two DC265-M2M ACQIRIS digitisers. The signals are split and fed to discriminators, which are set to 2/5 of single photoelectron level. The trigger is then a 5-fold coincidence of the direct signals, in anti-coincidence with a high-threshold discrimination on the central PMT signal. The latter is used to reject events, which are saturating the DAq, and so avoid the dead-time caused by reading them out. The DAq may be triggered by primary or secondary pulses; a 3-fold coincidence threshold is then applied off-line to all pulses. The pulses are sampled at 2 ns, with a waveform duration of 200 μs. A typical event is shown in Ref. 12.

The veto PMTs are summed and recorded on the last DAq channel. A 3-fold coincidence logic delayed signal is added to this sum, enabling the differentiation between large, single-site pulses (PMT noise or events) and proper gamma events in the scintillator volume.

General system parameters, like temperature, pressures or refrigerant load, are recorded every minute by a 48-channel slowcontrol system.

4. Detector performance

A daily calibration is performed using the 122 keV γ-line from ^{57}Co, introduced into the shielding and placed just below the target. A light yield of 0.55 photoelectrons/keV (with a drift field of 1 kV/cm) has been measured. The 122 keV line illuminates the bottom few mm of the active volume only, with a relative increase in statistics at the positions where the target vessel thickness has been reduced to 2 mm (from 15 mm). The latter feature is enabling to check the position reconstruction algorithms.

The scintillation and ionisation signals, from their production process, are anti-correlated.[9] Correlating the two signals improved by 20% the energy resolution, to reach ~17% at 122 keV. Moreover, a ~40 keV line from inelastic scattering in neutron calibration data, which was predicted by simulation but unresolved in neither of the two signals, can be clearly seen when analysing the data in anti-correlation. This enables to test the detector linearity to lower energy.

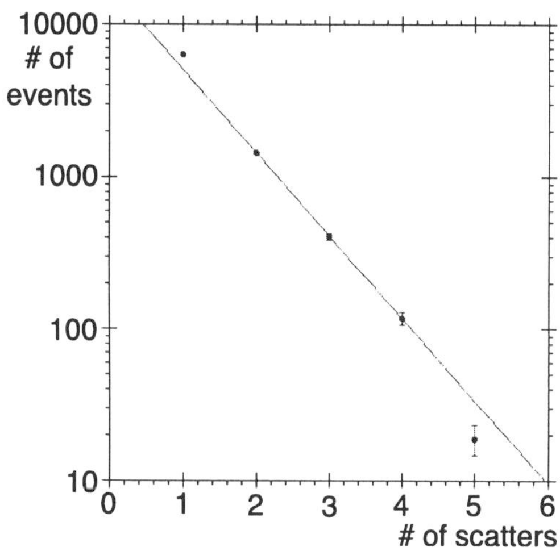

Fig. 2. Number of scatters from neutron event, extracted from AmBe calibrations.

Neutron calibrations are testing the detector response to Dark Matter: a neutron mimics a WIMP interaction by inducing nuclear recoils. However, contrary to WIMPs, neutron can suffer multiple scattering in the target volume, benefiting from larger mass. In a preliminary analysis, the evaluation of the number of scatters per event extracted from neutron calibrations (Fig. 2) shows the behaviour expected from simulations (Fig. 3).

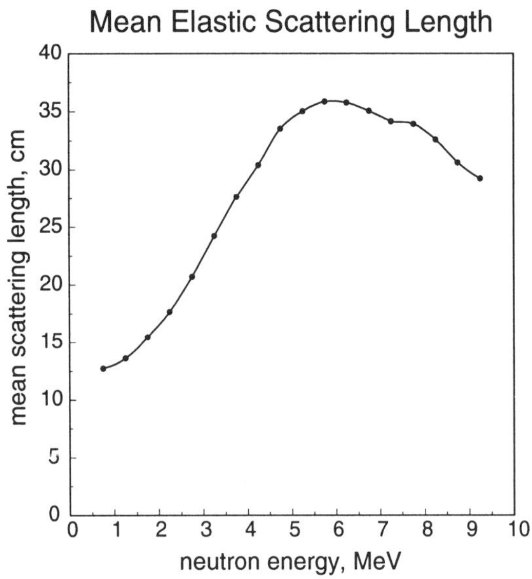

Fig. 3. Neutron mean scatter length as simulated by Geant4 for a Zeplin-II geometry (right).

5. Preliminary analysis

Recorded data are reduced off-line. Parameters like pulse height, area, width, time constant, ..., of all pulses are calculated prior to event selection. A single pulse responding to the criteria of a primary scintillation is requested, cutting out multiple events. One and only one pulse with the features of a secondary electroluminescence should be present, reducing multiple scattering events. More details on event reduction and selection can be found in Ref. 12.

At present, only 10% of the recorded data are being analysed and compared to calibration obtained with an AmBe source (figure 4). Work is in progress to understand the noise, estimate the efficiencies and evaluate the discrimination function.

6. Conclusion

The Zeplin-II detector has been installed in the underground laboratory of Boulby Mine. It is operational and 1 ton $\times$ day of background data has been recorded. The analysis of these data is in progress. The comparison of 10% of the background data with a neutron source calibration has been presented.

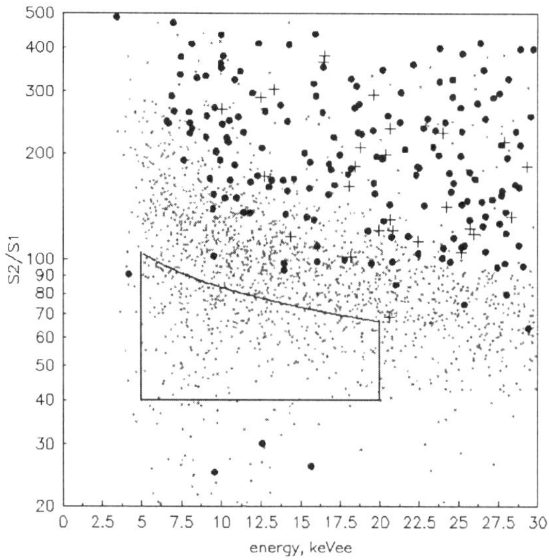

Fig. 4. A preliminary analysis of 10% of the data (black dots), in comparison with an AmBe calibration (grey dots). The area drawn is one example of an acceptance window being studied.

Acknowledgment

This work has been funded by PPARC, NSF and DOE. The authors would like to thank Cleveland Potash Ltd for their continuous support at Boulby Mine.

References

1. D. N. Spergel *et al.*, submitted to *ApJ* (2006); `http://lambda.gsfc.nasa.gov/product/map/current/map_bibliography.cfm`.
2. D. Clowe *et al.*, *astro-ph/0608407*; in press, *Astrophys J. Lett.* (2006).
3. G. J. Alner *et al.*, *Astropart. Phys.* **23**, 444 (2005).
4. H. Wang *et al.*, *Phys. Rep.* **307**, 263 (1998).
5. D. Y. Akimov *et al.*, in: *Proc. 4th Int. Workshop on the Identification of Dark Matter*, World Scientific, Singapore, 37 (2003).
6. D. Y. Akimov *et al.*, in: *3rd Intern Conf on Nonaccelerator New Physics* (NANP 01), Dubna (Russia), *Phys. Atom. Nucl.* **66**, 497 (2003).
7. A. C. Fonseca *et al.*, in: *IEEE Nuclear Science Symposium Conference Record 2004*, Vol. 1, 572 (2004).
8. E. M. Gushin *et al.*, *J. Exp. and Theor. Phys.* **76**, 5, 1685 (1979).
9. E. Conti *et al.*, *Phys. Rev. B* **68**, 054201 (2003).
10. T. J. Sumner *et al.*, this conference.
11. E. Daw *et al.*, this conference.
12. K Mavrokoridis *et al.*, this conference.

LIQUID RARE GASES AND INORGANIC SCINTILLATORS FOR WIMP SEARCHES

AKIRA HITACHI

Kochi Medical School, Nankoku, Kochi 783-8505, Japan
E-mail:jm-hitachia@kochi-u.ac.jp

Scintillation in liquid Ar and Xe, and NaI(Tl) and CsI(Tl) crystals is discussed in relation with dark matter detectors. The nuclear quenching factors for liquid Xe and CsI(Tl) are almost the same. However, measured recoil ion to γ ratios, RN/γ, show the opposite trend in energy dependence. The electronic quenching, q_{el}, is responsible for the difference. The electronic LET is obtained and used to estimate q_{el} and RN/γ in CsI(Tl). The results show good agreement with neutron scattering experiments.

1. Introduction

The direct detection of a dark matter candidate, WIMPs (weakly interacting massive particles) aims to observe recoil ions produced by elastic scattering with WIMPs. The knowledge of the signal produced by a few tens keV recoil nucleus is essential for the development of WIMP detectors and particle identification.

When ions are slow, $v < v_0 = e^2/\hbar$, the energy losses to electrons and atomic recoils are similar in magnitude. The secondaries, recoil atoms and electrons, may again go to the collision process and transfer the energy to new particles and so on. After this cascade process complete, the energy of the incident particle E is given to atomic motion ν and electronic excitation η [1]. Only a part of energy η can be used as scintillation and/or ionization signals. The nuclear recoil to gamma ratio, RN/γ, shows good agreement with the Lindhard factor (the nuclear quenching factor), $q_{nc} = \eta/E$, in semiconductors. However, for scintillators, in addition, electronic quenching and escaping electrons have to be considered [2]. The ratio is expressed as $RN/\gamma = q_{nc} \times q_{el} / S_\gamma$, where q_{el} is the electronic quenching factor and S_γ represents the scintillation efficiency for γs.

The electronic linear energy transfer (LET_{el}), the electronic energy deposition per unit length, plays an important role in considering the track structure, electronic quenching, decay shape, S/T ratio, etc.

2. Electronic quenching and LET

The scintillation efficiency for various particles in LXe [3] is shown in Fig. 1(a) as a function of LET. Nuclear quenching can be ignored for these particles (q_{nc}=1). The relativistic heavy ions (p, C, Si, Ar, Fe) lie in a flat-top region, suffer no quenching (q_{el} =1). However, The efficiency for αs and fission fragments are less (q_{el} <1) because of electronic quenching due to high-excitation density. The efficiency for low LET electrons and γs are also observed less because of inefficient recombination due to escaping electrons. The same trend has been reported for inorganic crystals [4], except that the flat-top region is missing for some scintillators as shown in Fig. 1(a) for CsI(Tl).

The electronic LET_{el} (= -dη/dR, R: range) for recoil ions are calculated and compared with the scintillation data mentioned above for evaluation of q_{el}. The Bragg-like curves, which show the ionization density along the track, for Ne, Ar and Xe were calculated using LET_{el} and shown in Fig. 1(b). The ions enter from the right hand side. Points are plotted at every 5 keV. The curves show an end region of the 'usual' Bragg curve. The vertical axis is plotted in ΔN_i, the number of ions produced per unit length. The area below the curve shows the number of ions produced. N_i was calculated using W-value in gas. The Bragg-like curve gives the directional information of the recoil ions in gas-TPC (time projection chamber). However it represents the ionization density also for liquids. The ionization density for discussion of electronic quenching as a function of energy in liquid is given by simply replacing W values in gas with W values liquids.

LET_{el} are also used to consider the track structure. The recoil ions of a few keV has similar LET as αs. This means that αs can be a good measure in considerir.g some characteristics, such as q_{el}, the S/T ratio and the charge yield, for recoil ions with energies of concern here. The track structure for heavy ion is characterized by a concentric cylinders consisting of a central core and a surrounding penumbra. The core is high excitation region mostly due to direct interaction, glancing collisions, and the penumbra is zone of low excitation density formed by δ–rays. The core radius is given by the Bohr-criterion, $r_0 = v/2E_1$, where E_1 is the energy of the lowest excitation state of medium [5].

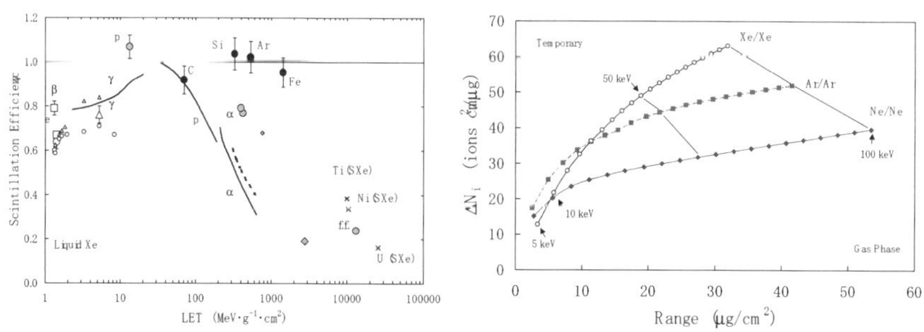

Fig. 1. (a). The scintillation efficiency as a function of LET in LXe (symbols) and CsI(Tl) (curves). (b) The Bragg-like curves for recoil ions in rare gases.

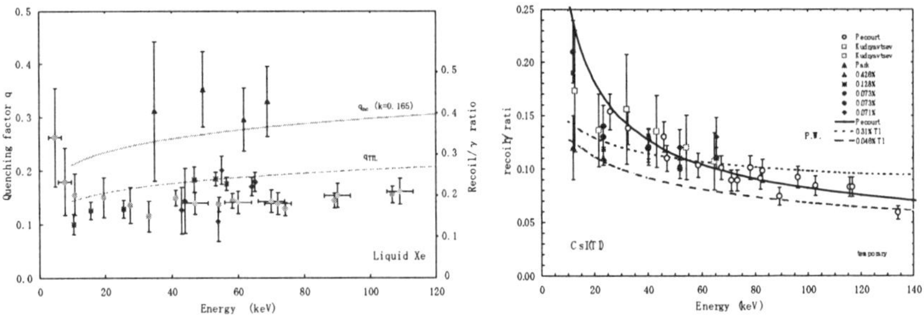

Fig. 2. (a) The nuclear, q_{nc}, and the total, $q_T = q_{nc} \times q_{el}$, quenching factors as a function of recoil ion energy in LXe [2,12]. (b) The recoil-ion to γ ratio reported in CsI(Tl) [8]. The broken curves are present estimates for Tl=0.31% (upper) and 0.046% (lower). The solid curve is a single parameter fitting by Pécour [7].

3. Quenching calculations

3.1. *Liquid Ar and Xe*

The electronic quenching mechanism in LAr and LXe has been discussed before [6]; therefore it is only briefly discussed here. The electronic energy η delivered to the liquid is divided in the core (T_c) and in the penumbra (T_p) of the ion track. Quenching takes place only in the high-ionization-density core. Then the energy T_s available for scintillation is, $T_s=q_T E=q_{el}\eta=q_c T_c+T_p$. Here q_{el} and q_c are the overall electronic quenching factor and that in the core, respectively.

The proposed quenching mechanism is biexcitonic collisions, $R^*+R^*\rightarrow R+R^++e^-$(K.E.), where, R^* is the "free" exciton [6]. The ejected electron e^-, may recombine with an ion to reform a free exciton and the overall result is that two excitons are required for one photon. The recombination component also undergoes this quenching processes. A diffusion-reaction model of free exciton with a specific reaction rate was used.

A simple method used here for estimation of the *RN*/γ ratio exploits the similarity in the track structure and assumes that q_c for the αs is the same as q_{el} for the recoil-ions. We assume $T_c/\eta=1$ for recoil ions since most δ-rays produced by recoil ions do not have sufficient energy to effectively penetrate the core, forming a fairly undifferentiated core. Using $q_T = q_{el}=0.71$ measured for αs, we have $T_s/E=0.71=0.72q_c+0.28$ then $q_c(\alpha)=0.6$, thus obtain $q_{el}=0.6$ for recoil-Ar ions. The total quenching factor for 60 keV Ar ions in LAr is $q_T = q_{nc} \times q_{el} \approx 0.4 \times 0.6 = 0.24$.

For LXe, we assume T_c/E for αs to be the same as in LAr. Using $q_T = q_{el} = 0.77$ for the αs in LXe [6], we have $0.77=0.72q_c +0.28$. We obtain $q_{el} = 0.68$

and $q_T = q_{nc} \times q_{el} \approx 0.28 \times 0.68 = 0.19$ for 60 keV recoil-Xe ions. With $q_{el}(\alpha) = 0.77$ and $\alpha/\gamma \sim 1$, we have $RN/\gamma = 0.25$. The results for LXe agree well with most of the nuclear recoil measurements as shown in Fig. 2(a).

3.2. *NaI(Tl) and CsI(Tl)*

Z and *A* for Cs and I atoms are very close and similar to Xe. These are treated similarly in the stopping calculations and q_{nc} values for LXe can be used for CsI(Tl) crystals.

The scintillation efficiency for slow ions in inorganic crystals is more or less a function of LET as shown for protons and αs in CsI(Tl) in Fig. 1(a) [4]. The range of recoil ions were obtained from the total stopping power for Xe ions in HMI tables. Then LET_{el} was calculated for recoil ions. The LET_{el} of recoil ions with energies of concern here are in the same range in LET as the αs. We use the scintillation efficiency for the αs at the same LET for that for the recoil ions. A scintillation efficiency for 122 keV γs was also taken [4]. Then, the RN/γ ratio was obtained for Tl concentrations of 0.046% and 0.31% as shown in Fig. 2(b) together with values for neutron recoil measurements [7, 8]. No parameter fittings has been made. The present estimations give good agreement with the experimental values. Pécourt [8] fitted their experimental values by the Birks-Lindhard model with single free parameter as shown with the solid line. The fitting shows a steeper curve than present calculations.

NaI(Tl) needs complex treatments since Z for Na and I atoms are quite different. The Bragg rule can be applied for the stopping calculation in compound. However, the evaluation of q_{nc} for $Z_1 \neq Z_2$ is quite hard. The recoil ions in α–decay in rare gases give one of the simplest cases. Lindhard [1] gave a power law approximation for very low energy. For Pb ions produced in α–decay in Ar, the power low approximation gives $\eta = 0.019E^{3/2}$ (*E*: keV).

Lindhard also gave a numerical calculation for this particular case in Ar. The results are shown in Fig. 3(a) together with numerical results for Ar ions in Ar. The RN/γ ratios measured by ionization in gas basically give the Lindhard factor q_{nc} and are shown for comparison [9]. The experimental points come in between the numerical result and the power low approximation.

The treatment for Na recoil ions in NaI is much complicated. The energy restriction in the power law approximation is rather sever and only applicable at a very low energy region for Na ions in NaI.

The simplest case, Na ions in Na, was used to obtain q_{nc} for Na ions in NaI(Tl). Then LET_{el} was calculated using the q_{nc} value and the range obtained from the total stopping power for Na ions in NaI by SRIM. The LET_{el} was compared with

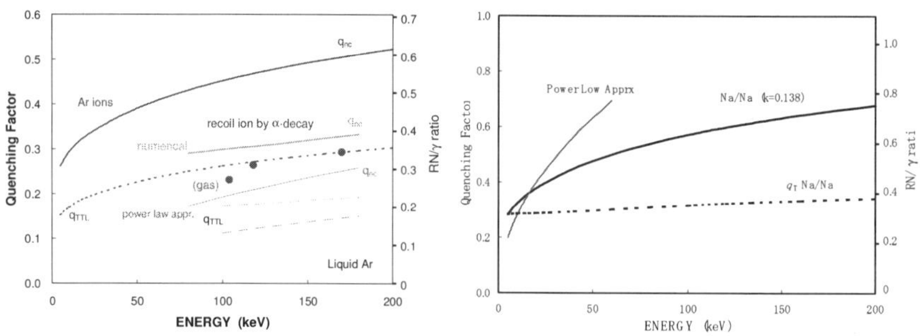

Fig. 3. Nuclear (solid curves) and total (broken curves) quenching factors for (a) Ar and Pb ions in LAr. •; q_{nc} measured for recoil ions in α–decay in gas Ar [9]. (b) Na ions in NaI(Tl). For the *RN*/γ ratio, refer the axis on the right.

the scintillation efficiency measurements as a function of LET in literature [4]. NaI(Tl) has a flat-tip region [10] and q_{el} =1 was taken there. Similar to the CsI(Tl) calculation, the q_{el} values for α at the same LET_{el} was taken for q_{el} for recoil Na ions in NaI(Tl). The electronic quenching is quite small at the low energy and increases considerably as the energy. The calculation give the total quenching factor q_T of 0.29 to 0.34 in 10-200 keV, slowly increasing with the energy, as shown in Fig. 3(b). Taking the scintillation efficiency for γ–rays (S_γ = 0.9), the *RN*/γ ratio (= q_T/S_γ) was compared with measured values [11]. The calculated value is considerably larger than the measured values.

4. Discussion

The *RN*/γ ratios for LXe and CsI(Tl) show the opposite trends in each other in spite of the fact that q_{nc} are almost the same for those scintillators. The *RN*/γ ratio for LXe increase slowly as the energy of the recoil ions whereas that for CsI(Tl) decreases (Fig. 2). The difference is due to q_{el}. LAr and LXe are the most strong among solid and liquid scintillators against high LET quenching and change in LET does not affect much on q_{el}. On the other hand, quenching in inorganic crystals is rather strong and changes as LET changes. The LET_{el} increases as the energy of recoil ions increases, as a result q_T and *RN*/γ ratio for CsI(Tl) decreases.

The recent results reported for LXe shows that the quenching factor rises at ≤ 10 keV [12] (Fig. 2(a)). The increasing trend can be explained by the behavior of LET_{el}. The Bragg-like curve for LXe decreases rapidly as the energy decreases at this energy range as shown in Fig. 1(b). This suggests that the electronic quenching, particularly for Xe, becomes less as the recoil energy decreases. Detailed calculation for LAr is underway.

The XENON group reported the charge yield for recoil Xe in LXe [13]. The increase in charge under the electric field is accompany by the reduction in scintillation. The charge and scintillation are complementary to each other and related by an equation with a quenching factor q [14]. The observed charge yield is much larger than that expected by the scintillation reduction. The increase can be a recovery of electronic quenching due to an applied field which is observed for heavy ions in LAr [14].

5. Summary

The electronic quenching in addition to nuclear quenching basically reproduces the RN/γ ratios in recoil measurements in liquid rare gases and in inorganic crystals. Three types of energy dependences; increasing (LXe), decreasing (CsI(Tl)) and nearly constant (NaI(Tl)) behaviors have been explained. The electronic LET plays important roles.

Acknowledgements

The author would like to thank Dr. P.A. Majewski for his help with preliminary calculations in NaI(Tl).

References

1. J. Lindhard *et al.* Mat. Fys. Medd. Dan. Vid. Selsk. **33**, no.10 (1963).
2. A. Hitachi, Astroparticle Physics, **24**, 247 (2005) and references therein.
3. T. Tanaka *et al.*, Nucl. Instr. Meth. A **457**, 454 (2001) .
4. R.B. Murray and A. Meyer, Phys. Rev. **122**, 815 (1961).
5. A. Mozumder et al. Advance in Chem. **81**, 27 (1968).
6. A. Hitachi, T. Doke and A. Mozumder, Phys. Rev. B **46**, 11463 (1992).
7. S. Pécour *et al.* Astroparticle Phys. **11**, 457 (1999).
8. H. Park *et al.* Nucl. Instr. Meth. A **491**, 460 (2002) and refs. therein.
9. B.S. Madsen, Mat. Fys. Medd. Dan. Vid. Selsk. **23**, no.8 (1945).
10. T. Doke, E. Shibamura and S. Kubota, J. Phys. Soc. Jpn. **68**, 2433 (1999).
11. H. Chagani *et al.*, to appear in Proc. 6th Int. Workshop on the Identification of Dark Matter.
12. V. Chepel *et al.*, Astroparticle Physics **26**, 58 (2006).
13. E. Aprile *et al.*, Phys. Rev. Lett. **97**, 0813202 (2006).
14. A. Hitachi *et al.*, Phys. Rev. A **35**, 3956 (1987): IEEE Trans. on Dielectric and Electrical Insulation, **9**, 45 (2002).

NEUTRALINO-NEUTRALINO ANNIHILATION TO γZ IN MSSM*

TH. DIAKONIDIS[a], G.J. GOUNARIS[a], J. LAYSSAC[b],
P.I. PORFYRIADIS[a] and F.M. RENARD[b]

[a] *Department of Theoretical Physics, Aristotle University of Thessaloniki, Gr-54124, Thessaloniki, Greece.*

[b] *Laboratoire de Physique Théorique et Astroparticules, UMR 5207, Université Montpellier II, F-34095 Montpellier Cedex 5.*

The complete 1-loop computation of the processes $\tilde{\chi}_i^0\tilde{\chi}_j^0 \to \gamma Z$ has been performed at an arbitrary c.m. energy for any pair of MSSM neutralinos. As an application suitable for Dark Matter (DM) searches, the neutralino-neutralino annihilation is studied at the limiting case $v_{ij} \to 0$ describing the present DM distribution in the galactic halo, and at $v_{ij} \simeq 0.5$, determining the neutralino relic density contributions. Our results led to the FORTRAN code PLATONd-mgZ, which may be used for any set of real MSSM parameters. A sample of 6 MSSM models taken, already used in literature and effectively describing the various possible neutralino properties. A comparison with other existing works has also been made.

1. Introduction

The nature of the Cold Dark Matter (DM), constituting almost 23% of the energy of the Universe, is one of the most exciting subjects of physics today.[1,2] Within the minimal R-parity conserving Supersymmetric (SUSY) framework, a most obvious candidate for such matter is of course the lightest neutralino $\tilde{\chi}_1^0$.[1,3] A striking signature for such DM would then be the detection of γ-rays obtained through the annihilation of two neutralinos.

In this paper we present an analytic calculation of the 1-loop process $\tilde{\chi}_i^0\tilde{\chi}_j^0 \to \gamma Z$, following the same philosophy as in Refs. 4 and 5. The results are based on analytically expressing the helicity amplitudes in terms of PV functions,[6] and they are valid for any set of real MSSM parameters

*Work supported by the Greek Ministry of Education and Religion and the EPEAEK program Pythagoras, and by the European Union program HPRN-CT-2000-00149.

where[a] $0 \leq v_{ij} \lesssim 0.7$. They can be obtained using the FORTRAN code PLATONdmZ, which is herewith released in Ref. 7.

2. Procedure

The process studied here is

$$\tilde{\chi}_i^0(l_1, \lambda_1) \ + \ \tilde{\chi}_j^0(l_2, \lambda_2) \to \gamma(p_1, \mu_1) + Z(p_2, \mu_2) \ , \tag{1}$$

The corresponding helicity amplitudes, which satisfy the standard Jacob-Wick conventions,[8] are denoted as $F^{ij}_{\lambda_1, \lambda_2; \mu_1, \mu_2}(\theta)$, where θ is the c.m. scattering angle of the outgoing photon with respect to the incoming neutralino $\tilde{\chi}_i^0$. According to,[8] the $\chi_i^0 \chi_j^0$ antisymmetry due to the fermionic nature of the neutralinos, obliges the helicity amplitudes to obey

$$F^{ij}_{\lambda_1, \lambda_2; \mu_1, \mu_2}(\theta) = (-1)^{\mu_1 - \mu_2} \ F^{ji}_{\lambda_2, \lambda_1; \mu_1, \mu_2}(\pi - \theta) \ , \tag{2}$$

while the CP symmetry for real SUSY-braking parameters implies

$$F^{ij}_{-\lambda_1, -\lambda_2; -\mu_1, -\mu_2}(\theta) = (-1)^{\lambda_1 - \lambda_2 - (\mu_1 - \mu_2)} \ \eta_i \eta_j F^{ij}_{\lambda_1, \lambda_2; \mu_1, \mu_2}(\theta) \ , \tag{3}$$

where $\eta_i = \pm 1$ is the CP eigenvalue of $\tilde{\chi}_i^0$.[9] Relation (2) is very useful, since it is used below to reduce the number of independent diagrams to be calculated.

In the calculation we use the 'tHooft-Feynman gauge $\xi_W = \xi_Z = \xi_\gamma = 1$, together with the standard linear gauge fixing. The complete set of independent diagrams, consists of: the box diagrams of Fig. 1, the bubbles and the initial and final triangles of Fig. 2, the t-channel triangle diagrams Fig. 3 and the $\gamma - Z$ self energy contributions in Fig. 4.

The first 10 boxes in Fig. 1 are called "direct". The contribution of the corresponding boxes with $\gamma \leftrightarrow Z$ exchanged, is then determined from them by enforcing (2), through the addition to

$$F^{ij}_{\lambda_1, \lambda_2; \mu_1, \mu_2}(\theta)^{\text{direct Box}} \ , \tag{4}$$

of the

$$(-1)^{(\mu_1 - \mu_2)} F^{ji}_{\lambda_2, \lambda_1; \mu_1, \mu_2}(\pi - \theta)^{\text{direct Box}} \tag{5}$$

contribution. Thus (2) greatly facilitates the computation.

The last 8 boxes in Fig. 1, called "twisted", have been checked to satisfy (2) by themselves. This is also true for the bubble and initial and final triangle diagrams of Fig. 2.

---

[a] As usual, the light velocity in vacuum c, is taken as the unit of velocities.

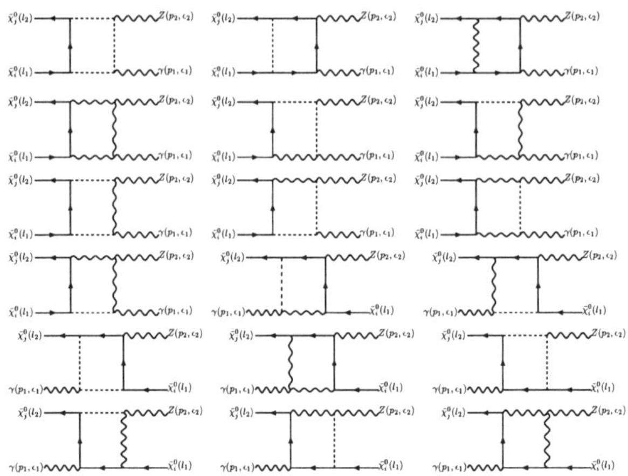

Fig. 1. Box Feynman diagrams needed for $\chi_i^0 \chi_j^0 \rightarrow \gamma Z$, full internal lines denote fermionic exchanges, broken lines scalars, and wavy lines gauge bosons.

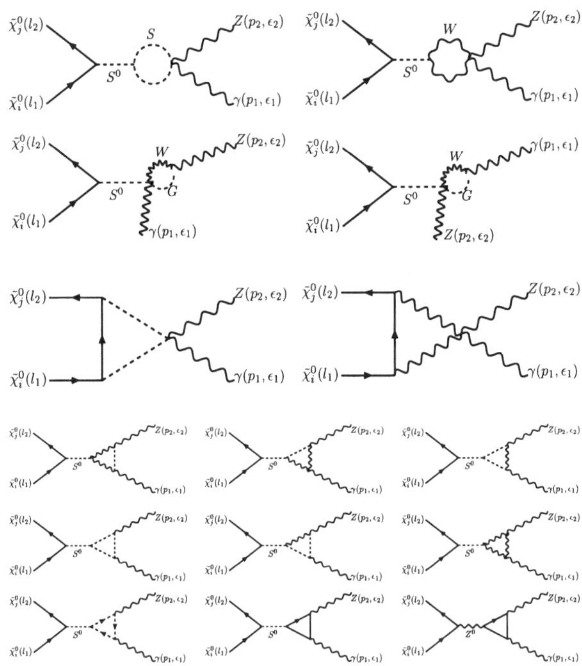

Fig. 2. The bubble, initial and final triangle diagrams for $\chi_i^0 \chi_j^0 \rightarrow \gamma Z$. All these contributions automatically satisfy (2).

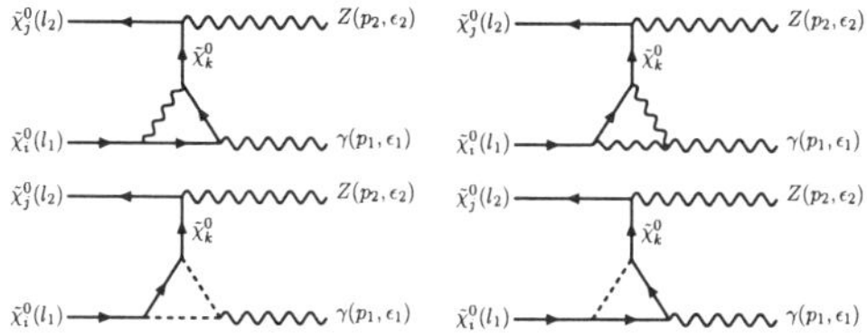

Fig. 3. t-channel triangles. By enforcing (2), the u-channel triangle contribution is obtained from it through (7); see text.

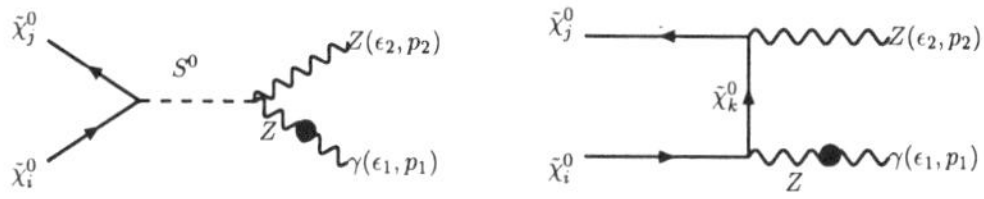

Fig. 4. The needed $\gamma - Z$ self energy diagrams. The left diagram describes the s-channel exchange of $S^0 = h^0, H^0$. The right the t-channel neutralino exchange contribution, while the corresponding u-channel contribution is obtained from it through (8).

The third set of needed diagrams consists of the t-channel triangles of Fig. 3, whose contribution we denote as

$$F^{ij}_{\lambda_1,\lambda_2;\mu_1,\mu_2}(\theta)^{\text{t-triangle}} , \tag{6}$$

to these we should add the contribution of the u-channel triangles, obtained by interchanging $\tilde{\chi}_i^0 \leftrightarrow \tilde{\chi}_j^0$, (with the arrows always running from $\tilde{\chi}_i^0$ to $\tilde{\chi}_j^0$). By enforcing (2), these u-channel triangle contribution is given by

$$F^{ij}_{\lambda_1,\lambda_2;\mu_1,\mu_2}(\theta)^{\text{u-triangle}} = (-1)^{(\mu_1-\mu_2)} F^{ji}_{\lambda_2,\lambda_1;\mu_1,\mu_2}(\pi-\theta)^{\text{ t-triangle}} , \tag{7}$$

so that only the t-channel triangles of Fig. 3 need to be calculated.

Finally, the needed $\gamma - Z$ self energy contribution is shown in Fig. 4, where the left diagram describes the contribution due to s-channel exchange of (H^0, h^0), while the right one the one induced by a t-channel neutralino exchange. Associated to this t-channel self-energy contribution, there exist a corresponding u-channel one obtained by enforcing (2)on it, through

$$F^{ij}_{\lambda_1,\lambda_2;\mu_1,\mu_2}(\theta)^{\text{u-self energy}} = (-1)^{(\mu_1-\mu_2)} F^{ji}_{\lambda_2,\lambda_1;\mu_1,\mu_2}(\pi-\theta)^{\text{ t-self energy}} . \tag{8}$$

We have made several tests on the results of $F^{ij}_{\lambda_1,\lambda_2;\mu_1,\mu_2}(\theta)$ amplitudes, in order to eliminate, as much as possible, the possibility of errors. The validity of (2) for the contribution of the 8 twisted boxes in Fig. 1 and those of Fig. 2, already provides a generally stringent test of their correctness. The calculations of the 10 direct boxes of Fig. 1, and t-channel triangles of Fig. 3, have been checked several times. Moreover, since these 10 direct boxes also contribute to the $\gamma\gamma$-amplitudes, we have checked that the γZ results smoothly go to the $\gamma\gamma$ ones, as $p_2^0 \to 0$.

Finally, we have also checked that our results respect the correct helicity conservation properties at high energy and fixed angles.[10] Such tests check stringently the correctness of the mass-independent contribution for both transverse-Z and longitudinal-Z amplitudes; since any seemingly innocuous misprint could easily transform the expected logarithmic energy-dependence of the amplitudes at high energy, to a linear or quadratic one. In addition to these, we have assured that all UV divergences,[4] as well as any scale dependence, cancel out exactly in the amplitudes and that (3) is respected for real MSSM parameters.

The observable quantities needed for DM studies of the process (1), are expressed in terms of the helicity amplitudes as

$$\Sigma_{ij} = v_{ij} \cdot \sigma_{ij} \simeq \frac{s - m_Z^2}{64\pi[s^2 - (m_i^2 - m_j^2)^2]} \left\{ \int_{-1}^{+1} d\cos\theta \sum_{\lambda_1 \lambda_2 \mu_1 \mu_2} |F^{ij}_{\lambda_1 \lambda_2 \mu_1 \mu_2}|^2 \right\}_s , \tag{9}$$

where v_{ij} describes the relative velocity of the $\tilde{\chi}_i^0 \tilde{\chi}_j^0$-pair. The transverse contribution Σ_{ij}^{TT} to (9) is obtained by discarding the $\mu_2 = 0$ contributions (longitudinal Z) in the helicity summation in (9).

3. Results and Comparisons

For understanding DM observations from neutralino-neutralino annihilation to γZ in *e.g.* the center of our Galaxy or in nearby galaxies like Draco,[11] the quantity Σ_{ij} (9) should be known at $v_{ij} \simeq 10^{-3}$.[1] At so small velocities, the relative orbital angular momentum of the $\tilde{\chi}_i^0 \tilde{\chi}_j^0$-pair must vanish, and the system must be in either an ${}^1S_0$ or an ${}^3S_1$ state. In such cases, the angular distribution of $d\Sigma_{ij}/d\cos\theta$ is flat. Angular momentum conservation implies that the ${}^1S_0$ state can only contribute to transverse helicities for both γ and Z; while the ${}^3S_1$ state can also give non-vanishing longitudinal-Z contributions. If the two neutralinos happen to be identical, Fermi statistics only allows the ${}^1S_0$-state at vanishing relative velocities, so that γ and Z, are both transverse. At higher velocities though, like *e.g.* $v_{ii} = 0.5$, longitudinal-Z amplitudes can also arise.

On the other hand, if $i \neq j$, longitudinal-Z amplitudes can also contribute, even for vanishing relative velocities.

We also note that for $i = j$, the angular structure of $d\Sigma_{ij}/d\cos\theta$, at non-negligible relative velocities, is always forward-backward symmetric. But for $i \neq j$, this is not true any more. Depending on the content of the two neutralinos, $d\Sigma_{ij}/d\cos\theta$ is sometimes peaked in the forward region, and others in the backward; compare (2). For $v_{ij} \lesssim 0.5$, $d\Sigma_{ij}/d\cos\theta$ was found to be flat, in all examples we have considered.

The next step is to compare our work, with that of other authors'. The only preexisting results in the literature apply to i=j=1 for $v_{11} = 0$ and $v_{11} = 0.5$, presented by Ref. 12, and for $v_{11} = 0$ presented by Ref. 13. We use the same sample of models as in Ref. 12.[b] These models, have been named Sugra, nSugra, higgsino-1, higgsino-2, wino-1 and wino-2 by Ref. 12. In Sugra and nSugra, the lightest neutralino (LSP) is a bino. In higgsino-1 and higgsino-2, the two lightest neutralinos are almost or exactly degenerate higgsinos. Finally, in wino-1 and wino-2, the lightest supersymmetric particle (LSP) is a wino, while the NLSP is a bino.

The PLATONdmZ results for i=j=1 and $v_{11} = 0$, perfectly agree with those of Ref. 12, while they deviate from those of Ref. 13. As expected, Z is completely transverse, in this case.

For $v_{11} = 0.5$ though, longitudinal-Z contributions, are also possible. Appreciable longitudinal Z contribution at $v_{11} = 0.5$, only appears for Sugra. At $v_{11} = 0.5$, important discrepancies between our predictions for Σ_{11} and those of Ref. 12, only appear for nSUGRA, reaching the 40% level. We also note that in nSugra and wino-2, Σ_{11} is very sensitive to the relative velocity v_{11}.[14] This can be inferred from the big difference between the $v_{11} = 0$ and $v_{11} = 0.5$ results, for $i = j = 1$, in these models. Is this sensitivity partly responsible for the discrepancy, of the present results, with respect to those of Ref. 12? And then, why it does not induce any discrepancy in wino-2?

We also produce results for Σ_{12} and Σ_{13} for the above 6 models,[14] at $v_{ij} = 0$ and $v_{ij} = 0.5$. Longitudinal Z production is often important in these cases, and sometimes it even dominates the transverse Z contribution, at both velocities.

4. Conclusions and Outlook

Through the present paper, an extensive analytical study of the 1-loop neutralino amplitudes in any unconstrained minimal supersymmetric model (MSSM) has been completed, and the related FORTRAN codes have been released.[7] The only restriction in these codes is that the MSSM parameters are assumed real.

More explicitly, the DM relevant process

$$\tilde{\chi}_i^0 \tilde{\chi}_j^0 \to \gamma Z \ , \tag{10}$$

has been studied analytically here, for any kinematic configuration.

[b] We also use $m_Z = 0.091187 TeV$ and $s_W^2 = 0.2319$, as in Ref. 12, and $m_t = 0.174 TeV$.

It determines also the "reverse" process

$$\gamma + e^{\mp} \to \gamma + (\gamma^*, Z^*) + e^{\mp} \to \tilde{\chi}_i^0 + \tilde{\chi}_j^0 + e^{\mp} \ , \tag{11}$$

where (γ^*, Z^*) denote photon and Z emission from the $e^{\mp}$-line. Such a process could be studied in a future $e^{\mp}\gamma$ Linear Collider. It is straightforward to get the amplitudes for (11), from those of the process (10) studied here. We hope to present results for this in the future.

Acknowledgments

GJG also gratefully acknowledges support from the European Union program RTN MRTN CT-2004-503369.

References

1. D.N. Spergel *et.al.*arXiv:astro-ph/0302209; G. Jungman, M. Kamionkowski and K. Griest, Phys. Rept. **267**:195 (1996); M. Kamionkowski, hep-ph/0210370; M. Drees, Pramana **51**,87(1998); M.S. Turner, J.A. Tyson, astro-ph/9901113, Rev. Mod. Phys. **71S**:145(1999); M.M. Nojiri, hep-ph/0305192; M. Drees hep-ph/0210142; D.P. Roy, Acta Phys. Polon. **B34**:3417 (2003) .
2. For a recent review see *e.g.* G. Lazarides, hep-ph/0601016.
3. G. Bertone, D. Hooper, J. Silk, Phys. Rept. **405**:279 (2005), hep-ph/0404175; J.L. Feng, hep-ph/0405215.
4. G.J. Gounaris, J. Layssac, P.I. Porfyriadis and F.M. Renard, Phys. Rev. **D69**:075007 (2004), hep-ph/0309032.
5. G.J. Gounaris, J. Layssac, P.I. Porfyriadis and F.M. Renard, Eur. Phys. J. **C32**:561 (2004), hep-ph/0311076.
6. G. Passarino and M. Veltman Nucl. Phys. **B160**:151 (1979).
7. PLATON codes can be downloaded from http://dtp.physics.auth.gr/platon/
8. M. Jacob and G.C. Wick, Annals of Phys. **7**:404 (1959), Annals of Phys. **281**:774 (2000).
9. See *e.g.* G.J. Gounaris, C. Le Mouël and P.I. Porfyriadis, Phys. Rev. **D65**:035002 (2002), hep-ph/0107249.
10. G.J. Gounaris and F.M. Renard, Phys. Rev. Lett. **94**:131601 (2005), hep-ph/0501046
11. W. de Boer, hep-ph/0508108; L. Bergström and D. Hooper, hep-ph/0512317; S. Profumo and M. Kamionkowski, astro-ph/0601249; Y. Mambrini and E. Nezri, hep-ph/0507263; Y. Mambrini, C. Muñoz, E. Nezri and F. Prada, hep-ph/0506204.
12. F. Boudjema, A. Semenov and D. Temes, Phys. Rev. **D72**:055024 (2005), hep-ph/0507127.
13. P. Ullio and L. Bergström, Phys. Rev. **D57**:1962 (1998).
14. Th. Diakonidis, G.J. Gounaris, J. Layssac, P.I. Porfyriadis, F.M. Renard hep-ph/0602049.

AXION: PAST, PRESENT AND FUTURE

JIHN E. KIM

Department of Physics and Astronomy and
Center for Theoretical Physics,
Seoul National University, Seoul 151-747, Korea
E-mail: jekim@phyp.snu.ac.kr

The current status of axion physics is presented. There still exists the axion window 10^9 GeV $\leq F_a \leq 10^{12}$ GeV. The recent CAST solar axion search experiment on the axion-photon-photon coupling strength has to be improved by a factor of 100 to reach down to the region of superstring axions. The calculable $\bar{\theta}$ and $m_u = 0$ cases for strong CP solutions, and axino cosmology in SUSY extension of axion are also commented.

Keywords: Axion, Axino, Strong CP Problem, Superstring Axion.

1. Introduction

Modern cosmology needs dark matter and dark energy in the universe: $\Omega_{\rm CDM} \simeq 0.23, \Omega_\Lambda \simeq 0.73$. There are several particle physics candidates for CDM: LSP, axion, axino, gravitino, LKP and other hypothetical heavy particles with some kind of Z_2 symmetries.

The old electroweak scale axion is the pseudo-Goldstone boson[1] arising from breaking the global Peccei-Quinn (PQ) symmetry.[2] The very light axion is the invention from the need to solve the strong CP problem through PQ symmetry with electroweak singlet field(s).[3,4] Superstring axions[13,14] may be in this very light axion category. The existence of instanton solutions in nonabelian gauge theories needs θ vacuum, introducing a CP violating interaction.[5] In the θ vacuum, the physically meaningful interaction is parametrized by $\bar{\theta}$

$$\frac{\bar{\theta}}{32\pi^2} F_{\mu\nu}\tilde{F}^{\mu\nu} \equiv \{F\tilde{F}\}; \quad \bar{\theta} = \theta_{\rm QCD} + \theta_{\rm weak}$$

where $\theta_{\rm QCD}$ is the value determined from high energy scale and $\theta_{\rm weak} =$ Arg. Det. M_q is the one contributed when the electroweak CP violation is introduced. Here $\bar{\theta}$ is the final value taking into account the electroweak

CP violation. For QCD to become a correct theory, this CP violation by $\bar{\theta}$ must be sufficiently suppressed. A nonvanishing value $\bar{\theta}$ contributes to the neutron electric dipole moment d_n. From the experimental limit,[6] we obtain the bound

$$|d_n| < 0.63 \times 10^{25}\ e\text{cm} \to |\bar{\theta}| < 10^{-9}.$$

Why is this so small? It is the strong CP problem. There are three types for the solution: (1) Calculable $\bar{\theta}$, (2) Massless up quark, and (3) Axion.

One may argue that there were no strong CP problem in the beginning. In particular in 5D extension, since the instanton solution is the one in 4D. I think this does not work or at best belongs to the calculable $\bar{\theta}$ type, because in the 4D effective theory one can always consider a 4D theory after integrating out the 5th coordinate. Let us briefly comment on two solutions first.

- The Nelson-Barr type:[7] CP violation is introduced spontaneously. So, original Yukawa couplings are real. Spontaneous CP violation is introduced at high energy by introducing vectorlike heavy quarks so that they mix with light quarks. If the *heavy* vectorlike quarks are not introduced, the CP violation of light quarks originated by the high energy scale CP violation will be tiny due to the decoupling theorem. Not to be affected by the decoupling theorem and to guarantee a tree level Arg. Det. $M_q = 0$, specific forms for Yukawa couplings are assumed: SU(2)xU(1) breaking real VEVs appear only between $F - F$ Yukawas, and CP violating phases in the VEVs appear only in $F - R$ Yukawas, where F are the SM fermions and R are the heavy fermions. If heavy vectorlike fermions are integrated out, the effective Yukawa coupling structure of the low energy sector is of the Kobayashi-Maskawa form.
- Massless up quark: Suppose that we chiral-transform a quark,

$$q \to e^{i\gamma_5 \alpha} q.$$

It is equivalent to changing $\theta \to \theta - 2\alpha$. Thus, if it is allowed to have such a symmetry then strong CP problem is not present. The massless quark case belongs here. This solution was known from the very beginning of the strong CP problem but was not taken seriously because the up quark seemed to be massive.[8] The problem is, "Is $m_u = 0$ allowed phenomenologically?" The famous up/down quark mass ratio from chiral perturbation theory (cPT) calculation is $\frac{m_u}{m_d} = \frac{5}{9}$. But physics below 100 GeV is more involved. There is the determinental interaction of 't Hooft, pictorially shown as

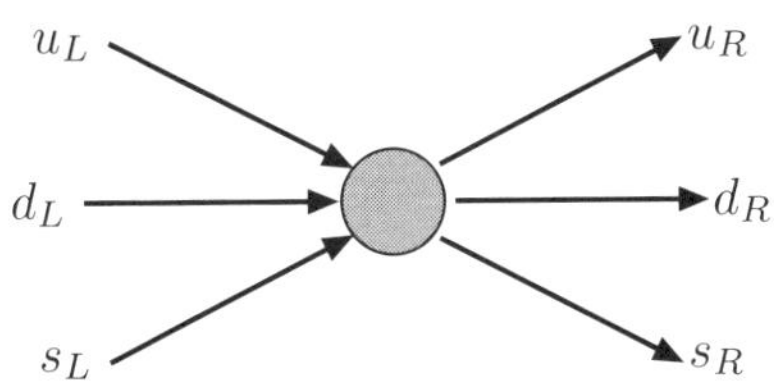

Below the electroweak scale quarks obtain mass. Suppose, the up quark is massless. Then, there is no strong CP problem. But chiral perturbation theory can be done with instanton generated up quark mass from the above 't Hooft interaction, $m_u = m_d m_s/\Lambda$, where Λ is at the QCD scale. So it is the problem whether the instanton calculus really gives the desired magnitude, in which case θ is still unphysical. In the community, still there is a disagreement on this issue: Kaplan and Manohar (KM), and Choi belongs to the positive group,[9,10] and Leutwyler (L) belongs to the negative group.[11] CP even observables do not see m_u. From the figure, for example, we have $m_{u,eff} = m_d m_s/\Lambda$, $m_{d,eff} = m_d + m_u m_s/\Lambda \simeq m_u m_s/\Lambda$, $m_{s,eff} \simeq m_s$. But CP odd observables see m_u. Is $Z = m_u/m_d$ small? KM shows from the 2nd order cPT $Z \simeq 0.2$, and they could not rule out the $m_u = 0$ case. Explicitly, cPT has the L_7 parameter in the term $L_7\langle M^\dagger U - MU^\dagger\rangle$, where $M = 3\times 3$ mass matrix and $U = 3\times 3$ matrix for meson fields. KM shows

$$m_u = 0 : \text{cPT} \Rightarrow \begin{cases} L_7 \sim +1.5\times 10^{-4} \text{ or} \\ (2L_8 - L_5) \simeq (-1.2 \sim -2.5)\times 10^{-3} \end{cases} \tag{1}$$

where L_8, L_5 are another parameters in the cPT. On the other hand L attempted to compute L_7, using the QCD sum rule for the SU(3) singlet pseudoscalar η' dominance (similarly to the vector meson dominance),

$$L_7 \simeq L_{7,\eta'} \simeq (-2\text{–}4)\times 10^{-4} (\text{Gasser} - \text{Leutwyler coefficients}) \tag{2}$$

with a notable sign difference from (1). If (2) were true, the case $m_u = 0$ is ruled out. But Choi[10] argues that if η' gets mass from instanton calculus, which is the modern wisdom on the U(1) problem resolution, he can change the sign of (2) to

$$L_7 \simeq (3 \sim 8)\times 10^{-4}. \tag{3}$$

So, we can have the possibility of $m_u = 0$.
In recent years, lattice calculation has been performed toward this issue.[12] In $16^3 \times 32$ lattice calculation, they obtain: $2L_8 - L_5 \sim 10^{-4}$ and $m_u/m_d = 0.484 \pm 0.027$. If true, $m_u = 0$ is ruled out.
I consider that the problem on $m_u = 0$ is not completely settled yet, even though $m_u \neq 0$ seems to be the majority opinion of the community.

- These show that the axion solution is the most compelling solution which is discussed in the subsequent section.

2. Axion

The axion potential is of the form where the vacuum is shown as a bullet.

The vacuum stays there for a long time, and oscillates when the Hubble time $(1/H)$ is larger than the oscillation period$(1/m_a)$: $H < m_a$. This occurs when the temperature is about 1 GeV. Axion is directly related to θ. Its birth was from the PQ symmetry whose spontaneous breaking introduced a dynamical degree, a pseudo-Goldstone boson called axion. But "pseudo-Goldstone" nature is specific in axion in that axion is a pseudoscalar a without any potential except that arising from,

$$\frac{1}{32\pi^2}\frac{a}{F_a}F\tilde{F} \equiv \frac{a}{F_a}\{F\tilde{F}\}. \tag{4}$$

This kind of nonrenormalizable term can arise in several ways. The first important scale is F_a, defining the strength of nonrenormalizable interaction. It can arise from higher dimensional fundamental interactions with the Planck scale F_a,[13,14] from composite models with the composite scale F_a,[15] from spontaneously broken renormalizable field theories. In the last case, the global symmetry must have the gluon anomaly and is called the PQ symmetry.[2] If this PQ symmetry is spontaneously broken, there arises a pseudo-Goldstone boson[1] coupling to the anomaly with the global symmetry breaking scale F_a.

In QFT, a very light axion is embedded in the phase of a complex $SU(2)_L$ singlet scalar field s, (it may contain very tiny components ($\lesssim 10^{-7}$) from $SU(2)_L$ doublet phases),[4]

$$s = \frac{V+\rho}{\sqrt{2}}e^{ia/F_a}, \quad a \equiv a + 2\pi N_{DW}F_a, \quad V = N_{DW}F_a. \tag{5}$$

So, F_a is in general smaller than $\langle s \rangle$. The potential arising from the anomaly term after integrating out the gluon field is the axion potential. Three properties of the axion potential are known:

(i) It is periodic with $2\pi F_a$ periodicity,
(ii) The minima are at $\langle a \rangle = 0,\ 2\pi F_a,\ 4\pi F_a, \cdots,$[2,16]
(iii) A set of minima is identical, leaving to a few (N_{DW}) distinct vacua.[17]

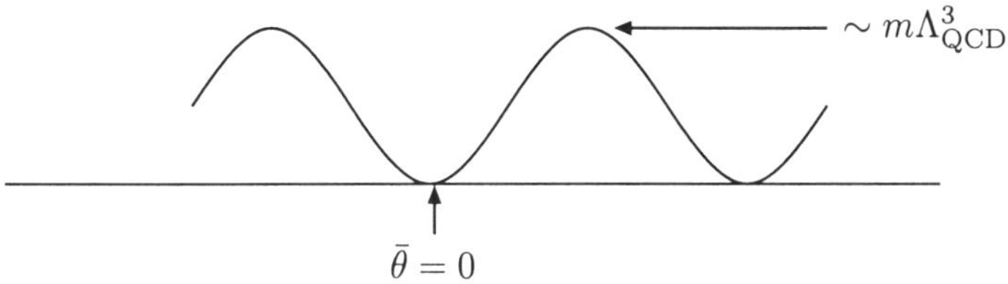

Fig. 1. Vacua are at $\bar\theta = 2n\pi$. The height of the axion potential is given by the instanton interaction and boson mixing.

The height of the axion potential is the scale Λ of the nonabelian gauge interaction and the boson mixing as shown in Fig. 1. We simply take this value as the QCD scale, but in fact it is $m\Lambda_{\rm QCD}^3$ where m is the light quark mass.[18] The dominant one $\Lambda_{\rm QCD}^4$ corresponds to the η' potential. If there are quarks, the height is adjusted since as we have seen before a massless quark makes it flat. The u and d quark phenomenology gives

$$V[a] = \frac{Z}{(1+Z)^2} f_\pi^2 m_\pi^2 \left(1 - \cos\frac{a}{F_a}\right). \tag{6}$$

The essence of the axion solution is that $\langle a \rangle$ seeks $\bar\theta = 0$ in the evolving universe whatever happened before. It is a cosmological solution[20] as shown in Fig. 1. The weak CP violation makes the minimum of the potential shifted a little bit at $\bar\theta = O(10^{-17})$. The axion mass is given by $m_a \simeq (10^7 {\rm GeV}/F_a)$ 0.6 eV.

There are several laboratory experiments, restricting the axion decay constants: (i) meson decays, $J/\Psi \to a + \gamma, \Upsilon \to a\gamma, K^+ \to \pi^+ + a$, (ii) beam dump experiments, p(or $e-$)$N \to aX, a \to \gamma\gamma$ and e^+e^-, (iii) and nuclear deexcitation, $N^* \to Na, a \to \gamma\gamma$ and e^+e^-. Thus, we obtain the inner space bound $F_a \geq 10^4$ GeV from the laboratory experiments. So, from the beginning, it was known that the PQWW axion, arising from the electroweak scale, is ruled out.[21] Thus, F_a has to be very large, having led to the so-called invisible axion. But, there is a possibility of detecting it,[22] and hence it should be called a *very light axion or sub-meV axion.*

3. Axion window to outer space

But the stringent lower bounds on the axion decay constant comes from the outer space observations. Firstly, stellar evolutions, if axion existed, are affected by axion emissions and the successful standard energy loss mechanism due to weak interactions restricts the axion mass toward a smaller region, or the axion decay constant to a larger region. The stringent bound comes from the study of supernova evolution,[23] especially from the SN1987A study to give $F_a \geq 10^9$ GeV.[24] On the other hand, the very interesting upper bound on F_a is obtained from the axionic contribution to dark energy in universe.[20]

3.1. *Stars*

The current supernova[24] (globular cluster[25]) limit on F_a is 10^9 GeV(10^{10} GeV). It uses primarily the Primakoff process with the following coupling,[3]

$$\mathcal{L}_{a\gamma\gamma} = -\, c_{a\gamma\gamma}\frac{a}{F_a}\frac{e^2}{32\pi^2}F_{\rm em}\tilde{F}_{\rm em} \Rightarrow \mathbf{E}\cdot\mathbf{B} \text{ interaction} \tag{7}$$

$$c_{a\gamma\gamma} = \tilde{c}_{a\gamma\gamma} + 6 \sum_{i=\text{light q}} \tilde{\alpha}_i Q^2_{\text{em},i} \simeq \tilde{c}_{a\gamma\gamma} - 1.93, \quad Z = \tfrac{5}{9} \tag{8}$$

$$\tilde{c}_{a\gamma\gamma} = \text{determined from high energy physics} \tag{9}$$

$$\tilde{\alpha}_u \simeq \tfrac{1}{1+Z}, \quad \tilde{\alpha}_d \simeq \tfrac{Z}{1+Z} \tag{10}$$

where the chiral symmetry breaking of u, d quarks are taken into account. The number 1.93 corresponds to $Z = \frac{5}{9}$. Since the instanton contribution to light quark masses is present,[9,10] we may take a band around 1.93.

In the hot plasma in stars, once produced, they most probably escape the core of the star and take out energy. This contributes to the energy loss mechanism of star and should not dominate the luminocity: (i) The Primakoff process: $\gamma \to a$ (present in any model): $g_{a\gamma\gamma} < 0.6\times 10^{-10}$ GeV^{-1} or $F_a > 10^7$ GeV, and 0.4 eV $< m_a <$ 200 keV ruled out because too heavy to produce,

(ii) Compton-like scattering: $\gamma e \to ae$ (DFSZ axion has aee coupling) $g_{aee} < 2.5 \times 10^{-13}$, 0.01 eV $< m_a <$ 200 keV, and

(iii) SN1987A, $NN \to NNa 3\times 10^{-10} < g_{aNN} < 3\times 10^{-7} \Longrightarrow F_a >$ 0.6×10^9 GeV.

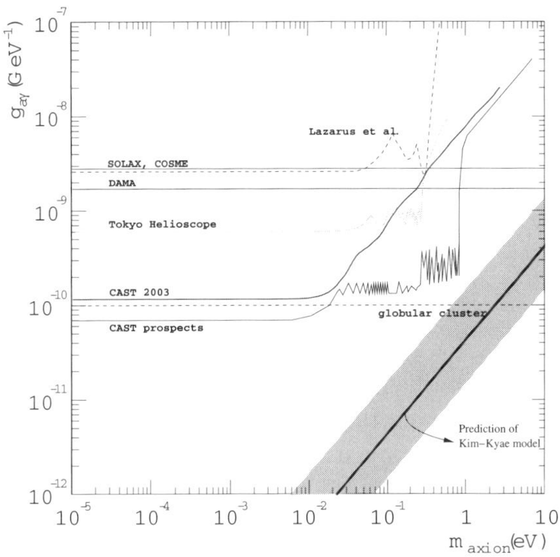

Fig. 2. Experimental bound from various experiments, especially including CAST 2003 data. This figure is taken from the result paper of CAST experiment.[26] Here,we show the prediction from Z_{12} model also.

Stellar evolution uses the energy loss mechanism, with the aforementioned lower bound on F_a. But laboratory experiments can offer a more effective bound on F_a than just the energy loss mechanism, as done by the CAST(CERN axion solar telescope) experiment of Fig. 2.[26]

3.2. *Universe*

In the standard big bang cosmology (SBB), there is a severe domain wall problem.[27] The SBB allows only the domain wall number $N_{DW} = 1$.[28] But the most interesting inflationary cosmology solves this domain wall problem at one stroke if the reheating temperature after inflation is lower than F_a. The inflationary cosmology seems to get support from the COBE and WMAP observations of density perturbations in the early universe, and NDW problem is not an issue since in SUSY models the reheating temperature is required to be smaller than 10^9 GeV.[29] In axion cosmology, the following items are important:

- The axion decay constant F_a,
- Axion couplings to γ, e, p, n,
- The domain wall number N_{DW}.

If a singlet scalar VEV $\langle s\rangle$ breaks the PQ symmetry, then $F_a = \langle s\rangle/N_{DW}$ defines the domain wall number. Reheating temperature after inflation is required to be below F_a if $N_{DW} > 1$.

Axions are created at $T \simeq F_a$, but the universe does not change $\langle a\rangle$ until $H \simeq m_a(T = 1\text{GeV})$. Then, the classical field $\langle a\rangle$ starts to oscillate. From the harmonic oscillator type energy density $m_a^2 F_a^2$, we have $m_a \times$ number density $\Rightarrow$ CDM-like energy:[30]

$$\rho_a(T_\gamma) = m_a(T_\gamma)n_a(T_\gamma) \simeq 2.5\frac{F_a}{M_P}\frac{F_a m_a}{T_1}T_\gamma^3\left(\frac{A(T_1)}{F_a}\right)^2 \tag{11}$$

where the oscillation start-up temperature T_1 is the strong interaction scale 1 GeV and T_γ is the present temperature. If F_a is large($> 10^{12}$ GeV), then the axion energy density dominates the energy density of the universe. Since the energy density is proportional to the number density, it behaves like a CDM.

3.3. *Axion window and search for cosmic axions*

The above astro- and cosmological-bounds on F_a are summarized as

$$10^9\text{GeV} \le F_a \le 10^{12}\text{GeV} \tag{12}$$

If axions are the CDM component of the universe, then they can be detected. The feeble coupling can be compensated by a huge number of axions. The number density $\sim F_a^2$, and the cross section $\sim 1/F_a^2$, and there is a hope of detecting it. Sikivies cavity detector[22] with dimension of tens of cm has been used to give coarse bounds[31] on axion parameters in the axion mass of order 10^{-5} eV.

4. Axions from superstring

Superstring tells us definite things about global symmetries. If axion is present, it is better to be realized in superstring. They are the bosonic degrees in B_{MN} (MI-axion[13] is $B_{\mu\nu}$ and MD-axion[14] is B_{ij}) and furthermore additional massless bosons from compactification are candidates. Superstring does not allow global symmetries. But there is an important exception to this claim: the shift symmetry of $H_{\mu\nu\rho}$, which gives the MI-axion. It is the only allowed global symmetry. B_{ij} are generally heavy;[32] but it is a model dependent statement.

The superstring axion decay constants are expected near the string scale which is too large:[33] $F_a > 10^{16}$ GeV. A key question in superstring models is "How can one obtain a low value of F_a?" An idea is the following:

> In some compactifications, anomalous U(1) results,[34] where U(1) gauge boson eats the MI-axion to become heavy. Earlier, this direction, even before discovering anomalous U(1) gauge boson, was pointed out by Barr.[35] It became a consistent theory after discovering the anomalous U(1). Then, a global symmetry survives down the string scale. F_a may be put in the axion window. It was stressed in several references.[36,37]

However, this idea does not work necessarily, as will be commented later.

Somehow MD-axion(s) may not develop a large superpotential terms. But the problem here is the magnitude of the decay constant. MD-axion decay constants were tried to be lowered by localizing them at fixed points.[19,38] It uses the flux compactification idea and it is possible to have a small F_a compared to the string scale as in the RS model. One needs the so-called throat as schematically shown in Fig. 3.

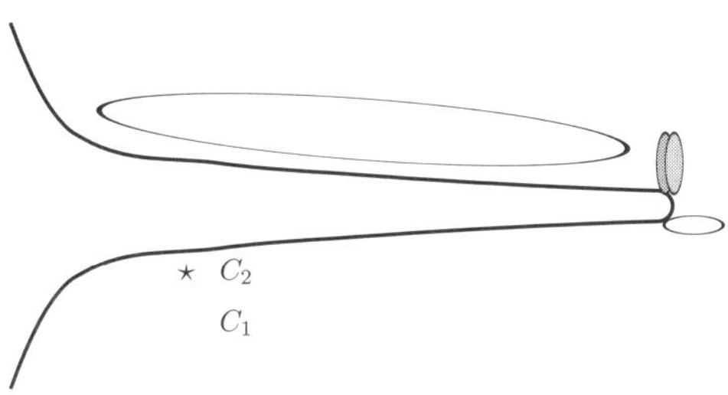

Fig. 3. A schematic view of the $S_2 \times S_2 \times S_1$ throat. At the tip, one elongated S_2 can shrink to a point shown as a star. The un-shrunk S_2 has the cycle C_2 where a MD-axion resides as a harmonic 2-form.

Axion mixing Even if we lowered some F_a, we must consider hidden sector also. In this case, axion mixing must be considered. There is an important theorem.

> *Cross theorem on potential heights and decay constants:*[39] Suppose two axions a_1 with F_1 and a_2 with F_2 ($F_1 \ll F_2$) couples to two

> nonabelian groups whose scales having a hierarchy, $\Lambda_1 \ll \Lambda_2$. The higher potential Λ_2 couples to both axions. Then, the diagonalization process chooses the larger potential Λ_2 corresponds to the smaller decay constant F_1, and the smaller potential Λ_1 corresponds to the larger decay constant F_2.

So, just obtaining a small decay constant is not enough. Hidden sector may steal the smaller decay constant. It is likely that the QCD axion chooses the larger decay constant. Recently, the mixing effect has been stressed by I.-W. Kim *et. al.*[19,40]

And most probably, our axion will couple to the μ term:

$$H_u H_d f(S_1, S_2, \cdots).$$

After all, the topologically attractive B_{MN} may not be the axion we want. Let us go back to earlier field theoretic very light axion. In string models, its effect toward phenomenology was not calculated before. Now we have an explicit model for MSSM,[41] and we can see here whether the idea of approximate global symmetry is realized. It must be that at sufficiently higher orders the PQ symmetry is broken. In the Z_{12-I} model, we calculate the axion-photon-photon coupling[40] whose result is shown in Fig. 2. But the decay constant is at the GUT scale. In this kind of calculation, there are so many Yukawa couplings to consider. For example, we encountered O(104) terms for d=7 superpotential terms and it is not a trivial task to find an approximate PQ symmetry direction.

In addition, we point out that the MI-axion with anomalous U(1) always has a large decay constant since most of the fields are charged under this anomalous U(1). Phenomenologically successful axion must need an approximate PQ symmetry.

An approximate PQ global symmetry with discrete symmetry in SUGRA was pointed out long time ago: given by Lazarides and Shafi[42] for a discrete $Z_3 \times Z_3$. But this field theoretic method does not guarantee that string models realize this idea. In this sense, an explicit demonstration of an approximate PQ symmetry is vital for a string matter axion.

5. SUSY extension and axino

The SUSY extension always introduces gravitino. Gravitinos produced thermally after inflation decay very late in cosmic time scale ($> 10^3$ s) and can dissociate the light nuclei by its decay products. Not to have too many gravitinos, the reheating temperature must be bounded,[29,43]

$T_r \leq 10^9$GeV (old value), $T_r \leq 10^7$GeV (new but model dependent value). Therefore, in SUSY theories we must consider the relatively small reheating temperature, and the domain wall problem does not matter in axion cosmology.

The SUSY extension with the strong CP solution via axion introduces its superpartner *axino*. Its cosmological significance is in that it can serve as keV range warm dark[44] matter or GeV range cold dark matter.[45] Let us comment on its CDM possibility.

For axino to be CDM, it must be stable or practically stable. Without the R-parity conservation, this can not happen. Thus, we require the practical R-parity conservation for the possibility of axino CDM. In addition, for axino to be LSP it must be lighter than the lightest neutralino whose mass is expected to be around 100 GeV. Thus, the estimation of the axino mass is of prime importance. The conclusion is that there is no theoretical upper bound on the axino mass and axino mass can be easily in the GeV range. Since axion is almost massless, one expects that its superpartners are massless in the first approximation. Its scalar partner, saxion, obtains the mass of order the soft terms after SUSY breaking. Its cosmological effect is relatively late decaying nature, adding more photons after its decay.[46] Regarding mass, saxion is like the other SM SUSY scalars.

However, the axino mass is intimately related to the SUSY breaking scenario and symmetries of the superpotential. The PQ symmetry allows the following superpotential[47]

$$W = fZ(S_1S_2 - F_a^2), \quad Z, S_1, S_2: \text{ singlets}$$

There also exist SUSY breaking soft terms. Thus, the following potential is obtained

$$V = |f|^2(|S_1|^2 + |S_2|^2)|Z|^2 + (A_1 fS_1S_2Z - A_2 fF_a^2 Z + \text{h.c.})$$

which determines the VEV of Z. Since S_1 and S_2 are of order F_a, $\langle Z\rangle$ is of order the A term. Thus, the fermion partners have the mass matrix of the form

$$\begin{pmatrix} 0 & m_{\tilde{a}} & fF_a \\ m_{\tilde{a}} & 0 & fF_a \\ fF_a & fF_a & 0 \end{pmatrix}, \quad m_{\tilde{a}} = f\langle Z\rangle$$

The lightest eigenvalue is the axino mass. The others are of order F_a. As shown above, the axino mass is basically a free parameter, and expected to be smaller than the naive SUSY breaking scale due to the small coupling.

Most probably, it is lighter than neutralino χ. But its mass can be much smaller than the SUSY scale as shown from a superpotential of the form,[48]

$$W' = fZ(S_1S_2 - X^2) + \frac{1}{3}(X - M)^3$$

where X carries the vanishing PQ charge. This potential is much more complicated to analyze. We show $m_{\tilde{a}} = O(A - 2B + C) + O(m_{3/2}^2/F_a)$. For the standard pattern of soft terms, we have $B = A - m_{3/2}$ and $C = A - 2m_{3/2}$.[49] Thus, the axino mass is of order keV. Even the tree level axino mass needs the knowledge on the full superpotential, we treat the axino as the LSP which is the most probable choice. Its mass is left as a free parameter. KeV axinos can be warm dark matter which is thermal relic. GeV axinos can be CDM. In this case, the reheating temperature must be low.[45]

If gravitino is the next lightest LSP (NLSP), $m_{\tilde{a}} < m_{3/2} < m_\chi$, the gravitino problem can be resolved,[50] since the thermally produced gravitinos would decay to axino and axion which do not affect BBN produced light elements.

Most probably, χ would be the NLSP, and the thermal production mechanism restricts the reheating temperature after inflation as summarized in Fig. 4. At high reheating temperature, thermal production contributes dominantly in the axino production. Even though the reheating temperature is below critical energy density line, there still exists the CDM possibility by the non-thermal production (NTP) axinos. Covi *et. al.* shows[45]

$$NTP: \quad \Omega_{\tilde{a}}h^2 = \frac{m_{\tilde{a}}}{m_\chi}\Omega_\chi h^2 \quad \text{for} \quad m_{\tilde{a}} < m_\chi < m_{3/2}$$

In the figure, NTP axinos can be CDM for relatively low reheating temperature < 10 TeV, in the region

$$10 \text{ MeV} < m_{\tilde{a}} < m_\chi, \quad \text{NTP axino as CDM possibility.}$$

The shaded region corresponds to the MSSM models with $\Omega_\chi h^2 < 10^4$, but a small axino mass renders the possibility of axino closing the universe or just 30% of the energy density. If all SUSY mass parameters are below 1 TeV, then $\Omega_\chi h^2 < 100$ and sufficient axino energy density requires

$$m_{\tilde{a}} > 1 \text{ GeV}, \quad \begin{cases} \text{Low reheating is good in view of} \\ \text{the recent gravitino problem.} \\ \text{But not good with leptogenesis.} \end{cases}$$

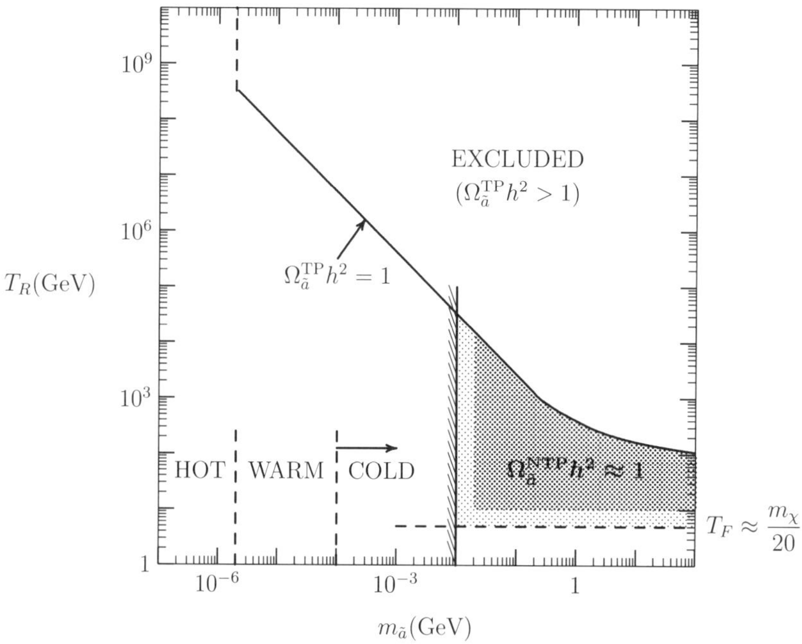

Fig. 4. The solid line gives the upper bound from thermal production on the reheating temperature as a function of the axino mass. The dark region is the region where non-thermal production can give cosmologically interesting results ($\Omega_{\tilde{a}}^{\rm NTP}h^2 \simeq 1$).

6. Conclusion

I reviewed strong CP and axion. In particular,

- Solutions of the strong CP problem: Nelson-Barr, $m_u = 0$, axion. Axion a is the most attractive and plausible solution.
- Axions can contribute to CDM. Maybe solar axions are easier to detect. Most exciting is, it confirms instanton physics by observation.
- Tried to present a superstring matter axion coupling for the first time. A QCD axion from superstring may be a window to string.
- With SUSY extension, O(GeV) axino can be CDM. It is difficult to detect this axino from the DM search, but possible to detect at LHC as missing energy.

Acknowledgments

This work is supported in part by the KRF ABRL Grant No. R14-2003-012-01001-0. J.E.K. is also supported in part by the KRF grants, No. R02-2004-000-10149-0 and No. KRF-2005-084-C00001.

References

1. S. Weinberg, Phys. Rev. Lett. **40** (1978) 223; F. Wilczek, Phys. Rev. Lett. **40** (1978) 279.
2. R. D. Peccei and H. R. Quinn, Phys. Rev. Lett. **38** (1977) 1440.
3. For a review, see, J. E. Kim, Phys. Rep. **150** (1987) 1.
4. J. E. Kim, Phys. Rev. Lett. **43** (1979) 103; M. A. Shifman, V. I. Vainstein, and V. I. Zakharov, Nucl. Phys. **B166** (1980) 4933; A. Zhitnisky, Sov. J. Nucl. Phys. **31** (1980) 260; M. Dine, W. Fischler, and M. Srednicki, Phys. Lett. **B104** (1981) 199.
5. C. G. Callan, R. F. Dashen and D. J. Gross, Phys. Lett. **B63** (1976) 334; R. Jackiw and C. Rebbi,Phys. Rev. Lett. **37** (1976) 172.
6. P. G. Harris et al., Phys. Rev. Lett. **82** (1999) 904.
7. A. Nelson, Phys. Lett. **B136** (1984) 387; S. Barr, Phys. Rev. Lett. **53** (1984) 329.
8. S. Weinberg, in *Festschrift for I. I. Rabi*, Trans. of New York Academy of Sciences, Vol. 38 (1977) 185.
9. D. B. Kaplan and A. Manohar, Phys. Rev. Lett. **56** (1986) 2004.
10. K. Choi, Nucl. Phys. **B383** (1992) 58; K. Choi, C. W. Kim and W. K. Sze, Phys. Rev. Lett. **61** (1988) 794.
11. H. Leutwyler, Nucl. Phys. **B337** (1990) 108.
12. D. R. Nelson, G. T. Flemming and G. W. Kilcup, Phys. Rev. Lett. **90** (2003) 021601.
13. E. Witten, Phys. Lett. **B149** (1984) 351.
14. E. Witten, Phys. Lett. **B153** (1985) 243.
15. J. E. Kim, Phys. Rev. **D31** (1985) 1733; K Choi and J. E. Kim, Phys. Rev. **D32** (1985) 1828.
16. C. Vafa and E. Witten, Phys. Rev. Lett. **53** (1984) 535.
17. P. Sikivie, Phys. Rev. Lett. **48** (1982) 1156.
18. V. Baluni, Phys. Rev. **D19** (1979) 2227. For a recent discussion, see, the subsequent reference.
19. I.-W. Kim and J. E. Kim, Phys. Lett. **B639** (2006) 342.
20. J. Preskill, M. B. Wise and F. Wilczek, Phys. Lett. **B120** (1983) 127; L. F. Abbott and P. Sikivie, Phys. Lett. **B120** (1983) 133; M. Dine and W. Fischler, Phys. Lett. **B120** (1983) 137.
21. R. D. Peccei, in *Proc. 19th ICHEP*, eds. S. Homma *et. al.* (Phys. Soc. of Japan, Tokyo, 1979) p.385.
22. P. Sikivie, Phys. Rev. Lett. **51** (1983) 1415.
23. A. Pantziris and K. Kang, Phys. Rev. **D33** (1986) 3509.

24. G. Raffelt and D. Seckel, Phys. Rev. Lett. **60** (1988) 1793; M. S. Turner, Phys. Rev. Lett. **60** (1988) 1797; R. Mayle, J. R. Wilson, J. R. Ellis, K. Olive, D. N. Schramm, and G. Steigman, Phys. Lett. **B203** (1988) 188; K. Choi, K. Kang and J. E. Kim, Phys. Rev. Lett. **62** (1989) 849. For reviews, see, M. S. Turner, Phys. Rep. **197** (1990) 67 and G. Raffelt, Phys. Rep. **198** (1990) 1.
25. G. G. Raffelt, Ann. Rev. Nucl. Part. Sci. **49** (1999) 163.
26. K. Zioutas *et al.* [CAST Collaboration], Phys. Rev. Lett. **94** (2005) 121301.
27. P. Sikivie, Phys. Rev. Lett. **48** (1982) 1156.
28. S. M. Barr, K. Choi and J. E. Kim, Nucl. Phys. **B283** (1987) 591.
29. J. Ellis, J. E. Kim and D. V. Nanopoulos, Phys. Lett. **B145** (1984) 181.
30. M. S. Turner, Phys. Rev. **D33** (1986) 889.
31. S. DePanfilis *et. al.*, Phys. Rev. Lett. **59** (1987) 839; C. Hargmann *et. al.*, Phys. Rev. **D42** (1990) 1297; C. Hargmann *et. al.*, Nucl. Phys. (Proc. Suppl.) **B51** (1996) 1415; S. Matsuki, talk presented at IDM-1998, Buxton, England, 7–11 Sep. 1997, Proc. "*The identification of Dark Matter*", ed. N. J. C. Spooner and V. Kudryavtsev (World Scientific Pub. Co., Singapore, 1999) p. 441.
32. X. G. Wen and E. Witten, Phys. Lett. **B166** (1986) 397.
33. K. Choi and J. E. Kim, Phys. Lett. **B154** (1985) 393.
34. M. Dine, N. Seiberg, and E. Witten, Nucl. Phys. **B289** (1987) 317; J. J. Atick, L. J. Dixon, and A. Sen, Nucl. Phys. **B292** (1987) 109; M. Dine, I. Ichinose, and N. Seiberg, Nucl. Phys. **B293** (1987) 253.
35. S. M. Barr, Phys. Lett. **B158** (1985) 397.
36. J. E. Kim, Phys. Lett. **B207** (1988) 434.
37. B. Svrcek and E. Witten, JHEP **0606** (2006) 051.
38. J. P. Conlon, JHEP **0605** (2006) 078.
39. J. E. Kim, JHEP **9905** (1999) 022; JHEP **0006** (2000) 016.
40. K.-S. Choi, I.-W. Kim and J. E. Kim, hep-ph/0612107.
41. J. E. Kim and B. Kyae, hep-th/0608085; hep-th/0608086.
42. G. Lazarides, C. Panagiotakopoulos and Q. Shafi, Phys. Rev. Lett. **56** (1986) 432.
43. M. Kawasaki, K. Khori and T. Moroi, Phys. Lett. **B625** (2005) 7; Phys. Rev. **D71** (2005) 083502.
44. K. Rajagopal, M. S. Turner and F. Wilczek, Nucl. Phys. **B358** (1991) 447.
45. L. Covi, J. E. Kim and L. Roszkowski, Phys. Rev. Lett. **82** (1999) 4180; L. Covi, H. B. Kim, J. E. Kim, and L. Roszkowski, JHEP **0105** (2001) 033.
46. J. E. Kim, Phys. Rev. Lett. **72** (1994) 1956; S. Chang and H. B. Kim, Phys. Rev. Lett. **77** (1996) 591.
47. J. E. Kim, Phys. Lett. **B136** (1984) 378.
48. E. J. Chun, J. E. Kim and H. P. Nilles, Phys. Lett. **B287** (1992) 123.
49. H. P. Nilles, Phys. Rep. **110** (1984) 1.
50. T. Asaka and T. Yanagida, Phys. Lett. **B494** (2000) 297.

OVERLOOKED ASTROPHYSICAL SIGNATURES OF AXION(-LIKE) PARTICLES

K. ZIOUTAS[1,2], Y. SEMERTZIDIS[1,3] and TH. PAPAEVANGELOU[2]

[1]*Physics Department, University of Patras, GR-26500 PATRAS, Greece*
[2]*CERN, CH-1211 Geneva 23, Switzerland*
[3]*Physics Department, Brookhaven National Laboratory, Upton, NY 11973-5000, USA*
Emails: zioutas@physics.upatras.gr; yannis@bnl.gov; Thomas.Papaevangelou@cern.ch

The working principle of axion helioscopes can be behind unexpected solar X-ray emission. Because this is associated with solar magnetic fields (~B^2), which become in this framework the catalyst and not the otherwise suspected / unspecified energy source of solar X-rays. In addition, the built–in fine tuning we may (not) be able to fully reconstruct, and, we may (not?) be able to copy. Solar axion signals are transient brightenings, or, continuous radiation violating the second law of thermodynamics and Planck's law of black body radiation. To understand the corona problem and other mysteries like flares, sunspots, etc., we arrive at two exotica: a) trapped, radiatively decaying, massive axions allow a continuous self-irradiation of the Sun, explaining the sudden temperature inversion ~2000 km above the surface and b) outstreaming light axions interact with local fields (~B^2), depending crucially on the plasma frequency which must match the axion rest mass, explaining the otherwise unpredictable transient solar phenomena. Then, the energy of a related phenomenon points at the birth place of involved axions. For example, this suggests that the ~2 MK solar corona has its axion roots at the top of the radiative zone. The predicted B ≈ 10–50 T make this place a coherent axion source, while the multiple photon scattering enhances the photon-to-axion conversion unilaterally, since axions escape. We conclude that the energy range below some 100 eV is a window of opportunity for axion searches, and that it coincides with a) the derived photon energies for an external self-irradiation of the Sun, which has to penetrate until the transition region, and b) with the bulk of the soft solar X-ray luminosity of unknown origin. Thus, indirect signatures support axions or the like as an explanation of enigmatic behaviour in the Sun and beyond. Axion antennas could take advantage of such a feed back.

1. Introduction

With the ongoing direct search for (solar) axions [1] one can ask whether the same detection principle is at work behind certain mysterious solar or other astrophysical observations. Indirect signatures for dark matter particle candidates might be strong, if, for example, the same process(es) can explain consistently more celestial phenomena. Here we mainly discuss striking solar observations of unknown origin which seem to require the involvement of axions or other exotica with similar properties. Previous work [2–4] addressed a steady X-ray emission due to very few gravitationally trapped massive axion-like particles (about 1 in 10^7, or only ~100 *kg*/s since 4.6 Gyears), giving rise to a self-irradiation of the whole Sun. In this work, we mainly argue in favour of a second solar X-ray component expected to originate from the widely accepted interaction of light axions or the like with the inner/outer solar magnetic fields, as they stream out of the Sun. In this work we emphasize the ubiquitous solar magnetic field, which is the dominating "element" in all axion experiments.

2. Astrophysical Signatures

The solar corona problem: Stellar observations and theory of stellar evolution cannot be reconciled with stars having atmospheres that emit X-rays [5], suggesting the question: Where do these X-rays come from? The mechanism that heats the solar corona to some MK has remained elusive since its discovery in 1939. The solar corona problem is one of the most important and challenging problems in astrophysics [6], since it violates, at first sight, the second law of thermodynamics and Planck's law of black body radiation, which is actually improbable. The radiative decay of gravitationally trapped massive particles like the generic Kaluza–Klein axions, being created by the Sun itself, could provide the invisible source that sustains the self-heating of the solar atmosphere [2], thus reconciling observation and thermodynamics. In fact, the expected decay hard X-rays from massive axions, accumulated around the Sun since cosmic times, fit only the energetic part (above ~2 keV) of the reconstructed analog photon spectrum from the quiet Sun during solar minimum (see Figure 8 and 9 in ref. [2]). However, this is only one X-ray component within the axion scenario, and probably the weakest one.

Interestingly, it was shown in ref. [3] that the radial distribution of the two off-pointing observations with the YOHKOH X-ray satellite fit the massive axion scenario also in the low–energy range (~0.5–4 keV), thus suggesting more measurements of this kind. In fact, the RHESSI mission has recently repeated several off-pointing observations above the solar limb during non-flaring, spotless and active-region-free Sun, i.e., during quiet Sun conditions [7], arriving apparently at a hard X-ray spectrum, which is rather similar to the reconstructed one from YOHKOH measurements in the previous solar cycle minimum (see Figure 9 in ref. [2]). The values obtained with RHESSI in the ~3–12 keV range, for such extremely quiet periods at the present solar minimum phase, correlate with those from another orbiting X-ray solar telescope (GOES), implying a real signal in this actually not so low energy band [7,8] for a cool Star like our Sun. This we consider as additional supporting evidence for the (massive) solar axion scenario [2]. Because, a conventional explanation with electrons raises a new problem: How is such a population of energetic electrons created in the quiet corona?

Magnetic field related observations: The long lived massive axion(-like) particles can not explain:

a) local and transient solar phenomena, and
b) the prevailing soft X-ray emission, which makes the bulk of the solar X-ray luminosity during periods of quiet (see Figure 9 in ref. [2]) as well as active [i.e. flaring] Sun.

Therefore, an as yet unnoticed additional X-ray source must be at work. The Primakoff–effect with "conventional" axions, e.g. like the QCD-motivated ones, can interact inside the solar magnetic fields, which is after all the most natural process to expect, as it resembles the working principle of an axion helioscope like CAST. Thus, near the solar surface X-rays should also be emitted due to magnetic-field-induced radiative "decay" of outstreaming axions (see below). Depending on the relative intensity of converted light axions, the resulting radial distribution of X-rays, coming from both components near the surface of the Sun, can be different from that expected for massive ones only [2,3]. Interestingly, the available two off-pointing observations with YOHKOH [3] show a radial distribution that agrees within ~30% with the massive axion simulation. Therefore, in our heuristic approach explaining the otherwise unexpected solar X-rays with axions or the like, this second magnetic–field–related component offers a possible explanation also for this rather small but significant discrepancy.

There is strong observational evidence that (transient) solar X-ray emission correlates with the local magnetic field strength squared ($\sim B^2$) [9], which is characteristic for light axion involvement, and, it determines the performance of an axion helioscope [1] à la Pierre Sikivie, and, à la Karl van Bibber *et al.*. Actually in all axion experiments, the B^2 - dependence of a potential signal is generally accepted as the ultimate method for axion identification. Large–scale magnetic fields of several kGauss exist in the enigmatic sunspots, which show enhanced and unpredictable activity. Interestingly, it is known that the magnetic field somehow heats the quiet solar corona, with the exact energy release mechanism still enigmatic in solar physics; within the axion picture, this is naturally expected. A few magnetic–field–related cases in favour of the axion(-like) scenario have already been discussed in ref. [9]. For completeness, we repeat some of them here briefly in an updated form (see also ref. [9]):

a) The X-ray emission from an isolated solar active region, measured with YOHKOH's X-ray telescope, during non-flaring periods, shows a striking $B^{1.94\pm0.12}$ dependence in the range ~0.5–4 keV; within the axion scenario, this points at a low energy solar axion spectrum (below ~ 1 keV), in contrast with the widely used solar axion spectrum that peaks at ~4.2 keV. Such a new component is of utmost interest for the design of direct solar axion searches. Similarly, a plethora of solar soft X-ray flux measurements correlates with the local magnetic field (~B^n), with the exponent n≈2 varying smoothly by ~20% during the solar cycle(!) [10], pointing possibly to a deeper axion implication in the dynamic Sun.

b) Remarkably, above sunspots the corona is hotter (e.g. ~2.1 MK instead of 1.3 MK) and the photosphere just underneath is cooler (e.g. ~4000 K instead of 5770 K) than near quiet Sun regions. Each of these observations is consistent with a Primakoff–effect taking place inside the extended strong surface magnetic fields with the plasma frequency matching the axion rest mass. The emerging picture is: 1) energetic axions streaming out of the hot inner Sun can be converted back to X-rays, further heating–up the preexisting ~1.3 MK quiet Sun corona, and 2) near the photosphere or even deeper into the Sun, photons with energy below ~10–100 eV can undergo the reverse Primakoff–process escaping into axions or the like, thus making the photosphere cooler.

c) Generally, there is strong evidence that magnetic elements in the Sun with higher magnetic flux are less bright, with the strongest magnetic field in a sunspot umbra being at its darkest position [11]. Low–energy solar axion production due to surface magnetic fields fits the observed ~B^2-dependent dimming effects of the sunspot brightness in the visible. Also recently [12], some 900 sunspots have shown an increase in brightness while the magnetic field strength was decreasing. Then, the dark sunspots can be low–energy axion sources, whose strength should correlate with their level of darkness, thus providing an axion trigger (see also [13]), which is exceptional. To be more quantitative, we work out here ***a numerical example***: We use the PVLAS derived parameters [14], i.e. $g_{a\gamma\gamma} \approx 2.5\ 10^{-6}\text{GeV}^{-1}$ and m ≈ 1 meV. We take as a mean free path length of visible photons in the photosphere equal to ~100 km, which is equal to the coherence length of the photon-to-axion conversion; the condition $\hbar\omega_{plasma} \approx mc^2 \approx 1$ meV seems reasonable. For a sunspot magnetic field B ≈ 3 kGauss (see ref. [22]), and, provided the PVLAS finding is correct, we estimate a 'photon-to-axion' conversion efficiency $P_{\gamma a} \approx 0.001$. Furthermore, assuming a surface of a single sunspot of a few ‰ of the solar surface, the expected solar axion-like luminosity is $L_{axion} \approx 10^{28}$erg/s. The derived signal rate in the visible, applying CAST performance [1], is S ≈ 300 Hz. Such a high signal rate along with the rather conservative values we used, allow to be sensitive to a much smaller coupling constant and check the PVLAS result.

In summary, axion–photon oscillations depend a) on the squared transverse magnetic field component along the axion/photon propagation, b) on the local plasma density, since at resonance $\hbar\omega_{plasma} \approx m_{axion}c^2$ the coherence length is limited by the photon absorption, and c) on the magnetic field configuration. Note that each of these parameters changes permanently near the solar surface and also deeper inside the turbulent (i.e. boiling) convective zone. For a quantitative calculation, since it implies restless density fluctuations for any place above the radiative zone ($R > 0.7 \bullet R_{solar}$), it is crucial to know this dynamical behaviour; resonance can be restored temporally repeatedly covering a large bandwidth in axion rest mass, e.g., ~0-10 eV/c^2, if we take as maximum density the one near the bottom of the convective zone. Then, axion (dis)appearance is not at all a static process; it can reflect the unpredictable dynamical behaviour of the Sun from the visible to X-rays. In fact, such a rapid change of the axion-relevant parameters defines instantly certain volume elements, where resonance-coherence effects can result in an enhanced axion production as well as in axion-related brightenings or light dimming. Note that a fine tuning is also exercised in CAST Phase II, by changing the density of the refractive gas in the magnetic volume [1].

Flares: What produces solar flares and Coronal Mass Ejections is one of the great solar mysteries. However, the energy that powers them is generally believed to be (connected to) the magnetic field [15]. Following recent work [16], these strong X-ray emitting events correlate (within 1.8σ) with the solar surface magnetic field-squared (see Figure 1). In fact, our fit to the X-ray data gives a ~$B^{1.6\pm0.22}$ dependence, while the authors in ref. [16] refer even to a B^2 correlation, fitting the assumed axion scenario.

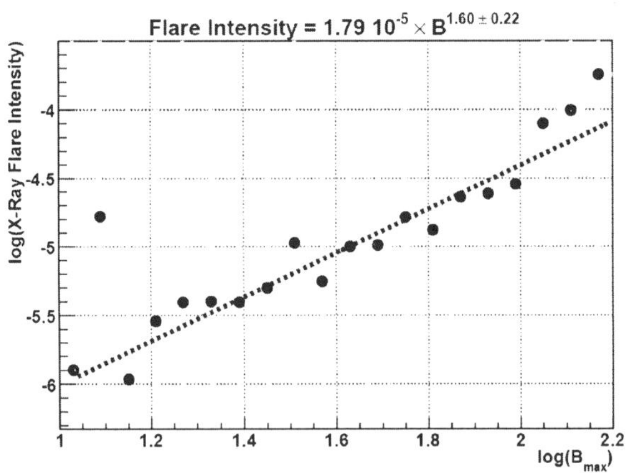

Figure 1. Rebinned peak flare X-ray intensity vs. the maximum magnetic field (Bmax), reconstructed from ref. [16]. The blue dashed line gives the derived fit to the data points (~$B^{1.6\pm0.22}$).

Note that the conventional solution for this radiation emission is that the hard X-rays are coming from electrons. However, this leads to the electron "problem" [17], since the energetic electron flux must be ~10^5 times higher than the X-rays.

We stress here that within the axion scenario the local surface magnetic field is "only" the required catalyst for the axion–photon reactions to take place, and not the otherwise suspected / unspecified energy source of solar X-rays. In this framework, the inner Sun is the actual energy source that creates the outstreaming axions.

3. Solar Magnetic Fields

We recall that it is the macroscopic coherence effects in magnetic axion helioscopes [1] which result in a better axion-to-photon conversion than in detectors, where the axions interact incoherently with the detector atoms. Therefore, helioscopes are more sensitive to axions. Surprisingly, solar magnetic fields [18] have not been included in the model calculations of the axion production rate. However, inside the magnetic Sun, when the local plasma frequency fits the axion rest mass, photon–axion conversion might even be the dominant axion production channel, since the coherence length can be maximum, i.e., equal to the photon mean free path. This is ~1 mm in the core and ~100 km (!) at the photosphere. Such an additional solar axion production channel can modify the previously expected solar axion spectrum (its intensity and its shape) without taken into consideration magnetic fields [19]. To know this is of utmost importance in interpreting solar observations as well as in adopting the appropriate parameter values for optimum performance of an axion antenna.

Interestingly, strong magnetic fields outside the hot core, like the simulated ones in Figure 2, can selectively enhance the production of low(est) energy axions. More precisely, in this scenario, the soft quiet Sun X-ray luminosity (from the celebrated ~2 MK corona) could originate from converted massive or light axions due to their spontaneous or "induced" decays, respectively, both kinds of axions stemming from a shell between the radiative and the convective zone ($R \sim 0.7 \cdot R_{solar}$); since it has the required temperature of ~2 MK [20] and is permeated with a predicted ~ 30-50 Tesla magnetic field (see Figure 2), this becomes a source of low–energy axions. The same reasoning applies to other potential candidates, e.g. microflares with a mean temperature of 12.6 MK [21], including large flares whose temperature distribution is, remarkably, around ~15-20 MK, i.e. equal to the solar core temperature. It is also interesting to note that usually, the power of the several MK hot flares, scaled up to the whole solar surface, does not actually exceed the total solar luminosity [21]. Each on its own, temperature and power, is not a signal, but it fits with the heuristic picture we follow with flares too, while in the opposite case, the reasoning would have been negative.

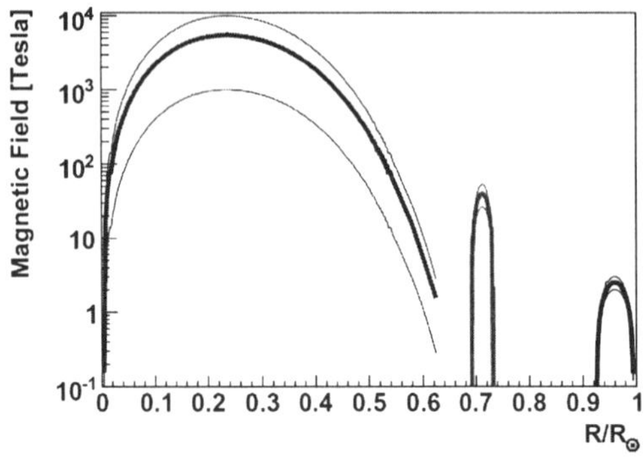

Figure 2. The reconstructed solar magnetic field simulation from ref. [18]: 10^3–10^4 Tesla (left), 30–50 Tesla (middle) and 2–3 Tesla (right), with a temperature of ~9 MK, ~2 MK and ~200 kK, respectively. The thin lines show the estimated range of values for each magnetic field component. Internal rotation was not included in the calculation. An additional axion production at those places can modify both intensity and shape of the solar axion spectrum (Courtesy Sylvaine Turck-Chièze).

We stress that a few 100 eV massive axions of the Kaluza–Klein type might also be created coherently, since the many axion mass states allow a quasi continuous resonance crossing throughout the magnetic Sun, which is a unique fine tuning, only if Kaluza–Klein-like axions exist. If these less massive axions (see ref. [2]) build up a large trapped component around the Sun, even their spontaneous decay alone could explain the low–energy X-rays from the quiet Sun. The induced "decay" in magnetic fields of light axions at the same lower energy range, created probably in the same place, comes in addition. In favour of this conclusion we recall the following findings:

a) the quiet Sun soft X-ray intensity is increasing at lower energies and its origin is unknown,
b) the measured radial distribution of the two off-pointing observations with YOHKOH (~0.5–4 keV) actually supports such a less massive axion scenario, which was not covered in ref. [2],
c) in order to reach the depth of the solar transition region (at ~2000 km) of the Sun being externally self-irradiated from decaying axions orbiting around the Sun, the estimated photon energies were ~50–350 eV [2], which is exactly the energy range of the ~2 MK solar corona radiation.

So far, the estimated X-ray luminosity due to the generic massive solar axions was even strongly suppressed below ~1 keV [2], since their assumed birth place was the ~16 MK hot solar core. Therefore, it seems that magnetic fields might be one of the key parameters for the solar axion scenario that accommodates light as well as massive axions or the like. Furthermore, local magnetic fields up to ~0.6 Tesla have been measured at the surface of sunspots [22], where axion-to-photon oscillations and vice versa can take place and cause otherwise unexplained phenomena. Then, the whole Sun is a multicomponent axion source.

4. Discussion – Conclusion

Challenging questions like the origin of the soft and hard solar X-ray emission remain elusive within conventional astrophysics, with the solar corona problem being present ~70 years. Also, it has been known for many years that the magnetic field plays a crucial role in heating the solar corona, with the exact energy storage and release mechanism(s) being still a nagging unsolved problem for solar physics [23]. In this work, supporting evidence has been presented in favour of a second, magnetic–field–related solar X-ray component, which can originate from converted axions streaming out of the Sun, explaining thus also transient solar X-ray emission.

We also mention briefly other unpredictable X-ray observations, which seem to be of potential relevance for axions or the like: class 0 protostars [24]; the galactic centre, the Inter Cluster Medium, the ubiquitous X-ray background radiation [4]; the Fourier analyzed data of solar neutrinos and soft X-rays result in surprisingly identical frequencies [25]; the solar metallicity problem [26], i.e., the disagreement of the Ne/O abundance between solar model prediction and observation, could be due to an axion surface effect. Following Martin Asplund, a pioneer of this discovery, "one possibility is rather the extra heating from the X-ray absorption of converted axions in the atmosphere which might increase the temperature in the spectral line formation region of the Sun" [27].

Finally, combining all these kinds of signatures, one might be able to constrain the appropriate parameter values in direct axion searches with Earth-bound experiments. One first practical conclusion is that axion helioscopes should lower their threshold energy below a few 100 eV, following the scenarios with (a few 100 eV) massive and light axions as being co-responsible for the dominant and otherwise unexplained X-ray emission in this energy range, from the quiet and the active Sun alike. The axion signatures discussed remained unnoticed before, probably because of their multifaced appearance like QCD-inspired axions, massive Kaluza–Klein axions, or other exotic forms which have not yet been predicted.

Acknowledgements

We thank Susan Leech O'Neale for reading the manuscript. The support by the greek funding agency GSRT is gratefully acknowledged. This research was partially supported by the ILIAS (Integrated Large Infrastructures for Astroparticle Science) project funded by the EU under contract EU-RII3-CT-2004-506222.

References

[1] K. Zioutas *et al.* [CAST Collaboration], *Phys. Rev. Lett.* **94** 121301 (2005), and references therein [hep-ex/0411033]; see also http://www.cern.ch/CAST and the contribution by M. Kuster in this conference.

[2] L. Di Lella, K. Zioutas, *Astroparticle Phys.* **13** 145 (2003) [astro-ph/0207073].

[3] K. Zioutas, K. Dennerl, L. Di Lella, D.H.H. Hoffmann, J. Jacoby, Th. Papaevangelou, *ApJ.* **607** 575 (2004) [astro-ph/0403176].

[4] K. Zioutas, K. Dennerl, D.H.H. Hoffmann, Th. Papaevangelou, *Science* **306** 1485 (2004), and references therein.

[5] LW Acton, Magnetodynamic Phenomena in the SolarAtmosphere (1996) 3.

[6] E.g., S.M. Jefferies, S.W. McIntosh, J.D. Armstrong, T.J. Bogdan, A. Cacciani, B. Fleck, *ApJ.* **648** L151 (2006); J.A. Klimchuk, *Solar Phys.* **234** 41 (2006); E.R. Priest, D.W. Longcope, J. Heyvaerts, *ApJ.* **624** 1057 (2005).

[7] I.G. Hannah, G.J. Hurford, H.S. Hudson, R.P. Lin, AAS, SPD meeting #37, #14.02 (June 2006); see also:
http://sprg.ssl.berkeley.edu/~hannah/presentations/pdf_spd_06.pdf;
http://www.phys.uu.nl/~rutten/rrtex/bibfiles/ads/abs/hudson.txt

[8] http://sprg.ssl.berkeley.edu/~hhudson/workshop/rhessiparis_II/wg4_wed.txt
http://sprg.ssl.berkeley.edu/~hannah/presentations/qs_spd_06.pdf

[9] K. Zioutas, K. Dennerl, M. Grande, D.H.H. Hoffmann, J. Huovelin, B. Lakic, S. Orlando, A. Ortiz, Th. Papaevangelou, Y. Semertzidis, Sp. Tzamarias, O. Vilhu, Proc. TAUP2005, *J. Phys. : Conf. Series* **39** 103 (2006) [astro-ph/0603507].

[10] D.H.H. Hoffmann, K. Zioutas, *Nucl. Phys. B Suppl.* **151** 359 (2006).

[11] N.A. Krivova, S.K. Solanki, M. Fligge, Y.C. Unruh, *Astron. Astrophys.* **399** L1 (2003); W. Livingston, *Solar Phys.* **207** 41 (2002).

[12] M.J. Penn, W. Livingston, *ApJ.* **649** L45 (2006).

[13] K. Rottler, contribution to this conference.

[14] E. Zavattini, G. Zavattini, G. Ruoso, E. Polacco, E. Milotti, M. Karuza, U. Gastaldi, G. Di Domenico, F. Della Valle, R. Cimino, S. Carusotto, G. Cantatore, M. Bregant (PVLAS Collaboration), *Phys. Rev. Lett.* **96** 110406 (2006) [hep-ex/0507107].

[15] G. Barnes, K.D. Leka, *ApJ.* **646** 1303 (2006); K.D. Leka, G. Barnes, *ApJ.* **595** 1277 (2003).

[16] D. Mason, R. Komm, F. Hill, R. Howe, D. Haber, B.W. Hindman, *ApJ.* **645** 1543 (2006).

[17] G. Emslie, http://www.astro.auth.gr/%7Evlahos/ascona/memberstalks/energeticsEmslie.ppt
[18] S. Couvidat, S. Turck-Chieze, A.G. Kosovichev. *ApJ.* **599** 1434 (2003).
[19] K. Van Bibber, P.M. McIntyre, D.E. Morris, G.G. Raffelt, *Phys. Rev.* **D39** 2089 (1989).
[20] J. Christensen-Dalsgaard, http://bigcat.ifa.au.dk/~jcd/solar_models/cptrho.l5bi.d.15c
[21] I.G. Hannah, S.D. Christe, S. Krucker, R.P. Lin, AGU, SH23A-0348 (2006). (http://sprg.ssl.berkeley.edu/~hannah/presentations/agu2006_mf_poster03.pdf)
[22] S.K. Solanki, *Astron. Astrophys. Rev.* **11** 153 (2003); W. Livingston, J.W. Harvey, O.V. Malanushenko, L. Webster, *Solar Phys.* **239** 41 (2006).
[23] K. Galsgaard, C.E. Parnell, *Astron. Astrophys.* **439** 335(2005); R.B. Dahlburg, J.A. Klimchuk, S.K. Antiochos, *ApJ.* **622** 1191 (2005).
[24] K. Hamaguchi, M.F. Corcoran, R. Petre, N.E. White, B. Stelzer, K. Nedachi, N. Kobayashi, A.T. Tokunaga, *ApJ.* **623** 291 (2005).
[25] D.O. Caldwell, P.A. Sturrock, *Astropart. Phys.* **23** 543 (2005).
[26] M. Asplund, N. Grevesse, M. Guedel, A.J. Sauval, *Nature* **436** 525 (2005); M. Asplund, IAU General Assembly Meeting, Prague, (August 2006).
[27] M. Asplund, private communication.

MICROMEGAS FOR RARE EVENT DETECTION

I. G. IRASTORZA[a,*], S. AUNE[b], P. COLAS[b], TH. DAFNI[b], T. ENQVIST[c], G. FANOURAKIS[d], E. FERRER-RIBAS[b], J. GALAN[a], T. GERALIS[d], I. GIOMATARIS[b], P. GORODETZKY[e], F. IGUAZ[a], K. KOUSOURIS[d], G. LUZON[a], J. MORALES[a], M. PIVOVAROFF[f], A. RODRIGUEZ[a], J. RUZ[a], P. SALIN[e], I. SAVVIDIS[g], R. SOUFLI[f], A. TOMAS[a], K. VAN BIBBER[f] and J. A. VILLAR[a]

[a] *Instituto de Física Nuclear y Altas Energías, Universidad de Zaragoza, Zaragoza, Spain*

[b] *DAPNIA, Centre d'Études Nucléaires de Saclay (CEA-Saclay), Gif-sur-Yvette, France*

[c] *CUPP, Pyhasalmi, Finland*

[d] *National Center for Scientific Research "Demokritos", Athens, Greece*

[e] *APC, Universit Paris 7 Denis Diderot, Paris, France*

[f] *Lawrence Livermore National Laboratory (LLNL), Livermore, CA, USA*

[g] *Aristotle University of Thessaloniki, Greece*

Micropattern detectors (MPD) and in particular MICROMEGAS are being developed very actively in recent years. While increasingly used now in high energy physics experiments, their application to rare event searches is relatively more recent. In this talk the status of three initiatives in this respect are presented: MICROMEGAS for axion searches in the CAST experiment at CERN, the use of MICROMEGAS technology to measure the recoil direction in WIMP searches, and the development of the spherical TPC concept with applications in low energy neutrino detection.

Keywords: Micromegas; gaseous detectors; axion; WIMP; neutrino; rare event searches.

1. Introduction

Micromesh gas structures (MICROMEGAS) is one of the most successful MPD concepts.[1] As other MPDs, it surpasses the conventional multiwire plane readouts in ease of construction, robustness, cost as well as in the readout granularity and spatial resolution potentially achievable. Comparatively to other MPDs, MICROMEGAS outstands by its stability, en-

*Attending speaker. Email: Igor.Irastorza@cern.ch

Fig. 1. Design of the new Micromegas line with the new x-ray optics, as will be installed in the CAST magnet (on the right, only the end cap is shown).

ergy and time resolution, among other experimental parameters.[2] Already widely used in high energy physics experiments, its interest in the field of rare events is being stressed since recently. The main motivation for MICROMEGAS in rare events is the accurate reconstruction achievable of the low energy events in gas, exploiting the high spatial and time resolution offered by this technology. Stable operation at high-gains and good energy resolution are key elements for achieving low energy threshold.[3] The group of CEA/Saclay, creator and main developer of the MICROMEGAS concept is involved –in collaboration with other groups– in several projects in which MICROMEGAS are used for rare event searches. The status of those projects are reviewed in the following sections.

2. MICROMEGAS and axions in the CAST experiment

The CAST experiment[4,5] is looking for solar axions since 2003, using a decommissioned LHC test magnet pointing to the Sun, to trigger their conversion to photons. One of the three X-ray detectors is a MICROMEGAS, which has shown high stability of operation, as well as a power of discrimination which has allowed to reach the lowest level of intrinsic background of the experiment, even with unshielded operation.[6] Presently a new detection line is being constructed, equipped with a new MICROMEGAS

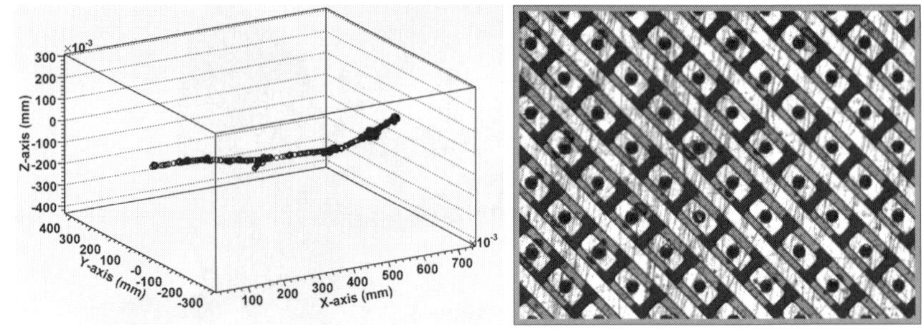

Fig. 2. **Left:** Example of WIMP-induced nuclear recoil, as simulated by SRIM (100 keV Ar nuclei recoiling in pure Ar gas at 0.5 bar). **Right:** 2D Readout pattern used in CAST.

detector coupled to a new X-ray optics. The new optics is manufactured for CAST using a novel low-cost technique, developed by Lawrence Livermore National Laboratory based on the use of plastic substrates. Its design follows that of a concentrator with a 1.3 m focal length and 47 mm diameter, and consists of 14 nested polycarbonate conic shells, each 125 mm long and coated with iridium. The optic is expected to transmit and focus $\sim$36% of the 0.5–10 keV flux emerging from the magnet bore, into a spot of about $\sim$ 3 mm diameter, therefore increasing the signal-to-background ratio considerably.

The new MICROMEGAS detector is optimized for operation with the X-ray optics. It will be smaller in size, and consequently it will allow to be shielded, further reducing its background. Its efficiency will be increased compared to the present detector, due to a more transparent X-ray window and to the use of Xe-based gas mixture as conversion gas (which offers an increased photon conversion by $\sim$10%). The detector's shield is composed of copper, lead, cadmium, nitrogen and polyethylene following the experience of the TPC detector of CAST, which achieved a relative background reduction factor of about $\sim$4 due to it.

The new detection line (which schematics is shown in Fig. 1), is expected to be installed in CAST for the first runs of the ^{3}He-phase by mid-2007.

3. MM to look for WIMP directionality

Dark Matter WIMP searches aim at the detection of the rare, low energy nuclear recoils produced by these particles. These events are in general very difficult to distinguish from the many other types of (background) events, so the experimental strategies of these searches focus on lowering the detector

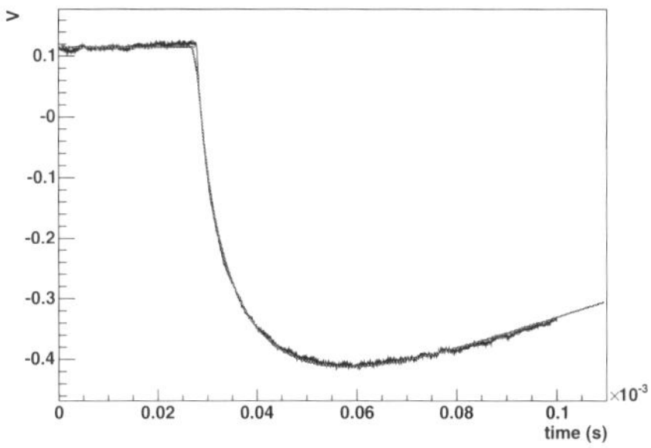

Fig. 3. **Left:** Photograph of the prototype of spherical TPC presently working at Saclay. **Right:** Example of ^{55}Fe 5.9 keV event in the spherical TPC equiped with the spherical electrode. Note the high signal-to-noise ratio achieved.

backgrounds and developing new rejection and identification capabilities. It is widely accepted that the measurement of the direction of the nuclear recoils would offer an unmistakable and ultimate signature of the WIMP.

Such a measurement is however a very important experimental challenge. Typical ∼100 keV recoiling nuclei tracks are in the ∼mm range in gaseous media, as for example the one shown on the left of Fig. 2. MPD detectors can achieve the readout granularity needed to perform the imaging of such small tracks. A relevant example of this is the first MICROMEGAS readout with a 2D pattern used successfully in CAST. Such readout is shown on the right of Fig. 2, and has a pitch of 350 μm. Latest developments on MICROMEGAS manufacturing offer the way to build large surfaces of readout easily, at low cost and with excellent performance. Present studies are addressing issues like the operation underground (in the Canfranc Underground Laboratory) of a MICROMEGAS CAST-like prototype, in order to assess the background rejection capabilities underground. Next issues in the experimental program include the construction of larger prototypes to study more complex readout design aspects, as well as quenching factor measuremente in neutron beam facilities.

4. The spherical TPC

The spherical TPC is a novel concept[7] with very promising features, among which is the possibility of easily instrumenting large target masses with very low energy threshold. This could open the way of detecting the tiny (few hundreds of eV) nuclear recoils produced by neutrino-nucleus coherent interaction, which, although a standard process, has never been within reach of current detectors' sensitivities. While this remains the main physics motivation for this development, a series of physics by-products have been identified. One example is supernova neutrino detection: a network of small spherical TPCs, distributed around the world has been proposed[12] as a very efficient supernova detector, with very simple operation and low cost. Other physics topics are neutrino charge radius[15] and other non-standard properties of the neutrinos,[16] low energy neutrino oscillation studies[7,13] using a tritium source, which could have sensitivity to θ_{13} and neutrino magnetic moment.[14] Another by-product of the spherical TPC is to use it as a neutron detector by partially (or totally) filling it with ^{3}He. That set-up would be able to do neutron spectrometry in very low neutron fluxes, being extremely interesting to characterize neutron backgrounds in underground laboratories.

The spherical TPC consists of a spherical vessel connected at ground, which tightly encloses the conversion gas, and a high voltage electrode placed at the center, so the electric field lines follow a radial distribution. The ionization charges produced in the interaction drift towards the center and are collected by an adequate gaseous readout, which covers the surface of the central electrode. In the simplest design, such amplification structure is just a small spherical electrode, around which the avalanche is produced, and which is read by a single channel electronic chain. More sophisticated options using Micromegas would profit of higher precision, fast response and excellent energy resolution. High efficiency for detecting single electrons have been proved with Micromegas[8] even at high pressures.[9]

A first prototype has been built in Saclay to demonstrate the concept. The external sphere is a 1.3 m diameter, 6 mm thick copper vessel, and the central amplification electrode is a sphere of 14 mm. First tests with Ar/CO_2 and Ar/Isobutane mixtures have proven issues like robustness, gain and stability of operation.[11] In spite of the simplicity of the readout, some spatial resolution is achieved in the radial coordinate by inspecting the time pattern of the charge pulse detected at the center of the sphere. Its temporal extension is determined by the longitudinal diffusion of the ionization cloud and therefore by the distance drifted. The $1/r^2$ dependence of the electric field enhances this effect with respect to plane or cylindrical TPCs. Preliminarily, a resolution of at least 10 cm has been achieved in the estimation of the radial coordinate, and a better value is expected for a faster readout structure like MICROMEGAS. This will also allow to have transversal dispersion information, by means of an appropriate pixelization of the MICROMEGAS readout. Nevertheless, the result already achieved by the current prototype is enough to perform rough fiducial cuts and to achieve some degree of event identification and background rejection.

Very recently the Saclay prototype has been upgraded and equipped with a small double (face-to-face) MICROMEGAS readout placed at the center. First runs have been performed in this configuration and tests are presently ongoing.

5. Conclusions

MICROMEGAS is a fruitful concept that opens new windows in the detection of rare events in gas. The status of three lines of research applying MICROMEGAS to rare event searches has been briefly reported. They are the new detection line in construction for solar axion search in CAST, the measurement of the directionality signal of WIMPs, and the spherical TPC concept with applications in low energy neutrino physics.

References

1. Y. Giomataris, P. Rebourgeard, J. P. Robert and G. Charpak, Nucl. Instrum. Meth. A **376**, 29 (1996).
2. Y. Giomataris, Nucl. Instrum. Meth. A **419** (1998) 239.
3. J. I. Collar and Y. Giomataris, Nucl. Instrum. Meth. A **469** (2001) 249 [arXiv:hep-ex/0009063].
4. K. Zioutas *et al.*, Nucl. Instrum. Meth. A **425** 480 (1999).
5. K. Zioutas *et al.* [CAST Collaboration] Phys. Rev. Lett. **94** (2005) 121301 [arXiv:hep-ex/0411033].
6. S. Aune *et al.*, arXiv:physics/0512094.
7. Giomataris Y and Vergados J D 2004 *Nucl. Instrum. Meth.* A **530** 330 (*Preprint* hep-ex/0303045).
8. Derre J, Giomataris Y, Rebourgeard P, Zaccone H, Perroud J P and Charpak G 2000 *Nucl. Instrum. Meth.* A **449** 314
9. Gorodetzky P *et al.* 1999 *Nucl. Instrum. Meth.* A **433** 554
10. Giomataris I *et al.* 2004 *Proc. Int. Workshop on Identification of Dark Matter (IDM2004) Edinburgh* ed N J C Spooner and V Kudryavtsev (World Scientific) pp 607-14.
11. S. Aune *et al.*, J. Phys. Conf. Ser. **39** (2006) 281 [arXiv:physics/0511221].
12. Giomataris I and Vergados J D 2005 *Preprint* hep-ph/0503029
13. Gounaris G J, Paschos E A and Porfyriadis P I 2004 *Preprint* hep-ph/0409053
14. McLaughlin G C and Volpe C 2004 *Phys. Lett.* B **591** 229 (*Preprint* hep-ph/0312156)
15. Papavassiliou J, Bernabeu J and Passera M 2005 *Preprint* hep-ph/0512029
16. Barranco J *et al.* 2005 *Preprint* hep-ph/0508299.

PHOTON REGENERATION PLANS

A. RINGWALD[†]

Deutsches Elektronen-Synchrotron DESY,
Notkestraße 85, D-22607 Hamburg, Germany
[†] *E-mail: andreas.ringwald@desy.de*
www.desy.de/~ringwald

Precision experiments exploiting low-energy photons may yield information on particle physics complementary to experiments at high-energy colliders, in particular on new very light and very weakly interacting particles, predicted in many extensions of the standard model. Such particles may be produced by laser photons send along a transverse magnetic field. The laser polarization experiment PVLAS may have seen the first indirect signal of such particles by observing an anomalously large rotation of the polarization plane of photons after the passage through a magnetic field. This can be interpreted as evidence for photon disappearance due to particle production. There are a number of experimental proposals to test independently the particle interpretation of PVLAS. Many of them are based on the search for photon reappearance or regeneration, i.e. for "light shining through a wall". At DESY, the Axion-Like Particle Search (ALPS) collaboration is currently setting up such an experiment.

1. Introduction

The standard model of particle physics is phenomenologically extremely successful. However, there are a number of hints which point to the possibility that there is new physics beyond it. Proposed extensions of the standard model aim at a solution of important problems such as the unification of all forces, including gravity, or at an explanation of the absence of CP violation in strong interactions, among many others.

Many attempts to embedd the standard model into a more general, unified framework, notably the ones based on string theory, predict, apart from new very heavy, $m \gg 100$ GeV, particles also a number of new very light, $m \ll 1$ eV particles, which are very weakly coupled to ordinary matter. Prominent candidates for such particles go under the names axions (arising in the course of a solution of the strong CP problem), dilatons, and millicharged particles.

2. Polarization Experiments

Laboratory experiments to search for such particles may be based on the possibility to produce them by shining laser photons along a transverse magnetic field. Searches for a change of the polarization state of initially linearly polarized photons after the passage through the magnetic field, in particular for a possible rotation (dichroism) and ellipticity (birefringence) (cf. Fig. 1), are particularly sensitive to new light particles [2,3].

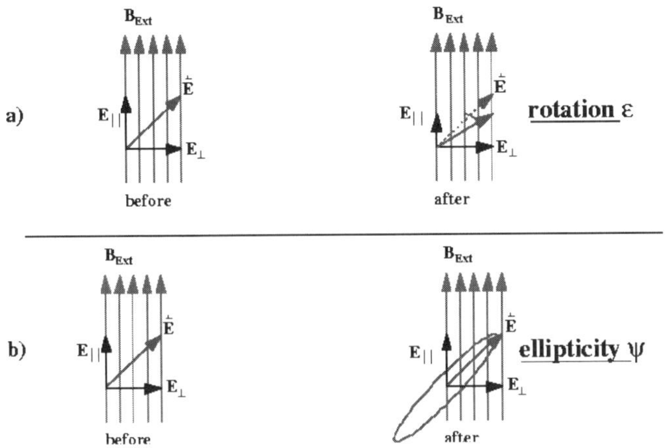

Fig. 1. Possible changes of the polarization state of initially linearly polarized photons after the passage through a magnetic field (adapted from Ref. [1]).

In a pioneering laser polarization experiment, the BFRT collaboration established an upper limit both on a possible vacuum magnetic (VM) dichroism and birefringence [4]. Recently, however, the PVLAS collaboration reported the observation of a VM dichroism [5]. Moreover, preliminary data seem to indicate also evidence for a non-vanishing VM birefringence. These observations have led to a number of theoretical and experimental activities, since the magnitude of the reported signals exceeds the standard model expectations by far.

Two viable particle physics explanations of the reported signals have been proposed: the real and virtual production
(i) of a neutral spin-0 (axion-like) particle (ALP) ϕ [2] with mass m_ϕ and a coupling to two photons via

$$\mathcal{L}_{\rm int}^{(+)} = -\frac{1}{4} g\phi^{(+)} F_{\mu\nu} F^{\mu\nu} = \frac{1}{2} g\phi^{(+)} (\vec{E}^2 - \vec{B}^2), \qquad (1)$$

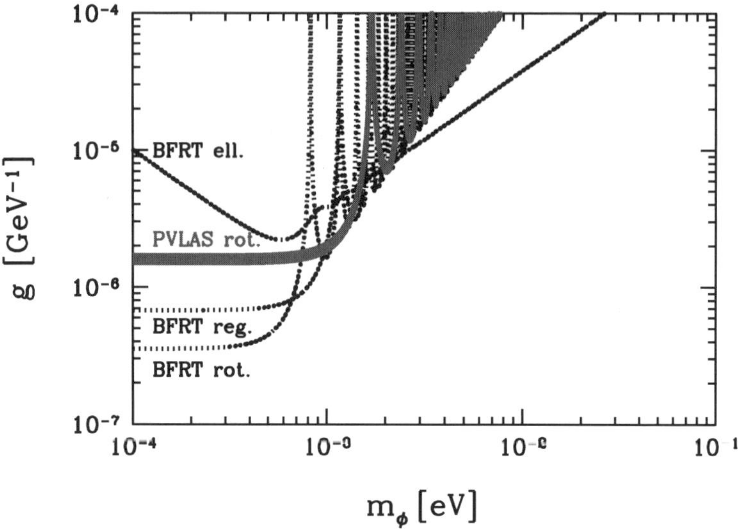

Fig. 2. ALP interpretation of BFRT and PVLAS data: two photon coupling g versus mass m_ϕ. The 95 % confidence level upper limits from BFRT data [4] on polarization (rotation and ellipticity data) and photon regeneration are displayed as dotted lines. The preferred values corresponding to the anomalous rotation signal observed by PVLAS [5] are shown as a thick solid line.

or

$$\mathcal{L}_{\rm int}^{(-)} = -\frac{1}{4} g \phi^{(-)} F_{\mu\nu} \widetilde{F}^{\mu\nu} = g \phi^{(-)} (\vec{E} \cdot \vec{B}), \qquad (2)$$

depending on its parity, denoted by the superscript $(\pm)$, or
(ii) of a pair of millicharged, $Q_\epsilon = \epsilon e$, particles (MCP) $\epsilon^+\epsilon^-$ [3] with mass m_ϵ, coupling to photons in the usual way via the minimal substitution $\partial_\mu \to D_\mu \equiv \partial_\mu - {\rm i}\epsilon e A_\mu$ in the Lagrangian.
Indeed, as apparent from Fig. 2, the rotation observed by PVLAS can be reconciled with the non-observation of a rotation and ellipticity by BFRT, if there is an ALP with a mass $m_\phi \sim$ meV and a coupling $g \sim 10^{-6}$ GeV^{-1} [5]. Alternatively, both experimental results are compatible with the existence of an MCP with $m_\epsilon \sim 0.1$ eV and $\epsilon \sim 10^{-6}$ [3] (cf. Fig. 3). These parameter values, however, seem to be in serious conflict with astrophysical bounds, arising for example from energy loss considerations of stars [6,7]. They may be evaded if the production of ALPs or MCPs is suppressed in astrophysical plasmas [8,9], as realized, for example, in extensions of the standard model involving extra U(1) gauge bosons which kinetically mix with our familiar hypercharge U(1) [10], such as in some realistic string compactifications [11].

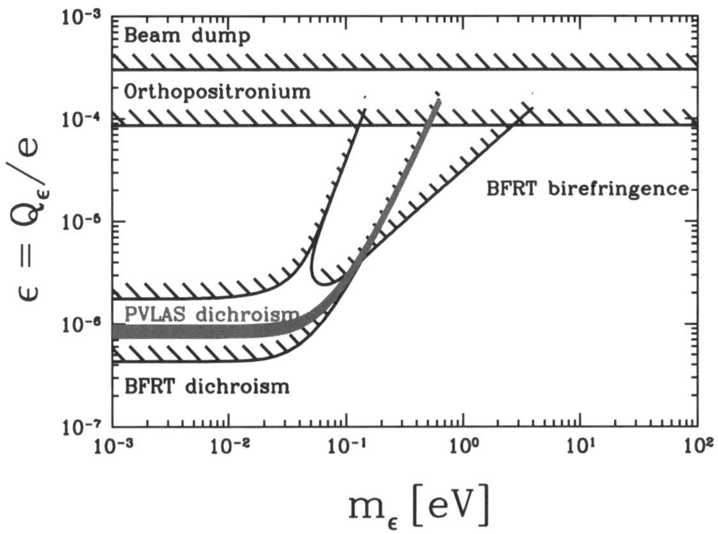

Fig. 3. MCP interpretation of BFRT and PVLAS data: fractional electric charge $\epsilon = Q_\epsilon/e$ versus mass m_ϵ. The preferred values corresponding to the anomalous rotation signal observed by PVLAS [5] are shown as a thick solid line.

It is very comforting that a number of laboratory-based low-energy tests of the ALP and MCP interpretation of the PVLAS anomaly are currently set up and expected to yield decisive results within the upcoming year. For example, in addition to PVLAS, the BMV [12] and Q&A [13] collaborations will run further polarization experiments with different experimental parameter values which finally may lead to a discrimination between the ALP and the MCP hypothesis [14].

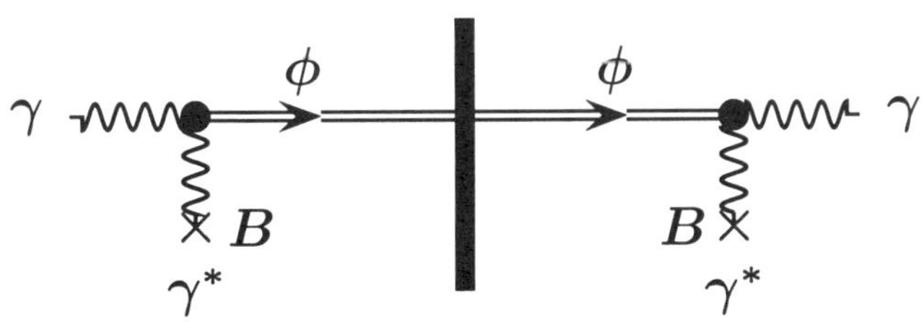

Fig. 4. Schematic view of ALP production through photon conversion in a magnetic field (left), subsequent travel through a wall, and final detection through photon regeneration (right).

3. Regeneration Experiments

The ALP interpretation of the PVLAS signal will crucially be tested by photon regeneration (sometimes called "light shining through walls") experiments [15–18], presently under construction or serious consideration [12,19–22] (cf. Table 1). In these experiments (cf. Fig. 4), a photon beam is directed across a magnetic field, where a fraction of them turns into ALPs. The ALP beam can then propagate freely through a wall or another obstruction without being absorbed, and finally another magnetic field located on the other side of the wall can transform some of these ALPs into photons — seemingly regenerating these photons out of nothing. A pioneering photon regeneration experiment has been done also by the BFRT collaboration [4]. No signal has been found and the corresponding upper limit on g vs. m_ϕ is included in Fig. 2.

Table 1. Experimental parameters of proposed photon regeneration experiments: magnetic fields B_i and their length ℓ_i on production ($i = 1$) and regeneration ($i = 2$) side (cf. Fig. 4); and the corresponding probability $P_{\gamma\phi\gamma}$, with $g \sim 2 \times 10^{-6}$ GeV^{-1} in the PVLAS preferred region (cf. Fig. 2).

Name	Laboratory	Magnets	$P_{\gamma\phi\gamma}$ for $g \sim 2 \times 10^{-6}$ GeV^{-1}
ALPS [19]	DESY/D	$B_1 = B_2 = 5$ T $\ell_1 = \ell_2 = 4.21$ m	$\sim 10^{-19}$
BMV [12]	LULI/F	$B_1 = B_2 = 11$ T $\ell_1 = \ell_2 = 0.25$ m	$\sim 10^{-21}$
LIPSS [20]	Jlab/USA	$B_1 = B_2 = 1.7$ T $\ell_1 = \ell_2 = 1$ m	$\sim 10^{-23.5}$
PVLAS [21]	Legnaro/I	$B_1 = 5$ T $\ell_1 = 1$ m $B_2 = 2.2$ T $\ell_2 = 0.5$ m	$\sim 10^{-23}$
— [22]	CERN/CH	$B_1 = B_2 = 11$ T $\ell_1 = \ell_2 = 7$ m	$\sim 10^{-17}$

Clearly, crucial experimental parameters for such an experiment are the magnetic fields B_i and their length ℓ_i on the production ($i = 1$) and the regeneration ($i = 2$) side of the apparatus. Indeed, the conversion probability for the process $\gamma \to \phi \to \gamma$ is given by

$$P_{\gamma\to\phi\to\gamma} = P_{\gamma\to\phi}(B_1, \ell_1, q_1)\, P_{\phi\to\gamma}(B_2, \ell_2, q_2)\,,$$
$$P_{\gamma\to\phi}(B, \ell, q) = P_{\phi\to\gamma}(B, \ell, q) = \frac{1}{4}\,(g\,B\,\ell)^2\; F(q\ell)\,, \tag{3}$$

where $F(q\ell) \leq 1$ is a form factor which equals unity, if the photons and the

ALPs act coherently over the whole length of the magnet, i.e. for

$$q\ell = \left| \frac{2\,(n_\gamma - 1)\,\omega^2 - m_\phi^2}{2\omega} \right| \ell \ll 1, \tag{4}$$

ω being the photon energy and n_γ being the refraction index of eventual buffer gas in the beam pipe [23]. At small momentum transfer, $q\ell \ll 1$, therefore, the regeneration probability scales as $(B_1\ell_1 B_2\ell_2)^2$. Correspondingly, strong and long dipole magnets from high-energy storage rings (e.g. HERA [18] or LHC) are particularly suited for a photon regeneration experiment, as can be seen in Table 1.

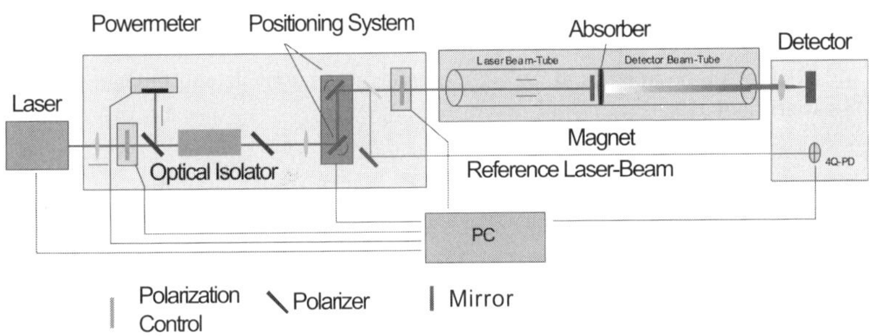

Fig. 5. Schematic view of the experimental setup of the ALPS experiment [19].

Importantly, one may optimize the sensitivity in certain mass regions by essentially tuning q towards small values by adjusting n_γ, i.e. by varying the gas pressure in the magnetic field regions [23]. This effect will be heavily exploited at the Axion-Like Particle Search (ALPS) experiment [19] at DESY (cf. Fig. 5), where the photon beam of a high-power, ~ 200 W, infrared laser, corresponding to an initial flux of $\sim 10^{21}$ photons/s, will be sent along the ~ 5 T, ~ 8.8 m transverse magnetic field of a superconducting HERA dipole magnet (cf. Fig. 6). Filling in a buffer gas with a refractive index $n_\gamma - 1 \sim 10^{-7}$, one will have the maximum sensitivity in the PVLAS preferred parameter region (cf. Fig. 7). ALPS, a collaboration of DESY, the Laserzentrum Hannover and the Sternwarte Bergedorf, has been approved in principle by the DESY directorate and is planning to take data in summer 2007.

Fig. 6. HERA dipole magnet at DESY's magnet test stand. It will be exploited for the ALPS experiment [19].

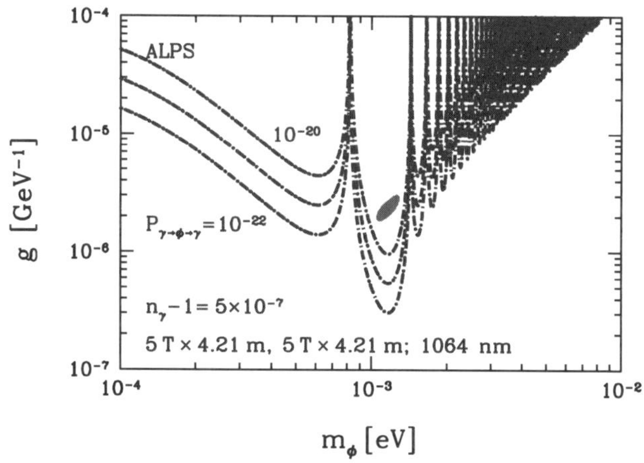

Fig. 7. Isocontours of the probability $P_{\gamma\to\phi\to\gamma}$, Eq. (3), for the parameters of ALPS experiment: $B_1 = B_2 = 5$ T, $\ell_1 = \ell_2 = 4.21$ m, exploiting an infrared photon beam ($\omega = 1.17$ eV) in buffer gas with a refractive index $n_\gamma - 1 = 5 \times 10^{-7}$ (from Ref. [19]).

4. Conclusions

The evidence for a vacuum magnetic dichroism found by PVLAS has triggered a lot of theoretical and experimental activities. Fortunately, in the upcoming year a number of decisive laboratory based tests both of the axion-like and millicharged particle interpretation will be done. In particular, the planned photon regeneration experiments (cf. Table 1) will firmly establish or exclude the axion-like particle interpretation. In summary, even in the LHC era, small experiments might have a big impact on particle physics!

References

1. F. Brandi *et al.* [PVLAS Collaboration], Nucl. Instrum. Meth. A **461**, 329 (2001).
2. L. Maiani, R. Petronzio and E. Zavattini, Phys. Lett. B **175**, 359 (1986).
3. H. Gies, J. Jaeckel and A. Ringwald, Phys. Rev. Lett. **97**, 140402 (2006).
4. R. Cameron *et al.* [BFRT Collaboration], Phys. Rev. D **47**, 3707 (1993).
5. E. Zavattini *et al.* [PVLAS Collaboration], Phys. Rev. Lett. **96**, 110406 (2006).
6. G. G. Raffelt, Stars As Laboratories For Fundamental Physics: The Astrophysics of Neutrinos, Axions, and other Weakly Interacting Particles, University of Chicago Press, Chicago, 1996.
7. S. Davidson, S. Hannestad and G. Raffelt, JHEP **0005**, 003 (2000).
8. E. Masso and J. Redondo, JCAP **0509**, 015 (2005).
9. J. Jaeckel, E. Masso, J. Redondo, A. Ringwald and F. Takahashi, hep-ph/0605313; hep-ph/0610203.
10. E. Masso and J. Redondo, Phys. Rev. Lett. **97**, 151802 (2006).
11. S. A. Abel, J. Jaeckel, V. V. Khoze and A. Ringwald, hep-ph/0608248.
12. C. Rizzo [BMV Collaboration], 2nd ILIAS-CERN-CAST Axion Academic Training 2006, http://cast.mppmu.mpg.de/
13. S. J. Chen, H. H. Mei and W. T. Ni [Q&A Collaboration], hep-ex/0611050.
14. M. Ahlers, H. Gies, J. Jaeckel and A. Ringwald, hep-ph/0612098.
15. P. Sikivie, Phys. Rev. Lett. **51**, 1415 (1983) [Erratum-ibid. **52**, 695 (1984)].
16. A. A. Anselm, Yad. Fiz. **42**, 1480 (1985).
17. K. Van Bibber, N. R. Dagdeviren, S. E. Koonin, A. Kerman and H. N. Nelson, Phys. Rev. Lett. **59**, 759 (1987).
18. A. Ringwald, Phys. Lett. B **569**, 51 (2003).
19. K. Ehret *et al.* [ALPS Collaboration], LoI subm. to DESY directorate.
20. K. Baker [LIPSS Collaboration], 2nd ILIAS-CERN-CAST Axion Academic Training 2006, http://cast.mppmu.mpg.de/
21. G. Cantatore [PVLAS Collaboration], 2nd ILIAS-CERN-CAST Axion Academic Training 2006, http://cast.mppmu.mpg.de/
22. P. Pugnat *et al.*, Czech. J. Phys. **55**, A389 (2005); **56**, C193 (2006).
23. K. van Bibber, P. M. McIntyre, D. E. Morris and G. G. Raffelt, Phys. Rev. D **39**, 2089 (1989).

AXINO DARK MATTER FROM Q-BALLS

LESZEK ROSZKOWSKI

Department of Physics and Astronomy, University of Sheffield
Sheffield, United Kingdom

OSAMU SETO*

Department of Physics and Astronomy, University of Sussex
Brighton BN1 9QJ, United Kingdom

We show that the $\Omega_b - \Omega_{DM}$ coincidence can naturally be explained in a framework where axino cold dark matter is predominantly produced in non-thermal processes involving decays of Q-balls formed in Affleck-Dine baryogenesis. In this approach the similarity between Ω_b and Ω_{DM} is a direct consequence of the (sub-)GeV scale of the mass of the axino.

Keywords: Axino; Q-ball; Affleck-Dine baryogenesis; Cold Dark Matter.

1. Introduction

The origin of non-baryonic cold dark matter (DM) and baryon asymmetry in the Universe are among the most long-standing puzzles in cosmology as well as in particle physics today. In particular, the question why the energy density of baryons Ω_b and that of dark matter Ω_{DM} are within an order of magnitude from each other remains a mystery.

A standard candidate for the non-baryonic cold dark matter is a weakly interacting massive particle (WIMP). Perhaps the most popular WIMP is the lightest supersymmetric particle (LSP) which remains stable owing to the conservation of R parity, in particular the lightest neutralino in the Minimal Supersymmetric Standard Model (MSSM). We often assume that the neutralino freezes out of thermal equilibrium in the early Universe, however this economical framework does not explain why Ω_b and Ω_{DM} are so close to each other. There are other well motivated candidates, gravitino,[1] axion which may solve the strong CP problem in QCD by Peccei-Quinn mechanism,[2] and its fermionic supersymmetric partner axino.[3,4]

*Speaker

Affleck-Dine (AD) mechanism[5] is a promising baryogenesis in supersymmetric models. Several years ago it was suggested that the question may be answered in the framework of Affleck-Dine (AD) baryogenesis because both baryon asymmetry and dark matter could be produced by a single mechanism. In particular, LSPs would be produced by the decay of Q-balls, non-topological solitons of an Affleck-Dine condensate.[6–8] Since then, however, it was realized that Q-ball decays typically causes an overproduction of neutralino LSPs.[9,10]

One should remember that the lightest neutralino is not only candidate for LSP. Gravitino or axino also is a good candidate for LSP and possibly dark matter. Although Affleck-Dine mechanism works well for gravitino LSP dark matter, the relation between Ω_b and Ω_{DM} is lost,[11] because most of gravitinos need to be produced by thermal processes.[12]

Here, we instead consider the axino LSP as a solution of the LSP overproduction problem by Q-balls and the simultaneous explanation of Ω_b and Ω_{DM}.[13]

2. Baryon asymmetry and dark matter abundance in Affleck-Dine mechanism

We now briefly review Affleck-Dine baryogenesis and LSP overproduction problem from Q-ball decays within the framework of gravity mediated supersymmetry breaking.

2.1. *Baryon asymmetry*

The potential of the AD flat direction field is, in general, lifted by soft supersymmetric (SUSY) breaking terms and non-renormalizable terms.[14,15] The full potential of the AD field is given by

$$V(\phi) = \left(m_\phi^2\left[1 + K\ln\left(\frac{|\phi|^2}{\Lambda^2}\right)\right] - c_1 H^2\right)|\phi|^2 + \left[\left(c_2 H + A m_{3/2}\right)\lambda\frac{\phi^n}{nM^{n-3}} + \text{H.c.}\right] + \lambda^2\frac{|\phi|^{2n-2}}{M^{2n-6}}. \tag{1}$$

Here, m_ϕ is the soft SUSY breaking scalar mass for the AD field with radiative correction $K\ln|\phi|^2$. A flat direction dependent constant, K, takes values from -0.01 to -0.1.[16] Λ denotes a renormalization scale and $-c_1H^2$ represents the negative mass squared induced by the SUSY breaking effect which comes from the energy density of the inflaton, with an order unity coefficient c_1.[15] λ is the coupling of nonrenormalizable term and M is some large scale acting as its cut-off. Terms proportional to A and c_2 are the A-terms from low energy SUSY breaking and those induced by the inflaton respectively, where $m_{3/2}$ denotes the gravitino mass. Here, we omitted possible terms appear by thermal effects.[17,18] These terms are negligible as long as we consider a sufficient low reheating temperature after inflation, as we will. Moreover, the model would face with "gravitino problem",[19] if the reheating temperature after inflation is so high that these thermal effect become effective, unless gravitino is LSP.[11]

The baryon number density for the AD field ϕ is given by $n_b = iq(\dot{\phi}^*\phi - \phi^*\dot{\phi})$ where q is the baryonic charge for the AD field. By using the equation of motion of the AD field, the charge density can be rewritten as

$$n_b(t) \simeq \frac{1}{a(t)^3}\int^t dt' a(t')^3 \frac{2qm_{3/2}}{M^{n-3}}\mathrm{Im}(A\phi^n), \tag{2}$$

with $a(t)$ being the scale factor. When the AD field starts to oscillate around the origin, the baryon number density is induced by the relative phase between A-terms. Since $s = 4\pi^2 g_* T^3/90$, after reheating the baryon to entropy ratio is

$$\frac{n_b}{s} = \left.\frac{T_R n_b}{4M_P^2 H^2}\right|_{t_{\rm os}} \simeq \frac{q|A|m_{3/2}}{2}\frac{T_R|\phi_{\rm os}|^n}{M_P^2 H_{\rm os}^3 M^{n-3}}\sin\delta. \tag{3}$$

Here, $M_P \simeq 2.4\times10^{18}$ GeV is the reduced Planck mass, $t_{\rm os}$ is the time of the start of the oscillation and $\sin\delta$ is the effective CP phase.

From now on, as an example, we consider the case of $n = 6$ because a promising AD field for our scenario, a $\bar{u}\bar{d}\bar{d}$ direction, belongs to this class. The baryon asymmetry Eq. (3) is estimated as

$$\frac{n_b}{s} \simeq 2\times10^{-10}\frac{q|A|\sin\delta}{2\lambda^{3/2}}\left(\frac{m_{3/2}}{100\mathrm{GeV}}\right)\left(\frac{10^3\mathrm{GeV}}{m_\phi}\right)^{3/2}\left(\frac{T_R}{100\mathrm{GeV}}\right)\left(\frac{M}{M_P}\right)^{3/2}, \tag{4}$$

where we assumed $M \sim M_P$. This assumption is crucial for our scenario, because Q-balls have too high decay temperature or can not survive the evaporation if M is too small, as we will show soon. Here one may find that a relatively low reheating temperature is required to explain the appropriate baryon asymmetry. In this case, Q-balls are formed for most of the flat directions.

2.2. *LSP production from Q-balls*

First, we briefly summarize relevant properties of Q-balls in gravity mediated SUSY breaking models. The radius of a Q-ball, R, is estimated as

$R^2 \simeq 2/(|K|m_\phi^2)$.[7] Numerical calculations provide a fitting formula for the Q-ball charge

$$Q \simeq \bar{\beta}\left(\frac{|\phi_{\rm os}|}{m_\phi}\right)^2 \times \begin{cases} \epsilon & \text{for} \quad \epsilon \gtrsim \epsilon_c \\ \epsilon_c & \text{for} \quad \epsilon < \epsilon_c \end{cases}, \tag{5}$$

where $\epsilon_c \simeq 10^{-2}$, $\bar{\beta} = 6 \times 10^{-3}$ and $\epsilon \equiv n_b/n_\phi|_{t_{\rm os}} \simeq (2q|A|/\lambda)(m_{3/2}/m_\phi)\sin\delta$.[20] The Q-ball charge can be evaluated as

$$Q \sim 1 \times 10^{20}\frac{q|A|\sin\delta}{\lambda^{3/2}}\left(\frac{m_{3/2}}{100\text{GeV}}\right)\left(\frac{1\text{TeV}}{m_\phi}\right)^{5/2}\left(\frac{M}{M_P}\right)^{3/2}. \tag{6}$$

A part of the charge of a Q-ball can evaporate by the interaction with particles in the thermal bath. The evaporated charge is estimated as $\Delta Q = \mathcal{O}(10^{18})$.[21] Thus, if the charge of a Q-ball is larger than $\mathcal{O}(10^{18})$, Q-balls can survive the evaporation. For a Q-ball charge of $M \sim M_P$ in Eq. (6), the decay temperature of Q-ball is $T_d \simeq 1$ GeV,[22] which is lower than the typical freeze out temperature of WIMPs, T_f. LSPs are then generated by Q-ball decay.

Unless supersymmetric particles produced by Q-ball decay annihilate, the final LSP abundance is given as

$$Y_{LSP} \simeq 3\left(\frac{N}{3}\right)f_B\frac{n_b}{s} = 3 \times 10^{-10}\left(\frac{N}{3}\right)\left(\frac{f_B}{1}\right)\left(\frac{n_b/s}{10^{-10}}\right), \tag{7}$$

where N is the number of produced supersymmetric particles per one baryonic charge and at least three, or more. f_B is a ratio of baryon number in Q-balls to the total baryon number and takes a value of order of unity.[10] Here, we simply neglect the contribution to the LSP density from freeze-out. If we include it and it is significant, the situation becomes only worse.

Recalling that

$$\frac{\rho_{DM}}{s} \simeq 3.9 \times 10^{-10}\left(\frac{\Omega_{DM}h^2}{0.11}\right)\text{GeV}, \tag{8}$$

Eq. (7) leads to the well known result that, in order to be consistent with WMAP, the mass of the dark matter WIMP must be about 1 GeV. This range is inconsistent with bounds from LEP and dark matter direct search for the neutralino LSP in the MSSM. Thus, the possibility that the Affleck-Dine baryogenesis accounts for both the dark matter and the baryon asymmetry due to the Q-ball decay which was originally explored in Ref. 7, does not quite work for the case of the lightest neutralino of the MSSM. This is the LSP overproduction problem in Q-ball decays.

So far, in order to avoid this problem, some ideas have been proposed. If supersymmetric particles produced by Q-ball decay can annihilate, the neutralino LSP with the mass of $\mathcal{O}(10^2)$ GeV becomes acceptable,[23] but the relation of the energy density between baryon and dark matter is lost. On the other hand, if one want to preserve the relation between the energy densities of baryons and dark matter, another choice of LSP is necessary. An example is the singlino-like neutralino in the Next-to-MSSM as the LSP[10] which can escape experimental bounds. Hereafter, we will investigate the axino LSP dark matter.

3. Axino dark matter from Q-balls

In general, axinos are produced by both thermal processes (TP) and non-thermal processes (NTP), e.g., the late decay of long-lived objects. The former consists of the scatterings and the decays of particles in the thermal bath. The latter is given by the decay of thermal relic NSP after its freeze-out, or the decay of Q-balls. The Boltzmann equations can be written as

$$\begin{aligned} \dot{n}_{NSP} + 3Hn_{NSP} &= -\langle\sigma v\rangle(n_{NSP}^2 - {n_{NSP}^{\rm eq}}^2) \\ &\quad + \gamma(\mathrm{Q-ball} \to \mathrm{NSPs}) - \Gamma_{NSP} n_{NSP}, \qquad (9) \\ \dot{n}_{\tilde{a}} + 3Hn_{\tilde{a}} &= \langle\sigma v(i + j \to \tilde{a} + ...)\rangle_{ij} n_i n_j + \langle\sigma v(i \to \tilde{a} + ...)\rangle_i n_i \\ &\quad + \Gamma_{NSP} n_{NSP}, \qquad (10) \end{aligned}$$

where $\gamma(\mathrm{Q-ball} \to \mathrm{NSPs})$ denotes the contribution by Q-balls decay, $\langle\sigma v\rangle_{ij}$ and $\langle\sigma v\rangle_i$ are the scattering cross section and the decay rate for the thermal production of axinos, and Γ_{NSP} is the decay rate of the NSP.

The total NSP abundance is given by

$$Y_{NSP} = N f_B \frac{n_b}{s} + Y_{NSP}^{TP} \qquad (11)$$

where

$$Y_{NSP}^{TP} \simeq \left.\frac{H}{s}\right|_{T=m_{NSP}} \frac{m_{NSP}/T_f}{\langle\sigma v\rangle_{ann}} = \frac{(\pi^2 g_*/90)^{-1/2}}{4 m_{NSP} M_P} \frac{m_{NSP}/T_f}{\langle\sigma v\rangle_{ann}}, \qquad (12)$$

is the thermal relic density as usual. Since $n_{NSP} = n_{\tilde{a}}$, the resulting number density of axino is given by

$$Y_{\tilde{a}} = Y_{\tilde{a}}^{NTP} + Y_{\tilde{a}}^{TP}, \qquad (13)$$

with

$$Y_{\tilde{a}}^{NTP} = 1 \times 10^{-10} N \left(\frac{f_B}{1}\right)\left(\frac{n_b/s}{1 \times 10^{-10}}\right) + Y_{NSP}^{TP}, \qquad (14)$$

where $Y_{\tilde{a}}^{TP}$ denotes the axinos produced by thermal processes. Since typically $Y_{NSP}^{TP} \sim 10^{-11}$, one can see that non-thermal production of axinos due to the thermal relic NSPs decay can easily be negligible compared to that from Q-ball production. The thermally produced axino $Y_{\tilde{a}}^{TP}$ also can be subdominant, say $Y_{\tilde{a}}^{TP} \lesssim 10^{-11}$, for the reheating temperature of $T_R \sim 100$ GeV.[4,24] Hence, the axino dark matter density is estimated as

$$\Omega_{\tilde{a}} h^2 \simeq 0.11 \left(\frac{m_{\tilde{a}} N f_B}{4.6\,\mathrm{GeV}} \right) \left(\frac{\Omega_b h^2}{0.02} \right). \tag{15}$$

One can see that the baryon asymmetry and the dark matter abundance are readily linked.

One may worry the constraints from the Big Bang Nucleosynthesis (BBN), because this model involves the late decay of long-lived particles. However, this problem can be easily avoided for axino LSP, because the interactions of axino are suppressed by the PQ scale $f_a \simeq 10^{11}$ GeV unlike those of gravitino which is suppressed by Planck scale.[3,4] In fact, for example, if the NSP is heavier than about 150 GeV, the model is almost free from this problem for any mass of axino.

4. Conclusions

We have shown that the framework with cold dark matter axino LSP produced in Q-ball decays can explain the abundance of dark matter and the baryon asymmetry simultaneously and may be an answer to the $\Omega_b \sim \Omega_{DM}$ coincidence. In this approach, the similarity between Ω_b and Ω_{DM} is explained by basically only the axino mass of order of (sub-)GeV.

Acknowledgments

The work of O.S. is supported by PPARC. We thank the European Network of Theoretical Astroparticle Physics ILIAS/N6 under contract number RII3-CT-2004-506222 for financial support.

References

1. J. R. Ellis, J. E. Kim and D. V. Nanopoulos, Phys. Lett. B **145**, 181 (1984).
2. R. D. Peccei and H. R. Quinn, Phys. Rev. Lett. **38**, 1440 (1977).
3. L. Covi, J. E. Kim and L. Roszkowski, Phys. Rev. Lett. **82**, 4180 (1999).
4. L. Covi, H. B. Kim, J. E. Kim and L. Roszkowski, JHEP **0105**, 033 (2001).
5. I. Affleck and M. Dine, Phys. Lett. B **145**, 181 (1984).
6. A. Kusenko and M. E. Shaposhnikov, Phys. Lett. B **418**, 46 (1998).

7. K. Enqvist and J. McDonald, Phys. Lett. B **425**, 309 (1998); K. Enqvist and J. McDonald, Nucl. Phys. B **538**, 321 (1999).
8. S. Kasuya and M. Kawasaki, Phys. Rev. D **61**, 041301 (2000).
9. K. Enqvist and J. McDonald, Phys. Lett. B **440**, 59 (1998).
10. K. Enqvist and J. McDonald, Nucl. Phys. B **570**, 407 (2000).
11. O. Seto, Phys. Rev. D **73**, 043509 (2006).
12. L. Roszkowski, R. Ruiz de Austri and K.-Y. Choi, JHEP **0508**, 080 (2005); D. G. Cerdeno, K. Y. Choi, K. Jedamzik, L. Roszkowski and R. Ruiz de Austri, JCAP **0606**, 005 (2006).
13. L. Roszkowski and O. Seto, hep-ph/0608013.
14. K. W. Ng, Nucl. Phys. B **321**, 528 (1989).
15. M. Dine, L. Randall and S. Thomas, Nucl. Phys. B **458**, 291 (1996).
16. K. Enqvist and J. McDonald, Phys. Lett. B **425**, 309 (1998); K. Enqvist, A. Jokinen and J. McDonald, Phys. Lett. B **483**, 191 (2000).
17. R. Allahverdi, B. A. Campbell and J. R. Ellis, Nucl. Phys. B **579**, 355 (2000).
18. A. Anisimov and M. Dine, Nucl. Phys. B **619**, 729 (2001).
19. M. Y. Khlopov and A. D. Linde, Phys. Lett. B **138**, 265 (1984); J. R. Ellis, J. E. Kim and D. V. Nanopoulos, Phys. Lett. B **145**, 181 (1984).
20. S. Kasuya and M. Kawasaki, Phys. Rev. D **62**, 023512 (2000).
21. R. Banerjee and K. Jedamzik, Phys. Lett. B **484**, 278 (2000).
22. A. G. Cohen, S. R. Coleman, H. Georgi and A. Manohar, Nucl. Phys. B **272**, 301 (1986).
23. M. Fujii and K. Hamaguchi, Phys. Lett. B **525**, 143 (2002); Phys. Rev. D **66**, 083501 (2002).
24. L. Covi, L. Roszkowski, R. Ruiz de Austri and M. Small, JHEP **0406**, 003 (2004).

STERILE NEUTRINO DARK MATTER

MIKHAIL SHAPOSHNIKOV

Institut de Théorie des Phénomènes Physiques, Ecole Polytechnique Fédérale de Lausanne, CH-1015 Lausanne, Switzerland

An extension of the Standard Model by three right-handed neutrinos with masses smaller than the electroweak scale (the νMSM) can explain simultaneously dark matter and baryon asymmetry of the Universe, being consistent with the data on neutrino oscillations. A dark matter candidate in this theory is the sterile neutrino with the mass in keV range. We discuss the constraints on the properties of this particle and mechanisms of their cosmological production. Baryon asymmetry generation in this model is reviewed. Crucial experiments that can confirm or rule out the νMSM are briefly discussed.

1. Introduction

There is compelling evidence that the Minimal Standard Model (MSM) of strong and electroweak interactions is not complete. There are several *experimental* facts that cannot be explained by the MSM. These are neutrino oscillations, the presence of dark matter in the Universe, the baryon asymmetry of the Universe, its flatness, and the existence of cosmological perturbations necessary for structure formation. Indeed, in the MSM neutrinos are strictly massless and do not oscillate. The MSM does not have any candidate for non-baryonic dark matter. Moreover, with the present experimental limit on the Higgs mass, the high-temperature phase transition, required for electroweak baryogenesis, is absent. Finally, the couplings of the single scalar field of the MSM are too large for the Higgs boson to play the role of the inflaton. This means that the MSM is unlikely to be a good effective field theory up to the Planck scale.

In Refs. 1–3 it was proposed that a simple extension of the MSM by three singlet right-handed neutrinos and by a real scalar field (inflaton) with masses smaller than the electroweak scale may happen to be a correct effective theory up to some high-energy scale, which may be as large as the Planck scale. This model was called "the νMSM", underlying the fact that it is the extension of the MSM in the neutrino sector. Contrary to Grand

Unified Theories, the νMSM does not have any internal hierarchy problem, simply because it is a theory with a single mass scale. As well as the MSM, the νMSM does not provide any explanation why the weak scale is much smaller than the Planck scale.

As we demonstrated in Refs. 1 and 2, the νMSM can explain simultaneously dark matter and baryon asymmetry of the Universe being consistent with neutrino masses and mixings observed experimentally. Moreover, in Ref. 3 we have shown that inclusion of an inflaton with scale-invariant couplings to the fields of the νMSM allows us to have inflation and provides a common source for electroweak symmetry breaking and Majorana neutrino masses of singlet fermions – sterile neutrinos. The role of the dark matter is played by the lightest sterile neutrino with mass m_s in the keV range. In addition, the coherent oscillations of two other, almost degenerate, sterile neutrinos lead to the creation of baryon asymmetry of the Universe[2] through the splitting of the lepton number between active and sterile neutrinos[4] and electroweak sphalerons.[5] For other astrophysical applications of sterile neutrinos see Refs. 6–8.

In this talk I review the structure of the νMSM and discuss its dark matter candidate – sterile neutrino. The baryogenesis in this model is briefly reviewed.

2. The νMSM

If three singlet right-handed fermions N_I are added to the Standard Model, the most general renormalizable Lagrangian describing all possible interactions has the form:

$$L_{\nu MSM} = L_{MSM} + \bar{N}_I i\partial_\mu \gamma^\mu N_I - F_{\alpha I}\,\bar{L}_\alpha N_I \Phi - \frac{M_I}{2}\,\bar{N}_I^c N_I + h.c., \quad (1)$$

where L_{MSM} is the Lagrangian of the MSM, Φ and L_α ($\alpha = e, \mu, \tau$) are the Higgs and lepton doublets, respectively, and both Dirac ($M^D = f^\nu \langle\Phi\rangle$) and Majorana ($M_I$) masses for neutrinos are introduced. In comparison with the MSM, the νMSM contains 18 new parameters: 3 Majorana masses of new neutral fermions N_i, and 15 new Yukawa couplings in the leptonic sector (corresponding to 3 Dirac neutrino masses, 6 mixing angles and 6 CP-violating phases).

Let us discuss in general terms what kind of scale for Majorana neutrino masses M_I one could expect. If Dirac neutrino masses $(M^D)_{\alpha I} = F_{\alpha I} v$ (where $v = 174$ GeV is the vacuum expectation value of the Higgs doublet) are much smaller than the Majorana masses M_I, the see-saw formula for active neutrino masses

$$m_\nu = -M^D \frac{1}{M_I}[M^D]^T \tag{2}$$

is valid. Though it is known that the masses of active neutrinos are smaller than $O(1)$ eV, it is clear that the scale of Majorana neutrino masses cannot be extracted. This is simply because the total number of physical parameters describing m_ν is equal to 9 (three absolute values of neutrino masses, three mixing angles and three CP-violating phases), which is two times smaller than the number of new parameters in the νMSM.

A most popular proposal[9] is to say that the Yukawa couplings F in the active-sterile interactions are of the same order of magnitude as those in the quark and charged lepton sector. Then one has to introduce a new energy scale, $M_I \sim 10^{10} - 10^{15}$ GeV, which may be related to grand unification. The model with this choice of M_I has several advantages in comparison with the MSM: it can explain neutrino masses and oscillations, and give rise to baryon asymmetry of the Universe through leptogenesis[10] and anomalous electroweak number non-conservation at high temperatures.[5] However, it cannot explain the dark matter as the low energy limit of this theory is simply the MSM with non-zero active neutrino masses coming from dimension five operators. On a theoretical side, as a model with two very distinct energy scales it suffers from a fine-tuning hierarchy problem $M_I \gg M_W$.

Another possibility is to say that no new energy scale is introduced, and $M_I < M_W$. In this case the Yukawa couplings must be much smaller than those in the quark sector, $F < 10^{-6}$. Clearly, no internal hierarchy problem appears for this choice. Also, the neutrino masses and mixing can be easily incorporated. In addition, all the parameters of the νMSM can be potentially determined experimentally since only accessible energy scales are present.

3. Dark Matter

Though the νMSM does not offer any stable particle besides those already present in the MSM, it contains a sterile neutrino with a life-time exceeding the age of the Universe, provided the corresponding Yukawa coupling is small enough. The decay rate of N_1 to three active neutrinos and antineutrinos (assuming that N_1 is the lightest sterile neutrino) is given by

$$\Gamma_{3\nu} = \frac{G_F^2 M_1^5 \theta^2}{96\,\pi^3}\,, \quad \theta = \frac{m_0}{M_1}\,, \quad m_0^2 = \sum_{\alpha=e,\mu,\tau} |M^D \alpha 1|^2\,, \tag{3}$$

where G_F is the Fermi constant. For example, a choice of $m_0 \sim O(1)$ eV and of $M_1 \sim O(1)$ keV leads to a sterile neutrino life-time $\sim 10^{17}$ years.[11]

The mass of the sterile dark matter neutrino cannot be too small. An application of the Tremaine-Gunn arguments to the dwarf spheroidal galaxies gives the lower bound[12] $M_1 > 0.3$ keV. If the sterile neutrino mass is in the keV region, it may play a role of warm dark matter.[13,14] Sterile neutrino free streaming length an matter-radiation equality is given by

$$\lambda_{FS} \simeq 1 \text{ Mps} \left(\frac{1 \text{ keV}}{M_I}\right) \left(\frac{\langle p_s/T\rangle}{3.15}\right) \tag{4}$$

and the mass inside λ_{FS} is

$$M_{FS} \simeq 3 \times 10^{10} M_\odot \left(\frac{1 \text{ keV}}{M_I}\right)^3 \left(\frac{\langle p_s/T\rangle}{3.15}\right)^3 ,$$

where $< p_s >$ $(< p_a >)$ is an average momentum of sterile (active) neutrino at the moment of structure formation, $M_\odot$ is the solar mass. One normally defines cold dark matter (CDM) as that corresponding to $M_{FS} < 10^5 M_\odot$, hot DM as the one with $M_{FS} > 10^{14} M_\odot$, and warm DM as anything in between. Potentially, WDM could solve some problems of the CDM scenario, such as the missing satellites problem[15,16] and the problem of cuspy profiles in the CDM distributions.[17,18]

Even stronger constraint on the mass of sterile neutrino comes from the analysis of the cosmic microwave background and the matter power spectrum inferred from Lyman-α forest data: $M_I > M_0 \left(\frac{<p_s>}{<p_a>}\right)$. According to Ref. 19, $M_0 = 15.4$ keV, whereas Ref. 20 gives $M_0 = 11$ keV, with a most conservative bound $M_0 \simeq 2$ keV.

Yet another constraint on the parameters of dark matter sterile neutrino comes from radiative decay $N_1 \to \nu\gamma$, suppressed in comparison with $N_1 \to 3\nu$ by a factor $O(\alpha)$ (α is a fine structure constant). This two body decay produces a line in the spectrum of X-rays coming from dark matter in the Universe; corresponding constraints are shown in Fig. 1 (see Ref. 21 and references therein).

4. Cosmological Production of Sterile Neutrinos

Let us discuss now cosmological production of sterile neutrinos. In the region of the parameter space admitted by X-ray observations sterile neutrinos were never in thermal equilibrium in the early Universe.[13] This means that their abundance cannot be predicted in the framework of the νMSM:[22]

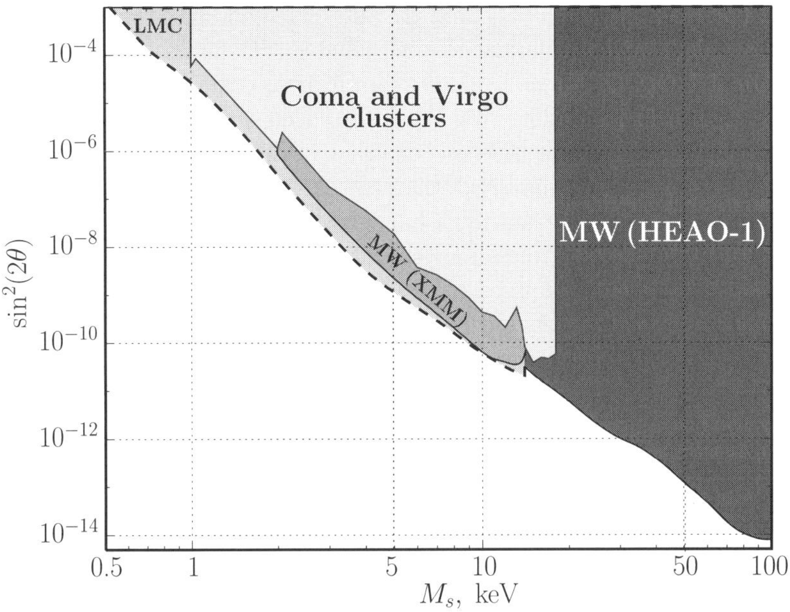

Fig. 1. Constraints on the mixing angle θ of sterile neutrino with active neutrino from different X-ray observations of XMM-Newton and HEAO-1 satellites of different astrophysical objects. MW = Milky Way, LMC = Large Magellanic Clouds.

one should either fix the concentration of sterile neutrinos at temperatures greater than 1 GeV, or specify the physics beyond the νMSM.

One can address the question how many sterile neutrinos are produced due to the νMSM interactions, Eq. (1), i.e. because of the mixing with active neutrino flavours characterized by parameter θ. In fact, this mixing is temperature dependent:[23]

$$\theta \to \theta_M \simeq \frac{\theta}{1 + 2.4(T/200\ \mathrm{MeV})^6(\mathrm{keV}/\mathrm{M_I})^2} ,$$

so that the rate Γ of sterile neutrino production is strong suppressed at $T > 100$ MeV, $\Gamma \propto T^{-7}$. The rate peaks roughly at[13] $T_{peak} \sim 130 \left(\frac{M_I}{1\ \mathrm{keV}}\right)^{1/3}$ MeV, which corresponds to the temperature of the QCD cross-over for keV scale sterile neutrinos. This fact makes an exact estimate of the number of produced sterile neutrinos to be a very difficult task, since T_{peak} happens to be exactly at the point where the quark-gluon plasma is strongly coupled and the dilute hadron gas picture is not valid. The chiral perturbation theory works only at $T < 50$ MeV. The perturbation theory in QCD works only at $T \gg \Lambda_{QCD}$, and the convergence is very slow. The lattice simulations work very well for pure gluodynamics. However, no results with three light quarks and with reliable extrapolation to continuum limit are available yet. Also, it is not clear how to deal with hadronic initial and final states in reactions $\nu + q \to \nu + q$, $q + \bar{q} \to \nu\bar{\nu}$. Nevertheless, in spite of all these uncertainties one can show[24] that if Lyman-α constraints of Ref. 19 are taken for granted, the production of sterile neutrinos due to active-sterile neutrino transitions happens to be too small to account for observed abundance of dark matter. In other words, physics beyond the νMSM is likely to be required to produce dark matter sterile neutrinos.

In Ref. 3 it was proposed the the νMSM may be extended by a light inflaton in order to accommodate inflation. To reduce the number of parameters and to have a common source for the Higgs and sterile neutrino masses the inflaton-νMSM couplings can be taken to be scale invariant on the classical level:

$$\mathcal{L}_{\nu\text{MSM}} \to \mathcal{L}_{\nu\text{MSM[M}\to 0]} + \frac{1}{2}(\partial_\mu \chi)^2 - \frac{f_I}{2}\,\bar{N_I}^c N_I \chi + \text{h.c.} - \text{V}(\Phi, \chi)\ , \quad (5)$$

where the Higgs-inflaton potential is given by:

$$V(\Phi, \chi) = \lambda\left(\Phi^\dagger \Phi - \frac{\alpha}{\lambda}\chi^2\right)^2 + \frac{\beta}{4}\chi^4 - \frac{1}{2}m_\chi^2 \chi^2.$$

The requirement that the chaotic inflation[25] produces the correct amplitude for scalar perturbations leads to the constraints:

$$\beta \simeq 10^{-13}, \quad \alpha \lesssim 10^{-7}, \quad f_I \lesssim 10^{-3}\ .$$

For $\alpha > \beta$ inflaton mass is smaller than the Higgs mass, $m_I < M_H$.

One can show[3] that the inflaton with mass $m_I > 300$ MeV is in thermal equilibrium thanks to reactions $\chi \leftrightarrow e^+e^-$, $\chi \leftrightarrow \mu^+\mu^-$ down to $T < m_I$. The sterile neutrino abundance due to inflaton decays: $\chi \to NN$ is given by

$$\Omega_s \simeq 0.26 \frac{\Gamma M_0 m_s}{m_I^2 \times 12\ \text{eV}} \frac{2\pi\zeta(5)}{\zeta(3)}\ .$$

So, for $m_I \sim 300$ MeV ($m_I \sim 100$ GeV) the correct Ω_s is obtained for $m_s \sim 16 - 20$ keV ($m_s \sim \mathcal{O}(10)$ MeV). A sterile neutrino in this mass range is perfectly consistent with all cosmological and astrophysical observations. As for the bounds on mass versus active–sterile mixing coming from X-ray observations of our galaxy and its dwarf satellites,[21,26] they are easily satisfied since the production mechanism of sterile neutrinos discussed above has nothing to do with the active–sterile neutrino mixing leading to the radiative mode of sterile neutrino decay.

5. Baryon Asymmetry of the Universe

The baryon (B) and lepton (L) numbers are not conserved in the νMSM. The lepton number is violated by the Majorana neutrino masses, while $B+L$ is broken by the electroweak anomaly. As a result, the sphaleron processes with baryon number non-conservation are in thermal equilibrium for $100\ \text{GeV} < T < 10^{12}\ \text{GeV}$. As for CP-breaking, the νMSM contains 6 CP-violating phases in the lepton sector and a Kobayashi-Maskawa phase in the quark sector. This makes two of the Sakharov conditions[27] for baryogenesis satisfied. Similarly to the MSM, this theory does not have an electroweak phase transition with allowed values for the Higgs mass,[28] making impossible the electroweak baryogenesis, associated with the non-equilibrium bubble expansion. However, the νMSM contains extra degrees of freedom - sterile neutrinos - which may be out of thermal equilibrium exactly because their Yukawa couplings to ordinary fermions are very small. The latter fact is a key point for the baryogenesis in the νMSM, ensuring the validity of the third Sakharov condition.

In Ref. 4 it was proposed that the baryon asymmetry can be generated through CP-violating sterile neutrino oscillations. For small Majorana masses the total lepton number of the system, defined as the lepton number of active neutrinos plus the total helicity of sterile neutrinos, is conserved and equal to zero during the Universe's evolution. However, because of oscillations the lepton number of active neutrinos becomes different from zero and gets transferred to the baryon number due to rapid sphaleron transitions. Roughly speaking, the resulting baryon asymmetry is equal to the lepton asymmetry at the sphaleron freeze-out.

The kinetics of sterile neutrino oscillations and of the transfers of leptonic number between active and sterile neutrino sectors has been worked out in Ref. 2. The effects to be taken into account include oscillations, creation and destruction of sterile and active neutrinos, coherence in sterile neutrino sector and its lost due to interaction with the medium, dynamical asymmetries in active neutrinos and charged leptons. The corresponding equations are written in terms of the density matrix for sterile neutrinos and concentrations of active neutrinos and are rather lengthy and will not be presented here due to the lack of space. They can be found in the original work.[2] The corresponding equations are to be solved with the choice of the νMSM parameters consistent with the experiments on neutrino oscillations and with the requirement that dark matter neutrino has the necessary properties.

The value of baryon to entropy ratio $\frac{n_B}{s}$ can be found from the solution of the kinetic equations and is given by[2]

$$\frac{n_B}{s} \simeq 1.7 \cdot 10^{-10} \, \delta_{\rm CP} \left(\frac{10^{-5}}{\Delta M_{32}^2 / M_3^2} \right)^{\frac{2}{3}} \left(\frac{M_3}{10 \text{ GeV}} \right)^{\frac{5}{3}},$$

where $M_{2,3}$ are the masses of the heavier sterile neutrinos, $\Delta M_{32}^2 = M_3^2 - M_2^2$, and the CP-breaking factor $\delta_{\rm CP}$ is expressed through the different mixing angles and CP-violating phases, parameterizing the Dirac neutrino masses, and can be $\mathcal{O}(1)$, given the present experimental data on neutrino oscillations. This shows that the correct baryon asymmetry of the Universe $\frac{n_B}{s} \simeq (8.8-9.8) \times 10^{-11}$ is generated when the heavier sterile neutrinos with the masses, say, 1 GeV are degenerate to one part in 10^5. This looks like a strong fine tuning but may also indicate that there exists some symmetry making the degeneracy automatic.[29]

It is interesting to note that for masses of sterile neutrinos > 100 GeV the mechanism does not work as the sterile neutrinos equilibrate. Also, the temperature of baryogenesis is rather low, $T_L \simeq (\Delta M^2 M_{Pl})^{\frac{1}{3}} > 10^2$ GeV, i.e. validity of νMSM is only required at the scales smaller than M_W or so.

6. Conclusions

The νMSM is, perhaps, the simplest and the most economical extension of the Minimal Standard Model. It shares with the MSM its advantages (renormalizability and agreement with most particle physics experiments) and its fine-tuning problems (the gauge hierarchy problem, flavour problem, etc). However, unlike the MSM, the νMSM can explain simultaneously three different phenomena, observed experimentally, namely neutrino oscillations, dark matter, and baryon asymmetry of the Universe. The parameter-space of the model is rather constrained: the dark matter neutrino should have a mass in the keV region and be much lighter than two heavier sterile neutrinos, which are required to be quite degenerate. The model has a number of testable predictions. In astrophysics, one should search for X-rays from decays of dark matter neutrinos, which could be achieved with a X-ray spectrometer in Space with good energy resolution $\delta E/E \sim 10^{-3} - 10^{-4}$ getting signals from our Galaxy and its Dwarf satellites.[21] In particle physics, the νMSM predicts the absolute values of active neutrino masses,[1] and existence of relatively light singlet fermions[29] which can be searched for in decays of charmed, beauty and even K or π-mesons in experiments similar to Refs. 30 and 31.

Acknowledgments

This work was supported in part by the Swiss National Science Foundation.

References

1. T. Asaka, S. Blanchet and M. Shaposhnikov, Phys. Lett. B **631** (2005) 151.
2. T. Asaka and M. Shaposhnikov, Phys. Lett. B **620** (2005) 17.
3. M. Shaposhnikov and I. Tkachev, Phys. Lett. B **639** (2006) 414.
4. E. Akhmedov, V. Rubakov and A. Smirnov, Phys. Rev. Lett. **81** (1998) 1359.
5. V. A. Kuzmin, V. A. Rubakov and M. E. Shaposhnikov, Phys. Lett. B **155** (1985) 36.
6. A. Kusenko and G. Segré, Phys. Lett. B **396** (1997) 197.
7. P. L. Biermann and A. Kusenko, Phys. Rev. Lett. **96** (2006) 091301.
8. J. Stasielak, P. L. Biermann and A. Kusenko, arXiv:astro-ph/0606435.
9. P. Minkowski, Phys. Lett. B **67** (1977) 421; T. Yanagida, Progr. Theor. Phys. **64** (1980) 1103 ; M. Gell-Mann, P. Ramond and R. Slansky, in *Supergravity*, North Holland, Amsterdam 1980.
10. M. Fukugita and T. Yanagida, Phys. Lett. B **174** (1986) 45.
11. A. D. Dolgov and S. H. Hansen, Astropart. Phys. **16** (2002) 339.
12. J. J. Dalcanton and C. J. Hogan, Astrophys. J. **561** (2001) 35.
13. S. Dodelson and L. M. Widrow, Phys. Rev. Lett. **72** (1994) 17 .
14. K. Abazajian, G. M. Fuller and M. Patel, Phys. Rev. D **64** (2001) 023501.
15. B. Moore et al., Astrophys. J. **524** (1999) L19.
16. P. Bode, J. P. Ostriker and N. Turok, Astrophys. J. **556** (2001) 93.
17. T. Goerdt et al., Mon. Not. Roy. Astron. Soc. **368** (2006) 1073.
18. G. Gilmore et al., arXiv:astro-ph/0608528.
19. U. Seljak et al., arXiv:astro-ph/0602430.
20. M. Viel et al., Phys. Rev. Lett. **97** (2006) 071301.
21. A. Boyarsky et al., arXiv:astro-ph/0603660.
22. T. Asaka, M. Laine and M. Shaposhnikov, JHEP **0606** (2006) 053.
23. R. Barbieri and A. Dolgov, Phys. Lett. B **237** (1990) 440.
24. T. Asaka, M. Laine and M. Shaposhnikov, in preparation.
25. A. D. Linde, Phys. Lett. B **129** (1983) 177.
26. S. Riemer-Sorensen, S. H. Hansen and K. Pedersen, Astrophys. J. **644** (2006) L33.
27. A. D. Sakharov, Pisma ZhETF **5** (1967) 32.
28. K. Kajantie et al., Phys. Rev. Lett. **77** (1996) 2887.
29. M. Shaposhnikov, arXiv:hep-ph/0605047.
30. G. Bernardi *et al.*, Phys. Lett. B **203** (1988) 332.
31. P. Astier *et al.* [NOMAD Collaboration], Phys. Lett. B **506** (2001) 27.

NEUTRINO-LESS DOUBLE BETA DECAYS AND NEUTRINO MASSES THE PRESENT AND FUTURE EXPERIMENTS

H. EJIRI

Research Center for Nuclear Physics, Osaka University,
Ibaraki, Osaka 567-0047, Japan, and
Japan Synchrotron Radiation Laboratory, SPring-8,
Sayou-machi, Hyogo 679-5198, Japan
E-mail: ejiri@rcnp.osaka-u.ac.jp

Neutrino-less double beta decays are of great interest for studying the Majorana nature of ν's, the ν mass spectrum, the absolute ν-mass scale, the Majorana CP phases and other properties of neutrinos and weak interactions. The present $\beta\beta$ detectors with sensitivities of $150 \sim 300$ meV are used to study the quasi-degenerate ν masses. Future higher-sensitivity experiments with mass sensitivities of the atmospheric and solar ν-masses are indispensable for studying ν-masses in the inverted and normal hierarchy spectra, respectively. Theoretical and experimental studies for evaluating nuclear matrix elements $M^{0\nu}$ are cruial for extracting the ν mass. This report is a brief review of double beta decays and neutrinos with emphasis on future $0\nu\beta\beta$ experiments for the Majorana ν mass in the 30-100 meV region.

1. Double Beta Decays and Neutrinos

Neutrino-less $\beta\beta$ decays $(0\nu\beta\beta)$ provide evidence for the Majorana nature of ν's, the ν mass spectrum, the absolute ν mass scale, the Majorana CP phases and others beyond the standard model (SM). Theoretical and experimental studies of $\beta\beta$ decays have extensively been made as given in recent reviews by [1], [2], [3], [4] and references therein.

Recent ν oscillation data show nonzero neutrino mixings and nonzero neutrino mass-square differences. They, however, do not provide the absolute mass scale nor the nature of the neutrinos, that can be studied by high-sensitivity $0\nu\beta\beta$ studies. Presently, $\beta\beta$ decays are unique and realistic probes to access such small ν masses as suggested by the ν oscillation.

On the basis of the recent ν oscillation studies, the effective mass sensitivity required for observing the $0\nu\beta\beta$ rate is of the order of the atmospheric

ν mass scale of $\delta m_A \sim 50$ meV in the case of the inverted mass hierarchy (IH) and of the order of the solar ν mass scale of $\delta m_S \sim 8$ meV in the case of the normal hierarchy (NH).

The present $\beta\beta$ detectors with sensitivities of $150 \sim 300$ meV are used to study the quasi-degenerate ν mass spectrum. Future detectors with higher sensitivities of the orders of δm_A, using different nuclei and methods (calorimetric, spectroscopic), are indispensable for establishing $0\nu\beta\beta$ in the IH region.

Theoretical and experimental studies for evaluating nuclear matrix elements $M^{0\nu}$ are important for extracting the sensible ν mass from the $0\nu\beta\beta$ rate. Charge exchange reactions by means of nuclear, electron and ν probes provide useful data for $M^{0\nu}$. Recent works on the nuclear matrix elements are given in recent reviews [1,4–6] and in references therein.

This is a brief review presented at IDM06 on recent $0\nu\beta\beta$ studies and the Majorana ν mass with emphasis on high-sensitivity $0\nu\beta\beta$ experiments in near future. The present review is partially given in the recent review article [4].

2. Double Beta Decays and Majorana Neutrino Masses

Double beta decays associated with neutrinos and weak interactions beyond SM are neutrino-less decays of $0\nu\beta\beta$, which violate the lepton number conservation by $\Delta L = 2$. They provide evidence for the Majorana nature of the neutrino and the absolute ν-mass scale.

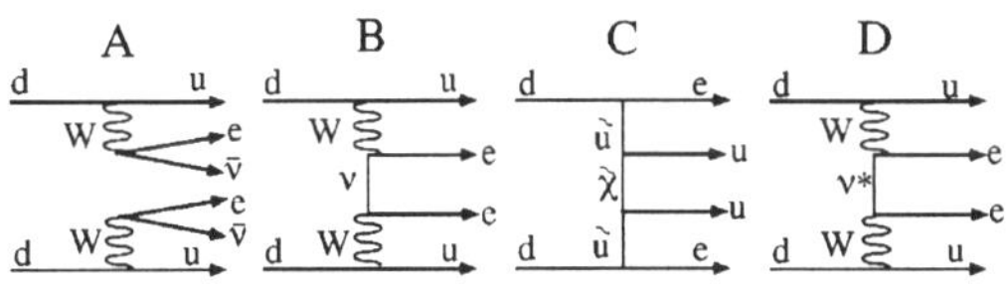

Fig. 1. Double beta decay processes. A: $2\nu\beta\beta$, B: $0\nu\beta\beta$ with the light-ν exchange, C: $0\nu\beta\beta$ with the SUSY exchange, D: $0\nu\beta\beta$ with the excited-ν exchange.

The $0\nu\beta\beta$ processes beyond SM include the light-ν exchange process, the SUSY particle exchange process with the R-parity-violating interaction, the coupling with the Majorana field followed by the Majoron emission ($0\nu\beta\beta$B) and others. Schematic diagrams of $2\nu\beta\beta$ and $0\nu\beta\beta$ processes are shown in Fig. 1. In the present report we discuss the case of Fig. 1B, where the light Majorana ν emitted from one neutron is reabsorbed by another neutron in the same nucleus. Details of the $\beta\beta$ theories are given in the review papers and refs therein. On the basis of the left and right symmetric model, the weak Hamiltonian for $0\nu\beta\beta$ is expressed as

$$H = G_F \cos(\theta_c)(2)^{-1/2}[j_{L\mu}J^{+}_{L\mu} + \kappa j_{L\mu}J^{+}_{R\mu} + \eta j_{R\mu}J^{+}_{L\mu} + \lambda j_{R\mu}J^{+}_{R\mu}], \quad (1)$$

where $j_{L\mu}$ and $J_{L\mu}$ are the left-handed lepton and hadron currents, and $j_{R\mu}$ and $J_{R\mu}$ are the right-handed lepton and hadron currents, respectively.

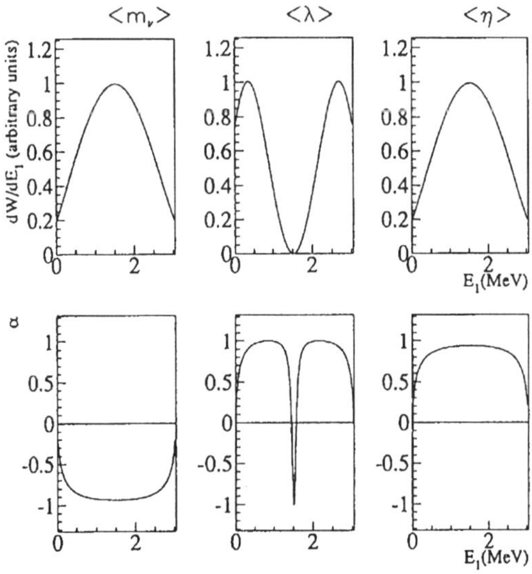

Fig. 2. Correlations for the ^{100}Mo $0\nu\beta\beta$ process caused by the mass and right-handed current terms of $\langle m \rangle$, $\langle \lambda \rangle$ and $\langle \eta \rangle$. Top: Single-β spectra. Bottom: Angular correlation coefficients α defined by $W(\theta_{12}) = 1 + \alpha \cos \theta_{12}$. [4]

The first term is the ν mass term and the others are associated with the right-handed weak interaction. The mass and the right-handed current terms are identified by observing the energy and angular correlations of the two β rays, as shown in Fig. 2.

The $0\nu\beta\beta$ decay rate due to the Majorana mass term is written as

$$T^{0\nu} = G^{0\nu} \left|M^{0\nu}\right|^2 \left|\langle m_\nu \rangle\right|^2. \quad (2)$$

The effective mass $\langle m_\nu \rangle$ is expressed using the mixing coefficients and the Majorana phases as

$$\begin{aligned} \langle m_\nu \rangle = & \ |\sum_i |U_{ei}|^2 m_i e^{i\alpha_i}|, \\ = & \ |C_{12}^2 C_{13}^2 m_1 \ + \ C_{13}^2 S_{12}^2 m_2 e^{i\phi_2} \ + \ S_{13}^2 m_3 e^{i\phi_3}|, \end{aligned} \tag{3}$$

where $\phi_2 = \alpha_2 - \alpha_1$ and $\phi_3 = -\alpha_1 - 2\delta$ are the phases for $|m_2\rangle$ and $|m_3\rangle$.

The ν masses m_1, m_2 and m_3 are rewritten using the mass-square differences $\delta m_{ij}^2 = m_i^2 - m_j^2$, which are derived from the ν oscillation data. We consider the three cases of the neutrino mass spectra; the normal mass hierarchy (NH) with $m_1 < m_2 \ll m_3$, the inverted hierarchy (IH) with $m_3 \ll m_1 < m_2$, and the quasi-degenerate spectrum (QD) with $m_1 \approx m_2 \approx m_3 \gg \sqrt{\delta m_{ij}^2}$.

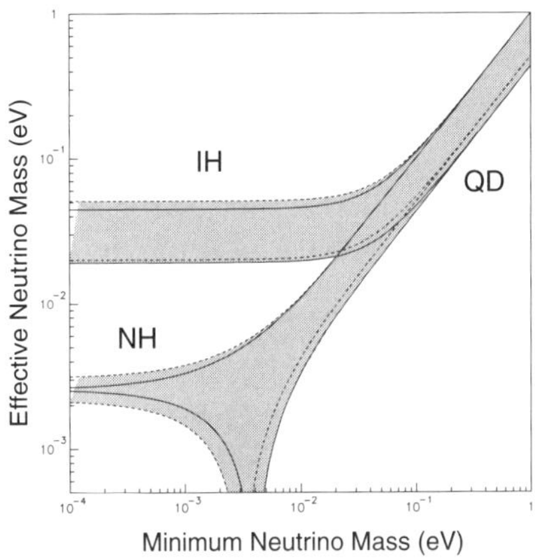

Fig. 3. Effective masses $\langle m^{NH} \rangle$ and $\langle m^{IH} \rangle$ for the NH and IH spectra as a function of the lightest mass of $m_1 (m_3)$. [4]

Then the effective masses to be studied by $0\nu\beta\beta$ are expressed for NH, IH, and QD spectra as

$$\langle m^{NH} \rangle = |C_{12}^2 C_{13}^2 m_1 + C_{13}^2 S_{12}^2 (\delta m_{21}^2 + m_1^2)^{1/2} e^{i\phi_2} + \tag{4}$$

$$S_{13}^2 (\delta m_{32}^2 + \delta m_{21}^2 + m_1^2)^{1/2} e^{i\phi_3}|, \tag{5}$$

$$\langle m^{IH} \rangle = |C_{12}^2 C_{13}^2 (\delta m_{23}^2 - \delta m_{21}^2 + m_3^2)^{1/2} + \tag{6}$$

$$C_{13}^2 S_{12}^2 (\delta m_{23}^2 + m_3^2)^{1/2} e^{i\phi_2} + S_{13}^2 m_3^2 e^{i\phi_3}|, \tag{7}$$

$$\langle m^{QD} \rangle \approx |C_{12}^2 C_{13}^2 + C_{13}^2 S_{12}^2 e^{i\phi_2} + S_{13}^2 e^{i\phi_3}| m. \tag{8}$$

The effective masses for the simple cases of the CP conserving phases are plotted as a function of the lightest mass (m_1 for NH and m_3 for IH) in Fig. 3 Next-generation DBD experiments with high mass sensitivity of the order of 20 meV is of great interest in establishing $0\nu\beta\beta$ and the neutrino mass spectrum for either QD or IH.

The electron neutrino mass derived from the analysis of the β-energy spectrum is the square root of the incoherent sum of the individual m_i^2,

$$m_\nu(\beta) = (\sum_i |U_{ei}|^2 m_i^2)^{1/2}. \tag{9}$$

Tritium β decays give the upper limit of around 2 eV. KATRIN aims at studying the QD mass region down to $m_\nu(\beta) = 300 \sim 200$ meV It will give constraints on the absolute masses of m_i.

Neutrinos constitute a fraction of hot dark matter (HDM), and the sum of the neutrino masses are given by cosmological observations [7,8]. The contribution of the massive neutrinos to the cosmological mass density is

$$\Omega_\nu h^2 = \sum m_\nu/(94eV), \tag{10}$$

where Ω_ν, h and $\sum m_\nu$ are the massive ν density, the Hubble constant in units of 100 km/s/Mpc, and the sum of the ν masses. For QD with $m_i^2 \gg \delta m_A^2$, the effective mass to be studied by $0\nu\beta\beta$ is

$$\langle m_\nu \rangle \ \leq \ \frac{1}{3}(\sum m_\nu)|C_{12}^2 + S_{12}^2 e^{i\phi_2}|. \tag{11}$$

The recent WMAP [8] analysis gives an upper limit of $\sum m_\nu \leq 0.7$ eV, which leads to the limit $\langle m_\nu \rangle \leq 230 \sim 120$ meV, depending on the phase. High-precision studies of the cosmological observation and the single-β spectrum may clarify a possible $\langle m_\nu \rangle$ in the QD region.

3. Nuclear Responses for Double Beta Decays

The nuclear matrix element for $0\nu\beta\beta$ is crucial for extracting the effective mas from the $0\nu\beta\beta$ transition rate. It includes the momentum-dependent terms and the tensor term, as discussed in the review [3]. Here, we consider the F- and GT-type matrix elements, for simplicity, as expressed as

$$M^{0\nu} = M^{0\nu}(\tau\sigma)[1 + R_F] \tag{12}$$

where R_F is the ratio of the Fermi and GT components. They are expressed in terms of the neutrino potentials $h_\pm$ as

$$M^{0\nu}(\tau\sigma) = \sum_c \langle 0_f | h_+(r_c, E_c) \ \tau\sigma\tau\sigma | 0_i \rangle \,, \tag{13}$$

$$M^{0\nu}(\tau) = \sum_c \langle 0_f | h_+(r_c, E_c)\ \tau\tau | 0_i \rangle\ . \qquad (14)$$

The nuclear matrix element $M^{0\nu}$ is very sensitive to the nuclear spin-isospin correlations and the nuclear interaction parameters. Actually, calculated matrix elements spread over $\pm 50\%$, depending on nuclear models and parameters used for the calculations.

Experimental study of nuclear structures relevant to $0\nu\beta\beta$ decays are also important [1,4]. Charge exchange nuclear reactions such as (^{3}He, t), (t,^{3}He), (d,^{2}He), IAS γ-rays, and μ^- capture reactions are used to obtain the spin-isospin responses for 1^+, 2^-, and so on. High-flux ν beams may be obtained from J-PARC and NuSNS [9,10].

$M^{0\nu}$ may be expressed as a separable form of $M^{0\nu}(k) \approx \Sigma_J \left[M_S(J) M_{S'}(J) / \Delta_S(J) \right]$, where $M_S(J)$ and $M_{S'}(J)$ are nuclear matrix elements for single-β decays through the intermediate states J [11]. The single-β matrix elements can be obtained experimentally by using nuclear, μ and ν beams, as discussed above.

4. Present Status of $\beta\beta$ Experiments

Experimental studies of neutrino-less double beta decays have been carried out extensively so far and are currently in progress by using various types of high-sensitivity detectors as given in reviews [2–4,12].

The ν mass limits depend on the nuclear matrix elements. Calorimetric experiments with the high E-resolution ^{76}Ge and TeO_2 detectors give stringent upper limits on the $0\nu\beta\beta$ rates. The IGEX and HM data with the ^{76}Ge semiconductors gave half-life limits of 1.6 and 1.9×10^{25}y, corresponding to around 0.33 -1.3 eV, respectively [13,14]. A claim for the $0\nu\beta\beta$ peak in the ^{76}Ge will be discussed by the next speaker(H.K). CUORICINO with 41 kg TeO_2 gives the half-life limit of 1.8 $\times 10^{24}$ y for ^{130}Te, and the mass limit of 0.4 $\sim$ 1.3 eV [15].

Spectroscopic methods have been used for $\beta\beta$ decays from ^{100}Mo, ^{82}Se, ^{150}Nd and other isotopes. ELEGANT V with drift chambers was used to study $\beta\beta$ decays from ^{100}Mo and ^{116}Cd [16,17]. NEMO III with large tracking chambers is being used in $\beta\beta$ experiments of ^{100}Mo, ^{82}Se, ^{150}Nd and other isotopes [18]. Quite stringent lower half-life limits and upper mass limits were obtained as 4.6 10^{23} y and 0.66-2.81 eV for ^{100}Mo and 1.0 10^{23} y and 1.75-4.86 eV for ^{82}Se. The mass limits depend on the nuclear matrix elements.

It is noted that the neutrino masses to be studied using the present detectors are limited by their mass sensitivity of 0.2–1 eV. Among them, CUORICINO and NEMO III are expected to reach the mass region of 0.2–0.5 eV in the near future.

5. Perspectives of $\beta\beta$ Experiments

Experimental proposals for future $\beta\beta$ experiments with the mass sensitivity of δm_A have been made on $\beta\beta$ isotopes [4,12], as listed in Table 1. Brief comments are given below on some of the future experiments.

Table 1. Isotopes and detectors to be used for future $\beta\beta$ experiments. S_N: nuclear sensitivity [4] in units of 10^{-24} y^{-1} $(eV)^{-2}$.

Isotope	A %	$Q_{\beta\beta}$ MeV	S_N	Experiment
^{48}Ca	0.187	4.276	0.11	CANDLES[a]
^{76}Ge	7.8	2.039	0.22	MAJORANA[b] GERDA[c],
^{82}Se	9.2	2.992	0.86	Super-NEMO[d]
^{100}Mo	9.6	3.034	2.02	MOON[e],
^{116}Cd	7.5	2.804	0.90	COBRA[f] CAMEO[g]
^{130}Te	34.5	2.529	0.73	CUORE[h], COBRA[f]
^{136}Xe	8.9	2.467	0.13	EXO[i], XMASS[j]
^{150}Nd	5.6	3.368	11.3	DCBA[k]

a: ref. [19] *b*: ref. [20] *c*: ref. [21] *d*: ref. [28] *e*: ref. [23] *f*: ref. [24] *g*: ref. [25] *h*: ref. [26] *i*: ref. [28] *j*: ref. [27] *k*: ref. [29].

CANDLES for ^{48}Ca $\beta\beta$ decays

CANDLES (CAlcium fluoride for studies of Neutrinos and Dark matter by Low Energy Spectrometers) is an array of CaF_2 crystals to study $\beta\beta$ decays of ^{48}Ca and dark matter [19]. Itis based on ELEGANT VI with CaF_2 crystals surrounded by CsI active shields [30]. The $Q_{\beta\beta}$ is large, but the natural abundance of ^{48}Ca is only 0.2%. CANDLES aims at studying the QD ν mass region with 3.2 ton of CaF_2 crystals.

MAJORANA and GERDA for ^{76}Ge $\beta\beta$ decays

MAJORANA aims at studies of ^{76}Ge $\beta\beta$ decays by a high-purity ^{76}Ge detector array [20], being based on IGEX. The detector consists of segmented intrinsic Ge detectors with a total mass of $M \approx 120$ kg. The expected sensitivity is 5.5 10^{26} y with 4.5 y run, which corresponds to the mass sensitivity around 100 meV. Pulse shape discrimination (PSD) and segmentation of detector (SED) are powerful for reducing various kinds of BG's. Natural RI impurities are reduced by electro-forming of the copper used for the detectors. Cosmogenic ^{68}Ga BG's can be reduced by SSTC, i.e., by measuring the preceding 10.4 keV X-rays following K-EC [4].

GERDA uses enriched ^{76}Ge detectors imensed in liquid argon [21]. Here the liquid argon serves as an active shield. GERDA collects the HM and IGEX enriched Ge crystals with the total mass of 18 kg to study the KKDC claim in Phase 1, and plan to add 20 kg to them in Phase II to get the sensitivity of 2 10^{26} y , corresponding to 90 - 290 meV.

MOON for $^{\mathbf{100}}$Mo $\boldsymbol{\beta\beta}$ decays

MOON (Molybdenum Observatory Of Neutrinos) [23] for ^{100}Mo $\beta\beta$ decays is a "hybrid" $\beta\beta$ and solar ν experiment with ^{100}Mo to study the Majorana ν masses with a sensitivity of 20 $\sim$ 30 meV and the pp and ^{7}Be solar ν's. MOON is based on the ELEGANT V [17]. The large $Q_{\beta\beta}$ of 3.034 and 1.903 MeV for the ground and excited states make it realistic to study the $\beta\beta$ decays to them.

MOON is a spectroscopic correlation study of two β-rays (charged particles) to identify the ν-mass term for $0\nu\beta\beta$. Selection of $0\nu\beta\beta$ and reduction of correlated and accidental BG's are made by means of SSSC and SSTC.

MOON has ^{100}Mo isotopes of an order of 1 ton , the energy resolution of $\sigma \approx$2.5% at $E = Q_{\beta\beta}$, and the position resolution of 10^{-8} per ton. Enriched ^{100}Mo isotopes are obtained by centrifugal separation of MoF_6 gas. It gives the sensitivity of $T^{0\nu}_{1/2} \approx 7 \times 10^{26}$ y , corresponding to $\sim$30 meV with $M^{0\nu} =$ 3. Currently, a prototype MOON (MOON 1) with 6-layer PL plates, each 0.53 m square and 1 cm thick, is set in the Cu-Pb shield container used for ELEGANT V at the Oto Cosmo Observatory. It gives a good overall E-resolution better than σ <3% as required for the 30 meV sensitivity.

MOON with $\beta\beta$ source $\neq$ detector is used to study other $\beta\beta$ isotopes such as ^{82}Se, ^{150}Nd and ^{116}Cd as well as ^{100}Mo.

NEMO III and SuperNEMO.

NEMO III will be improved by reducing Rn contributions. The expected sensitivity for 5 y is 4×10^{24} y for ^{100}Mo and 8×10^{23} y for ^{82}Se, corresponding to the effective masses of 0.2$\sim$0.35 eV and 0.65$\sim$1.8 eV [18]. A much larger project with 100 kg of $\beta\beta$ source is under consideration to search for the smaller ν mass below 0.1 eV [22]. The collaboration aims at the sensitivity of 2 10^{26} y, corresponding to 35 - 105 meV, for ^{82}Se.

COBRA COBRA uses an array of 1 cm^3 CdZnTe semiconductor detectors to study $\beta\beta$ decays of ^{116}Cd and others with high E-reslution [24]. A array of 64-detectors with mass of about 0.5 kg is now at LNGS. The sensitivity with 64 K detector for 5 y will be around 1-2 10^{26} y. corresponding to 50-30 meV.

CUORICINO and CUORE

CUORE (Cryogenic Underground Observatory for Rare Events) [26] is a calorimetric detector to measure the $0\nu\beta\beta$ of ^{130}Te with $Q = 2.529$ MeV. CUORICINO, which is a prototype for CUORE, is now running [26]. It consists of 64 TeO_2 crystals with the total mass of 41 kg.

Merits of ^{130}Te are the large natural abundance of 34% and the large ratio of Te in the TeO_2 crystal. Then the expected sensitivity for 3 y is around $T_{1/2} \approx 1.5 \times 10^{25}$ y, and the corresponding mass sensitivityis 100~300 meV, depending on the nuclear matrix element.

CUORE is a scaled-up version of CUORICINO. It consists of 988 TeO_2 crystals with the Te mass of 600kg . The net ^{130}Te mass is 203 kg. The CUORE group expects to reduce the BG rate to 0.01-0.001 per keV kg y and improve the energy resolution to 5 keV. Then CUORE aims at the sensitivity of $T_{1/2} \approx (2 - 6.6) \times 10^{26}$ y for a 5 y run with the BG rate of 0.01 - 0.001 / kev kg y. The mass sensitivity of $(35 - 120) \sim (20 - 65)$ meV, depending on the nuclear matrix element [26].

EXO and XMASS for ^{136}Xe $\beta\beta$ decays

EXO (Enriched Xenon Observatory) is a $\beta\beta$ experiment of ^{136}Xe with $Q = 2.467$ MeV [28]. It uses a large scale enriched ^{136}Xe detector.

The unique feature of EXO is the identification of the decay product of ^{136}Ba by means of a laser spectroscopy technique to select ^{136}Ba and suppress all kinds of BG's. Assuming no BG other than the $2\nu\beta\beta$ tail in the $0\nu\beta\beta$ window and the large efficiency of 70%, the 1 ton enriched Xe detector with an energy resolution of σ =1.6% gives a sensitivity of $T_{1/2} \approx 2 \times 10^{27}$ y for a 5 y run. The 10 ton Xe detector with the improved energy resolution of σ= 1% will give the sensitivity of 4.1×10^{28} y for 10 y run. The ν mass sensitivities are 50~68 meV and 11~15 meV for the one ton and 10 ton detectors, respectively, depending on the nuclear matrix elements. The 200 kg detector is under construction to be used at WIPP.

XMASS is a large liquid Xe scintillator to study dark matter and solar neutrinos as well as ^{136}Xe double beta decays [27]. A prototype detector is now under progress. The E-resolution is around a few percent at 1 MeV.

DCBA for ^{150}Nd $\beta\beta$ decays

DCBA (Drift Chamber Beta-ray Analyzer) uses a tracking chamber in a magnetic field to study ^{150}Nd $\beta\beta$ decays [29]. The β energy is obtained by the β-ray trajectory analysis. The energy resolution is crucial to reduce the BG contribution from $2\nu\beta\beta$. The goal is to achieve the half-life sensitivities of 10^{25} y and 10^{26} y with natural and enriched ^{150}Nd sources, respectively.

6. Concluding Remarks and Discussions

Neutrino-less double beta decays ($0\nu\beta\beta$) are of great interest for studying fundamental ν properties such as the Majorana nature and the absolute mass scale. High-sensitivity studies of $0\nu\beta\beta$ decays with the ν mass sensitivity of the order of 10 meV and reliable evaluations of the $0\nu\beta\beta$ matrix elements are crucial. The region of the effective mass to be studied by $\beta\beta$ decays is constrained from the ν oscillation experiments and the cosmological observations. Then $\langle m_\nu \rangle$ is of the order of 2 meV and 20 meV in case of the NH (normal mass hierarchy) and IH (inverted one), respectively, while it is of the order of 100 meV if the ν masses are quasi degenerate (QD).

Nuclear matrix element $M^{0\nu}$ for $0\nu\beta\beta$ is necessary for extracting the effective ν mass from the $0\nu\beta\beta$ rate. Theoretical calculations of the matrix elements and experimental studies of nuclear structures relevant to $0\nu\beta\beta$ decays are inportant.

The ν masses to be studied with the presently running experiments are limited, by the detector sensitivities, to be around $\langle m_\nu \rangle = 0.2 \sim 0.5$ eV depending on the nuclear matrix element. High-sensitivity experiments for the IH ν-mass of around 30 meV can be carried out by means of next-generation detectors with ton-scale $\beta\beta$ iotopes.

New-generation $\beta\beta$ experiments with the mass sensitivities of 100-25 meV are of vital importance in studies of neutrinos and weak interactions beyond SM. Proposals for high-sensitivity $\beta\beta$ experiments have been made for several $\beta\beta$ isotopes.

In view of the importance and the scale of new-generation $\beta\beta$ experiments, internationally corporative works in both experiments and theories is important in order to ensure successfully executing high-sensitivity future experiments for at least a few isotopes. It is noted that high-sensitivity $\beta\beta$ detectors are of potential interest for other rare decay experiments such as low-energy solar neutrinos, search for WIMPs, nucleon decays and others.

References

1. H. Ejiri, *Phys. Rep. C* **338**, 265 (2000).
   H. Ejiri, *Prog. Particle and Nuclear Physics* **48**, 185 (2002).
2. S.R. Elliott and P. Vogel, *Annu. Rev. Nucl. Part. Sci.* **52**, 115 (2002).
3. J.D. Vergados, *Phys. Rep. C* **361**, 1 (2002).
4. H. Ejiri, *J. Phys. Soc. Japan* **74**, 2101 (2005).
5. J. D. Suhonen and O. Civitarese, *Phys. Rep.* **300**, 123 (1998).
6. F. Simkovic and A. Faessler, *Prog. Part. Nucl. Phys.* **48**, 201 (2002).
7. M.J. Santos et al.*Phys. Rev. Lett.* **88**, 24302 (2002).
   S. Mosi et al.,*Prog. Nucl. Part. Phys.* **48**. 243 (2002).
   N.W. Halverson et al., *Astrophys. J.* **568**, 38 (2002).
8. D.N. Spergel et al., *Astrophys. J. Suppl.* **148**, 175 (2003).
9. H. Ejiri, *Nucl. Instr. Meth. Phys. Research* **503**, 276 (2003).
10. E. Hungerfoer, *NNR workshop proceedings, Dec. 2005, ed H.Ejiri,*, http://www.spring8.or.jp/ext/en/appeal/nnr05/.
11. H. Ejiri and H. Toki, *J. Phys. Soc. Japan Lett.* **65**. 7 (1996).
12. F. Avignone, *Proc. Neutrino Cconference, ν 2004* (2004).
13. C.E. Aalseth et al., *Phys. Rev. C* **59** (1999) 2108; D **65** 092007 (2002).
14. L. Baudis et al., *Phys. Rev. Lett.* **83**, 41 (1999).
15. C. Arnaboldi et al., *Phys. Lett. B* **584**, 260 (2004).
16. H. Ejiri et al., *J. Phys. Soc. Japan* **64**, 339 (1995).
17. H. Ejiri, *Phys. Rev. C* **63**, 065501 (2001).
18. X. Sarazin et al., *Proc. Neutrino Conference ν2004* (2004).
    R. Arnold et al., *Phys. Rev. Lett.* **95**, 182302 (2005).
19. S. Umehara et al., *Proc. NDM03 eds. H.Ejiri and I.Ogawa*, http://ndm03.phys.sci.osaka-u.ac.jp/proc/index.htm (2003).
20. C.E. Aalseth et al., *Yad. Fiz.* **67**, No. 11 (2004).
21. S. Scheonert, *Proc. Neutrino 2006, Santa Fe*, (2006).
22. D. Lallene, et al., *Pivate communication, in A. Pipke, Proc. Neutrion 2006, Santa Fe* (2006).
23. H. Ejiri, et al., *Phys. Rev. Lett.* **85** 2917 (2000).
    H. Ejiri: Czechoslovak Journal of Physics **54 B**, 317 (2004).
24. K. Zuber, *Phys. Lett. B* **519**, 1 (2001).
25. G. Bellini et al., *Eur. Phys. J. C* **19** 43 (2001).
26. C. Arnaboldi et al., *Nucl. Instr. Meth. A* **518**, 775 (2004).
    R. Maruyama et al., *Proc. Neutrino 2006, Santa Fe* (2006).
27. S. Moriyama et al., *Proc. Workshop XENON01, Tokyo* (2001).
28. A. Piepke, *Proc. Neutrion 2006, Santa Fe* (2006).
29. N. Ishihara et al., *Nucl. Instr. Meth. A* **443**, 101 (2000).
30. I. Ogawa et al., *Nucl. Phys. A* **730**, 215 (2004).

A NEW VIEW ON THE FUTURE OF NUCLEAR DOUBLE BETA DECAY

H. V. KLAPDOR-KLEINGROTHAUS
Max-Planck-Institut für Kernphysik,
P.O. Box 10 39 80, D-69029 Heidelberg, Germany
Spokesman of HEIDELBERG-MOSCOW, HDMS and GENIUS Collaborations
E-mail: H.Klapdor@mpi-hd.mpg,
Home-page: http://www.mpi-hd.mpg.de.non_acc/

This paper describes the lessons we have to draw after the observation of neutrinoless $\beta\beta$ decay by the enriched ^{76}Ge experiment, for present and future experiments (a) to fulfill the task to confirm the present result (b) to deliver additional information on the main contributions, of effective neutrino mass and right-handed weak currents etc. to the $0\nu\beta\beta$ amplitude. It is pointed out that presently running and planned experiments are mostly not sensitive enough to check the present evidence on a reasonable time scale. More important, the only way to get more information on the individual contributions of m,η,λ etc. to the $0\nu\beta\beta$ amplitude is to go to completely different types of experiments, e.g. mixed-mode $\beta^{+}EC$ decay experiments, such as ^{124}Xe decay.

1. Introduction

Nuclear double beta decay provides an extraordinarily broad potential to search for beyond standard model physics[2,3]. Its occurrence has enormeous consequences: it means that *total lepton number is not conserved*, and second that the *neutrino is a Majorana particle.* Furthermore it can provide, *under some assumptions*, an absolute scale of the neutrino mass, and yields sharp restrictions for SUSY models, leptoquarks, compositeness, left-right symmetric models, test of special relativity and equivalence principle in the neutrino sector and others[2,3].

Among the many existing efforts in search for $0\nu\beta\beta$ decay, since 13 years now the HEIDELBERG-MOSCOW experiment[1] which operated 11 kg of enriched ^{76}Ge in the period 1990-2003 in the GRAN SASSO underground laboratory, is the by far most sensitive experiment.

2. Status of Search for $0\nu\beta\beta$

What is the main result from the HEIDELBERG-MOSCOW experiment (see Refs. 8, 7, 9, 12, 10, 11).

- There is now a $>6\sigma$ signal for $0\nu\beta\beta$ decay.
- The neutrino is a Majorana particle.
- *Total* lepton number is violated (neutrino oscillations prove only violation of *family* lepton number).
- The neutrinos are (if we assume vanishing contributions of right-handed weak currents and of other contributions to the $0\nu\beta\beta$ amplitude, see below) degenerate in mass or (if the LSND result will be confirmed) allow existence of a *sterile* neutrino[7,12]).
- The $0\nu\beta\beta$ process yields very strong limits for other fields of Beyond SM physics often well competitive to high energy accelerators (see Refs. 2, 13, 3).

To put the experiment into historical perspective, the first (non-geochemical) discovery of $2\nu\beta\beta$ decay, in 1987 with a half-life of 1.1×10^{20} y for ^{82}Se, relied on a 2.2σ signal (35 events)[21].

Now we see this decay process, for ^{76}Ge, in the HEIDELBERG-MOSCOW experiment with **160 000 events** ($T_{1/2}^{2\nu} = 1.74 \times 10^{21}$ y)[4], i.e. the experimental sensitivity has been increased by a factor of **50 000 !!** This is what *allowed us* to see the $0\nu\beta\beta$ signal, with a half-life of $\mathbf{1.19 \times 10^{25}}$ **y** on a $>\mathbf{6}\sigma$ c.l.[7,10,11] (without methods to reduce the γ -background, on a 4.2σ level). The experiment is the by far most sensitive $0\nu\beta\beta$ experiment since 13 years and its sensitivity will be unfortunately not reached by other experiments in very near future (see below).

It has the largest source strength ever in operation (11.0 kg), the lowest background in such type of experiment (0.11 counts kg y keV *without* pulse shape analysis), the highest efficiency for detection of $\beta\beta$ events ($\sim$95%), the highest energy resolution ($\sim$3.3 keV), the highest duty cycle and the highest collected statistics (71.7 kg y, i.e by a factor of 8.2 more than one of the other ^{76}Ge experiments (IGEX) which finished operation in 1999 (see e.g. Ref. 6)). Further, the background of the experiment is very well understood from extensive Monte Carlo simulations[4] and from independent analysis by Ref. 5.

Of decisive importance for the reduction of the γ-background in the range of the $Q_{\beta\beta}$ value of the $0\nu\beta\beta$ process was, to develop methods of pulse shape analysis which were able to separate $0\nu\beta\beta$ from γ background events. We have developed two independent methods allowing to project out $0\nu\beta\beta$ events with practically no background from surrounding γ-rays.

One is based on application of a neuronal net[18,19], the other on calculated libraries of pulse shapes of $\beta\beta$-like events, starting from Monte Carlo simulated time history and spatial distribution of $0\nu\beta\beta$ events as function of location in the detector (including the dependence on the spectral angular correlation of the emitted electrons)[9,19] (see Fig. 1).

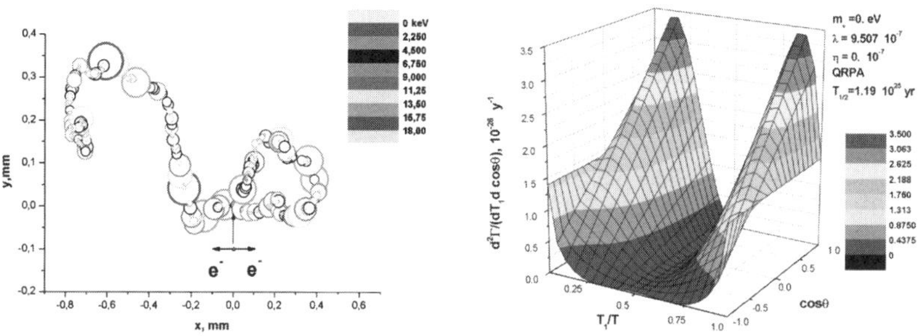

Figure 1. Typical calculated event for $0\nu\beta\beta$ decay without photon emission (bremstrahlung). Right: Calculated spectral angular correlation for the $\langle\lambda\rangle^2$ - term for $0\nu\beta\beta$ decay of ^{76}Ge (see Ref. 9).

Both methods *fulfill the criteria* required to prove observation of neutrinoless $\beta\beta$ decay: 1-select $0\nu\beta\beta$ events at $Q_{\beta\beta}$; 2-reduce strongly surrounding γ-events. In the period 1995-2003, which delivered the main set of data, the time structure of all events has been measured, using 250 MHz flash ADCs.

We show here in Figs. 2 the spectrum selected by the neuronal net around $Q_{\beta\beta}$ – compared with the measured full spectrum (see Refs. 7, 15 and 11). The signal at $Q_{\beta\beta}$ has a confidence level of 6.4σ (7.05 ± 1.11 events). The other method gives similar results.[10]

The energy of the line observed, (see Ref. 10) seems to be slightly below the 'best' value reported for $Q_{\beta\beta}$[28] of 2039.006 ± 0.050 keV (which may be understood as a result of the non-corrected ballistic deficit, see Ref. 10). Other measurements report $Q_{\beta\beta}$ =2040.71 ± 0.52 keV, 2038.56 ± 0.32 keV and 2038.7 ± 2.2 keV[28].

3. Lessons for Present and Future

3.1. *To 'confirmation' of the ^{76}Ge experiment*

The actual experimental status is in 2006 similar as it was in 2001. There is an observed signal and various experiments trying to check this result

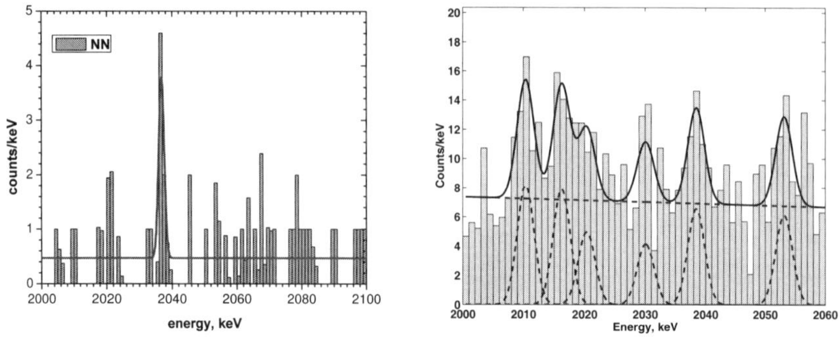

Figure 2. Upper part: The pulse shape selected spectrum in the range 2000-2100 taken with detectors 2,3,4,5 and the corresponding full spectrum of all five detectors in the range 2000-2060 keV (bottom part), in the period 1995-2003 (see Refs. 10, 7).

(some of them meanwhile stopped operation). What is required: - *Very good energy resolution.* Not fulfilled by NEMO III[32] and EXO[30] which have 400 and 100 keV, respectively, to be compared to 3.3 keV in the HEIDELBERG-MOSCOW experiment[7,15,11]. - *Large efficiency.* Nemo III has only 14%, i.e. a 10 kg experiment is effectively only a 1.4 kg experiment. - *The measured spectrum should be shown and analysed over the full energy range* to show that the background is fully understood. - the $2\nu\beta\beta$ spectrum should be measured as well as a help for normalization of the $0\nu\beta\beta$ matrix element. This is at present not possible with sufficient precision for CUORICINO/CUORE.

3.1.1. *Problems*

The main problem is that present and future 'confirmation' experiments partly because of the reasons mentioned are mostly not sensitive enough: A good example is the NEMO III experiment. The half-life limits reached (at a 1.5σ level) of $T^{0\nu}_{1/2}$ =1.0×10^{23} and 4.6×10^{23} years for ^{100}Mo and ^{82}Se (see Ref. 32) after 389 days of effective measurement are a factor of $\sim$ 20 away from the half-lives required to check the HEIDELBERG-MOSCOW ^{76}Ge result on a 1.5σ level. Since the half-life is connected with the measuring time by $T^{0\nu}_{1/2} = \sqrt{\frac{t \cdot M}{\delta E \cdot B}}$, this means that NEMO III would have to measure *more than 400 years, to see the signal on a 1.5σ level,* and correspondingly longer, to see it on a higher c.l.[15]. The future project SuperNEMO, with a by a factor of ten larger source strength, but same background and efficiency, would still require 40 years for the same purpose.

CUORICINO - which has the general problem, that it *cannot* distinguish between β and γ-events, and because of its high background cannot see the $2\nu\beta\beta$ spectrum of ^{130}Te, could see the HEIDELBERG-MOSCOW ^{76}Ge signal, assuming an uncertainty in the knowledge of the nuclear matrix element[25] of a factor of only 2, within 1 and 30 years - on a 1.5σ c.l.[15]. It can thus *never disprove* the HEIDELBERG-MOSCOW ^{76}Ge result (see also Ref. 29).

The large version CUORE with a by a factor of 16 larger mass, also would need many years for a statement on a 6σ level.

EXO - the main problem is that *no tracks* are visible in a liquid ^{136}Xe experiment[30]. This kills the main idea of the experiment to separate $\beta\beta$ from γ events, and just reduces it to complicated calorimeter. Since the other main idea - laser identification of the daughter nucleus, is not (yet) working, the present rather modest aim is to reach a background level as reached in the HEIDELBERG-MOSCOW ^{76}Ge experiment, instead of the factor of 1000 less, projected earlier[31].

GERDA/MAJORANA - (the first of them being essentially a copy of the GENIUS project proposed in 1997[22], planning to operate naked ^{76}Ge crystals in liquid nitrogen). From our earlier Monte Carlo calculations we expected to get a large potential of GENIUS for $\beta\beta$ research. The only long-term experience with naked detectors in liquid nitrogen has been collected since then with our GENIUS-Test-Facility[23,24]. The main problem found[14,24] is that over longer operation time the leakage current of the detectors raised which led to *serious restrictions of the high voltages applicable* and finally to *destruction of the detectors.* MAJORANA, if performed with conventional detector mounting in Cu cups, will have serious difficulties - as already visible by the IGEX experiment stopped in 1999[6] - to improve the background beyond the level reached by the HEIDELBERG-MOSCOW -experiment. If MAJORANA will also take up the GENIUS idea, the remarks given for GERDA should be considered.

COBRA - using or planning to use CdTe pixelized semiconductor detectors which may have in principle the potential of looking for $\beta^+\beta^+$ and β^+EC decay, still *is almost 10 orders of magnitude* away (see Ref. 33) from the required sensitivity, to become useful for double beta decay research (see Refs. 9, 26), e.g. for the points discussed in section 3.3. The increase in sensitivity of about one order of magnitude in the last four years shows that rather long time scales may be expected for this technique to come to fruitful applicability for $\beta\beta$ decay.

3.2. *A problem of present and future experiments concerning the neutrino mass and right-handed weak currents*

Concerning expected information on the ν mass, there is another problem in present and future experimental approaches. Even if one of the above or other $\beta^-\beta^-$ experiments would be able to confirm the HEIDELBERG-MOSCOW ^{76}Ge result, no *new* information would be obtained.

It is known for 20 years - but surprisingly often overlooked - that a $\beta^-\beta^-$ experiment can give information on the effective neutrino mass *only* under some *assumption* on the contribution of right-handed weak currents (parameters η, λ) or of others like SUSY etc. to the $\beta\beta$-amplitude (see e.g. Ref. 2). The half-life for the neutrinoless decay mode is given, if we neglect for simplicity contributions from SUSY, leptoquarks, etc. (for Refs. see Ref. 2) by[16,17]

$$[T^{0\nu}_{1/2}(0^+_i \rightarrow 0^+_f)]^{-1} =$$

$$= C_{mm}\frac{\langle m\rangle^2}{m_e^2} + C_{\eta\eta}\langle\eta\rangle^2 + C_{\lambda\lambda}\langle\lambda\rangle^2 + C_{m\eta}\langle\eta\rangle\frac{\langle m\rangle}{m_e} + C_{m\lambda}\langle\lambda\rangle\frac{\langle m_\nu\rangle}{m_e} + C_{\eta\lambda}\langle\eta\rangle\langle\lambda\rangle,$$

$$\langle m\rangle = |m^{(1)}_{ee}| + e^{i\phi_2}|m^{(2)}_{ee}| + e^{i\phi_3}|m^{(3)}_{ee}|\ ,$$

where $m^{(i)}_{ee} \equiv |m^{(i)}_{ee}|\exp{(i\phi_i)}$ $(i = 1, 2, 3)$ are the contributions to the effective mass $\langle m\rangle$ from individual mass eigenstates, with ϕ_i denoting relative Majorana phases connected with CP violation, and $C_{mm}, C_{\eta\eta}, ...$ denote nuclear matrix elements squared. Ignoring contributions from right-handed weak currents on the right-hand side of the above equation, only the first term remains.

Thus, *in general* one obtains *only an upper limit on* $\langle m\rangle$. So if neutrino masses are deduced from $0\nu\beta\beta$ experiments, this is always done under the *assumption* of vanishing η, λ etc. In that sense it is highly premature to compare as often done such number with numbers deduced e.g. from cosmological experiments, such as WMAP or other experiments, or to use it as a landmark for future tritium experiments.

It is unfortunate that even an additional high-sensitive $\beta^-\beta^-$ experiment (e.g. ^{136}Xe) *together* with the ^{76}Ge result can give *no* information to decide the individual contribution of $\langle m\rangle$, $\langle\eta\rangle, \langle\lambda\rangle$ to the $0\nu\beta\beta$ decay rate. This has been shown already in 1994[9,26] (see Fig. 3).

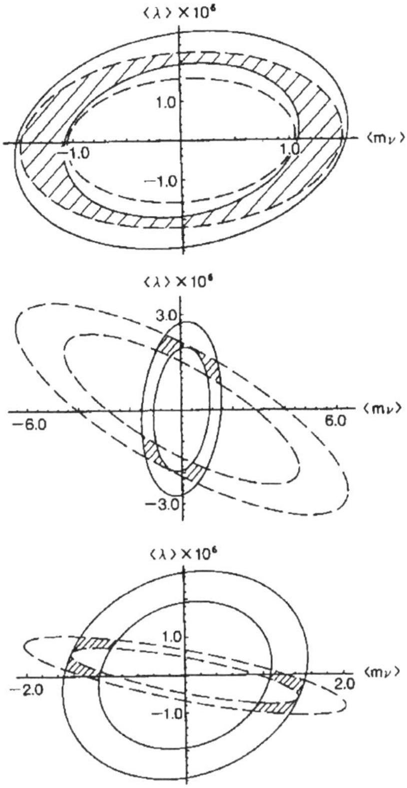

Figure 3. Allowed region of parameter space if $0\nu\beta\beta$ decay would be observed in two $0\nu\beta\beta$ decay experiments. Shown is only the $\langle\lambda\rangle - \langle m_\nu\rangle$ plane for simplicity. Upper part: For the plot schematically a $0\nu\beta\beta$ decay half life of (1.5±0.50 × 10^{24} y has been assumed for ^{76}Ge (full lines) and ^{136}Xe (dashed lines). The dashed area is consistent with both experiments (for discussions see Ref. 26). Middle part: As upper part, but for one $0\nu\beta^-\beta^-$ decay experiment (^{76}Ge) and one experiment measuring the neutrinoless mixed mode β^+EC^- decay of ^{124}Xe. For ^{124}Xe a $0\nu\beta\beta$ decay half life of $T_{1/2}^{0\nu\beta\beta}$=(1.5±0.50 × 10^{25} y has been assumed. Bottom part: As in middle part, but for a $0\nu\beta^+$EC decay half life of $T_{1/2}^{0\nu\beta\beta}$=(1.5±0.50 × 10^{26} y) for ^{124}Xe (see Ref. 26).

3.3. *A proposed way out*

In the same paper[26] it has been shown that the only realistic way to get this information on the individual contributions of m, η, λ is to combine the $\beta^-\beta^-$ result from ^{76}Ge, with a very high-sensitivity mixed mode β^+EC decay experiment (e.g. of ^{124}Xe) on a half-life level of 10^{27}years (see Ref. 9 and Fig. 3). The essential point is that the $0\nu\beta^+$EC decay is relatively more sensitive to the $\langle\lambda\rangle$ mechanism, than to the mass mechanism. The reasons for this are described in Ref. 26. While in Fig. 3 the *principle* behaviour is shown, recent calculations, with the *measured* $0\nu\beta\beta$ half-life of ^{76}Ge, and with ^{124}Xe and some other $\beta\beta$ sources confirm this general behavior[20].

Another theoretical possibility is to look for the $0\nu\beta\beta$ half-life for the $0^+ \to 2^+$ transition, e.g. in ^{76}Ge, for which the mass mechanism vanishes in first order and the transition is driven mainly by the λ and η mechanism. Since the half-life to be expected[26,27] for these modes would lie, however, with $\sim 10^{30}$ and $\sim 5 \times 10^{32}$ years almost in the range of that of proton decay, such experiment may be only of academic interest.

So it might be wise to combine future efforts to confirm the ^{76}Ge result with a possibility to pin down the various contributions to the $0\nu\beta\beta$ decay amplitude.

4. Summary and Outlook

We reached with the HEIDELBERG-MOSCOW ^{76}Ge experiment[1,7,10,11], what we wanted to learn from our large GENIUS project, proposed in 1997[22] - at a time where a signal was not yet seen - namely observation of $0\nu\beta\beta$ decay. There is now a $>6\sigma$ signal for $0\nu\beta\beta$ decay. The neutrino is a Majorana particle. Total lepton number is violated. Presently running and planned experiments are mostly not sensitive enough to check the ^{76}Ge result on a reasonable time scale. They further cannot give *any* information on the absolute (effective) neutrino mass and on the contribution of other processes to the total $0\nu\beta\beta$ amplitude. A way out of this dilemma is outlined.

Acknowledgements

The author acknowledges the invaluable support from DFG, for this project.

References

1. H.V. Klapdor, *Proposal, Internal Report*, MPI-1987-**V17**, September 1987.
2. H.V. Klapdor-Kleingrothaus, "60 Years of Double Beta Decay - From Nuclear Physics to Beyond the Standard Model", (World Scientific), Singapore (2001).
3. H.V. Klapdor-Kleingrothaus, *Int. Journ. of Mod. Phys. D* **13** (2004) 2107-2126
4. Ch. Dörr, H.V. Klapdor-Kleingrothaus, *Nucl. Instr. Meth. A* **513** (2003) 596-621.

5. K.Ya. Gromov et al., *J. Part. Nucl. Lett.* **3** (2006) 30-41.
6. H.V. Klapdor-Kleingrothaus, A. Dietz, I.V. Krivosheina, *Phys. Rev. D* **70** (2004) 078301.
7. H.V. Klapdor-Kleingrothaus, I.V. Krivosheina et al., *Phys. Lett. B* **586** (2004) 198-212 and H.V. Klapdor-Kleingrothaus, A. Dietz, I.V. Krivosheina et al., *Nucl. Instr. & Methods A* **522** (2004) 371-406.
8. H.V. Klapdor-Kleingrothaus et al. *Mod. Phys. Lett. A* **16** (2001) 2409-2420; *Found. Phys.* **31** (2002) 1181-1123 (Corrigenda 2003).
9. H.V. Klapdor-Kleingrothaus, I.V. Krivosheina, I.V. Titkova, *Phys. Lett. B* **632** (2006) 623-631, and *Phys. Rev. D* **73** (2006) 013010, and *Intern. J. of Modern Phys. A* **21** (2006) 1159-1188.
10. H.V. Klapdor-Kleingrothaus, I.V. Krivosheina, *Modern Phys. Lett. A* **21** (2006) 1547-1566.
11. H.V. Klapdor-Kleingrothaus, *Phys. Scripta T* **127** (2006) 40-42.
12. H.V. Klapdor-Kleingrothaus, H. Päs, A.Yu. Smirnov, *Phys. Rev. D* **63** (2001) 073005.
13. H.V. Klapdor-Kleingrothaus, In Proc. of Int. Conf. "Neutrinos and Implications for Physics Beyond the Standard Model", Stony Brook, Oct., 2002, ed. R. Shrock, World Scientific (2003) 367-382; and hep-ph/0303217; Proc. of Int. Conf. "Beyond the Desert 2003", BEYOND03, Tegernsee, Germany, June 2003, Springer, (2004), ed. H.V. Klapdor-Kleingrothaus, 307-364.
14. H.V. Klapdor-Kleingrothaus, I.V. Krivosheina, *Phys. Scripta T* **127** (2006) 52.
15. H.V. Klapdor-Kleingrothaus, in Proc. of Int. Conf. "Neutrino Telescopes", Febr. 2005, Venice, Italy, ed. M. Baldo-Ceolin, p. 215, hep-ph/0512263.
16. K. Muto, H.V. Klapdor, in "Neutrinos", Graduate Texts in Contemporary Physics", ed. H.V. Klapdor, Berlin, Germany: Springer (1988) 183-238.
17. K. Muto, E. Bender and H.V. Klapdor, *Z. Phys. A - Atomic Nuclei* **334** (1989) 187 and 177.
18. J. Hellmig, H.V. Klapdor-Kleingrothaus, *Nucl. Instrum. Meth. A* **455** (2000) 638-644; and B. Majorovits, H.V. Klapdor-Kleingrothaus. *Eur. Phys. J. A* **6** (1999) 463.
19. H.V. Klapdor-Kleingrothaus, I.V. Krivosheina et al., *Phys. Lett. B* **636** (2006) 235-247.
20. H.V. Klapdor-Kleingrothaus et al., in preparation
21. M. Moe et al., *Phys. Rev. Letters*, **59** (1987) 989.
22. H.V. Klapdor-Kleingrothaus, J. Hellmig and M. Hirsch, *GENIUS-Proposal* 20 Nov. 1997; J. Hellmig and H.V. Klapdor-Kleingrothaus, *Z. Phys. A* **359** (1997) 351-359 and nucl-ex/9801004; H.V. Klapdor-Kleingrothaus, M. Hirsch, *Z. Phys. A* **359** (1997) 361-372; H.V. Klapdor-Kleingrothaus, J. Hellmig, M. Hirsch, *J. Phys. G* 24 (1998) 483-516.
23. H.V. Klapdor-Kleingrothaus et al., *Nucl. Instr. Meth. A* **511** (2003) 341; H.V. Klapdor-Kleingrothaus, CERN Courier **43 Nr.6** (2003) 9; H.V. Klapdor-Kleingrothaus et al., Proc. of the 3nd Intern. Conf. on Particle Physics Beyond the Standard Model, BEYOND02, Castle Ringberg, Germany, 2002, IOP 2003, ed. H.V. Klapdor-Kleingrothaus, 499.

24. H.V. Klapdor-Kleingrothaus et al., *Nucl. Instr. Meth. A* **530** (2004) 410-418; H.V. Klapdor-Kleingrothaus, I.V. Krivosheina, *Nucl. Instr. Meth. A* **566** (2006) 472-476.
25. A. Staudt, K. Muto, H.V. Klapdor-Kleingrothaus, *Eur. Lett.* **13** (1990) 31.
26. M. Hirsch, K. Muto, T. Oda, H.V. Klapdor-Kleingrothaus, *Z. Phys. A* **347** (1994) 151.
27. T. Tomoda, *Nucl. Phys. A* **484** (1998) 635.
28. R.J. Ellis et al., *Nucl. Phys. A* **435** (1985) 34-42; J.G. Hykawy et al., *Phys. Rev. Lett.* **67** (1991) 1708; G. Audi and A.H. Wapstra, *Nucl. Phys. A* **595** (1995) 409-480; G. Doysset et al., *Phys. Rev. Lett.* **86** (2001) 4259- 4262.
29. For CUORICINO and CUORO Collab. C. Brofferio, in Proc. of Int. Conf. "Neutrino Telescopes", Febr. 2005, Venice, Italy, ed. M. Baldo-Ceolin, p. 239; and in Proc. Int. Conf. "NEUTRINO", June 2006, Santa Fe, USA.
30. J. Vuilleumier for the EXO coll., Proc. idm2004, Edinburg, Scotland 2004, World Scientific, Singapore (2005) 635.
31. A. Piepke, Talk at Heidelberg, ν Workshop, 2005.
32. NEMO Collabor., *Phys. Rev. Lett.* **95** (2005) 182302.
33. H. Kiel et al., Nucl. Phys. A 723 (2003) 499-514.

THE GERDA NEUTRINOLESS-DOUBLE-BETA DECAY EXPERIMENT

BÉLA MAJOROVITS FOR THE GERDA COLLABORATION

Max-Planck-Institut für Physik, Föhringer Ring 6, 80805 München, Germany
email: bela@mppmu.mpg.de

Neutrinoless double beta ($0\nu\beta\beta$)-decay could be the key to understanding the nature of the neutrino. The GErmanium Detector Array (GERDA) is designed to search for $0\nu\beta\beta$-decay of the isotope ^{76}Ge. Germanium crystals enriched in ^{76}Ge, acting as source and detector simultaneously, will be submerged directly into their ultra pure cooling medium that also serves as a radiation shield. This concept will allow for a reduction of the background by up to two orders of magnitudes with respect to earlier experiments.

1. Introduction

Neutrino accompanied double beta-decay is an allowed second order weak process. If the neutrino is a massive Majorana–particle the decay can occur without the emission of a neutrino [1]. The GERmanium Detector Array, GERDA, [2] is designed to search for $0\nu\beta\beta$-decay of ^{76}Ge. The importance of such a search is emphasized by the fact that the observation of a non-zero neutrino mass from flavor oscillations [3] does not give any information on its absolute value. The neutrino-oscillation observations do allow for two different mass scenarios with a normal or an inverted hierarchy [4]. These can be disentangled by $0\nu\beta\beta$-decay if a sensitivity for the effective Majorana neutrino-mass of 10 meV can be achieved [5].

The most sensitive $0\nu\beta\beta$ experiments are based on High-Purity-Germanium, HPGe,T detector technology. This is due to the combination of a very good energy resolution of the detectors at the $Q_{\beta\beta}$-value of ^{76}Ge, the very high purity of the detectors (very low intrinsic background) and the high signal detection efficiency of an experiment with detector being equal the source.

Currently the Heidelberg-Moscow (HdMo) and IGEX experiments give lower limits for $0\nu\beta\beta$-decay of $1.9{\cdot}10^{25}y$ and $1.6{\cdot}10^{25}y$, respectively [6,7].

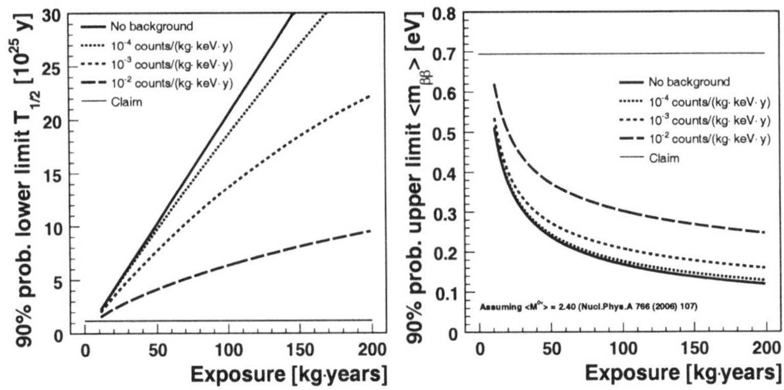

Fig. 1. Left: 90% probability to set lower limit on $T_{1/2}^{0\nu\beta\beta}$ higher than displayed as a function of exposure for different background indices. Right: 90% probability to set upper limit on effective Majorana neutrino mass lower than displayed using matrix elements from [10] as a function of exposure.

These upper limits can be translated into upper limits for the effective Majorana neutrino-mass of 0.35 keV-1.2 keV and 0.3 keV-1.5 keV for the two experiments, respectively. The large range is due to uncertainties in the matrix-element calculations [8]

A part of the HdMo collaboration claims to have observed a peak at $Q_{\beta\beta}$ with 4.2σ confidence level which can be attributed to $0\nu\beta\beta$-decay with a half-life of $T_{1/2}^{0\nu\beta\beta}$=1.29·$10^{25}y$. [9].

2. Sensitivity and the Principles of GERDA

The sensitivity obtainable for double beta experiments with a given exposure and background index is displayed in Fig. 1. The 90 % probability for obtaining lower limits higher than the displayed values as a function of exposure for given background indices are given in the left panel. The right panel shows the upper bounds that can be put on the effective Majorana-neutrino mass using matrix elements from [10] with 90 % probability. The values were calculated using Monte-Carlo ensemble test on the basis of Bayesian statistics [11].

For a given exposure the background limits the sensitivity. Therefore the goal is to minimize the background. This can be achieved by using an ultra-pure cryogenic liquid as the cooling medium and as shield against gamma radiation simultaneously [12]. The cryo-tank will be made out of carefully selected stainless steel. Additionally it will contain a low-background copper inlet as a radiation shield against the steel of the cryo-tank. The cryogenic

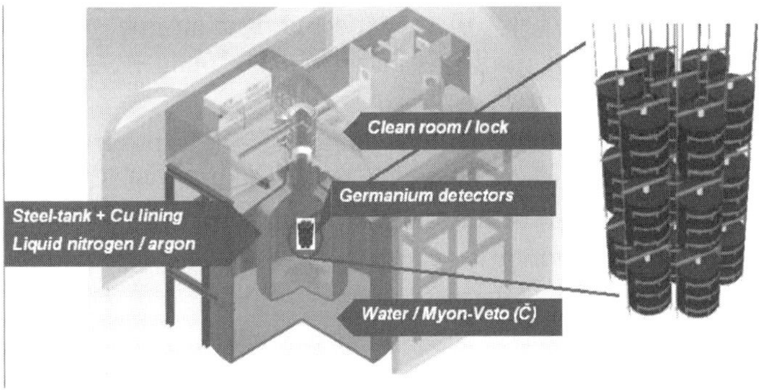

Fig. 2. Schematic view of the GERDA setup. The detector array is sitting in the center of the cryo tank. The cryo tank is surrounded by a water buffer serving as additional shield and as a muon-Čerenkov veto system. The detector array shown on the right is loaded through a lock system from the top of the tank. The lock will be installed in a clean-room

volume is surrounded by a buffer of ultra-pure water acting as an additional gamma and neutron shield. The water buffer is additionally used as a muon-Čerenkov veto. The setup is schematically depicted in Fig. 2.

GERDA will be installed in Hall A of the Gran Sasso underground Laboratory (LNGS), Italy. The experiment is foreseen to proceed in two phases. In the first phase, enriched detectors which were previously operated by the HdMo- and IGEX- collaborations will be redeployed. The aim is to take 15 $kg \cdot y$ of data with a background level of 10^{-2} counts/$(kg \cdot y \cdot keV)$ at the $Q_{\beta\beta}$–value of 2039 keV. As can be seen in Fig. 1 this will be enough to either confirm or refute the claim from the Heidelberg-Moscow-Experiment. In the second phase custom made detectors will be installed which have a true coaxial geometry and are 18-fold segmented. An exposure of 100 $kg \cdot y$ with a background level of 10^{-3} counts/$(kg \cdot y \cdot keV)$ is foreseen. A lower limit on the half-life of $0\nu\beta\beta$-decay of higher than $1.5 \cdot 10^{26}$ y corresponding to an upper limit of 100 meV (using the matrix elements from [10]) can be set with 90 % probability for the case of a null signal (Fig. 1).

3. Main Background Sources

To estimate the background expected for Phase II of the GERDA experiment Monte Carlo studies were performed using the GEANT 4 implementation MaGe [13]. The geometries assumed for the detectors are according to the design of the first segmented prototype detector that has been developed and successfully operated [14].

$5\cdot 10^{-4}$ counts/$(kg\cdot\ keV\cdot\ y)$ are expected from internal contaminations of the detector. The origin of this background is mainly cosmogenic ^{68}Ge and ^{60}Co.

This background is reduced by minimizing the exposure of the material to cosmic rays. Since ^{68}Ge decays with a half-life of $T_{1/2}$=271 d this component will decrease within the life-time of the experiment.

The largest contribution to the background is expected from the detector infrastructure, i.e. detector support cabling and electronics with $2.1\cdot 10^{-3}$ counts/$(kg\cdot\ keV\cdot\ y)$. This component could already be considerably decreased by reducing the mass of the signal cables around the HPGe-crystal by a factor of four. Further material selection and material minimization are under way.

From the external infrastructure (cryo-tank, water shield, etc.) roughly $4\cdot 10^{-4}$ counts/$(kg\cdot keV\cdot y)$ are expected for liquid nitrogen as the cryogenic liquid. Using liquid argon this is reduced to $3\cdot 10^{-5}$ counts/$(kg\cdot keV\cdot y)$ being negligible compared to the dominant background sources.

Neutrons and muons are expected to yield roughly $2\cdot 10^{-4}$ counts/$(kg\cdot keV\cdot y)$ the most important contribution being due to the delayed decay of ^{77}Ge produced by the neutron capture on ^{76}Ge [15].

If the background resulting from the detector infrastructure is reduced by a further factor of two, the goal of a total background index of less than 10^{-3}Counts/$(kg\cdot\ keV\cdot\ y)$ can be reached.

4. Phase I and Phase II Status

All IGEX and HdMo detectors are presently underground at LNGS. The first of them have been taken out of their cryostats without any technical problems. The left panel of Fig. 3 shows the dismounting of the first enriched HdMo detector. The detectors are presently being checked for their properties and will then be installed into the phase I suspension. A prototype detector has been operated in liquid argon in the holder at LNGS since beginning of 2006. It has gone through more than twenty cooling and warming up cycles without showing major deterioration.

The enriched material for phase II of the experiment has been procured. 35.5 kg of germanium enriched to 87%-88% in ^{76}Ge in form of GeO were transported from Krasnoyarsk, Siberia to Munich, Germany in a steel cylinder designed to reduce cosmogenic activation. The powder was unloaded, weighed and inspected (see Fig. 3) and subsequently transported to a 500 *mwe* underground cite.

The phase II detectors will be 18-fold segmented true coaxial n-type.

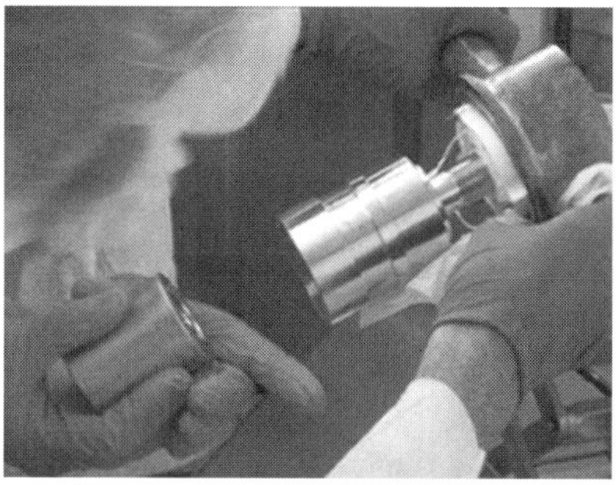

Fig. 3. Left: Dismounting of first enriched detector of the HdMo experiment from its copper vacuum-cryostat. Right: Unloading of the 35 kg enriched germanium. The green cylinder was built to shield the material against cosmogenic activation while transport from Siberia to Germany.

The segmentation will help to identify multiple Compton-scatter events in the region of interest. As shown in [16] the Compton background can be identified with high efficiency depending on its source and location.

A prototype detector was produced by Canberra-France. The 18 segments are read out using a novel contacting scheme [14]. The copper contact pads of a Kapton printed circuit board are pressed directly onto the contact areas of the segmented detector. This is shown in the left panel of Fig. 4. The prototype detector was extensively tested in a conventional test cryostat. The energy resolution of all signals and the core were around 3 keV at 1.3 MeV. The right panel of Fig. 4 shows a measurement taken with a ^{60}Co calibration source. The black histogram denotes the full spectrum from the core signal. The red histogram shows the spectrum requiring that only one segment had an energy deposit above a threshold of 20 keV. The Compton background discrimination was a factor of 10 for this measurement. The data could be reproduced very well by Monte Carlo simulations [16].

5. Conclusions

The Gerda neutrinoless double beta-decay experiment will be installed in the LNGS underground laboratory. For the first phase 15 $kg \cdot y$ data taking with a background index of 10^{-2} counts/$(kg \cdot keV \cdot y)$ are planned. This will allow to check the claim of a positive evidence from the HdMo-experiment. The goal of the second phase of the experiment is to collect 100 kg y of data with a background index of not more than 10^{-3} counts/$(kg \cdot keV \cdot y)$. This

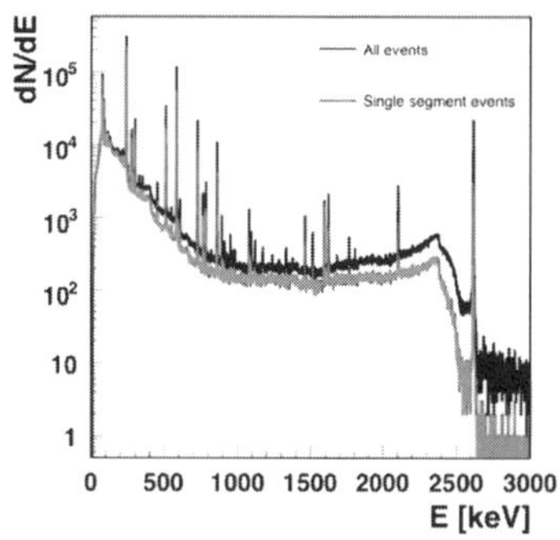

Fig. 4. Left: Germanium crystal with the new contacting scheme. The 18 segments are contacted by pressing the contact pads of Kapton printed circuit boards onto the segment contact area. Right: ^{60}Co calibration spectrum taken with the first 18 fold segmented true coax n-type HPGe detector. For more details see the text.

will allow to set a lower limit for the $0\nu\beta\beta$ half life of $1.5{\cdot}10^{26}$ y corresponding to an upper limit of the effective Majorana neutrino-mass of 100 meV. According to the Monte Carlo calculations this goal can be achieved. The detectors for phase I of the experiment are at the experimental site. The dismounting of the first detectors from the IGEX and the HdMo experiments was successful. A first prototype 18-fold segmented n-type HPGe detector has been produced and checked. Its performance is as expected.

References

1. S. Elliott and P. Vogel, Ann. Rev. Nucl. Part. Sci **52**(2002)115
2. I. Abt et al., hep-exp/0404039
3. S. Fakuda et al., Phys. Rev. Lett. **82**(1999)2644 and C. Ahmad et al., Phys. Rev. Lett **92**(2004)181301
4. S. Bilenky et al., Phys. Rev. C **64**(2001)053010
5. F. Feruglio et al., Nucl. Phys. B **637**(2002)345
6. L. Baudis et al., Phys. Rev. Lett. **83**(1999)41
7. Aalseth et al., Phys. Rev. D **65**(2002)092007
8. A. Faessler and F. Simkovic, J. Phys. G **24**(1998)2139
9. Klapdor-Kleingrothaus et al., Phys. Lett. B **586**(2004)198
10. V. Rodin et al., Nucl. Phys. A **766**(2006)107
11. A. Caldwell and K. Kröninger, physics/0608249
12. G. Heusser, Ann. Rev. Nucl. Part. Sci. **45** (1995)543
13. M. Bauer et al., J. Phys. Conf. Series **39**(2006)362
14. I. Abt et al., accepted for publication in Nucl.Instr.Meth. A
15. L. Pandola et al., accepted for publication in NIM A
16. I. Abt et al., to be published

MAJORANA DARK MATTER

ALEXEY ANISIMOV

Institut de theorie des phenomenes physiques,
Ecole polytechnique federale de Lausanne,
Lausanne, 1015, Switzerland
**E-mail: alexey.anisimov@epfl.ch*

In this letter in the framework of a simple see-saw scenario with three quasi degenerate Majorana neutrinos we propose* that one of these neutrinos can be very weakly coupled, yet there is a mechanism of the generation of the abundance of such "dark" neutrino in the early universe. The mechanism of production is due to oscillations between "dark" Majorana neutrino and one of the Majorana's which has relatively large Yukawa couplings ("bright" Majorana neutrino). The transition of "bright" Majorana into a "dark" one is nonadiabatic. We point out on the similarity with the Landau-Zener transition regime. In our model one can explain the observed dark matter density, present matter-antimatter asymmetry and active neutrino data all at the same time.

Keywords: Dark matter; Neutrino oscillations; Landau-Zener transition; See-saw models.

1. Introduction

The observed mater-antimatter asymmetry and the presence of yet unknown form of matter in the universe known as Dark Matter (DM) both require the extension of the Standard Model which does not contain neither a DM candidate nor a mechanism to produce large enough baryon asymmetry in order to match the observed value. On the other hand, the observed neutrino mass scales and mixings can not be incorporated within the Standard Model (SM) as well. The sew-saw mechanism[2] provide probably the most natural and simple explanation why neutrinos are massive, yet so light compared to all other Standard Model particles. At the same time, neutrino mixing data represent a positive test for leptogenesis,[3] an attractive explanation of the the observed matter-antimatter asymmetry of the Universe as

*This letter is based on the original work.[1]

a direct consequence of the see-saw mechanism. Can DM be accommodated within the see-saw model as well? It is known, that it is enough to have just two Majorana's to account for both leptogenesis and active neutrino data (see, e.g.[4]). Assuming that the number of Majorana neutrinos is the same as the number of lepton flavors in Standard Model one can ask the question if one of the Majorana neutrino can be made almost decoupled and, thus, stable at present but being produced efficiently enough in the early universe. However, if a particle is almost stable at present there is typically quite hard to produce it during earlier epoch. One clear possibility is the DM production from the inflaton decay along with the rest of particles. There is no argument why not to assume, for example, that inflaton couplings to the DM particle and to the SM matter are not of the same order. In that case DM particle can be produced as efficiently as ordinary matter. However, such mechanism, due to the present lack of the detailed picture of inflation and most importantly the exit from it, is strongly model dependent. Besides, nothing would require one of the Majorana neutrinos to be a DM particle. Any other particle decoupled from the SM fields would be as plausible. It is appealing, therefore, to look for another more constrained mechanism of the production of the "dark" Majorana neutrino. The perturbative production out of the thermal bath can be a solution. However, typically this would require the reheating temperature after inflation to be quite high.

In this letter we propose that "dark" Majorana neutrino is produced *via* oscillation mechanism which is similar to the Landau-Zener type transition. The production starts and terminates near some critical temperature T_c due to the oscillations between "dark" Majorana neutrino and one of the strongly coupled (or "bright") Majorana neutrino. The critical temperature can be as low as the Majorana neutrino mass scale which, in turn, can be as low as $\mathcal{O}(\text{TeV})$. Contrary to the perturbative production mechanism in our model high reheating temperature is not required. In order for oscillation mechanism to be successful both neutrinos have to be quasi degenerate in mass.[a] We will discuss two scenarios. The one, called "minimal", has only the see-saw type couplings being added to the SM Lagrangian. In this case the production rate turns out to be not efficient enough. In the "next-to-minimal" scenario we will introduce a simple dimension five operator which couples Higgs field to Majorana neutrinos and is suppressed by the high

[a]In the case of quasi degenerate neutrinos the Majorana mass scale can be made quite low as the consequence of resonant enhancement of the kinematic part of the expression for the CP-asymmetry, which compensates for the smallness of the Yukawa couplings.[5]

energy physics scale Λ. Such operator could result from GUT extensions of the SM, supergravity etc. This operator is nearly irrelevant for any physics at present. However, at relatively high temperatures ($T > T_{\rm EW}$) it will turn out to be an effective source for the "dark" Majorana production.[b]

2. The Model

The see-saw extension of the Standard Model is described by the Lagrangian

$$\mathcal{L} = \mathcal{L}_{\rm SM} + i\bar{N}_i\hat{\partial}N_i - F^{ik}\bar{L}_iN_k\Phi^c - \frac{M_i}{2}\bar{N}_i^cN_i + \text{h.c.} \tag{1}$$

For simplicity we will assume that only one of the "bright" Majorana neutrino serves as a source for the production of the "dark" Majorana neutrino. We will employ the density matrix formalism.[10,11] The evolution of the number densities of the Majorana and of the active neutrinos is described by the density matrix which obeys quantum kinetic equation:

$$i\dot{\rho}_{\mathbf{k}} = [H_{\mathbf{k}}, \rho_{\mathbf{k}}] - \frac{i}{2}\{\Gamma^d_{\mathbf{k}} + \Gamma^p_{\mathbf{k}}, \rho\} + i\Gamma^p_{\mathbf{k}}, \tag{2}$$

where $\Gamma^{d,p}_{\mathbf{k}}$ are destruction and production rates correspondingly (see, e.g.[12]). In general case, this is $2 \times (3 + N_M)$ matrix equation, where N_M is the number of the Majorana neutrinos. In practice, one can always neglect the off-diagonal elements ρ_{NL} of this matrix because the latter undergo rapid oscillations. Because the helicity flip and lepton number violation processes are suppressed the opposite helicity parts of this matrix evolve independently. This simplifies the Eq. (2) further. Finally, in our model the oscillations of "bright" neutrino into a "dark" one is effective only during a brief period, so that light lepton part of the density matrix does not have any effect on the final abundance of the "dark" Majorana neutrino. Therefore, it is enough to consider only the part of this matrix which describes two quasi degenerate Majorana neutrinos: one "dark" and one "bright". This is described by 2×2 matrix the $\rho = (\rho)_{AB}$ is (A, B are flavor indices). This equation, in general, has to be solved for each mode $\mathbf{k}$ and the resulting spectral matrix $\rho_{\mathbf{k}}$ then has to be integrated over the phase volume to

[b]A simultaneous picture of baryogenesis and DM has been already proposed.[6,7] Here the matter-antimatter asymmetry is generated through RH neutrino oscillations[8] and the lightest $\mathcal{O}$(KeV) mass RH neutrino is a warm DM. The other two Majorana neutrinos in this scenario are much heavier but still have to be lighter then the electroweak scale as well. Previously, some other models were already proposed where the light sterile neutrino is produced by the mixing with the ordinary light ones and plays the role of Dark Matter.[9] In this letter we explore the case when Majorana neutrinos mass scale is well above the electroweak scale.

obtain the number density matrix ρ:

$$\rho = \int \frac{d^3\mathbf{k}}{(2\pi)^3} \rho_{\mathbf{k}}. \tag{3}$$

The diagonal elements of the matrix ρ give the physical number densities of the two neutrinos after being multiplied by the massless fermion number density (or, equivalently, by the photon number density n_γ).

In our model one of the neutrinos is decoupled and the other is in thermal equilibrium.[c] In that case the Eq. (2) simplifies to

$$i\dot{\rho}_{\mathbf{k}} = [H_{\mathbf{k}}, \rho_{\mathbf{k}}]. \tag{4}$$

Furthermore, instead of solving the Eq. (4) for each mode we will adopt a more simplified approach choosing $k \sim T$ in order to facilitate further discussion. A more quantitative analysis will be given elsewhere.[1] In this letter we aim on a qualitative discussion so that the latter simplification is justified. Because the abundance of the "dark" Majorana neutrino is expected to be much smaller then thermal abundance of the "bright" Majorana neutrino at the moment of production, we can set $\rho_{BB} = \rho_{BB}^{\rm eq}$. In addition we will assume that there is no CP violation in the $A - B$ sector. With all these assumptions the Eq. (4) for a given momentum $k \sim T$ leads to the following set of equations:

$$\begin{aligned}
\dot{\rho}_{AA} &= 2H_{AB}\rho_-, \\
\dot{\rho}_- &= \rho_+(H_{BB} - H_{AA}) - H_{AB}\rho_{BB}^{\rm eq}, \\
\dot{\rho}_+ &= \rho_-(H_{AA} - H_{BB}),
\end{aligned}$$

where we have defined $\rho_+ = (\rho_{AB} + \rho_{BA})/2$, $\rho_- = -i\,(\rho_{BA} - \rho_{AB})/2$. The Hamiltonian can be computed from the real part of the Majorana neutrino two-point Green function on Fig. (1). Below we will consider two cases with two different effective temperature dependent Hamiltonians result from self-energy corrections shown on Fig. (2-3).

3. The Minimal Scenario

The Lagrangian of the model (1) is written in the Majorana neutrino mass eigenstate basis. The Yukawa matrix in the case of two generations is given

[c] We assume that the critical temperature T_c satisfies the condition $M < T_c < T^{\rm eq}$, where M is the Majorana neutrino mass scale and $T^{\rm eq}$ is the temperature at which the second neutrino come to the thermal equilibrium. This condition put some constraints on the model, that will be discussed later.

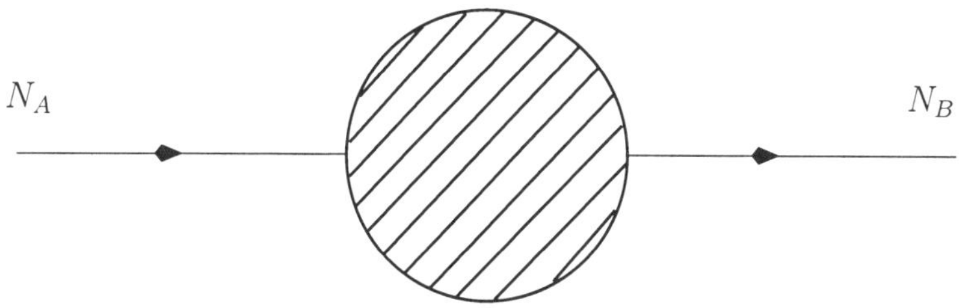

Fig. 1. Majorana neutrino two-point Greens function.

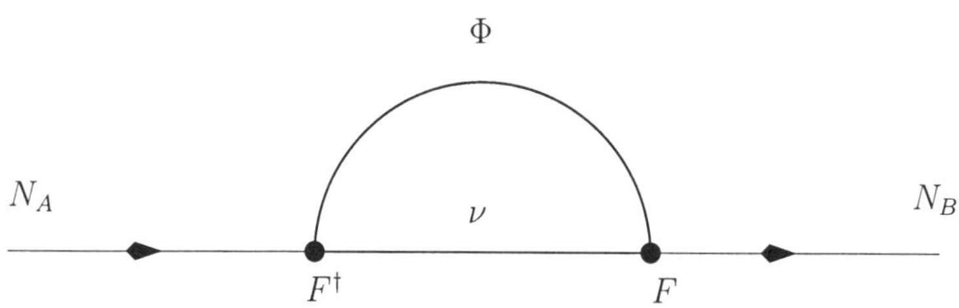

Fig. 2. Correction to the Majorana neutrino self-energy: minimal scenario.

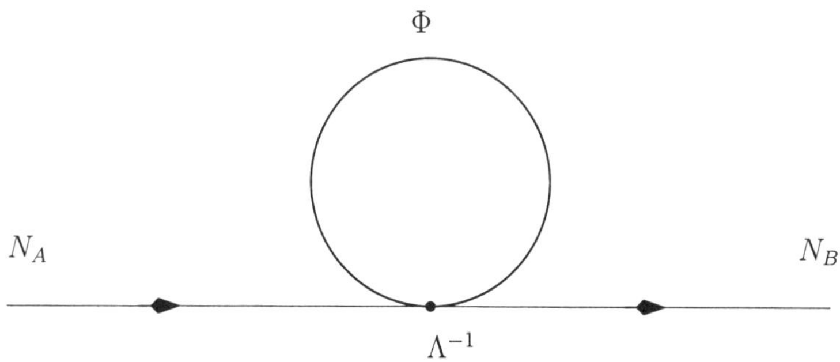

Fig. 3. Correction to the Majorana neutrino self-energy: next-to-minimal scenario.

by[d]

$$(F)_{ij} = \begin{pmatrix} F_{11} & F_{12} \\ F_{21} & F_{22} \end{pmatrix} \approx \begin{pmatrix} 0 & F_{12} \\ F_{21} & F_{22} \end{pmatrix}, \tag{5}$$

where by assumption $F_{11} \ll F_{12,21} \ll F_{22}$, therefore we neglect F_{11} keeping only $F_{12,21}$ elements. The Hamiltonian can be computed following[13] and is given by[e]

$$(H)_{ij} = \sqrt{M_1^2 + k^2}\, I + \frac{T}{8} \begin{pmatrix} K_{11} & K_{12} \\ K_{21} & K_{22} + \frac{8\,\Delta M_{21}^2}{T\sqrt{M_1^2+k^2}}, \end{pmatrix} \tag{6}$$

where I is the identity matrix and the matrix K is defined as

$$K = F^\dagger F. \tag{7}$$

The first term, proportional to the identity matrix, cancels out in the evolution equation so that we will not keep it further. The computation of the "dark" Majorana neutrino production can be made in the mass eigenstate basis by computing ρ_{11} and noting that $\rho_{11} \approx \rho_{AA}$. Instead, we will make a rotation into the flavor basis and compute ρ_{AA} directly. The diagonalization of the matrix F_{ij} can be made with the help of two unitary matrices U (rotation of the Majorana neutrino basis) and V (rotation of the light lepton basis). Introducing two different angles θ_U and θ_V and diagonalizing matrix F_{ij} one obtains

$$(F)_{AB} \approx \begin{pmatrix} -\frac{F_{12}F_{21}}{F_{22}} & 0 \\ 0 & F_{22} \end{pmatrix}, \tag{8}$$

where it was used that both angles $\theta_{U,V}$ are small and given by

$$\theta_{U,V} \approx \frac{F_{21,12}}{F_{22}}. \tag{9}$$

The Hamiltonian is rotated according to

$$H' = U^\dagger H U, \tag{10}$$

and one obtains the Hamiltonian familiar from the active neutrino oscillations and MSW effect:

$$(H')_{AB} \simeq \begin{pmatrix} 0 & \frac{M^2\theta_U\Delta}{E} \\ \frac{M^2\theta_U\Delta}{E} & -\frac{M^2\Delta}{E} + \frac{1}{8}TK_{22} \end{pmatrix}, \tag{11}$$

---

[d] The indices i, j will be used for the mass eigenstate basis, while letter indices will represent flavor basis. The index A is always reserved for the "dark" Majorana neutrino.
[e] Assuming that the two RH neutrinos are strongly degenerate, such that $\Delta \equiv |M_2 - M_1|/M_1 \ll 1$.

where $E = \sqrt{M^2 + k^2}$, $M = M_1 \simeq M_2$. Assuming that $E \approx k \sim T$, $\Delta M_{21}^2 \equiv -2M^2\Delta < 0$, where $\Delta \equiv (M_1 - M_2)/M_2 \ll 1$ one obtains solving the Eqs. (5)

$$\rho_{AA} \approx \frac{m_{pl}}{M} \left(\frac{K_{22}}{8\Delta} \right)^{\frac{1}{2}} K_{11} \leq \frac{m_{pl}}{2M} K_{11}, \tag{12}$$

where $m_{pl} = M_{pl}\sqrt{90/8\pi^3 g_*} \sim 10^{18}$GeV. The last inequality follows from the requirement that the critical temperature $T_c = M(8\Delta/K_{22})^{1/2}$ at which H'_{BB} becomes equal to zero satisfies $T_c > M$.

It is instructive to compare (12) to the value of the Landau-Zener parameter Q that controls the regime in which the transition occurs. Its value is given by (see, e.g.[14])

$$Q = \frac{M^2 \Delta \sin^2 2\theta_U}{E \cos 2\theta_U h_R}, \tag{13}$$

where

$$h_R = \left| \frac{2\dot{T}}{T} \right| = \frac{2T^2}{m_{pl}}. \tag{14}$$

For MSW effect[15] it is essential that $Q \gg 1$ while in the Landau-Zener regime $Q \leq 1$. Taking into account account that $\theta_U \ll 1$ and estimating for $k \sim T$ the value of (13) one obtains at $T = T_c$

$$Q = \frac{m_{pl}}{M} \left(\frac{K_{22}}{8\Delta} \right)^{\frac{1}{2}} K_{11}. \tag{15}$$

One can see that $\rho_{11} \approx Q$, as expected since both represent the probability of the transition in the nonadiabatic regime. The upper bound on the DM energy density can be estimated as follows

$$\frac{\Omega_{DM} h^2}{0.1} \approx 10^{25} K_{11}. \tag{16}$$

Given that the lifetime of the flavor state N_A has to exceed the age of the universe in order to be a DM candidate one can estimate that $K_{11} \leq 10^{-42}(1\ \text{TeV}/M_1)$ and the corresponding density (16) is by far too small.

4. The Next-to-Minimal Scenario

As it was shown in the previous section the minimal model fails to produce the correct amount of the "dark" Majorana neutrino abundance. In this section we propose an extension to the previous setup by introducing (explicitly) no new particles but allowing a dimension five effective operator

$$\mathcal{L}_{\text{eff}} = \frac{(\lambda)_{AB}}{\Lambda} |\Phi|^2 \bar{N}_A^c N_B, \tag{17}$$

where Φ is usual Higgs field, Λ is some high energy scale and the elements of the matrix $(\lambda)_{AB} \sim \mathcal{O}(1)$. The correction of the type (17) is generic if one assumes a certain additional physics at energies much higher then the electroweak scale. For example the operator (17) generically will result from supergravity with Λ being M_{pl} or from tadpole type graphs in GUT theories *via* heavy Higgs exchange. There are other possibilities to generate (17). In what follows we will not address any specific model which provides the relation of the scale Λ to any other scales like M_{pl}, M_{GUT}, M_{SUSY} etc. Instead we will treat Λ as phenomenological parameter and impose the constraints on its value from the requirement that the operator (17) is responsible for the correct DM abundance at present. Additionally, we assume that

$$\frac{F_{12}\Delta M}{F_{22}} \ll \frac{T_{\text{EW}}^2}{\Lambda}. \tag{18}$$

The latter conditions simply implies that the the operator (17) in the Majorana neutrino flavor basis gives a contribution to the off-diagonal entrees of the Hamiltonian which dominate over the term

$$\frac{\Delta M^2 \sin 2\theta}{2E} \tag{19}$$

at all temperature above the electroweak phase transition. In that case we may disregard (19) in the computation below. The validity (18) can be easily inforced by choosing θ_U small enough (or even setting it to zero).

The resummed Majorana propagator will contain the nondiagonal entrees which are given by

$$-\Sigma_{AB} \approx \frac{1}{\Lambda} \int_0^\infty \frac{d^4 q}{(2\pi)^3} \delta(q^2 - m_\Phi^2) n_{\text{b}}(q), \tag{20}$$

where

$$n_{\text{b}}(q) = \frac{1}{e^{\frac{|q\cdot u|}{T}} - 1}, \tag{21}$$

with u being a four velocity of the thermal bath. The resulting expression for the real finite part of Σ_{AB} in Boltzmann approximation is

$$-\text{Re}[\Sigma_{AB} - \Sigma_{AB}(T=0)] \approx \frac{T^2}{2\pi^2 \Lambda} z_\Phi K_1(z_\Phi) u_\mu \gamma^\mu, \tag{22}$$

where $z_\Phi = (m_\Phi/T) \sim \mathcal{O}(1)$, and $K_1(z)$ is a modified Bessel function. Alternatively, we can neglect thermal mass of the Higgs. In that case one obtains

$$-\text{Re}[\Sigma_{AB} - \Sigma_{AB}(T=0)] \approx \frac{T^2}{12\Lambda} u_\mu \gamma^\mu. \tag{23}$$

In what follows the exact coefficient in front of T^2/Λ, which is of order of $\mathcal{O}(10^{-1})$ is not particularly important and we will use (23) for simplicity. Assuming as well that[f]

$$\frac{1}{8} K_{22} \gg \frac{T_c}{\Lambda}. \tag{24}$$

results in the following effective Hamiltonian:

$$H \approx \begin{pmatrix} 0 & -\frac{T^2}{12\Lambda} \\ -\frac{T^2}{12\Lambda} & -\frac{M^2\Delta}{E_{\mathbf{k}}} + \frac{1}{8} T K_{22} \end{pmatrix}. \tag{25}$$

Solving Eq. (4) one obtains that

$$\rho_{AA} \sim \frac{m_{pl} M}{z_c \Lambda^2 K_{22}}, \tag{26}$$

where z_c is given by

$$z_c = \left(\frac{K_{22}}{8\Delta} \right)^{\frac{1}{2}}. \tag{27}$$

The above estimate transfers into the DM abundance which is at present

$$\frac{\Omega_{DM} h^2}{0.1} \approx \frac{1}{z_c} \left(\frac{10^{24}\text{GeV}}{\Lambda} \right)^2 \left(\frac{M}{10^9\text{GeV}} \right). \tag{28}$$

The operator (17) after electroweak symmetry breaking leads to the mixing term $(v_{\text{EW}}^2/\Lambda)\bar{N}_A N_B$ between "dark" and "bright" neutrinos. This results in the decay of the "dark" neutrino *via* oscillations into the "bright" one and its consequent decay. The rate of this process is

$$\Gamma(N_A \to N_B \to ...) = \left(\frac{v_{\text{EW}}^2}{\Lambda} \right)^2 \frac{\Gamma_B}{\left(\frac{\Gamma_B}{2}\right)^2 + M^2\Delta^2}. \tag{29}$$

The decay rate of the "bright" neutrino is dominated by $\Gamma(N_B \to \Phi + \nu)$ and one estimate that

$$\Gamma(N_A \to N_B \to ...) \sim 10^{-35} z_c^4 \left(\frac{10^{24}\text{GeV}}{\Lambda} \right)^2 \left(\frac{10^9\text{GeV}}{M} \right)^2 \text{eV}. \tag{30}$$

[f]This assumption is not crucial but simplifies analytics and turns out to be satisfied for a wide range of parameters after restrictions coming from the "dark" neutrino abundance at present and its decay rate are imposed.

There is phase space of parameters $\{M, \Lambda, z_c\}$ which leads to the correct DM abundance and the lifetime of the "dark" neutrino larger then the age of the Universe. For example, one can choose $z_c = 10^{-2}$, $M = 10^7$ GeV and $\Lambda = 10^{24}$GeV. For this choice of parameters the condition (24) is satisfied as well. The "extreme" case would be to take $z_c = 10^{-3}$ (because, at smaller values it is hard to have the "bright" neutrino being thermally produced), $M = 1$ TeV and $\Lambda \sim 10^{22.5}$GeV. In this case the "dark" neutrino has the mass near electroweak scale and a lifetime close to the present age of the Universe. Of course, a more stringent bounds may come from astrophysical observations and the lower bound on the neutrino mass scale quoted here may go up substantially.

There is an another constraint which comes from the four-body decay $N_A \to 2\Phi + N_B \to 3\Phi + \nu$. The rate of such decay can be estimated as

$$\Gamma_4 \sim 10^{-7} \frac{M^3}{\Lambda^2} K_{22} \sim 10^{-24} \left(\frac{M}{10^9 \text{GeV}} \right)^4 \left(\frac{10^{24} \text{GeV}}{\Lambda} \right)^2 \text{eV}. \quad (31)$$

Requiring that this rate is smaller then the value of the Hubble parameter at present leads to the constraint

$$\left(\frac{M}{10^9 \text{GeV}} \right)^2 \left(\frac{10^{24} \text{GeV}}{\Lambda} \right) \leq 10^{-4.5}. \quad (32)$$

This implies that M can not be higher then about $10^8 - 10^9$GeV depending on the value of z_c.

5. Conclusions

In this letter we proposed the see-saw model in which one of three Majorana neutrinos ("dark" neutrino) is weakly coupled and is a DM candidate. The mechanism of the generation of a suitable abundance was suggested. At least one of the strongly coupled Majorana neutrino ("bright" neutrino) have to be quasi degenerate with the "dark" one. The allowed quasi degenerate Majoranas mass scale M can, in principle, be as low as $\mathcal{O}$(TeV) and up to as high as usually suggested in GUT scenarios, i.e. $M \sim 10^8 - 10^9$GeV. The mass of the third neutrino is adjusted correspondingly: if M is below 10^8GeV the third neutrino has to be also quasi degenerate with the other two in order to have lepton number produced efficiently. Otherwise it remains a relatively free parameter as in the standard hierarchical leptogenesis scenario. This letter is based on the results which will be reported in much more details in.[1]

References

1. A. Anisimov, P. Di Bari, *in preparation*
2. P. Minkowski, Phys. Lett. B **67** (1977) 421; T. Yanagida, in *Workshop on Unified Theories*, KEK report 79-18 (1979) p. 95; M. Gell-Mann, P. Ramond, R. Slansky, in *Supergravity* (North Holland, Amsterdam, 1979) eds. P. van Nieuwenhuizen, D. Freedman, p. 315; S.L. Glashow, in *1979 Cargese Summer Institute on Quarks and Leptons* (Plenum Press, New York, 1980) eds. M. Levy, J.-L. Basdevant, D. Speiser, J. Weyers, R. Gastmans and M. Jacobs, p. 687; R. Barbieri, D. V. Nanopoulos, G. Morchio and F. Strocchi, Phys. Lett. B **90** (1980) 91; R. N. Mohapatra and G. Senjanovic, Phys. Rev. Lett. **44** (1980) 912.
3. M. Fukugita, T. Yanagida, Phys. Lett. **B 174** (1986) 45.
4. P. Frampton, S. Glashow and T. Yanagida, Phys. Lett. **B548** (2002) 119
5. A. Pilaftsis, T. Underwood, Nucl. Phys. **B692** (2004)303; A. Anisimov, A. Broncano and M. Plumacher, Nucl. Phys. **B737** (2006) 176.
6. K. S. Babu, D. Eichler and R. N. Mohapatra, Phys. Lett. B **226** (1989) 347; L. M. Krauss, S. Nasri and M. Trodden, Phys. Rev. D **67** (2003) 085002; K. Cheung and O. Seto, Phys. Rev. D **69** (2004) 113009; T. Asaka, S. Blanchet and M. Shaposhnikov, Phys. Lett. **B631** (2005) 151
7. T. Asaka and M. Shaposhnikov, Phys. Lett. B **620** (2005) 17; N. Sahu and U. A. Yajnik, Phys. Lett. B **635**, 11 (2006)
8. E. K. Akhmedov, V. A. Rubakov and A. Y. Smirnov, Phys. Rev. Lett. **81** (1998) 1359
9. S. Colombi, S. Dodelson and L. Widrow, Astrophys. J. **458** (1996) 1; X. Shi and G. Fuller, Phys. Rev. Lett. **82** (1999) 2832; A. Dolgov and S. Hansen, Astropart. Phys. **16** (2002) 339
10. R. Barbieri and A. Dolgov, Nucl. Phys. **B 349** (1991) 743
11. G. Sigl and G. Raffelt, Nucl. Phys. **B406** (1993) 423
12. H. A. Weldon, Phys. Rev. **D26** (1982) 2789
13. H. A. Weldon, Phys. Rev. **D28** (1983) 2007
14. C. W. Kim and A. Pevzner, "*Neutrinos in Physics and Astrophysics*", Contemporary Concepts in Physics, Volume 8.
15. L. Wolfenstein, Phys. Rev. **D17** (1978) 2369; S. Mikheev and A. Smirnov, Nuovo Cim. **C9** (1986) 17

ICECUBE, THE WORLD'S LARGEST DARK MATTER DETECTOR

HAGAR LANDSMAN FOR THE ICECUBE COLLABORATION

Physics department, University of Wisconsin,
Madison, WI 53703, USA
** E-mail: hagar@icecube.wisc.edu*
http://icecube.wisc.edu

IceCube is a kilometer scale high-energy neutrino observatory, currently under construction at the South Pole. It is a photo-detector, using the deep Antarctic ice as detection medium for the Cherenkov photons induced by relativistic charged particles. These charged particles may be atmospheric muons or reaction products from neutrino interactions in the vicinity of the instrumented volume. The experiment searches for neutrinos originating in astrophysical sources, and can also detect neutrinos from WIMP interaction in the Sun or Earth. In the last two austral summers, 9 in-ice strings and 16 surface IceTop stations (out of up to 80 planned) were successfully deployed, and the detector has been taking data ever since. In this proceedings, IceCube design, present status, performance and dark matter detection sensitivities will be discussed.

Keywords: IceCube; Neutrino; AMANDA; WIMP; Neutralino; Earth; Sun

1. Introduction

High energy neutrino detectors consist of an array of light detectors ("optical modules") spread evenly in deep and clear media, like the Antarctic ice. They are designed to detect Cherenkov photons radiated by charged particles. The particles that are most likely to penetrate through the matter above the fiducial volume of the detector are muons (from interactions of cosmic rays in the atmosphere), and neutrinos (atmospheric or from any other source). Based on measurements of the number of photons arriving at different light detectors and their arrival times, a track or a cascade can be reconstructed.

The IceCube detector is a neutrino detector currently under construction at the South Pole [1]. The main scientific goal of IceCube is to map the neutrino sky, searching for high energy neutrinos that may be created in hadronic processes, for example in Gamma-Ray Bursts or Active Galactic Nuclei. IceCube will also look for high energy GZK neutrinos [2] that are expected to be created by interaction of the cosmic ray protons with the background radiation. It will study air showers, and high energy atmospheric neutrinos [3]. In a special data acquisition mode, IceCube will detect supernovae neutrinos, where the detection probability depends on the distance and neutrino flux from the explosion [4]. To some extent, IceCube measurements can be used for neutrino mass hierarchy and CP phase measurements [5].

Any model predicting a certain neutrino flux, can be tested by IceCube. These sources include, but are not limited to, dark matter, super-symmetry, magnetic monopoles and quantum gravity.

The neutralino is the lightest Weakly Interacting Massive Particle (WIMP) candidate of the Minimal Super-Symmetric extension of the Standard Model (MSSM). It is stable due to R-parity conservation. In direct searches, the WIMPs are detected by the scattered nuclei in the rare elastic scattering of WIMPs on a suitable target [6]. In indirect searches the annihilation products of two neutralinos are detected. If the decay products include neutrinos, neutrino telescopes like IceCube and AMANDA can be used for detection [7].

Relic neutralinos from the Galactic halo can lose energy through scattering with matter and become trapped in the gravitational field of massive objects, such as the Sun and the Earth, where they accumulate over time and annihilate with each other. The decay products may vary, and most of them will interact and decay in the massive body. If neutrinos are created from those secondaries, they will escape and create a neutrino flux.

The number of detected neutrinos, from neutralino interaction, depends on the source characteristic (like capture rate, neutralino annihilation cross section and mass) and on the detector response.

2. WIMPs Searches in AMANDA

AMANDA, the predecessor experiment to IceCube, consists of 677 optical modules deployed on 19 strings at depth between 1.5 km and 2 km. A search for an excess of muon neutrinos from the center of the Earth or the Sun was preformed using the AMANDA detector [8,9].

Muons created by cosmic-ray interactions in the atmosphere, can penterate through the 1-2 km of ice and trigger the detector. Their abundance will overpower smaller fluxes. They are filtered using the matter of Earth, blocking muons from the northen hemisphere, but allowing neutrinos to pass through. Therefore only events transversing through Earth, "upgoing events" are used. For neutrinos with energies above one PeV, the interaction cross section is high enough and Earth starts to become opaque, depending on the angle the neutrino comes in. For WIMP studies, we are intrested in lower energies since it is predicted that the mass of the neutralino will be less than 340 TeV, based on unitarity conditions.

Since the position of the Sun in the South Pole, reaches no more than 23° below the horizon, it is not possible to exploit the full advantage of the upgoing events and the separation between signal and atmospheric muons is harder. In the case of searches for WIMPs from the center of the Earth, the expected signal is veritically upgoing, and the atmospheric muon background is easily removed, with no signal loss.

In both cases, tracks are reconstructed using times of hits in optical modules at known positions in the ice. The reconstructed zenith angle is used to separate background events from signal in the Earth analysis, and the space angle between the Sun and the reconstructed track is used to separate signal from background for the Sun analysis.

The neutralino induced events were simulated using the DARKSUSY [10] program, with neutralino masses in the range of 100 GeV to 5 TeV. Two decay channels were analyzed: A 'soft' channel where the neutralinos annihilate to a pair of b-quarks, and a 'hard' channel where they decay to W-bosons.

The background consists of atmospheric neutrinos that arrive to the detector from all directions (upgoing and downgoing events) and from mis-reconstructed downgoing muon. The background is estimated using off source measurements and simulation. No excess of signal was observed in either the Sun or the Earth analysis.

Based on theoretical models and simulation this result is translated to upper limits on the flux of neutrinos originating from the Sun or from the Earth, and further extrapolated to limits on models. These predictions are model dependent. Not only they depend on SUSY models, that determine branching ratios, but to some extent also on astrophysical assumptions. The threshold energy of the detector plays an important role in the limit extrapolation. As can be seen in figures 1 and 2, once completed, based on the current simulations, IceCube will contribute to the limits of WIMPs from the Sun.

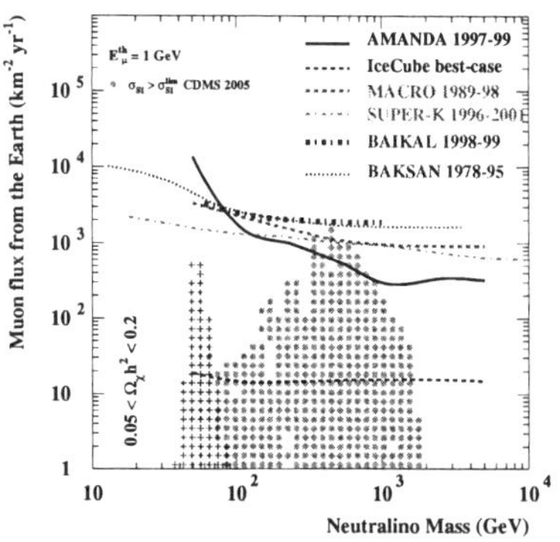

Fig. 1. Upper limits on the muon flux from the Earth, from neutralino annihilation in the 'hard' channel, as a function of the neutralino mass. IceCube best case refers to simulation of 80 strings with 3 years worth of data.

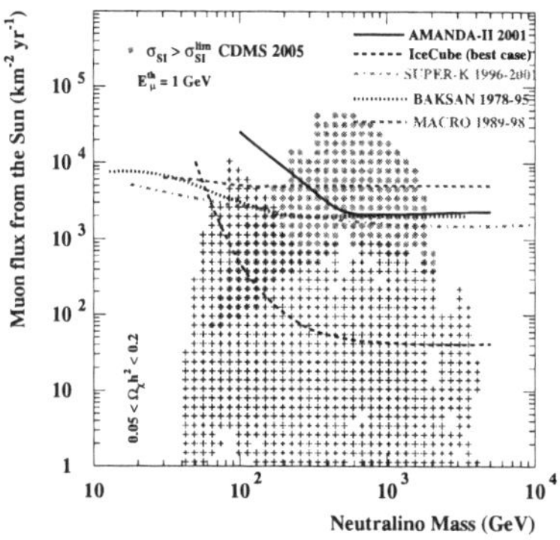

Fig. 2. Upper limits on the muon flux from the Sun, from neutralino annihilation in the 'hard' channel, as a function of the neutralino mass. The crosses show predictions based on DARKSUSY. The dots show models disfavored by direct searches. The limits were rescaled to muon threshold of 1 GeV.

3. Current Status and Verification of IceCube

A larger and improved detector will increase the number of events that can be measured and improve the track reconstruction resolution allowing better separation between signal and background. This will increase the sensitivity for WIMPs, and allow us to measure more neutralinos, in case of detection, or set better limits on models.

The IceCube detector will consist of up to 4800 Digital Optical Modules (DOMs) covering a fiducial volume of 1 cubic km [1]. The DOMs will be equally spaced on up to 80 strings at depths from 1.5 to 2.5 km in the clear Antarctic ice. An array of surface stations, called IceTop, enhances the ability to trigger on, or veto, down-going showers. Each IceTop station consists of two clear ice tanks, each instrumented with 2 DOMs. An IceTop station is located roughly 10 meters from each bore hole of the array.

Each DOM is an autonomous data collecting and digitizing unit consisting of a 10" Hamamatsu PMT in a 13" pressure sphere. A main board inside the DOM can digitize up to 300 Mega Samples per Second (MSPS) for 400 ns and 40 MSPS for $6.4\mu s$. A flasher board, equipped with 12 LEDs, produces pulses used for optical and timing calibration.

In the winter of 2004-2005 a single IceCube In-Ice string and 4 IceTop stations were deployed. At the end of the 2006 Austral summer IceCube consisted of 9 strings and 16 IceTop stations. A set of measurements was performed after the first year to confirm the design goal of the detector and check its performance [11].

3.1. *Timing*

In order to reconstruct tracks over the entire array, a timing resolution of a few nanoseconds is needed. The time calibration of the entire array is done by sending a fast bipolar pulse at known intervals from the surface to every DOM. Each DOM then sends an identical fast bipolar pulse back to the surface after a known delay. Based on the time measured on the surface and on the DOM, and based on the shape of the measured pulse, the round trip time of the signal is calculated. Figure 3 shows the round trip times of the calibration pulse. The time increases with the depth of a DOM, and is constant for the IceTop array. The variation of the round trip time from one calibration to the next provides a measurement of the time calibration accuracy (figure 4).

The time resolution of each DOM was verified in two independent ways. First, a LED on a DOM was flashed and an adjacent DOMs triggered on it. The time delay between the flashing and the triggering was measured multiple times. This procedure was repeated for all DOMs, and the maximum time RMS resolution was found to be less than 2ns. A different way to estimate the time resolution is by reconstructing down-going muon tracks excluding one DOM, and calculating the time difference between the measured hit time and the expected hit time. Figure 5 shows the distribution of those time residuals for a single DOM using multiple events. The process is repeated for all DOMs. The resolution was found to be less than 3 ns, after accounting for ice properties.

3.2. *Muon track reconstruction*

In order to point back to the origin of the neutrinos, a muon track is reconstructed and the zenith and azimuth angles are calculated. Reconstruction of muon tracks, originating from downward going cosmic rays, or from neutrino interactions in the detector volume, is done by analyzing the waveform measured on each DOM.

A comparison of the measured atmospheric neutrinos and downgoing muons flux to the predicted and known fluxes is an important consistency check on the detector response and simulation. The quality of the reconstruction depends on the number of DOMs hit (determined by the geome-

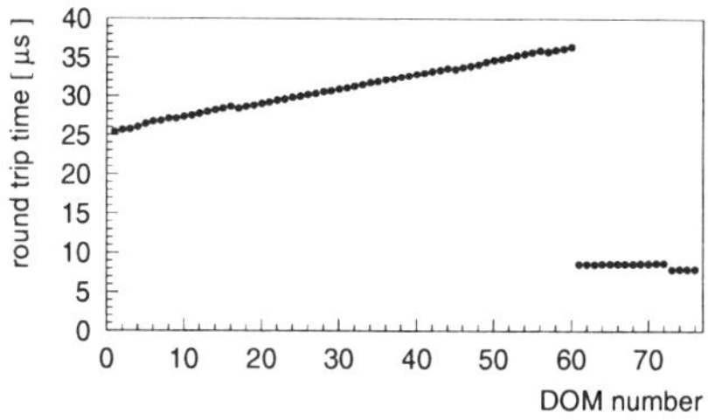

Fig. 3. Round trip time of the time calibration pulse. DOM number 1 is at the top of the string, and DOM 60 is at the bottom. DOMs 60-76 are part of the surface array.

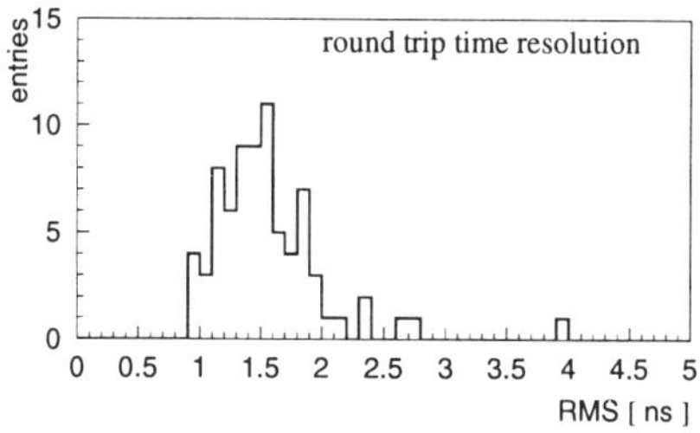

Fig. 4. The rms variation of the round trip time of the time calibration pulse for 76 DOMs on IceCube's first string and 4 IceTop tanks.

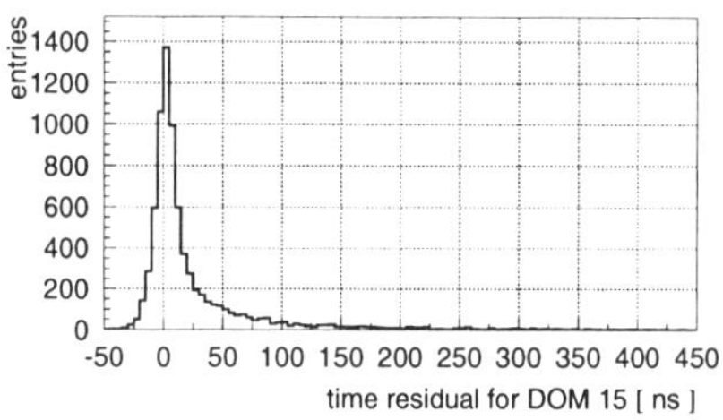

Fig. 5. Distribution of time residuals of photons arriving at a DOM from nearby tracks reconstructed with the rest of the string.

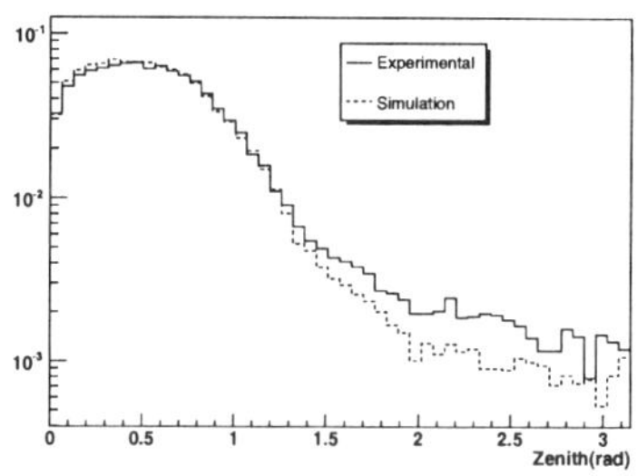

Fig. 6. Distribution of reconstructed and measured zenith angle for downgoing events.

try and the energy of the event), the ice quality (impurities lead to photon scattering and absorption) and timing resolution. Figure 6 compares the measured downgoing events reconstructed zenith angle to the predicted one using simulation of downgoing muons. Figure 7 shows the predicted angular resolution of the IceCube detector with 9 and 80 strings.

3.3. *Summary*

A neutrino flux from the center of the Earth or the Sun, may indicate neutralino capture inside those heavy bodies. No excess of events from the Sun or the Earth, above the estimated atmospheric neutrino background, was measured by the AMANDA detector. Limits on the fluxes have been set for different neutralino masses.

The IceCube detector, under construction in the South Pole will be a powerful tool for high energy neutrino searches. Among other results, it

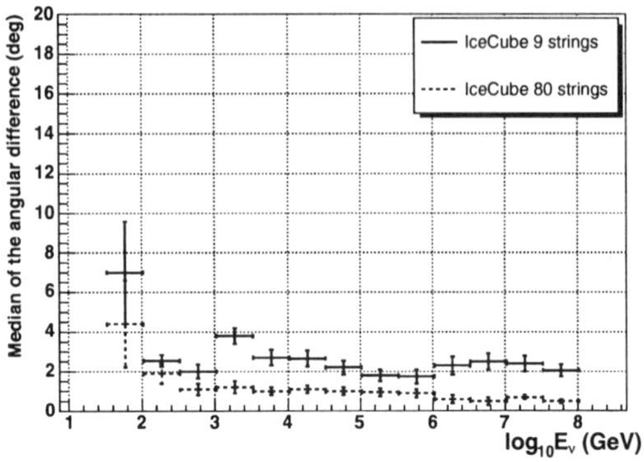

Fig. 7. The angular resolution for 9 strings and for the full IceCube array as function of event energy (preliminary).

will also improve the limits on neutrinos from neutralino annihilation, in the case of no signal. The higher threshold energy will decrease its sensitivity to lower mass neutralinos, but its larger size and better track reconstruction abilities, together with data collected on AMANDA, will improve the limits, especially for higher mass. Until December 2007, IceCube consisted of 9 strings and 16 surface stations. The detector will rapidly grow as more strings and stations will be deployed. Measurements done on IceCube show good timing and reconstruction ability throughout the detector.

References

1. IceCube Project Preliminary Design Document, Ahrens et al. (IceCube Collaboration), http://icecube.wisc.edu.
2. R.Engel, D.Seckel and T.Stanev, Phys.Rev. D64 093010 (2001)
3. M.C. Gonzalez-Garcia, F.Halzen and M.Maltoni, Phys.Rev. D71, 092010 (2005)
4. J.Ahrens et al. (AMANDA Collaboration), Astropart. Phys. 16 , 345 (2002)
5. W. Winter, Phys.Rev. D74, 033015 (2006)
6. CDMS Collaboration, Phys.Rev.Lett. 93 211301 (2004)
7. F. Halzen and D.Hooper, Phys. Rev. D73 1233507 (2006)
8. A. Achterberg et al. (IceCube collaboration), Astropart. phys. 24 (2006) 459-466
9. A. Achterberg et al. (IceCube collaboration), Astropart. Phys. 26 (2006) 129-139
10. P.Gondolo et al., Journ. of Cosm. & Astropart. Phys. 0407, 008, 2004
11. A. Achterberg et al. (IceCube collaboration), Astropar. Phys. 26, (2006) 155

THE ANTARES DETECTOR

A. MARGIOTTA ON BEHALF OF THE ANTARES COLLABORATION*

Dipartimento di Fisica, Università di Bologna and Sezione INFN di Bologna viale C. Berti-Pichat, 6/2
Bologna, 40127, Italy
e-mail: margiotta@bo.infn.it

The ANTARES Collaboration is building an underwater neutrino telescope in the Mediterranean Sea, at a depth of 2500 m, 40 km offshore from Toulon (France). The main goal of the detector is the search for high energy neutrinos from possible Galactic and extra-Galactic sources as well as from the decay of neutralinos and other exotic particles. The present status of the detector, with the first two lines in acquisition since Mar and Sep 2006, is presented.

1. Introduction

A new window in the exploration of the Universe could be opened by the detection of high energy astrophysical neutrinos. Several observations support the hypothesis that the core of astrophysical sources of high energy cosmic rays, such as Active Galactic Nuclei (AGNs), Gamma-Ray Bursters (GRB), SuperNova Remnants (SNR) or micro-quasar binary systems, host powerful cosmic accelerators where hadronic mechanisms are active. High energy neutrinos could be produced in the interaction of accelerated protons with the surrounding matter. Unlike charged cosmic rays and high energy photons, neutrinos carry information directly from their production sites, being essentially unaffected by Galactic/Intergalactic matter, photons or magnetic fields.

In the last decades several detectors have been producing physics results (Baikal [1] and AMANDA [2]), others are fully funded and taking data in their initial configuration (ANTARES [3,4,5] and IceCube [6]), others are under test and R&D phase, such as NEMO [7] and NESTOR [8].

The ANTARES Collaboration, formed by about 200 physicists, engineers and technicians from several European countries, is building a medium sized neutrino telescope at about 40 km off the Southern French coast, near Toulon, at 2500 m below the sea surface.

---

*For a complete list of ANTARES Collaborators see http://antares.in2p3.fr/

ANTARES is a multidisciplinary experiment, whose main goal is the detection of high energy neutrinos from Galactic and extra-Galactic sources. Moreover it will search for possible signals due to the annihilation of neutralinos, gravitationally trapped inside the core of massive objects like the Earth, the Sun or the Galactic centre. In parallel a number of researches in the fields of marine biology, oceanography and seismology are being carried out. It is also meant as a first step towards a bigger detector in the Mediterranean Sea. The ANTARES Collaboration is involved in the KM3NeT consortium, funded by the European Community FP6 and by participating institutions over the period 2006-2009, with the objective of producing a design study for a Mediterranean km^3 neutrino detector [9].

2. Detector design

The small neutrino cross section and the low expected fluxes require enormous volume detectors. Deep sea water or polar ice provide very large amount of Cherenkov radiators at a reasonable cost and represent themselves an efficient shield from the background due to atmospheric muons. The detection principle is shown in fig. 1. Upgoing neutrinos interact in or in the proximity of the sensitive volume, in the Earth's crust or in the sea water and produce relativistic muons emitting Cherenkov photons. A three-dimensional array of photomultipliers (PMT) registers their position and arrival time, thus allowing a precise reconstruction of the muon direction.

The final detector will consist of 12 lines, each holding 75 10" Hamamatsu PMTs [3], arranged in triplets (storeys), oriented at 45 degrees to the downward vertical. Each PMT is hosted in a pressure resistant glass sphere. The front-end electronics of ANTARES optical modules (OM) includes the Analog Ring Sampler (ARS) ASIC, developed to process PMT signals and to measure the arrival time and charge for single photoelectrons and the full waveform, at a sampling rate of 1 GHz, for signals above a preset amplitude. The storeys are separated by 14.5 m, starting at 100 m from the seabed, while the distance between pairs of lines is about 60-70 m. Each PMT triplet is held by a titanium frame attached to a vertical electro-optical cable used both for data and clock signals and for power transmission. A titanium cylinder encloses the readout/control electronics and compasses/tilt meter boards. A system of LED and laser beacons, distributed in some of the storeys and at the bottom of the line, provides a tool for timing calibrations. An acoustic system is used to perform a precise positioning (~10-20cm). Each line is connected, via an electronics module located at its bottom (the String Control Module, SCM) to a Junction Box (JB) linked to the shore station through an electro-optical cable.

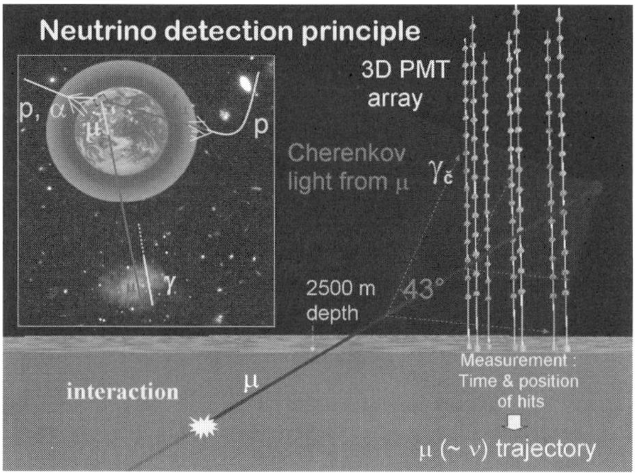

Figure 1. Neutrino detection principle for an underwater telescope.

Figure 2. Artist's impression of the ANTARES neutrino telescope. The different items are not on scale (© F. Montanet, CNRS/IN2P3 and UFJ).

All data reaching the shore are sent to a computer farm and processed to search for events satisfying predetermined trigger requirements [4]. Timing is distributed via the cables to the electronics modules using a 20 MHz high accuracy on-shore clock synchronised with GPS. In fig. 2 a pictorial view of the apparatus is shown. An additional line, the Instrumented Line (IL), will host a number of instruments monitoring continuously environmental parameters like temperature, pressure, salinity, light attenuation length and sound velocity.

3. Expected performances

An intensive study to simulate the detector response to neutrino fluxes and to reconstruct events has been performed. In fig. 3 (left) the effective area for neutrinos is shown. It is larger than 10 m^2 above 10^6 GeV and presents a strong decrease for almost vertical high energy neutrinos.

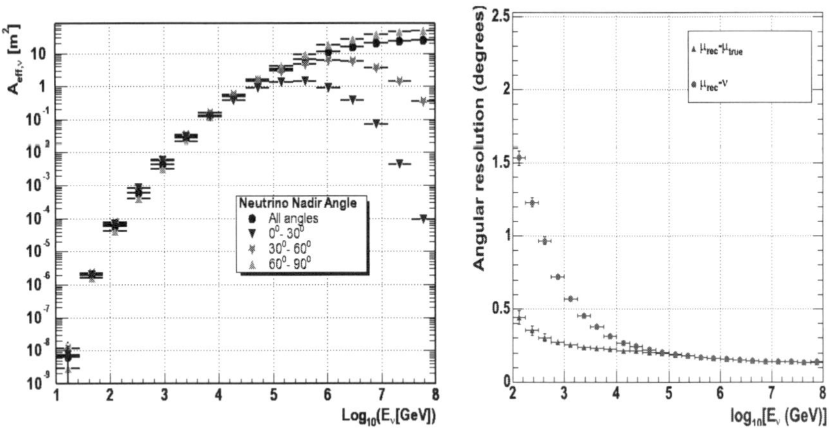

Figure 3. (left) neutrino effective area as a function of energy; (right) neutrino (dots) and muon (triangle) angular resolutions (defined as the median of the distributions of simulated-reconstructed muon angle and of the simulated neutrino - reconstructed muon angle) vs. neutrino energy.

The neutrino angular resolution, fig. 3 (right), is dominated by electronics, transit time spread of PMTs and light scattering, at high energies, where reaches a value of ~ 0.2°÷0.3°, while at lower energies the kinematics of neutrino interactions dominates. Though the main goal of the ANTARES experiment is the detection of high energy astrophysical neutrinos, it can give a contribution to indirect searches for Dark Matter (DM) [10]. Weakly Interacting Massive Particles (WIMPs) are considered among the most promising DM candidates. In a supersymmetric framework, assuming R-parity conservation, a theoretically-favored candidate is the lightest supersymmetric particle, which, in many regions of the SUSY parameter space, is the lightest neutralino, χ. Neutralinos are expected to scatter elastically inside celestial bodies, becoming gravitationally trapped in their core, where they would annihilate producing standard model particles among which only neutrinos could reach us. Though direct production is suppressed, neutrinos can be produced in the decay of the primary products, carrying ~1/10 ÷1/2 of the neutralino energy.

Detection of neutralinos from the Sun, which represents the most promising source, is favoured by ANTARES location at intermediate latitudes. The detection of a neutrino flux from the Earth core is more difficult because of the lower Earth mass and of the depletion effect of the gravitational field of the Sun. The possibility of neutralinos detection from the Galactic Centre is uncertain and is still under study.

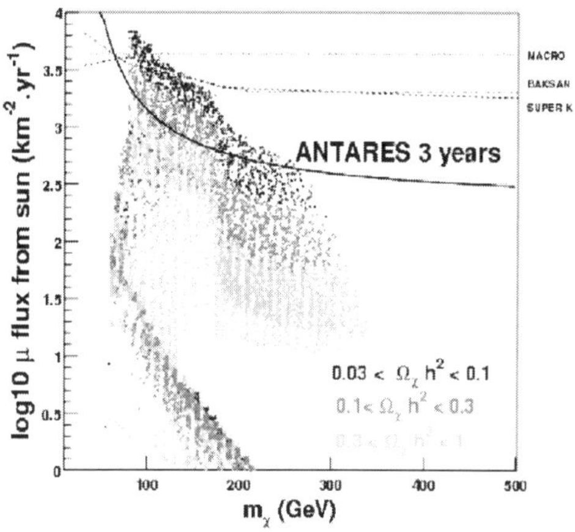

Figure 4. Indirect detection experiment sensitivities in the (μ flux from the Sun, m_χ) plane with E_μ = 5 GeV threshold. Dotted, dash-dotted and dashed curves are respectively the MACRO, Baksan and Super-Kamiokande upper limits. Solid line indicates the expected ANTARES sensitivity (for colour code of models see [10].

4. ANTARES construction milestones and present status

Since 1996 the ANTARES Collaboration started an intense R&D and site evaluation activity that included the deployment of the 40 km electro-optical cable, of the JB and of several prototype lines used to test specific sub-systems of the detector. In 2005 the so-called MILOM [5] (a line with 4 PMTs and several calibration and monitoring instruments) was installed and connected to the JB and is still running and taking data. At the beginning of 2006, the first complete line of the ANTARES detector was deployed and connected to the JB, using a Remotely Operated Vehicle (ROV). The second line was deployed in July and connected at the end of September. The 3rd line was deployed on the

29th of November and will be connected in December 2006; the full detector will be operative by the end of 2007.

5. First results

After the connection of the first line, the time resolution of its PMTs was measured using the optical beacon located on the first floor of the MILOM. Results are in figure 5: the left plots show the time differences between hits in a pair of OMs in storeys at different distances from the light source. The top distribution refers to OMs at 150 m from emitter and the effect of the light scattering is visible. The bottom plot corresponds to a 70 m distant storey. The distribution is Gaussian and has a width of 0.7 ns.

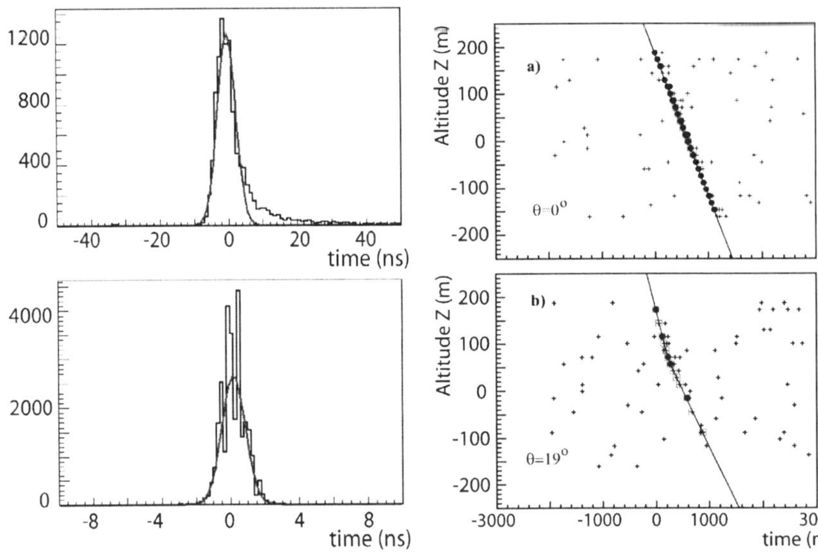

Figure 5. (left) time resolution of ANTARES PMTs; (right) examples of reconstructed tracks.

At 2000-2500 m under the sea surface the atmospheric muon flux is strongly reduced, nevertheless they represent the most abundant signal in the ANTARES telescope. Immediately after the line 1 connection the first muons were registered by the PMTs and data processed to search for events satisfying trigger conditions. The trigger condition currently applied requires at least 5 space-time correlated hits within 20 ns on 1 storey or one hit with an amplitude > 2.5 pe. This helps to reject the hits due to the optical background. The background rate in an underwater neutrino telescope consists of different components: a constant rate due to the β-decay of ^{40}K present in sea water and a slow varying (time scale of some hours) rate due to biological activity of microrganisms whose global effect results in an almost constant counting rate showing a significant seasonal dependence. At present, it is in the range 70 – 100 kHz. Furthermore, bursts of bioluminescence have been measured with a rise time of few milliseconds and a few seconds duration. These bursts may induce a counting rate of some MHz in a PMT.

Though the track reconstruction and angular resolution suffers from the geometrical ambiguity derived by the use of a single line, a simplified reconstruction algorithm is used based on a least-square minimization of hit time and altitude. In figure 5 (right) the reconstructions of two events are shown. The display shows heights (Z) along the line and the measured times wrt an arbitrary zero. Crosses correspond to all the hits within a 1 μs time window, black dots to hits in coincidence while those used in the event fit are surrounded by a red square.

Conclusions

During 2006 the ANTARES Collaboration reported a great success with the deployment and connection of the first 3 complete lines of the detector (the third one will be connected in December 2006). By the end of 2006 the deployment of a fourth line is foreseen. First measurements confirm the expected performances in terms of time resolution. By the end of 2007 the full detector will be operative.

References

1. V. Bolkanov *et al., Nucl. Phys. Proc. Suppl.* **B438,** 91 (2001).
2. E. Andres *et al., Astrop. Phys.* **13**, 1 (2000).
3. J. A. Aguilar *et al., Nucl. Instr. and Methods* **A555** 132 (2005).
4. J. A. Aguilar *et al., Astropart. Phys.* accepted, astro-ph/0610029.
5. J. A. Aguilar *et al., Astropart. Phys.* **26,** 314 (2006).
6. T. Montaruli, astro-ph/0608140.
7. A.Margiotta, *Phys. Scripta,* **127,** 107(2006).
8. G.Aggouras *et al., Nucl. Phys. Proc. Suppl.* **151,** 279(2006).
9. http://www.km3net.org/
10. J. Carr *et al., Rept.Prog.Phys.* **69** 2475 (2006).

HELYCON: TOWARDS A SEA-TOP INFRASTRUCTURE*

SPYROS EUST. TZAMARIAS

School of Science and Technology, Hellenic Open University, Tsamadou 13-15
Patras, 26223, Greece

The "HEllenic LYceum Cosmic Observatories Network" - HELYCON collaboration is constructing a network of detector stations distributed over three extended geographical areas. The goal of HELYCON is to observe Extensive Air Showers and to collect data corresponding to the flux, the direction and possible correlations between very energetic cosmic rays. In this report the design, the construction and the performance of a prototype detector array is presented. The results of a feasibility study on the use of HELYCON detectors for the calibration of the Mediterranean neutrino telescope, KM3NeT, are also presented.

1. Introduction

The HELYCON: HEllenic LYceum Cosmic Observatories Network, collaboration is constructing an Extensive Air Shower (EAS) detector array. HELYCON will consist of several detector stations distributed over extended geographical areas in Western Greece, in the island of Chios and in the area of Nicosia, Cyprus. Each station is equipped with three or four large plastic scintillation counters, each having an effective area of 1 m^2. It also includes a Global Positioning System (GPS) device, trigger and digitization electronics, slow control electronics and a computer based data acquisition system. The stations will be mounted on the roofs of high-school and university buildings whilst the time synchronization between the detector stations will rely on the timing signals from the GPS navigation satellites.

The main goal of the project is to contribute to the research in the field of astroparticle physics. A particularly interesting prospect is the study of time and directional correlations between cosmic ray activities on station clusters

*This work was support in part by the European Social Fund and Greek National Resources EPEAEK II (PYTHAGORAS).

separated by large distances[a]. Such correlations would indicate that the cosmic ray particles have some common history. The creation mechanism of cosmic ray events arriving in time coincidence could be, for example, the photodisintegration of heavy nuclei by solar photons [1] or transient cosmic gamma-ray activity [2].

In parallel, HELYCON provides the workbench for the development of a calibration system for a deep sea neutrino telescope, as the Mediterranean KM3NeT [3]. A floating array of HELYCON detectors can be used as a sea-top calibration infrastructure, on top of the KM3NeT neutrino telescope. Such an array can detect the copiously produced, low energy and small zenith angle, atmospheric showers and the collected data can be used for the reconstruction of the direction and of the impact parameter of the shower axis. According to Monte Carlo studies, presented in this paper, 35% of the cosmic showers in the energy range of 10^{14}-$5{\cdot}10^{15}$eV contain energetic muons able to penetrate the 4000m deep sea water and reach the KM3NeT detector. These muons are detectable by the deep sea telescope and the muon track parameters can be estimated with high accuracy. The comparison of the reconstructed muon track parameters with the direction and the impact of the shower axis can be used in order to: a) reveal systematic angular errors in the determination of the track parameters by the neutrino telescope and b) provide an absolute positioning of the undersea detectors.

The design, the construction and the operating characteristics of a prototype HELYCON array are presented in Section 2 of this paper. Section 3 describes the performance of prototype HELYCON stations in detecting and reconstructing the direction of cosmic air showers. The results of a feasibility study on the use of floating arrays of HELYCON detectors to calibrate the deep sea Mediterranean neutrino telescope - KM3NeT, are contained in Section 4, whilst the conclusions are in Section 5.

2. The HELYCON Detector

2.1. *Description*

Each HELYCON station consists of four (or three) charged particle detectors, arranged on a rectangular (or triangular) formation, of a side of about 20 meters. Each station includes also a GPS antenna, digitization and control

---

[a]The distance between the HELYCON stations in the same site will vary from 500m to 30 km. However there are 390 km between the HELYCON sites of Patras and Chios whilst the Patras - Nicosia distance is about 1100 km.

electronics, as well as a data acquisition system controlled by a personal computer. The collected data are broadcasted through the internet to a main counting room. A single station is able to detect atmospheric showers initiated by cosmic particles of energy more than 10^{14} – 10^{15} eV. However, in order to monitor the low energy cosmic ray flux, in some of the 4-detector stations, two of the detectors are placed in coincidence, either on top of each other or side by side.

The HELYCON charged particle detector is a scintillation counter of 1 m^2 active area. It is made of two layers of plastic scintillator tiles (Protvino SC-301[4]), adjusted on a wooden frame. Each layer consists of 80 tiles (tile dimensions: 12x10x0.5 cm3) wrapped in reflective paper (Dupont Tyvek 2460B[5]). The light is collected by wave shifting fibers (Bicron BCF91A[6]) embedded inside the grooves of the scintillating tiles and it is detected by a fast 19mm Photonis XP1912 [7] photomultiplier tube (PMT).

A programmable high voltage dc-dc converter (EMCO CA20N[8]), mounted on the wooden frame of the detector, supplies high voltage (HV) to the photomultiplier tube whilst a single, external power supply provides power (12V dc) to all the detectors of the station. The scintillator, the PMT and the dc-dc converter are shielded against the ambient electromagnetic noise by means of a thick (0.3 mm) aluminium foil, wrapped around the wooden frame. Finally, the whole construction is enclosed inside a water proof wooden box wrapped with Tyvek. Photos during successive construction stages can be found in [9].

The electrical pulses of each PMT are curried to the digitization electronics by a long (~30m) RG58 cable. The control and monitor signals are transmitted through standard UTP cables: these include the dc-dc converter controls and monitors as well as the readings of temperature sensors mounted on the wooden frame near the PMTs. The ReadOut system is based on a High Precision Time to Digital Converter (HPTDC) [10] chip, designed at CERN, whilst a 12-bit multifunction USB data acquisition card (National Instruments USB-6008 [11]) is used for the slow controls. The synchronization between the HELYCON stations relies on the GPS (Motorola M12+ [12]) time-signal.

The HELYCON ReadOut [13] card offers up to 5 analogue inputs, each one for a scintillation detector. The input signals are amplified and compared to six predefined (remotely adjustable) thresholds. The corresponding times of the PMT waveform-threshold crossings are digitized with an accuracy of 100ps by the HPTDC, along with the 1 PPS output of the GPS receiver. A first level trigger is created as the fast hardware coincidence between two of the PMT inputs. Second level triggers are realized in the Field Programmable Gate

Array (FPGA) of the ReadOut card, requiring a certain time relation between the input signals. Both the first and the second level triggers are provided as NIM and as TTL card outputs. The FPGA is also responsible for formatting the data and for communicating with the station (local) PC through the USB or/and the Ethernet ports of the card. The data are saved on the hard disk of the local computer and transmitted on request, via the Internet, to the central server of the HELYCON telescope.

2.2. *Calibration and Test Results*

The charged particle detectors, before their commission, undergo several evaluation tests and calibration procedures, including the calibration of the photomultiplier tubes, measurements of the response of each detector to a minimum ionizing particle (MIP), evaluation of the uniformity of each detector response with respect to the MIP incident point, as well as the synchronization of all detectors belonging to the same HELYCON station.

The photomultipliers, before integration to the detector, have been illuminated by LED light pulses (at 420nm wave length) in order to measure the gain variation as a function of the high voltage, as well as to evaluate the PMT response characteristics at a single photoelectron emission. At the operating voltage, the typical PMT response to one photoelectron emitted from the photocathode corresponds to a pulse amplitude of 1.0 mV and to a total charge of 0.06 pCb.

The response of each detector to a MIP and the uniformity of the detector's response have been studied by using a small muon telescope, in order to identify relativistic muons. This is a hodoscope consisting of two plastic scintillators, of 5cm × 10cm effective area and 1cm thickness each and a lead absorber, 10cm thick, between the scintillators, as shown in Figure 1. It was found that the typical response of a HELYCON detector to a MIP corresponds to 21 photoelectrons (or 1.2 pCb) with a variation of less than 10%, depending on the position of the muon hit. The uniformity of the response of a typical detector to a MIP is demonstrated in Figure 2, where the response to a MIP is shown as a function of the coordinates of the muon hit. However, due to the finite time duration of the scintillation (decay constant 2.5 ns) and the wave shifting (decay constant 12 ns) processes, the mean PMT pulse height corresponding to a MIP is only 5 mV.

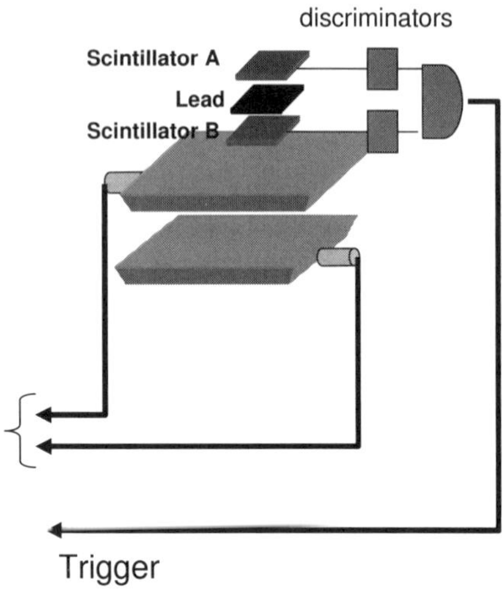

Figure 1. The small telescope is used to identify muons and to locate the muon hit on the detector surface.

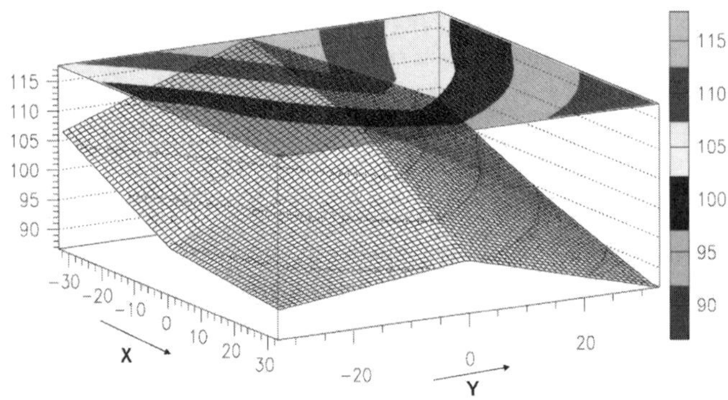

Figure 2. The variation of a typical HELYCON detector's response to a MIP as a function of the muon hit coordinates with respect to the centre of the counter. The response refers to the charge of the detector signal, in arbitrary unites, whilst the distance is measured in cm.

The experimental set up shown in Figure 1, was also used to study the resolution of the HELYCON detectors in recording the arrival time[b] of single muons. Using the small muon telescope to identify relativistic muons and to restrict the muon hit position, the time difference between the electrical pulses

[b]The arrival time is defined as the first time that the PMT waveform crosses a certain voltage threshold, typically 1.5 mV.

of two HELYCON detectors was measured. Distributions of this time difference are shown in Figure 3 for nine different positions of the hodoscope. The mean value of this distribution varies as a function of the hodoscope position due to the fact that, in this experimental set up the light paths towards the two PMTs are complimentary. It was found that this variation is consistent with the corresponding difference in the light path lengths towards the PMTs and the refractive index (n=1.6) of the wave length shifting fibers. Consequently, when the muon hit coordinates are unknown, there is a systematic uncertainty in the estimation of a single muon arrival time of the order of 1.5ns for 1m long detectors.

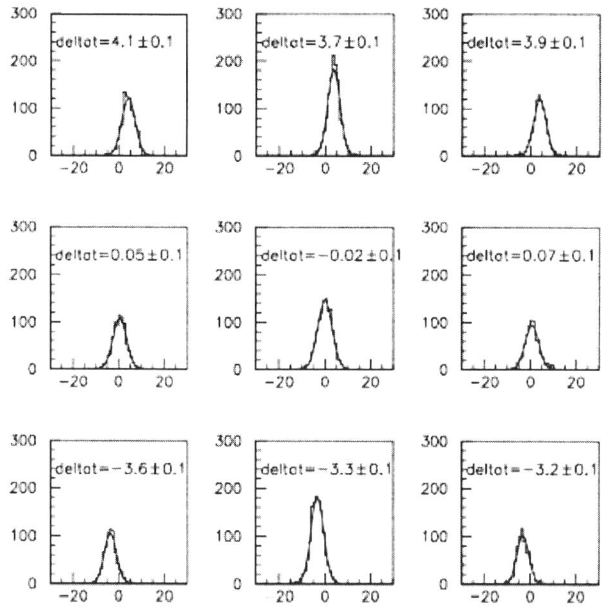

Figure 3. Distributions of the time difference between the two HELYCON detectors in recording the arrival time of the same muon, for several hodoscope positions. The solid lines correspond to Gaussian fits, whilst the parameter "deltat" refers to the deviation of the mean value from zero.

However, the variance of this time difference was found to remain practically constant, independent of the hit position and the light path length. The histogram in Figure 4 represents the time difference distribution when the hodoscope is at the center of the detectors, whilst the solid line represents a Gaussian fit to the experimental distribution, resulting to a mean value of 0 ns and a sigma of 2.8 ns. Assuming that the two detectors have identical local time resolutions, the local time resolution of each of the detectors to a MIP is estimated to be $2.8\text{ns}\cdot(2)^{-1/2}=2$ ns.

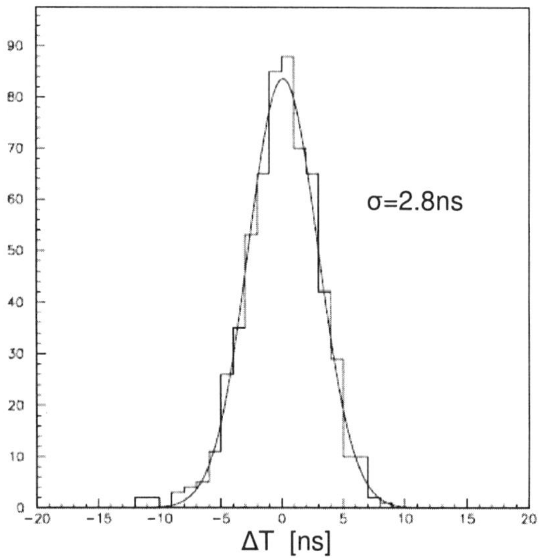

Figure 4. Distribution of the time difference between the two HELYCON detectors in recording the arrival time of the same muon, when the muon telescope is located at the centre of the detectors.

The detector signals produced by EAS are much larger than the pulses corresponding to a single MIP. The systematic and the statistical errors in the estimation of the arrival time of these signals were studied by using the experimental set-up shown in Figure 5. The two outer detectors operated in coincidence, in order to establish the event selection trigger, whilst the waveforms of the inner two detectors were digitized. By studying the distribution of the arrival time[c] difference for the two inner detectors, the global time resolution and the slewing (systematic time offset) were determined as a function of the charge of the PMT signal. These dependences are shown in Figures 5(b) and 5(c). For pulses larger than the signal corresponding to 4 MIPs, the slewing effect is insignificant, whilst the resolution in estimating the arrival time is typically of the order of 1.5 ns, depending on the particular detector.

[c]The arrival time of each detector signal was defined as the first time when the digitized waveform crosses the threshold of 3 mV.

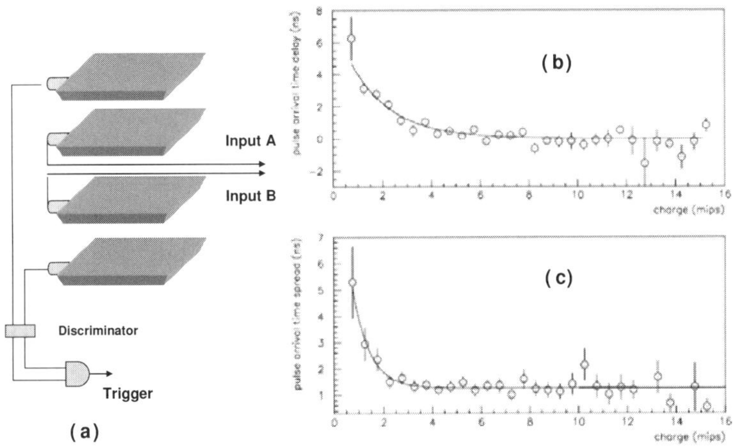

Figure 5. The experimental set-up (a) and the measurements of the corrections needed due to the slewing effect (b) and the time resolution of a typical HELYCON detector (c), as functions of signals charge.

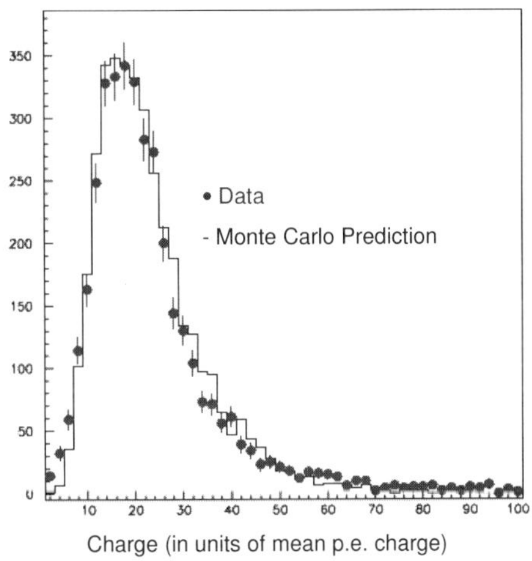

Figure 6. The distribution of the charge of PMT signals produced by muons. The points with error bars represent the experimental data whilst the histogram corresponds to the simulation prediction.

The calibration results presented above have been compared with the predictions of a detailed Monte Carlo (MC) description of the HELYCON detectors, which was developed by utilizing the GEANT4 simulation framework [14]. As an example, in Figure 6 the measured response of a HELYCON detector to a MIP is presented, in comparison to the Monte Carlo prediction.

The very good agreement between the MC description and the experimental data verifies that the detector operation characteristics are well understood.

3. Observations of Extensive Air Showers

The performance of HELYCON in detecting and reconstructing showers has been studied by operating a system of eight HELYCON detectors in the laboratory. The detectors were arranged in the formation shown in Figure 7, under a light wooden roof with brick tiles. The detectors T_1 and T_2 were used in coincidence, within a time window of 40 ns, in order to define an event selection trigger. The trigger was formed by requiring the signals to exceed in pulse height a certain threshold, equivalent to the mean pulse amplitude of 1.5 MIPs for every one of the trigger detectors. A NIM pulse, formed as a result of the above coincidence, was used to trigger synchronously two fast oscilloscopes (Tektronix TDS3052B and DPO4054 [15]) which sample and digitize the waveforms of the other six detectors. The active time window was defined as an interval of 150 ns around the trigger arrival time. The digitized waveforms were considered as an event and were saved on hard disk for further offline analysis.

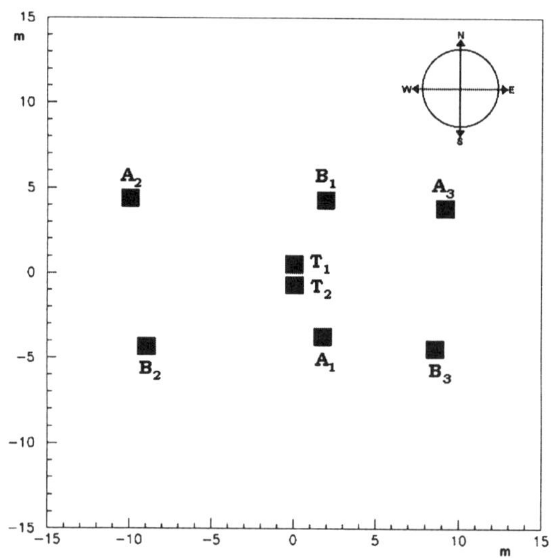

Figure 7. The formation of 8 HELYCON detectors in the laboratory, in order to cross check the simulation performance and to measure the angular resolution of a single HELYCON station.

The collected experimental data and the results from the offline analysis were compared to simulation predictions. In particular, the CORSIKA[d] air shower simulation software [16] was used to produce air showers, initiated by cosmic protons entering isotropically the upper atmosphere, whilst the detector response was simulated by the specific HELYCON MC package. The events produced by the simulation were stored using the same format as the experimental data and were analyzed as real events.

In the offline analysis procedure a detector was considered to be active if its signal exceeded in charge the signal corresponding to 4 MIPs. Subsequently, an event was consider as a candidate of a "shower event" if at least one of the detector groups $\{A_1, A_2, A_3\}$ or $\{B_1, B_2, B_3\}$ was active. The choice of these specific detector groups guaranteed that there were at least three active detectors in the event, separated by more than 15 m. Figure 8 demonstrates the success of the simulation package to describe the charge distribution of an active detector in a shower event.

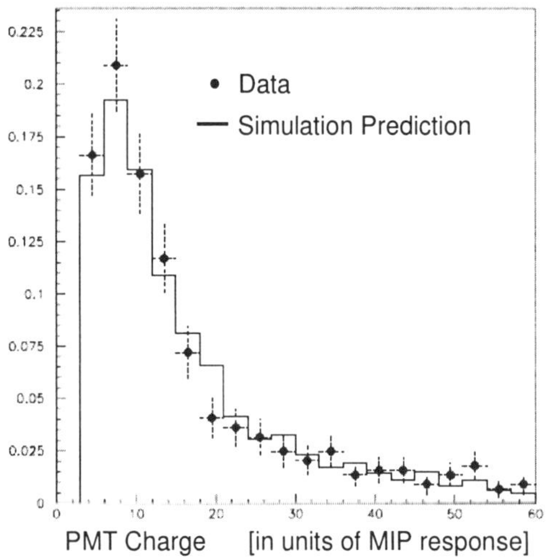

Figure 8. The charge distribution of an active HELYCON detector in a shower event.

The demand for detector signals larger than 4 MIPs rejected the majority of accidental background and reduced the systematic errors in estimating the

---

[d]In CORSIKA simulation the QGSJET II model was used for the high energy hadronic interactions and the GHEISHA model for the low energy hadronic interactions. EGS4 was used for the electromagnetic interactions.

arrival time of the shower front due to the slewing effect. It also reduced the systematic and statistical inaccuracies[e] due to the spherical shape and the thickness of the shower's front. These uncertainties were evaluated and parameterized as functions of the signal's charge[f], using the detailed detector simulation. Figure 9 shows the parameterizations of the systematic correction needed to be applied to the detector signal and of the statistical error in measuring the arrival times.

The direction of the shower axis was reconstructed by using the relative arrival times of the detector signals. For this reconstruction, a simple plane-wave hypothesis was used for the shower front whilst the systematic correction shown in Figure 9 was applied to the data. The zenith (θ) and azimuth (φ) angle of the shower axis were estimated by a χ^2 fit or as an exact solution of the triangulation technique, when only three detectors were active. The errors, $\delta\theta$ and $\delta\varphi$, in reconstructing the direction of every single shower, were estimated by taking into account the statistical inaccuracies for the arrival times discussed above.

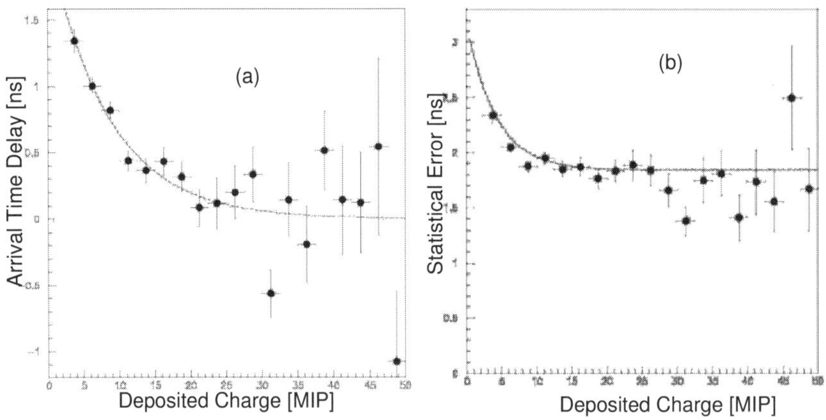

Figure 9. The parameterization of the systematic time delay due to the shape (a) and of the systematic error in determining the arrival time (b) of the shower particle front.

---

[e]The systematic error is due to the fact that the particles far away from the shower centre are delayed with respect to the expected arrival time, assuming a plane particle front. The statistical error corresponds to the thickness of the shower front.

[f]In these parameterizations, the charge of the detector signal was chosen due to the lack of knowledge concerning the distance from the shower centre, before the reconstruction of the shower axis.

In order to evaluate the performance of a typical HELYCON station, the analysis focused on the subset of events in which all six detectors were active. The axis direction of every single one of the above events was reconstructed by using, a) all six detectors, b) only the group of detectors A: $\{A_1, A_2, A_3\}$ and c) only the group of detectors B :$\{B_1, B_2, B_3\}$. Figure 10 presents the zenith angle distributions of these shower events, as reconstructed using each of the detector groups (A and B) or all six detectors. The solid lines represent the corresponding MC predictions, in very good agreement with the experimental data points.

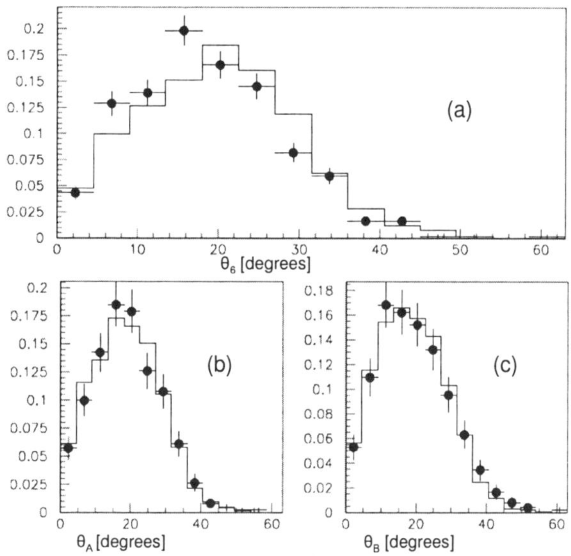

Figure 10. Distributions of the reconstructed zenith angle by: a) all 6 detectors, b) only the group of detectors A and c) only the group of detectors B. The points represent the experimental data whilst the histograms correspond to the simulation predictions.

In the case of MC analyzed events, the possibility exists for a direct comparison of the reconstructed to the generated zenith angle values revealing the angular resolution of each detector set-up. Figure 11 shows the distribution of these differences fitted to Gaussian shapes, with the corresponding resolution values for the detector group A, detector group B and all 6 detectors to be:

$$\sigma_A^{MC}=4.5^o \pm 0.5^o , \sigma_B^{MC}=5.2^o \pm 0.6^o , \sigma_6^{MC}=3.5^o \pm 0.3^o \ .$$

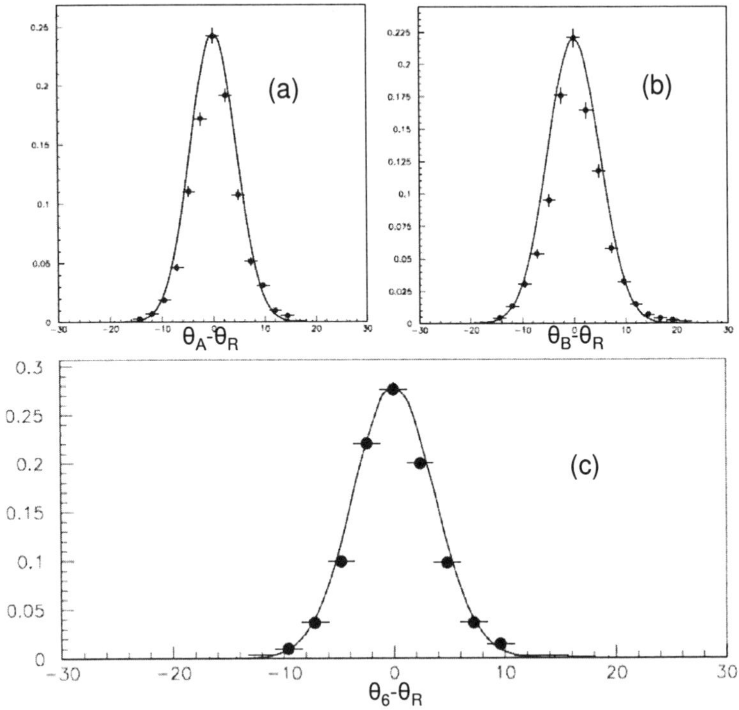

Figure 11. Te deviation of the estimated zenith angle from the generated value: a) using the detector group A, b) using the detector group B and c) using all 6 detectors.

It must also be noticed that: a) the mean value of each of the above distributions equals to zero, proving an unbias estimation procedure and b) the deviation of the reconstructed from the generated zenith angle value, when normalized to the corresponding reconstruction error, follows a normal distribution with σ=1, proving, an unbias error evaluation.

One can also evaluate the HELYCON station resolution, solely from the real data, by comparing the results obtained by the two detector groups (A and B) on an event by event basis. Figure 12 presents the difference ($\Delta\theta=\theta_A-\theta_B$) of these two estimated values of the zenith angle. The histogram corresponds to the MC prediction whilst the solid curve represents a Gaussian fit to the data points, with a sigma parameter, $\sigma^{DATA}=7.2^{o}\pm0.2^{o}$. This spread is consistent with the MC prediction for the resolution of each detector groups shown in Figure 10:

$$\sigma^{MC}=\sqrt{\left(\sigma_A^{MC}\right)^2+\left(\sigma_B^{MC}\right)^2}=6.9^{o}\left(\pm0.5^{o}\right).$$

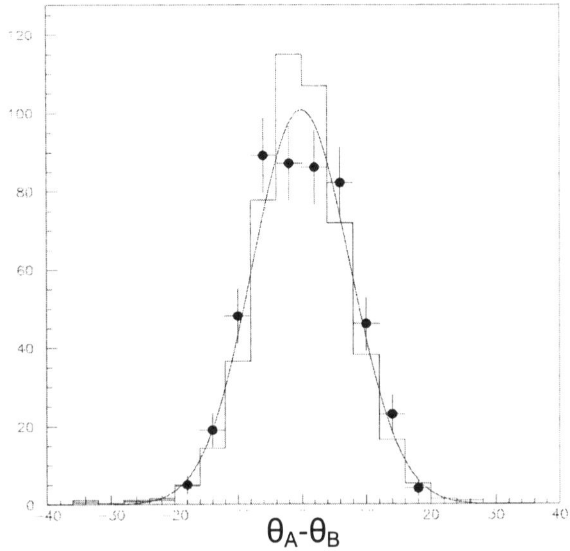

Figure 12. The distribution of the difference of the two detector-group estimations. The solid curve represents a Gaussian fit to the data points, whilst the histogram correspond to the simulation prediction.

Finally, the above difference ($\Delta\theta$) normalized to the relevant[g] reconstruction errors follows a normal distribution with a sigma equal to 1, indicating that the estimated reconstruction errors are evaluated correctly.

4. KM3NeT Calibration with HELYCON Detector Arrays

4.1. *Calibration Set-Up*

The analysis of atmospheric showers observed in coincidence by the Antarctic Muon and Neutrino Detector Array (AMANDA) [17] and the South Pole Air Shower Experiment (SPASE)[18] resulted in the calibration and survey [19] of the deep AMANDA detector, as well as the response of AMANDA to muon bundles. The IceTop [20], a larger cosmic shower telescope, being built on top of the IceCube [21] neutrino telescope, will provide calibration data for the under-ice detector, information for the cosmic ray composition and will veto the atmospheric muon background.

---

[g]This is the squared root of the quadratic sum of the two statistical errors, corresponding to the zenith angle reconstructions by the two detector groups.

In the case of a deep sea neutrino telescope, as the Mediterranean KM3NeT, a cosmic shower array on the nearby shore cannot be used for calibration purposes. Indeed, even if the neutrino telescope site is close (20 km) to the shore and the telescope is installed at 4 km under the sea surface, only very energetic showers with a zenith angle greater than 80° can trigger both the KM3NeT and an on-shore EAS array. It is estimated, using results from [22], that the total detection rate for such showers cannot exceed 2 showers per year.

However, a floating HELYCON detector array on top of the neutrino telescope can be used to calibrate the deep-sea detector. This array employs the copiously produced, low energy and small zenith angle, atmospheric showers. According to Monte Carlo studies, 35% of the cosmic showers in the energy range of 10^{14}-$5\cdot10^{15}$eV contain energetic muons (E > 2 TeV) able to penetrate the 4000m deep sea water and reach the KM3NeT detector. Moreover, the two thirds of these muons can be reconstructed by the neutrino telescope and their direction can be evaluated with an accuracy of 0.1 degrees.

The calibration capabilities of three autonomous HELYCON detector arrays on floating platforms, were quantified by a Monte Carlo study using: a) CORSICA to simulate EAS, b) the HELYCON detector simulation package and c) KM3Sim [23], a GEANT4 based simulation package to describe the passage (energy losses, electromagnetic shower production, multiple scattering, Cherenkov light emission) of muons through the water and the operation of a large underwater neutrino telescope. In this study the platforms were assumed to float 4000 m above the neutrino telescope, 150 m apart of each other, around the vertical symmetry axis of the telescope. Each platform contained 16 HELYCON detectors, arranged on a two dimensional grid (5m x 5m cell size) covering an area of about 360 m^2. It was assumed that every single floating detector array was operated independently from the others, as an isolated detector.

In this study, the neutrino telescope was assumed to consist of 80 strings, 125 m apart, in hexagonal geometry as in the IceCube detector [21]. Each string was carrying 60 optical modules (OM), with a vertical distance of 17 m between successive storeys. Each OM consisted of 40 cylindrical PMTs (XP53X2 Photonis, 3'' in diameter [24]) inside a 17'' benthos sphere, covering 4π in solid angle. The simulated response[h] of the neutrino telescope to down-coming muons, produced in the EAS, was used [25] to estimate the muon track parameters. Figure 13 demonstrates the resolution of this neutrino telescope in reconstructing the zenith angle (a), the azimuth angle (b) and the coordinates

[h]Including background signals from the decays of K^{40} contained in the sea water.

(c and d) of the impact point where the muon track intersects the horizontal plane on the sea surface.

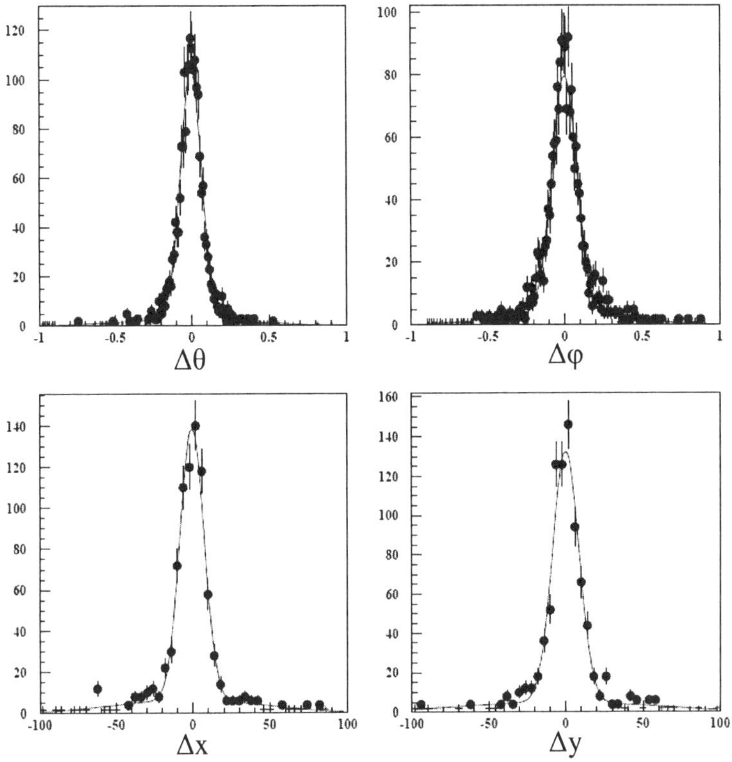

Figure 13. The deviation of the reconstructed muon track parameters from the generated values. The solid curves represent Gaussian fits to the distributions. The resulted resolutions are: a) 0.070 ±0.010 in the zenith angle (θ) estimation, b) 0.080 ±0.020 in the azimuth angle (φ) estimation, c) 7.8±0.5 m in evaluating the X coordinate and d) 8.1±0.5 m in evaluating the Y coordinate of the intersection point with the sea surface.

4.2. *Investigation for a Systematic Angular Offset*

The simulated response of each HELYCON detector array to EAS was further analyzed in order to reconstruct the direction of the shower axis, as described in Section 3, employing only detector signals from the same detector array. A minimum of 3 active detectors, each with a signal exceeding 4 MIPs, was required to define an event as a "shower event" candidate. Any shower event that contained at least one energetic muon[i] reconstructed by the neutrino telescope, was used in the calibration.

The estimated zenith angles of the shower axis and of the muon track were compared on an event by event basis. The difference between these two angles, shown in Figure 14, follows a normal distribution with mean value equal to zero. A statistical significant deviation of this mean value from zero indicates that the estimations of the neutrino telescope suffer from a systematic angular offset[j]. The sigma parameter of the Gaussian fit, shown in Figure 14, expresses the resolution[k] of the proposed calibration technique. The calibration resolution

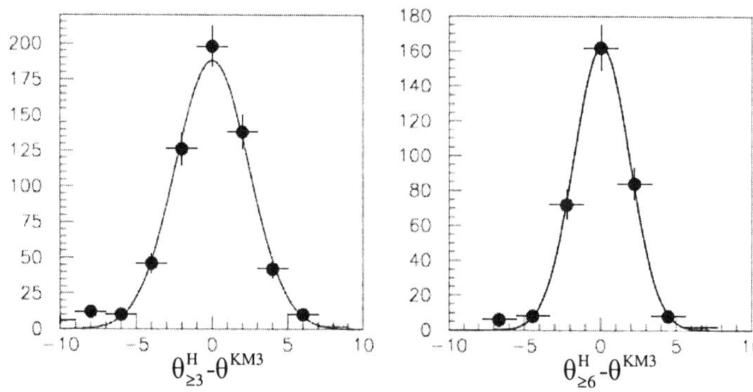

Figure 14. Distributions of the difference between the zenith angles the muon track (θ^{KM3}) and of the reconstructed shower axis when there more than 3 active ($\theta^{H}_{\geq 3}$) and more than 6 active ($\theta^{H}_{\geq 6}$) HELYCON detectors. The lines represent Gaussian fits with sigma parameters as: $\sigma_{\geq 3}=2.4^{\circ}\pm 0.1^{\circ}$ and $\sigma_{\geq 6}=1.8^{\circ}\pm 0.1^{\circ}$.

---

[i]A significant fraction (~ 33%) of these events contained more than one energetic muon. These muons are almost collinear within 0.01 degrees, very close to each other (1.8 m is the averaged separation with an RMS of 0.9 m). This muon bundle was reconstructed as a single track. The reconstruction resolution of muon bundles, with respect to the generated track parameters of the most energetic muon, is by a little worst than the reconstruction resolution of single muons, e.g. 0.11° (instead of 0.07°) in the zenith angle determination and 9 m (instead of 8. m) in the determination of the coordinates of the impact point on the sea surface. This effect was taken into account in the following analysis.

[j]The possibility that a found offset is due to systematic errors in estimating the shower axis direction can be easily overruled by bringing the platforms close together and use simultaneously two or three detector arrays to reconstruct the same shower, as described in Section 3.

[k]In the following, this parameter will be called "calibration resolution per single shower". It depends on the reconstruction resolution of the shower's zenith angle and on the angular misalignment of the muon track relative to the shower axis (typically 1.2°). The inaccuracy in estimating the muon zenith angle is smaller (by more than an order of magnitude) of the above factors and practically does not affect the calibration resolution per single shower. Moreover, the resolution of the calibration technique in evaluating the systematic angular offsets depends also on the number of reconstructed shower events.

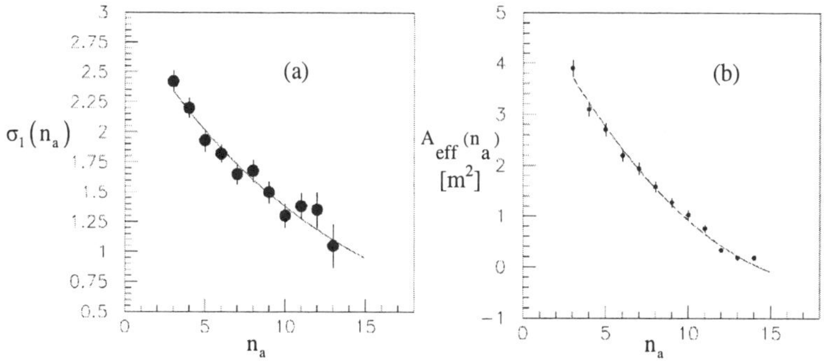

Figure 15. The calibration resolution per single shower (a) and the effective area of a floating detector array (b) as functions of the number of active detectors in the shower event.

per single shower decreases when events with more active detectors are selected, as shown in Figure 15a, because the reconstruction accuracy of the shower's direction improves. However, the requirement of more active detectors per event results to a reduction of the effective area[1] of the floating detector array, as shown in Figure 15b.

The calibration resolution, $\sigma_c(n_a)$, in identifying a possible angular offset in the neutrino telescope estimations using the three floating detector arrays, is approximated by the following equation:

$$\sigma_c(n_a) = \frac{\sigma_1(n_a)}{\sqrt{N}} = \frac{\sigma_1(n_a)}{\sqrt{3 \cdot \langle A_{eff}(n_a) \rangle_E \cdot 2\pi \cdot \int_{\Delta E} \Phi(E) dE \cdot \Delta T}} \quad (1)$$

where n_a is the minimum number of active detectors per shower event, $\sigma_1(n_a)$ is the calibration resolution per single shower, N is the total number of shower events containing energetic muons reconstructed by the HELYCON detector arrays and the neutrino telescope, $< A_{eff}(n_a)>_E$ is the averaged over energy effective area of one detector array, $\Phi(E)$ is the cosmic ray differential flux, E is the cosmic ray energy and ΔT is the time of operating the 3 floating detector arrays.

Figure 16 shows this calibration resolution for 10 days of operation, for different event selection criteria, demonstrating that the proposed calibration

[1]This corresponds to reconstructed shower events by both the HELYCON detector array and the neutrino telescope. For the evaluation of the effective area, $< A_{eff}(n_a)>_E$, a cosmic ray energy spectrum following a power law with index equal to -2.7 was assumed.

system will be able to measure a possible zenith angle offset with an accuracy of 0.05°. Furthermore, shower events with less than 5 active detectors do not contribute significantly in the calibration performance of the floating arrays.

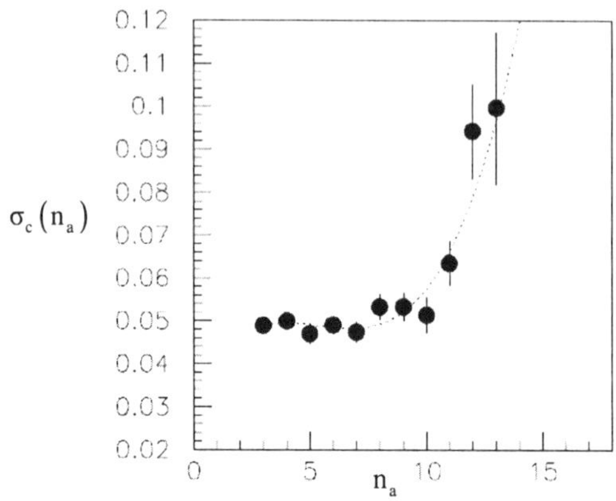

Figure 16. The calibration resolution of 3 HELYCON detector arrays, for 10 days of operation, as a function of the minimum number of active detectors per event.

4.3. *Determination of the KM3NeT Absolute Position*

The absolute position of the underwater telescope can be also measured using the floating HELYCON detector arrays. The technique is based on measuring the distance, on an event by event basis, between the impact points[m] of the reconstructed muon track and the shower axis. The muon's impact point is evaluated using the reconstructed track parameters with respect to the centre of the neutrino telescope (local system of coordinates: A). In parallel the impact point of the shower axis is estimated by a χ^2 fit [26], with respect to a fixed point on the floating detector array (local system of coordinates: B). This fit involves the signals of the active HELYCON detectors and a parameterization[n] of the expected averaged detector signal and its statistical variance as functions of the perpendicular distance from the shower axis. The projections of the

[m]These are the points of intersection with the horizontal (XY) plane at the sea surface.

[n]This is a parameterization of the simulation prediction concerning the density of particles at a certain distance from the shower core. In general, this parameterization depends also on the type, on the energy and on the zenith angle of the primary cosmic particle. In this study, only proton showers have been simulated, with energies according to the measured cosmic ray spectrum, and directions that can trigger both the floating and the underwater detectors. Consequently the parameterization corresponds to the lateral profile of an "averaged" detectable shower containing a reconstructed energetic muon.

distance between the two impact points on the X and Y axis follow symmetric, non-Gaussian, distributions[o]. The distribution of these projections, shown in Figure 17, can be approximated by the sum of two Gaussian functions, having equal mean values. The mean values of both X and Y projections found to be statistically consistent with zero, due to the fact that there was no displacement between the two local systems. In general, a measurement of these mean values defines the absolute position of the neutrino telescope, assuming that the position of the floating array is known with high accuracy using a large number of repeated GPS measurements.

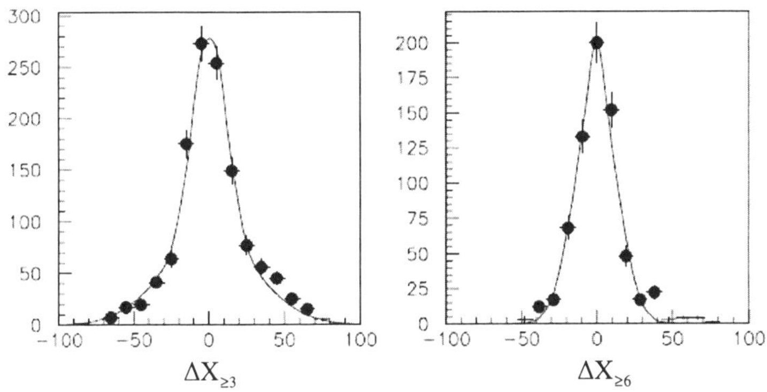

Figure 17. The X-projection of the displacement between the two impact points, when the selected shower events contain more than 3 (left plot) or more than 6 (right plot) active detectors. The standard deviations of the above distributions are 21m±0.5m and 16m±0.5m.

The number of the reconstructed events and the spread of the above, ΔX and ΔY, distributions determine the resolution in measuring the telescope's position. The distribution spread depends strongly on the shower axis reconstruction resolution, shown in Figure 18a, and less on the neutrino telescope resolution. The accuracy of this calibration technique can be quantified by a similar expression as Equation 1. Figure 18b presents the position resolution which can be achieved by 3 floating detector arrays, collecting data for 10 days. The proposed technique can estimate the absolute position of the neutrino telescope with an accuracy[p] of less than 0.5m.

[o]This is due to the fact that the showers have different zenith angles and that the muons are not collinear to the shower axis. However, the contribution of the broader Gaussian function to the fit is getting smaller when there are more active detectors in the shower event.

[p]In this analysis was assumed that the position of the floating arrays was known with infinite accuracy by averaging many GPS measurements.

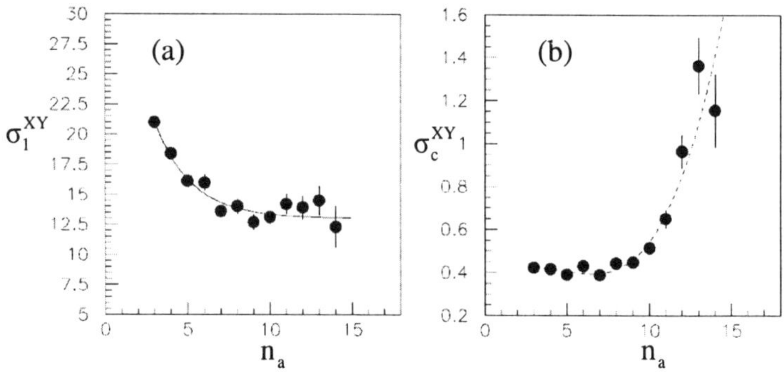

Figure 18. The resolution in estimating the (X-Y) coordinates of the under-water detector, as a function of the number of active detectors per event, using: a) a single reconstructed EAS and b) the shower events collected by three floating arrays during 10 days of operation.

4.4. *Estimation of the KM3NeT Angular Resolution*

The floating detectors arrays cannot be used for measuring the angular resolution of the neutrino telescope, due to the fact that the resolution of the neutrino telescope is by an order of magnitude better than the accuracy of the floating detector arrays in reconstructing the direction of the shower axis. However, it is possible to inter-calibrate the neutrino telescope by dividing the under-water detector in sub-detectors. The same muon track is then reconstructed by each sub-detector and the resolution of the neutrino telescope is evaluated by comparing the reconstruction results.

This MC study used the simulated response of the neutrino telescope to 1 TeV muons, distributed isotropicaly in the detector volume. Only simulated events with at least 14 active OMs, after filtering out the background hits, were used in the calibration analysis. The selected sample consisted, in average, of 24 active OMs per event, whilst the remaining contamination from K^{40} background hits was less than 0.5 OM per event. Each muon track was reconstructed using the information from the whole set of the active OMs as well as using the data from the two sub-groups, each containing the half of the selected OMs.

In a MC study there is the possibility of knowing the reconstruction resolution by comparing the estimated to the corresponding generated values. The accuracy in estimating the zenith angle, shown in Figure 19, depends on the number of active OMs which are used in the reconstruction. The information provided by the whole set of OM's resulted to a zenith angle resolution of $0.062^{\circ}\pm0.003^{\circ}$, whilst the angular resolution using the half of the data-points is $0.095^{\circ}\pm0.005^{\circ}$.

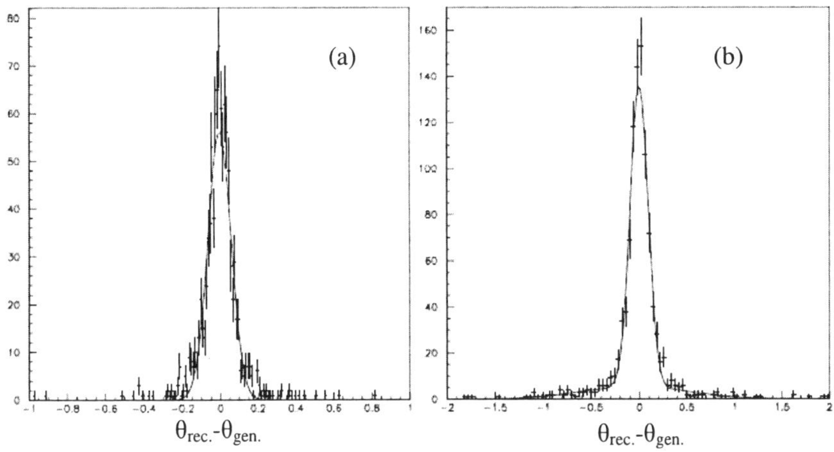

Figure 19. The deviation of the reconstructed zenith angles from the corresponding generated values, a) using the whole set and b) using the half of the active OMs. The sigma parameters of the Gaussian fit, represented by the solid curves, express the resolution in estimating the zenith angle of the muons.

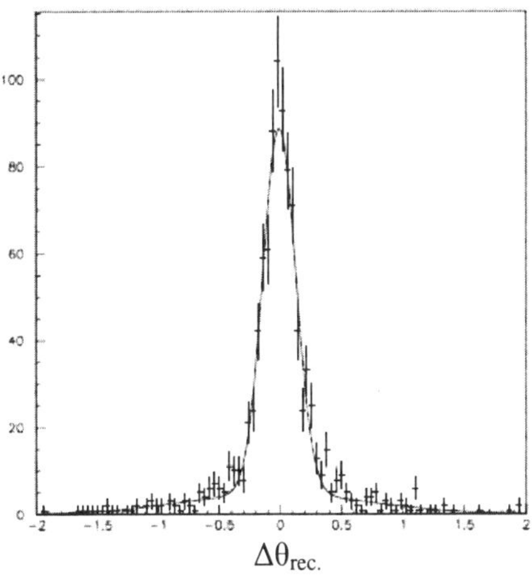

Figure 20. The distribution of the difference between the estimated muon zenith angles by the two sub-groups.

The difference between the two zenith angles, reconstructed by the two sub-groups of OMs, was evaluated on a track by track basis and was used to estimate the neutrino telescope's angular resolution. This zenith angle difference, shown in Figure 20, was found to follow a Gaussian distribution with a sigma parameter equal to 0.14°±0.01°. Assuming the same resolution for each of the two zenith angle estimations, the resolution of each sub-group can be estimated as:

$$0.14^\circ\left(\pm 0.01^\circ\right)/\sqrt{2}=0.099^\circ\left(\pm 0.007^\circ\right)$$

which is in very good agreement with the sub-group's resolution of found above.

4.5. *Conclusion*

The HELYCON collaboration has constructed, tested and calibrated 15 detectors in order to facilitate and operate four of the telescope stations in Patras, by the end of 2007. Furthermore a detailed Monte Carlo description of the detectors has been developed and it has been found in excellent agreement with the test and calibration results.

In parallel with the pilot operation of the first HELYCON station, more detector stations will be constructed and they will be deployed at high school and university buildings in the Patras area. During 2008 more detector stations will be commissioned in nearby cities, as well as in the island of Chios and around Nicosia in Cyprus. The geographical dispersion of the detector stations will make it possible to study long range correlations between air showers.

Furthermore, HELYCON provides the test bench for the development of the calibration techniques and instrumentation for an under-water neutrino telescope, as the Mediterranean KM3NeT. Monte Carlo studies, presented in this paper, have shown that three floating HELYCON detector arrays can measure a possible angular offset in the neutrino telescope estimations with an accuracy of 0.05°, whilst the absolute position of the underwater detector can be estimated with an accuracy of less than half a meter. Finally, a technique of measuring with high accuracy the angular resolution of the neutrino telescope has been proposed and demonstrated using simulation.

Acknowledgments

The author also wishes to acknowledge the help and support of all the members of the HELYCON collaboration, especially Dr. A. G. Tsirigotis, Dr. A. Leisos and Dr. G. Bourlis.

References

1. G. A. Medina-Tonco, A. A. Watson, Astroparticle Physics, 10, 157, (1999).
2. N. Ochi et al, J. Phys. G,. 29, 1169, (2003).
3. Ulrich F. Katz, "KM3NeT, Proceedings of the 2nd VLVNT Workshop on Very Large Neutrino Telescope (VLVNT2), Catania, Italy, 8-11 Nov 2005, to be published ; e-Print Archive: astro-ph/0606068.
4. http://www.ihep.su/scint/mold/product-e.htm
5. http://www.tyvek.com/
6. http://www.detectors.saint-gobain.com/
7. http://www.photonis.com/data/cms-resources/File/Photomultiplier_tubes/upload_pdf/19mm.pdf
8. http://www.emcohighvoltage.com/
9. http://helycon.eap.gr
10. http://micdigital.web.cern.ch/micdigital/hptdc.htm
11. http://www.ni.com/pdf/manuals/371303e.pdf
12. http://www.synergy-gps.com/images/stories/guides/m12+userguide.pdf
13. D. Loukas et al, "HELYCON Readout Electronics", to be published.
14. S. Agostinelli et al, Nuclear Instruments and Methods in Physics Research Section A, 506, 250, (2003).
15. http://www.tek.com/site/ps/41-12482/pdfs/41W_12482.pdf, http://www.tek.com/site/ps/48-19032/pdfs/48W_19032.pdf
16. J. Knapp, D. Heck, Nachr. Forsch. zentr. Karlsruhe, 30, 27 (1998); http://www-ik.fzk.de/corsika/.
17. http://amanda.uci.edu/
18. http://www.bartol.udel.edu/spase/
19. J. Ahrens et al, Nuclear Instruments and Methods in Physics Research Section A, 522, 347, (2004).
20. http://icecube.bartol.udel.edu/
21. http://icecube.wisc.edu/
22. M. Are et al. Ast.Part. 14: 109-120 2000 and http://www.ast.leeds.ac.uk/haverah/havpark.shtml
23. HOU simulation package:
    http://www.physicslab.eap.gr/PAS/PAS_index.htm
    A. Tsirigotis, PhD Thesis, Hellenic Open University, 2004
    http://www.eap.gr/tsirigotis/PhD

24. http://www.photonis.com/data/cms-resources/File/Photomultiplier_tubes/spec/XP53X2.PDF
25. NESTOR Collaboration, , Astropart.Phys.23, 377-392, (2005).
26. e.g. J. A. Hinton, PhD Thesis, University of Leeds.

SOLVING COSMOLOGICAL PROBLEM IN UNIVERSAL EXTRA DIMENSION MODELS BY INTRODUCING DIRAC NEUTRINO

SHIGEKI MATSUMOTO[a], JOE SATO[b], MASATO SENAMI[c] and MASATO YAMANAKA

[a] *Theory Group, KEK, Oho 1-1, Tsukuba, Ibaraki 305-0801, Japan*
[b] *Department of Physics, Saitama University, Shimo-okubo, Sakura-ku, Saitama, 338-8570, Japan*
[c] *ICRR, University of Tokyo, Kashiwa 277-8582, Japan*

Universal extra dimension (UED) models with right-handed neutrinos are studied. The introduction of the neutrinos makes us possible not only to describe Dirac neutrino masses but also to solve the cosmological problem called the KK graviton problem. This problem is essentially caused by the late time decay of a KK photon into a KK graviton and a photon, and it distorts the spectrum of the cosmic microwave background or the diffuse photon. We point out that, once we introduce right-handed neutrinos to UED models, the KK photon decays dominantly into neutrinos and does not emit a photon. We also discuss sub-dominant modes with a photon in the decay quantitatively, and show that their branching ratios are so small that the spectra are not distorted.

1. Introduction

Investigating the nature of dark matter in our universe is important for particle physics, cosmology and astronomy. Many candidates for dark matter has been proposed so far in the context of physics beyond the standard model (SM). Among those, universal extra dimension (UED) models [1] are one of the attractive scenario which provide a good candidate for non-baryonic cold dark matter naturally. In this scenario, all particles in the SM can propagate in those extra dimensions. Furthermore, UED models have Kaluza-Klein (KK) parity conservation, which is a remnant of the KK number conservation originated in the momentum conservation along an extra dimension. Under this parity, particles with odd KK number have odd charge, while the others have even ones. As a result, the lightest KK particle (LKP) is stabilized by this parity and is a good candidate for dark matter.

The model most extensively studied in this scenario is the minimal UED (MUED) model, which is defined in five dimensions. The extra dimension is compactified on an S^1/Z_2 orbifold for producing the SM chiral fermions.The mass spectrum of KK particles has been precisely calculated in Ref. [2] including radiative corrections. They have found that the LKP is the first KK particle of the photon, which is dominantly composed of the KK particle of the hypercharge gauge boson. In spite of the simple framework, this model is phenomenologically successful. In fact, it is consistent with the recent results of $b \to s\gamma$ [3], the anomalous muon magnetic moment [4, 5], $Z \to b\bar{b}$ [1,6], $B - \bar{B}$ $Z \to b\bar{b}$ [1,6], $B - \bar{B}$ oscillation [7], B and K meson decays [8], and electroweak precision measurements [1,9] when the compactification scale $1/R$ is large enough, $1/R \gtrsim 400$ GeV.

In addition, cosmological aspects such as the thermal relic abundance of the LKP dark matter has also been studied extensively [10–12]. The recent work [11] gives the most precise calculation about the abundance by including all second KK resonance processes in all coannihilation modes. From the analysis in the work, it is shown that the compactification scale, which is consistent with the WMAP observation [13], is 600 GeV $\lesssim 1/R \lesssim$ 1400 GeV. Furthermore, the upper bound of the Higgs mass has been obtained as $m_h \lesssim 230$ GeV.

Although phenomenologically successful, this model has two shortcomings. First one is the absence of neutrino masses. As well known, neutrino oscillationexperiments indicate the existence of neutrino masses [14]. However, neutrinos are massless in the MUED model, because this model is the simple extension of the SM to the extra dimension. Hence, the model is required to extend to describe neutrino masses. Second issue concerns the cosmological aspect of this model. For $1/R \lesssim 800$ GeV, it has been pointed out that the KK particle of the graviton is the LKP [11,12]. This is a serious problem for the model, because the lifetime of the next LKP (NLKP), that is the KK photon, is very long, and its decay in the early universe leads to an inconsistency with the observation of the cosmic microwave background (CMB) [15,16] or the diffuse photon [16].

In this letter, we show that these two problems can be solved simultaneously by introducing right-handed neutrinos. In UED models, neutrino masses are inevitably introduced by Dirac mass terms, because Majorana mass terms are not allowed due to the Lorentz invariance in higher dimensions. As a result, the UED models have tiny Yukawa couplings as $\sim m_\nu/v$, where m_ν is a neutrino mass and $v \sim 246$ GeV is the vacuum expectation value of the Higgs field. Then the first KK particles of right-handed neutrinos are the NLKPsin the region where the KK graviton is the LKP, because the radiative corrections to the KK right-handed neutrinos are negligibly small due to tiny Yukawa couplings. Therefore, a KK photon decays dominantly into a KK right-handed neutrino and an ordinary neutrino, and photons are not emitted in the leading process of its decay. Thus, the CMB and diffuse photon spectra are not distorted.

In this letter, we discuss the above solution quantitatively. First, we briefly summarize the KK graviton problem in the next section. Next, the right-handed neutrinos are introduced to provide neutrino masses, and some decay widths of the KK photon into the first KK right handed neutrino are calculated in Sec. 3. Then, we discuss that the decay of the first KK photon does not affect spectra of the CMB and diffuse photon, and also does not spoil the successful prediction of the big bang nucleosynthesis (BBN) in Sec.4. Finally, Sec. 5 is devoted to summary and discussion.

2. KK graviton problem in the MUED model

First, we consider the mass spectrum of first KK particles, which is of importance for the KK graviton problem. At tree level, the mass of a KK particleis determined by the compactification scale $1/R$ and the corresponding SM particle one as $m^{(n)} = (n^2/R^2 + m_{\rm SM}^2)^{1/2}$. Since the compactification scale is much larger than the SM particle masses, all KK particles at each KK mode are highly degenerated in mass around n/R. Mass differences among KK particles at each KK mode dominantly come from radiative corrections, which are calculated in Ref. [2]. Thus, for instance, colored KK particles are heavier than other KK particles, while the masses of KK U(1) gauge boson and right-handed leptons still remain $\sim n/R$.

The candidates for the LKP are the KK photon ($\gamma^{(1)}$), the KK charged Higgs ($H^{\pm(1)}$), and the KK graviton ($G^{(1)}$) depending on $1/R$ and the Higgs mass (m_h). Here $H^{\pm(1)}$ is the KK particle of the Goldstone boson. Since KK gauge bosons acquire mass term from the fifth components in gauge fields, KK particles of Goldstone bosons remain as physical states.

First, the mass of the KK photon is obtained by diagonalizing following mass squared matrix,

$$\begin{pmatrix} 1/R^2 + \delta m_{B^{(1)}}^2 + g'^2 v^2/4 & g' g v^2/4 \\ g' g v^2/4 & 1/R^2 + \delta m_{W^{(1)}}^2 + g^2 v^2/4 \end{pmatrix}, \tag{1}$$

which is written in the ($B^{(1)}$, $W_3^{(1)}$) basis. Here, $g(g')$ is $\mathrm{SU}(2)_L$ ($\mathrm{U}(1)_Y$) gauge coupling constant. Radiative corrections, $\delta m_{B^{(1)}}^2$ and $\delta m_{W^{(1)}}^2$, are

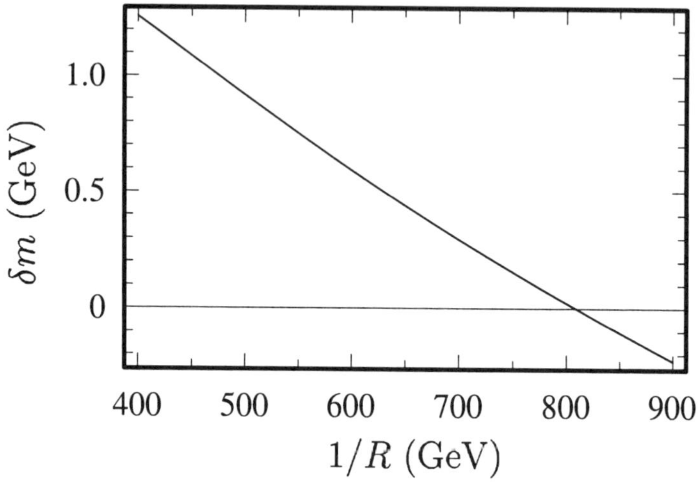

Fig. 1. The difference, $\delta m \equiv m_{\gamma^{(1)}} - 1/R$, is shown as a function of $1/R$. The cutoff scale Λ is set to be $\Lambda R = 20$.

given by

$$\delta m^2_{B^{(1)}} = -\frac{39}{2}\frac{g'^2\zeta(3)}{16\pi^4R^2} - \frac{1}{6}\frac{g'^2}{16\pi^2R^2}\ln(\Lambda^2R^2) ,$$
$$\delta m^2_{W^{(1)}} = -\frac{5}{2}\frac{g^2\zeta(3)}{16\pi^4R^2} + \frac{15}{2}\frac{g^2}{16\pi^2R^2}\ln(\Lambda^2R^2) , \tag{2}$$

where Λ is the cutoff scale of the MUED model, it is usually taken to be $\Lambda R \sim \mathcal{O}(10)$. In this work, we adopt the value, $\Lambda R = 20$. Even if different value of this cutoff scale was adopted, our conclusion is not changed significantly. The mass of the KK photon is smaller than $1/R$ for $1/R \gtrsim 800$ GeV. The difference, $\delta m \equiv m_{\gamma^{(1)}} - 1/R$, is shown as a function of $1/R$ in Fig. 1. As seen in this figure, δm is typically on the order of 1 GeV and small for larger $1/R$.

Next, the mass of the KK charged Higgs is given as

$$m^2_{H^{\pm(1)}} = 1/R^2 + m_W^2 + \delta m^2_{H^{(1)}} ,$$
$$\delta m^2_{H^{(1)}} = \left(\frac{3}{2}g^2 + \frac{3}{4}g'^2 - \lambda_h\right)\frac{1}{R^2}\frac{\ln\left(\Lambda^2R^2\right)}{16\pi^2} , \tag{3}$$

where λ_h is the Higgs self-coupling defined by $\lambda_h \equiv m_h^2/v^2$. As m_h is increased, the negative contribution in the radiative correction, $\delta m^2_{H^{(1)}}$, increases. Hence, for large λ_h, the mass difference between the KK photon and $H^{\pm(1)}$ can be negative.

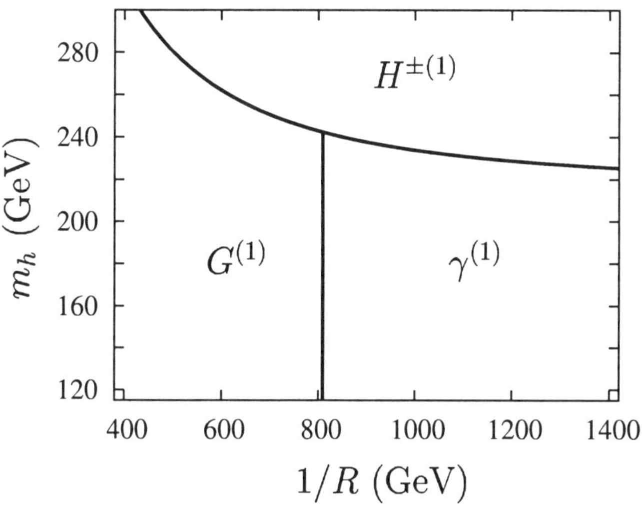

Fig. 2. The constituent of the LKP in the MUED model is shown. The cutoff scale Λ is set to be $\Lambda R = 20$.

Finally, we consider the mass of the KK graviton. Since all interactions relevant to the KK graviton are suppressed by the Planck scale, radiative corrections are negligible. Therefore its mass is simply given by $1/R$. By the comparison with the mass of the KK photon or KK charged Higgs, it turns out that the KK graviton is the LKP when $1/R$ and m_h are not so large. We summarize the result in Fig. 2. The parameter space in the MUED model is divided into three regions, where the LKP is $\gamma^{(1)}$, $H^{\pm(1)}$, and $G^{(1)}$, respectively. Obviously, the $H^{\pm(1)}$ LKP region is ruled out from the viewpoint of dark matter.

Moreover, the $G^{(1)}$ LKP region is also excluded by considering cosmological implication from NLKP decay. When the temperature of the universe is around $1/R$, the KK photon, which is the NLKP in this region, is in thermal equilibrium. On the contrary, the KK graviton is not in thermal equilibrium and its abundance is tiny at that time due to extremely weak interactions[a]. After the temperature of the universe becomes low enough, the annihilation process of the KK photon is frozen out, and its density per co-moving volume is fixed. The KK photon decays into the KK graviton long after the freeze-out through the process, $\gamma^{(1)} \to G^{(1)} + \gamma$, which

[a]We assume that the reheating temperature of the universe after inflation is so low that the KK graviton is not produced abundantly from the thermal bath.

occurs typically in or after the recombination era as shown in the next section. Since the observation of the CMB spectrum is very accurate and any deviation from the black-body spectrum has not been observed, this decay in the recombination era is severely constrained[b]. Furthermore, even if the decay takes place after the recombination era, it produces observable peak in the diffuse photon spectrum. As a result, the KK graviton decay is severely constrained from these measurements, and in fact, the entire $G^{(1)}$ LKP region is ruled out.

Fortunately, in the $\gamma^{(1)}$ in Fig. 2 LKP region, where the NLKP is $G^{(1)}$, there is no KK graviton problem when the reheating temperature is low enough [17]. However, most of the parameter region consistent with the dark matter abundance observed by WMAP is in the $G^{(1)}$ LKP region [11]. Furthermore, this region is more attractive than others for physics at colliders such as the LHC.

The KK graviton problem is not the intrinsic one in the MUED model. It will appear in various UED extensions of the SM. Thus, it is important to consider a mechanism in order to avoid the problem. The mechanism we present in this letter is very simple and can be applied to other UED models.

3. Introduction of right-handed neutrinos to the MUED model

In addition to the KK graviton problem, the MUED model has another problem, which is the absence of neutrino masses. Since the MUED model is the simple extension of the SM to the higher dimensions, neutrinos are treated as massless particles. On the other hand, the existence of neutrino masses and lepton flavor mixing have been established by neutrino oscillation experiments. Thus, we have to extend the MUED model in order to construct a more realistic one.

According to the algebraic structure of the five dimensional space-time, $SO(1,4)$, a spinor field in UED models is represented by a four-component field. Thus, Majorana mass terms can not be introduced as long as the Lorentz invariance is preserved in the five dimensional space-time[c]. Hence, we have to introduce Dirac mass terms to neutrinos in UED models. As a

[b] In fact, the constraint is more stringent than that from the BBN for the decay in the recombination era [16].

[c] Though the Lorentz invariance is violated at the fixed points, the Majorana mass term can not be introduced in a natural way as long as the KK parity is imposed.

result, tiny Yukawa couplings for neutrinos are also introduced. However, due to the existence of these tiny couplings, the KK graviton problem is resolved as discussed in the next section.

Once we introduce right-handed neutrinos in the MUED model, their KK particles automatically appear in the spectrum. Since these particles are the SM gauge singlet and have only small Yukawa interactions, radiative corrections to their masses are very small. The mass of the KK particle of a right-handed neutrino, $N^{(1)}$, is estimated as

$$m_{N^{(1)}} \simeq \frac{1}{R} + O\left(\frac{m_\nu^2}{1/R}\right) . \tag{4}$$

The KK right handed neutrino is the NLKP in the $G^{(1)}$ LKP region, since the mass is smaller than that of the KK photon. The existence of the $N^{(1)}$ NLKP changes the late time decay of the KK photon. In this section, we consider several decay modes of the KK photon, which are important for the detailed study of the thermal history of the universe.

The KK photon decays dominantly into $N^{(1)}$ and left-handed neutrino at tree level. This decay takes place through the diagram in which the KK left-handed neutrino is flipped into $N^{(1)}$, and its decay width is given by

$$\Gamma_{N^{(1)}\bar{\nu}} = \frac{\alpha'}{8}\frac{m_\nu^2 \delta m^2}{(1/R)^3} \simeq 2\times 10^{-7}\ [\mathrm{sec}^{-1}] \left(\frac{500\mathrm{GeV}}{1/R}\right)^3 \left(\frac{m_\nu}{0.1\mathrm{eV}}\right)^2 \left(\frac{\delta m}{1\mathrm{GeV}}\right)^2 , \tag{5}$$

where $\alpha' = g'^2/4\pi$ and the mass difference between $\gamma^{(1)}$ and $N^{(1)}$ is denoted by δm. We use the approximation $m_{N^{(1)}} = m_{G^{(1)}} = 1/R$ unless the mass difference appears. Here, m_ν is the heaviest neutrino mass. The lower limit of m_ν is obtained byatmospheric neutrino experiments, $m_{\nu_{\mathrm{atm}}} \simeq 0.05$ eV [14,18] . We adopt the upper bound from the WMAP observation, $\sum m_\nu <$ 2.0 eV [13,19]. In Fig. 3, the lifetime of $\gamma^{(1)}$ is shown. In two solid lines, the mass of $\gamma^{(1)}$ is obtained by the MUED model. The upper (lower) solid line is the result for $m_\nu = m_{\nu_{\mathrm{atm}}}(m_{\nu_{\mathrm{WMAP}}})$, where $m_{\nu_{\mathrm{WMAP}}} = 0.67$ eV. Dotted lines are depicted by regarding δm as a free parameter and fixing it on 1 (0.1) GeV for the lower(upper) line. The neutrino mass m_ν is set to be $m_{\nu_{\mathrm{atm}}}$.

The KK photon can also decay into $G^{(1)}$ through the Planck scale suppressed interaction. This process may be important for cosmological implication, because a photon is produced in the final state. Its decay width is given in Ref. [16] as

$$\Gamma_{G^{(1)}\gamma} = \frac{5\cos^2\theta_W \delta m^3}{9\pi M_{\mathrm{Pl}}^2} \simeq 7\times 10^{-14}\ [\mathrm{sec}^{-1}] \left(\frac{\delta m}{1\mathrm{GeV}}\right)^3 , \tag{6}$$

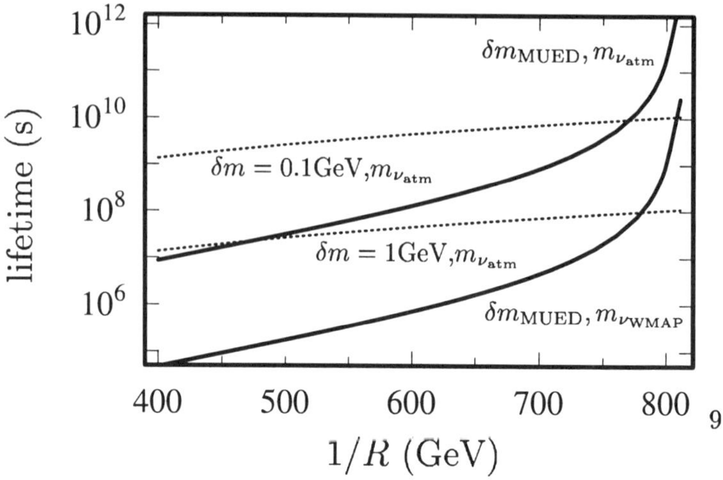

Fig. 3. The lifetime of $\gamma^{(1)}$ is shown as a function of $1/R$. In two solid lines, the mass of $\gamma^{(1)}$ is obtained by the MUED model. The upper line is for $m_\nu = m_{\nu_{\rm atm}}$, while the lower one is for $m_{\nu_{\rm WMAP}}$. In dotted lines, δm is regarded as a free parameter and fixing it on 1 (0.1) GeV for the lower(upper) line. The neutrino mass m_ν is set to be $m_{\nu_{\rm atm}}$.

where $M_{\rm Pl} \simeq 2.4 \times 10^{18}$ GeV is the reduced Planck mass.

In addition to these decay processes, there are two other modes in which a photon is emitted in the final state. First one is the γ associated $\gamma^{(1)}$ decay into $N^{(1)}$ ($\gamma^{(1)} \to N^{(1)}\bar{\nu}\gamma$), which occurs through one-loop diagrams. Its width is estimated as

$$\begin{aligned}\Gamma_{N^{(1)}\bar{\nu}\gamma} &\sim \frac{\alpha^3\alpha' \sin^2\theta_W}{8\pi^3} \frac{m_\nu^2 \delta m^4}{m_{\gamma^{(1)}} m_W^4} \times 10^{-4} \\ &\sim 10^{-20}\,[\sec^{-1}] \left(\frac{500{\rm GeV}}{1/R}\right) \left(\frac{m_\nu}{0.1{\rm eV}}\right)^2 \left(\frac{\delta m}{1{\rm GeV}}\right)^4 , \end{aligned} \qquad (7)$$

where m_W is the W boson mass and the factor 10^{-4} arises from loop integrals and phase space. Thus, this process is suppressed by one-loop integrals, tiny neutrino mass, and small phase space. As a result, its width turns out to be much smaller than the two bodies decay, $\gamma^{(1)} \to G^{(1)}\gamma$, and we can neglect it even for the maximum m_ν. Another mode is the decay process associated with a neutral pion, $\gamma^{(1)} \to N^{(1)}\bar{\nu}\pi^0$,in which photons are emitted in the final state through $\pi^0 \to 2\gamma$ decay. However, this process is also suppressed by f_π^2/m_Z^2, neutrino mass, and three bodies phase space.

Hence, this decay width is much smaller than the two bodies decay, and we can also neglect the process.

After all, the decay of the KK photon is governed by the process, $\gamma^{(1)} \to N^{(1)}\bar{\nu}$. On the other hand, the dominant decay mode associated with a photon comes from the Planck suppressed process, $\gamma^{(1)} \to G^{(1)}\gamma$, and its branching ratio is

$$\mathrm{Br}_{X\gamma} = \frac{\Gamma_{G^{(1)}\gamma}}{\Gamma_{N^{(1)}\bar{\nu}}} = 5 \times 10^{-7} \left(\frac{1/R}{500\mathrm{GeV}}\right)^3 \left(\frac{0.1\mathrm{eV}}{m_\nu}\right)^2 \left(\frac{\delta m}{1\mathrm{GeV}}\right) . \tag{8}$$

Hence, the decay associated with a photon is very suppressed.

4. $N^{(1)}$ dark matter and the KK graviton problem

In the early universe ($T \sim 1/R$), $\gamma^{(1)}$ is in thermal equilibrium, while $N^{(1)}$ and $G^{(1)}$ are out of equilibrium due to the tiny couplings. Thus, their abundances are negligibly small at that time, and they can be produced only by $\gamma^{(1)}$ decay. The KK photon decays dominantly into $N^{(1)}$ at $T \sim (\Gamma_{N^{(1)}\bar{\nu}} M_{\mathrm{Pl}})^{1/2}$. However, $N^{(1)}$ can not decay into $G^{(1)}$, since it is kinematically forbidden. As a result, $N^{(1)}$ remains as a non-baryonic cold dark matter in the present universe. Since the mass difference between $N^{(1)}$ and $\gamma^{(1)}$ is negligibly small, the allowed region consistent with the WMAP observation is the same as that of the $\gamma^{(1)}$ LKP.

As discussed in the previous section, $\gamma^{(1)}$ can decay into the final state with a photon, though its branching ratio is small. Since constraints from observation of background photons to the late time decay is very stringent, we should quantitatively check that the decay does not really affect the photon spectrum. In the following, we consider cosmological implications of the MUED model with the right-handed neutrinos. The decay of $\gamma^{(1)}$ occurs at $t = 10^5 - 10^{12}$ s as shown in Fig. 3, it corresponds to the period between the BBN and the recombination. As a result, the constraint from measurements of the diffuse photon spectrum can be neglected. On the other hand, emitted photons may spoil the successful prediction of the BBN by destroying light nuclei. Moreover the photons may distort the spectrum of the CMB. Thus, it is important to consider constraints from measurements of the BBN and CMB to this model.

The distortion of the CMB spectrum is parametrized by chemical potential μ when a process changing the energy of a background photon are effective, or the Compton y-parameter after the process become ineffective. These values are constrained as $|\mu| < 9\times 10^{-5}$ and $|y| < 1.2\times 10^{-5}$ [20]. The total injection energy from the decay is used as the quantity constrained from the CMB measurement. In fact, μ and y are proportional to the energy.Furthermore, the constraint from the BBN is also represented using the energy. Since the energy of a emitted photon is rapidly redistributed through inverse Compton scattering ($\gamma e^- \to \gamma e^-$) and e^+e^- creation with background photons ($\gamma\gamma_{\rm BG} \to e^+e^-$), the important quantity is the total injection energy and not its spectrum.

The injection energy is given by $\epsilon{\rm Br}_{X\gamma}Y_{\gamma^{(1)}}$,where ϵ is the typical energy of emitted photon and $Y_{\gamma^{(1)}} = Y_{N^{(1)}} = n_{N^{(1)}}/n_{\gamma_{\rm BG}}$ is the number density of $\gamma^{(1)}$ normalized by that of background photons. The branching ratio is given in Eq. (8) and ϵ is less than δm. The yield $Y_{N^{(1)}}$ is estimated by requiring that the relic abundance of $N^{(1)}$ accounts for the observed abundance of dark matter,

$$Y_{N^{(1)}} = \frac{\Omega_{\rm DM}\rho_c}{1/R}\frac{1}{n_0} \simeq 5\times 10^{-12}\left(\frac{\Omega_{\rm DM}h^2}{0.10}\right)\left(\frac{500{\rm GeV}}{1/R}\right), \tag{9}$$

where $\rho_c = 1.1\times 10^{-5}h^2$ GeVcm^{-3} is the critical density of the universe, and $n_0 = 410$ cm^{-3} is the number density of background photons in the present universe. Therefore, the total injection energy is estimated as

$$\epsilon{\rm Br}_{X\gamma}Y_{N^{(1)}} \lesssim 3\times 10^{-18}\ {\rm GeV}\left(\frac{1/R}{500{\rm GeV}}\right)^2\left(\frac{0.1{\rm eV}}{m_\nu}\right)^2\left(\frac{\delta m}{1{\rm GeV}}\right)^2\left(\frac{\Omega_{\rm DM}h^2}{0.10}\right) \tag{10}$$

The successful BBN and CMB scenarios are not disturbed unless this value exceeds $10^{-9} - 10^{-13}$ GeV [16]. The prediction of the MUED model with right-handed neutrinos is several order of magnitude smaller than the bound. Therefore, the KK graviton problem is avoided in the MUED model with right-handed neutrinos. Moreover, this results is expected to hold for other UED models with right-handed neutrinos.

5. Summary and discussion

In this letter, we have pointed out that the introduction of right-handed neutrinos resolves two shortcomings of the MUED model simultaneously, which are the absence of neutrino masses and the KK graviton problem. A KK photon decays dominantly into a KK right-handed neutrino and an ordinary neutrino. In other words, no photon is emitted by the KK photon decay, and hence this model are free from constraints on the late time decay from the BBN, CMB, and diffuse photon measurements. With the calculation of the relic abundance of the KK photon, this fact allows us to consider a small compactification scale in the MUED model ($1/R \sim 600$ GeV), which is consistent with all results in particle physics experiments and observed abundance of dark matter.

The dark matter in this region is the KK right-handed neutrino and difficult to observe a signal in detection measurements for dark matter. However, smaller value of $1/R$ has a great advantage for collider experiments. For instance, not only first KK particles but also second KK ones can be produced at the LHC, and we can easily find signals of second KK gauge bosons using energetic dilepton channels [21]. Furthermore, the MUED model has only a few undetermined parameters, and these values will be observed in a good accuracy even at the LHC. Once the values are determined, we can predict the relic abundance of dark matter theoretically, and can discuss the connection between collider physics and cosmology by comparing the prediction with WMAP and Planck results [22]. The comparison will be a good tool to make sure that the UED model provides dark matter in our universe.

References

1. T. Appelquist, H. C. Cheng and B. A. Dobrescu, Phys. Rev. D **64** (2001) 035002.
2. H. C. Cheng, K. T. Matchev and M. Schmaltz, Phys. Rev. D **66**, 036005 (2002).
3. K. Agashe, N. G. Deshpande and G. H. Wu, Phys. Lett. B **514**, 309 (2001).
4. K. Agashe, N. G. Deshpande and G. H. Wu, Phys. Lett. B **511**, 85 (2001).
5. T. Appelquist and B. A. Dobrescu, Phys. Lett. B **516**, 85 (2001).
6. J. F. Oliver, J. Papavassiliou and A. Santamaria, Phys. Rev. D **67** (2003) 056002.
7. D. Chakraverty, K. Huitu and A. Kundu, Phys. Lett. B **558**, 173 (2003); P. Colangelo, F. De Fazio, R. Ferrandes and T. N. Pham, Phys. Rev. D **73**, 115006 (2006).
8. A. J. Buras, M. Spranger and A. Weiler, Nucl. Phys. B **660**, 225 (2003).
9. T. Appelquist and H. U. Yee, Phys. Rev. D **67**, 055002 (2003); I. Gogoladze and C. Macesanu, arXiv:hep-ph/0605207.
10. G. Servant and T. M. P. Tait, Nucl. Phys. B **650**, 391 (2003); M. Kakizaki, S. Matsumoto, Y. Sato and M. Senami, Phys. Rev. D **71**, 123522 (2005); M. Kakizaki, S. Matsumoto, Y. Sato and M. Senami, Nucl. Phys. B **735**, 84 (2006); K. Kong and K. T. Matchev, JHEP **0601**, 038 (2006); F. Burnell and G. D. Kribs, Phys. Rev. D **73**, 015001 (2006).
11. M. Kakizaki, S. Matsumoto and M. Senami, Phys. Rev. D **74**, 023504 (2006).
12. S. Matsumoto and M. Senami, Phys. Lett. B **633**, 671 (2006).
13. C. L. Bennett *et al.*, Astrophys. J. Suppl. **148**, 1 (2003); D. N. Spergel *et al.* [WMAP Collaboration], Astrophys. J. Suppl. **148**, 175 (2003); D. N. Spergel *et al.*, arXiv:astro-ph/0603449.
14. G. L. Fogli, E. Lisi, A. Marrone and A. Palazzo, Prog. Part. Nucl. Phys. **57**, 742 (2006), and reference therein.

15. W. Hu and J. Silk, Phys. Rev. Lett. **70**, 2661 (1993); W. Hu and J. Silk, Phys. Rev. D **48**, 485 (1993).
16. J. L. Feng, A. Rajaraman and F. Takayama, Phys. Rev. Lett. **91**, 011302 (2003); J. L. Feng, A. Rajaraman and F. Takayama, Phys. Rev. D **68**, 063504 (2003).
17. J. L. Feng, A. Rajaraman and F. Takayama, Phys. Rev. D **68**, 085018 (2003).
18. J. Hosaka *et al.* [Super-Kamiokande Collaboration], arXiv:hep-ex/0604011; K. abe *et al.* [Super-Kamiokande Collaboration], arXiv:hep-ex/0607059; E. Aliu *et al.* [K2K Collaboration], Phys. Rev. Lett. **94** (2005) 081802; M. H. Ahn [K2K Collaboration], arXiv:hep-ex/0606032.
19. K. Ichikawa, M. Fukugita and M. Kawasaki, Phys. Rev. D **71**, 043001 (2005); M. Fukugita, K. Ichikawa, M. Kawasaki and O. Lahav, Phys. Rev. D **74**, 027302 (2006).
20. D. J. Fixsen, E. S. Cheng, J. M. Gales, J. C. Mather, R. A. Shafer and E. L. Wright, Astrophys. J. **473**, 576 (1996); G. Smoot and D. Scott, arXiv:astro-ph/9711069.
21. A. Datta, K. Kong and K. T. Matchev, Phys. Rev. D **72**, 096006 (2005) [Erratum-ibid. D **72**, 119901 (2005)].
22. [Planck Collaboration], arXiv:astro-ph/0604069.

NEUTRON BACKGROUND IN LARGE-SCALE DARK MATTER DETECTORS

V. A. KUDRYAVTSEV ON BEHALF OF ILIAS AND UKDMC

Department of Physics and Astronomy, University of Sheffield,
Sheffield S3 7RH, United Kingdom
E-mail: v.kudryavtsev@sheffield.ac.uk
www.shef.ac.uk/physics

Neutron background for the high-sensitivity dark matter experiments is discussed. Neutron production via spontaneous fission and (α,n) reactions from U and Th, and by cosmic-ray muons is considered.

Keywords: dark matter, WIMPs, neutron background

1. Introduction

The sensitivity of future large-scale underground detectors for direct WIMP searches can be restricted by various types of backgrounds. Knowledge of background fluxes and ability to suppress or reject background events are essential for estimating the detector sensitivity, interpreting experimental results and designing future experiments. Background studies have been initiated by several collaborations to evaluate the potential of future large-scale experiments for dark matter searches, design their shielding and active veto systems. A coordinated approach for such studies, a possibility to share the Monte Carlo codes, test them and exchange all relevant information have been offered by ILIAS – Integrated Large Infrastructures for Astroparticle Science – a European Programme within Framework 6.

In this paper we present a short overview of neutron background that may limit the sensitivity of dark matter experiments. Neutrons underground arise from two sources: i) local radioactivity, and ii) cosmic-ray muons. Neutrons associated with local radioactivity are produced via spontaneous fission of ^{238}U and (α,n) reactions initiated by α-particles from U/Th traces in the rock and detector components. The muon-induced neutron flux depends strongly on the muon flux, i.e. the depth of the laboratory site.

2. Neutron background from radioactivity

To achieve the neutron-induced background rate of a few events/tonne/year in a dark matter detector, a suppression of neutron background from rock by a factor of 10^6 is required. To formulate the requirements for shielding and veto systems, this background should be accurately calculated.

The neutron yields and energy spectra from (α,n) reactions and spontaneous fission in the U/Th decay chains can be calculated using the code SOURCES.[1] A number of changes has recently been made to the original code. More cross-sections have been added to the code library either from experimental data[2] or from calculations[3] using the EMPIRE software package (version 2.19).[4] The upper limit for α energies has been extended to 10 MeV. The modified code SOURCES has been used to simulate neutron production in NaCl[2,5] and in the Modane rock.[3] Neutron spectra in both types of rock have peaks at 1-2 MeV and are softer than suggested in some other publications. This is due to the more accurate calculations of the transitions of the final nucleus to excited states in SOURCES that leaves less energy to neutrons. Despite higher radioactivity levels for Modane rock compared to NaCl, the neutron production rates were found to be very similar in both types of rock.[3] This is due to the higher (α,n) contribution per unit U/Th concentration in NaCl compared to Modane rock, because of low energy thresholds for (α,n) reactions on ^{23}Na (3.5 MeV) and ^{37}Cl (4.1 MeV). This results in 98% contribution from (α,n) reactions in NaCl (only 2% is coming from spontaneous fission) and only 77% contribution from (α,n) reactions in the Modane rock. The code SOURCES is now subject to further modifications and extensions which are briefly described in Ref. 6.

The neutron propagation can be done with MCNP(X)[7] or GEANT4[8] simulation codes. Recently a comparison between MCNPX-2.5 and GEANT4.7.0.p01 (with corrected inelastic cross-section on chlorine) has been carried out.[3] Starting from the neutron production spectra provided by SOURCES, neutrons have first been propagated to the rock boundary and then through different configurations of shielding that included lead and hydrocarbon in a simple geometry. The neutron flux at the rock/cavern boundary in the Modane rock is smaller than in NaCl despite higher U/Th concentrations and similar neutron production rates (discussed above). This is due to the presence of a small amount of hydrogen (in water) in the Modane rock and its absence in NaCl. Hydrogen serves as a good neutron moderator and the flux is very sensitive to the hydrogen abundance. Hydrogen reduces the neutron flux above 100 keV (1 MeV) at Modane by a factor 4.7 (1.8). Hence it is important to take into account small concentration of some isotopes (for instance, hydrogen) that can affect significantly neutron transport through moderation, inelastic scattering or capture.

The agreement between the two codes is satisfactory for the purpose of the shielding design. After 50 g/cm^2 of CH_2, MCNPX gives the differential flux 50% higher than the GEANT4 flux, translating into an additional 1-2 cm thick CH_2 layer. About 20 cm of lead and 50 g/cm^2 of CH_2 suppress the neutron flux below an observable level in a tonne-scale experiment. The thickness of CH_2 should be bigger if no lead is used. In practice, since lead is needed to shield a detector from rock gammas, a combination of lead and paraffin is probably the best option, unless very large thickness of paraffin or water is chosen to substitute lead as a gamma shield.

After suppression of rock neutrons is achieved through the installation of passive shielding, neutron background from detector components may become the limiting factor. For liquid noble gases PMTs may give the dominant contribution. Even ultra-low background PMTs can produce a rate of a few tens of events/tonne/year in liquid xenon.[2] For cryogenic detectors without PMTs the neutron background from detector components can be significantly smaller. The code SOURCES[1] can be used to calculate neutron yields and spectra from detector components (including shielding), and GEANT4 or MCNP(X) – to transport neutrons to the detector and estimate the nuclear recoil rate.

A factor of 2-3 suppression of neutron rate from internal detector components can be achieved by using an active veto system just around the main target.[9,10] Neutrons can hit the main target first and then give a signal (either proton recoils or gammas from neutron capture) in the veto.

3. Muon-induced neutrons

At deep underground sites ($\geq$ 3 km w. e.), the neutron production rate from muons is at least 3 orders of magnitude lower than the rate of neutrons from U/Th traces in rock. Muon-induced neutron flux can be important, however, for experiments intending to reach high sensitivity to WIMPs or other rare events. This flux is proportional to the muon flux but also increases with the mean muon energy underground (see, for example, Ref. 11 for discussion). As neutrons are produced not only in rock but also in all materials in and around a dark matter detector, the total neutron flux is strongly affected by the composition of the detector and its surroundings. It can be enhanced significantly by the presence of high-A (high-Z) target close to the detector. It can also be significantly reduced at energies below 10 MeV by the presence of hydrogen in the shielding. All these effects complicate simulations of the muon-induced neutron background. To make accurate predictions full Monte Carlo of the experimental setup is needed.

The muon flux has been measured for most underground laboratories giving enough of information for testing Monte Carlo codes for muon simulations. Muons can be transported through large thickness of rock using general particle physics codes such as GEANT4 or FLUKA,[12] or specially developed, simple, fast and widely used muon propagation codes such as MUSIC.[13] In most cases the muon transport through the rock is not needed and can be substituted with a sampling of muons according to the known muon energy spectrum and angular distribution at a particular depth. Such a sampling can be done with a code MUSUN[11] that has already been used as a starting point for neutron simulations.[2,5,11,14]

Two general-purpose codes GEANT4 and FLUKA have been used so far for production, transport and detection of muon-induced neutrons. The validation of the codes has been done through the comparison of them with each other[15,16] and with available experimental data.[11,15,17] Several models for neutron production in GEANT4 have been tested and compared to FLUKA predictions[16] allowing one to choose the most accurate model. The results of the simulations with the most accurate model agree reasonably well with most available experimental data.[15] Agreement within 50% has been found for neutron yields, spectra and lateral distributions in light materials but a possibility for a significantly larger neutron yield in heavy targets, such as lead, compared to the GEANT4 or FLUKA predictions, still remains as discussed in Refs. 11 and 15. However, most experiments that measured muon-induced neutrons, did not present accurate Monte Carlo simulations which would include their set-ups together with production, transport and detection of all particles produced by muons. This makes the interpretation of experimental results difficult.

Most available experimental data involved detection of delayed pulses from neutron capture on various isotopes: hydrogen, gadolinium and others. This leads to two important consequences. Firstly, simulations of neutron production by muons in a simple geometry[11,15,17] cannot reproduce the complexity of experiments and thus, accurate Monte Carlo of all physical processes, geometry and other experimental conditions and comparison with data should be done by the groups carrying out the experiments, to allow any definite conclusion to be derived. Secondly, detectors measured the rate of already thermalised neutrons, not the flux of fast neutrons which can potentially mimic WIMP interactions. So whether there is or not an agreement between the neutron measurements and GEANT4 or FLUKA simulations of neutron yields, this is hard to extrapolate to the flux of fast (> 1 MeV) neutrons dangerous for direct dark matter searches.

For high-energy muons (>100 GeV) and most targets the neutron yield from GEANT4 has been found to be less (by up to a factor of 2) than that from FLUKA.[15,16] Direct neutron production in muon spallation reactions is similar in both codes.[15] The neutron production in electromagnetic cascades always dominates in GEANT4 over that in hadronic cascades. In FLUKA hadronic cascades play more important role at high muon energies (>10 GeV) in most materials.[11] GEANT4 shows enhanced neutron photoproduction in electromagnetic cascades in light materials compared to FLUKA. For heavy elements, FLUKA overproduces neutrons in both electromagnetic and hadronic cascades initiated by muons compared to GEANT4. Despite these differences, the simulated total fluxes of fast neutrons (>1 MeV) from salt rock entering the underground cavern were found to agree within 20%.[15] The neutron fluxes generated by FLUKA and GEANT4 behind the lead and hydrocarbon shielding also agree.

A few important points to note about simulations of muon-induced neutrons. Total neutron yields in light and heavy materials are very much different and the spectra of neutrons are different too. In heavy materials the enhancement of the neutron flux occurs mainly at low and intermediate energies (< 20 MeV), whereas the spectra above 20 MeV are not much different (in shape and absolute fluxes) from those in light targets.[11,15,17] Considering the energy range of interest to dark matter searches (>1 MeV), any extrapolation from one material to another one without accurate Monte Carlo may result in a significant error in the neutron flux.

Table 1. Neutron fluxes in Boulby rock above specified energies as calculated in Ref. 18 and in Ref. 2 (in units 10^{-9} cm^{-2} s^{-1}).

	Total	> 1 keV	> 1 MeV	> 10 MeV	> 100 MeV
Ref. 18	4.86		1.34	1.11	0.277
Ref. 2		5.61	0.87	0.25	0.064

Recently new simulations of the muon-induced neutrons have been carried out for several underground sites[18] leading to the conclusion of higher neutron-induced rates in dark matter detectors than calculated by Carson et al.[2] We have found, however, that the neutron fluxes coming out of the Boulby rock, calculated in these two papers using the same code FLUKA (but most likely different versions) are quite different. The difference reaches a factor of 4 or higher for neutron energies above 10 MeV (see Table 1).

In Ref. 18 a wrong rock composition for Boulby was used: instead of using NaCl as specified in Ref. 2, the authors used the average values for Z and A for all rock between the surface and the Boulby Underground Laboratory. This should not cause, however, the reported large difference between the neutron fluxes at high energies.

It is attractive to simplify simulations and save the CPU time by doing this in two stages and separating neutron production from neutron transport and detection. Such an approach may lead to a big error in the neutron event rate estimate due to the following reasons. Neutrons, and particularly fast neutrons, are not emitted isotropically with respect to the muon direction. They are emitted preferentially along the muon path.[17] If the second stage of simulation starts with sampling neutrons (not muons) according to the pre-defined energy spectrum, the angular distribution of these neutrons is not known since the information about direction, energy etc. of the muon (or other particle that produced this neutron) was lost after the first step of simulation (neutron production by muons) was completed. Whatever assumption is made about the neutron angular distribution, it will not be accurate since muon angular distribution is not isotropic. Also, separating neutron transport and detection from neutron production means that an event which may have neutron-induced recoil AND other energy deposition associated with a muon or other particles produced by the muon, could be seen as an event with nuclear recoil only, thus increasing the rate of background mimicking WIMP signal.

The case of 250 kg xenon detector has been considered in Refs. 2 and 15. The spectra of energy depositions have been found to be very similar in GEANT4 and FLUKA.[15] Both codes predict similar rate of nuclear recoils and similar rejection factor for most events with nuclear recoils due to the presence of energy deposition from other particles in muon-induced cascades or muons themselves. In fact, only a few percent (5-8%) of nuclear recoils in a large-scale xenon detector are 'pure' nuclear recoils without other energy depositions. Such coincidences will be missed if the simulation of production, propagation and detection are done for neutrons only but not for all other particles.

The rate of single nuclear recoils from muon-induced neutrons in a large xenon detector is expected to be about (9 ± 3) events/tonne/year at 10-50 keV Xe recoil energies. This is the rate of events not accompanied by any other energy deposition due to muons or muon-induced secondaries (electrons, photons, hadrons). With an active veto around the detector this rate can be reduced to below 1 event/tonne/year.

4. Conclusions

A good agreement between MCNPX and GEANT4 is found for propagation of low-energy neutrons in various materials. A shielding equivalent to about 55 g/cm^2 of CH_2 (for instance 20 cm of Pb plus 45 g/cm^2 of CH_2) would be suitable for one tonne experiment. The simulations show that muon-induced neutrons at the depth of $\sim$3 km w. e. will not limit the sensitivity of tonne-scale xenon dark matter detectors down to $(2-3)\times 10^{-10}$ pb at the minimum of the sensitivity curve even without a veto system.

Acknowledgments

This work was supported by the ILIAS integrating activity (Contract RII3-CT-2004-506222). The author is grateful to all members of the working group on background studies within the JRA1 and N3 activities of ILIAS for their contributions to this work.

References

1. W. B. Wilson et al., SOURCES-4A, Technical Report LA-13639-MS, Los Alamos (1999).
2. M. J. Carson et al., Astrop. Phys. **21** (2004) 667.
3. R. Lemrani et al., Nucl. Instrum. & Meth. in Phys. Res. **A560** (2006) 454.
4. M. Herman et al., EMPIRE-2.19 (2005), http://www.nndc.bnl.gov/empire219.
5. M. J. Carson et al., Nucl. Instrum. & Meth. in Phys. Res. **A546** (2005) 509.
6. V. Tomasello and V. A. Kudryavtsev, this proceedings.
7. MCNP-4B, Technical Report LA-12625-M (Ed. J. F. Briesmeister), Los Alamos, 1997.
8. S. Agostinelli et al., Nucl. Instrum. & Meth. in Phys. Res. **A506** (2003) 250.
9. M. J. Carson et al., Nucl. Instrum. & Meth. in Phys. Res. **A548** (2005) 418.
10. C. Bungau et al., Astrop. Phys. **23** (2005) 97.
11. V.A.Kudryavtsev et al., Nucl.Instrum. & Meth. Phys.Res. **A505** (2003) 688.
12. A. Fassò, A. Ferrari and P. R. Sala, Proceedings of the MonteCarlo 2000 Conference, ed. A.Kling, F.Barao, M.Nakagawa, L.Tavora, P.Vaz (Springer-Verlag, 2001), p. 159; A.Fassò, A.Ferrari, A.Ranft and P.R.Sala, *ibid.* p. 995.
13. P. Antonioli et al., Astrop. Phys. **7** (1997) 357; V. A. Kudryavtsev et al., Phys. Lett. **B471** (1999) 251.
14. H.Wulandari et al., Proc. 5th Intern. Workshop on the Identification of Dark Matter, ed. N.Spooner and V.Kudryavtsev (World Scientific, 2005), p. 477.
15. H. M. Araújo et al., Nucl. Instrum. & Meth. in Phys. Res. **A545** (2005) 398.
16. M. Bauer et al., Proc. 5th Intern. Workshop on the Identification of Dark Matter, ed. N.Spooner and V.Kudryavtsev (World Scientific, 2005), p. 494.
17. Y. F. Wang et al., Phys. Rev. D 64 (2001) 013012.
18. D.-M. Mei and A. Hime, Phys. Rev. D 73 (2006) 053004.

BACKGROUND STUDIES FOR A TON-SCALE ARGON DARK MATTER DETECTOR (ArDM)

L. KAUFMANN and A. RUBBIA

Institute for Particle Physics, ETH Zurich, Switzerland
E-mail: lilian.kaufmann@cern.ch, andre.rubbia@cern.ch

The ArDM project aims at operating a large noble liquid detector to search for direct evidence of Weakly Interacting Massive Particles (WIMP) as Dark Matter in the universe. Background sources relevant to ton-scale liquid and gaseous argon detectors, such as neutrons from detector components, muon-induced neutrons and neutrons caused by radioactivity of rock, as well as the internal ^{39}Ar background, are studied with simulations. These background radiations are addressed with the design of an appropriate shielding as well as with different background rejection potentialities. Among them the project relies on event topology recognition, event localization, density ionization discrimination and pulse shape discrimination. Background rates, energy spectra, characteristics of the background-induced nuclear recoils in liquid argon, as well as the shielding performance and rejection performance of the detector are described.

1. Introduction

Astronomical observations suggest that over 80% of the matter contained in the universe is Dark Matter, which presumably consists of unknown and invisible WIMPs (Weakly Interacting Massive Particles). The most common candidate is the lightest supersymmetric particle. Direct detection experiments aim at measuring elastic scattering of WIMPs with nuclei. The energy transferred to the nuclei is typically below 100 keV. Due to this low energy range and the small cross section, such a signal is difficult to measure and signal sensitivity is limited by ordinary backgrounds. Background radiation arises from several sources and is of different significance. One dominant background source is the contamination with radioactive elements, namely uranium and thorium chains, and potassium-40 resulting in α, β and γ emissions as well as neutron production processes, of the materials used in and around the detector. Another source of neutron background in underground environments comes from interactions of cosmic muons with matter, resulting in neutron production. A third background source can arise from internal contamination of the target material, e.g. with radioactive isotopes.

The experiment for which the present study has been carried out is ArDM [1–4], a ton-scale liquid argon detector. The technical concept of the experiment relies on the independent readout of ionization charge by LEM and of scintillation light by photo-detectors [4]. This paper summarizes the results of Monte Carlo studies carried out to assess the impact of different background radiation in this project.

2. Neutron background

An important background is neutron-induced nuclear recoils which are hardly distinguable from WIMP events. The neutrons from radioactivity arise mainly from contaminations of materials or rock by the two radioactive elements uranium and thorium, and their decay-chain daughters. The decay chains of ^{238}U and ^{232}Th contain α decays with α energies of 3.5 to 11 MeV. These α undergo (α,n) reactions, thereby producing neutrons with energies in the MeV range. The cross section depends on the material and on the α-energy [5]. Typical yields are 10^{-8} to 10^{-5} neutrons per α. Besides (α,n) reactions, neutrons can arise from spontaneous fission. The number of neutrons arising from these two processes is typically of the same order of magnitude, but depends on the material. Muon-induced neutrons arise from cosmic muon interactions with surrounding materials. Highly energetic muons are able to penetrate deep underground. Neutrons are produced by spallation or photonuclear processes, or by secondary interactions of muon-induced hadronic showers. The shielding of the detector optimized for natural radioactivity is less efficient in this case and faster neutrons can penetrate inside the detector fiducial region. Shielding and detector components can also act as a target for muons, however the expected production rates are low.

2.1. *Neutrons from radioactivity in detector components*

The ArDM detector parts consist mainly of the following materials: stainless steel, vetronite, polyethylene, borosilicate glass, ceramics and copper (see Ref. [4] for details). In general, metallic materials contain much less U/Th contamination, i.e. typically a few ppb. Minerals and glasses generally have higher contaminations of the order of a few hundred ppb. We have estimated the number of emitted neutrons using the data from Ref. [5]. Furthermore, the emission numbers have been simulated with SOURCES4a for cross-check. The numbers obtained with the two different approaches agree well. The biggest contribution for standard materials comes from photomultiplier tubes located on the bottom of the detector and large electron multiplier (LEM) plates located on the top of the detector, since these two parts contain glass. Low background versions of these two components should reduce the rate by about two orders of magnitude. The resulting neutron production rates and the precise assumptions for component masses and contaminations are summarized in Table 1.

Table 1. Assumptions for detector components and estimated neutron production based on [5] and SOURCES simulation. A description of the detector geometry can be found in [2–4].

Component	Mass (kg)	ppb U	ppb Th	n/year
Dewar (steel)	1000	0.6	0.7	380
LEM Vetronite	2	1000	1000	10000
LEM low bg.	4	<2	<2	<40
PMT std. version	2.4 (80 tubes)	400	400	12000
PMT low bg.	9.8 (14 tubes)	30	30	1400
Pillars	13 (8 Pillars)	20	20	310

2.1.1. *Neutron spectra*

The energy spectra of neutrons coming from U and Th decay chains have two contributions, namely the one from spontaneous fission and the other from (α,n) reactions. The spontaneous fission spectrum is described by $dN/dE \propto \sqrt{E}\exp(-E/1.29)$. About 2 neutrons are emitted per spontaneous fission. The spectrum of neutrons coming from (α,n) reactions is more involved, since the (α,n) cross section is material-dependent. For the computation of the spectra in detector component materials, the simulation program SOURCES4a was used. The code was extended to α energies above 6 MeV by [6]. The resulting spectra for stainless steel (dewar) and Vetronite (LEM) are shown in Fig. 1.

2.2. *Neutrons from radioactivity in rock and concrete*

The minerals constituting the rock overburden in an underground laboratory also contain small amounts of U and Th, causing neutron radiation in the same way as in the detector components. The level of the contamination

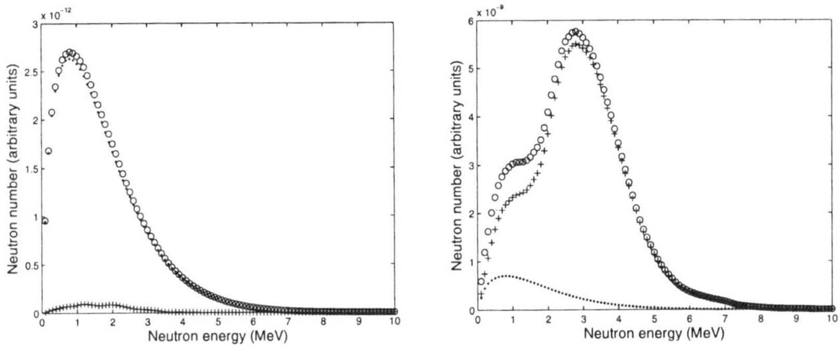

Fig. 1. Simulated neutron energy spectra for the dewar (left) and LEM (right). The dotted line is the contribution from spontaneous fission, the crossed line the contribution from (α,n.) reactions and the circled line is the sum of the two.

strongly depends on the location and the elemental composition of the rock. Usually the U/Th contamination is on the order of a few hundred ppb. The studies described here have been carried out for the Canfranc Underground Laboratory. It is located in the Pyrenees at a depth of 2450 m.w.e. A concrete layer covers the walls of the laboratory, which also affects the neutron radiation flux.

CH_2 is the typical material used as shield against these ambient neutrons. The number of neutrons coming from rock and concrete and reaching different depths of the shielding was simulated with GEANT-4 [8]. The neutron flux decreases by approximately one order of magnitude after every 10 cm of CH_2 at energies up to 6 MeV, for higher energies the suppression is slightly less. The suppression for concrete neutrons is slightly lower, especially for higher energies. On the average, 60 cm of CH_2 reduce the flux of neutrons by about six orders of magnitude.

2.3. *Detector response to neutrons*

Background neutrons interact with the detector via elastic scattering with argon nuclei. The imparted recoil energy on the nucleus caused by a neutron with energy E_n and a scattering angle θ is $E_R \simeq 2E_n \frac{M_n M_{Ar}}{(M_n+M_{Ar})^2}(1-\cos\theta)$. Neutrons from the various sources mentioned above have been simulated with the proper energy spectrum and fully propagated through a detailed geometry of the detector with GEANT-4. As an example, the resulting argon recoil spectrum for neutrons from detector components is shown in Fig. 2.

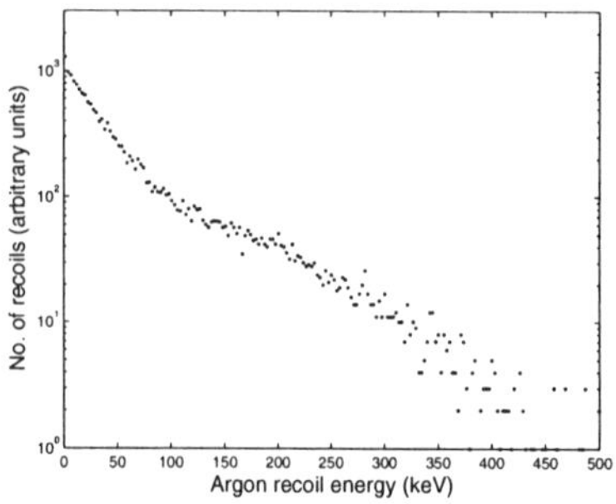

Fig. 2. Spectrum of argon recoils caused by neutrons from detector components.

Of the neutrons entering the fiducial volume, more than 50% can be rejected because of multiple scattering, since a WIMP would never scatter more than once within the detector active volume. The number of neutron scatters depends on the size of the fiducial volume, the distance of the neutron emitting components to the fiducial volume and on the energy threshold. Assuming that a WIMP-like event is a single recoil with an energy between 30 keV and 100 keV, approximately 5% of the neutrons from detector components produce WIMP-like events. The number of remaining events per year is 12 for the neutrons from the dewar, 800 for the standard material LEM, below 2 for the low background LEM, 480 for the standard PMTs, 56 for the low background PMTs and 19 for the pillars; to be compared with 3500 WIMP events at $\sigma_{\mathrm{WIMP-nucleon}} = 10^{-7}$ pb.

3. ^{39}Ar electron background

Commercially available Argon is procured by liquefaction of air and contains radioactive isotopes. ^{39}Ar decays via β-disintegration into ^{39}K with a half-life of 269 years and Q=565 keV. The concentration of ^{39}Ar in atmospheric argon is $(7.9\pm0.3)\cdot 10^{-16}$g/g [7], causing a decay rate of 1 kHz in one ton of argon. γs from U/Th of the detector components produce an interaction rate which is about three orders of magnitude smaller.

The rejection of electron and γ events is facilitated by two means: charge/light ratio discrimination [1] and pulse shape discrimination [9]. The first uses the fact that the ionization yield of nuclear recoils is highly quenched compared to that of electron/γ, while the scintillation yield is similar. The pulse shape analysis relies on different populations of the fast and slow components of scintillation. In order to overcome the internal ^{39}Ar background, a combined rejection power of about 10^8—depending on the WIMP-nucleon cross section—is required.

4. Conclusion

Three main background sources for the ArDM experiment—a liquid argon ton-scale detector—have been studied. Neutron radiation is the most important background, since neutron events have the same signature inside the detector as WIMP events. The main neutron background comes from contaminations of detector components with radioactive elements. Neutrons from rock and concrete of the underground laboratory can be shielded by a thick enough CH_2 shield. The internal ^{39}Ar background is strongly suppressed if the light/charge ratio and the scintillation light time distribution are measured precisely enough.

References

1. A. Benetti *et al.*, Nucl. Instrum. Meth. A 327 (1993) 203
2. A. Rubbia, Talk given at APP UK 2006, First Meeting of the Astroparticle Physics Group of the IOP, Sheffield (UK), May 2006
3. L. Kaufmann and A. Rubbia, arXiv:hep-ph/0611288.
4. A. Rubbia, J. Phys. Conf. Ser. **39**, 129 (2006) [arXiv:hep-ph/0510320].
5. R. Heaton *et al.*, Nucl. Instrum. Meth. A 276 (1989) 529
6. M. J. Carson *et al.*, Astropart. Phys. 21 (2004) 667
7. H. H. Loosli, Earth and Planetary Science Letters, 63 (1983) 51 and "Nachweis von ^{39}Ar in atmosphärischem Argon", PhD thesis University Bern 1968
8. S. Agostinelli *et al.* [GEANT4 Collaboration], Nucl. Instrum. Meth. A **506**, 250 (2003).
9. M. G. Boulay and A. Hime, Astropart. Phys. **25**, 179 (2006).

CHARACTERIZATION OF THE CANFRANC UNDERGROUND LABORATORY: STATUS AND FUTURE PLANS

G. LUZÓN*, J.M. CARMONA, S. CEBRIÁN, F. IGUAZ, I.G. IRASTORZA, H. GÓMEZ, J. MORALES, A. ORTIZ de SOLÓRZANO, A. RODRÍGUEZ, J. RUZ, A. TOMÁS and J.A. VILLAR

Laboratory of Nuclear and High Energy Physics,
University of Zaragoza, E-50009 Zaragoza, SPAIN
**E-mail: luzon@unizar.es*

The new *Laboratorio Subterráneo de Canfranc* (LSC) was inaugurated on March 2006. We report the first measurements of radon concentration, gamma background and muon fluxes performed in the Laboratory.

Keywords: Enviromental radioactivity; Underground laboratory; Low background

1. Introduction

The scientific interest in underground physics is growing worldwide. Underground laboratories shield against cosmic rays and provide a low radioactive background environment essential to look for dark matter candidates, investigate neutrino properties, detect gravitational waves and, in general, study rare event physics.

Since March 2006, European Underground Science has at its disposal a new large international installation located under the Aragonese Pyrenees. This facility is conceived as a Consortium of the Spanish Ministry of Education and Science, the Aragon Regional Goverment and the University of Zaragoza. Constructed 100 m away from the old Laboratory, the vertical overburden at the site provides 2500 Meters Water Equivalent (MWE) of shielding from cosmic rays and offers a low background environment for the next generation of experiments exploring the frontiers of particle and astroparticle physics. By the beginning of 2007, the first experiments hosted by the new Laboratory are expected to be installed in the LSC.

The new experimental hall ($40\times15\times12$ m^3) is oriented towards CERN to allow experiments using neutrino beams coming from there. A smaller hall used as an ultra low background facility ($15\times10\times8$ m^3), offices, clean room and workshops for a total of 1,000 m^2 completes the installation. The air conditioning system operates continuously with an air inlet flow of 11,000 m^3/h from outside of the tunnel. The ventilation system circulates and filters also a flow of 25,000 m^3/h. In standard conditions, the temperature is 20° and the humidity level is stable around 30% in winter and 40% in summer. Fig. 1 shows a view of the new LSC.

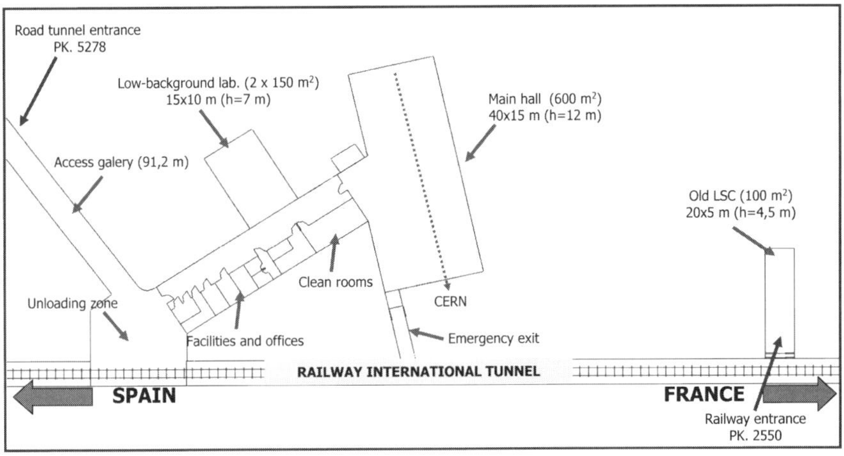

Fig. 1. Squeme of the new installation.

As a new facility, it is required to know the background levels of the Laboratory. The first characterisation measurements began in January 2006. Preliminary data corresponding to radon concentration, gamma background and muon fluxes are presented in this paper.

2. Radon concentration

Two ALPHAGUARD PQ2000PRO detectors, with an active volume of 0.56 l and 2 Bq/m^3 as lower limit of Rn concentration, have been used for measurements in real time. These detectors allow also to determine selected climatic parameters (air temperature, pressure and humidity) connected with radon concentration.

The first and second sets of data have been registered every 30 min. (integration time) for periods of 6-7 days at different points of Hall A (main experimental Hall) and Hall B, always 1 meter above floor level, and also at the air entrance. Table 1 summarises the radon average values for the period January–June of 2006. A third set of measurements taken from July 2006 until now is under analysis showing similar results.

Table 1. Average values of radon concentration (Bq/m^3).

Hall A	Hall B	Air entrance
62.2±0.4	71.2±0.4	72.0±1.9

The values of the radon concentration at Halls A and B and at the air entrance tell us that our ventilation system works properly eliminating the radon emanation from the laboratory walls. The quite high value of radon concentration registered at the air entrance is due to the around 22 minutes of transit time of the outer air through concrete tubes.

Radon detectors show us also that temperature is under control around 20° up to 0.1°C and that humidity levels are stable below 30% in winter and 40% in summer. Also a slow control of parameters as temperature, humidity, pressure and radon is underway.

3. Gamma background

Gamma ray measurements have been performed using a standard high efficiency 3″ × 3″ NaI(Tl) detector manufactured by Saint-Gobain. Its resolution is of 6.6% at the 662 keV peak of ^{137}Cs. The detector has been supplied by CANBERRA in a hermetically sealed assembly which includes the crystal, a photomultiplier tube, an internal magnetic/light shield, an aluminium housing, and a 14-pin connector. This assembly plugs directly into a uniSpec Tube-Base MCA. The whole system is 2.0–USB–connected to a laptop which controls both the power supplied and the acquisition using Genie 2000 software.

The identification of a set of peaks in the background spectra coming from potassium and the radioactive chains of uranium and thorium (see Fig. 2), allowed us to estimate the number of photons. The detector efficiency has been calculated thanks to a calibrated source of ^{152}Eu and Geant4 simulations.[1]

A first set of measurements was performed in June 2006 (three hours), with the detector placed at the center of the Low Background Laboratory (Hall B) one meter above floor surface. Table 2 collects results for the

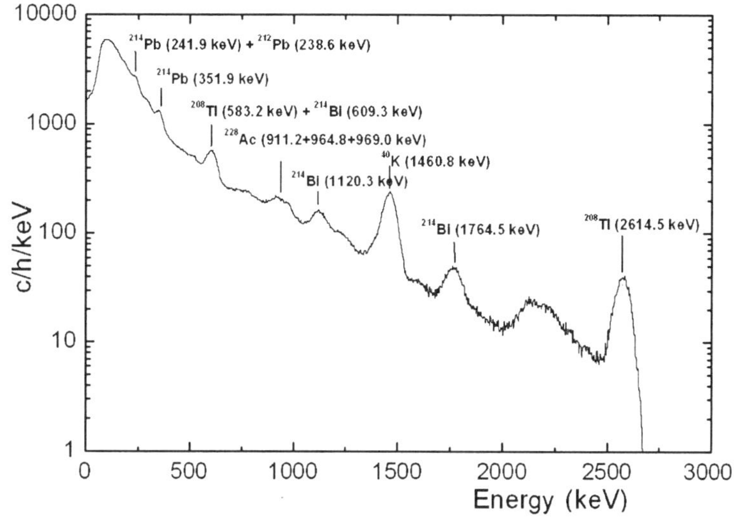

Fig. 2. Typical background spectrum, including identified peaks, measured at the center of the Hall B.

radioactive chains and potassium. Mean activity values of the surrounding walls have been also estimated assuming these walls to be the main sources of the gamma background.

Table 2. Estimated gamma values at the center of the Hall B.

Nuclide	γ ($cm^{-2}s^{-1}$)	Activity (Bq/m^3)
^{40}K	0.33±0.01	169±5
^{238}U	0.71±0.12	41.4±7.0
^{232}Th	0.85±0.07	34.4±2.8

The total number of photons, 1.9±0.2 $cm^{-2}s^{-1}$, is in good agreement with LNGS reported values[2] (1 γ $cm^{-2}s^{-1}$). At present we are involved in more detailed measurements of walls radioactivity in both, Hall A and Hall B. A more precise qualitative and quantitative determination of the activity of the walls using High purity Germanium detectors inside the ILIAS (Integrated Large Infrastructures for Astroparticle Science)[3] project is foreseen for the near future.

4. Muon fluxes

Muon measurements have been performed with two $80\times40\times5\,\mathrm{cm}^3$ BICRON BC408 plastic scintillator detectors using a standard coincidence system. Such a device has been checked at our laboratory in Zaragoza, 200 meters above sea level. With this aim we have studied two different setups for the detectors: joined both detectors together and with 1 meter of separation. Results are in good agreement with Geant4 simulations. The flux of muons obtained after a run of 12 hours is $(1.26 \pm 0.07) \times 10^2$ m^{-2} s^{-1}, in good agreement to the number calculated from the review of particle data[4] for our detector configuration, 1.25×10^2 m^{-2} s^{-1}.

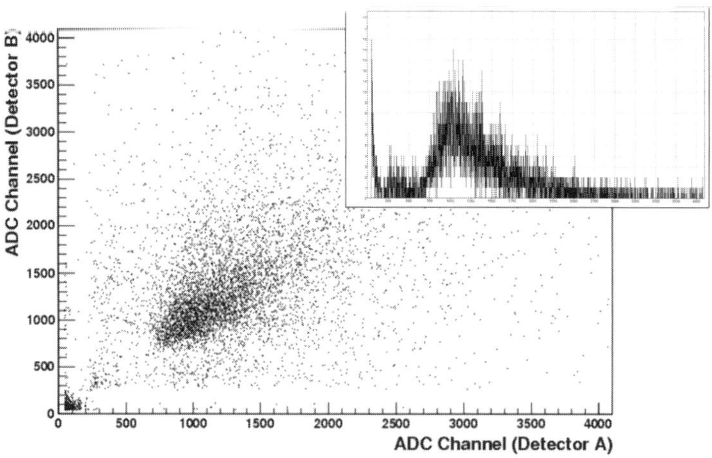

Fig. 3. Coincidence plot obtained in the LSC (Hall B) after 50 days of data. Right on the corner, muon spectrum collected by one of the detectors.

Then, we moved the whole system to the Canfranc Underground Laboratory where two 50–days–measurements have been performed in both Hall B and Hall A. Fig. 3 shows the coincidence plot corresponding to the first measurement, taken at Hall B. The two set of data have allowed us to calculate a preliminary number for muon fluxes at the LSC main Hall (Hall A), $(4.71 \pm 0.06(\mathrm{stat.}) \pm 0.24(\mathrm{sys.})) \times 10^{-3}$ m^{-2} s^{-1}, and also at the Hall B, $(5.04 \pm 0.06(\mathrm{stat.}) \pm 0.25(\mathrm{sys.})) \times 10^{-3}$ m^{-2} s^{-1}, values about 2 times higher than a previous measurement at the old laboratory. The difference could be due to the profile of the mountain whose maximum overburden is just on top of the old laboratory, almost 100 meters far away from the new installations. To check this hypothesis, new measurements in the old laboratory and a complete program of simulations to study the influence of the mountain profile are underway.

5. Neutrons

Up to the moment, we can just give a total number for neutrons at the old laboratory, 3.8×10^{-6} cm^{-2}s^{-1}, inferred from IGEX-DM experiment data.[5] However, in the near future neutron measurements are planned in the framework of the ILIAS collaboration.

6. Conclusion

Preliminary data about background sources at the new Canfranc Laboratory have been reported here. They point out to an average radon concentration of about 70 Bq m^{-3}, a number of photons of 1.9 ± 0.2 cm^{-2}s^{-1} and a flux of muons of around 5×10^{-3} m^{-2} s^{-1}. The characterisation program will continue during next year with more detailed measurements and simulations. We expect also in this time to be able to get data about neutrons in the Laboratory.

Acknowledgements

This research was partially funded by the Spanish Ministry of Education and Science (MEC) under contract No. FPA2004-00974. We also acknowledge the funding from the EU FP6 project ILIAS (contract number RII3-CT-2004-506222).

References

1. S. Agostinelli *et al.* [GEANT4 Collaboration], *NIM A* **506** 250 (2003).
2. C. Arpesella, *Nucl. Phys.B* (Proc. Suppl) **28** 420 (1992).
3. `http://ilias.in2p3.fr/`
4. W.M. Yao *et al.*, *J. Phys. G: Nucl. Part. Phys.* **33** 1 (2006).
5. J.M. Carmona *ct al.*, *Astr. Phys.* **21** 523 (2004).

FIRST MEASUREMENT OF LOW INTENSITY FAST NEUTRON BACKGROUND FROM ROCK AT THE BOULBY UNDERGROUND LABORATORY

E. TZIAFERI*, M. J. CARSON, V. A. KUDRYAVTSEV, R. LERNER,
P. K. LIGHTFOOT, S. M. PALING, M. ROBINSON and N. J. C. SPOONER

Department of Physics and Astronomy, University of Sheffield, Sheffield S3 7RH, UK
** E-mail: e.tziaferi@sheffield.ac.uk*

A technique to measure low intensity fast neutron flux has been developed in order to measure the neutron background from rock at the Boulby Underground Laboratory (2800 m.w.e.). The experiment was performed using a liquid scintillation detector loaded with Gd. A flux of fast neutrons from rock was measured as $(1.72 \pm 0.61(stat.) \pm 0.38(syst.)) \times 10^{-6}$ cm^{-2} s^{-1} above 0.5 MeV.

1. Introduction

Neutrons are an important background for a large variety of underground experiments. It is crucial that the neutron flux is suppressed by shielding, active veto systems and the use of ultra-pure materials. To design shielding and veto systems for sensitive detectors, the neutron flux from rock and other materials should be known. We present the first measurements of fast neutron flux from rock at the Boulby Underground Laboratory. A two-pulse signature (proton recoils followed by gammas from neutron capture) was used to identify the neutron events. The experiment was performed with a liquid scintillation detector (6.1 litre) loaded with Gd to increase the probability of neutron capture. The detector was viewed by two PMTs and it was installed inside a 'castle' made of ultra-pure copper and lead to ensure the suppression of the gamma flux from the rock.

2. Calibrations

To characterise the detector and its response to various radiations the following calibrations were carried out:

(i) Energy calibration using gamma-rays from several radioactive sources. Accurate calibration was possible by comparing the measured energy distribution with simulations which were carried out using the GEANT4 toolkit [1] taking into account the configuration of the detector, shielding and the position of the source. (ii) Calibration with neutrons from a ^{252}Cf source with the detection of delayed coincidences to demonstrate the sensitivity of the detector to neutrons and to estimate efficiency. (iii) Calibration with gamma-rays from ^{60}Co source with the detection of delayed coincidences.

The DAQ was configured to detect two pulses in delayed coincidence within a time window of 200 μs – the first pulse due to proton recoils from neutron elastic scattering and the second due to gammas from neutron capture. As the incident neutron flux from the ^{252}Cf source was very high, there was a non-zero probability of detecting random coincidences within 200 μs between two pulses produced by proton recoils from two neutrons or two pulses produced by gammas from two captured neutrons. Fig. 1(a) shows the time delay distribution of the two pulses. The presence of random coincidences described above adds a flat distribution superimposed on the exponential. From the fitting of the time delay distribution, the index of the exponential was $\tau = 84.21 \pm 5.50$ μs and the rate, having the source 1 m from the detector, was 0.53 ± 0.04 s^{-1}. The time delay distribution from ^{60}Co calibration is shown in Fig. 1(b). The presence of the two pulses in an event in this case is due to random concidences of two gammas. The time delay distribution is flat proving that the exponential shape observed with the neutron source is not an effect of DAQ but is due to neutron events.

In addition, a ^{252}Cf run was carried out recording single pulses (mainly proton recoils with some admixture of gammas), without delayed coincidences. The energy spectrum of proton recoils as measured with the neutron source on top of the castle is shown in Fig. 2a (dashed histogram). Also the corresponding spectrum (without any normalisation) calculated with GEANT4 (solid histogram) is shown, taking into account typical quenching of scintillations for protons in organic liquid scintillators [2]. The small difference, at high energies between the measured and simulated spectra is likely due to the presence of gammas from neutron capture in the calibration data. The agreement in the rates allows us to conclude that GEANT4 simulates neutron transport and production of proton recoils with reasonable accuracy.

The calibration runs described above proved that the detector is sensitive to neutrons and that the Monte Carlo simulations using the GEANT4 toolkit were able to reproduce the energy spectrum and absolute rate of proton recoils.

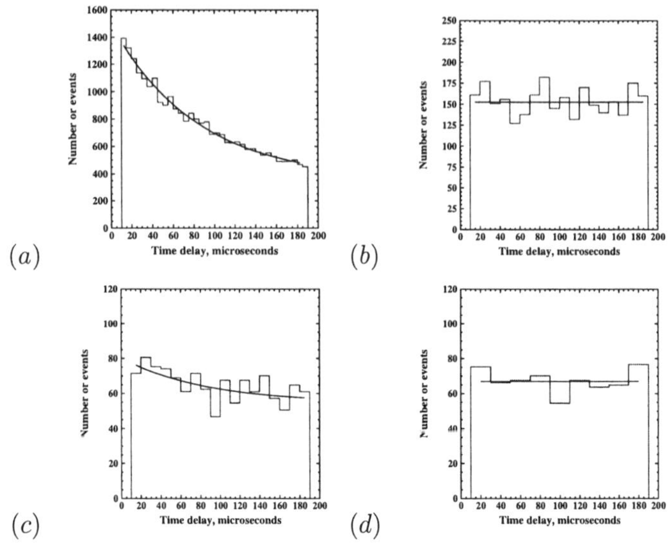

Fig. 1. Distribution of time delay between the 1st and 2nd pulses in events from (a) ^{252}Cf source, (b) ^{60}Co source, (c) data with unshielded detector; (d) data with shielded (CH_2) detector.

3. Measurements of the neutron flux from rock

The experiment to measure the neutron flux at the Boulby Undergound Laboratory was carried out from November 2004 to September 2005. The first three months of the experiment were dedicated to the study of the behaviour of the detector and its background.

An important part of the experiment was the rejection of background events. Two sources of background events producing delayed coincidences were found in this experiment. The first are afterpulses from the PMTs and the second source of background arises from two consecutive decays in the ^{238}U decay chain: ^{214}Bi$\rightarrow^{214}$Po $\rightarrow^{210}$Pb. The first decay goes via beta emission whereas the second one is an alpha-decay. The half-life of ^{214}Po is 164 μs which makes the two pulses from this decay chain capable of mimicking the signal from neutrons. The distribution of the time delay between the first and second pulses in the events from this correlated background is well fitted with an exponential of 237 μs consistent with the origin of this background being due to the Bi$\rightarrow$Po$\rightarrow$Pb decays. These background events were rejected using pulse shape analysis: the second pulses are due to alphas in the correlated background events whereas for neutron events they are due to Compton electrons.

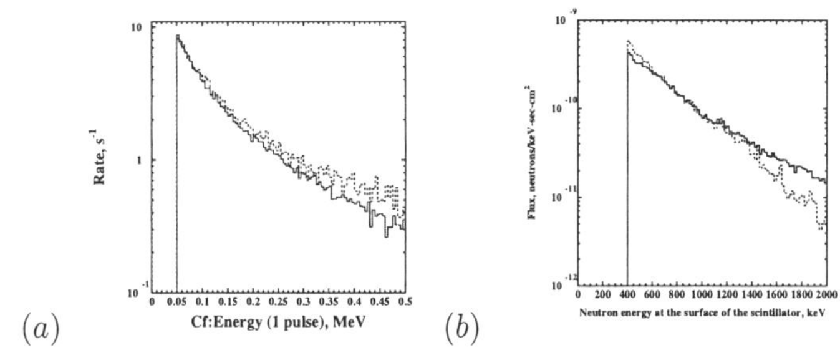

Fig. 2. (a) Proton recoil spectra (rates in absolute units per 5 keV without any normalisation) from data (dashed histogram) and simulations (solid histogram) of ^{252}Cf run. (b) Flux of neutrons, which produced proton recoils, from the background (dashed histogram) and the Cf source (solid histogram), at the outer surface of the scintillator.

Data included in the analysis of neutrons from rock were collected during 123 days of live time. After rejection of events from these two sources of background, there remained two types of events in the data: neutron events caused by radioactivity in rock and random coincidences of background pulses within a 200 μs time window. The distribution of time delayes between the two pulses are shown for all events included in the analysis (after cuts) in Fig. 1c. To obtain the number of neutron events, the measured distribution was fitted to the same formula as used for the neutron calibration analysis, having fixed the exponential to 84.21 μs. The result of the fit gave a rate of 1.82 ± 0.64 events per day, in the energy range for proton recoils of 50-500 keV.

A second long run was carried out for 3 months, with the same settings as for the neutron measurements, but with a neutron shielding (CH_2) in place. The aim of this run was to show that the neutrons observed without neutron shielding are indeed coming from the rock. Fig. 1d shows the resulting time delay distribution for this run, a nearly flat distribution, consistent with the detection of mainly random coincidences.

To convert the measured rate into neutron flux, we need to know the neutron detection efficiency. To determine this efficiency, a Monte Carlo simulation of neutron production, transport and detection can be used. The Monte Carlo simulations were carried out in several stages, using GEANT4. Neutrons generated from ^{238}U and ^{232}Th decay chains (calculated using the modified code SOURCES [3]), were propagated to the rock-laboratory boundary, then to the outer surface of the shielding, then to the scintillator and, finally, inside the sensitive volume of the detector where they produced proton recoils. The same procedure was applied to neutrons from the ^{252}Cf source.

As a result of the simulations the absolute rate and energy spectra of neutrons at different surfaces and those of proton recoils in the detector were obtained. The simulated proton recoil rate from ^{252}Cf source placed at a distance of 1 m from the detector (same position as in the calibration run) was 14.51 s $^{-1}$ at 50-500 keV. Typical statistical error for all simulation results is less than 1% and will be neglected further on. The efficiency of detecting delayed coincidences was calculated as the ratio of the measured rate from Cf source to the calculated rate of single proton recoils at 50-500 keV and was found to be 0.037 $\pm$ 0.003. The same efficiency was then used to convert the calculated rate of proton recoils from background neutrons into the expected rate of delayed coincidences. Fig. 2b depicts the fluxes of neutrons from the background and Cf source at the outer surface of the scintillator, which produced proton recoils in the scintillator. The spectrum from Cf was normalised to the overall area of the spectrum from the background. The shapes of the two spectra are quite similar with an estimated difference in total rate of 11% in the energy range of 0.4-1 MeV. Neutrons from this region contribute the most to the proton recoils at 50-500 keV and also are more likely to be captured in the scintillator.

4. Results

As mentioned before, pulse shape analysis used to reject background events from Bi→Po→Pb decay. The efficiency of this cut was 61%. Taking into account this cut efficiency, the efficiency of detecting delayed coincidences was reduced to 0.023$\pm$0.002.

The energy spectrum of neutrons at the outer surface of shielding which produced proton recoils in the detector (within 50-500 keV energy range) indicates a threshold of 0.5 MeV for the measured neutron flux. For a flux of 9.11×10^{-8} cm^{-2} s^{-1} at the rock-lab boundary, assuming 10 ppb U and Th, a rate of 6.40 d^{-1} for proton recoils at 50-500 keV was obtained. Taking into account the efficiency of detecting delayed coincidences we get a predicted rate of events in the background run of 0.147 ± 0.013 per day. As the detector in fact measured a rate of 1.82 ± 0.64 per day, the neutron flux was $(1.72 \pm 0.61) \times 10^{-6}$ cm^{-2} s^{-1} above 0.5 MeV at shielding, and $(1.14 \pm 0.40) \times 10^{-6}$ cm^{-2} s^{-1} above 0.5 MeV at the entrance of the cavern. There is a 51% increase in the flux at shielding which is in agreement with Lemrani et al. [4], who claimed a 50% increment to the flux above 1 MeV due to the contribution of the back-scattered neutrons. Our measured neutron flux agrees well with the simulation result 1.21×10^{-6} cm^{-2} s^{-1} assuming measured concentrations of 67 ppb of U and 127 ppb of Th [5].

The same re-scaling, as was used to derive the neutron flux, should be applied to the original concentrations of U and Th resulting in 127 ± 45 ppb of U and Th assuming their equal concentrations. If we assume that Th is twice as abundant as U, as measured by Smith et al. [5], then the resulting concentrations are 95 ± 34 ppb of U and 190 ± 69 ppb of Th. These values agree within errors with the recent measurements of the U and Th concentrations at Boulby [5]. They are also consistent with most of the previous measurements of the U and Th concentrations in Boulby rock using different techniques keeping in mind the large spread of the measured values [6].

The errors given above are purely statistical. In addition we can estimate systematic uncertainties. These are associated with the following parameters used to evaluate the neutron flux: (i) conversion of the charge to the energy scale for proton recoils, gives a possible error in the flux intensity of 10%; (ii) uncertainties in the efficiency estimate comes from the statistical error of the fit to the time delay distribution obtained with the neutron source (8%) and from the different shape of the energy spectra of neutrons from the ^{252}Cf source and background at the outer surface of the scintillator (11%), see Fig. 2a; (iii) accuracy of simulations, 14%. The total systematic uncertainty is 22%.

5. Conclusions

A technique to measure low intensity fast neutron flux has been developed. The neutron flux in the Boulby Underground Laboratory was measured as $(1.72 \pm 0.61(stat.) \pm 0.38(syst.)) \times 10^{-6}$ cm^{-2} s^{-1} above 0.5 MeV. This flux corresponds to the concentrations of $95 \pm 34(stat.) \pm 21(syst.)$ ppb of U and $190 + 69(stat.) \pm 42(syst.)$ ppb of Th, assuming Th is twice as abundant as U.

References

1. S. Agostinelli et al., GEANT4 Collaboration, NIM A 506 (2003) 250.
2. M. Anghinolfi et al., NIM A 165 (1979) 217224.
3. W.B. Wilson et al., American Nuclei Society/Radiation Protection and Shielding Division (2002); M.J. Carson et al., Astropar.Phys. 21 (2004) 667.
4. R. Lemrani et al., NIM A, 560 (2006), 454.
5. P.F. Smith et al., Astrop.Phys. 22 (2005) 409.
6. http://hepwww.rl.ac.uk/ukdmc/Radioactivity/uk.html.

SIMULATION OF SHIELDING DESIGNS FOR FUTURE LARGE-SCALE DARK MATTER SEARCH EXPERIMENTS

M. BAUER*, J. JOCHUM and S. SCHOLL

Physikalisches Institut I, Eberhard Karls Universität Tübingen,
Auf der Morgenstelle 14,
72076 Tübingen, Germany
**E-mail: bauer@pit.physik.uni-tuebingen.de*

The propagation of radiation in several possible shielding designs for future large-scale dark matter search experiments has been simulated with respect to the muon-induced background using the GEANT4 Monte Carlo toolkit. Results indicate that it seems to be advisable to have as few high-Z materials as possible in the vicinity of the detector.

Keywords: Shielding design; Monte Carlo simulations; GEANT4; neutron background

1. Motivation

One of the most important steps to increase the sensitivity of future dark matter search experiments is to reduce the background by several orders of magnitude. To reach this goal it is necessary to shield the detector as good as possible. In this work we compare several generic shielding designs with respect to the muon-induced background.

2. Shielding designs

The conventional shielding design for low background experiments consists — from inside to outside — of lead or copper, polyethylene and a muon veto made of liquid or plastic scintillator. A possible disadvantage is that it contains a large amount of high-Z material near the detector, which acts itself as a target for cosmic-ray muons, in which for example neutrons can be produced, which then enter the detector. A solution is to use also *internal* polyethylene inside the lead shielding to moderate and stop these neutrons. Another possibility could be to replace the whole shielding by several meters of water or even liquid scintillator. This design avoids high-Z materials completely, while through its large thickness it should nevertheless be sufficient to stop gammas and moderate neutrons, and it can also act as a muon veto by itself.

3. Simulations

We carried out simulations to compare the different shielding strategies. The simulations were done using GEANT4 version 8.1 [1]. We used a modified version of the QGSP_GN physics list [2]. The geometry used in the simulations is a "toy" geometry consisting of concentric cubes of material. The detector is a cube of natural germanium with dimensions 1.6 m x 1.6 m x 1.6 m, which is about the total size of the expected detection volume in future experiments. It is segmented into 3x3x3 pieces during analysis to consider anti-coincidences between the separate detectors which will be used in the real experiment. Muons were shot to the detector from a circle with a radius of 9.9 m at a height of 9.9 m above the center of the detector, with an energy spectrum and angular distribution as found in the Gran Sasso laboratory.

4. Simulations without rock

First we simulate the different shielding designs in air. We vary the thicknesses of the internal polyethylene, lead and external polyethylene separately while keeping the other two values constant. The thickness of the plastic scintillator used as a muon veto was fixed to 5 cm in all simulations. Results are shown in tables 1, 2, 3 and 4. Values are given for numbers of events where

- something is seen in the detector (*seen in det.*)
- something is seen in only one segment of the detector (*only one seg.*)
- a "hadronic" particle is seen in the detector, i. e. one which is no electron, positron, muon or gamma (*had. in det.*). This will potentially produce a nuclear recoil.
- a "hadronic" particle is seen in only one segment of the detector (*had. in one seg.*)

The last category is the most dangerous one, since it closely resembles a WIMP signal; however, in our simulations with one million primary muons hadronic particles are always accompanied by "electromagnetic" particles (electron, positron, muon or gamma) and all detected events are also seen in the muon veto. Results indicate that it is advisable to have about 20 cm of internal polyethylene to minimize the number of "dangerous" events and that the thickness of the lead should be kept as small as possible, while the thickness of the external polyethene has almost no influence. Alternative shielding designs without high-Z materials give comparable or even better results.

Table 1. Numbers of events for different thicknesses of internal PE, all with 30 cm Pb and 30 cm external PE, for 1 million primary muons

Internal PE [cm]	seen in det.	only one seg.	had. in det.	had. in one seg.
0	12282	964	5788	506
10	13647	1493	3429	367
20	15580	2216	3118	321
30	17368	3025	3283	374
40	18566	3852	3167	409
50	19280	4407	3370	418

Table 2. Numbers of events for different thicknesses of lead, all with 30 cm internal and 10 cm external PE, for 1 million primary muons

Lead [cm]	seen in det.	only one seg.	had. in det.	had. in one seg.
0	13840	2893	1553	1
10	16029	2498	2126	38
20	16637	2832	2595	184
30	17281	3009	3215	385
40	17469	3128	3630	546
50	18000	3257	4202	708

5. Simulations with rock

Now we simulate the shielding designs in a hall with side length of 13 m, surrounded by 50 cm of Gran Sasso concrete and 2.95 m of Gran Sasso rock. The composition of the concrete and the rock is taken from Ref. [3]. The muon source is at the same position above the detector as in the previous simulation, so that muons have to go through the rock and concrete in order to enter the hall. Results are shown in tables 5, 6 and 7. Here the thickness

Table 3. Numbers of events for different thicknesses of external PE, all with 30 cm internal PE and 30 cm Pb, for 1 million primary muons

External PE [cm]	seen in det.	only one seg.	had. in det.	had. in one seg.
0	16878	2934	3336	313
10	17281	3009	3215	385
20	17313	2971	3237	380
30	17368	3025	3283	374
40	17213	2926	3189	381
50	17205	2985	3361	355

Table 4. Numbers of events for alternative shielding designs, for 1 million primary muons. The value without any shielding is of course only given for reference; "no shielding" is not a good option if also radioactivity is taken into account and not only muon-induced background

Shielding	seen in det.	only one seg.	had. in det.	had. in one seg.
100 cm PE	15381	3809	1640	2
5 m Water	16556	4104	1980	126
5 m Liq. Scint.	17742	4741	1894	86
No shielding	9569	790	1454	3

of the external polyethylene is not varied since it does not seem to have a noticeable effect. Results are similar to the situation without rock, but favor a little larger amount of internal polyethylene of about 30 cm.

Table 5. Numbers of events for different thicknesses of internal PE, all with 30 cm Pb and 30 cm external PE, for 1 million primary muons; setup with rock

Internal PE [cm]	seen in det.	only one seg.	had. in det.	had. in one seg.
0	12767	1160	6311	719
10	14302	1782	5531	798
20	16005	2385	4594	647
30	17149	3112	3949	566
40	18891	3946	3911	605
50	19550	4564	3782	573

Table 6. Numbers of events for different thicknesses of lead, all with 30 cm internal and 10 cm external PE, for 1 million primary muons; setup with rock

Lead [cm]	seen in det.	only one seg.	had. in det.	had. in one seg.
0	19134	5971	2041	155
10	16505	2873	2972	231
20	17072	3012	3775	446
30	17231	3123	3902	593
40	17869	3393	4058	726
50	18212	3488	5149	966

Table 7. Numbers of events for alternative shielding designs, for 1 million primary muons; setup with rock. The value without any shielding is again only given for reference

Shielding	seen in det.	only one seg.	had. in det.	had. in one seg.
100 cm PE	18647	5750	1948	145
5 m Water	16266	4088	1810	212
No shielding	63402	35610	2451	159

6. Outlook

The simulations presented here take only muon-induced background directly from above into account. Future work will also be concerned with muon-induced showers where the muon does not pass the detector. For a complete assessment of the shielding also background from local radioactivity has to be taken into account.

Acknowledgements

We acknowledge the funding from the Network on Direct Dark Matter Detection N3 of the EU FP6 project ILIAS.

References

1. S. Agostinelli *et al.*, GEANT4 — a simulation toolkit, NIM A 506 (2003) 250
2. M. Bauer *et al.*, Simulations of muon-induced neutron background with GEANT4, IDM2004 proceedings, World Scientific 2005
3. H. Wulandari *et al.*, Neutron Flux Underground Revisited, Astropart. Phys. 22 (2004) 313 and hep-ex/0312050

SIMULATION OF INELASTIC NEUTRON SCATTERING WITH GEANT4

S. SCHOLL, M. BAUER and J. JOCHUM

Physikalisches Institut I, Eberhard-Karls Universität,
72076 Tübingen, Auf der Morgenstelle 14, Germany
**E-mail: scholl@pit.physik.uni-tuebingen.de*

Since neutrons provide the principal background in direct dark matter detection experiments, the simulation of neutron induced nuclear recoils is crucial for background reduction. GEANT4 provides a framework for such simulations. While the simulation of the elastic scattering reactions agrees very well with the theoretical expectations, the simulation of the inelastic scatterings generally fails, i.e. energy conservation per reaction and angular–energy correlation of the recoiling nucleus are not respected. An improved code is used to correct these shortcomings. Further necessary improvements are addressed as well.

1. Introduction

Nowadays the existence of non–baryonic dark matter is clearly supported by several astrophysical observations, e.g. the CMB anisotropy,[1] high–z supernova searches,[2] the analysis of galaxy redshifts[3] and microlensing of the bullet cluster.[4] However the dark matter still eludes direct detection in a laboratory experiment, which is crucial for the determination of its nature. Among the various theoretical candidates for a dark matter particle, a weakly interacting massive particle (WIMP) seems most promising. The aim of a WIMP direct detection dark matter experiment is to detect nuclear recoils from the scattering of WIMPs on the nuclei of the detector. Due to the small cross–section for WIMP nucleon scattering, the expected signal rate is much smaller than the background signals from ambient and cosmogenic radioactivity. Thus, a competitive experiment must be able to discriminate nuclear recoils from electron recoils in the energetic region of interest. Once this goal is achieved, the remaining background consists of neutrons which scatter off the nuclei in the detector.

The neutrons cannot be eliminated by active background rejection techniques and thus limit the sensitivity of the experiments. Three possible neutron reactions are able to produce a fake WIMP signature in the detector; elastic scattering, inelastic scattering in which the de–excitation gamma escapes the detector and neutron capture where the de–excitation gamma escapes detection. Inelastic scattering in which a charged particle is emitted or spallation of the nucleus will generally deposit too much energy to be confused with WIMP scattering.

A reliable simulation of the neutron–induced reactions in a given detector is necessary for the construction of a direct detection experiment and the analysis of the collected data.

2. Neutron Simulation in GEANT4

The GEANT4 simulation toolkit is the object–oriented, C++ based toolkit to simulate the passage of particles through matter developed at CERN.[5,6] It is originally designed for simulations regarding the LHC, but several optional packages for low–energy applications are also available.

For our simulations, the important optional physics process class in GEANT4 is the G4NDL class and it is used for the treatment of low energy neutron reactions below 20 MeV based on the ENDL B-VI evaluated neutron scattering data.[7] In this code, four general classes of processes are defined for neutrons; capture processes, elastic scattering, inelastic scattering and fission reactions. For our simulation of the neutron background, elastic and inelastic scattering reactions are the most important. Capture processes could also generate a fake signal if the gamma escapes the detector, but with the setup described below, only very few capture events were recorded.

The setup of the simulations was kept simple; one million monoenergetic 10 MeV neutrons were shot on a cube with 10 cm edge length made of Germanium 76 *. The test data was generated by a simulation based on the GEANT4.7.0.p01 version using the high precision neutron data library G4NDL3.7. In addition, a modified code correcting the issues with energy conservation and the angular depency of the recoil energy in inelastic scattering reactions was used. Since the energy given in the data files of the G4NDL is always the sum of the kinetic energy of the ejected hadron and the recoiling nucleus, this sum must be disentangled. One way is to calculate the length of the momentum vector in the center of mass system using the invariant mass of the reaction. In relativistic kinematics the invariant

*This choice of material was made to achieve comparabilty with other simulations regarding questions of double beta decay done in our institute.

mass–square can be calculated from the mass of the incident particle and the mass of the target particle:

$$s^2 = (m_{neutron} + m_{target})^2 \tag{1}$$

In addition, the length of three–momentum vector in the center of mass system can be calculated, if the invariant mass s, the mass of the recoil nucleus m_{rec}, its excitation energy E_{ex} and the mass of emitted particle m_{em} are known:

$$\vec{p}_f = \frac{1}{2s}\sqrt{\left[s^2 - (m_{em} - (m_{rec} + E_{ex}))^2\right]\left[s^2 - (m_{em} + (m_{rec} + E_{ex}))^2\right]} \tag{2}$$

The direction of the momentum vector of the scattered neutron in the laboratory system is extracted from the data files. With both information, the resulting neutron momentum and recoil nucleus momentum can be reconstructed.

3. Results

One problem is found with the choice of Germanium 76 as detector material: The data files for the inelastic scattering are given for the several isotopes as total cross–section, but for the various final states, only the cross–section for the natural abundancy of isotopes is given. If this occurs, the G4NeutronHP class uses the final state cross–section for the next lighter element†. We resolved this problem by removing the single isotope cross–sections and used the cross–section data for the the natural abundancy of Germanium.

Since now the used cross–section is the sum of the cross–section of all isotopes, the simulation will produce gammas even when inelastic scattering for this particular isotope is not allowed. The energy of the gammas is given by the energy of the first transition gamma for the particular isotope. Thus energy conservation in the reaction is violated and the kinematically improved code can yield an imaginary length of the momentum vector.

This problem occurs for virtually all elements with more than one stable isotope where the natural abundancy cross–section data files are used, limiting the predictive power of GEANT4 for both elastic and inelastic neutron scatterings. For the elastic scatterings done by the G4NeutronHPElastic process, one can see in fig. 1 and fig. 2 that the energy conservation per scattering is respected and the the energy–angular correlation of the recoiling nucleus behaves as expected.

†This feature is known as the Gallium Bug for quite a time.

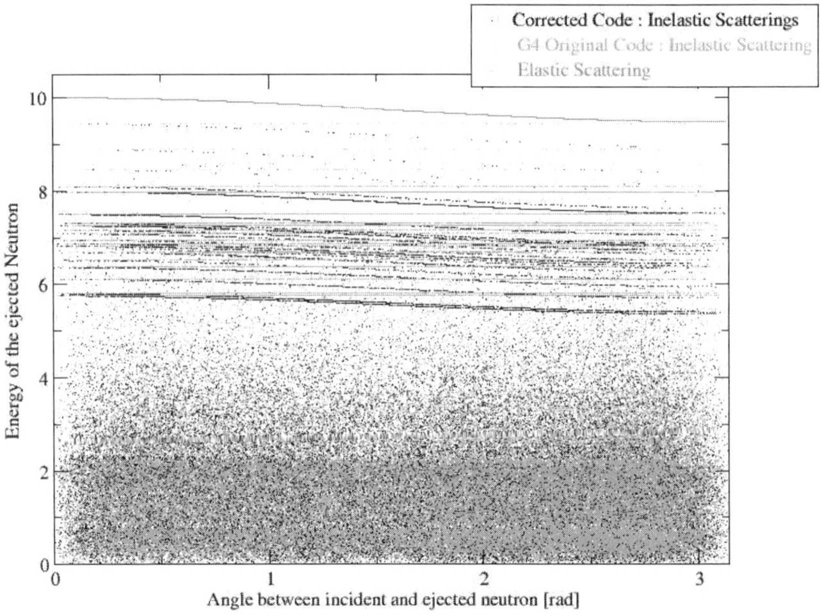

Fig. 1. The energy of the ejected neutron in dependency of the scattering angle as scatterplot; the dark dots are calculated with the improved code showing the expected energy–angular dependency, the lighter dots representing the result from the orginal code do not show any variation with the scattering angle. The medium shaded dots forming the prominent line at 10 MeV show the elastic scattering events.

The energy conservation in inelastic scatterings is flawed as can be seen in fig. 2. Additionally, fig. 2 shows that no angular–energy correlation of the ejected neutron exists. The energy excess on the statistical level and the missing angular–energy correlation of the reocil nucleus are due to a programming glitch in the class responsible for inelastic scattering, were the wrong frame of reference is used for the calculation of the kinematics.

4. Discussion

A look at fig. 3 shows significant difference between the original code and the corrected code for the low energy end of the spectrum. The improved code predicts a lot less inelastic events with very low energy deposition than the original code. For higher energies, peaks are smeared out due to the corrected kinematics. Especially for simulations of direct dark matter searches, fig. 3 emphasizes the importance of correct the kinematics of inelastic scatterings in GEANT4.

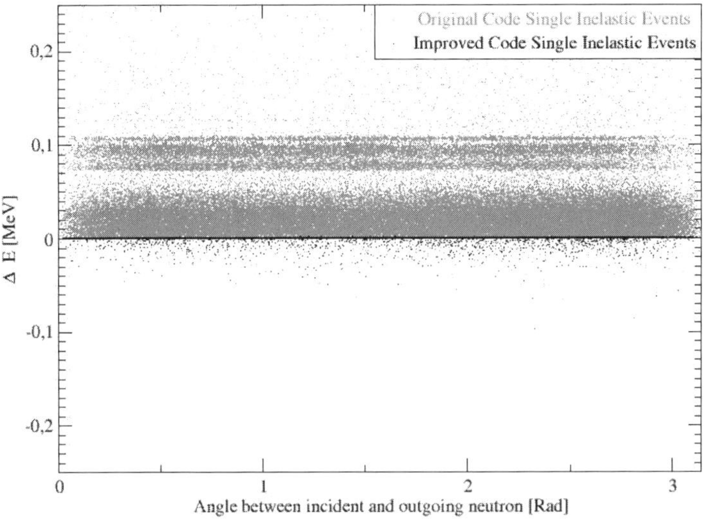

Fig. 2. The energy balance of single inelastic reactions of incident 10 MeV neutrons on Germanium in depency of the scattering angle between incident and ejected neutron, the lighter shaded points are data generated with the original code, the dark points are from the simulation with the improved code.

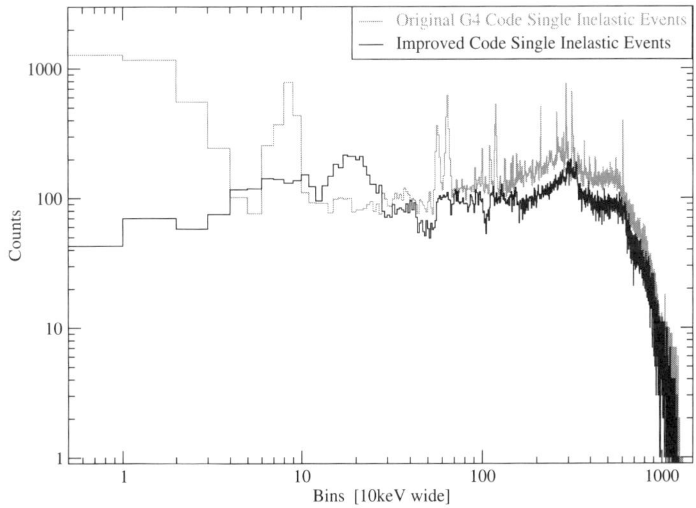

Fig. 3. Double logarithmic plot of the total deposited energy in the germanium crystal of single inelastic events only. The light curve shows the data obtained with the original code, the dark curve shows the data from the improved code.

If one plans to do a simulation in GEANt4, a close look to available cross–section data files is recommended since a lot of elements has no cross–section data.

Noteable examples of elements without any cross–section data given in GNDL3.7 are Neon, Scandium, Zinc and virtually all elements heavier than Arsen and lighter than Lead. A lot of the lighter elements have only data for the natural abundancy of isotopes given. The newer G4NDL3.9 suffers still the same problem, however, data for a few more elements can now be found.

Acknowledgements

We also acknowledge the funding from the Network on Direct Dark Matter Detection N3 of the EU FP6 project ILIAS.

References

1. E. L. Wright, *astro-ph/0306132* (2003).
2. S. Perlmutter *et al.*,*Astrophys. J.* **517**, 517 (1999).
3. W. J. Percival *et al.*,*MNRAS* **327**, 129 (2001).
4. D. Clowe, M. Bradac, A. H. Gonzalez, M. Markevitch, S. W. Randall, C. Jones, D. Zaritsky, *astro-ph/0608407*.
5. S. Agostinelli, J. Allison, K. Amako, J. Apostolakis, H. Araujo, P. Arce, M. Asai, D. Axen, S. Banerjee, G. Barrand *et al.*,*NIM A* **506**, 250 (2003).
6. J. Allison, K. Amako, J. Apostolakis, H. Araujo, P. Arce Dubois, M. Asai, G. Barrand, R. Capra *et al.*,*IEEE Transactions on Nuclear Science* **53**, 270 (2006).
7. Cross Section Evaluation Working Group, http://www.nndc.bnl.gov/csewg .

CALCULATION OF NEUTRON YIELD FROM RADIOACTIVITY IN MATERIALS RELEVANT TO DARK MATTER SEARCHES

VITO TOMASELLO* and VITALY A. KUDRYAVTSEV

Department of Physics and Astronomy, University of Sheffield
Hicks Building, Hounsfield road, Sheffield S3 7RH, UK
**E-mail: V.Tomasello@sheffield.ac.uk*

The sensitivity of future large-scale dark matter detectors may be limited by neutrons produced in detector components. For this reason it is important to use pure materials and estimate the background from this source. Neutrons are produced via spontaneous fission and (α, n) reactions from U and Th traces in the materials. To estimate neutron yields and spectra, the cross-sections of (α, n) reactions and probabilities of transitions to different excited states should be known. In this work these cross-sections and probabilities are calculated using EMPIRE-2.19 for some isotopes and compared to the experimental data. The results have been used to calculate neutron spectra from materials with the code SOURCES4. We plan to create a database of the (α, n) cross sections and neutron spectra for materials used in dark matter detectors. This work is on-going as part of the EU FP6 programme ILIAS.

1. Introduction

The underground physics deals with some extremely rare phenomena, which are very hard to distinguish from others more ordinary that come from cosmic rays and natural radioactivity. The best way to screen from cosmic rays is to carry out the experiments deep underground, because in this way the surrounding layers of rock reduce their flux by several orders of magnitude. Nevertheless, many radiation backgrounds are still present; the main ones are those originating from the cosmic-rays interactions (secondary reactions). There are many experiments, existing or in the planning phase, located deep underground. They are conceived to detect particles coming from the space like the WIMPs - Weakly Interacting Massive Particles, or astrophysical neutrinos, or to search for weak processes such neutrinoless double beta decay, predicted by new theories beyond the Standard Model. The radiation backgrounds are crucial for the sensitivity of these experiments to expected events. ILIAS - Integrated Large Infrastructure for Astroparticle Science - is a EU programme which includes working groups on background simulations and measurements. The aim is to understand the background which will help to design future large-scale detectors, their shielding and active veto systems, in particular for dark matter searches. In this work we show results of the (α, n) cross-sections, obtained with EMPIRE 2.19 [1], and neutron yield and spectra, calculated using SOURCES4A [2].

2. Direct WIMP detection

The main requirements for direct dark matter detection are: low intrinsic radioactive background from detector and surrounding components; good discrimination between e^- produced by gamma background and nuclear recoils expected from WIMP interactions; low energy threshold to achieve maximum sensitivity to WIMP-induced nuclear recoils. We focus on the first point, in particular on the neutron background.

We are able to find a way of rejecting other types of background, but for neutrons it is more difficult, since neutron interaction can mimic nuclear recoils produced by WIMPs. It is crucial to reduce neutron background below the sensitivity of the experiments. There are two main sources of neutrons: local radioactivity and cosmic-ray muons. In this work we do not consider cosmic-ray muons leaving this for future studies. Here we consider neutrons produced by local radioactivity: spontaneous fission and (α, n) reactions. The (α, n) reactions are initiated by an α-particle from U/Th traces in the rock and detector elements. Significant contribution to the neutron flux from spontaneous fission is present only for 238U.

3. Objectives, tools and method

Our aim is to run simulations of neutron background for underground experiments, so neutron yields and spectra are required. To calculate them, an input such as cross-section of (α, n) reactions and the transition probabilities to the different excited states (branching ratios), are needed. The code SOURCES4A [2] was used to calculate neutron yield and spectra (see [2,3] for tests of code reliability).

For this work a modified version of SOURCES4A [4] has been used; the original code provides a treatment of (α, n) reactions only up to 6.5 MeV limit. These restricted significantly the reliability of the results, because the cross section of (α, n) reactions rises with energy, and the average neutron energy also increases with the parent α energy. In the modified version this

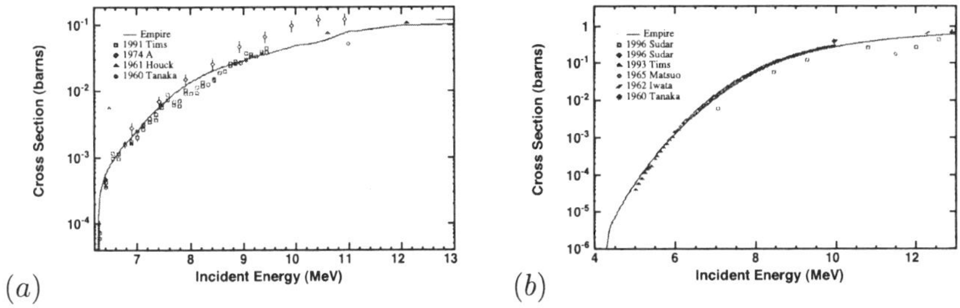

Fig. 1. Comparison of results with experimental data (a) ^{54}Fe, (b) ^{55}Mn.

limit was moved to 10 MeV, and some cross-sections were extended to 10 MeV using experimental data.

Another problem with original SOURCES4A was the absence of cross-sections and branching ratios for some isotopes on the libraries of the code.

Following previous work [5], we have chosen the EMPIRE 2.19 [1] code to calculate cross-sections and branching ratios, but we have made two significant improvements. The first concerns the cross-section and branching ratio data. We have realized that the results are sensitive to the values of incident α energy present in the input file of EMPIRE 2.19, for which we wanted to calculate the cross-sections and branching ratios. Choosing smaller step in energy (bigger number of input energies) results in an improvement in the accuracy. We have reduced the step in energy until the variation became insignificant.

The second improvement concerns the excited states: we have changed the code SOURCES4A to allow much larger number of energetic levels to be stored in the library.

Comparison between EMPIRE 2.19 calculations and experimental data is shown in Fig. 1 for ^{54}Fe and ^{55}Mn. There is a good agreement between data and simulations especially for energies below 10 MeV, important for dark matter experiments. This gave us more confidence in EMPIRE 2.19 reliability in calculating cross-sections for others isotopes, for which comparison is not possible because the absence of measurements.

4. Result

Cross-sections and branching ratios for several isotopes have been calculated with EMPIRE 2.19 and inserted in the libraries of SOURCES4A. We have ran SOURCES4A for some of the most used materials in detectors constructed for dark matter searches. A concentration of 10 ppb of U or 10 ppb of Th in all calculations is assumed.

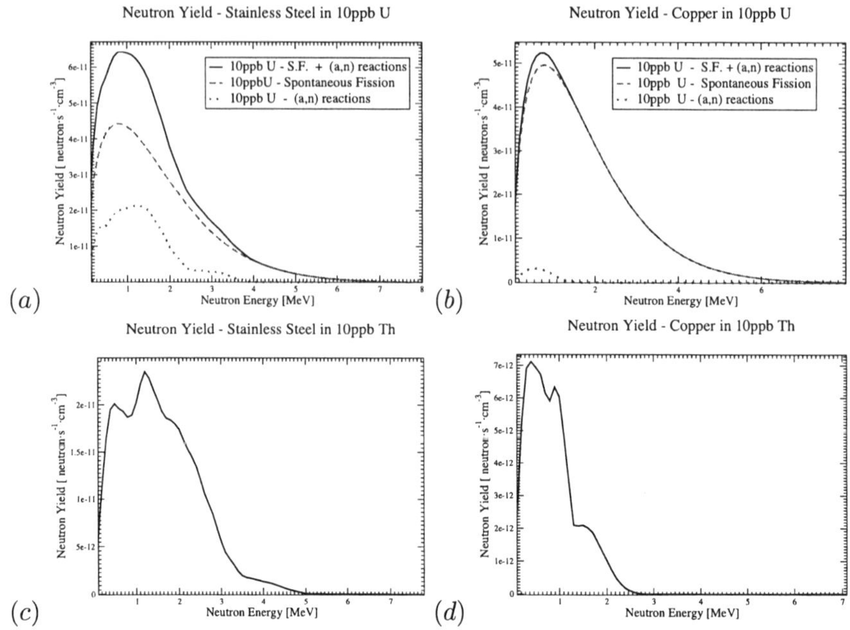

Fig. 2. Neutron spectra from stainless steel (a,c) and copper (b,d) for 10 ppb of U (a,b) and 10 ppb of Th (c,d).

The spectra from stainless steel and copper are presented in Fig. 2, showing a significant contribution from spontaneous fission from 238U. Fig. 3 shows comparison of our calculations for stainless steel, using 10 ppb of U and Th, with a previous work [4]. In that work the neutron yield was calculated with SOURCES4 too, but using a different library, that included only transitions to the ground state. This assumption lead to an overestimation of the neutron energy and underestimation of the neutron flux.

5. Future plans and conclusions

The work presented here is aimed at developing a tool for calculating neutron background with various radioactive isotope concentrations and materials. This will be a very useful input for running dark matter experiment simulations. In order to reach our objective, we plan to complete the SOURCES4A libraries with cross-sections for (α, n) reactions and branching ratios for all isotopes relevant to dark matter searches. We also plan

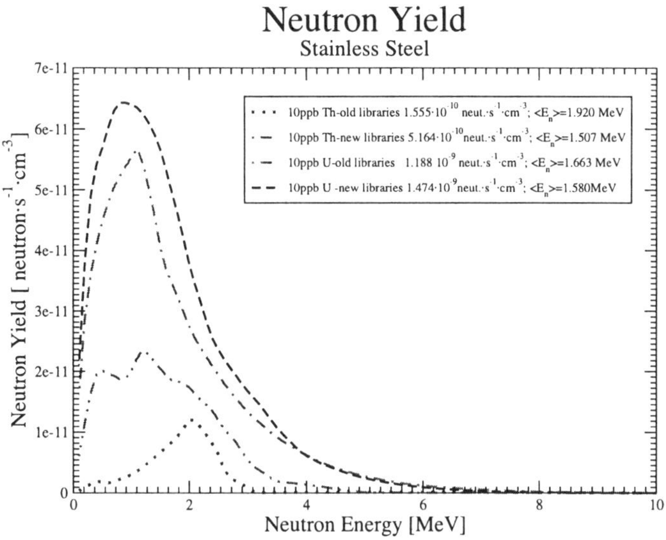

Fig. 3. Neutron yield from stainless steel with 10 ppb U and 10 ppb Th calculated using SOURCES4A, with different libraries. The "new" library refers to our present results, "old" libraries refers to [4]. The main difference between the two libraries is that in the old one only transitions to the ground state were considered.

to obtain neutron yields and spectra from materials used in dark matter detectors and from rocks surrounding the experimental sites.

Acknowledgments

This work has been supported by the ILIAS integrating activity (Contract No. RII3-CT-2004-506222) as part of the EU FP6 programme in Astroparticle Physics. We thank Drs. R. Lemrani and M. Robinson for their kind help.

References

1. EMPIRE 2.19: *www.nndc.bnl.gov/empire219*
2. W.B. Wilson et. al., *SOURCES4A, Technical Report LA-13639-MS, Los Alamos (1999).*
3. V.A.Kudryavtsev, *Proc. of the Topical Workshop on Low Radioactivity Techniques (Sudbury, Canada), AIP Conference Proceedings vol 785, ed B Cleveland, R Ford and M Chen (New York: Melville) p 238.*
4. M.J.Carson et. al., *Astropart. Phys. 21 (2004) 667.*
5. R.Lemrani et al., *Nucl. Instr. and Meth. A 560 (2006) 454.*

BACKGROUNDS AND THRESHOLD OF XENON10 DARK MATTER EXPERIMENT

P. F. SORENSEN
ON BEHALF OF THE XENON COLLABORATION
Department of Physics, Brown University, Providence, RI 02912, USA
E-mail: peter_sorensen@brown.edu
http://xenon.brown.edu

The XENON10 Experiment is a dual (liquid/gas) phase time-projection chamber designed for direct detection of dark matter via elastic scattering of the Xe nuclei. It was deployed in March 2006 at LNGS ($\sim$ 3100 mwe) and is presently operational. The ratio of prompt (primary) scintillation to secondary scintillation (via electro-luminesence caused by ionization electrons extracted into the gas phase) allows event-by-event discrimination of electron and nuclear recoil events. A threshold $<$ 10 keV nuclear recoil energy (keVr) is demonstrated, as well as a background rate of $<$ 1 cts/kg/keVee/day (dru) before rejection. With 3-D position reconstruction (to define a fiducial region) and >99% discrimination of gamma background (at 50% acceptance for nuclear recoils), XENON10 expects to be sensitive to a WIMP-nucleon interaction cross section of $< 10^{-43}$ cm^2 for WIMP particles with mass $>$ 50 GeV.

Keywords: dark matter, liquid xenon

1. Introduction

There is compelling evidence for a significant cold dark matter (CDM) component in the universe,[1] and perhaps the best-motivated candidate is the lightest super-particle from super-symmetry (SUSY).[2] Such a particle is more generally classified as a Weakly Interacting Massive Particle (WIMP), and the open question of the nature of such WIMPs is being addressed by numerous direct and indirect detection experiments.[3] A majority of theoretical models predict that current direct-detection experiments have only begun to probe the available phase-space for SUSY dark matter,[4] and much larger-scale experiments will be necessary to do so. XENON10 is a first step towards development of a tonne-scale detector with a reach of $\sim 10^{-46}$ cm^2.

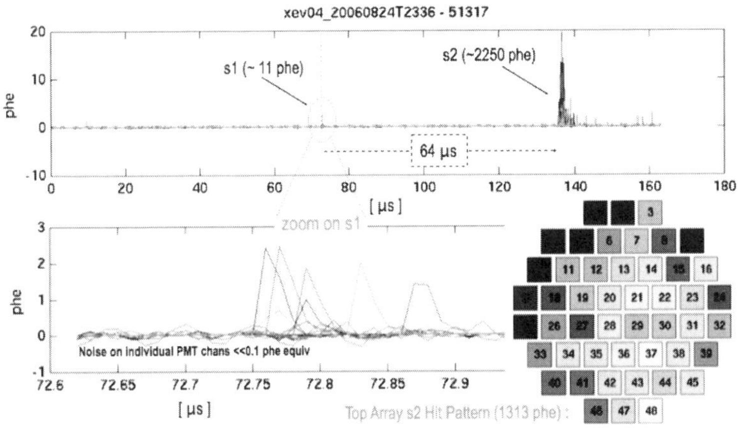

Fig. 1. A typical 5 keVee event (2.2 S1 phe/keVee). **(top)** the primary scintillation (S1) signal at 73 μs, followed by the proportional scintillation (S2) signal at 138 μs. **(bottom)** A zoom on the S1 shows individual photoelectrons from the PMTs. **(inset)** the S2 hit-pattern on the top PMT array reconstructs $(x, y) \sim (2.6, -2.2)$ cm. (each PMT is 2.5x2.5 cm)

2. Detector Response

XENON10 has an active liquid region 20 cm ⌀ by 15 cm deep, with mesh to define electric fields in the liquid and gas regions. A particle interaction creates primary scintillation (175 nm) and ionization; the former are detected by 48 (top) / 41 (bottom) Hamamatsu R8520 UV-sensitive PMTs (the "S1" signal), and the latter (those electrons which escape recombination) are drifted (E ~1 kV/cm) through the bulk liquid (electron lifetime > 1 ms). They are extracted into the gas Xe, where they are accelerated (E > 10 kV/cm) and undergo proportional scintillation, which is then also detected by the PMTs (the "S2" signal). A typical 5 keV gamma event (referred to as electron-equivalent energy, keVee) is shown in Fig. 1. The S1 signal consists of 11 photo-electrons (phe). The 65 μs delay between the signals is due to the drift time of the electrons through the liquid, and indicates the z coordinate of the interaction (here, $\sim$ 12 cm below liquid surface). The $x - y$ coordinates are re-constructed from the S2 PMT hit pattern.[5]

2.1. *Discrimination*

The stopping power and ionization density in liquid Xe differ for charged particles (electron recoil) versus neutral particles (nuclear recoil), with a larger fraction of ionization escaping recombination in the case of an electron recoil (relative to an equivalent energy deposition in the detector). The ratio of $S2/S1$ is therefore a key aspect of background rejection. The rejection power of XENON10 is $\sim 99\%$ with 50% acceptance for nuclear recoil events.[6]

2.2. *Trigger Threshold*

Trigger systems based on the S1 and S2 signals were implemented and studied. The S1 trigger required $n \geq 6$ coincidence in an 80 ns window (the Xe scintillation decay time is 27 ns),[9] on any of the bottom 41 PMTs. The threshold levels were optimized for each individual PMT and correspond to $\sim$ 80% acceptance for a single photo-electron (phe) pulse in a given PMT. Due to the detector geometry, the bottom PMT array receives 80% of the available scintillation photons; the PMTs have 20 – 25% quantum efficiency and a 60% – 70% first-strike collection efficiency, and an average $\sigma \sim 56\%$ for the spread in the PMT single photo-electron peak. A Monte Carlo simulation accounting for these factors, and also Poisson fluctuations in the number of phe's generated by the PMT photo-cathode, is compared with the measured S1 trigger efficiency over about 20 live days of WIMP-search data (Fig. 2).

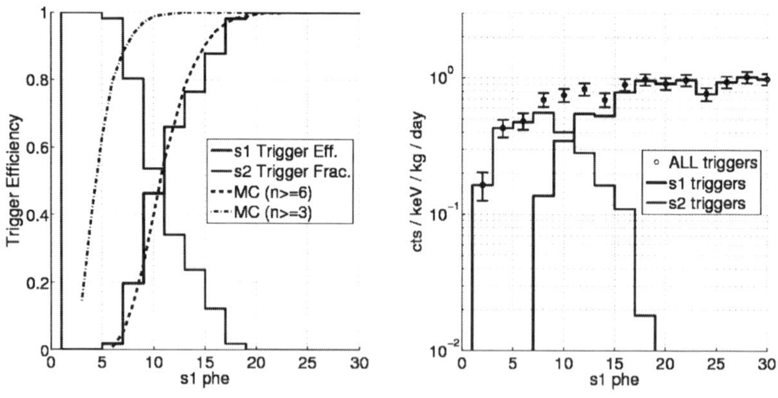

Fig. 2. **(left)** Measured S1 trigger efficiency (blue) and fraction of S2 triggers (magenta); Monte Carlo simulation results for S1 coincidence $n \geq 6$ (black dashed) and $n \geq 3$ (black dotted). **(right)** The S1 spectrum for the same data, showing ~0.90 cts / keVee / kg / day in the Fiducial Volume ($r < 8$ cm, $1 < z < 13$ cm).

The S2 trigger consists of time-integrating (1μs, the FWHM of a typical S2 pulse) the summed top PMT array, and passing the output to a threshold discriminator. It was found to have a threshold low enough to catch the ~ 24 phe S2 signal from a single electron extracted into the gas Xe. The S2 trigger offers a lower threshold than the S1 trigger (even if the S1 trigger is operated with a lower n-fold coincidence, as indicated in Fig. 2, left); additionally, operating the S1 trigger below $n \geq 6$ was found to have a high (> 5 Hz) rate of S1-only events (no associated S2), which do not correspond to events in the active Xe target region (and would therefore be rejected).

2.3. *Detector Threshold*

Due to nuclear form factor suppression, the expected differential event rate for a WIMP in Xe falls by 2 orders of magnitude between 0-60 keV nuclear recoil energy (keVr),[10] so it is advantageous to push the threshold as low as possible. Gamma calibration data (662 keV from ^{137}Cs source) indicate a primary (S1) light yield of 2.2 phe/keVee for events at the center of the Xe target, and recent results from nuclear recoil quenching factor measurements for scintillation[7] indicate that for nuclear recoils the light yield should be ~ 0.45 this value, or about 1 phe/keVr .

3. Backgrounds

The XENON10 detector is shielded by 33 tonnes (20 cm thick) of Pb outside of 1.5 tonnes (20 cm thick) polyethylene. Prior to fiducial volume cuts or discrimination, the gamma background rate < 30 keVee in the full Xe target volume is > 1 dru, as shown in Fig. 3. The external neutron flux ($1 - 10$ MeV) is reduced by $> \times 200$; because the Pb reduces the neutron energy before it reaches the polyethylene,[11] the raw efficacy of the polyethylene shielding is increased by $\times 10$. Additionally, XENON10 benefits from self-shielding: excluding the outer ~ 2 cm of Xe (via $x - y$ and drift-time fiducialization) further reduces the event rate to about 0.90 dru, as shown in Fig. 2 (right) and Fig. 3.

3.1. *Gamma Background*

The primary sources of gamma background are (1) U/Th/K contamination in the PMTs, (2) ^{60}Co activity from the Stainless Steel of the outer cryostat and inner detector vessel, (3) U/Th/K contamination in ceramic HV feed-throughs (replacement with lower activity components will be done

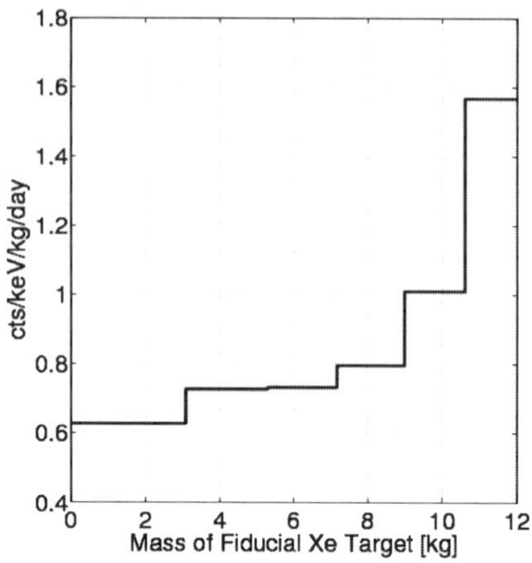

Fig. 3. Event rate in Xe target vs. mass of fiducial Xe target; self-shielding cylindrical shells defined via fiducial cuts in $x - y - z$, as described in Sec. 2. Total mass of inner Xe is 13.7 kg.

shortly), (4) beta decay of ^{85}Kr in the Xe (using Xe with < 10 ppb Kr, purchased from Spectra Gases; CWRU have also produced 25 kg LXe with < 10 ppt Kr using charcoal column separation[12]), (5) ^{210}Pb activity in the Pb shield, and (6) ^{222}Rn inside the shield cavity (the ambient level at LNGS of $\sim$100 Bq/m^3 is reduced to < 4 Bq/m^3 by LN boil-off purge). The contributions are summarized in Table 1 (event rates were estimated from Monte Carlo). Because XENON10 was conceived primarily as a proof-of-principle step towards a larger-scale detector, there is still room for improvement in (2), for example using Cu instead of Steel.

3.2. *Neutron Background*

The primary sources of neutron background are (1) from (α,n) reactions in the PMTs, (2) from (α,n) and fission due to U/Th contamination in the surrounding rock, and (3) neutrons generated by muons interacting in the Rock and Pb shield.

Neutron production in the PMTs is dominated by the Si, O and B used in the window and feed-throughs in the back-plane, from which one expects 0.1 n/10^6 α's (averaged over the U/Th emission spectrum).[13] An activity of 1 mBq U/Th per PMT yields $\sim$0.02 neutrons/PMT/year (into 4π); considering only single-scatter events in the full Xe target, based on GEANT4 simulation of the detector geometry, we find < 2 cts/year $(0 - 30$

Table 1. Summary of Primary γ-Background Sources [3 – 16 keVee]. Event Rates are based on MC Simulation.

Material	Event Rate in Xe [10^{-3} dru] (8 cm Radius Fiducial)	Radio-Impurities [10^{-3}Bq/kg] ^{238}U/^{232}Th/^{40}K/^{60}Co
R8520 PMT	28 ± 10	0.17/0.20/10/0.51 [a]
PMT base	130 ± 60	3.85/3.62/6.81/0.12 [b]
Vessel Stainless Steel	200 ± 60	21/61/12/101 [c]
Cryostat Stainless Steel	135 ± 48	21/61/12/25 [d]
HV Feed-Through	100 ± 50	4800/500/2100/ $\sim$ [e]
^{85}Kr in Xe	< 220	[< 10 ppb Kr in Xe]
^{222}Rn in Shield Cavity	< 20	[< 4 Bq/m^3 ^{222}Rn] [f]
^{210}Pb in Pb Shield	80 ± 20	[33 ± 4 Bq/kg ^{210}Pb] [g]
Polyethylene (Shield)	*N/A*	[being assessed]
Total (measured)	**~900**	

Note:
[a] based on counting 7 R8520-06-AL PMTs [LNGS counting facility]. 5 of the 89 PMTs in XENON10 are type R8520-M4F, with activities ×10 higher (scheduled for replacement). Radio-Impurities per PMT; Event Rate per 89 PMTs.
[b] includes resistors, capacitors and Cirlex boards (not solder). Radio-Impurities per Base; Event Rate per 89 Bases.
[c] based on counting scrap from vessel walls [Soudan SOLO facility]; used as baseline activity for ^{60}Co in SS pieces which were not counted.
[d] estimated from knowledge that cryostat was > 12 years old and assumed ^{60}Co activity ×3 above baseline.
[e] based on counting equivalent item from same vendor [Soudan SOLO facility].
[f] measured in-situ with LN purge operational, using Durridge RAD7
[g] vendor: Fonderie Gentilly

keVr) for the 89 PMTs. Recent measurements suggest that the activity per PMT is even lower (see Table 1).

The neutron flux (1 – 10 MeV) from cavern rock at LNGS is 0.4×10^{-6} – 3×10^{-6} n/cm^2/s,[14] with the lower values coming from more recent results. GEANT4 simulation using 2×10^{-6} n/cm^2/s shows < 5 cts/year (0 – 30 keVr) in the full Xe target. The muon-induced component extends well above the effective range of the polyethylene shielding (20 MeV), however the total muon-induced flux above 1 MeV is $< 1 \times 10^{-9}$ n/cm^2/s,[16] thus the flux of high energy neutrons (20 – 200 MeV) is sub-dominant. The rate of neutron events due to muons in the Pb shield was also simulated, considering the muon flux/spectrum at LNGS and a neutron yield in Pb of $\sim 4 \times 10^{-3}$ n/μ/g/cm^2. The resulting rate in the full Xe target was also < 5 cts/year (0-30 keVr), hence a muon veto is unnecessary for XENON10.

4. Conclusions

XENON10 has established a trigger system based on the S2 signal which maintains full efficiency for events which generate less than 5 S1 photoelectrons; this is expected to correspond to < 3 keVee. The gamma background count rate is 0.87 dru ($3-16$ keVee) for events in the fiducial volume ($r < 8$ cm, $1 < z < 13$ cm), and the major radio-impurity contributions from each piece of the detector have been studied. The background count rate due to neutrons is not expected to affect the sensitivity of XENON10.

References

1. W. Freedman and M. Turner, *Rev. Mod. Phys.*, **75**, (2003) 1433
2. M.W. Goodman and E. Witten, *Phys. Rev. D*, **31**, (1985) 3059
3. R. J. Gaitskell, *Ann. Rev. Nucl. Part. Sci.*, **54**, (2004) 315
4. J. Ellis *et al.*, *Phys. Rev. D*, **71** (2005) 095007
5. K. Ni *et al.*, *Proc. 7th UCLA Symp.* (2006) [arXiv:astro-ph/0609714]
6. M. Yamashita, *Proc. 6th IDM Workshop* (2006) [current proceedings]
7. E. Aprile *et al.*, *Phys. Rev. D*, **72**, (2005)
8. T. Shutt *et al.*, (2006) [arXiv:astro-ph/0608137]
9. T. Doke *et al.*, *NIM A*, **420** (1999) 62
10. J. D. Lewin & P. F. Smith, *Astropart. Phys.*, **6** (1996) 87
11. M. J. Carson *et al.*, *Astropart. Phys.*, **21** (2004) 667
12. A. I. Bolozdyna *et al.*, Nucl. Instr. Meth. A (2006) [to be published]
13. R. Heaton *et al.*, *NIM A*, **276** (1989) 529
14. H. Wulundari *et al.*, *Astropart. Phys.*, **22** (2004) 313
15. M. J. Carson *et al.*, *NIM A*, **548** (2005) 418
16. D.-M. Mei and A. Hime, *Phys. Rev. D*, **73** (2006) 053004

BACKGROUND REDUCTION FOR GERMANIUM DOUBLE BETA DECAY DETECTORS

H. GÓMEZ, S. CEBRIÁN, J. MORALES and J. A. VILLAR
Laboratory of Nuclear and High Energy Physics,
University of Zaragoza,
50009 Zaragoza, Spain
E-mail: hgomez@unizar.es

Several techniques must be combined to reduce the background level close to the value needed in the new generation experiments to search for the neutrinoless double beta decay of the ^{76}Ge, we have to combine several techniques. Three of the most important points to study are: granularity, segmentation of the crystals and spatial resolution of the detector directly correlated with an offline Pulse Shape Analysis (PSA). Preliminary studies about these strategies for background reduction were developed during last months, obtaining some promising results.

1. Introduction

To explore effective neutrino masses around 40 meV, a background level $\sim 10^{-3}$ c keV^{-1} kg^{-1} y^{-1} in the region of interest must be achieved in germanium double beta decay experiments. The two main sources of the background registered in the region of interest for the neutrinoless double beta decay of ^{76}Ge (RoI: 2-2.1 MeV) have been identified in previous experiments like IGEX to be cosmogenic activation of germanium detectors (mainly ^{60}Co and ^{68}Ge)[1] and external gamma background above 2 MeV coming from ^{232}Th and ^{238}U chains.

The recent advances in high-purity germanium detector technologies[2], together with the analysis of pulse shapes in highly segmented detectors[3], can help to differentiate between real double beta decay events, that leave only one energy deposit ("monosite" events), and background events, that leave in many cases more than one ("multisite" events). This could allow us to reject all the background events producing more than one energy deposit.

This work in particular is focused on quantifying the reduction of the background coming from the main sources identified, reached by the granularity of the detector system, the segmentation of each individual germanium detector and the application of Pulse Shape Analysis (PSA) techniques. With this purpose a set of Monte Carlo simulations using GEANT4[4] package has been developed. The precise conditions necessary to achieve a background level $\sim 10^{-3}$ c keV^{-1} kg^{-1} y^{-1} in the region of interest will be also discussed.

2. Granularity

Cylindrical crystals with the same value for diameter and height and masses between 0.1 and 4 kg have been simulated in order to determine the optimal mass distribution to have a background level as low as possible in a hypothetical final set up of some tens of kilograms.

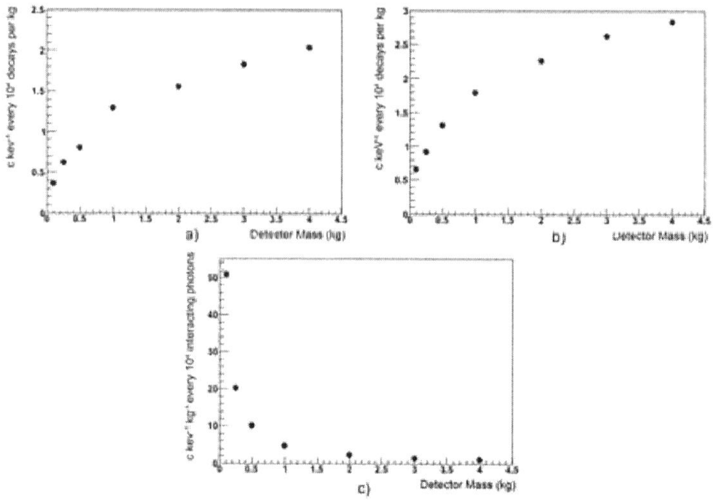

Figure 1. Background level in the 2-2.1 MeV region of interest depending on the component detector mass. For internal contaminations ^{60}Co (a) and ^{68}Ge (b), it is represented in counts per keV every 10^4 decays per kg. For external photons of 2614.5 keV (c), it is represented in counts per keV per kg every 10^4 interacting photons.

We can see how for internal contaminations and a given specific activity (see Figs. 1a, b), 4-kg detectors register 28% (26%) more events for ^{60}Co (^{68}Ge) than 2-kg ones in the RoI. Whereas for external contamination and a given activity (see Fig. 1c), 4-kg detectors register 54% less events coming from 2614.5 keV photons than 2-kg detectors.

3. Segmentation

We can quantify the maximum reduction factor obtained from the anticoincidences applied between the segments in each individual detector. The segmentation schemes studied are a combination of longitudinal and transversal segmentation. For these schemes we have to take into account the difficulty associated to the reduction of the width of the transversal segments or to have a segment size big enough not to lose efficiency in double beta decay events detection.

As Table 1 shows, for a 4 kg detector with 11 by 6 segments, approximately only 2 (5) out of 100 events will not be rejected for ^{60}Co (^{68}Ge) contamination. For 2614.5 keV external gammas, the ratio is less than a half of total events. Simulation provides an "ideal" segmentation, for that reason, the rejection factors obtained must be considered as limits.

Table 1. Percentage of rejected events by crystal segmentation and anticoincidence techniques for registered events between 2 and 2.1 MeV, for 2 and 4-kg detectors and different segmentation schemes.

	2 kg				4 kg			
	7 planes	9 planes	7 by 6 segments	9 by 6 segments	9 planes	11 planes	9 by 6 segments	11 by 6 segments
^{60}Co	92.4	95.6	97.0	98.2	94.9	96.7	97.6	98.4
^{68}Ge	86.1	90.7	93.2	95.1	89.8	92.7	94.2	95.7
2614.5 keV photons	40.6	44.4	48.6	51.0	45.4	49.4	52.4	55.1

4. PSA

The obtained data in the simulations can be reanalyzed for a given spatial resolution, grouping all the partial energy deposits with a separation lower than the resolution and considering these groups like indivisible energy deposits. Then, it is possible to determine how many of these indivisible deposits each background event has and reject the "multisite" events.

Assuming 3 and 5 mm for the spatial resolution, which seem to be reachable today according to the work developed in Ref. [5], we can estimate the rejection efficiency. Assuming a 3 (5) mm spatial resolution, it is possible to reject 99.9 (99.5)% background events coming from ^{60}Co, 99.2 (97.8)% from ^{68}Ge and 60.2 (56.5)% from 2614.5 keV photons.

5. Expected background

To estimate a total background after application of these techniques is necessary to assume the raw background coming from the three sources considered in this work.

Cosmogenic activation in germanium double beta decay experiments has been specifically studied in Ref. [6] and the expected counting rate due to this effect in the neutrinoless DBD region has been evaluated using these and previous results. For ^{60}Co, an exposure time of 30 days and a production rate of 5 kg^{-1} d^{-1} have been considered. For ^{68}Ge, two hypothesis were formulated.

For the external 2614.5 keV, an environmental flux of 0.1 cm^{-2} s^{-1} has been assumed taking into account the recent measurements made in the Canfranc Underground Laboratory[7] and the use of 30 and 40 cm thick lead shield considered. For intrinsic ^{232}Th impurities in the lead shielding, the activity supposed, just as a reference, is 1 μBq kg^{-1}.

Table 2. Estimates of counting rates R (c keV^{-1} kg^{-1} y^{-1}) in the 2-2.1 MeV region of interest for cosmogenic contaminations and 2614.5 keV gamma emissions from ^{208}Tl in 2 and 4-kg germanium detectors, assuming maximum reduction factors deduced for segmented crystals and for segmented crystals using PSA. (Conditions for cosmogenic production and 2614.5 keV emissions are explained in Sec. 5)

	2 kg			4 kg		
	raw	segmentation	segmented PSA	raw	segmentation	segmented PSA
^{60}Co	2.9x10^{-3}	5.2x10^{-5}	8.7x10^{-6}	3.7x10^{-3}	6.0x10^{-5}	3.7x10^{-6}
^{68}Ge^{a}	1.2x10^{-2}	6.1x10^{-4}	1.4x10^{-4}	1.6x10^{-2}	6.8x10^{-4}	1.3x10^{-4}
^{68}Ge^{b}	3.1x10^{-2}	1.5x10^{-3}	3.4x10^{-4}	3.9x10^{-2}	1.7x10^{-3}	3.1x10^{-4}
2614.5 keV photons 30 cm of lead	3.0x10^{-2}	1.5x10^{-2}	1.3x10^{-2}	2.4x10^{-2}	1.1x10^{-2}	9.4x10^{-3}
2614.5 keV photons 40 cm of lead	2.6x10^{-4}	1.3x10^{-4}	1.1x10^{-4}	2.1x10^{-4}	9.3x10^{-5}	8.3x10^{-5}
Intrinsic 2614.5 keV in lead	2.8x10^{-3}	1.4x10^{-3}	1.2x10^{-3}	2.2x10^{-3}	9.9x10^{-4}	8.8x10^{-4}
BEST TOTAL	1.8x10^{-2}	2.1x10^{-3}	1.5x10^{-3}	2.2x10^{-2}	1.8x10^{-3}	1.1x10^{-3}

[a] Production rate 1 kg^{-1} d^{-1}. Exposure time 180 d. Cooling time 180 d.
[b] Production rate 10 kg^{-1} d^{-1}. Exposure time 180 d. Cooling time 730 d.

As Table 2 shows, a background level of ~10^{-3} c keV^{-1} kg^{-1} y^{-1} is reachable applying PSA in segmented detectors and considering the best background features (less cosmogenic activation for ^{68}Ge and 40 cm thick lead shielding).

6. Conclusions

The study shows that heavier crystals are better to reduce the contribution of external radioimpurities, but worse to reduce background coming from internal contaminations. By applying the most powerful segmentation techniques taken here into consideration in a 4 kg detector, 2 (5) out of 100 events due to internal impurities from ^{60}Co (^{68}Ge) would remain in the RoI, while for external contaminations, about half of the events would be rejected. A 3-dimensional spatial resolution of 3 mm, obtained by means of PSA in segmented detectors, would allow to reject more than 99% of background events due to cosmogenic isotopes induced in the crystal, and around 60% of those coming from external 2614.5 keV photons.

The use of large crystals, seems more adequate since background rejection by anticoincidence between segments or by PSA is efficient enough in the reduction of contributions from internal radioimpurities.

In summary, contribution from dominant background sources in previous germanium double beta decay experiments could be reduced down to 10^{-3} c keV^{-1} kg^{-1} y^{-1} using present detectors technologies.

Acknowledgments

This work has been funded by Spanish MCyT/MEC contract FPA2004-0974 and the ILIAS integrating activity (Contract number: EU-RII3-CT2003-506222).

References

1. C. E. Aalseth *et al.*, *Phys. Rev.* **C 59** (1999) 2108.
2. P. Sangsingkeow *et al.*, *Nucl. Instrum. and Meth.* **A 505** (2003) 183-186.
3. Th. Kröll, D. Bazzacco, *Nucl. Instrum. and Meth.* **A 463** (2001) 227-249.
4. S. Agostinelli *et al.*, *Nucl. Instrum. and Meth.* **A 506** (2003) 250. Web page: http://www.cern.ch/geant4.
5. Th. Kröll, D. Bazzacco, *Nucl. Ins. and Meth.* **A 565** (2006) 691-703.
6. S. Cebrián *et al.*, *Journal of Physiscs: Coneference Series* **39** (2006) 344-346. Institute of Physiscs, (Proceedings of the TAUP2005 Conference, Zaragoza, Spain, September 2005) http://www.iop.org/EJ/toc/1742-6596/39/1.
7. G. Luzón *et al,*. See these Proceedings.

BACKGROUND STUDIES OF THE PN-CCD DETECTOR OF CAST

A. RODRIGUEZ[†], B. BELTRAN, S. CEBRIAN, H. GOMEZ, I.G. IRASTORZA, G. LUZON, J. MORALES, J. RUZ and J.A. VILLAR
Laboratorio de Física Nuclear y Altas Energías
Facultad de Ciencias, Pedro Cerbuna 12, 50009, Zaragoza, Spain

M. KUSTER
Technische Universität Darmstadt, IKP
Schlossgartenstrasse 9, 64289 Darmstadt, Germany
Max-Planck-Institut für extraterrestrische Physik, Giessenbachstrasse, D-85748 Garching, Germany
Gesellschaft für Schwerionenforschung, GSI-Darmstadt, Plasmaphysik, Planckstr. 1,64291 Darmstadt, Germany

C. KLOSE
Technische Universität Darmstadt, IKP
Schlossgartenstrasse 9, 64289 Darmstadt, Germany

R. HARTMANN, R. KOTTHAUS and L. STRÜDER
MPI Halbleiterlabor Otto-Hahn-Ring 6, 81739 München, Germany
Max-Planck-Institut für extraterrestrische Physik, Giessenbachstrasse, D-85748 Garching, Germany

A backside illuminated pn-CCD detector in conjunction with an X-ray mirror optics is one of the three detectors used in the CERN Axion Solar Telescope (CAST) to register the expected photon signal. A background study performed for this detector shows that the level $(8.00\pm0.07)\times10^{-5}$ counts cm^{-2} s^{-1} keV^{-1} between 1 and 7 keV is to be dominated by the external gamma background due to natural activities at the experimental site, while radioactive impurities in the detector itself and cosmic neutrons contribute a much smaller fraction.

1. Introduction

The CERN Axion Solar Telescope (CAST) is intended to search for solar axions using a decommissioned 9.26 m long Large Hadron Collider (LHC) superconducting magnet providing a 9 T magnetic field. The magnet is installed

[†]Email address: mara@unizar.es

on a platform that permits horizontal movement from azimuthal angle 46° to 133° and a vertical movement of ±8°. As a consequence, it can follow the Sun three hours per day, 1.5 hours each during sunset and sunrise. The rest of the day is dedicated to background measurements.

Three X-ray detectors are installed on the ends of the magnet to search for an excess of X-rays coming from axion conversions inside the magnet during alignment with the Sun. The detectors are a Time Projection Chamber (TPC), a Micromesh Gaseous Structure (MICROMEGAS), and a Charge Coupled Device (CCD) in combination with an X-ray mirror telescope. The CCD and MICROMEGAS detectors observe the sun at sunrise, while the TPC detector, covering two magnet bores, observe at sunset.

During 2003 and 2004 the experiment operated with vacuum inside the magnet pipes (CAST Phase I). No signal above background was observed in 2003 data, implying an upper limit to the axion-photon coupling[1] of $g_{a\gamma\gamma} \leq 1.16 \times 10^{-10}$ GeV^{-1}. The CAST experimental setup was transformed to be able to fill the axion conversion volume with a buffer gas in 2005 (CAST Phase II) in order to increase the sensitivity to higher axion mass.

2. The CCD detector and the X-ray telescope

The X-ray telescope of CAST consists of a Wolter I type X-ray mirror optics[2] focusing a potential axion signal on a small area of the CCD detector which is located in the focal plane of the optics. The detector is a fully depleted back side illuminated pn-CCD with a depletion depth of 280 μm and a pixel size of 150 x 150 μm^2, optimized for the 0.2-10 keV energy range. Further enhancement of the sensitivity is achieved by adding a passive shield consisting of a combination of internal and external lead-copper components.

3. Background sources

The most important background contribution is expected from external gamma rays, produced mainly by primordial radio-nuclides like ^{40}K and the radioactive natural chains from ^{238}U, ^{235}U and ^{232}Th in laboratory soil, building materials and experimental set-up as well as by ^{222}Rn in air.

Intrinsic radioactive impurities (either primordial or cosmogenically induced) in the detector materials are important in experiments looking for rare event signals because of their alpha, beta and gamma emissions.

Cosmic rays on the Earth's surface are dominated by muons and neutrons. While muon interactions (as those of other charged particles) can be rejected by off-line analysis, signals from neutrons are an irreducible.

4. Background simulations

4.1 *The code*

The GEANT4 package[3] has been used for simulations with the G4NDL 3.7 library for neutrons. In the first simulations for neutrons and external gamma backgrounds a simplified description of the detector was implemented. A much more detailed geometry for the detector was defined to carry out simulations of the radioactive impurities in the detector components (see Figure 1).

An energy calibration with a ^{55}Fe source was first simulated and compared to the corresponding experimental spectrum where the photopeaks have been reproduced quite well.

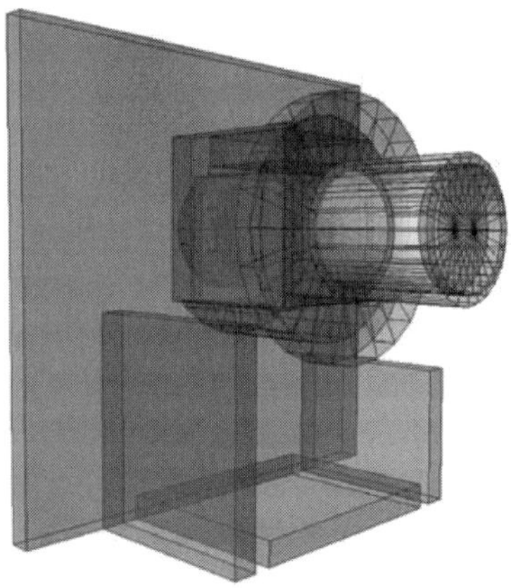

Figure 1. View of all simulated components, including shields, aluminum vessel and connection to telescope.

4.2 *External gamma background*

In order to evaluate the contribution of the environmental gamma background to the CCD counting rate and the effect of the four different shielding configurations (no shield, internal copper shield, internal copper shield and external lead shield and finally internal lead-copper shields and external lead shield) the response of the detector to photons has been studied.

Radon level measurements of the experimental site using a Ge spectrometer has been used as input for determing the expected detector response. A mean radon concentration of 10 Bq m^{-3} has been assumed. The total contribution from radon up to 7 keV is found to be ~10^{-6} counts cm^{-2} s^{-1} keV^{-1}, almost negligible compared to the measured background rate in the CCD detector. A counting rate of a 3-4x10^{-5} counts cm^{-2} s^{-1} keV^{-1} is then attributed to the measured activities of the experimental site walls.

4.3 *Neutrons*

Simulations have estimated the counting rate due to cosmic neutron background in the region up to 10 keV as 6x10^{-6} counts cm^{-2} s^{-1} keV^{-1}, other neutron background contribution due to radioactivity or muon induced neutron are at least one order of magnitude lower than the experimental background levels of the CCD detector, and therefore, not very significant at the present level of sensitivity.

Table 1. Total contribution to CCD counting rate between 1 and 7 keV from natural radioactivity of the individual components in units of counts cm^{-2} s^{-1} keV^{-1}

Detector Component	Differential Flux
CCD Board	(5.83±0.41)x10^{-7}
CCD Chip	< 2.2x10^{-5}
Ceramics	(1.17±0.14)x10^{-6}
Sockets	(1.34±0.15)x10^{-6}
Front Cooling Mask	< 3.6x10^{-7}
Back Cooling Mask	< 2.6x10^{-7}

4.4 *Intrinsic radioactive impurities*

The levels of radioactive impurities in the main components of the CCD detector were measured in the Canfranc Underground Laboratory in Spain, using an ultra-low background germanium detector, can be found in a database of radiopurity of materials[4] inside the ILIAS program (Integrated Large Infraestructures for Astroparticle Science). Activity comes mainly from the radioactive chains ^{235}U, ^{238}U, ^{232}Th and the isotope ^{40}K.

The total contribution to the background from the radioactivities measured in Table 1 is 2.6x10^{-5} counts cm^{-2} s^{-1} keV^{-1}. Compared to the measured background level in the CCD detector impurities from detector components account for up to ~33% of the counting rate according with the simulations.

5. Conclusions

Using the simulated response of the CCD detector to the external gamma background, an estimate of the contribution of this background component has been attempted, finding that measured radon levels in the air of the CAST site could produce on average ~1% of the registered counting rate, while just the measured activities from ^{238}U, ^{232}Th radioactive chains and the isotope ^{40}K in the walls of the CAST hall could justify more than 50% of this counting rate.

The response of the CCD to neutrons of different energies has also been simulated. They do not seem to be a very relevant source of the CCD background, producing just a few per cent of the observed counting rate.

Finally, the contribution to the CCD counting rate of the internal radioactive impurities of the main detector components measured with a Ge spectrometer at the Canfranc Underground Laboratory could justify at most 33% of the measured according to simulations.

Taking into account each of these contributions, a quite complete model for the background measured by the CCD detector has been obtained.

Acknowledgments

This work is supported by Spanish Ministry of Education and Science under contract FPA2004-00973, and by the Bundesministerium für Bildung und Forschung (BMBF) under the grant number 05 CC2EEA/9 and 05 CC1RD1/0, by the Virtuelles Institut für Dunkle Materie und Neutrinos - VIDMAN. The gratitude is also to the group of the Canfranc Underground Laboratory for material radiopurity measurements and the CAST collaboration and the ILIAS integrating activity (Contract number: EU-RII3-CT-2003-506222).

References

1. K.Zioutas et al, Phys. Rev. Lett. 94 121301-1 (2005).
2. L.Strüder et al, Astron. Astrophys. 365 L18-L26 (2001).
3. S. Agostinelli et al, Nucl. Instrum. Methods Phys.Res., Sect. A 506 250-303 (2003).
4. http://radiopurity.in2p3.fr

ZEPLIN II EVENT RECONSTRUCTION

K. MAVROKORIDIS ON BEHALF OF THE ZEPLIN II COLLABORATION*,†

Department of Physics and Astronomy, University of Sheffield,
Hicks Building, Hounsfield Road, Sheffield S3 7RH, United Kingdom
† *E-mail: k.mavrokoridis@shef.ac.uk*

The ZEPLIN-II detector has completed two months production run. The data acquisition system is described. Parameterisation of events and data reduction are discussed.

Keywords: WIMP direct detection, data acquisition

1. Introduction

The nature of dark matter remains one of the most significant problems in our present understanding of the universe [1]. This non luminous matter is expected to be non baryonic and "cold" (non relativistic at the time of decoupling). Weakly Interacting Massive Particles (WIMPs) are an excellent candidate for non baryonic, cold dark matter. These particles could be detected directly, via their interactions with nuclei in the target. ZEPLIN-II (ZonEd Proportional scintillation in LIquid Noble gases) is a 30 kg two phase liquid/gaseous xenon scintillation/ionisation WIMP detector. The detector design, shielding and veto are described elsewere in these proceedings [2,5]. ZEPLIN-II has been operating since December 2005 at Boulby mine. The Boulby underground laboratory is located at a depth of 2800 m.w.e. and the measured muon flux is 4.09 $\times 10^{-8}$ cm^{-2} s^{-1} [3].

2. Detection Principle

For a WIMP to be detected, it must first scatter off a nucleus of the target material. The initial interaction occurs in the liquid phase and causes a flash of UV light (primary pulse S1). Thc intensity of this light depends on

*CCLRC/Rutherfore Appleton Laboratory, Imperial College, Edinburgh, Sheffield, Coimbra, ITEP, Texas A&M, UCLA, Rochester.

the number of excited atoms and on a contribution from fast recombination of some fraction of the simultaneous ionisation produced. The remaining ionisation electrons which did not recombine are then drifted in an applied electric field to the surface of the liquid where a high field is applied to extract them into the gas phase. When the electrons enter the gas phase they are accelerated by another electric field which is sufficiently high to induce electroluminescence from excitation in the gas itself. This is seen as a second flash (secondary pulse S2 of duration around 1 μs) of UV light by the same PMTs. The ratio of the secondary to primary pulses is a measure of the relative amounts of escaping ionisation and excitation/recombination. This ratio depends on the nature of the interacting species and allows for a high level of discrimination between nuclear recoils produced by neutrons or WIMPs and electron recoils produced by gammas.

3. Data Acquisition

The data acquisition system (DAQ) consists of 2×4 channel Acqiris digitisers based on CompactPCI technology. The trigger setup can be seen in Fig. 1. The signal from each of the 7 PMTs splits at a 50 Ω splitter. Subsequently, one line is fed to a $\times 10$ amplifier and the other into the DAQ. From the amplifier the signals are fed to discriminator D1 which outputs 50 mV/channel for input signals above about 2/5 of a single photoelectron pulse. The logic sum of the signals is then fed to discriminator D2. When 5 out of 7 PMTs detect a signal above the threshold level, the discriminator D2 outputs a NIM pulse to dual timer T2. Meanwhile, T2 sends a 200 μs square-wave pulse to the veto input of D1. This prevents further triggers until the whole waveform is read by the DAQ. The amplified signal from the central PMT (PMT 1) is also fed to an attenuator before going into discriminator D3. On average the central PMT sees a larger signal than the outer PMTs. This is due to the relative placing of the PMTs and the PTFE support/reflector structure. Signals of a large amplitude can cause optical feedback in the target resulting in many noise pulses of long duration. A maximun cut-off 250 mV is set and those signals exceeding this cut-off are vetoed. Therefore large signals of an energy well above the range of interest for dark matter searches are prevented from triggering the system.

4. Data Reduction

Raw background data consist of digitised waveforms 200 μs long. Reduction of the raw data is made with a linux based application called UNZAP2

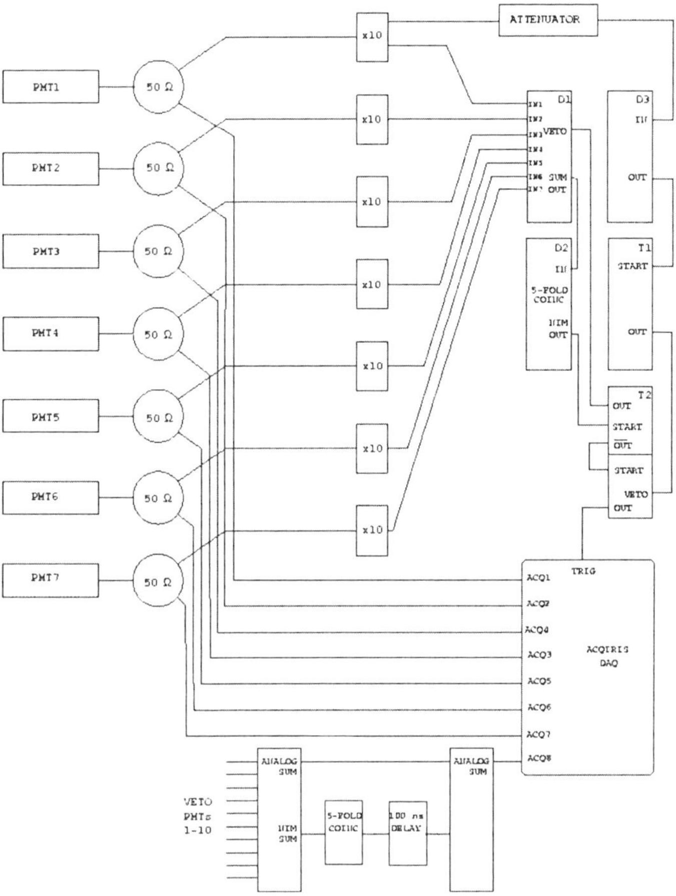

Fig. 1. Signal path from target and veto PMTs to DAQ via trigger electronics.

(Fig. 2). UNZAP2 consists of an event viewer which allows you to view raw traces in each PMT. It also contains a reduction algorithm which can find and parameterise the scintillation/ionisation pulses in the waveforms. The pulses are parameterised mainly according to their area, width, time constant and arrival time. They are then written in an hbook ntuple structure for the purpose of further analysis (Fig. 2). The user has to enter input variables in order for the reduction algorithm to identify correctly the pulses in the trace. These variables are clustering, smoothing, threshold and cable delays. Clustering allows the user to set a maximum time separation of two distinct signals associated with a single scintillation/ionisation event inside the detector. Smoothing is a boxcar smoothing algorithm which suppresses the high frequency noise in the waveform and increases the signal to noise ratio. This is used to identify single photoelectrons (SPEs) in the sum channel. Threshold allows the user to define the minimum amplitude of the SPE. More detailed information about these variables can be found in [4]. From December 2005 to the present day more than 6 TBytes of data have been collected. All the data are stored on magnetic digital tapes and on the analysis computers.

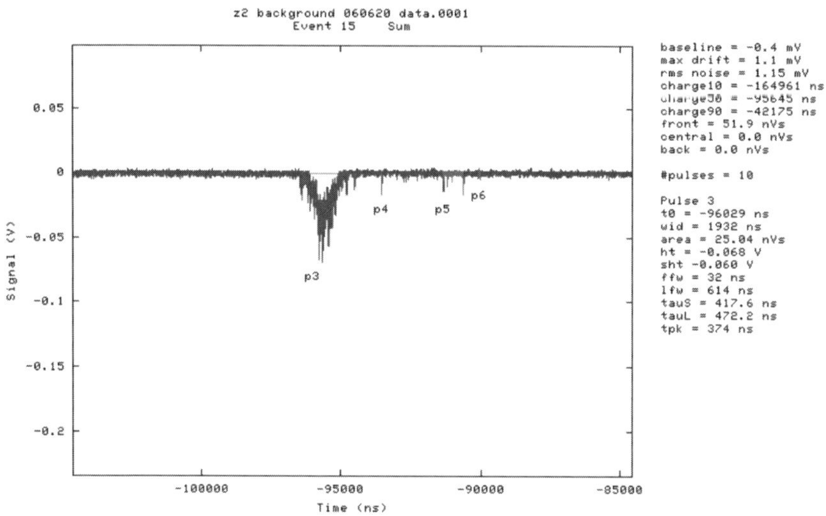

Fig. 2. A typical S2 signal for dark matter data. In the SUM channel the reduction algorithm finds all pulses that pass the set threshold. Then all the identified pulses are parametrised according to pulse height, area, width, arrival time and decay constant. These parameters are used to identify S1 and S2 signals.

5. Position Reconstruction

The secondary electroluminescence pulses occur in the vapour phase, close to the PMT array. This allows the reconstruction of the primary vertex position in the horizontal plane. Reconstruction of Z coordinate is done using the time delay between the primary and the secondary.

Fig. 3 shows a scatter plot of reconstructed positions of pulses in the horizontal plane. These pulses are due to 122 keV gammas from a ^{57}Co source positioned below the copper bottom of the target volume. The clusters of pulses forming two annular rings correspond to thin regions deliberately incorporated into the metal vessel bottom to allow calibration gammas to penetrate into the target. There is also a background gamma count which is fairly uniform due to gammas penetrating through the vessel between the holes.

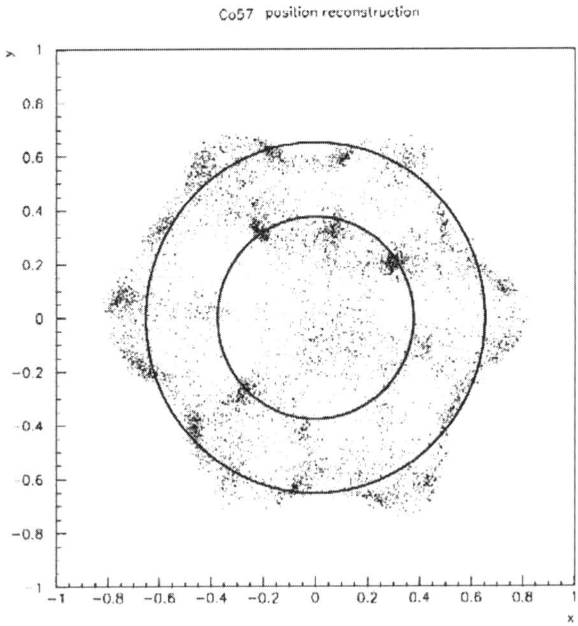

Fig. 3. x-y position reconstruction of the calibration holes.

6. Event Selection

The main selection cuts for valid dark matter events are the following: All selected events must have a single identifiable secondary (S2) pulse preceded by a single identifiable primary (S1) pulse. The secondaries need to have a 5-fold PMT coincidence (hardware trigger). A time duration and minimum area for S2 cut is also applied. The primaries need to have 3-fold PMT coincidence (software cut). A time duration cut and cut on time window for single PMT signals are applied as well. Furthermore, a fiducial cut rejects events from the bottom 12 mm of the target and outer 3/7 of the target radius, restricting the active volume to 7.2 kg. Finally, events that also trigger the veto are rejected (this has a measured efficiency of 49% for neutron events in the target [5]).

Fig. 4 shows that the calculated efficiencies are consistent with the relative efficiencies for Am-Be and ^{60}Co data. The relative efficiencies are obtained by dividing Am-Be and ^{60}Co spectra by the simulated single scatter differential spectra.

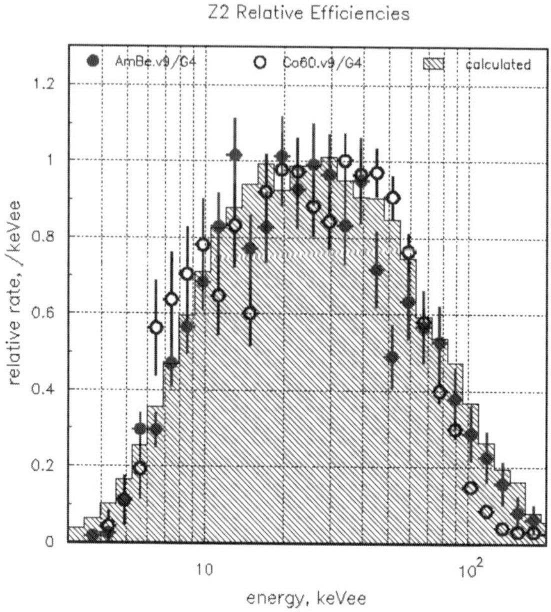

Fig. 4. Comparison between calculated and measured relative efficiencies.

7. Summary

The ZEPLIN II hardware trigger, the data reduction procedure and event selection cuts have been described. Preliminary results from this experiment are discussed elsewhere in these proceedings [2].

References

1. G. Jungman, M. Kamionkowski, K. Griest, *Phys. Rept.* **267**, 195-273 (1996)
2. R. Lüscher *et al.* (ZEPLIN II collaboration), *These proceedings.*
3. M. Robinson, Nucl. Instrum. Meth. **511**, 347 (2003) [arXiv:hep-ex/030601]
4. M. Carson *et al.*, (ZEPLIN II collaboration), *"The ZEPLIN-II dark matter detector: data acquisition system and data reduction"*, in preparation.
5. E. J. Daw *et al.*, (ZEPLIN II collaboration), *"Zeplin II Active Veto"*, *These proceedings.*

THREE DIMENSIONAL TRACK RECONSTRUCTION IN DRIFT-II

D. N. MUNA ON BEHALF OF THE DRIFT COLLABORATION

Department of Physics and Astronomy, University of Sheffield
Sheffield S3 7RH, United Kingdom
Email: d.muna@sheffield.ac.uk

DRIFT-II is a dark matter detector designed to measure WIMP recoils in three dimensions. It uses this information for background discrimination as well as a means to seek an isotropy as a positive signal for dark matter detection. Methods for three dimensional track reconstruction of DRIFT-II event data are presented.

Keywords: Dark Matter; Track Reconstruction; DRIFT.

1. Overview of DRIFT-II

The DRIFT-II experiment is unique as the only directionally sensitive dark matter detector currently in operation. It employs a time projection chamber (similar to those in high energy particle physics detectors), though rather than drifting electrons it drifts negative ions to the multiwire proportional chamber (MWPC) endcaps to spatially reconstruct tracks. The target mass is CS_2 held at low pressure to minimise track diffusion. While most detectors look for a signal above a zero background as a signal for dark matter, the three dimensional directional information opens the window to be able to look for a daily modulation that would follow the rotation of the Earth (and thus, the orientation of the detector). There are several sources for more in-depth information about the design of the detector including Alner, et al. 2005[1] as well as studies on the statistics pertaining to detecting a WIMP signal in a directional dark matter detector (Morgan, et al. 2005[2] and Morgan, Green, 2005[3]). See also the paper by D. Muna and N. Spooner in this volume for an introduction and status of DRIFT-II.

2. Motivation for Track Reconstruction

One of the difficulties in the analysis of data from a dark matter experiment is the accurate identification of events caused by any background signal from the lab. The background in the DRIFT-II detector is primarily from radon decays within the fiducial volume plus a small contribution of alphas from the inner surface of the shielding surrounding the detector. The range of a neutron recoil in low pressure CS_2 is typically 2–6 mm (corresponding to ~20–60 keV for a C recoil, ~100–325 keV for S)[4] while a ~5 MeV alpha will produce a track on the order of ~300 mm or more. Track reconstruction, even as a projection in two dimensions, thus provides a powerful and often clear-cut background discrimination.

The primary motivation for reconstructing tracks in three dimensions is to look for a directional signal that is anisotropic. Any background signal that is terrestrial in origin (i.e. contamination within the lab) will be isotropic when viewed in a galactic frame. However, a true WIMP signal will have a preferred direction in the direction of solar motion. The DRIFT-II detector cannot measure the original or recoil direction of the incoming WIMP, but instead measures the direction of the recoil of the sulphur nucleus in the target gas. (The WIMP-sulphur cross section is expected to be considerably higher than for carbon nucleons, although simulations show that the reverse is true for neutron sources.) This recoil distribution has been shown to correlate to the WIMP flux,[2] even accounting for diffusion as the recoil track travels to the MWPC readout.

3. Two Dimensional Track Reconstruction

The process for two dimensional reconstruction is as follows. First, all pulses above a set software threshold within the event record are collected and ordered by their voltage-weighted time. Beginning with the first pulse, a second pulse is sought within a given time window, a free parameter usually chosen to be between 200–500 μs. Pulses are added to the track as long as another is found within the time beginning from the previous pulse. A few criteria are placed on the track, and if it fails any of these it is rejected altogether. The track must traverse adjacent wires, i.e. have no gaps (missing wires) along its length, and two sequential pulses cannot fall on the same wire. The track is rejected if any pulse has a full width half-max that is less than the shaping time (usually an indication that the event occurred within the MWPC itself). Any pulses that have a pulse of opposite polarity, an area >60% of the main pulse, and that precede the main signal are also discarded as these are most likely caused by induction rather than an actual deposit of charge.

Several parameters are calculated from the track. The (x, z) co-ordinates for each anode pulse are taken from the wire hit and the time at which the

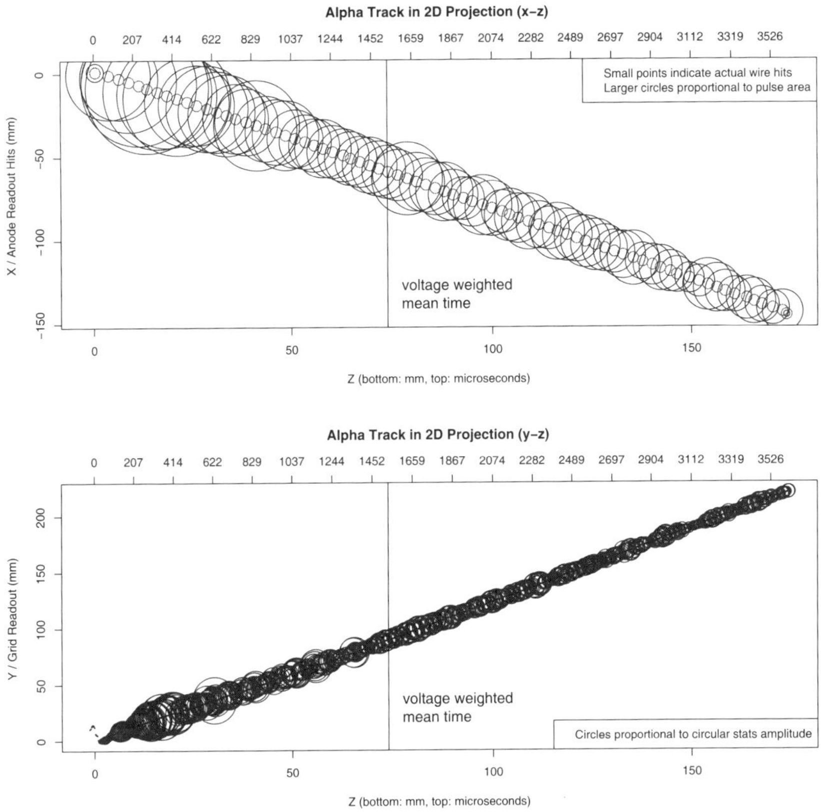

Fig. 1. The alpha track reconstructed here in two dimensions. Data in the upper plot are calculated by direct anode hits at a resolution of 2 mm. The lower plot shows induced grid (cathode) data which has a resolution equal to the DAQ sampling rate (here, 1000 kHz).

pulse reached the maximum value. It is impossible to know the absolute x location where the track occurred within the fiducial volume, but it is irrelevant as only the direction is important. The distance in the drift direction, dz, is calculated as the total time of the track times the drift velocity. The total time is taken as the start time of FWHM of the first pulse to the end time of the FWHM of the last pulse, minus 10 μs to account for the shaping time. The sampling rate of the DAQ (usually 1 μs) determines the resolution in this direction. The minimum track length across the anode wires, dx, is defined as the number of wires crossed minus 1 times the pitch (2 mm). The maximum length (without triggering another wire) is found by adding 1 mm to each end of the track. Thus, dx is the number of wires hit plus 1 mm $\pm$ 1 mm. Finally, the length of the 2D projection on the dx/dz plane is simply $R_2 = \sqrt{dx^2 + dz^2}$.

An example of an event record of an alpha and a neutron is presented in Figure 5 of the paper "The DRIFT-II Dark Matter Detector" in this volume, and it is these two events that will be further analysed. The long, roughly straight track illustrates the cyclical nature of the grouped readouts. The upper plot in Figure 1 shows the reconstruction of the anode wires and time (x–z), showing that the track crossed 74 anode wires (146 mm in the x direction). The small circles indicate the peak positions of each pulse on the readouts, while the larger circles show the relative charge deposited (area of each pulse). For higher energy events such as this one, the difference between the amount of charge deposited at the beginning of the track and the end reveals the sense of the track — the particle travelled from the left to the right based on the Bragg curve. It's likely that just noting which half of the track in time the voltage-weighted mean time falls is enough to determine the sense for alphas. The upper plot in Figure 2 is the reconstruction of the neutron event, where here only two points are available for reconstruction in x.

4. Three Dimensional Track Reconstruction

The third dimension is extracted from the signals on the cathode wires, labelled G1–G8 on the event plots. Reconstructing this axis is not as straightforward as the others. Charge deposited on a single anode wire will induce a charge on several cathode wires, and this is further entangled when multiple anode wires are hit. The cathode readouts over the time period where the track is recorded is divided into 5 μs bins. For each bin, the readout is averaged to give one number per wire. The y co-ordinate corresponding to the time z at the midpoint of the bin is taken as the point where the largest "weight" of the charge falls. The spacing of the cathode wires is 2 mm, but by performing a fit using the available values a resolution smaller than this can be attained. It is worth noting that "single" charges are not deposited in the same way as on the anode wires, rather, charge is induced on many wires simultaneously on the perpendicular plane.

Matters are slightly complicated by the grouping of the 512 cathode wires down to eight readouts. Since wire 1 is technically adjacent to wire 8, the fit must be cyclical in nature. Circular statistics, common in the field of biostatistics, are employed here. It is best visualised as placing each of the eight values on equidistant positions on a circle. There is no preferred order in this arrangement as opposed to a normal linear ordering which has a start

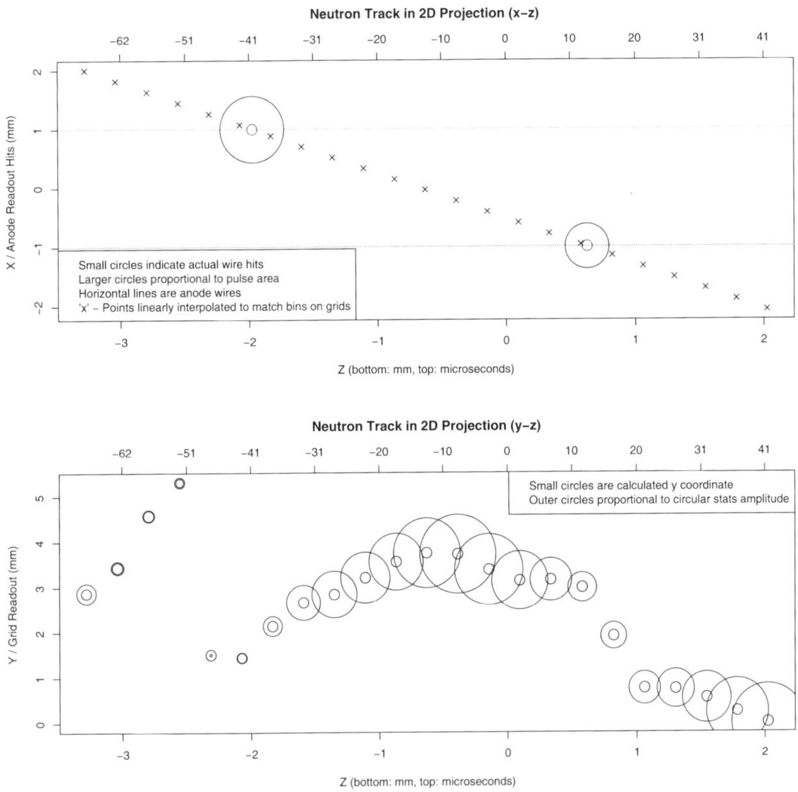

Fig. 2. The neutron track reconstructed in 2D projections. To produce a 3D plot, additional values of x are linearly interpolated between anode wire hits, indicated as x's on the plots. These points are the midpoints of the time bins in the lower plot.

and an end. The statistics produce an angle where the greatest charge falls and a magnitude. The angle is converted back to a wire number and gives the x co-ordinate while the magnitude can be used as a "goodness of fit" parameter. A magnitude that is close to zero is what would be expected from random noise; although an angle is always returned, a magnitude approaching zero is equivalent to no preferred direction. Other statistical means are also valid here are long as the cyclical nature of the points is observed and the function being fit is even.

The lower plot in Figure 1 shows the y–z reconstruction of the alpha track. Here again the Bragg curve can be seen. The neutron reconstruction is the lower plot of Figure 2. The proportional magnitude returned by the circular statistics is again indicated in the plot. The very small circles on the

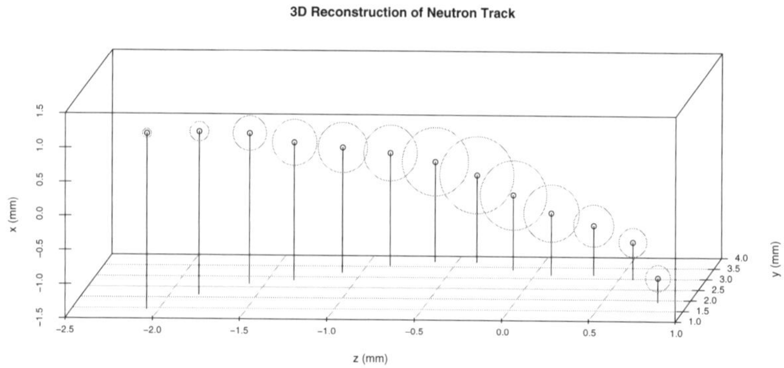

Fig. 3. 3D Reconstruction. The grey circles indicate the magnitude on the grids.

left are taken to mean no signal, where after that a clear segment appears that is coincident with the two anode hits above. Another signal appears after a break (the end of the track), and this is likely due to the undershoots of the signals induced by the charges on the anodes. Together, these plots provide x, y, and z co-ordinates which are used to reproduce the track in three dimensions (Figure 3).

5. Future Work

Much of this work is in the early stages, but the indicators are very promising. Here we demonstrate that low energy (i.e. WIMP-like) events can be reconstructed in three dimensions. From this, the next step will be to plot an "all-sky" distribution of the angles of reconstructed tracks and compare between directed neutron runs and background runs (where the isotropic radon progeny background that exists in this data can at least be used to some advantage for comparison). Finally, determining the sense of low energy tracks will be the top priority. Details will be presented in an upcoming publication.

References

1. G. J. Alner and et al., *Nuclear Instruments & Methods A* **555**, 173 (2005).
2. B. Morgan, A. M. Green and N. J. C. Spooner, *Physical Review D* **71**, p. 103507(May 2005).
3. B. Morgan and A. M. Green, *Physical Review D* **72**, 123501(December 2005).
4. D. P. Snowden-Ifft, T. Ohnuki, E. S. Rykoff and C. J. Martoff, *Nuclear Instruments & Methods A* **498**, 155(February 2003).

ON THE SENSITIVITY OF SUPERFLUID ^{3}He BOLOMETERS FOR ULTIMA

J. ELBS, C. WINKELMANN, YU. M. BUNKOV, E. COLLIN and H. GODFRIN

CNRS-CRTBT

25, av. des Martyrs, BP 166, 38042 Grenoble cedex 9, France

** E-mail: henri.godfrin@grenoble.cnrs.fr*

http://www-crtbt.grenoble.cnrs.fr/

The keV energy sensitivity of superfluid ^{3}He bolometers is due to the extremely small heat capacity of this fluid. At ultra-low temperatures, however, their thermal response may be dominated by a few atomic layers of adsorbed ^{3}He. We report on direct measurements of this effect, as well as on the techniques that allow to suppress it to achieve the highest detection sensitivity.

Keywords: bolometer; superfluid ^{3}He; dark matter; detector.

1. Introduction

Bolometric particle detectors are devices sensitive to the energy deposited by incoming particles in a well chosen material, causing a change in its temperature. Very low heat capacity materials are therefore used for this purpose. Superfluid ^{3}He has clear advantages, owing to its ordering temperature of about 1 mK, and the feasability of thermal measurements at temperatures as low as 100 μK. This spin 1/2 ordered system is highly sensitive to the axial interaction with WIMPS. ULTIMA is therefore complementary with other experiments, like CDMS, Edelweiss, etc.

An absolute energy calibration of superfluid ^{3}He bolometers can be made. The calibration coefficient depends only, in principle, from geometrical factors and the heat capacity of superfluid ^{3}He. We found, however, variations of this coefficient at the lowest temperatures from run to run. These were expected, since at ultra-low temperatures, the heat capacity of the solid ^{3}He monolayers adsorbed on the surface of the cell walls is of the same order that that of the bulk superfluid ^{3}He. This should result in an additional thermal load. The puzzle, in fact, was that this contribution was not always present in our calibration measurements. We describe here the results of several experiments which allowed us to determine the adsorbed ^{3}He heat capacity, to understand the origin of the calibration coefficient changes, and to optimize the sensitivity of the ^{3}He bolometer.

2. Superfluid bolometers

A detector consists of a cell containing superfluid ^{3}He cooled down to 100 μK by a nuclear demagnetization stage.[1,2] Many cells are arranged to form a matrix of bolometers which allow, by correlation of their signals, to reject background events. Each cell is a box (typically a 0.13 cm^3) which is in contact with the ^{3}He bath through a small hole. When a particle-^{3}He collision occurs, the temperature of the bolometer increases, and this is measured with a sensitive Vibrating Wire Resonator (VWR) thermometer.[3,4] It consists of a superconducting wire loop, excited in the presence of an external magnetic field at its mechanical resonance frequency by an a.c. current, and which thus vibrates inside the cell. Its damping, measuring directly the scattering of thermal excitations, is determined from the frequency response of the signal. The motion of the wire in the magnetic field induces a voltage, which is measured using lock-in techniques.

This type of bolometers can be calibrated by direct transfer of energy to quasiparticles, using a second VWR.[2,5] Thermometry is then achieved by measuring continuously the damping coefficient W of a VWR driven at resonance (W is measured as the FWHH of the oscillator's lorentzian resonance in frequency space). In practice, we monitor the amplitude at the resonance, and the width, expressed in Hz, is calculated. The effect of a pulse appears as an abrupt increase in the resonance width. The height of the resulting peak (H, in Hz), for small pulses, is proportional to the energy (E) injected in the cell. The calibration coefficient $\sigma = H/E$, inversely proportional to the heat capacity, has a weak dependence on temperature. It can be fitted with only one adjustable parameter, a geometric factor characterising the VWR.

3. Calibration measurements

We have measured the calibration coefficient at different temperatures in many experimental runs. Each of the data points correspond to at least 15 pulses of different energies. The calibration coefficient has then been obtained by a linear fit of the peak height vs. energy plot.

We have used different condensation procedures for the ^{3}He gas to test the sensitivity of the calibration coefficient to the amount of ^{4}He reaching the bolometric cell. The usual procedure consists in condensing the gas at relatively high temperatures (≈ 2 K). Under these conditions, ^{4}He impurities (typically less than 100 ppm) are dissolved in the ^{3}He, and can reach the bolometric cell (2 ppm concentration of ^{4}He in the bolometrical cell is sufficient to cover the cell with 2 monolayers of solid ^{4}He). This can be avoided by condensing the gas when the cryostat is at low temperatures (≈ 10 mK). The equilibrium concentration of ^{4}He in ^{3}He is then extremely small. ^{4}He is preferentially adsorbed in high surface area parts of the set-up, and the rest remains in a guard cell as droplets of ^{4}He. Under these conditions, the bolometric cell and the adsorbed monolayers contain essentially pure ^{3}He.

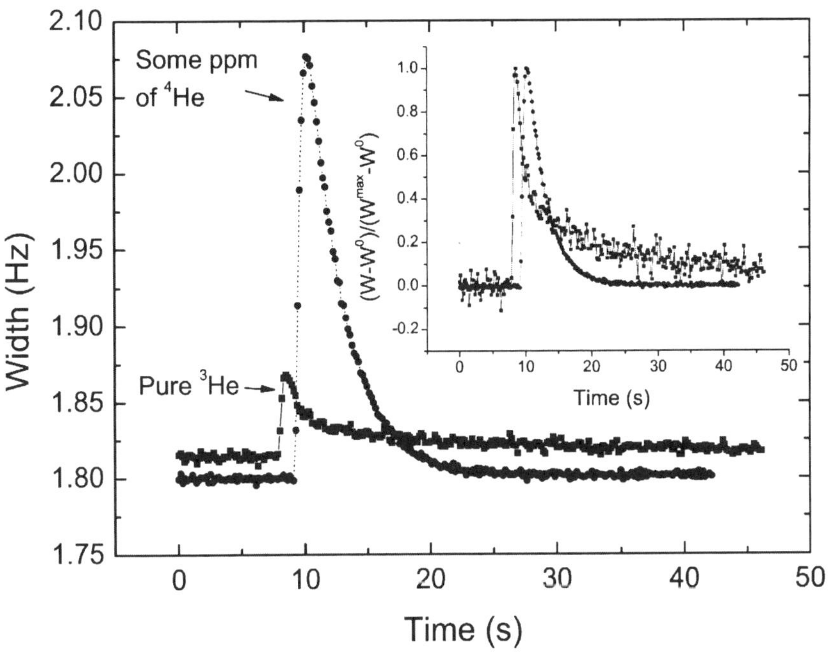

Fig. 1. Response of the bolometer submitted to a calibration pulse of 750 keV, for the cell filled by Helium in cold and warm conditions. Insert : normalized signals.

We first filled very slowly the cell with ^{3}He when the cryostat was already at ≈ 10 mK. The heater pulses done in the subsequent 7 demagnetization runs all showed unusually low response on the thermometer and hence low calibration coefficients (Fig. 1). Between these runs, the temperature in the cell never went above 100 mK. After the seventh demagnetization, the system was heated to about 10 K in order to evaporate the Helium and force a redistribution of the adsorbed layers. We then cooled the system down to ultra low temperatures, and we observed much larger signals after the calibration heat pulses.

The calibration coefficient determined with pure ^{3}He (cell filled at low temperatures) is strongly reduced, and displays a strong temperature dependence. This can be explained when one considers the additional heat capacity of the solid ^{3}He adsorbed layers: $C = C_{bulk} + C_{surface}$.

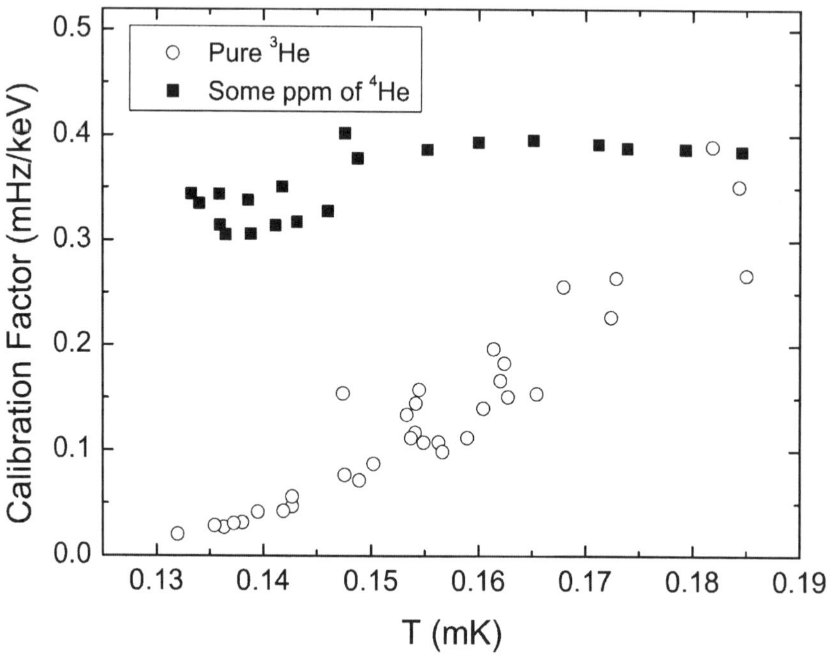

Fig. 2. Calibration coefficient as a function of temperature, before and after rebaking. The curve after rebaking corresponds to ^{4}He-coated walls. In the pure ^{3}He case, a strong reduction of the bolometer sensitivity is observed.

We have estimated the heat capacity of the adsorbed ^{3}He layers considering a surface area of $1.5*10^{-4}\text{m}^2$, calculated from the geometric dimensions of the cell, regardless of the rugosity of the copper walls. We find a surface heat capacity of approximately 1000 μJK^{-1}m^{-2} at 150 μK, a value about 2 orders of magnitude higher than what has been found in the mK range. Our experiments are performed in magnetic fields ranging from 60 mT to 150 mT. At our working temperatures of about 150μK we are thus in the vicinity of the maximum of the Schottky anomaly of the adsorbed ^{3}He. The addendum specific heat is thus of magnetic nature. The scattering observed in the results is partly due to the influence of the magnetic field. For the calibrations done without the surface heat capacity, we observed that the calibration coefficient decreases with increasing field, as expected from the magnetic dependence of the superfluid gap. Detailed measurements on this effect are in progress.

4. Conclusions

The calibration coefficient of a superfluid ^{3}He detector has been carefully measured using different filling procedures of the ^{3}He bolometer. We have observed a substantial difference between "warm" and "cold" fillings, showing than in the last case the strong reduction of the calibration coefficient can be ascribed to the formation of adsorbed ^{3}He monolayers on the copper walls of the bolometric cell. The heat capacity of these adsorbed layers is large, and magnetic field dependent. This effect, which can reduce significantly the detector sensitivity, can be suppressed by filling the cell at relatively high temperatures, and using an appropriate cooling procedure.

Acknowledgements

The ULTIMA project is funded by the ANR, France. We are also thankful to the ESF Program COSLAB for many stimulating discussions.

References

1. D. I. Bradley, Yu. M. Bunkov, D. J. Cousins, M. P. Enrico, S. N. Fisher, M. R. Follows, A. M. Guénault, W. M. Hayes, G. R. Pickett and T. Sloan, *Phys. Rev. Lett.* **75** 1887 (1995)
2. C. Bäuerle, Yu. M. Bunkov, S. N. Fisher and H. Godfrin, *Phys. Rev. B* **57**, 14381 (1998).
3. A. M. Guénault, V. Keith, C. J. Kennedy S. G. Mussett, and G. R. Pickett, *J. Low Temp. Phys.* **62**, 511 (1986)
4. C. B. Winkelmann, E. Collin, Yu. M. Bunkov and H. Godfrin, *J. Low Temp. Phys.* **135**, 3 (2004)
5. S. N. Fisher, A. M. Guénault, C. J. Kennedy and G. R. Pickett, *Phys. Rev. Lett.* **69**, 1073 (1992).

ZEPLIN II ACTIVE VETO

E. J. DAW ON BEHALF OF THE ZEPLIN II COLLABORATION*

Department of Physics and Astronomy, University of Sheffield, Hicks Building, Hounsfield Road, Sheffield S3 7RH, United Kingdom
**E-mail: e.daw@shef.ac.uk*

The ZEPLIN II experiment utilizes an active neutron veto below and to the sides of the 35 kg liquid xenon target. We describe the veto hardware and present some preliminary results of performance measurements.

Keywords: WIMP direct detection active veto shielding organic scintillator

1. Hardware

The ZEPLIN II detector has been described elsewhere.[1,2] Figure 1 shows the active veto and passive shielding surrounding the target. The veto consists of a 30 cm layer of organic liquid scintillator consisting of 80% mineral oil and 20% phenylxenylethane. The liquid is housed in a stainless steel roughly hemispherical shell. The inner surface of this shell is coated with 2 mm of a resin impregnated with 0.5% gadolinium by mass. Readout is through 10 photomultiplier tubes (PMTs).

Figure 2 shows the readout electronics. Each PMT output feeds a discriminator, and the 10 discriminator outputs feed a majority logic unit triggering on 3-fold coincidence. In parallel, the 10 PMT signals are summed. Finally the 3-fold coincident trigger is delayed by 100ns and added to the analog sum. The resultant signal feeds channel 8 of the ZEPLIN II DAQ, an Acqiris digitizer having a 2 ns sampling period.

Figure 3 shows an event display of an event from a neutron calibration having a veto pulse around 30ns after the primary scintillation light from the xenon target. This pulse is due to a neutron that scatters in the target, and subsequently scatters in the veto.

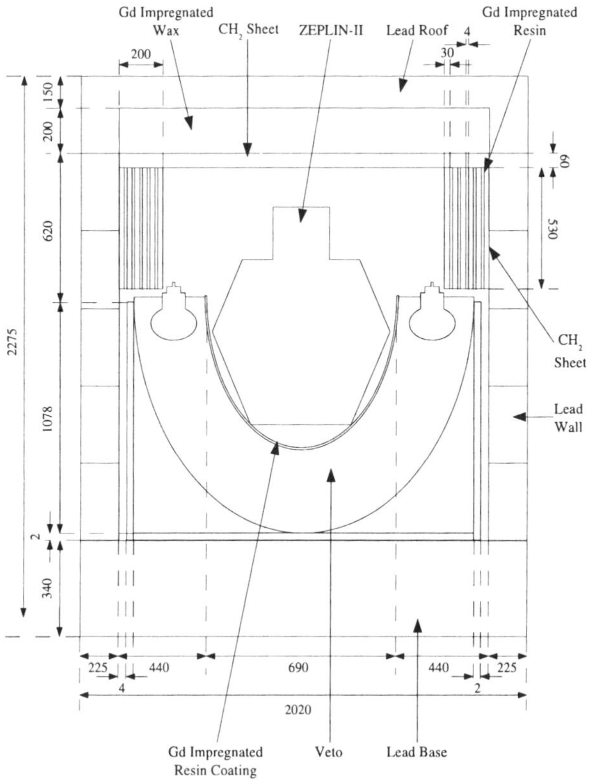

Fig. 1. Schematic diagram of the passive shielding and active veto surrounding the ZEPLIN II target. The active veto consists of organic liquid scintillator read out from above by 10 PMTs.

2. Relative Location of Veto and Target Primary Scintillation Pulses

For events where there was both a target event and a veto triple coincidence, the distribution of time delays between target primary scintillation and veto scintillation pulses was studied for dark matter data and AmBe data. Figure 4 shows histograms of the time delay on three timescales. There are three main features visible, a flat background in the interval $[-100\mu s, +100\mu s]$, a sharp peak in the interval $[-100ns, +100ns]$, and indications of an exponential decay in the rate between 0ns and $+100\mu s$. The event rate drops off for intervals of more than ± 100 μs as the length of the DAQ time records is $200\mu s$.

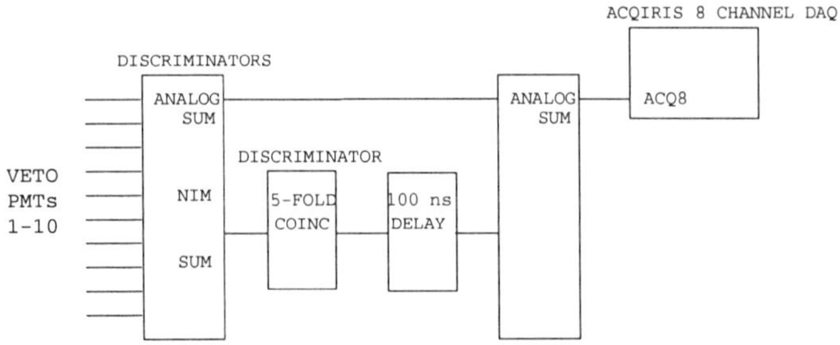

Fig. 2. Schematic diagram of the ZEPLIN II veto readout electronics. The 10 PMT outputs feed 10 channels of a discriminator/buffer NIM module. The sum of the discriminator outputs is passed into a second discriminator whose threshold is set to output a pulse if at least 3 of the PMTs fire in coincidence. This pulse is delayed by 100ns in a delay line, and the output is added to the analog sum of the PMT outputs.

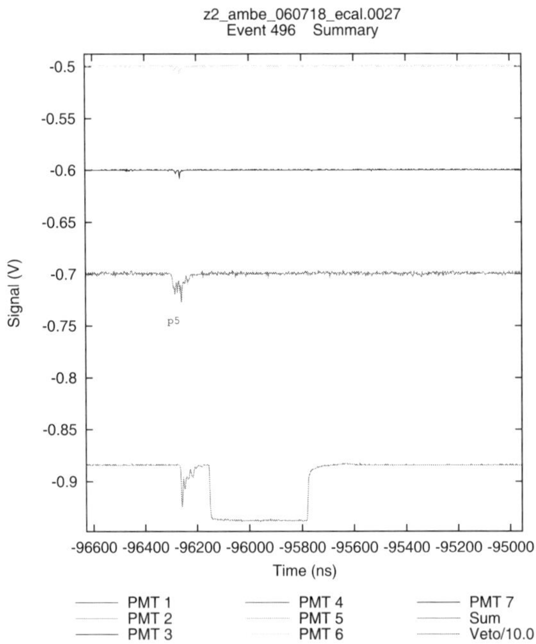

Fig. 3. Data from a single event from an Americium-Beryllium source calibration run. The upper two traces are from two of the seven target photomultiplier tubes (PMTs). The third trace down is the sum of all seven target PMTs. The bottom trace is the veto channel divided by 10.

Figure 5 is a fit to the portion of the AmBe data between 0ns and $+80\mu s$. The fit is a flat background plus an exponential decay with a mean life of

$14\mu s$. This exponential is evidence that a portion of the events in this run are due to neutrons scattering in the target, then being captured onto some nucleus, perhaps Gd. The flat background is accidental coincidences, and the peak at zero is double scattering, once off a target nucleus and once of a veto nucleus.

Figure 6 shows the same time delay distribution for a dark matter run. Events are γ double Compton scatters in the target and veto. The time delay between the two scatters is too short to be resolved by our PMTs.

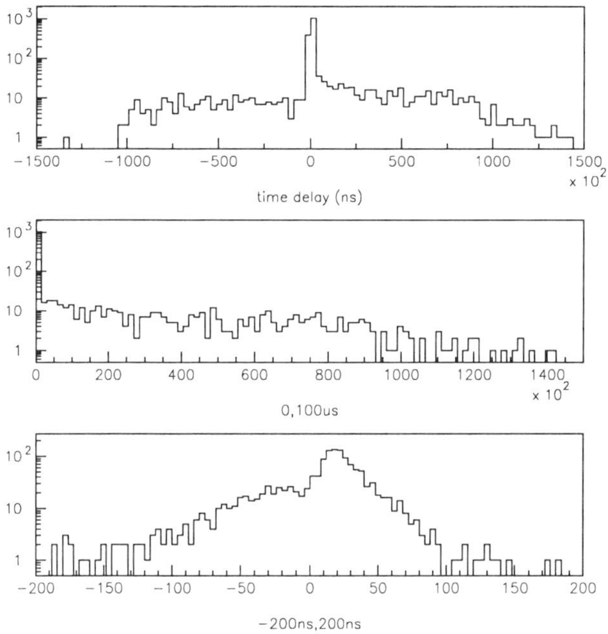

Fig. 4. Histograms of time delay between target scintillation and veto triple coincident pulses for AmBe data on three timescales.

3. Veto Background Rate and False Coincident Rate

Tests were conducted to determine the veto false coincidence rate, due to accidental coincidence between noise pulses in the veto and target hits. The data acquisition system triggers on pulses in the ZEPLIN II target. The background rate was determined by counting veto triggers in data

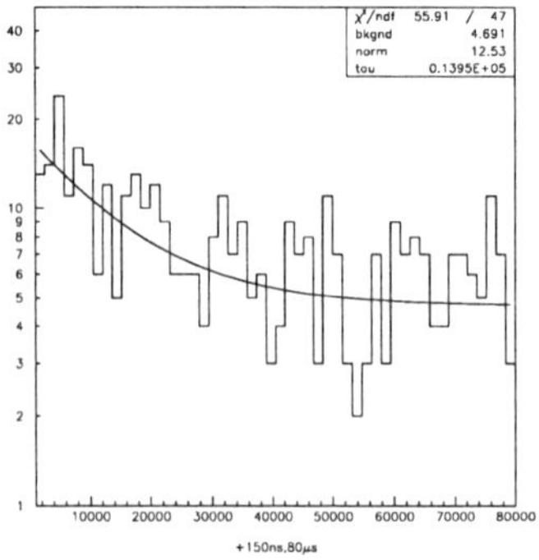

Fig. 5. Fit to time delay data in the interval $[0, +80\mu s]$ for AmBe data. The fit function is a flat background plus an exponential decay.

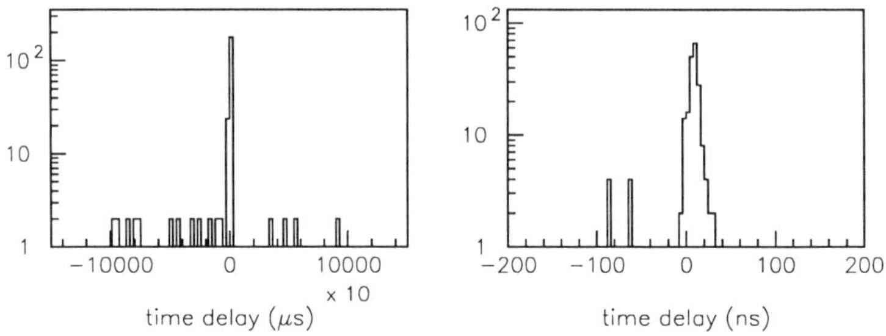

Fig. 6. Histograms of the distribution of time delays between the identified primary scintillation pulse in the target and the veto pulse in background data. The left hand histogram is over the domain $[-150, +150]\mu s$ in delay, the right hand histogram is over the much narrower domain $[-200, +200]$ns.

preceding primary signals in the target, and dividing by the sum of the pretrigger times.

Three run types were considered. Dark matter runs, where there was no installed calibration source, ^{60}Co runs, and AmBe runs, where data was taken during calibration with gamma and neutron sources. Measured background rates are shown in the sixth row of Table 1. The veto cut applied to target data rejects events within a time interval $[-500\text{ns}, +\Gamma_V]$ of the primary scintillation pulse for the event. The false coincidence rate is the fraction of target events vetoed due to accidental coincidence with a veto trigger. For a background veto trigger rate R, the false coincidence rate is $1 - \exp\left[-R(500ns + \Gamma_V)\right]$. Table 1 shows the false coincidence rate for three run types where $\Gamma_V = 100\mu s$. The veto efficiency, defined as the probability of the acceptance of a WIMP signal, is 1 minus the false coincidence rate for background runs, or 99.2%.

For each run, Table 1 also shows the total number of events selected, the number of these events that were vetoed, and the ratio. We subtract from the ratio the false coincident rate to obtain the efficiency of the veto. We conclude that the nuclear recoil rejection efficiency is 49%, consistent with simulation studies.[3–5] For gamma recoil rejection, the background run and the ^{60}Co run give rejection efficiencies at the 13.6% and 19.7% levels respectively. The 6% higher veto efficiency in the ^{60}Co run may be due to its harder spectrum of gammas, or to the position of the source which is such that all straight trajectories from the source pass through the veto.

Table 1. Results of tests on the veto trigger for background and source data.

	bgnd	AmBe	^{60}Co
number of target events selected (fiducial cut and energy < 100 keVee)	821	3434	2212
number of these events vetoed	118	1980	1138
% of events vetoed	14.4%	57.6%	51.4%
number of vetoed events with veto pulse within [-100ns,+100ns] of primary	114	1417	640
% of events vetoed having veto pulse within [-100ns,+100ns] of primary	96.6%	71.6%	56.2%
background rate	83 Hz	850 Hz	3.8 kHz
false coincidence rate	0.83%	8.2%	31.7%
efficiency of veto beyond expectation from random coincidence	13.6%	49.1%	19.7%

4. Conclusions

The ZEPLIN II target and active veto coincidence cut has an efficiency of 49% for the rejection of nuclear recoils having recoil energies in the target of up to 100keVee. The cost of this efficiency is the rejection of 0.2% of WIMP nuclear recoils due to accidental coincidence with the veto.

References

1. K. Mavrokoridis *et al.* (ZEPLIN II collaboration) *These proceedings.*
2. R. Lüscher *et al.* (ZEPLIN II collaboration) *These proceedings.*
3. C. Bungau *et al.*, Astropart. Phys., **23** 97-115 (2005)
4. P. F. Smith *et al.*, Astropart. Phys., **22** 409-420 (2005)
5. M. J. Carson *et al.* Nucl. Instrum. Meth. A **548** (2005) 418.

BEKENSTEIN-SANDERS TENSOR–VECTOR–SCALAR THEORY AND COSMOLOGY

C. SKORDIS*

Perimeter Institute,
Waterloo, Ontario N2L 2Y5, Canada
**E-mail: cskordis@perimeterinstitute.ca*
www.perimeterinstitute.ca

The missing mass problem is the longest standing problem of modern cosmology. This is readily solved by positing the existence of an unknown form of matter, called the dark matter. Still, a particle with the properties required is yet to be found experimentally. An alternative explanation is that gravity deviates from General Relativity for small enough accelerations. One such a theory was proposed by Bekenstein, building on earlier work of Sanders and Milgrom. The theory has a tensor, a vector and a scalar field as dynamical agents of gravity and is thus called Tensor–Vector–Scalar (TeVeS). In this talk I give a short overview of TeVeS theory and discuss its cosmology.

Keywords: TeVeS; Modified gravity; Dark Matter; Tensor–Vector–Scalar theory; Cosmology.

1. Introduction

Bekenstein has recently proposed a relativistic theory of gravity where gravity is mediated by a tensor, a vector and a scalar field, thus called TeVeS,[1] aiming at explaining the missing mass problem.

The missing mass problem is the longest standing problem of modern cosmology: the observed mass coming from all visible matter at the scales of interest, cannot account for the Newtonian (or General Relativistic) gravitational force observed acting on the same objects. The problem has a long history,[2] and manifests as discrepancies in the rotation curves of galaxies,[3] motions of clusters of galaxies, gravitational lensing, and the absence of strong damping of linear perturbations on very large scales, to name a few.

The popular approach to solving the missing mass problem, is to posit a dark matter component which does not interact with electromagnetic radiation but whose presence is evident from the pull of gravity. Dark matter is also required to be cold, if it is to explain galaxies. A plethora of such particles generically arises in particle physics models beyond the standard model quite naturally (see Ref. 4 for a recent review). Still, the actual nature of cold dark matter is left to speculation at the present time.

The alternative approach is to alter to the laws of motion, or the gravitational field. In the Modified Newtonian Dynamics (MOND) scenario, Milgrom[5] proposed that for accelerations smaller than an acceleration scale a_0, Newton's 2nd law of motion is altered (which is still valid for large accelerations), in such a way as to explain the flat rotation curves of galaxies.

In the following I review the problem and the motivation for MOND. Then I discuss the passing to modified gravity theories and eventually TeVeS. I finally conlude with the discussion of cosmology in this theory.

2. From MOND to TeVeS

A generic non–relativistic gravitational theory can be built as follows. We define two potentials, namely a Poisson potential Φ_P such that it obeys the Poisson equation, and a Newtonian potential Φ_N which obeys Newton's 2nd law of motion. More explicitely we have

$$\nabla^2\Phi_P = 4\pi G_N\rho, \tag{1}$$

where G_N is Newton's constant and ρ is the total matter density, and

$$\vec{a} = -\nabla\Phi_N \tag{2}$$

where $\vec{a}$ is the acceleration of a test body due to the Newtonian potential. To complete the theory, a relation between Φ_N and Φ_P must be given. In the simplest case, Newtonian gravity, the two potentials are equal : $\Phi_N = \Phi_P$.

Lets now discuss the motivation behind MOND. Consider a spherical object. If gravity is Newtonian then Φ_N at distance r from the galactic center falls as $\Phi_N \propto 1/r$ (outside the mass distribution) and using (2) we have that $|\vec{a}| \propto 1/r^2$. However for circular orbits we have that the speed of a test body e.g. a star, is $v^2 \propto |\vec{a}|r$. Therefore according to Newtonian gravity the speed of stars outside a galaxy should be falling as $v \propto r^{-1/2}$.

Observations show the contrary; the velocity of a star, stays roughly constant[3] with r in many cases and even in the ones that it does not, it is still very different than the Newtonian prediction. The problem is readily alleviated if there is dark matter. This adds an extra density profile to the RHS of (1) s.t. the solution to Φ_N is now an interior rather than the exterior solution and the star is moving within a halo of dark matter.

On the other hand one may try to modify either the LHS of (1) or (2) hence in this case one is dealing with modified gravity (these two modifications are to some extend equivalent[6]). The 2nd approach was in fact the one used by Milgrom,[5] i.e modifying (2) while keeping (1) intact. Milgrom noticed that the peculiarity with the rotation curves was kicking in, not when a certain scale was reached, but rather when an *acceleration* scale was reached. He then modified (2) to

$$f(|\vec{a}|/a_0)\vec{a} = -\nabla\Phi_P \tag{3}$$

where $f(x) \to 1$ as $x >> 1$ where one recovers the Newtonian limit, and $f(x) \to x$ as $x << 1$ which is the ultra MOND limit. In the ultra MOND limit we then have $|\vec{a}|^2 = a_0|\nabla\Phi_P|$, which in the spherically symmetric case (with $\Phi_P \propto 1/r$ and $v^2 \propto |\vec{a}|r$) gives constant velocity curves! The function f is left free with only the two limits completely specified.

One can cast MOND into a proper theory : Aquadradic Lagrangian gravity[6] (AQUAL). Start from (3) together with Poisson equation (1). Then use definition of the Newtonian potential to get the AQUAL equation

$$\vec{\nabla}\cdot\left[f(|\vec{\nabla}\Phi_N|/a_0)\vec{\nabla}\Phi_N\right] = 4\pi G_N\rho \tag{4}$$

This equation is derivable from a Lagrangian with aquadratic kinetic term. The theory considered this way will therefore satisfy all conservation laws like conservation of energy, momentum and angular momentum.

2.1. *Relativistic MOND*

The relativistic analogue of the Newtonian/Poisson potentials must be encoded in a metric tensor. To do that we can use two metrics. The matter frame metric $g_{\mu\nu}$ is defined such that matter will follow its geodesics i.e. if v^μ is the matter velocity and ∇_μ is the connection compatible with $g_{\mu\nu}$ then the geodesic equation is obeyed i.e. $v^\mu\nabla_\mu v^\nu = 0$. In the non–relativistic weak field limit we have that $g_{00} = -1 + 2\Phi_N$ and $g_{ij} = (1 + 2\Phi_N)\delta_{ij}$ and the geodesic equation reduces to Newton's 2nd law (2). This also means that the action for generic matter fields χ^A is built from $g_{\mu\nu}$ alone, i.e. $S_m = S_m[g_{\mu\nu}, \chi^A, \nabla_\mu\chi^A]$. Variation of S_m wrt $g^{\mu\nu}$ defines the stress-energy tensor of matter, $T_{\mu\nu}$ as

$$\delta S_m = -\frac{1}{2}\int d^4x\sqrt{-g}\,T_{\mu\nu}\,\delta g^{\mu\nu} \tag{5}$$

The analogue of the Poisson equation cannot come from the same metric. Rather, one postulates a 2nd metric tensor $\tilde{g}_{\mu\nu}$, called the Einstein–Hilbert frame metric tensor since it is required that its action is

$$S_{\tilde{g}} = \frac{1}{16\pi G}\int d^4x\sqrt{-\tilde{g}}\tilde{R} \tag{6}$$

where G is a "bare" gravitational constant and $\tilde{g}$ and $\tilde{R}$ are the metric determinant and scalar curvature of $\tilde{g}_{\mu\nu}$ The above action ensures that the non–relativistic weak field limit of the field equations of $\tilde{g}_{\mu\nu}$ will be the Poisson equation for a potential defined by $\tilde{g}_{00} = -1 + 2\Phi_P$ and $\tilde{g}_{ij} = (1 + 2\Phi_P)\delta_{ij}$.

The theory is completed by providing a relation between $\tilde{g}_{\mu\nu}$ and $g_{\mu\nu}$, which ultimately induces a relation between Φ_N and Φ_P.

In the simplest relativistic MOND theory considered in Ref. 6, Bekenstein and Milgrom assume the existence of a scalar field ϕ which conformally relates the two metrics : $\tilde{g}_{\mu\nu} = e^{2\phi} g_{\mu\nu}$. This completes the theory by providing an action for ϕ which to ensure an AQUAL-behaviour in the non-relativistic limit, it must be an aquadratic function of the kinetic term of ϕ.

Relativistic AQUAL and its descendants fail in many respects but most notably gravitational lensing. In fact it can be shown that any relativistic theory of MOND built out of the metric and a scalar field even with a disformal relation between the two metrics, will give the wrong gravitational lensing if it is to be causal.[7] Generic local or non-local theories based on only one metric also fail.[8]

3. TeVeS in a nutshell

TeVeS and the BSTV variant[9] differ from the theories above, in that a unit–timelike vector field A^μ is posited in addition to the scalar. The vector field was introduced by Sanders,[10] who showed that it can solve the lensing problem. Sanders only considered the vector field in a fixed frame. Bekenstein included an action for the vector field to make the theory covariant.[1]

Unlike other relativistic MOND theories, the metric transformation in TeVeS is no longer conformal but rather

$$g_{\mu\nu} = e^{-2\phi}(\tilde{g}_{\mu\nu} + A_\mu A_\nu) - e^{2\phi} A_\mu A_\nu \tag{7}$$

Bekenstein assumes the following action for the vector field.

$$S_A = -\frac{1}{32\pi G} \int d^4x \sqrt{-\tilde{g}} \left[K F_{\mu\nu} F^{\mu\nu} - 2\lambda(A_\mu A^\mu + 1)\right] \tag{8}$$

where $F_{\mu\nu} = \tilde{\nabla}_\mu A_\nu - \tilde{\nabla}_\nu A_\mu$, K is a constant and λ is a lagrange multiplier ensuring that the vector field is unit–timelike. This action is by no means generic and could be modified in future versions of TeVeS if warranted.

The scalar field action is very similar to the relativistic AQUAL action with the additional coupling of the scalar to the vector field. It is

$$S_\phi = -\frac{1}{16\pi G} \int d^4x \left[\mu \left(\tilde{g}^{\mu\nu} - A^\mu A^\nu\right) \tilde{\nabla}_\mu \phi \tilde{\nabla}_\nu \phi + V(\mu)\right] \tag{9}$$

where μ is a non–dynamical field with a free function $V(\mu)$. This free function is not completely arbitrary but must obey certain requitements in order that the theory has MOND and Newtonian limits (see below). In the BSTV variant μ is made dynamical by including an extra action for it.

The field equations are easily found and the reader is refered to Ref. 11.

Although the theory was designed to provide for a relativistic theory of MOND, taken at face value this need not be the case. The requirement that both exact MOND and exact Newtonian limits exist within the theory puts some constraint on the possible forms of $V(\mu)$. In particular these two requirements fix the derivative of $V(\mu)$ to be of the form[1,12]

$$V'(\mu) \equiv \frac{dV}{d\mu}(\mu) = -\frac{1}{16\pi\ell_B^2}\frac{\mu^2}{(1-\mu/\mu_0)^m}f(\mu) \tag{10}$$

where μ_0 is a dimensionless constant, ℓ_B is a scale and $f(\mu)$ is an arbitrary function, not related to the MOND or Newtonian limits, whose only constraint is that $f(0)$ is non–vanishing. The μ^2 in the numerator is what is precisely required to have an exact MOND limit (as $\mu \to 0$), whereas the factor $1 - \mu/\mu_0$ is what is required to have an exact Newtonian limit (as $\mu \to \mu_0$). We follow Bekenstein by choosing the simplest case, $m = 1$.

The above completely fix G_N and a_0 in terms of the other parameters[12] in the theory regardless of $f(\mu)$ or the power m. In particular we have

$$G_N = \frac{\mu_0(1-K/2)}{\mu_0+2-K}G \tag{11}$$

and

$$a_0 = \frac{e^{\phi_c}}{\ell_B\sqrt{\pi}}\left(\frac{1}{2-K}+\frac{1}{\mu_0}\right) \tag{12}$$

where ϕ_c is the cosmological value of ϕ when the quasistatic system in question broke away from the cosmological expansion.

4. Cosmology

As Bekenstein assumes, cosmology lives on a different branch of μ than the quasistatic branch, i.e. in the branch $\hat{\mu} \geq \mu_a$. Different solutions can be studied after fixing the form of $f(\mu)$. Consider a power series expansion[12] $f(\mu) = \sum_n c_n(\mu/\mu_0 - \mu_a)^n$ where c_n are constants. Then for individual powers the solutions are as follows. For $n \geq 1$ the scalar field is tracking the background fluid (which has equation of state w) with relative density[13]

$$\Omega_\phi = \frac{(1+3w)^2}{3\mu_a\mu_0(1-w)^2}. \tag{13}$$

This kind of evolution, shown in the left panel of Fig. 1 , is akin to tracking quintessence[14] fields.

The individual cases $-2 \leq n \leq 0$ suffer from singularities and are pathological, while the case $n = -3$ is ok only if $w = -1$ but otherwise problematic. Finally for $n \leq -4$, the fluid dominates completely (no tracking).

Cosmological perturbation theory for TeVeS is given in Ref. 11 while numerical studies have been carried out[13] for the special case $c_n = \delta_{n,2}$ and $\mu_a = 2$. The important result is a growth of density fluctuations even in the absence of dark matter. Indeed it is possible (Right panel of Fig. 1) in this theory to form structure without dark matter. The precice reason has been uncovered[15] by Dodelson and Liguori. It turns out that at least in this case, fluctuations in ϕ are unimportant while for small enough values of $K\mu_0$, the scalar mode of the fluctuation of the vector field, has a growing solution. This is what helps baryonic density perturbations to grow after recombination but comes at a price: at the same time the difference of the two metric potentials (which is tiny in standard cosmology) is of $O(1)$. This can have dramatic effects, e.g. on ISW, and can be potential model discriminator.

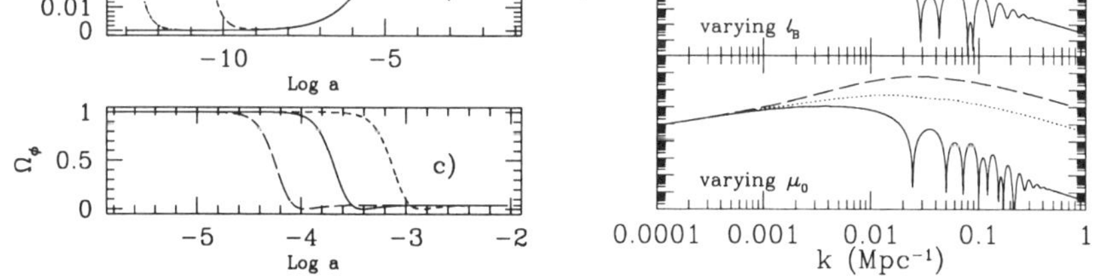

Fig. 1. LEFT : Ω_ϕ as a function of the scale factor (arbitrary units). ϕ has an attractor which corresponds to tracking the three constituents independently of the initial conditions. We illustrate this here for: *a*) radiation era, *b*) matter era and *c*) the Λ era. RIGHT : The effect of the TeVeS parameters on the power of spectrum of the baryonic density fluctuations for a range of parameters for each case (see Ref. 13).

5. Beyond TeVeS?

It turns out[16] that one can eliminate the Einstein–Hilbert metric and write the theory entirely in the matter frame. In other words the total action is $S = S_g[g] + S_A[g, A] + S_m[g, \chi^A]$ where $S_g[g]$ is the Einstein-Hilbert action of $g_{\mu\nu}$. This comes with a twist : ϕ is also eliminated and is absorbed into the vector field. Moreover the matter action and the vector field action still remain completely decoupled! This opens the way to new generalizations and interpretations of TeVeS (see for example[17]).

6. Conclusion

I gave a short review of Bekenstein-Sanders TeVeS theory,[1,9] a relativistic formulation of Milgrom's MOND.[5] The theory is the first one to have been shown to form structure without dark matter.[13] It forms a platform for future studies and generalizations to answer the question : is the solution to the missing mass problem a modified gravity theory or dark matter?

References

1. J. D. Bekenstein, Phys. Rev. D**70**, 083509 (2004).
2. J. Oort, Bull. Astron. Inst. Neth. **6**, 249 (1932); ibid. **15**, 45 (1960); F. Zwicky, Helv. Phys. Acta **6** , 110 (1933); S. Smith Astrophys. J.**83**, 23 (1936).
3. V. C. Rubin and W. K. Ford Astrophys. J.**159**, 379 (1970); V. C. Rubin *et al.*, Astrophys. J.**289**, 81 (1985).
4. G. Bertone, D. Hooper and J. Silk, Phys.Rept.**405**, 279 (2005).
5. M. Milgrom, Astrophys. J.**270**, 365 (1983); ibid. Astrophys. J.**270**, 371 (1983); ibid. Astrophys. J.**270**, 384 (1983).
6. J. D. Bekenstein and M. Milgrom, Astrophys. J.**286**, 7 (1984).
7. J. D. Bekenstein & R. H. Sanders, Astrophys. J.**429**, 480 (1994).
8. M. E. Soussa and R. P. Woodard, Phys. Lett. B**578**, 253 (2004).
9. R. H. Sanders, Mon.Not.Roy.Astron.Soc.**363**, 459 (2005).
10. R. H. Sanders, Astrophys. J.**480**, 492 (1997).
11. C. Skordis, Phys. Rev. D**74**, 103513 (2006).
12. F. Bourliot *et al.*, astro-ph/0611255.
13. C. Skordis *et al.*, Phys. Rev. Lett.**96**, 011301 (2006).
14. B.Ratra and P.J Peebles, Phys. Rev. D **37**, 3406 (1988); P.G. Ferreira and M.Joyce, Phys. Rev. Lett., **79**, 4740 (1997); A.Albrecht and C.Skordis, Phys. Rev. Lett. **84**, 2076 (2000), C.Skordis and A.Albrecht, Phys. Rev. D **66** 043523 (2002).
15. S.Dodelson and M.Liguori, Phys. Rev. Lett.**97**, 231301 (2006).
16. T.G. Zlosnik *et al.*, Phys. Rev. D**74**, 044037 (2006).
17. T.G. Zlosnik *et al.*, astro-ph/0607411.

COSMOLOGY FROM XMM HIGH-Z AGN CLUSTERING

MANOLIS PLIONIS[1,2] and SPYROS BASILAKOS[3]

[1] *Institute of Astronomy & Astrophysics, National Observatory of Athens, Greece*
[2] *INAOE, Puebla, Mexico*
[3] *Academy of Athens, Athens, Greece*

We review the high-redshift X-ray selected AGN clustering, based on the XMM/2dF survey, and compare it with other recent XMM-based results. Using the recent Hasinger et al. (2005) and La Franca et al. (2005) luminosity functions we find that the spatial clustering lengths, derived using Limber's inversion equation, are ~ 17 and $20\ h^{-1}$ Mpc for the soft and hard band sources while their median redshifts are $\bar{z} \sim 1.2$ and 0.8, respectively. The corresponding bias factors at $z = \bar{z}$ are $\sim 5.3 \pm 0.9$ and $\sim 5.1 \pm 1.1$, respectively. Within the framework of a flat cosmological model we find that our results support a model with $\Omega_m \simeq 0.26$, $\sigma_8 \simeq 0.75$, $h \simeq 0.72$, w$\simeq -0.9$ (in excellent agreement with the 3 year WMAP results). We also find the present day bias of X-ray AGNs to be $b_o \sim 2$.

1. Introduction

Active Galactic Nuclei (AGN) can be detected out to high redshifts and thus their clustering properties can provide information on the large scale structure, the underlying matter distribution and the evolution with redshift of the AGN phenomenon. Optically selected AGN surveys miss large numbers of dusty systems and therefore provide a biased census of the AGN phenomenon. X-ray surveys, are least affected by dust providing an efficient tool for compiling uncensored AGN samples over a wide redshift range.

Early studies of the X-ray AGN clustering, using *Einstein* and *ROSAT* data, produced contradictory results (eg. [9]; [43]; [11]; [1]; [31]). Recently, there has been an effort to address this confusing issue and determine the clustering properties of both soft and hard X-ray selected AGNs, based on the new XMM and Chandra missions (eg. [46]; [47]; [7]; [8], [21]; [36]; [17]; [30]). Most of these studies find a relatively large correlation length for the high-z X-ray AGNs (eg., [7]; [8]; [36]). However, the recent XMM-COSMOS survey results [30] provided significantly smaller clustering lengths, opening again the contradiction realm.

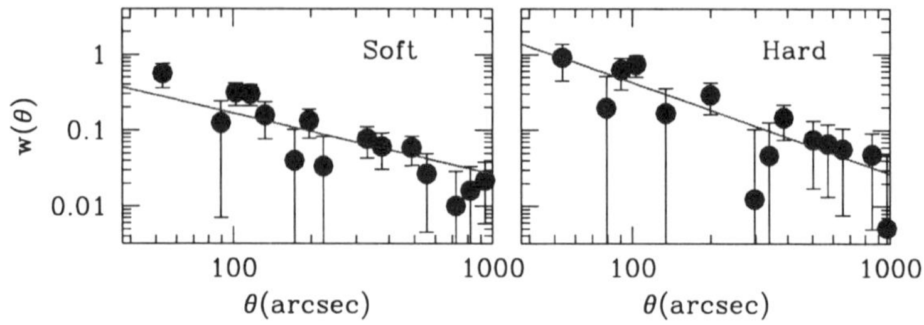

Fig. 1. The XMM/2dF survey angular correlation function for the 2 bands considered.

2. The XMM/2df Survey Angular and Spatial Clustering

The XMM-*Newton*/2dF survey is a shallow (2-10 ksec per pointing) survey comprising of 18 XMM-*Newton* pointings equally split between a Northern and Southern Galactic region.Due to elevated particle background we analysed a total of 13 pointings. A full description of the data reduction, source detection and flux estimation are presented in Georgakakis et al. [19], [18].

We derive the source $\log N$-$\log S$ after constructing sensitivity maps in order to estimate the area of the survey accessible to point sources above a given flux limit (see [7], [8]). In Table 1 we also provide the effective flux-limit of the different surveys, estimated using the corresponding area curves, by: $S_{\rm lim}^{\rm eff} \simeq \sum \Omega_i S_{\rm lim,i} / \sum \Omega_i$, where Ω_i is the survey solid angle of which the flux limit is $S_{\rm lim,i}$.

We calculate the angular correlation function using the estimator: $\mathrm{w}(\theta) = f(N_{DD}/N_{DR}) - 1$, of which the uncertainty is: $\sigma_w = \sqrt{(1+\mathrm{w}(\theta))/N_{DR}}$, where N_{DD} and N_{DR} are the number of data-data and data-random pairs, respectively, in the interval $[\theta - \Delta\theta, \theta + \Delta\theta]$. The normalization factor is $f = 2N_R/(N_D - 1)$, with N_D and N_R the total number of data and random points, respectively. For each XMM pointing we produce 100 Monte Carlo random catalogues having the same number of points as the real data accounting also for the sensitivity variations across the surveyed area (for details see [7], [8]). We use a standard χ^2 minimization procedure to fit the measured correlation function assuming a power-law form: $\mathrm{w}(\theta) = (\theta_\circ/\theta)^{\gamma-1}$ and fixing γ to 1.8. The fitting is performed for angular separations in the range $40'' - 1000''$ The resulting raw values of $\theta_\circ$ are corrected for the integral constraint and the amplification bias (for details see [43] and [8]), although such corrections are quite small. The final results are presented in Table 1. Now, the spatial correlation function can be modeled as (eg. [14]): $\xi(r,z) = (r/r_\circ)^{-\gamma} \times (1+z)^{-(3+\epsilon)}$, where ϵ

Table 1. Correlation function analysis results with their 1σ uncertainties for the different surveys and X-ray bands. The correlation lengths assume a slope $\gamma = 1.8$.

Band (keV)	Survey	#	$S_{\rm lim}^{\rm eff}$	$\theta_\circ ('')$	$r_\circ$ (h^{-1} Mpc)	$\bar{z}$
0.5-2	XMM/2dF	432	1.3×10^{-14}	10.4 ± 3	17 ± 2.2	1.25
0.5-2	XMM-COSMOS	1037	1.9×10^{-15}	1.9 ± 0.3	9.3 ± 0.7	1.33
0.5-2	XMM-ELAIS	395	4.5×10^{-15}	5.2 ± 3.8	13.8 ± 4.5	1.32
2-8	XMM/2dF	177	4.6×10^{-14}	22.2 ± 9	20 ± 3.7	0.80
4.5-10	XMM-COSMOS	151	1.6×10^{-14}	6 ± 2	12 ± 1.8	0.89
2-10	XMM-ELAIS	205	1.8×10^{-14}	12.8 ± 7.8	17 ± 4.6	0.88

parametrizes the clustering evolution model. For $\epsilon = \gamma - 3$ (ie., $\epsilon = -1.2$ for $\gamma = 1.8$), the clustering is constant in comoving coordinates, a model which appears to be appropriate for active galaxies (eg. [26]).

In order to invert the angular correlation function to three dimensions we utilize Limber's integral equation (eg. [34]). For a spatially flat Universe, Limber equation can be written as:

$$\mathrm{w}(\theta) = 2\frac{H_\circ}{c}\int_0^\infty \left(\frac{1}{N}\frac{\mathrm{d}N}{\mathrm{d}z}\right)^2 E(z)\mathrm{d}z \int_0^\infty \xi(r,z)\mathrm{d}u \qquad (1)$$

where the number of objects in a survey of a solid angle Ω_s and within the shell $(z, z + \mathrm{d}z)$, is given by:

$$\frac{\mathrm{d}N}{\mathrm{d}z} = \Omega_s x^2 \phi(x) \left(\frac{c}{H_\circ}\right) E^{-1}(z) \qquad (2)$$

with $\phi(x)$ the selection function and x the proper distance related to the redshift through (see [34]):

$$x(z) = \frac{c}{H_\circ}\int_0^z \frac{\mathrm{d}u}{E(u)} \quad \text{with} \quad E(z) = [\Omega_{\rm m}(1+z)^3 + \Omega_\Lambda]^{1/2} \; , \qquad (3)$$

Since we do not have complete redshift information for our sources we estimate $\mathrm{d}N/\mathrm{d}z$ using the X-ray source luminosity function and folding in the area curve, via the relation: $\phi(x) = \int_{L_{\rm min}(z)}^\infty \Phi(L_x, z)\mathrm{d}L$, where $\Phi(L_x, z)$ is the luminosity dependent density evolution luminosity (LDDE) function. In the present analysis we use that of Hasinger et al. [23] for the soft-band and that of La Franca et al. [28] for the hard-band.

In Fig. 2 we present the expected redshift distributions of the soft and hard X-ray sources for all three recent XMM based surveys together with the histogram of some limited spectroscopic and photo-z data (see caption for details). It is evident that all three surveys trace similar depths although

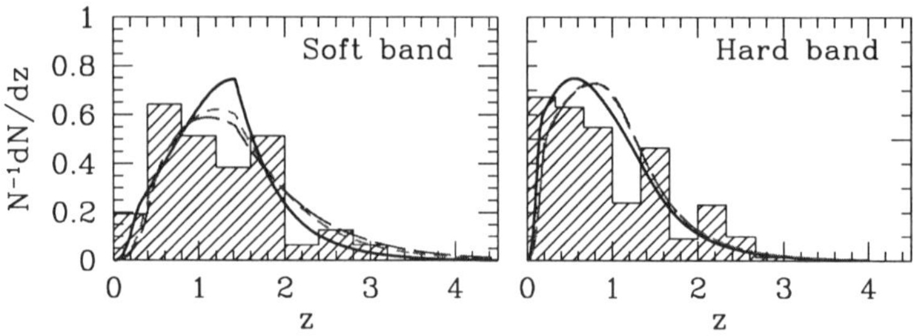

Fig. 2. *Left Panel:* The expected z-distributions using the Hasinger et al [23] soft-band luminosity function for the XMM/2dF survey (thick line), the XMM-COSMOS survey (blue long-dashed line) and the XMM-ELAIS-S1 survey (red short-dashed line). *Right Panel:* The corresponding redshift distributions using the LDDE hard-band luminosity function of La Franca et al. [28]. The histogram in the left panel corresponds to the distribution of the X-ray sources in the ROSAT Lochman Deep Field [39], while that of the right panel on limited spectroscopic and photo-z data of our XMM/2dF survey.

the XMM-COSMO and XMM-ELAIS surveys have a slightly larger contribution from $z \gtrsim 2$ in the soft-band. The predicted median redshift of the three XMM surveys for the soft and hard sources are shown in Table 1.

Then the inversion of eq. (2), using the LDDE luminosity function, $\epsilon = -1.2$, the concordance cosmological model, and integrating out to $z = 4.5$, provides a correlation length of $r_\circ \simeq 17 \pm 2.2\ h^{-1}$ Mpc and $\simeq 20 \pm 3.7\ h^{-1}$ Mpc, for the soft and hard bands, respectively (the slight differences with respect to the Basilakos et al. [7], [8] results are due to the different luminosity functions used and a better treatment of the errors). These results are in good agreement with the XMM-ELAIS results [36] but significantly larger than the XMM-COSMOS [30] results (see Table 1).

If we assume that all three XMM based surveys provide correct results, and that their differences are due to the cosmic variance, we can volume weight them to obtain an average estimate of the clustering length of the XMM point sources. The relative volume weights for the XMM-COSMOS, XMM-ELAIS and XMM/2dF surveys are 1:0.9687:0.8527 and 1:0.9947:0.701 for the soft and hard bands, respectively (we have not corrected for the slight differences of the hard-bands used in the three surveys). Doing so we obtain: $\langle r_\circ \rangle \simeq 13.2$ and $16.0\ h^{-1}$ Mpc, for the soft and hard bands, respectively. These large correlation lengths are comparable to those of Extremely Red Objects [20], of luminous radio sources ([37]; [32]; [38]) and of bright distant red galaxies [15]. They are however, significantly larger than those derived from optical AGN surveys: $r_\circ \simeq 5.4 - 8.6\ h^{-1}$ Mpc (eg. [12]; [27]; [13]; [22]; [35]; [44]).

3. Cosmological Constraints

It is well known [24] that according to linear biasing the correlation function of the mass-tracer ($\xi_{\rm obj}$) and dark-matter one ($\xi_{\rm DM}$), are related by:

$$\xi_{\rm obj}(r,z) = b^2(z)\xi_{\rm DM}(r,z) \ , \tag{4}$$

where $b(z)$ is the bias evolution function. In this study we use the bias model of Basilakos & Plionis [3], [4]. We quantify the underlying matter distribution clustering by presenting the spatial correlation function of the mass $\xi_{\rm DM}(r,z)$ as the Fourier transform of the spatial power spectrum $P(k)$:

$$\xi_{\rm DM}(r,z) = \frac{(1+z)^{-(3+\epsilon)}}{2\pi^2}\int_0^\infty k^2 P(k)\frac{\sin(kr)}{kr}{\rm d}k \ , \tag{5}$$

where k is the comoving wavenumber and $\epsilon = -1.2$, according to the constant in comoving coordinates clustering evolution model. As for the power spectrum, we consider that of CDM models, where $P(k) = P_0 k^n T^2(k)$ with scale-invariant ($n = 1$) primeval inflationary fluctuations. In particular, we use the transfer function parameterization as in [2], with the corrections given approximately by Sugiyama [41]. Note that we also use the non-linear corrections introduced by Peacock & Dodds [33]. We have chosen to use either the standard normalization given by: $\sigma_8 \simeq 0.5\Omega_{\rm m}^{-\gamma}$ with $\gamma \simeq 0.21 - 0.22{\rm w} + 0.33\Omega_{\rm m}$ [45], or to leave σ_8 a free parameter.

Firstly, using equations (1), (4) and (5) and evaluating at $z = \bar{z}$ we derive the bias factor of our sources at the corresponding redshift. We find $b(\bar{z}) \simeq 5.1 \pm 1.1$ and $\simeq 5.3 \pm 0.9$ for the hard and soft bands respectively, which are significantly smaller than the values attributed to our work in Miyaji et al. [30].

Next, we utilize a χ^2 likelihood procedure to compare the measured XMM soft source angular correlation function with the prediction of different spatially flat cosmological models (see [5], [6]). In particular, we define the likelihood estimator as: $\mathcal{L}^{\rm AGN}(\mathbf{c}) \propto \exp[-\chi^2_{\rm AGN}(\mathbf{c})/2]$ with: $\chi^2_{\rm AGN}(\mathbf{c}) = \sum_{i=1}^{n}\left[w_{\rm th}(\theta_i,\mathbf{c}) - w_{\rm obs}(\theta_i)/\sigma_i\right]^2$, where $\mathbf{c}$ is a vector containing the cosmological parameters that we want to fit and σ_i the observed angular correlation function uncertainty. We assume a flat ($\Omega_{\rm tot} = 1$) cosmology with primordial adiabatic fluctuations and baryonic density of $\Omega_{\rm b}h^2 \simeq 0.022$ (eg. [25]; [40]). In this case the corresponding vector is $\mathbf{c} \equiv (\Omega_{\rm m}, {\rm w}, \sigma_8, h, b_\circ)$ and we densely sample the various parameters. Note that in order to investigate possible equations of state, we allow the parameter w to take values below -1, corresponding to the so called *phantom* cosmologies (eg. [10]).

The resulting best fit parameters for $\epsilon = -1.2$ and the Miyaji et al [29] luminosity function are presented in Table 2. In the first two rows we

Table 2. Cosmological parameters from the likelihood analysis. Errors of the fitted parameters represent 1σ uncertainties. Note that the fitted parameters correspond to results marginalized over the parameters that do not have errorbars, for which we use the values indicated.

Data	$\Omega_{\rm m}$	σ_8	w	h	$b_\circ$
XMM	$0.31^{+0.16}_{-0.08}$	0.93	$w=-1$	$0.72^{+0.02}_{-0.18}$	$2.30^{+0.70}_{-0.20}$
XMM/SNIa	0.28 ± 0.02	0.95	$-1.05^{+0.10}_{-0.20}$	0.72	2.30
XMM	0.28 ± 0.03	0.75 ± 0.03	$w=-1$	0.72	$2.0^{+0.20}_{-0.25}$
XMM/SNIa	0.26 ± 0.04	0.75	$-0.9^{+0.10}_{-0.05}$	0.72	2.0

present results based on the traditional [45] σ_8 normalization. Note that our estimate of the Hubble parameter h is in very good agreement with those derived by the HST key project [16]. In the last two rows of Table 2 we leave σ_8 free but fix the Hubble constant to $h = 0.72$. In this case our fit provides a σ_8 value which is in excellent agreement with the recent 3-years WMAP results [40] Therefore, allowing values $w < -1$ we can derive a $(\Omega_{\rm m}, \sigma_8)$ relation, a good fit of which is provided by :

$$\sigma_8 = 0.34(\pm 0.01)\, \Omega_{\rm m}^{-\gamma(\Omega_{\rm m},w)} \tag{6}$$

with $\gamma(\Omega_{\rm m}, w) = 0.22(\pm 0.04) - 0.40(\pm 0.05)w - 0.052(\pm 0.040)\Omega_{\rm m}$. Note that w is degenerate, within the 1σ uncertainty, with respect to $\Omega_{\rm m}$. Therefore, in order to put further constraints on w we additionally use a sample of 172 supernovae SNIa of Tonry et al. [42].

The joined likelihood analysis is performed by marginalizing the X-ray clustering results over σ_8, h and b_0. The vector $\mathbf{c}$ now becomes: $\mathbf{c} \equiv (\Omega_{\rm m}, w)$. The SNIa likelihood function can be written as: $\mathcal{L}^{\rm SNIa}(\mathbf{c}) \propto \exp[-\chi^2_{\rm SNIa}(\mathbf{c})/2]$, with: $\chi^2_{\rm SNIa}(\mathbf{c}) = \sum_{i=1}^{172} \left[\log D_{\rm L}^{\rm th}(z_i, \mathbf{c}) - \log D_{\rm L}^{\rm obs}(z_i)/\sigma_i\right]^2$, where $D_{\rm L}(z)$ is the dimensionless luminosity distance, $D_{\rm L}(z) = H_\circ(1+z)x(z)$ and z_i is the observed redshift. The joint likelihood function peaks at: $\Omega_{\rm m} = 0.26 \pm 0.04$ with $w = -0.90^{+0.1}_{-0.05}$. Using eq. (6) we find that the normalization of the power spectrum that corresponds to these cosmological parameters is $\sigma_8 \simeq 0.73$, in excellent agreement with the recent 3-year WMAP results [40].

Acknowledgments

Many thanks are due to I. Georgantopoulos and A. Georgakakis.

References

1. Akylas, A., Georgantopoulos, I., Plionis, M., 2000, MNRAS, 318, 1036

2. Bardeen, J.M., Bond, J.R., Kaiser, N. & Szalay, A.S., 1986, ApJ, 304, 15
3. Basilakos, S. & Plionis, M., 2001, ApJ, 550, 522
4. Basilakos, S. & Plionis, M., 2003, ApJ, 593, L61
5. Basilakos, S. & Plionis, M., 2005, MNRAS, 360, L35
6. Basilakos, S. & Plionis, M., 2006, ApJ, 650, L1
7. Basilakos, S., et al., 2004, ApJL, 607, L79
8. Basilakos, S., et al., 2005, MNRAS, 356, 183
9. Boyle, B. J., & Mo, H. J., 1993, MNRAS, 260, 925
10. Caldell, R. R., 2002, Physics Letters B, 545, 23
11. Carrera, F.J., et al., 1998, MNRAS, 299, 229
12. Croom, S. M., & Shanks, T., 1996, MNRAS, 281, 893
13. Croom, S.M., et al., 2002, MNRAS, 335, 459
14. de Zotti, G., et al., 1990, ApJ, 351, 22
15. Foucaud, S., et al., 2006, MNRAS, *submitted*, astro-ph/0606386
16. Freedman, W., L., et al., 2001, ApJ, 553, 47
17. Gandhi, P., et al., 2006, A&A, 457, 393
18. Georgakakis, A., et al., 2004, MNRAS, 349, 135
19. Georgakakis, A., et al., 2003, MNRAS, 344, 161
20. Georgakakis, A., et al., 2005, ApJ, 620, 584
21. Gilli, R., et al. 2005, A&A, 430, 811
22. Grazian, A., et al., 2004, AJ, 127, 592
23. Hasinger, G., Miyaji, T., Schmidt, M., 2005, A&A, 441, 417
24. Kaiser N., 1984, ApJ, 284, L9
25. Kirkman, D., et al., 2003, ApJS, 149, 1
26. Kundić, T., 1997, ApJ, 482, 631
27. La Franca F., Andreani, P., Cristiani, S., 1998, ApJ, 497, 529
28. La Franca, F. et al., 2005, ApJ, 635, 864
29. Miyaji, T., Hasinger, G., Schmidt, M., 2000, A&A, 353, 25
30. Miyaji, T., et al., 2006, astro-ph/0612369
31. Mullis C.R., et al., 2004, ApJ, 617, 192
32. Overzier, R.A., et al., 2003, A&A, 405, 53
33. Peacock, A. J., &, Dodds, S. J., 1994, MNRAS, 267, 1020
34. Peebles P.J.E., 1993. *Principles of Physical Cosmology*, Princeton Univ.
35. Porciani, C., Magliocchetti, M. & Norberg, P., 2004, MNRAS,
36. Puccetti, S., et al., 2006, AA *submitted*, astro-ph/0607107
37. Roche, N. D., Dunlop, J., Almaini, O., 2003, MNRAS, 346, 803
38. Röttgering, H., et al., 2003, New Astr. Rev., 47, 309
39. Schmidt, M., et al., 1998, A&A, 329, 495
40. Spergel D. N., et al., ApJ, 2006, *submitted*, astro-ph/0603449
41. Sugiyama, N., 1995, ApJS, 100, 281
42. Tonry, et al. , 2003, ApJ, 594, 1
43. Vikhlinin, A. & Forman, W., 1995, ApJ, 455, 109
44. Wake, D.A. et al., 2004, ApJ, 610, L85
45. Wang, L. & Steinhardt, P.J., 1998, ApJ, 508, 483
46. Yang, Y., et al. 2003, ApJ, 585, L85
47. Yang, Y., Mushotzky, R.F., Barger, A.J., Cowie, L.L., 2006, ApJ, 645, 68

MODELING THE GALAXY/LIGHT-MASS CONNECTION WITH COSMOLOGICAL SIMULATIONS

ARGYRO TASITSIOMI

Lyman Spitzer Jr. Fellow, Dept. of Astrophysical Sciences, Princeton University, Princeton, NJ 08544, USA
E-mail: iro@astro.princeton.edu

I review some results on the galaxy/light-mass connection obtained by dissipationless simulations in combination with a simple, non-parametric model to connect halo circular velocity to the luminosity of the galaxy they would host. Ia focus on the galaxy-mass correlation and mass-to-light ratios obtained from galaxy up to cluster scales. The predictions of this simple scheme are shown to be in very good agreement with SDSS observations.

Keywords: galaxy-mass correlation; mass-to-light ratio; cosmological simulations

1. Introduction

During the last decade, large observational galaxy surveys have greatly improved the information we have about the relation of galaxies and/or light with the underlying mass distribution. However, our understanding of these relations is at best incomplete, largely because they are shaped by processes of galaxy formation that are too numerous and complicated to be included in cosmological simulations. As a result, often other either semi-analytic or empirical ways are adopted to investigate these connections.

In this paper I summarize some results on the galaxy/light-dark matter relation predicted by a simple scheme used to assign luminosities to dark matter halos formed in cosmological simulations. In this scheme, simulated dark matter halos of a certain number density are assigned the luminosity of observed galaxies with the same number density. When calculating these number densities, we rank galaxies with respect to their SDSS r-band luminosity — i.e., we use the observed SDSS r-band luminosity function — while we rank the simulated dark matter halos using as tag the maximum of their rotation velocity, $V_{\max}$. Thus, the assumption we make is that there is a one-to-one relation between the luminosity of a galaxy and the gravitational potential of the halo within which the galaxy is formed, with the latter being described by $V_{\rm max}$. This is a reasonable assumption since we expect the ability of a halo to make baryons cool and form a galaxy to increase with the halo gravitational potential depth. Furthermore, V_{max} is particularly good as a potential depth indicator in the case of subhalos. Assuming that the initial — prior to accretion potential — is the one relevant for galaxy formation in subhalos, V_{max}, being a property of the inner parts of halos is not affected as much by tidal processes as other halo properties (e.g., the mass). Thus, it is a fair indicator of the initial potential within which the galaxy formed. In addition, we assign magnitudes in the remaining four SDSS bands (u, g, i, and z) using the observed relation between local galaxy density and color. More specifically, we use as a measure of density the distance to the 10th nearest neighbor above a certain luminosity. We measure this quantity for both our mock galaxies and actual observed SDSS galaxies. For each mock galaxy, we then choose a real SDSS galaxy which has a similar r-band luminosity and nearest neighbor distance, and assign the colors of this galaxy to the mock galaxy.

In what follows I present results obtained after assigning luminosities in halos simulated in collisionless simulations using the above scheme. More details and references can be found in Refs. 1–4.

2. Galaxy-mass clustering

The most recent observational study of galaxy-galaxy lensing by the SDSS collaboration [5] (hereafter S04), has significantly improved the accuracy of galaxy correlation measurements. The observable is $\Delta\Sigma$ defined as $\Delta\Sigma = \bar{\Sigma}(\leq R) - \bar{\Sigma}(R)$, where $\bar{\Sigma}(\leq R)$ is the mean surface density within the projected radius R, $\bar{\Sigma}(R)$ is the azimuthally averaged surface density at R, and $\Sigma_{\rm crit}$ is the critical density for lensing which depends on the angular diameter distances of the lens and the source. S04 also deprojected their $\Delta\Sigma$ to obtain the actual 3D galaxy–mass correlation function, ξ_{gm}.

We calculate $\Delta\Sigma$ and ξ_{gm} in simulations by selecting and weighting objects in accordance with S04. A comparison between our results and the S04 results is shown in Fig. 1. Our faintest bin is not exactly the same as that of S04, but the faintest galaxies that are missing from our sample have only a small contribution to the total signal. Given these small differences and the simplicity of our luminosity assignment scheme the agreement between simulations and observations is quite good. From the left panel we see that, in agreement with S04, the amplitude of the correlation function increases

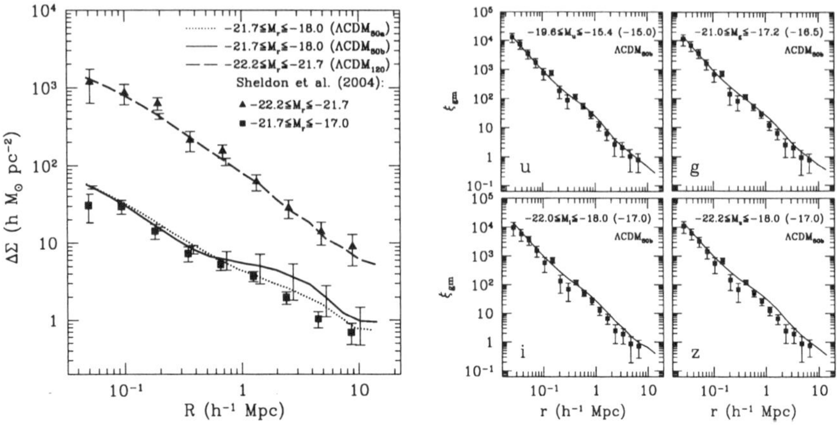

Fig. 1. *Left panel:* $\Delta\Sigma$ as a function of projected separation R as measured by SDSS [5] (*points*) and as calculated from the simulations (*lines*). Results are shown for two r-band luminosity bins: $-22.2 \leq M_r \leq -21.7$ (*triangles/dashed line*) and $-21.7 \leq M_r \leq -17.0(-18.0)$ (*squares* and *solid line*, respectively). The effect of σ_8 can be seen for the fainter sample for which we plot the results for the $\sigma_8 = 0.75$ (*dotted* line) and $\sigma_8 = 0.9$ (*solid* line) runs. *Right panel:* Comparison of the simulation galaxy–mass correlation function (*lines*) with the observations of [5] (*points*) in the u, g, i, and z SDSS bands for the fainter sample. The numbers in parentheses denote the actual faint magnitude limit of Ref. 5 sample. Reproduced from Ref. 1.

with luminosity on intermediate $(0.1 - 1h^{-1}\text{Mpc})$ scales, while it is nearly independent of luminosity on larger scales. This is indicative of an increase of the effective slope with luminosity. The left panel of the figure also shows the difference in $\Delta\Sigma(R)$ for two simulations with different values of σ_8. In the right panel we show the corresponding ξ_{gm} comparisons for the u, g, i and z bands for the faintest S04 sample. The overall agreement of our simple luminosity scheme, even in the bluer bandpasses, is surprisingly good. These results indicate that assigning colors based on luminosity and environment is sufficient at least for the purposes of galaxy-mass clustering.

3. Mass-light relation

Our luminosity assignment scheme does not distinguish between isolated halos and subhalos, i.e. halos within larger (host) halos. Furthermore, there is scatter between mass and V_{max}. Thus, it is not obvious that our scheme can give the observed mass-light relation. In Fig. 2 we present a comparison of the mass-to-light ratios as a function of virial mass obtained for several of our simulations with results obtained by a more recent version of the M/L analysis of Ref. 6. These authors constrained the parameters of the

Halo Occupation Distribution (HOD) by fitting the space density and the projected SDSS galaxy-galaxy correlation function. With this HOD they then calculated the mass-to-light ratios. The agreement between our results

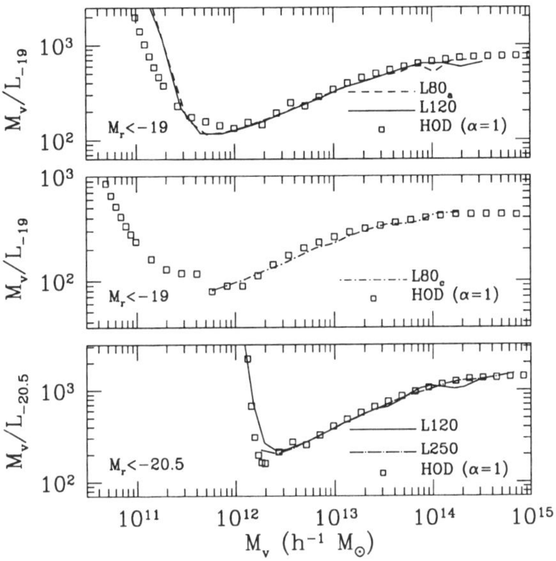

Fig. 2. Mass-to-light ratio against virial mass. For the top two panels the total luminosity of an object is calculated by summing up the luminosities of all galaxies with SDSS r-band magnitude $M_r < -19$. For the bottom panel this magnitude threshold is equal to -20.5. Simulations $L80_a, L120$ and $L250$ are for $\Omega_m = 0.3, \Omega_\Lambda = 0.3, \sigma_8 = 0.9$ and $h = 0.7$. Simulation $L80_c$ was run using the 3 year WMAP best fit parameters. The *squares* show the corresponding HOD predictions of a more recent version of the analysis of Ref. 6. In this analysis the authors fixed the power of the power law giving the mean number of satellite galaxies as a function of host mass, $\langle N_{sat} \rangle \propto M^\alpha$, to unity ($\alpha = 1$). Reproduced from Ref. 4.

and those from the HOD analysis is astonishing. Furthermore, the mass-to-light ratio varies as expected, namely it has a minimum at some mass scale, and increases towards larger and smaller masses. It increases more rapidly towards smaller masses, and this increase is expected since below a certain mass one expects that galaxies cannot form given that halos are not massive enough to be able to cool the baryons. The increase towards high masses is not as steep and is expected on the basis that galaxy formation efficiency goes down with mass at large masses possibly because of feedback effects, e.g. from AGNs. Moreover, there seems to be a flattening of the mass-to-light ratio above a mass of $\sim 10^{14} h^{-1} M_\odot$.

4. Going a step further

Since our simple scheme seems to be reproducing very well different kinds of observations (also see, e.g., [7] for comparisons with the galaxy auto-correlation function), we can use it to gain some insight, e.g., into interpreting/analyzing the observations. For example, the left panel of Fig. 3

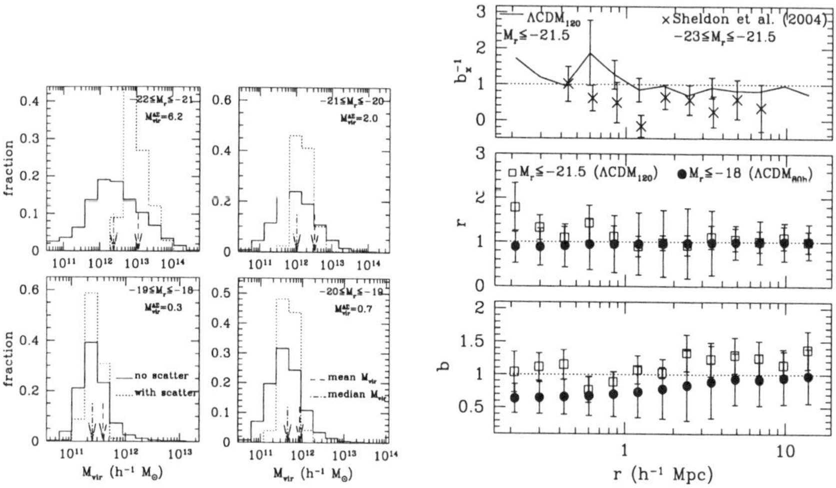

Fig. 3. *Left panel:* Halo virial mass distributions (defined using an overdensity of 180 with respect to the mean density of the universe) for central galaxies in four luminosity bins with and without scatter in the V_{max}-luminosity relation (*solid* and *dotted* histograms, respectively). $M_{\rm vir}^{\Delta\Sigma}$ is the virial mass obtained by fitting the $\Delta\Sigma$ of the corresponding subsample of galaxies, in units of $10^{12}h^{-1}M_{\odot}$. For the distributions with scatter, the mean and median virial masses are indicated with *dashed* and *dot-dashed* arrows, respectively. *Right panel:* Bias for the volume-limited samples. *Top panel:* Inverse cross-bias, b_x^{-1}, as estimated by [5] (*crosses*) for a sample with $-23 \leq M_r \leq -21.5$ and as measured in our simulations (*solid line*) for objects with $M_r \leq -21.5$. For clarity, error bars in the simulation results are plotted only for scales at which there appears to be some discrepancy between simulations and observations. *Middle panel:* Correlation coefficient as a function of scale for two volume-limited simulation samples with $M_r \leq -18$ (*solid circles*) and $M_r \leq -21.5$ (*open squares*). *Bottom panel:* Bias as a function of scale for the same samples as in the middle panel. Reproduced from Ref. 1.

addresses the question of how well weak lensing observations will be able to infer the typical mass of halos in a luminosity bin. Clearly, for finite luminosity bins as used in observations one has a distribution of halos masses. This may bias the estimated 'mean mass', especially if the distribution is asymmetric. As shown in the figure and for our assumed mass distribution, we find that the mass inferred via weak lensing $M_{\rm vir}^{\Delta\Sigma}$ lies somewhere between the median and the mean of the mass distribution. Thus, the mass derived cannot be interpreted in a straightforward way as the mass for galaxies of a given luminosity, unless one knows the mass distribution. Therefore, the mass–luminosity relations should be quoted and interpreted with caution. More discussion on the information we will be able to extract from weak lensing observations using the halo model formalism can be found in Ref. 2.

In addition, we can gain insight for quantities not directly observable. The right panel of Fig. 3 shows the correlation coefficient r and bias b for two volume-limited simulation samples of different luminosity. The top panel also shows comparisons of the cross-bias, b_x^{-1}, for the volume-limited sample of S04. If ξ_{mm}, ξ_{gg} and ξ_{gm} are the mass–mass, galaxy–galaxy and galaxy–mass correlation functions, respectively, then the above quantities are defined as: $b^2 = \xi_{\rm gg}/\xi_{\rm mm}$, $r = \xi_{\rm gm}/(\xi_{\rm mm}\xi_{\rm gg})^{1/2}$ and $b_x = \xi_{\rm gg}/\xi_{\rm gm} = b/r$. The simulation results agree with observations within the error bars. A small offset of the simulation results toward smaller values of b_x may be due to objects with $-23.0 < M_r < -22.5$, which are not present in our simulations due to the small box size and which could enhance the clustering signal somewhat. The correlation coefficient is approximately unity on scales $\geq 1h^{-1}$Mpc in our simulations. This means that on these scales the cross bias b_x which is observable is a fair measure of the standard bias b which is not directly observable.

Acknowledgments

I want to thank A.V. Kravtsov, R.H. Wechsler and J.R. Primack for allowing me to present here some of the results of our work. Also, I want to especially thank J. Tinker for making available the predictions of his HOD formalism for comparison with the simulation results.

References

1. A. Tasitsiomi, A. V. Kravtsov, R. H. Wechsler and J. R. Primack, *ApJ* **614**, 533 (2004).
2. R. Mandelbaum, A. Tasitsiomi, U. Seljak, A. V. Kravtsov and R. H. Wechsler, *MNRAS* **362**, 1451 (2005).
3. A. Slosar, U. Seljak and A. Tasitsiomi, *MNRAS* **366**, 1455 (2006).
4. A. Tasitsiomi *et al.*, *in preparation* (2006).
5. E. S. Sheldon *et al.*, *AJ* **127**, 2544 (2004).
6. J. L. Tinker, D. H. Weinberg, Z. Zheng and I. Zehavi, *ApJ* **631**, p. 41 (2005).
7. C. Conroy, R. H. Wechsler and A. V. Kravtsov, *ApJ* **647**, p. 201 (2006).

CDM ABUNDANCE IN NON-STANDARD COSMOLOGIES

C. PALLIS

School of Physics and Astronomy, The University of Manchester, Manchester M13 9PL, UNITED KINGDOM

E-mail: Constantinos.Pallis@manchester.ac.uk

The relic density of a cold dark matter (CDM) candidate is calculated in the context of three non-standard cosmological scenaria and its value is compared with the one obtained in the standard regime. In particular, we consider the decoupling of the CDM particle during: (i) A decaying-particle dominated phase or (ii) a kinetic-energy dominated phase or (iii) the decay of a massive particle under the complete or partial domination of kination. We use plausible values (from the viewpoint of supersymmetric models) for the mass and the thermal averaged cross section times the velocity of the cold relic and we investigate scenaria of equilibrium and non-equilibrium production. In the case (i) a low reheat temperature, in the range $(1-20)$ GeV, significantly facilitates the achievement of an acceptable CDM abundance. On the other hand, the presence of kination in the case (ii) can lead to an enhancement of the CDM abundance up to three orders of magnitude. The latter enhancement can be avoided, in the case (iii). In a such situation, the temperature turns out to be frozen to a plateau value which is, mostly, lower than about 40 GeV.

1. Introduction

This review is based on Refs. [1,2,3]. We recall the calculation of the CDM abundance (Sec. 2) in the context of the Standard scenario (SC) (Sec. 3) and we show how this calculation is modified in three different cases: When we consider a Low Reheating Scenario (LRS)[1] (Sec. 4), a Quintessential Scenario (QKS)[2] (Sec. 5) or a Kination-Dominated Reheating (KRS)[3] (Sec. 6). A comparison between the various scenarios is displayed in Table 1 whereas in Table 2 we present some representative combinations of the parameters which produce the central observational value of the CDM abundance. Throughout the subscript or superscript 0 [I] is referred to present-day values (except for the coefficient V_0) [to the onset of each scenario] and $\bar\rho_i = \rho_i/\rho_{\rm c}^0$ where $\rho_{\rm c}^0 = 8.099\times10^{-47}h^2$ GeV^4 with $h = 0.72$.

2. The CDM Abundance

In light of the recent WMAP3 results[4], the relic density of any CDM candidate χ, $\Omega_\chi h^2$, is to satisfy the following range of values:

$$\Omega_\chi h^2 = 0.1045^{+0.0075}_{-0.0095} \;\Rightarrow\; 0.08 \lesssim \Omega_\chi h^2 \lesssim 0.12 \;\text{ at } 95\% \text{ c.l.} \quad (1)$$

We concentrate our presentation mainly on the Lightest SUSY Particle (LSP) which is stable within the SUSY models with R-parity conservation[5] and consists the most popular, promising and natural CDM candidate[6].

The calculation of $\Omega_\chi h^2$ is based on the formula:

$$\Omega_\chi h^2 = 2.741 \times 10^8 \; Y_0 \; m_\chi/\text{GeV}, \quad \text{where } \; Y_0 = n_\chi^0/s_0, \quad (2)$$

$s \propto T^3$ is the entropy density, m_χ is the mass of χ and n_χ is the number density of χ's, which satisfies a Boltzmann Equation (BE), provided that χ's achieve *kinetic* equilibrium with plasma. The form of the BE depends on the scenario under consideration. In general, it depends on: (i) The Hubble parameter, $H = \sqrt{\rho_{\rm BG}}/\sqrt{3}m_{\rm P}$ ($m_{\rm P} = M_{\rm P}/\sqrt{8\pi}$, with $M_{\rm P}$, the Planck scale) with $\rho_{\rm BG}$, the background energy density, (ii) the equilibrium number density of χ's, $n_\chi^{\rm eq}$, which obeys the Maxwell-Boltzmann statistics:

$$n_\chi^{\rm eq}(x) = \frac{g}{(2\pi)^{3/2}} m_\chi^3 \, x^{3/2} \, e^{-1/x} P_2(1/x), \quad \text{where} \quad x = T/m_\chi < 1, \quad (3)$$

with $g = 2$ the number of degrees of freedom of χ and $P_n(z) = 1 + (4n^2 - 1)/8z$, (iii) the thermal-averaged cross section times velocity of χ's, $\langle\sigma v\rangle$, which can be mostly expanded as: $\langle\sigma v\rangle = a + bx$. We focus on the case $\langle\sigma v\rangle = a$ with $10^{-15}\ \text{GeV}^{-2} \leq \langle\sigma v\rangle \leq 10^{-7}\ \text{GeV}^{-2}$ which can be naturally produced within SUSY models[1,2,3].

Moreover, two fundamental cases of χ-production can be singled out[7]: χ's do or do not maintain *chemical* equilibrium with plasma. In the first case (EP) the current value of n_χ/s follows $n_\chi^{\rm eq}/s$ and at some $T = T_{\rm F}$, n_χ/s becomes larger than $n_\chi^{\rm eq}/s$. On the other hand, in the case of non-EP, $n_\chi/s \gg n_\chi^{\rm eq}/s$ (non-EPI) or $n_\chi/s \ll n_\chi^{\rm eq}/s$ (non-EPII) at least at the point of the maximal χ production[3,7].

Table 1. Standard vs non-Standard Scenaria.

SC	LRS	QKS	KRS
$\bar\rho_q - \bar\rho_\phi = 0$	$\bar\rho_{\phi{\rm I}} \gg \bar\rho_{\rm RI},\ \bar\rho_q = 0$	$\bar\rho_{q{\rm I}} \gg \bar\rho_{\rm RI},\ \bar\rho_\phi = 0$	$\bar\rho_{q{\rm I}} \gg \bar\rho_{\phi{\rm I}} \gg \bar\rho_{\rm RI}$
$H \propto T^2$	$H \propto T^4$	$H \propto T^3$	$H \propto T^3$
$T \propto R^{-1}$	$T \propto R^{-3/8}$	$T \propto R^{-1}$	$T = \text{cst}$
$sR^3 = \text{cst}$	$sR^3 \neq \text{cst}$	$sR^3 = \text{cst}$	$sR^3 \neq \text{cst}$
$N_\chi = 0$	$N_\chi \neq 0$	$N_\chi = 0$	$N_\chi \neq 0$

3. The Standard Cosmological Scenario (SC)

According to the SC[5], χ's (i) are produced through thermal scatterings, (ii) reach *chemical* equilibrium with plasma and (iii) decouple from the cosmic fluid at a temperature $T = T_{\rm F} \sim 15$ GeV during the radiation-dominated (RD) era. The consequences of the assumptions above are: (i) The form of the relevant BE is (dot denotes derivative w.r.t the cosmic time),

$$\dot{n}_\chi + 3Hn_\chi + \langle\sigma v\rangle \left(n_\chi^2 - n_\chi^{\rm eq2}\right) = 0 \tag{4}$$

which can be solved numerically (or semi-analytically using the freeze-out procedure[1,2,3]) with initial condition $n_\chi(x = 1) = n_\chi^{\rm eq}(x = 1)$ or $n_\chi(x = 1) = 0$, (ii) the required $\langle\sigma v\rangle$ is $\langle\sigma v\rangle \gtrsim 10^{-20}$ GeV^{-2} (note that with $\langle\sigma v\rangle \simeq 2.9 \times 10^{-29}$ GeV^{-2} and $n_\chi(x = 1) = 0$, we can obtain $\Omega_\chi h^2 = 0.1$ if we allow for non-EPII), (iii) the cosmological evolution during the χ decoupling is RD and so, $\rho_{\rm BG} \simeq \rho_{\rm R} \propto T^4$. Therefore, $H \propto T^2$ and $T \propto R^{-1}$, where R is the scale factor of the universe (see Table 1).

In this context, the $\Omega_\chi h^2$ calculation depends only on two parameters: m_χ and $\langle\sigma v\rangle$. As shown in Table 2 (1st column), $\langle\sigma v\rangle \sim 10^{-9}$ GeV^{-2} can ensure acceptable $\Omega_\chi h^2$'s. Such a requirement strongly restricts the parameter space of many particle models. However, this picture can drastically change if one or more assumptions of the SC are lifted.

4. The Low Reheating Scenario (LRS)

The modern cosmo-particles theories are abundant in scalar massive particles (e.g. moduli, PQ-flatons, dilatons) which can decay when H becomes equal to their mass creating episodes of reheating. In the LRS, we assume that such a scalar particle ϕ with mass m_ϕ decays with a rate Γ_ϕ into radiation, producing an average number N_χ of χ's, rapidly thermalized. The energy densities of ϕ and the produced radiation and $\rho_{\rm R}$ and ρ_ϕ and $n_{\tilde\chi}$, satisfy the following BEs ($\Delta_\phi = (m_\phi - N_{\tilde\chi} m_{\tilde\chi})/m_\phi$):

$$\dot\rho_\phi + 3H\rho_\phi + \Gamma_\phi\rho_\phi = 0, \tag{5}$$

$$\dot\rho_{\rm R} + 4H\rho_{\rm R} - \Gamma_\phi\Delta_\phi\rho_\phi - 2m_\chi\langle\sigma v\rangle\left(n_\chi^2 - n_\chi^{\rm eq2}\right) = 0, \tag{6}$$

$$\dot{n}_\chi + 3Hn_\chi + \langle\sigma v\rangle\left(n_\chi^2 - n_\chi^{\rm eq2}\right) - \Gamma_\phi N_\chi n_\phi = 0, \tag{7}$$

The system above can be solved, imposing the following initial conditions:

$$H_{\rm I} = m_\phi \;\Rightarrow\; \bar\rho_{\phi{\rm I}} = m_\phi^2/H_0^2 \;\text{ and }\; \bar\rho_{\rm RI} = \bar\rho_{\chi{\rm I}} = 0. \tag{8}$$

Our investigation verifies that the reheating process is not instantaneous[8]. Until its completion – i.e., $\rho_\phi(T_{\rm RH}) = \rho_{\rm R}(T_{\rm RH})$ –, the maximal temperature, $T_{\max} = f(\Gamma_\phi, \bar\rho_{\phi{\rm I}})$, can become much larger than the so-called reheat

Table 2. Combinations of parameters leading to $\Omega_\chi h^2 = 0.1$ in the Standard and non-Standard Scenaria.

	SC	LRS			QKS		KRS		
χ-Production:	EP	EP	non-EPII	EP	EP	EP	non-EPI	non-EPII	EP
$\langle\sigma v\rangle$ (GeV^{-2})	2×10^{-9}	10^{-10}	10^{-10}	10^{-8}	3×10^{-6}	3.6×10^{-7}	10^{-10}	10^{-10}	10^{-10}
T_ϕ (GeV)	–	5.5	0.001	5.5	–	–	30	30	5.5
N_χ	–	0	10^{-3}	5×10^{-5}	0	0	10^{-6}	0	0
$\Omega_q^{\rm NS}$	–	–	–	–	0.1	0.001	0.02	3×10^{-9}	2.4×10^{-15}
$\log\bar\rho_{\phi_{\rm I}}$	–	95.62	95.62	95.62	–	–	69.3	73.815	76.5
$T_{\rm PL}$ (GeV)	–	–	–	–	–	–	1.2	16	32.5
$\Omega_\chi h^2\vert_{\rm SC}$	0.1	1.9	1.9	0.023	9×10^{-5}	7×10^{-4}	1.9	1.9	1.9

Note: We fix $m_\chi = 350$ GeV for the results presented in this table. We also use $m_\phi = 10^6$ GeV in the LRS and KRS and $\lambda = 0.5$ in the QKS and KRS (note, however, that the results on $\Omega_\chi h^2$ are λ-independent). Recall that in the LRS, $T_\phi = T_{\rm RH}$ and the results on $\Omega_\chi h^2$ are $\bar\rho_{\phi_{\rm I}}$-independent and remain invariant for fixed $N_\chi m_\phi^{-1}$ (and, obviously, T_ϕ, m_χ, $\langle\sigma v\rangle$) – however, the type of the χ-production does depend separately on N_χ and m_ϕ. In each case, shown is the type of the χ-production and the obtained value of $\Omega_\chi h^2$ in the SC, $\Omega_\chi h^2|_{\rm SC}$.

temperature[9], $T_{\rm RH}$. Also, for $T > T_{\rm RH}$, $\rho_{_{\rm BG}} \simeq \rho_\phi \Rightarrow H \propto T^4$ with $T \propto R^{-3/8}$ and an entropy production occurs (see Table 1).

The free parameters of the LRS are: m_χ, $\langle\sigma v\rangle$, Γ_ϕ, m_ϕ, N_χ, $\bar\rho_{\phi{\rm I}}$. However, the results on $\Omega_\chi h^2$ do not depend on the explicit value of $\bar\rho_{\phi{\rm I}}$ as long as $T_{\rm RH} < T_{\rm F} < T_{\rm max}$, and are invariant[1,10] for fixed $N_\chi m_\phi^{-1}$ (and T_ϕ, m_χ, $\langle\sigma v\rangle$). Moreover, Γ_ϕ can be replaced by $T_\phi \simeq T_{\rm RH}$ through the relation[3]:

$$\Gamma_\phi = 5\sqrt{\frac{\pi^3 g_{\rho*}(T_\phi)}{45}}\frac{T_\phi^2}{M_{\rm P}} = \sqrt{\frac{5\pi^3 g_{\rho*}(T_\phi)}{72}}\frac{T_\phi^2}{m_{\rm P}}. \tag{9}$$

As shown in Table 2, we can obtain acceptable $\Omega_\chi h^2$'s (i) for relatively low $\langle\sigma v\rangle$'s (2nd, 3rd columns) with low $T_{\rm RH} = 5.5$ GeV and $N_\chi = 0$[1,9] (EP) or much lower $T_{\rm RH} = 1$ MeV and $N_\chi \neq 0$ (non-EPII) and (ii) for larger $\langle\sigma v\rangle$'s (4th column) with larger N_χ's (EP).

5. The Quintessential Scenario (QKS)

Another role that a scalar field could play when it does not couple to matter (contrary to ϕ) is this of quintessence. Such a field, q, satisfies the equation:

$$\ddot q + 3H\dot q + dV/dq = 0, \quad \text{where} \quad V = V_0 e^{-\lambda q/m_{\rm P}} \tag{10}$$

is the adopted potential[11] and undergoes three phases during its cosmological evolution[2]: (i) The kination[12] dominated (KD) phase, where the energy density of q is $\rho_q = \dot q^2/2 + V \simeq \dot q^2/2$, (ii) the frozen-field dominated (FD) phase, where ρ_q is constant and (iii) the attractor dominated (AD) phase. Today we obtain a transition from the FD to the AD phase[2,13]. Although this is not a satisfactory solution to the coincidence problem, the observational data from WMAP3[4]:

$$\Omega_q^0 = \Omega_{\rm DE} = 0.74 \pm 0.12 \;\; \text{and} \;\; w_q < -0.83. \tag{11}$$

can be reproduced with $\lambda \leq 1.1$ and by conveniently adjusting V_0[2,13].

For a reasonable region of initial conditions ($q_{\rm I} = 0$ and $\Omega_q^{\rm I} = 1$[2]), $\rho_q \simeq \dot q^2/2$ can dominate over radiation, creating a totally KD era in conjunction with the satisfaction of the nucleosynthesis (NS) constraint[15], $\Omega_q^{\rm NS} \leq 0.21$. As a consequence, during the KD era we obtain: $\rho_{_{\rm BG}} \simeq \rho_q \Rightarrow H \propto T^3$ with $T \propto R^{-1}$ (see Table 1). Combining eqs. (4) and (10) we can show that if the χ-decoupling occurs during the KD era, $\Omega_\chi h^2$ increases[14] w.r.t $\Omega_\chi h^2|_{\rm SC}$ (see 5th and 6th column in Table 2). This enhancement of $\Omega_\chi h^2$ turns out to be a single-valued function of the quintessential parameter at the eve of NS[2], $\Omega_q^{\rm NS}$, for fixed m_χ and $\langle\sigma v\rangle$. Therefore, the $\Omega_\chi h^2$ calculation depends only on the parameters: m_χ, $\langle\sigma v\rangle$, $\Omega_q^{\rm NS}$ $(\lambda \leq 1.1)$.

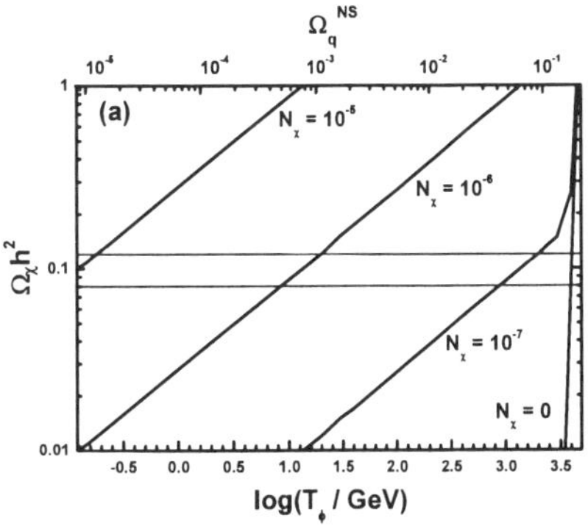

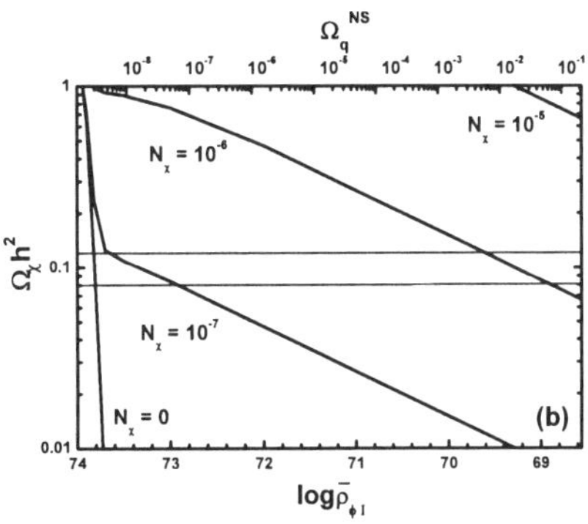

Figure 1. $\Omega_{\tilde{\chi}}h^2$ versus $\log T_\phi$ (or $\Omega_q^{\rm NS}$) for $\log\bar{\rho}_{\phi_{\rm I}} = 70$ (a) and $\log\bar{\rho}_{\phi_{\rm I}}$ (or $\Omega_q^{\rm NS}$) for $T_\phi = 30$ GeV (b). We take $m_\phi = 10^6$ GeV, $m_{\tilde{\chi}} = 350$ GeV, $\langle\sigma v\rangle = 10^{-10}$ GeV^{-2} and various N_χ's, indicated on the curves. The CDM bounds of Eq. (1) are, also, depicted by the two thin lines.

6. The Kination-Dominated Reheating (KRS)

In view of the two previous situations, the obvious question would be: What happens if we have both quintessence and low reheating? Or a low $T_{\rm RH}$ could assist us to the reduction of $\Omega_\chi h^2$, in the presence of a KD phase? This novel[16] cosmological set-up can be analyzed by solving the system of Eqs. (5)-(7) and (10) with constraint $\Omega_q^{\rm I} = 1$ and initial conditions ($q_{\rm I} = 0$):

$$q_{\rm I} = 0, \quad H_{\rm I} = m_\phi \;\Rightarrow\; \dot{q}_{\rm I} = \sqrt{2\rho_{\rm c}^0}(m_\phi/H_0) \;\; \text{and} \;\; \bar{\rho}_{_{\rm RI}} = \bar{\rho}_{\chi{\rm I}} = 0. \qquad (12)$$

We can distinguish two types of q-domination, depending whether ϕ decays before or after it becomes the dominant component of the universe. In both cases, $\rho_{_{\rm BG}} \simeq \rho_q \;\Rightarrow\; H \propto T^3$, entropy production occurs and a prominent period of constant maximal temperature, $T_{\rm PL} = f(\rho_{\phi_{\rm I}}, \Gamma_\phi/m_\phi)$, arises[3] (see Table 1). The free parameters of this scenario are: λ, $\bar{\rho}_{\phi_{\rm I}}$, m_ϕ, T_ϕ, N_χ, m_χ, $\langle\sigma v\rangle$. As in the QKS the $\Omega_\chi h^2$ calculation is λ-independent for $\lambda \leq 1.1$ but unlike the LRS it severely depends on $\bar{\rho}_{\phi_{\rm I}}$, and T_ϕ does not coincide with the maximal temperature of the RD era.

The crucial difference between the KRS and the QKS is that $\Omega_\chi h^2$ does not exclusively increase with $\Omega_q^{\rm NS}$. Indeed, when an increase of $\Omega_q^{\rm NS}$ is generated by an increase of T_ϕ (which results to an increase of $T_{\rm PL}$), $\Omega_\chi h^2$ increases with $\Omega_q^{\rm NS}$ – see Fig. 1-(a). On the contrary, when the increase of $\Omega_q^{\rm NS}$ is due to the decrease of $\bar{\rho}_{\phi_{\rm I}}$ (which results to a decrease of $T_{\rm PL}$), $\Omega_\chi h^2$ decreases as $\Omega_q^{\rm NS}$ increases – see Fig. 1-(b). This is, because n_χ decreases rapidly with $T_{\rm PL}$ due to the exponential suppression of $n_\chi^{\rm eq}$ in Eq. (3).

As shown in Table 2 (7th and 8th columns), $\Omega_\chi h^2$ reaches the range of Eq. (1) with $N_\chi \sim (10^{-7} - 10^{-5})$ when $T_{\rm PL} \ll T_{\rm F}$ (non-EPI) and with $N_\chi \sim 0$ when $T_{\rm PL} \sim T_{\rm F}$ (non-EPII). As $\Omega_q^{\rm NS}$ decreases, $(T_{\rm PL} - T_{\rm F})$ increases and $\Omega_\chi h^2$ approaches its value in the LRS (9th column in Table 2).

7. Conclusions

We considered three deviations from the SC and we showed[1,2,3] that: (i) In the LRS with $T_{\rm RH} < 20$ GeV, $\Omega_\chi h^2$ decreases w.r.t its value in the SC for low N_χ's and increases for larger N_χ's. Both EP and non-EP are possible for commonly obtainable $\langle\sigma v\rangle$'s, (ii) in the QKS, $\Omega_\chi h^2$ increases drastically (almost 3 orders of magnitude for $\Omega_q^{\rm NS}$ close to its upper bound) (iii) in the KRS, $\Omega_\chi h^2$ becomes cosmologically interesting for $N_\chi \sim (10^{-7} - 10^{-5})$ when $T_{\rm PL} \ll T_{\rm F}$ (non-EPI), and for $N_\chi \sim 0$ when $T_{\rm PL} \sim T_{\rm F}$ (non-EPII); EP is activated for $T_{\rm PL} > T_{\rm F}$ and the results on $\Omega_\chi h^2$ approach their values in the LRS or SC as $T_{\rm PL}$ increases well beyond $T_{\rm F}$.

Acknowledgments

The author would like to thank the organizers of IDM2006 for their invitation and the PPARC research grand PP/D000157/1 for financial support.

References

1. C. Pallis, *Astropart. Phys.* **21**, 689 (2004) [hep-ph/0402033].
2. C. Pallis, *J. Cosmol. Astropart. Phys.* **10**, 015 (2005) [hep-ph/0503080].
3. C. Pallis, *Nucl. Phys.* **B751**, 129 (2006) [hep-ph/0510234].
4. D.N. Spergel *et al.*, astro-ph/0603449.
5. H. Goldberg, *Phys. Rev. Lett.* **50**, 1419 (1983);
   J.R. Ellis *et al.*, *Nucl. Phys.* **B238**, 453 (1984).
6. For other candidates see, e.g., G. Lazarides, hep-ph/0601016.
7. E.W. Kolb, hep-ph/9910311.
8. R.J. Scherrer and M.S. Turner, *Phys. Rev.* D **31**, 681 (1985).
9. J. McDonald, *Phys. Rev.* D **43**, 1063 (1991);
   G.F. Giudice *et al.*, *Phys. Rev.* D **64**, 023508 (2001) [hep-ph/0005123];
   N. Fornengo *et al.*, *Phys. Rev.* D **67**, 023514 (2003) [hep-ph/0208072].
10. G. Gelmini and P. Gondolo, *Phys. Rev.* D **74**, 023510 (2006) [hep-ph/0602230].
11. C. Wetterich, *Nucl. Phys.* **B302**, 668 (1988).
12. B. Spokoiny, *Phys. Lett.* B **315**, 40 (1993) [gr-qc/9306008];
    M. Joyce, *Phys. Rev.* D **55**, 1875 (1997) [hep-ph/9606223].
13. U. França *et al.*, *J. High Energy Phys.* **10**, 015 (2002) [astro-ph/0206194];
    C.L. Gardner, *Nucl. Phys.* **B707**, 278 (2005) [astro-ph/0407604].
14. P. Salati, *Phys. Lett.* B **571**, 121 (2003) [astro-ph/0207396].
15. R.H. Cyburt *et al.*, *Astropart. Phys.* **23**, 313 (2005) [astro-ph/0408033].
16. See, also, A. Liddle *et al.*, *Phys. Rev.* D **68**, 043517 (2003) [astro-ph/0302054];
    B. Feng and M. Li, *Phys. Lett.* B **564**, 169 (2003) [hep-ph/0212233].

PROBING DARK ENERGY WITH TYPE Ia SUPERNOVAE

R. PAIN*

LPNHE, CNRS/IN2P3 and Universités Paris 6 and 7
Paris, 75005, France
**E-mail: reynald.pain@in2p3.fr*

The energy density of the universe seems to be dominated by a mysterious "dark-energy" which drives the acceleration of the expansion. This phenomenon was discovered by mapping out the history of cosmic expansion using Type Ia supernovae (SNe Ia) as distance indicators. SNe Ia currently provide one of the best way to characterize the nature of dark energy, by placing tight limits on its equation of state $w = p/\rho$. Current high-redshift supernova surveys, such as the 5-year Supernova Legacy Survey will deliver $\sim$ 1000 SN Ia detections with well-sampled *griz* light curves. Using this unprecedented dataset, a statistical precision of about 5% will be obtained on Ω_Λ, and w. In this contribution, I review the ongoing Type Ia surveys and present the recent constraints obtained on the dark energy equation of state.

Keywords: Cosmology, Supernovae

1. Introduction

Observations of cosmologically distant type Ia supernovae (SNe Ia) at redshifts $z < 1$[1,2] indicate that the expansion of the universe is currently accelerating. This behavior is attributed to a mysterious *dark-energy* component, with negative pressure. Combining the SNe Ia results with the study of the cosmic microwave background (CMB) anisotropies,[3] cosmic shear[4] and studies of galaxy clusters, one can show that the Universe is nearly flat, with about 70% of the energy contained in the form of dark energy.

Todays' projects aim at directly probe the nature of dark energy by constraining its equation of state: $w = p/\rho$. The historical cosmological constant (Λ) model predicts $w = -1$. The cosmological constant is however 10^{60} to 10^{120} times smaller than the vacuum energy predicted by standard theories of particle physics. The model cannot explain either why dark energy and dark matter have comparable densities today and over the last decade, many alternate dynamical-Λ models have been proposed, such as quintessence, k-essence, phantoms, tachyon fields.[5] These models allow values of w above -0.8 (quintessence) and below -1 (phantom fields), with a time-varying equation of state. The most recent observations constrain w to be close to that of a cosmological constant. They say little about the time evolution of w.

In a matter and dark-energy dominated universe, the expansion rate, $H(z)$ is a function of the matter density $\Omega_{\rm M}$, the dark energy density Ω_X and w_X:

$$H^2(z) = H_0^2\left(\Omega_{\rm M}(1+z)^3 + \Omega_X \exp\left[3\int_0^z (1+w(x))\, d\ln(1+x)\right]\right) \quad (1)$$

$H(z)$ can be related to various observables. To date, the best constraints on dark energy come from the luminosity distance $d_L(z) = (1+z)\times\int dz/H(z)$, using type Ia Supernovae as "standardizable" candles, *i.e.* objects of uniform luminosity.[6,11] Another promising observable is the angular distance versus redshift relation, $d_A(z) = 1/(1+z)\times\int dz/H(z)$. $d_A(z)$ can be constrained using clusters for which the angular distance can be derived from SZ and X observations. Constraints on $d_A(z)$ can also be derived from the detection of the baryon acoustic peak (BAO) at various redshifts and number counts of galaxy clusters, detected with the Sunayev-Zel'dovitch effect, X-ray emission, or weak-lensing surveys allow one to map out the redshift variations of the comoving volume element $dV/dzd\Omega = r^2(z)H(z)$. In this last case, the constrains are not purely geometrical. They also comes from how dark energy affects mass clustering.

In this contribution, we concentrate on the study of $d_L(z)$ in the redshift range $0.1 < z < 1.0$ using SNe Ia as cosmological distance indicators. In the next section, we will recall the measurement principles, and the progress made over the last decade. We also review second generation SN Ia surveys currently in progress. As an example of such surveys we then present the analysis of the SuperNova Legacy Survey (SNLS) first year dataset, which currently gives, when combined with BAO or CMB recent results, the most precise measurement of w.

2. Measuring the Cosmological Parameters Using Type Ia Supernovae

The first evidence for an accelerating Universe was published in 1998 by the Hi-z Supernova Team[1] and the Supernova Cosmology Survey.[2] Both teams had gathered distinct $z \sim 0.5$ supernova samples, but shared the same low redshift sample made of objects discovered in the early and mid 90s by various supernova projects. The constraints on the cosmological parameters comes from the comparison of the nearby and distant samples. The primary observation made by the two teams, is that distant SNe Ia appear fainter (by $\sim$20% at $z \sim 0.5$) than expected in a decelerating Einstein-de Sitter universe.

The acceleration of the expansion had not been anticipated, and was questioned during several years. The conclusions had such a potential impact on fundamental physics, that projects were started to verify the initial results and study the possible sources of systematic errors. Indeed, several astrophysical effects could have possibly biased the cosmological conclusions. First, some intergalactic "gray" dust, could be present that would dim SNe Ia without altering their colors, thus being undetectable. Second, evolution of the luminosity of SNe Ia could bias the cosmological conclusions, if SNe Ia at $z \sim 0.5$ were intrinsically fainter than nearby supernovae. Finally, detection biases in the distant and nearby supernova sample could alter the average supernova brightness in a redshift dependent way.

In the early 2000, experiments expanded the distant SN samples to $z \sim 1$[7–9] using ground based telescopes and the Hubble Space Telescope. Sullivan et al[10] studied the dependence of the Hubble diagram with the host galaxy types, and showed that the cosmological conclusion did not depend on the supernova host galaxy type. Finally, Riess et al[11] extended the Hubble diagram to a redshift of 1.7 with 16 very distant SNe Ia, discovered with the HST during the course of the GOODS ACS Treasury program. Using this combined sample, they showed that around $z \sim 0.45$ the Universe underwent a transition from a decelerated (matter dominated) era to an accelerated era, inconsistent with a simple gray dust model as an alternative to dark energy. This remarkable data sample also allowed one to place improved constraints on w and dw/dz, compatible with the cosmological constant model ($w = -1$, $dw/dz = 0$).

Improving significantly over current SN constraints on the dark energy requires (1) an improved nearby sample of several hundreds supernovae, discovered with a completeness controlled search and (2) a ten-fold larger sample (i.e. o(1000) in the redshift range $0.2 < z < 1.0$, where w is best measured). Such a dataset is needed in order to significantly improve on statistical errors but also, most importantly, on systematic uncertainties. These two requirements have motivated a second generation of surveys. The Nearby Supernova Factory is one of these new large programs. It aims at discovering several hundred SNe Ia at a redshift of $z \sim 0.05$, observed spectroscopically and photometrically with a good time sampling, using a dedicated integral field spectrometer. The SDSS Supernova Survey uses a "rolling search" strategy covering a large area of 300 square degrees every day, during 3 months each year. The goal of the project is to discover and follow-up in the *ugriz*-bands about 200 SNe Ia in the redshift range $0.1 < z < 0.3$. The ESSENCE and SNLS surveys are also based on the rolling search technique. They target the intermediate redshift range. ESSENCE covers 8 square degrees in the R- and I-passbands and aims at characterizing about 200 SNe Ia at redshifts between 0.2 and 0.8. The SNLS is more ambitious and targets about 1000 supernovae up to redshift unity. We will described SNLS in more details in the next section. Two very distant supernova surveys relying on HST observations are targeting the $1 < z < 1.5$ redshift range: the PANS Project[12] and the Supernova Cosmology Project High Redshift Cluster Search.[13] Due to the relatively small field of view of the ACS and the difficulty to secure large amount of time on the HST, these programs yield, however, limited numbers of objects.

Finally, proposed large wide field imaging projects, some already in preparation, plan to significantly improve on both the number of supernovae and the quality of the measurements. Pan-Starrs[14] is a collection of four 1.8-m telescopes equipped with 7 square degree cameras. It will cover the entire sky visible from Hawaii with a time-sampling rate of about 4 days and is expected to discover a large collection of Type Ia Supernovae up to a redshift $z \sim 0.6$. The Large Synoptic Survey Telescope (LSST)[15] is an 8-m telescope, equipped with a 9 square degree (3.2 billion pixel) camera. LSST will cover the entire visible sky every three days, and will discover of the order of 250,000 SNe Ia per year. Finally, several satellite missions are being proposed. SNAP[16,17] and DESTINY are candidates for the NASA Joint Dark Energy Mission (JDEM). DUNE is a European project currently under discussion. SNAP will discover and identify about 2000 SNe Ia up to redshifts of 1.7, allowing to map out the transition between the accelerated and decelerated eras. Such a redshift range should allow one to place tight constraints on dw/dz. The observing strategy developed for DUNE is somewhat different. The satellite will cover a wider area, thus discovering a larger number of supernovae, down to very low redshift. Such an option will drastically reduce the cross calibration uncertainties between the low- and high-redshift samples. On the other hand, DUNE will rely on an external - ground based - spectroscopic survey to perform the host galaxy typing and redshift determination. Expected launch dates for these projects range in the 2015-2020 time period.

3. An Example of 2nd Generation SN Survey: The SNLS

The goal of the Supernova Legacy Survey is to build a sample of several hundreds high-z type Ia supernovae in the redshift range $0.2 < z < 1.0$, detected before maximum light, identified spectroscopically, and with well sampled lightcurves in the four g, r, i and z-bands ($< \lambda_g > \sim 480$ nm, $< \lambda_z > \sim 880$ nm). The project was designed to improve significantly over the traditional strategy as follows: (1) discovery and photometric follow-up are performed with a wide field imager used in "rolling search" mode, where a given field is observed every third to fourth night as long as it remains visible; (2) service observing is exploited for both spectroscopy and imaging, reducing the impact of bad weather. Using a single imaging instrument to observe the same fields reduces photometric systematic uncertainties; service observing optimizes both the yield of spectroscopic observing time, and the light-curve sampling.

The project is comprised of two components: an imaging survey, to detect the supernovae and monitor their lightcurves, and a spectroscopic program, to confirm the nature of the candidates, measure their redshift and identify their host galaxy.

3.1. *The Imaging Survey: Finding the Supernova Candidates*

The imaging is taken as part of the deep component of the CFHT Legacy Survey[19] using the one square degree imager MegaCam.[20] In total, CFHTLS has been allocated about 500 nights over 5 years and consists of 3 surveys: a very wide shallow survey (1300 square degrees), a wide survey (120 square degrees) and a deep survey (4 square degrees). The 4 pointings of the deep survey are evenly distributed in right ascension. The observations for the deep survey are sequenced in a way suitable for detecting supernovae and measuring their light-curves: in every lunation in which a field is visible, it is imaged at five equally spaced epochs during a MegaCam run (which lasts about 18 nights, centered on new moon). Observations are taken in a combination of r, i plus g or z filters depending on the phase of the moon. Each field is observed for 5 to 7 consecutive lunations, as long as it remains visible. Observations lost to weather on any one night remain in the queue until the next clear observing opportunity, or until a new observation in the same filter is scheduled.

Observations and real-time pre-processing are performed by the CFHT staff using the Elixir reduction pipeline,[21] with the data products immediately available to the SN search teams. Two independent real-time pipelines have been set up, which analyze these pre-processed images to detect supernova candidates.

The detection of candidates is performed by subtracting a "past" image to the current images, where the past-image is constructed by stacking previous observations of the same field. The key element of these pipelines is matching the point spread function of a new exposure to the past-image. New candidates are detected and measured on the subtraction images; detections are matched to other detections in the field, if any. One of the pipelines processes all bands on an equal footing, the other detects in the i band (which is deep enough for trigger purposes) and measures fluxes in the other bands. The two candidate lists are merged after each epoch and typically have an overlap greater than 90% for $i(AB) < 24.0$ after two epochs in a dark run. The reasons for one candidate being found by only one pipeline are usually traced to different masking strategies or different handling of the CCD overlap regions.

3.2. *The Spectroscopic Survey: Identifying the Supernova Candidates*

Spectroscopy is critical to obtain SN redshifts and to determine the nature of each SN candidate. This requires observations on 8-10 meter class telescopes, due to the faintness of the distant supernovae. Spectroscopic follow-up time for the first year dataset was obtained at a variety of telescopes, mainly at the European Southern Observatory Very Large Telescope (60 hours per semester) and at Gemini-North and South (60 hours per semester). Spectroscopic time was also obtained at Keck-I and Keck-II (3 nights during each Spring semester).

The imaging survey still delivers more variable candidates than can be observed spectroscopically. Hence, ranking of these candidates is performed to optimize the SN Ia yield. The method uses both a photometric selection tool[18] which performs real-time light-curve fits to reduce the contamination of core-collapse SNe, and a history database of every variable object ever detected by our pipelines to remove AGN and variable stars which are seen to vary repeatedly in long-timescale data sets (more than one year). SN Ia candidates fainter than $i_{AB} = 24.5$ (likely at $z > 1$) and those with very low relative luminosity w.r.t. their host galaxy are usually not observed. With the real-time light-curve fit technique, approximately 70% of the candidates turn out to be SNe Ia. The possible biases associated with this selection were studied in Ref. 18 and found to be negligible.

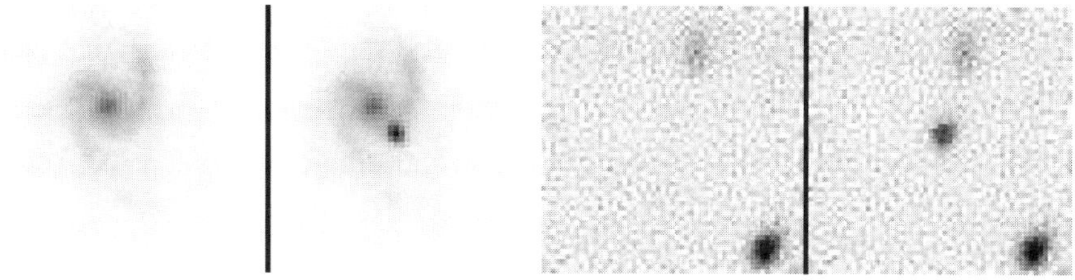

Fig. 1. Photometry model fit on the follow-up images. Left: host galaxy model without the supernova and host galaxy model with the supernova for a SNLS SN Ia at redshift $z = 0.28$. Right: host galaxy model without the supernova and host galaxy model with the supernova for a distant SNLS SN Ia at $z = 0.95$. The galaxy is barely visible.

3.3. *Building Supernova Light Curves*

For each supernova candidate, the image with the best image quality (subsequently called "reference") is identified, and all other images are resampled to the pixel grid defined by this reference. Convolution kernels that would match the Point Spread Function (PSF) of the reference image to the PSF of the other resampled science images are then derived. These convolution kernels not only match the PSFs, but also contain the photometric ratios of each image to the reference.

To measure the supernova flux evolution, all images in a given filter with a model that includes (1) a spatially variable galaxy (constant with time), and (2) a time variable point source (the supernova) are fitted. The shape of the galaxy and positions of both galaxy and supernova are fit globally. The non parametric galaxy "model" is made of independent pixels which represent the galaxy in the best image. Figure 1 show the photometry model fit on the images for an intermediate and a distant SNLS supernova.

All fluxes (f_i) are expressed in units of the reference image flux. The fit parameters are: the supernova position and the galaxy pixel values (common to all images), the supernova fluxes, and a constant sky background (different for each image). In some images in the series, the supernova flux is known to be absent or negligible; these frames enter the fit as "zero flux images" and are thus used to determine the values of the galaxy pixels.

3.3.1. *Photometric Calibration*

The SN flux measurement techniques described in the previous section deliver fluxes on the same photometric scale as the reference image. These fluxes have to be converted on the photometric scale of the nearby supernovae used in the Hubble diagram to supplement the SNLS dataset.

This is a two step process. For each field, a set of $\sim$ 1200 isolated field stars (tertiary standards) is selected. Calibrated magnitudes for these stars are determined using auxiliary observations of standard stars. Since the nearby SNe used in the cosmological fits were extracted from the literature and are typically calibrated using the standard star catalog of (Landolt, 1992),[22] the same calibration source was adopted for the high-redshift sample. This avoids introducing additional systematic uncertainties between the distant and nearby SN fluxes, which are used to determine the cosmological parameters. 5 to 20 calibrated observations performed during photometric nights were combined to compute the calibrated magnitudes of the tertiary standards.

The supernova lightcurves are then calibrated relative to the tertiary standards. For each tertiary standard, one flux for each image is obtained, using the same photometry code used for the supernovae, without the "zero-flux" images and without an underlying galaxy.

The observed standard deviation of the calibration residuals, which sets the upper bound to the repeatability of the photometric measurements, is about or below 0.01 mag in g, r and i, and about 0.016 mag in z. The systematic calibration uncertainties are dominated by the precision of the photometric alignment on the Landolt catalog, ($\sim$ 1%) and the determination of the broadband magnitudes in the Landolt system of Vega, the fundamental flux standard ($\sim$ 1%).

3.3.2. *Fitting Multicolor Light-curves*

The SN Ia light-curves are fitted in two or more bands using the SALT light-curve model[23] which returns the supernova rest-frame B-band magnitude m_B^*, a single shape parameter s and a single color parameter c. The stretch factor s is similar to that described in Ref. 24: it parameterizes the brighter-slower relation by stretching the time axis. The rest-frame color c is defined by $c = (B - V)_{B\,max} + 0.057$: it is a color excess (or deficit) with respect to a fiducial SN Ia (for which $B - V = -0.057$ at B-band maximum light). The light-curve model was trained on very nearby supernovae (mostly at $z < 0.015$) published in the literature. Figure 2 shows the points of a nearby and a distant SNLS object, along with the multi-color lightcurve models.

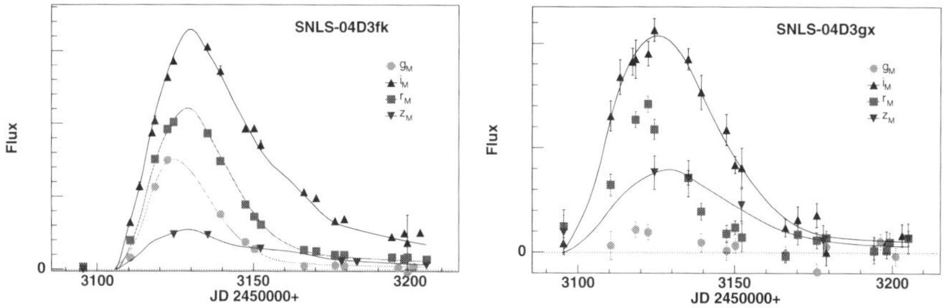

Fig. 2. SALT lightcurve model fit on a intermediate redshift SN Ia ($z = 0.36$) and on a distant SN Ia ($z = 0.91$). Both objects are SNLS discoveries. With a SN redshift of 0.36, the four measured passbands lie in the wavelength range of the lightcurve model, defined by rest-frame U and R bands, and all lightcurve points are therefore fitted simultaneously with four free parameters only (photometric normalization, date of maximum, a stretch and a color parameter). For a distant supernova, the bluest bands, g and r lie outside the wavelength coverage of SALT. Note the excellent time sampling of the SNLS observations.

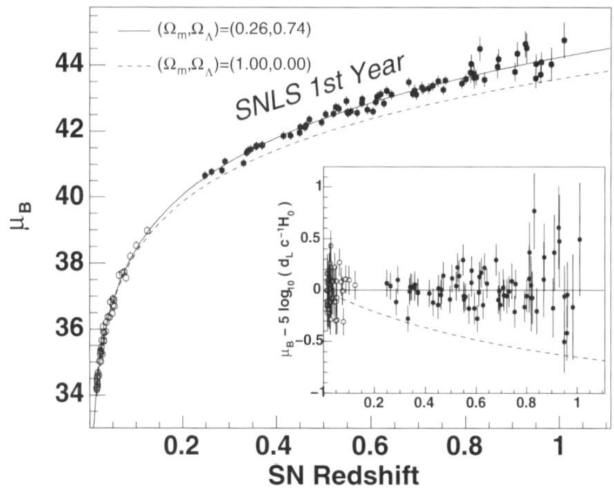

Fig. 3. Hubble diagram of SNLS and nearby SNe Ia with various cosmologies superimposed. The inset plot shows the residuals for the best fit to a flat Λ cosmology.

3.4. *Cosmological Results from the First Year Data Set*

3.4.1. *Fitting the Cosmology*

From the fits to the light-curves a rest-frame-B magnitude is computed, which, for perfect standard candles, should vary with redshift according to the luminosity distance. This rest-frame-B magnitude refers to *observed*

Table 1

fit	parameters (stat only)
$(\Omega_{\rm M}, \Omega_\Lambda)$	$(0.31 \pm 0.21, 0.80 \pm 0.31)$
$(\Omega_{\rm M} - \Omega_\Lambda, \Omega_{\rm M} + \Omega_\Lambda)$	$(-0.49 \pm 0.12, 1.11 \pm 0.52)$
$(\Omega_{\rm M}, \Omega_\Lambda)$ flat	$\Omega_{\rm M} = 0.263 \pm 0.037$
$(\Omega_{\rm M}, \Omega_\Lambda)$ + BAO	$(0.271 \pm 0.020, 0.751 \pm 0.082)$
$(\Omega_{\rm M}, w)$+BAO	$(0.271 \pm 0.021, -1.023 \pm 0.087)$

brightness, and therefore does not account for brighter-slower and brighter-bluer correlations (see Ref. 23 and references therein). As a distance estimator, the following expression was used:

$$\mu_B = m_B^* - M + \alpha(s-1) - \beta c$$

where m_B^*, s and c are derived from the fit to the light curves, and α, β and the absolute magnitude M are parameters which are fitted by minimizing the residuals in the Hubble diagram. The cosmological fit is actually performed by minimizing:

$$\chi^2 = \sum_{objects} \frac{(\mu_B - 5\log_{10}(d_L(\theta, z)/10pc))^2}{\sigma^2(\mu_B) + \sigma_{int}^2},$$

where θ stands for the cosmological parameters that define the fitted model (with the exception of H_0), d_L is the luminosity distance, and σ_{int} is the intrinsic dispersion of SN absolute magnitudes. The minimization is performed with respect to θ, α, β and M. The parameters α, β and M are treated as nuisance parameters in the cosmological fit. Since d_L scales as $1/H_0$, only M depends on H_0.

The Hubble diagram of SNLS SNe and nearby data is shown in Figure 3, together with the best fit Λ cosmology for a flat Universe. Table 1 summarizes the results of the cosmological fit. Figure 4 show the resulting confidence contours, together with the product of these confidence estimates with the probability distribution from baryon acoustic oscillations (BAO) measured in the SDSS (Eq. 4 in Ref. 25). Note that the constraints from BAO and SNe Ia are quite complementary.

3.4.2. *Systematic Uncertainties*

The systematic uncertainties are fully discussed in Astier et al, 2006.[6] The dominant systematic errors which affect the SNLS first year measurements come from the photometric calibration uncertainties, and the Malmquist bias (detection bias). Other systematic effects come from (1) imperfections

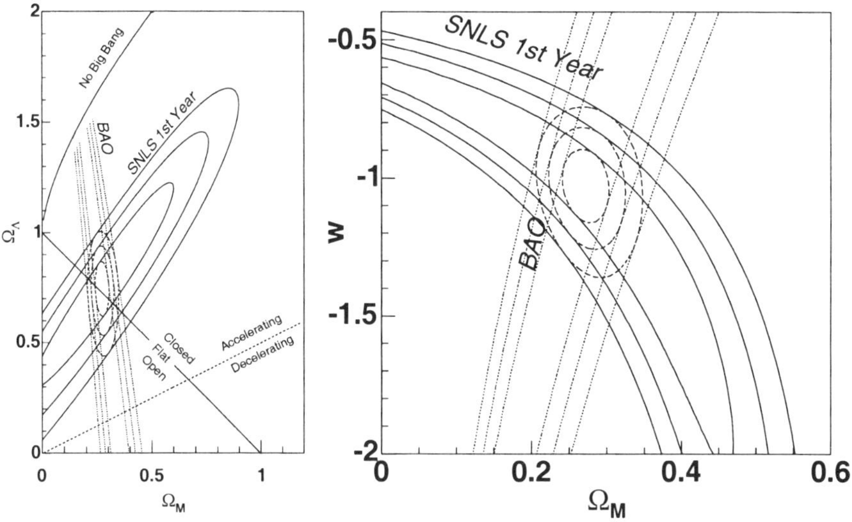

Fig. 4. Left: Contours at 68.3%, 95.5% and 99.7% confidence levels for the fit to an (Ω_M, ol) cosmology from the SNLS Hubble diagram (solid contours), the SDSS baryon acoustic oscillations (dotted lines), and the joint confidence contours (dashed lines). Right: Same contours to a flat (Ω_M, w) cosmology.

in the modeling of the supernova lightcurve and spectra (2) contamination of the Hubble diagram by non-SN Ia (3) gravitational lensing by mass inhomogeneities which deamplifies most supernovae and amplify only a few SNe and (4) gray dust. The measurement and isolation of systematic errors is a major goal of the SNLS. Some of these uncertainties will decrease as more data is acquired and future papers will examine a wider range of issues.

Table 2 summarizes the uncertainties affecting the cosmological parameter measurements. The table includes the impact of uncertainties in several parameter directions: the Ω_M direction for a flat $(\Omega_M, \Omega_\Lambda)$ (i.e. $w = -1$) cosmology, the Ω_{tot} direction for a general $(\Omega_M, \Omega_\Lambda)$ cosmology, and the w direction at fixed Ω_M for a (Ω_M, w) cosmology. Also reported here are the observed shifts when the BAO prior is applied to a flat (Ω_M, w) cosmology.

Combining tables 1 and 2, the following results was obtained:

$$\Omega_M = 0.263 \pm 0.042\ (stat) \pm 0.032\ (syst)$$

for a flat Λ cosmology, and

Table 2

Source	$\sigma(\Omega_M)$ (flat)	$\sigma(\Omega_{tot})$	$\sigma(w)$	$\sigma(\Omega_M)$ (with BAO)	$\sigma(w)$
Zero-points	0.024	0.51	0.05	0.004	0.040
Vega spectrum	0.012	0.02	0.03	0.003	0.024
Filter bandpasses	0.007	0.01	0.02	0.002	0.013
Malmquist bias	0.016	0.22	0.03	0.004	0.025
Sum (sys)	0.032	0.55	0.07	0.007	0.054
Meas. errors	0.037	0.52	0.09	0.020	0.087
U-B color(stat)	0.020	0.10	0.05	0.003	0.021
Sum (stat)	0.042	0.53	0.10	0.021	0.090

$$\begin{aligned} \Omega_M &= 0.271 \pm 0.021\ (stat) \pm 0.007\ (syst) \\ w &= -1.023 \pm 0.090\ (stat) \pm 0.054\ (syst) \\ w &< -0.85\ \ (95\%\ \ CL) \end{aligned}$$

for a flat cosmology with constant equation of state, when the constraints are combined with the BAO SDSS results. Assuming $w > -1$ brings the upper limit to -0.83 (at 95% CL). As visible on figure 4, supernovae alone give a marginal constraint: $w < -0.5$ at 95% CL.

4. Discussion

The determination of cosmological parameters from Type Ia supernovae has turned out to be a very powerful method. Today's best constraints are derived for a constant w from combining supernova with constraints derived either from BAO measurements or CMB data or both. Current precisions on w are in the 10% range, with a value close to the cosmological constant prediction of $w = -1$. Only marginal constraints can been obtained on a varying w.

The "rolling search" observation strategy has proven to be very efficient and successful. As of September 2006, about 300 identified supernovae have been detected and followed-up by SNLS, and the goal of 500 spectroscopically identified SNe Ia is within reach. The current statistical error budget is expected to improve by a factor two as shown in Table 3 where the expected precisions on $\sigma(\Omega_M)$ and $\sigma(w)$ were calculated for various scenarios. Assuming for example an improved constraint on BAO by a factor 2, 500 high-redshift supernovae and 3 times more nearby SNe will result in an improved statistical constraints on w by a factor 2 to δw(stat.)=0.044.

Table 3

Sample	1-yr	3-yr	5 yr
Nearby SNe	44	44	132
Distant SNe	71	213	500
$\sigma(\Omega_M)$ (BAO)	0.023	0.019	0.018
$\sigma(w)$ (BAO)	0.088	0.064	0.055
$\sigma(\Omega_M)$ (BAOx2)	0.016	0.014	0.013
$\sigma(w)$ (BAOx2)	0.081	0.054	0.044

More importantly, the SNLS dataset will be large and redundant enough to carefully test the various sources of possible systematic errors.

Future wide field imaging projects, whether in space or from the ground, will provide large samples of extremely carefully measured supernovae that will allow one to improve further on the measurement of cosmological parameters. These projects will make it possible to put significant constraints on a varying w, therefore testing dynamical-Λ models and improving our understanding of Dark Energy.

References

1. A. G. Riess et al, *Astronomical Journal* **116**, 1009 (1998)
2. S. Perlmutter et al, *Astrophysical Journal* **116**, 1009 (1998)
3. D. N. Spergel et al (WMAP Collaboration), Submitted to *Astrophysical Journal*
4. E. Semboloni et al, *Astronomy and Astrophysics* **452**, 51 (2006)
5. E. J. Copeland et al, `hep-th/0603057`, to be submitted to *Journal of Modern Physics D.*
6. P. Astier et al, *Astronomy and Astrophysics* **447**, 31 (2006)
7. R. A. Knop et al, *Astrophysical Journal* **598**, 102 (2003)
8. J. L. Tonry et al, *Astrophysical Journal* **594**, 1 (2003)
9. B. J. Barris et al, *Astrophysical Journal* **602**, 571 (2004)
10. M. Sullivan et al, *Monthly Notices of the Royal Atronomical Society* **340**, 1057 (2003)
11. A. G. Riess et al, *Astrophysical Journal* **607**, 665 (2004)
12. L. Strolger et al, *ASP Conf. Ser. 342*
13. `http://panisse.lbl.gov/ACSclustersearch`
14. `http://pan-starrs.ifa.hawaii.edu/public`
15. `http://www.lsst.org`
16. `http://snap.lbl.gov`
17. M. J. Sholl et al *Proceedings of SPIE* , 1473 (2004)
18. M. Sullivan et al, *Astronomical Journal* **131**, 960 (2006)
19. `http://cfht.hawaii.edu/Science/CFHTLS`

20. O. Boulade et al, *Proceedings of SPIE* **4081**, 72 (2003)
21. E. Magnier *Publications of the Astronomical Society of the Pacific* **116**, 449 (2004)
22. A. U. Landolt *Astronomical Journal* **104**, 340 (1992)
23. J. Guy *Astrophysical Journal* **443**, 781 (2005)
24. S. Perlmutter *Astrophysical Journal* **483**, 565 (1997)
25. D. J. Eisenstein *Astrophysical Journal* **633**, 560 (2005)
26. D. S. Hayes *IAU. Symp. Calibration of Fundamental Stellar Quantities* **111**, 225 (1985)

RELAXATION DARK ENERGY IN NON-CRITICAL STRING COSMOLOGIES AND ASTROPHYSICAL DATA

NIKOLAOS E. MAVROMATOS*

Physics Department, Theoretical Physics, King's College London, Strand, London WC2R 2LS, UK
E-mail: nikolaos.mavromatos@kcl.ac.uk

VASILIKI A. MITSOU

Instituto de Física Corpuscular (IFIC), CSIC – Universitat de València, Edificio Institutos de Paterna, P.O. Box 22085, E-46071 Valencia, Spain
and
CERN, Physics Department, CH-1211 Geneva 23, Switzerland
E-mail: vasiliki.mitsou@cern.ch

In this talk we review briefly the basic features of non-critical (dissipative) String Cosmologies, and we confront some of these models with supernova data. We pay particular attention to the off-shell and dilaton contributions to the dynamical evolution equations of the non-critical string Universe, as well as the Boltzmann equation for species abundances. The latter could have important consequences for the modification of astrophysical constraints on physically appealing particle physics models, such as supersymmetry. The data fits show that non-critical string cosmologies may be viable alternatives to ΛCDM model.

Keywords: Relaxation Dark Energy; Non-Critical String Cosmology; Astrophysical Data.

1. Introduction

There is a plethora of astrophysical evidence today, from supernovae measurements,[1,2] the spectrum of fluctuations in the cosmic microwave background,[3] baryon oscillations[4] and other cosmological data, indicating that the expansion of the Universe is currently accelerating. The energy budget of the Universe seems to be dominated (by more than 70%) at the present epoch by a mysterious dark energy component. Many theoretical

*Conference Speaker

models provide possible explanations for the dark energy, ranging from a cosmological constant[5] to super-horizon perturbations[6] and time-varying quintessence scenarios,[7] in which the dark energy is due to a smoothly varying (scalar) field which dominates cosmology in the present era. In the context of string theory,[8,9] the most successful and mathematically complete theory of quantum gravity available to date, such a time-dependent "quintessence" field is provided by the scalar dilaton field of the gravitational string multiple.[10–12] The current astrophysical data are capable of placing severe constraints on the nature of the dark energy, whose equation of state may be determined by means of an appropriate global fit. Most of the analyses so far are based on effective four-dimensional Robertson-Walker Universes, which satisfy on-shell dynamical equations of motion of the Einstein-Friedman form. Even in modern approaches to brane cosmology, which are described by equations that deviate during early eras of the Universe from the standard Friedman equation (which is linear in the energy density), the underlying dynamics is assumed to be of classical equilibrium (on-shell) nature, in the sense that it satisfies a set of equations of motion derived from the appropriate minimization of an effective space-time Lagrangian.

However, cosmology may not be an entirely classical equilibrium situation. The initial Big Bang or other catastrophic cosmic event, such as a collision of two brane worlds in the modern approach to strings,[9] which led to the initial rapid expansion of the Universe, may have caused a significant departure from classical equilibrium dynamics in the early Universe, whose signatures may still be present at later epochs including the present era. One specific model for the cosmological dark energy which is of this type, being associated with a rolling dilaton field that is a remnant of this non-equilibrium phase, was formulated[12–14] in the framework of non-critical string theory.[10,15,16] We call this scenario 'Q-cosmology'. It is of outmost importance to confront the currently available precision astrophysical data with such non equilibrium stringy cosmologies. The central purpose of this talk is to present a first step towards this direction, namely a confrontation of cosmological data on high-redshift supernovae[17] with Q-cosmologies and compare the results with the predictions of the conventional ΛCDM model[5] and the super-horizon model.[6] Care must be taken in interpreting the Q-cosmology scenario. Since such a non-equilibrium, non-classical theory *is not described by the equations of motion derived by extremising an effective space-time Lagrangian*, one must use a more general formalism to make predictions that can be confronted with the current data.

2. Dissipative Q-Cosmology Basics

2.1. *Off-shell (non-critical-string) terms and dissipative cosmological equations*

Q-cosmology are stringy cosmological models, based on the (perturbative) formulation of σ-models propagating mostly on dilaton and graviton backgrounds, which, however, are not world-sheet conformal invariant. The deviation from conformal invariance may be due to a number of reasons, for instance cosmically catastrophic events at an early stage of the Universe history, such as Big Bang, the collision of two (or more) brane worlds *etc.* An important formal ingredient of this approach is the identification[16] of target time with the zero mode ρ of the Liouville field,[15] which is an extra world-sheet field, introduced in order to restore the conformal invariance of the world-sheet theory. The dynamics of this latter identification is encoded in the solution of the generalized conformal invariance conditions, after Liouville dressing, which read in the σ-model frame:[15,16]

$$-\tilde{\beta}^i = g^{i''} + Qg^{i'}, \tag{1}$$

where the prime denotes differentiation with respect to ρ, and the overall minus sign on the left-hand side pertains to supercritical strings,[10] with a timelike signature of the Liouville mode, for which the central charge deficit $Q^2 > 0$ by convention. In physical terms, Q depends on the microscopic details of the specific non-equilibrium string theory model at hand. For instance, in the case of adiabatic brane world collisions with bouncing (recoiling) worlds,[12,18,19] $Q^2 \propto v^4$, where $v \ll 1$ is the (relative) recoil velocity of the brane worlds after the collision. Notice the dissipation, proportional to the (square root) of the central charge deficit Q, on the right-hand side of Eq. (1), which heralds the adjective *Dissipative* to the associate non-critical-string-inspired Cosmological model. Moreover, the Weyl anomaly coefficients $\tilde{\beta}^i$, $i = \{\Phi, g_{\mu\nu}\}$, whose vanishing would guarantee local conformal invariance of the string-cosmology background, are associated with off-shell variations of a low-energy effective string-inspired action, $\mathcal{S}[g^j]$:

$$\frac{\delta \mathcal{S}[g]}{\delta g^i} = \mathcal{G}_{ij}\tilde{\beta}^j, \qquad \mathcal{G}_{ij} = \mathrm{Lim}_{z\to 0} z^2\bar{z}^2 \langle V_i(z,\bar{z})V_j(0,0)\rangle, \tag{2}$$

with $z, \bar{z}$ (complex) world-sheet coordinates, $\mathcal{G}_{ij}$ the Zamolodchikov metric in theory space of strings,[8,16] and V_i the σ-model vertex operators associated with the σ-model background field g^i. It is this off-shell relation that characterises the entire non-critical (Q) Cosmology framework, associated physically with a non-equilibrium situation as a result of an initial cosmically catastrophic event, at the beginning of the (irreversible) Liouville/cosmic-time flow.

The detailed dynamics of Eq. (1) are encoded in the solution for the scale factor $a(t)$ and the dilaton $\Phi(t)$ in the simplified model considered in Ref. 14, after the identification of the Liouville mode with the target time. In fact, upon the inclusion of matter backgrounds, the associated equations, after compactification to four target-space dimensions, read in the so-called Einstein frame (i.e., an appropriate redefinition of the σ-model graviton, such that the graviton-dilaton effective action has a canonically normalised Einstein curvature scalar term[10]):

$$
\begin{aligned}
&3H^2 - \tilde{\varrho}_m - \varrho_\Phi = \frac{e^{2\Phi}}{2}\tilde{\mathcal{G}}_\Phi, \quad 2\dot{H} + \tilde{\varrho}_m + \varrho_\Phi + \tilde{p}_m + p_\Phi = \frac{\tilde{\mathcal{G}}_{ii}}{a^2}, \\
&\ddot{\Phi} + 3H\dot{\Phi} + \frac{1}{4}\frac{\partial \hat{V}_{all}}{\partial \Phi} + \frac{1}{2}\left(\tilde{\varrho}_m - 3\tilde{p}_m\right) = -\frac{3}{2}\frac{\tilde{\mathcal{G}}_{ii}}{a^2} - \frac{e^{2\Phi}}{2}\tilde{\mathcal{G}}_\Phi, \\
&\tilde{\mathcal{G}}_\Phi = e^{-2\Phi}\left(\ddot{\Phi} - \dot{\Phi}^2 + Qe^\Phi\dot{\Phi}\right), \\
&\tilde{\mathcal{G}}_{ii} = 2a^2(\ddot{\Phi} + 3H\dot{\Phi} + \dot{\Phi}^2 + (1-q)H^2 + Qe^\Phi(\dot{\Phi} + H))\ ,
\end{aligned}
\tag{3}
$$

where $\tilde{\varrho}_m$ ($\tilde{p}_m$) denotes the matter energy density (pressure), including dark matter contributions, and ϱ_Φ (p_Φ) are the corresponding quantities for the dilaton dark-energy fluid. The overdots in the above equations denote derivatives with respect to the Einstein time, which is related to the Robertson-Walker cosmic time t_{RW} by $t = \omega\ t_{RW}$. Without loss of generality we have taken $\omega = \sqrt{3}H_0$ where H_0 is the present-day Hubble parameter. With this choice for ω the densities appearing in Eqs. (3) are given in units of the critical density. In the above equations, $H = \dot{a}/a$ is the Hubble parameter and q is the deceleration parameter of the Universe $q \equiv -\ddot{a}a/\dot{a}^2$, and are both functions of (Einstein frame) cosmic time. The potential $\hat{V}_{all}$ appearing above is defined by $\hat{V}_{all} = 2Q^2\exp(2\Phi) + V$ where, in order to cover more general cases, we have also allowed for a (e.g., string-loop-induced) potential term in the four-dimensional action, $-\int d^4y\sqrt{-G}V$. We have also assumed a (spatially) flat Universe. The dilaton energy density and pressure are given in this class of models by: $\varrho_\Phi = \frac{1}{2}(2\dot{\Phi}^2 + \hat{V}_{all})$, $p_\Phi = \frac{1}{2}(2\dot{\Phi}^2 - \hat{V}_{all})$. The dependence of the central charge deficit $Q(t)$ on the cosmic time stems from the running of the latter with the world-sheet RG scale,[16,19] and is provided by the Curci - Paffuti equation[20] expressing the renormalisability of the world-sheet theory. To leading order in an α' expansion, which we restrict ourselves here, this equation in the Einstein frame reads:[14] $\frac{d\tilde{\mathcal{G}}_\Phi}{dt_E} = -6e^{-2\Phi}(H + \dot{\Phi})\frac{\tilde{\mathcal{G}}_{ii}}{a^2}$. For completeness we also state here the continuity equation of the matter stress tensor, which is not an independent equation, but can be obtained from Eq. (3) by appropriate algebraic manipulations: $\frac{d\tilde{\varrho}_m}{dt_E} + 3H(\tilde{\varrho}_m + \tilde{p}_m) + \frac{\dot{Q}}{2}\frac{\partial \hat{V}_{all}}{\partial Q} - \dot{\Phi}(\tilde{\varrho}_m - 3\tilde{p}_m) = 6(H + \dot{\Phi})\frac{\tilde{\mathcal{G}}_{ii}}{a^2}$.

A consistent numerical solution of $a(t)$, $\Phi(t)$, and the various densities, including back reaction of matter onto the space-time geometry, has been discussed in Ref. 14, where we refer the interested reader for further study. We note the existence of exotic matter scaling in this approach today, which no longer scales as dust, but it includes different scaling components, e.g. a^δ, with δ close to 4. We quote below the final result of our parametrisation for $H(z)$ in the Q-cosmology framework, at late eras, such as the ones pertinent to the supernova and other data ($0 < z < 2$), where some analytic approximations are allowed:[17]

$$\frac{H(z)}{H_0} = \sqrt{\Omega_3(1+z)^3 + \Omega_\delta(1+z)^\delta + \Omega_2(1+z)^2}\ , \quad \Omega_3 + \Omega_\delta + \Omega_2 = 1\ , \tag{4}$$

with the densities $\Omega_{2,3,\delta}$ corresponding to present-day values ($z = 0$). We stress, however, that a complete analysis of the non-critical and dilaton effects, which turn out to be important in the present era after the inclusion of matter, requires numerical treatment.[14] In view of this, the exponent δ is treated from now on as a fitting parameter, and in fact it *may even be* z*-dependent* to cover more general cases, and get agreement with the numerical treatment of Ref. 14. It is important to note that the various Ω_i contain contributions from *both* dark energy and matter energy densities. As explained in detail in Ref. 17, Ω_3 does not merely represent ordinary matter effects, but also receives contributions from the dilaton dark energy. In fact, the sign of Ω_3 depends on details of the underlying theory, and it could even be *negative.* For instance, Kaluza-Klein graviton modes in certain brane-inspired models[21] yield negative dust contributions. In a similar vein, the exotic contributions scaling as $(1+z)^\delta$ are affected by the off-shell Liouville terms of Q-cosmology. It is because of the similar scaling behaviours of dark matter and dilaton dark energy that we reverted to the notation Ω_i, $i = 2, 3, \delta$, in Eq. (4). To disentangle the ordinary matter and dilaton contributions one may have to resort to further studies on the equation of state of the various components, which we do not study here or in.[17] More generally, one could have included a cosmological constant Ω_Λ contribution in Eq. (4), which may be induced in certain brane-world inspired models. We do not do so in this work, as our primary interest is to fit Q-cosmology models,[12–14] which are characterised by dark energy densities that relax to zero.

2.2. *Dilaton and off-shell modifications to cosmic evolution of species abundances*

The above formalism can be used to derive the dilaton and non-critical-string effects on Boltzmann-type evolution equations[22] of a number of densities for certain particle physics species, playing the rôle of dominant dark matter candidates. In the presence of (time-dependent) dilaton and off-shell graviton fields, that couple to matter, the phase-space density of species depends on sources, $f(|\vec{p}|, t, \Phi(t,\rho), g_{\mu\nu}(t,\rho))$, where p is the momentum, and ρ is the Liouville mode, which eventually is going to be identified with a function of the target time t,[16] whose form is dictated by both conformal-field-theory and target-space dynamical considerations, such as the minimisation of the effective potential.[18] A detailed analysis[23] shows that the Boltzmann equation for the density of species X, with mass m_X, assumed to be a dominant dark matter candidate, $n = \int d^3p\, f$, can be written in a compact form that represents collectively the dilaton-dissipative-source and non-critical-string contributions as external-source $\Gamma(t)n$ terms:

$$\frac{dn}{dt} + 3\frac{\dot{a}}{a}n = \Gamma(t)n - \langle v\sigma\rangle\left(n^2 - n_{eq}^2\right), \quad \Gamma(t) \equiv \dot{\Phi} + \frac{1}{2}\eta e^{-\Phi} g^{\mu\nu}\tilde{\beta}^{\rm Grav}_{\mu\nu}, \tag{5}$$

where the last term on the right-hand-side of the Boltzman equation is the collision term, expressed in terms of the thermal average of the cross section σ times the Möller velocity v of the annihilated particles;[22] η is a world-sheet renormalization-scheme dependent parameter;[16,23] for our purposes here we work in the physical scheme $\eta = -1$. To find an explicit expression for $\Gamma(t)$ in our case requires the full solution of the Q-cosmological equations.[14] Depending on the sign of $\Gamma(t)$ one has different effects on the relic abundance of the species X with density n, which we now proceed to analyse briefly. Before the decoupling ('freeze out') time t_f, $t < t_f$, equilibrium is maintained and thus $n = n_{eq}$ for such an era. However, it is crucial to observe that, as a result of the presence of the source Γ terms, n_{eq} no longer scales with the inverse of the cubic power of the expansion radius a, which was the case in conventional (on-shell) cosmological models. Under some plausible assumptions, for instance that the entropy remains approximately constant after the freeze out temperature, and that one can define a temperature T for the ordinary matter degrees of freedom, but *not* for the ones pertaining to (dilatonic) dark energy, we can solve the Boltzmann equation and arrive at the dilaton and non-critical string corrections to the freeze out temperature of, say, the lightest supersymmetric particle, $\tilde{\chi}$, assumed to be the dominant dark matter candidate (we define $x \equiv T/m_{\tilde{\chi}}$):

$$x_f^{-1} = \ln\left[\left(0.03824 g_s \frac{M_{Planck}\; m_{\tilde{\chi}}}{\sqrt{g_*}} x_f^{1/2} \langle v\sigma\rangle_f\right)\left(\frac{g_*}{\tilde{g}_*}\right)^{1/2}\right] + \int_{x_f}^{x_{in}} \frac{\Gamma dx}{Hx}, \quad (6)$$

whereby $\tilde{g}_{eff}$ is simply defined by[23] $\varrho + \Delta\varrho \equiv \frac{\pi^2}{30}T^4\tilde{g}_{eff}$. $\Delta\varrho$ incorporates the effects of the additional contributions due to the non-critical (off-shell) terms and the dilaton dissipative source, which are not accounted for by the ordinary matter degrees of freedom g_{eff} of conventional Cosmology.[22] Recalling that only the degrees of freedom involved in ordinary matter energy density ρ are thermal,[22] $\rho = \frac{\pi^2}{30}T^4 g_{eff}(T)$, whilst from the dynamical equations of Q-cosmology $H^2 = \frac{8\pi G_N}{3}(\rho + \Delta\rho)$, we obtain[23] $\tilde{g}_{eff} = g_{eff} + \frac{30}{\pi^2}T^{-4}\Delta\rho$. We can also assume that the $\tilde{\chi}$'s decoupled before neutrinos. Then, for the relic abundance, we derive the following approximate result[23]

$$\Omega_{\tilde{\chi}} h_0^2 = \left(\Omega_{\tilde{\chi}} h_0^2\right)_{no-source} \times \left(\frac{\tilde{g}_*}{g_*}\right)^{1/2} \exp\left(\int_{x_0}^{x_f} \frac{\Gamma dx}{Hx}\right), \quad (7)$$

with $\left(\Omega_{\tilde{\chi}} h_0^2\right)_{no-source} = \frac{1.066\times 10^9 \mathrm{GeV}^{-1}}{M_{Planck}\sqrt{g_*}J}$, $J \equiv \int_{x_0}^{x_f} \langle v\sigma\rangle\, dx$, and x_f determined by Eq. (6). The merit of casting the relic density in such a form is that it clearly exhibits the effect of the presence of the source. It is immediately seen from Eq. (7) that, depending on the sign of the source Γ, one may have increase or reduction of the relic density as compared with the corresponding value in the absence of the source. In this way, the constraints imposed on supersymmetric extensions of the Standard Model from astrophysical data,[24] such as those from WMAP-satellite measurements,[3] *etc.*, based on the available bounds of, say, cold dark matter relic densities, need to be revisited. This is left for the future.

3. Confronting Q-Cosmologies with Supernova Data

We use the type-Ia supernovae (SN) data reported in Refs. 1,2, which are given in terms of the distance modulus $\mu = 5\log d_L + 25$, where the luminosity distance d_L (in megaparsecs) is related to the redshift z via the Hubble rate H:

$$d_L = c(1+z)\int_0^z \frac{dz'}{H(z')}. \quad (8)$$

We note that this observable depends on the expansion history of the Universe from z to the present epoch, and recall that most of the available supernovae have $z < 1$, although there is a handful with larger values of z.

In the analysis that follows, the predictions of the following three cosmological models are investigated:

ΛCDM model assuming a flat universe, where the Hubble rate is

$$H(z) = H_0 \left(\Omega_{\rm M}(1+z)^3 + \Omega_\Lambda (1+z)^{3(1+w_0)}\right)^{1/2} . \tag{9}$$

Super-horizon model in which the Hubble parameter is given by[6]

$$H(z) = \overline{a}^{-1}\frac{d\overline{a}(t)}{dt} = \frac{\overline{H}_0}{1-\Psi_{\ell 0}}\left(a^{-3/2} - a^{-1/2}\Psi_{\ell 0}\right), \tag{10}$$

where $(1+z)^{-1} = \overline{a}(t)$ and $\Psi_{\ell 0}$ is a free parameter.

Q-cosmology discussed in the previous sections, with H given by Eq. (4) and three parameters to be determined by the fit: Ω_3, Ω_δ and δ.

Measurements are available of the distance moduli of 157 supernovae in a so-called 'gold' sample,[1] observed by ground-based facilities and the Hubble Space Telescope (HST). In addition, a so-called 'silver' dataset of 29 SN 1a is also available, of a slightly lesser spectrometric and photometric quality. The analysis was also repeated including the 'silver' supernovae data — with a total of 186 SN— with results comparable to those with the 'gold' dataset, proving the robustness of the analysis. Furthermore, we analysed the measurements of 71 other high-redshift supernovae by the Supernova Legacy Survey (SNLS),[2] which are accompanied by a reference sample of 44 nearby SN Ia, yielding a total of 115 data points. Since the fits to the two datasets are quite compatible,[17] a combined analysis was finally performed.

For illustration purposes, both data and predictions of cosmological models are expressed in the following as residuals, $\Delta\mu$, from the empty-Universe ($\Omega_{\rm M} = 0$) prediction. The combined sample of the 'gold' (157 SN) plus the SNLS sample (71 SN), yielding a sample of 228 supernovae in total, is shown in Fig. 1, where the predictions of the cosmological models under study are also displayed for the best-fit parameter values.

The analysis involves minimisation of the standard χ^2 function with respect to the cosmological model parameters. The best-fit parameter values acquired by combining the 'gold' sample and the SNLS supernovae datasets, the 1σ errors and the corresponding χ^2 values are listed in Table 1 for the three cosmological models.

It is evident from Fig. 1 and Table 1 that the standard ΛCDM model fits the supernova data *very well*, as expected from earlier analyses. The super-horizon dark matter model also fits the supernova data *quite well.* Both of these models are *on-shell*, i.e., they satisfy the pertinent Einstein's

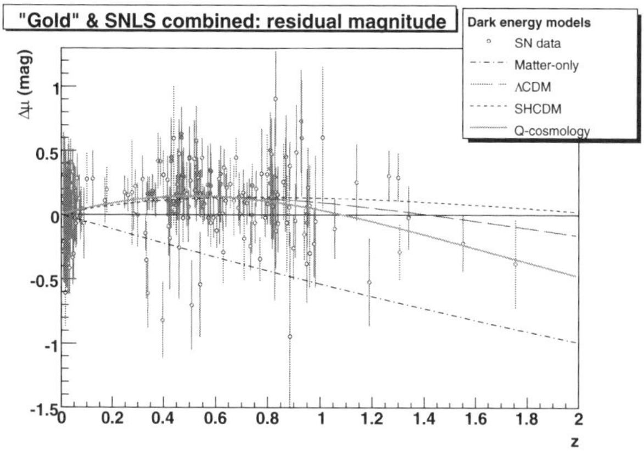

Fig. 1. Residual magnitude versus redshift for supernovae from the 'gold' and the SNLS datasets. Predictions of cosmological models for the best-fit parameter values are superimposed: (i) Empty Universe; (ii) Universe with matter only, $\Omega_M = 1$; (iii) ΛCDM model; (iv) super-horizon model; and (v) off-shell Q-cosmology model.

Table 1. Parameter values favoured by the 'gold' + SNLS combined data for various cosmological models.

Models	Best-fit parameters	χ^2	χ^2/dof
ΛCDM	$\Omega_M = 0.274 \pm 0.017$	239	1.05
Super-horizon	$\Psi_{\ell 0} = -0.87 \pm 0.06$	245	1.09
	$\Omega_3 = -3.7 \pm 1.1$		
Q-cosmology	$\Omega_\delta = 1.3 \pm 0.7$	237	1.05
	$\delta = 3.9 \pm 0.3$		

equations. Moreover, *off-shell* cosmology models are also compatible with the data. As we discussed above, off-shell effects are important in our Q-cosmology model. Introducing the appropriate parametrisation (4) to allow for these off-shell effects, we find that *the Q-cosmology model may fit the supernovae data as well as the standard ΛCDM model.*

In Fig. 2, the Hubble parameter predictions of the various models are compared with the data available from high-z red galaxies[25,26] and from sets of type Ia SN.[27] The bands correspond to the 68% confidence intervals deduced by the aforementioned analysis of 228 supernovae (cf. Table 1). The SN data constrain H only up to $z \simeq 1.3$ with a 50% uncertainty, in contrast to the red galaxies data, which provide stringent measurement of H until $z \simeq 1.8$. The ΛCDM model is compatible with the data, however the super-horizon model deviates from both galaxies and supernovae data.

Q-cosmology, on the other hand, predict higher values of H than the galaxies data measurements, though fully compatible with the SN data, as expected. Nevertheless, this discrepancy is hoped to be remedied by repeating the analysis of Ref. 17 with the inclusion of a sample of 21 type-Ia supernovae —including 13 with $z > 1$— recently released by HST.[27] Moreover, as remarked earlier, a z-dependent-δ parametrisation of $H(z)$ (4) may be necessary, as becomes clear from the highly non-linear form of the numerical solution of some Q-cosmology models presented in Ref. 14. In fact, fitting the numerical solution itself with the data is probably the most complete treatment. We hope to tackle such issues in the future.

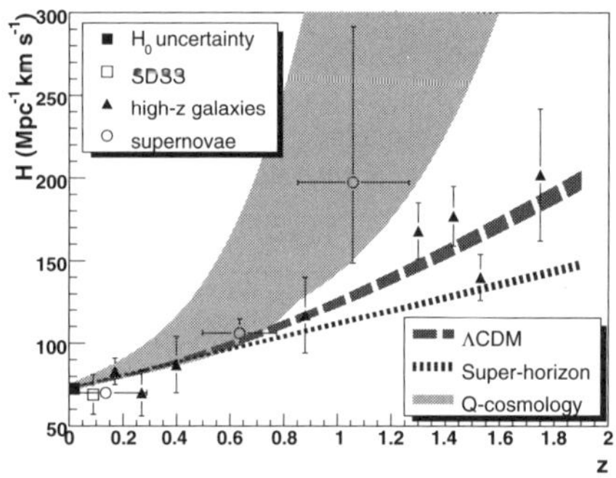

Fig. 2. The Hubble parameter redshift relation for various cosmological models and observational data. The bands represent 68% confidence intervals derived by the SN analysis of Ref. 17 for the standard ΛCDM, the super-horizon and the Q-cosmology models. The black rectangle shows the WMAP3[3] estimate for H_0, the squares show the measurements from SDSS galaxies,[25] the triangles result from high-z red galaxies,[26] and the circles correspond to a combined analysis[27] of supernovae data from Refs. 1 and 2.

For standard on-shell cosmologies, complementary information is provided by the data on baryon acoustic oscillations,[4] which show up in the galaxy-galaxy correlation function at $z \sim 0.35$. However, for reasons discussed in Ref. 17, the application of such an analysis to the off-shell Q-cosmology model is an open issue, since the underlying theoretical framework needs to be re-evaluated.

4. Conclusions

In this talk, we briefly reviewed the main predictions of Q-cosmologies, including a brief discussion about the time-dependent-dilaton and non-critical-string effects on the Boltzmann equation for species abundances,[23] which could lead to important modifications of the constraints imposed by astro-physical data on interesting particle physics models, such as supersymmetric extensions of the standard model. Then, we discussed some initial steps towards demonstrating[17] that the available supernova data are compatible with such non-critical-string-based cosmologies,[12,14] thus implying the possibility that such models may be viable alternatives to the Standard ΛCDM model. As more precision astrophysical data are coming into play, more stringent constraints can be imposed on our non-critical string Q-cosmologies.

Acknowledgements

We thank G. Diamandis, J. Ellis, B. Georgalas, A. Lahanas and D. Nanopoulos for the enjoyable collaboration and discussions. The work of N.E.M. was supported in part by the European Union through the Marie Curie Research and Training Network *UniverseNet* (MRTN-CT-2006-035863).

References

1. A. G. Riess *et al.* [Supernova Search Team Collaboration], Astrophys. J. **607**, 665 (2004) [arXiv:astro-ph/0402512].
2. P. Astier *et al.*, Astron. Astrophys. **447** (2006) 31 [arXiv:astro-ph/0510447].
3. D. N. Spergel *et al.*, arXiv:astro-ph/0603449.
4. D. J. Eisenstein *et al.*, Astrophys. J. **633** (2005) 560 [arXiv:astro-ph/0501171]; S. Cole *et al.* [The 2dFGRS Collaboration], Mon. Not. Roy. Astron. Soc. **362**, 505 (2005) [arXiv:astro-ph/0501174].
5. N. A. Bahcall, J. P. Ostriker, S. Perlmutter and P. J. Steinhardt, Science **284**, 1481 (1999) [arXiv:astro-ph/9906463].
6. E. W. Kolb, S. Matarrese, A. Notari and A. Riotto, arXiv:hep-th/0503117; E. W. Kolb, S. Matarrese and A. Riotto, arXiv:astro-ph/0506534.
7. A. Upadhye, M. Ishak and P. J. Steinhardt, Phys. Rev. D **72** (2005) 063501 [arXiv:astro-ph/0411803].
8. M.B. Green, J.H. Schwarz and E. Witten, *Superstring Theory*, Vols. I & II (Cambridge University Press, 1987).
9. J. Polchinski, *String theory*, Vols. I & II (Cambridge University Press, 1998).
10. I. Antoniadis, C. Bachas, J. R. Ellis and D. V. Nanopoulos, Phys. Lett. B **211**, 393 (1988); Nucl. Phys. B **328**, 117 (1989); Phys. Lett. B **257**, 278 (1991).
11. M. Gasperini, F. Piazza and G. Veneziano, Phys. Rev. D **65**, 023508 (2002) [arXiv:gr-qc/0108016].
12. J. R. Ellis, N. E. Mavromatos and D. V. Nanopoulos, Phys. Lett. B **619**, 17 (2005) [arXiv:hep-th/0412240]; J. R. Ellis, N. E. Mavromatos,

D. V. Nanopoulos and M. Westmuckett, Int. J. Mod. Phys. A **21** (2006) 1379 [arXiv:gr-qc/0508105].

13. G. A. Diamandis, B. C. Georgalas, N. E. Mavromatos and E. Papantonopoulos, Int. J. Mod. Phys. A **17**, 4567 (2002) [arXiv:hep-th/0203241]; G. A. Diamandis, B. C. Georgalas, N. E. Mavromatos, E. Papantonopoulos and I. Pappa, Int. J. Mod. Phys. A **17**, 2241 (2002) [arXiv:hep-th/0107124].
14. G. A. Diamandis, B. C. Georgalas, A. B. Lahanas, N. E. Mavromatos and D. V. Nanopoulos, Phys. Lett. B **642**, 179 (2006) [arXiv:hep-th/0605181].
15. F. David, Mod. Phys. Lett. A **3**, 1651 (1988); J. Distler and H. Kawai, Nucl. Phys. B **321**, 509 (1989); J. Distler, Z. Hlousek and H. Kawai, Int. J. Mod. Phys. A **5**, 391 (1990); see also: N. E. Mavromatos and J. L. Miramontes, Mod. Phys. Lett. A **4**, 1847 (1989); E. D'Hoker and P. S. Kurzepa, Mod. Phys. Lett. A **5**, 1411 (1990).
16. J. R. Ellis, N. E. Mavromatos and D. V. Nanopoulos, Phys. Lett. B **293**, 37 (1992) [arXiv:hep-th/9207103]; Mod. Phys. Lett. A **10**, 1685 (1995) [arXiv:hep-th/9503162]. *Invited review for the special Issue of J. Chaos Solitons Fractals*, Vol. 10, (eds. C. Castro amd M.S. El Naschie, Elsevier Science, Pergamon 1999) 345 [arXiv:hep-th/9805120];
17. J. R. Ellis, N. E. Mavromatos, V. A. Mitsou and D. V. Nanopoulos, arXiv:astro-ph/0604272, Astroparticle Physics in press.
18. E. Gravanis and N. E. Mavromatos, Phys. Lett. B **547**, 117 (2002) [arXiv:hep-th/0205298]; N. E. Mavromatos, arXiv:hep-ph/0309221.
19. J. R. Ellis, N. E. Mavromatos, D. V. Nanopoulos and A. Sakharov, New J. Phys. **6**, 171 (2004) [arXiv:gr-qc/0407089];
20. G. Curci and G. Paffuti, Nucl. Phys. B **286**, 399 (1987).
21. M. Minamitsuji, M. Sasaki and D. Langlois, Phys. Rev. D **71**, 084019 (2005) [arXiv:gr-qc/0501086].
22. E. W. Kolb and M. S. Turner, *The Early Universe,* Front. Phys. **69** (Redwood City, USA: Addison-Wesley 1990).
23. A. B. Lahanas, N. E. Mavromatos and D. V. Nanopoulos, arXiv:hep-ph/0608153.
24. A. B. Lahanas, D. V. Nanopoulos and V. C. Spanos, Phys. Lett. B **518**, 94 (2001) [arXiv:hep-ph/0107151]; J. R. Ellis, K. A. Olive, Y. Santoso and V. C. Spanos, Phys. Lett. B **565**, 176 (2003) [arXiv:hep-ph/0303043]; U. Chattopadhyay, A. Corsetti and P. Nath, Phys. Rev. D **68**, 035005 (2003) [arXiv:hep-ph/0303201]; for a review see: A. B. Lahanas, N. E. Mavromatos and D. V. Nanopoulos, Int. J. Mod. Phys. D **12**, 1529 (2003) [arXiv:hep-ph/0308251] and references therein.
25. R. Jimenez, L. Verde, T. Treu and D. Stern, Astrophys. J. **593** (2003) 622 [arXiv:astro-ph/0302560].
26. J. Simon, L. Verde and R. Jimenez, Phys. Rev. D **71** (2005) 123001 [arXiv:astro-ph/0412269].
27. A. G. Riess *et al.*, arXiv:astro-ph/0611572.

TRAPPED QUINTESSENTIAL INFLATION FROM FLUX COMPACTIFICATIONS

K. DIMOPOULOS*

Physics Department, Lancaster University,
Lancaster LA1 4YB, U.K.
**E-mail: k.dimopoulos1@lancaster.ac.uk*

Quintessential inflation is studied using a string modulus as the inflaton–quintessence field. It is assumed that the modulus crosses an enhanced symmetry point (ESP) in field space. Particle production at the ESP temporarily traps the modulus resulting in a period of inflation. After reheating, the modulus freezes due to cosmological friction at a large value, such that its scalar potential is dominated by contributions due to fluxes in the extra dimensions. The modulus remains frozen until the present, when it can become quintessence.

Keywords: String moduli, inflation, quintessence

1. Introduction

A plethora of observations concur that the Universe at present enters a phase of accelerated expansion. In fact, most cosmologists accept that over 70% of the Universe content at present corresponds to the elusive dark energy; a substance with pressure negative enough to cause the observed acceleration [1]. The simplest form of dark energy is a positive cosmological constant Λ, which however, needs to be incredibly fine-tuned to explain the observations [2]. This is why theorists have looked for alternatives, which could explain the observations while setting $\Lambda = 0$, as was originally assumed. A promising idea is to consider that the Universe at present is entering a late-time inflationary period [3]. The credibility of this option is supported also by the fact that the generic predictions of inflation in the early Universe are in excellent agreement with the observations. The scalar field responsible for this late-inflation period is called quintessence because it is the fifth element after baryons, photons, CDM and neutrinos [4].

Since they are based on the same idea, it is natural to attempt to unify early Universe inflation with quintessence. Quintessential inflation was thus born [5–8]. This attempt has many advantages. Firstly, quintessential inflation models allow the treatment of both inflation and quintessence within a single theoretical framework. Also, quintessential inflation dispenses with the tuning problem of the initial conditions for quintessence. Finally, unified models for inflation and quintessence are more economic because they avoid introducing yet another unobserved scalar field.

For quintessential inflation to work one needs a scalar field with a runaway potential, such that the minimum has not been reached until today and, therefore, there is residual potential density, which can cause the observed accelerated expansion. String moduli fields are suitable because they are typically characterised by such runaway potentials. The problem with such fields, however, is how to stabilise them temporarily, in order to use them as inflatons in the early Universe. In this work (see also Ref. [9]) we achieve this by considering that, during its early evolution our modulus crosses an enhanced symmetry point (ESP) in field space. When this occurs the modulus is trapped temporarily at the ESP [10], which leads to a period of inflation. After inflation the modulus picks up speed again in field space resulting into a period of kinetic density domination (kination) [11]. Kination ends when the thermal bath of the hot big bang (HBB) takes over. During the HBB, due to cosmological friction [12], the modulus freezes at some large value and remains there until the present, when its potential density dominates and drives the late-time accelerated expansion [8].

Is is evident that, in order for the modulus to become quintessence, it should not decay after the end of inflation. Reheating, therefore should be achieved by other means. We assume that the thermal bath of the HBB is due to the decay of some curvaton field [13] as suggested in Refs. [8,14]. By considering a curvaton we do not add an *ad hoc* degree of freedom, because the curvaton can be a realistic field, already present in simple extensions of the standard model (e.g. a right-handed sneutrino [15], a flat direction of the (N)MSSM [16] or a pseudo Nambu-Goldstone boson [17,18] possibly associated with the Peccei-Quinn symmetry [19]). Apart from reheating, the curvaton can provide the correct amplitude of curvature perturbations in the Universe. Consequently, the energy scale of inflation can be much lower than the grand unified scale [20]. In fact, in certain curvaton models, the Hubble scale during inflation can be as low as the electroweak scale [18,21].

2. The Runaway Scalar Potential

String theories contain a number of flat directions which are parametrised by the so-called moduli fields, which correspond to the size and shape of the compactified extra dimensions. Many such flat directions are lifted by non-perturbative effects, such as gaugino condensation or D-brane instantons [22]. The superpotential, then, is of the form

$$W = W_0 + W_{\rm np} \quad \text{with} \quad W_{\rm np} = Ae^{-cT}\,, \tag{1}$$

where $W_0 \approx$ const. is the tree level contribution from fluxes, A and c are constants and T is a Kähler modulus in units of m_P. Hence, the non-perturbative superpotential $W_{\rm np}$ results in a runaway scalar potential characteristic of string compactifications. For example, in type IIB compactifications with a single Kähler modulus, $\sigma \equiv {\rm Re}(T)$ is the so-called volume modulus, which parametrises the volume of the compactified space. In this case, the runaway behaviour leads to decompactification of the internal manifold. The tree level Kähler potential for a modulus, in units of m_P^2, is

$$K = -3\ln(T + \bar{T}) \equiv -3\ln(2\sigma)\,, \tag{2}$$

and the corresponding supergravity potential is[a]

$$V_{\rm np}(\sigma) \simeq \frac{cAe^{-c\sigma}}{2\sigma^2 m_P^2}\left(\frac{c\sigma}{3}Ae^{-c\sigma} - W_0\right)\,. \tag{3}$$

To study the cosmology, we turn to the canonically normalised modulus ϕ which, due to Eq. (2), is associated with σ as

$$\sigma(\phi) = \exp\left(\sqrt{\tfrac{2}{3}}\,\phi/m_P\right)\,. \tag{4}$$

Suppose that the Universe is initially dominated by the above modulus. The non-perturbative scalar potential in Eq. (3) is very steep (exponential of an exponential), which means that the field soon becomes dominated by its kinetic density. Once this is so, the particular form of the potential ceases to be of importance. To achieve inflation we assume that, while rolling, the modulus crosses an ESP and becomes temporarily trapped at it.

3. At the Enhanced Symmetry Point

In string compactifications there are distinguished points in moduli space at which there is enhancement of the gauge symmetries [23]. This results in some massive states of the theory becoming massless at these points.

Even though from the classical point of view an ESP is not a special point, as the modulus approaches it certain states in the string spectrum

[a]We considered $c\sigma > 1$ to secure the validity of the supergravity approximation and we have assumed that the ESP lies at a minimum in the direction of ${\rm Im}(T)$.

become massless [24]. In turn, these massless modes create an interaction potential that may drive the field back to the symmetry point. In that way a modulus can become trapped at an ESP [10]. The strength of the symmetry point depends on the degree of enhancement of the symmetry.

Such modulus trapping can lead to a period of so-called 'trapped inflation' [10], when the trapping is strong enough to make the kinetic density of the modulus fall below the potential density at the ESP. However, it turns out that the number of e-foldings of trapped inflation cannot be very large. Therefore, with respect to cosmology, the main virtue of the ESPs relies on their ability to trap the field and hold it there, at least temporarily.

Because ESPs are fixed points of the symmetries we have

$$V'(\phi_0) = 0\,, \tag{5}$$

where the prime denotes derivative with respect to ϕ and ϕ_0 is the value of the modulus at the ESP. The above means that the ESP is located either at a local extremum (maximum or minimum) or at a flat inflection point of the scalar potential, where $V'(\phi_0) = V''(\phi_0) = 0$. This means that the presence of an ESP deforms the non-perturbative scalar potential (see Fig. 1). This deformation may be enough so that, after trapped inflation, the field undergoes slow-roll inflation over the flat region of the scalar potential at the vicinity of the ESP. The total duration of inflation may, thus, be enough to solve the flatness and horizon problems of the HBB.

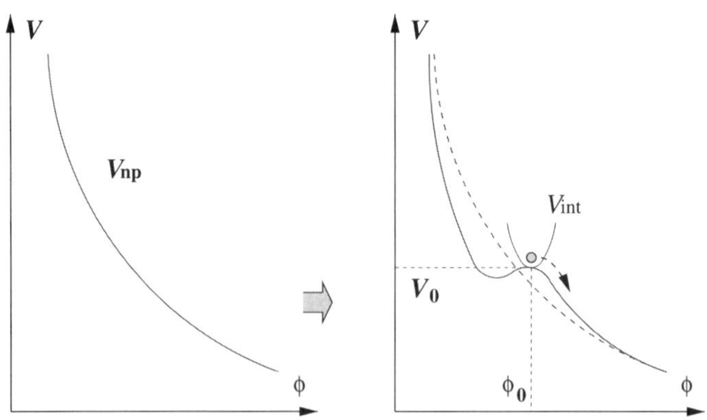

Fig. 1. Illustration of how the appearance of an ESP at $\phi = \phi_0$ deforms the non-perturbative scalar potential $V_{\rm np}$ to generate, for example, a local maximum at potential density V_0. The crossing modulus is temporarily trapped by the emergence of an interaction potential $V_{\rm int}$ due to its enhanced interaction with other fields. After released from trapping, the modulus may drive slow-roll inflation while sliding over the potential hill.

3.1. *Trapped Inflation*

Let us briefly study the trapping of the modulus at the ESP. We assume that around the ESP there is a contribution to the scalar potential due to the enhanced interaction between the modulus ϕ and another field χ, which we take to be also a scalar field [10]. The interaction potential is

$$V_{\rm int}(\phi,\chi) = \frac{1}{2}g^2\chi^2\bar{\phi}^2\,, \tag{6}$$

where $\bar{\phi} \equiv \phi - \phi_0$ with g being a dimensionless coupling constant.

Thus, at the ESP the χ particles are massless. The time dependence of the effective (mass)^2 of the χ field results in the creation of χ-particles. This takes place when the field is within the production window $|\phi| < \Delta\phi \sim (\dot{\phi}_0/g)^{1/2}$, where $\frac{1}{2}\dot{\phi}_0^2$ is the kinetic density of the modulus when crossing the ESP and the dot denotes derivative with respect to the cosmic time t.

The effective scalar potential near the ESP is $V_{\rm eff}(\phi) \approx V_0 + \frac{1}{2}g^2\langle\chi^2\rangle\bar{\phi}^2$ where $V_0 \equiv V(\phi_0)$ with $V(\phi)$ being the 'background' scalar potential. Following Ref. [10] we have $\langle\chi^2\rangle \simeq n_\chi/g|\phi|$, where n_χ denotes the number density of χ particles produced after the crossing of the ESP. This means that $V_{\rm eff}(\phi) \sim V_0 + gn_\chi|\phi|$ and the field climbs a *linear* potential since n_χ is constant outside the production window.

After the first crossing, the field reaches the amplitude Φ_1, determined by its initial kinetic density. To avoid overshooting the ESP we require $\Phi_1 \lesssim m_P$ since for larger values the coupling softens [25]. After reaching Φ_1, the field reverses direction and crosses the production window again, generating more χ particles and, therefore, increasing n_χ. Thus, it now has to climb a steeper potential reaching an amplitude $\Phi_2 < \Phi_1$. The process continues until the ever decreasing amplitude becomes comparable to the production window (see Fig. 2). At this moment particle production stops.

After the end of particle production, $\langle\chi^2\rangle$ remains roughly constant during an oscillation and the modulus continues oscillating in the quadratic interaction potential. Studying this oscillation, we found that, due to the Universe expansion, the amplitude and frequency decrease as $\Phi \sim \Delta\phi/a$ and $\langle\chi^2\rangle \propto a^{-2}$ [9], where the scale factor $a(t)$ is normalised to unity at the end of particle production. Hence, the quadratic potential becomes gradually "diluted" due to the Universe expansion (see Fig. 3). The above mean that the kinetic density of the oscillating modulus scales as $\rho_{\rm osc} \propto a^{-4}$. When $\rho_{\rm osc}$ becomes redshifted below V_0, trapped inflation begins.

The above process assimilates a multitude of initial conditions (provided overshooting the ESP is avoided) because any kinetic density in excess of V_0 is depleted before the onset of trapped inflation. Trapped inflation dilutes

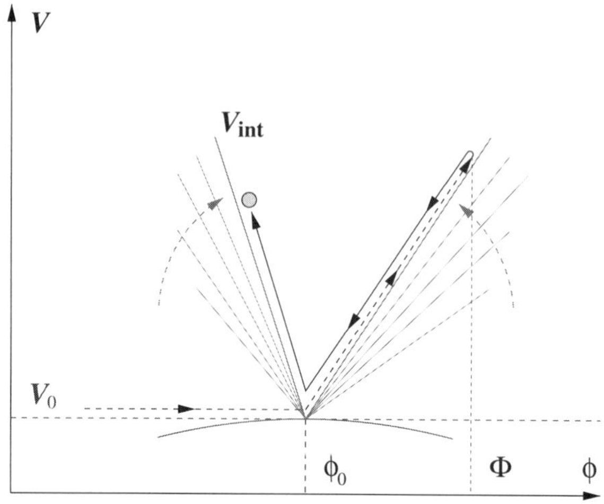

Fig. 2. Illustration of the trapping of a modulus crossing the ESP during particle production. Outside the production window, the modulus oscillates in a linear interaction potential, which steepens progressively due to the production of more χ-particles every time the modulus crosses the ESP.

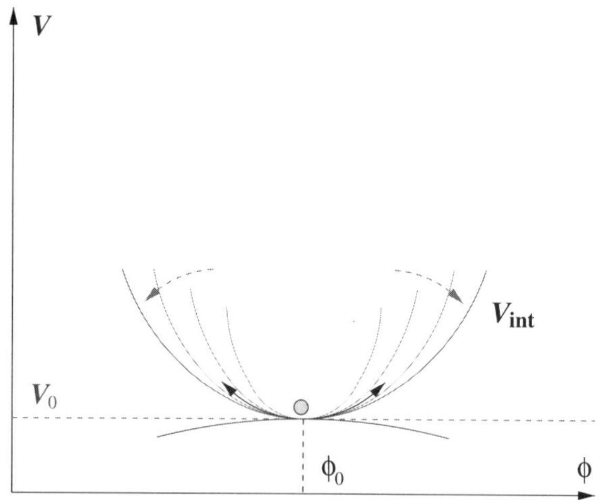

Fig. 3. Illustration of the trapping of a modulus crossing the ESP after particle production. Inside the production window, the modulus oscillates in a quadratic interaction potential, which becomes gradually diluted due to the Universe expansion.

exponentially the density of the χ–particles, which quickly redshifts $V_{\rm int}$. Therefore, after a rather limited number of e-foldings of trapped inflation, the modulus is released from the ESP.

3.2. *Slow-Roll Inflation*

Since the ESP is located at a locally flat region of the potential there is a chance that, after $V_{\rm int}$ becomes negligible, the modulus drives a period of slow-roll inflation while sliding away from the ESP. To study this period we need to quantify the deformation of the scalar potential due to an ESP.

The appearance of an ESP generates either a local extremum or a flat inflection point at ϕ_0. In all cases, in the vicinity of the ESP, the scalar potential can be approximated by a cubic polynomial [9]. Hence, the characteristics of the potential depend only on $m_\phi^2 \equiv V''(\phi_0)$ and $V_0^{(3)} \equiv V'''(\phi_0)$. In fact, we can parametrise the deformation of the scalar potential using [9]

$$|V_0^{(3)}| \sim \xi^2 \sigma_0^3 H_*^2 / m_P \,, \tag{7}$$

where $\sigma_0 \equiv \sigma(\phi_0)$ and H_* is the Hubble parameter during inflation: $H_*^2 \approx V_0/3m_P^2$. The ξ parameter accounts for the strength of the symmetry point; the smaller the ξ, the stronger the deformation and the wider the inflationary plateau. The requirement that the deformation becomes negligible at distances larger than m_P results in the lower bound $\xi > 1$, which also guarantees that the modulus does not overshoot the ESP [9].

By studying inflation after the modulus escapes trapping, we have obtained the following results, depending on the ESP morphology [9]. In each case, one has to achieve enough inflationary e-folds to solve the horizon and flatness problems, while also taking care that the curvature perturbations due to the modulus are not excessive compared to observations.

Consider first the case of a flat inflection point. In this case, we can have enough e-foldings of slow roll inflation if $|V_0^{(3)}| < g^2 H_* \ll H_*$. The case of a local minimum is indistinguishable from the above if $m_\phi^2 < g^2 H_* |V_0^{(3)}|$. If the opposite is true then the modulus becomes trapped in the local minimum and must escape through tunnelling. Afterwards the modulus can drive a period of slow-roll inflation with total number of e-foldings given by $N \sim (H_*/m_\phi)^2$. Hence, to solve the horizon and flatness problems we need $m_\phi \ll H_*$. Finally, in the case of a local maximum, after the end of trapping, one can have a phase of fast/slow roll inflation provided $|m_\phi| \lesssim H_*$.

Thus, we have found that, in all cases, *enough slow-roll inflation to solve the horizon and flatness problems of the HBB is attainable provided* $|m_\phi|, |V_0^{(3)}| < H_*$. Choosing for illustrative purposes an intermediate value for the Hubble scale: $H_* \sim 1$ TeV we have found that one can achieve enough inflationary e-foldings (up to $N_{\rm max} \sim 10^4$) without producing excessive curvature perturbation if $1 < \xi^2 < 10^4$. Thus, *there is ample parameter space for slow-roll inflation to occur after the modulus escapes trapping* *at the ESP*. Note also, that, while H_* is determined by the location of the ESP in field space (by the vacuum density V_0), the values of $|m_\phi|$ and $|V_0^{(3)}|$ are due to the deformation of the scalar potential at the vicinity of the ESP which is not directly related to V_0. Hence, the requirement that the latter are smaller than H_* does not necessarily imply fine-tunning.

4. After the End of Inflation

After inflation, the field rolls away from the ESP. Soon the influence of the ESP on the scalar potential diminishes and $V(\phi) \approx V_{\rm np}(\phi)$. The steepness of $V_{\rm np}$ results in the kinetic domination of the modulus density. As a result a period of kination occurs, during which the field equation is: $\ddot{\phi}+3H\dot{\phi} \simeq 0$. Hence, the density of the Universe scales as $\rho \simeq \frac{1}{2}\dot{\phi}^2 \propto a^{-6}$ [11]. During kination, the scalar field is oblivious of the particular form of the potential.

Kination is terminated when the density of the decay products of a curvaton field dominates the kinetic density of the modulus [14]. Thus, the end of kination corresponds to reheating, with reheating temperature $T_{\rm reh} \sim \sqrt{H_{\rm reh} m_P}$, where $H_{\rm reh}$ is the Hubble parameter at reheating.

After the onset of the HBB, the rolling scalar field is subject to cosmological friction [8,12], which asymptotically freezes the field at the value $\phi_F/m_P \simeq \frac{1}{\sqrt{6}} \ln(V_0/T_{\rm reh}^4)$. Note that this value depends on $T_{\rm reh}$ which, in turn, is determined by curvaton physics. The modulus remains frozen until the present when it plays the role of quintessence. This guarantees that there is no dangerous variation of fundamental constants during the HBB. The evolution of the modulus until today is depicted in Fig. 4.

5. Quintessence

Since $\sigma_F \equiv \sigma(\phi_F) \sim (V_0/T_{\rm reh}^4)^{1/3} > 1$, the modulus rolls to large values before freezing. At such values we can assume that the scalar potential is

$$V(\sigma) \simeq \frac{C_n}{\sigma^n} \quad \Rightarrow \quad V(\phi) \simeq C_n e^{-b\phi/m_P}\,, \tag{8}$$

where C_n is a density scale and $b = \sqrt{\frac{2}{3}}\, n$. The above is a typical uplift potential introduced by flux compactifications as discussed below.

If the modulus is to account for the required dark energy, it must satisfy the coincidence requirement: $V(\sigma_F) \simeq \Omega_\Lambda \rho_0$, where $\Omega_\Lambda \simeq 0.73$ is the dark energy density parameter and ρ_0 is the critical density at present. Hence,

$$T_{\rm reh} \sim V_0^{1/4}\,(\rho_0/C_n)^{\sqrt{3/8n^2}}\,. \tag{9}$$

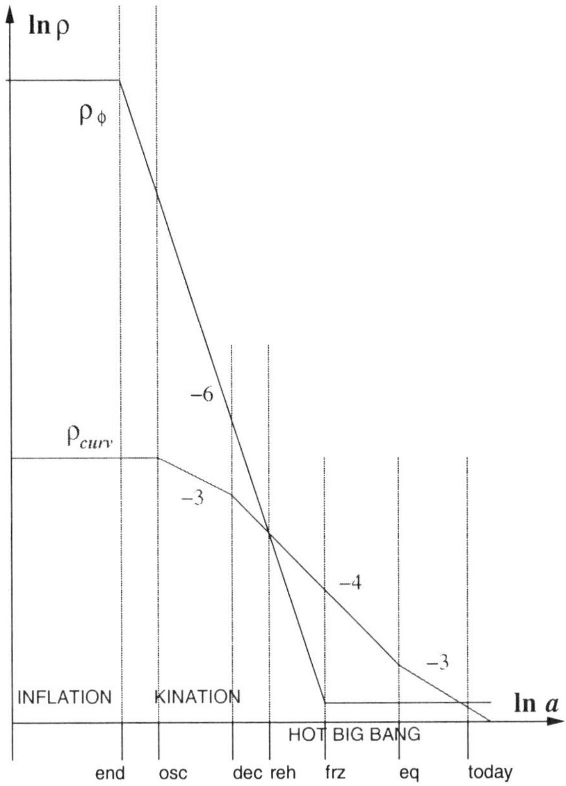

Fig. 4. Illustration of the evolution of the modulus density ρ_ϕ and the density of the curvaton and its decay products $\rho_{\rm curv}$ with respect to the scale factor of the Universe a. In inflation, $\rho_{\rm curv}$ is subdominant and remains constant until, after the end of inflation (denoted by 'end') the curvaton begins oscillating (at time denoted by 'osc'). During the oscillations, $\rho_{\rm curv}$ scales as pressureless matter. Sometime afterwards (denoted by 'dec') the curvaton decays into the thermal bath of the HBB. This thermal bath dominates the Universe at reheating (denoted by 'reh') soon after which the modulus freezes (at time denoted by 'frz') assuming constant potential density comparable to the density today.

Thus, the density scale C_n is determined by $T_{\rm reh}$ which, in turn, is determined by curvaton physics. An upper bound on C_n is obtained by demanding that reheating occurs before big bang nucleosynthesis (BBN):

$$C_n \lesssim \rho_0 \left(V_0^{1/4}/T_{\rm BBN}\right)^{2n\sqrt{2/3}}, \tag{10}$$

where $T_{\rm BBN} \sim 1$ MeV is the temperature at BBN.

The scalar potential in Eq. (8) may have a multitude of origins. For example, using the volume modulus, we may consider a stack of $\overline{D3}-$branes located at the tip of a Klebanov-Strassler throat. The uplift potential is [26]

$$\delta V \sim \exp(-8\pi K/3Mg_s)\, m_P^4/\sigma^2 \equiv C_2/\sigma^2\,, \tag{11}$$

where M and K, in the warp factor, are the units of RR and NS three-form fluxes. To satisfy Eq. (10) we must have $C_2^{1/4} \lesssim 10^{-20} m_P$. This can be attained by choosing the ratio of fluxes as $K/Mg_s \gtrsim 22$. Taking $g_s = 0.1$, only twice as many units of K flux as those of M flux are needed.

It is also possible to consider fluxes of gauge fields on $D7-$branes [27]. In this case, the scalar potential obtains a contribution

$$\delta V \sim 2\pi E^2/\sigma^3 \equiv C_3/\sigma^3\,, \tag{12}$$

where E depends on the strength of the gauge fields considered. The constraint in Eq. (10) requires now $C_3^{1/4} \lesssim 10^{-15} m_P \sim 1$ TeV.

The future of the modulus after unfreezing depends on the steepness of the scalar potential, or equivalently the value of b in Eq. (8).

• For $b \leq \sqrt{2}$, the modulus dominates the Universe for ever, leading to eternal acceleration. This results in future horizons, which pose a problem for the formulation of the S-matrix in string theory [28].
• For $\sqrt{2} < b \leq \sqrt{3}$, the modulus dominates the Universe but results only in a brief accelerated expansion period. Such is the fate of the $n = 2$ case.
• For $\sqrt{3} < b \leq \sqrt{6}$, the modulus does not dominate the Universe, albeit causing a brief period of accelerated expansion. Afterwards the modulus density remains at a constant ratio with the background matter density. This is the fate of the $n = 3$ case.
• For $b > \sqrt{6}$, the modulus does not cause any accelerated expansion and so cannot be used as quintessence. After unfreezing, the modulus rolls fast down the quintessential tail of the scalar potential with its density approaching asymptotically kinetic domination (and subsequently freezing at a value larger than σ_F [8]). This case corresponds to $n > 3$.

The brief acceleration period caused by the unfreezing modulus is due to the fact that the modulus oscillates around an attractor solution [29], which in itself does not result to acceleration [7]. In Ref. [30] it was claimed that brief acceleration occurs if $\sqrt{2} < b \leq 2\sqrt{6}$, which corresponds to the range $\sqrt{3} < n \leq 6$. More recent studies, however, have reduced this range. Brief acceleration in the range $\sqrt{2} < b \leq \sqrt{3}$ has been confirmed by Ref. [31]. This includes the $n = 2$ case which corresponds to the most popular uplift potential. The range for b was expanded further in Ref. [32], where it is shown that brief acceleration can explain the observations at least up to $b \simeq \frac{3}{4}\sqrt{6} \approx 1.837$. Since the data are interpreted using a number of priors, we believe that the $n = 3$ case is still marginally acceptable.

6. Conclusions

Quintessential inflation is possible to achieve in string theory with flux compactifications using as inflaton a string modulus which rolls down its runaway potential. Inflation is due to the presence of an enhanced symmetry point (ESP), which traps the modulus and creates a locally flat region over which the modulus can slow-roll. There is ample parameter space for successful inflation provided: $m_\phi, |V_0^{(3)}| \ll H_*$. Trapping assimilates a multitude of initial conditions provided overshooting the ESP is avoided.

After inflation, the modulus becomes (again) kinetically dominated causing a period of kination. Reheating is due to the Universe domination by the decay products of a curvaton field, which also accounts for the correct amplitude of the curvature perturbations in the Universe. During the Hot Big Bang, the modulus freezes and remains constant until the present.

At the frozen value, the potential is dominated by an uplift term of exponential form. This residual potential density begins to dominate today, when the modulus unfreezes, leading to a brief period of late inflation.

Curvaton physics fixes the reheating temperature and determines the value of the frozen modulus σ_F and the density scale C_n in the uplift potential. Coincidence and BBN constrains on C_n allow realistic values much less fine-tunned than the cosmological constant Λ in the ΛCDM model.

Acknowledgements

This work was done with Juan C. Bueno Sánchez and was supported by the E.U. Marie Curie Research and Training Network *"UniverseNet"* (MRTN-CT-2006-035863) and by PPARC (PP/D000394/1). I am grateful to the Royal Society for supporting my participation to the IDM2006 conference.

References

1. P. J. E. Peebles and B. Ratra, Rev. Mod. Phys. **75** (2003) 559; E. J. Copeland, M. Sami and S. Tsujikawa, Int. J. Mod. Phys. D **15** (2006) 1753.
2. S. Weinberg, Rev. Mod. Phys. **61** (1989) 1.
3. B. Ratra and P. J. E. Peebles, Phys. Rev. D **37** (1988) 3406.
4. R. Caldwell, R. Dave and P. J. Steinhardt, Phys. Rev. Lett. **80** (1998) 1582.
5. P. J. E. Peebles and A. Vilenkin, Phys. Rev. D **59** (1999) 063505.
6. R. A. Frewin and J. E. Lidsey, Int. J. Mod. Phys. D **2** (1993) 323; M. Peloso and F. Rosati, JHEP **9912** (1999) 026; K. Dimopoulos, Nucl. Phys. Proc. Suppl. **95** (2001) 70; A. S. Majumdar, Phys. Rev. D **64** (2001) 083503; N. J. Nunes and E. J. Copeland, Phys. Rev. D **66** (2002) 043524; A. Gonzalez, T. Matos and I. Quiros, Phys. Rev. D **71** (2005) 084029.
7. K. Dimopoulos and J. W. F. Valle, Astropart. Phys. **18** (2002) 287.

8. K. Dimopoulos, Phys. Rev. D **68** (2003) 123506.
9. J. C. Bueno Sanchez and K. Dimopoulos, Phys. Lett. B **642** (2006) 294; hep-th/0606223.
10. L. Kofman, A. Linde, X. Liu, A. Maloney, L. McAllister and E. Silverstein, JHEP **0405** (2004) 030.
11. M. Joyce and T. Prokopec, Phys. Rev. D **57** (1998) 6022.
12. R. Brustein, S. P. de Alwis and P. Martens, Phys. Rev. D **70** (2004) 126012; T. Barreiro, B. de Carlos, E. Copeland and N. J. Nunes, Phys. Rev. D **72** (2005) 106004.
13. D. H. Lyth and D. Wands, Phys. Lett. B **524** (2002) 5; K. Enqvist and M. S. Sloth, Nucl. Phys. B **626** (2002) 395.
14. B. Feng and M. z. Li, Phys. Lett. B **564** (2003) 169; A. R. Liddle and L. A. Urena-Lopez, Phys. Rev. D **68** (2003) 043517.
15. J. McDonald, Phys. Rev. D **68** (2003) 043505; Phys. Rev. D **70** (2004) 063520; A. Mazumdar and A. Perez-Lorenzana, Phys. Rev. Lett. **92** (2004) 251301.
16. K. Enqvist, S. Kasuya and A. Mazumdar, Phys. Rev. Lett. **90** (2003) 091302; K. Enqvist, A. Jokinen, S. Kasuya and A. Mazumdar, Phys. Rev. D **68** (2003) 103507; M. Bastero-Gil, V. Di Clemente and S. F. King, Phys. Rev. D **67** (2003) 103516; M. Postma, Phys. Rev. D **67** (2003) 063518; J. McDonald, Phys. Rev. D **69** (2004) 103511. S. Kasuya, M. Kawasaki and F. Takahashi, Phys. Lett. B **578** (2004) 259; K. Enqvist, Mod. Phys. Lett. A **19** (2004) 1421.
17. K. Dimopoulos, D. H. Lyth, A. Notari and A. Riotto, JHEP **0307** (2003) 053; E. J. Chun, K. Dimopoulos and D. Lyth, Phys. Rev. D **70** (2004) 103510; R. Hofmann, Nucl. Phys. B **740** (2006) 195.
18. K. Dimopoulos and G. Lazarides, Phys. Rev. D **73** (2006) 023525.
19. K. Dimopoulos, G. Lazarides, D. Lyth and R. Ruiz de Austri, JHEP **0305** (2003) 057.
20. K. Dimopoulos and D. H. Lyth, Phys. Rev. D **69** (2004) 123509.
21. K. Dimopoulos, D. H. Lyth and Y. Rodriguez, JHEP **0502** (2005) 055; K. Dimopoulos, Phys. Lett. B **634** (2006) 331.
22. J. P. Derendinger, L. E. Ibanez and H. P. Nilles, Phys. Lett. B **155** (1985) 65; E. Witten, Nucl. Phys. B **474** (1996) 343.
23. C. M. Hull and P. K. Townsend, Nucl. Phys. B **451** (1995) 525.
24. S. Watson, Phys. Rev. D **70** (2004) 066005.
25. R. Brustein, S. P. De Alwis and E. G. Novak, Phys. Rev. D **68** (2003) 023517.
26. S. B. Giddings, S. Kachru and J. Polchinski, Phys. Rev. D **66** (2002) 106006.
27. C. P. Burgess, R. Kallosh and F. Quevedo, JHEP **0310** (2003) 056.
28. S. Hellerman, N. Kaloper and L. Susskind, JHEP **0106** (2001) 003; W. Fischler, A. Kashani-Poor, R. McNees and S. Paban, JHEP **0107** (2001) 003;
29. E. J. Copeland, A. R. Liddle and D. Wands, Phys. Rev. D **57** (1998) 4686.
30. J. M. Cline, JHEP **0108** (2001) 035.
31. U. J. Lopes Franca and R. Rosenfeld, JHEP **0210** (2002) 015; R. Kallosh, A. Linde, S. Prokushkin and M. Shmakova, Phys. Rev. D **66** (2002) 123503; R. Kallosh and A. Linde, JCAP **0302** (2003) 002; A. Kehagias and G. Kofinas, Class. Quant. Grav. **21** (2004) 3871.
32. D. Blais and D. Polarski, Phys. Rev. D **70** (2004) 084008.

OBSERVATIONAL SIGNATURE OF EXTENDED GRAVITY THEORIES

L. PERIVOLAROPOULOS*

Department of Physics, University of Ioannina,
Ioannina, 451 10, Greece
**E-mail: ab_leandros@uoi.gr*
http://leandros.physics.uoi.gr

The observed current accelerating expansion of the universe can be attributed to either the existence of a negative pressure energy component (dark energy) in the universe or to modifications of general relativity. The discovery of observational signatures that may distinguish between the two classes of theories is a major challenge for modern cosmology. Here I briefly discuss one such potential signature: The change of sign of the expression $\frac{d(H^2(z))}{dz} - 3\Omega_{0m}H_0^2(1+z)^2$ (where $H(z$ is the Hubble parameter at redshift z) which is equivalent to an equation of state parameter $w(z)$ crossing the $w = -1$ line. I also review the current information provided by observational probes with respect to the above cosmological test.

1. Introduction

A wide range of cosmological observations have indicated that the universe has entered a phase of accelerating expansion starting at a recent redshift ($z \simeq 0.5$). This acceleration is described by the expansion rate as a function of redshift $H(z) = \frac{\dot{a}}{a}$ which can be probed by several observational tools. Such tools include the use of Type Ia supernovae (SnIa) standard candles[1,2] to construct the Hubble diagram at recent redshifts as well as the use of standard rulers obtained from the physics of the Cosmic Microwave Background[3] (CMB) fluctuations spectrum (size of the horizon at recombination), the Baryon Acoustic Oscillations[4] (BAO) in the Large Scale Structure (LSS) correlation function and the Cluster Gas Mass Fraction.[5] The observed expansion rate $H(z)$ at recent redshifts can be combined with other cosmological observations (CMB or LSS) probing the density of matter in the universe to test the Friedman equation as predicted by general relativity which in a flat matter dominated universe takes the form

$$H^2(a) = \left(\frac{\dot{a}}{a}\right)^2 = \frac{8\pi G}{3}\rho_m(a) \tag{1}$$

where $a(t)$ is the scale factor of the universe and $\rho_m \sim a(t)^{-3}$ its average matter density. The above observational probes have indicated that the two sides of the Friedman equation (1) can not be equal even if the (theoretically disfavored) possibility of a non-flat universe is allowed. There are two possible resolutions to this puzzle: Either modify the right side of the Friedman equation (1) introducing a new form of *'dark' energy* ideal fluid component ($\rho(a) = \rho_m(a) + \rho_X(a)$) with suitable evolution in order to restore the equality or modify both sides by changing the way energy density affects geometry thus modifying the Einstein equations.

In the first class of approaches the required gravitational properties of dark energy (see Refs. 6–8 for recent reviews) needed to induce the accelerating expansion are well described by its equation of state $w(z) = \frac{p_X(z)}{\rho_X(z)}$ which enters in the second Friedman equation as

$$\frac{\ddot{a}}{a} = -\frac{4\pi G}{3}(\rho_m + \rho_X(1+3w)) \tag{2}$$

implying that a negative pressure ($w < -1/3$) is necessary in order to induce accelerating expansion. The simplest viable example of dark energy is the cosmological constant[9] ($w = -1$). This example however even though consistent with present data lacks physical motivation. Questions like 'What is the origin of the cosmological constant?' or 'Why is the cosmological constant 10^{120} times smaller than its natural scale so that it starts dominating at recent cosmological times (coincidence problem)?' remain unanswered.[10] Attempts to replace the cosmological constant by a dynamical scalar field (quintessence[11]) have created a new problem regarding the initial conditions of quintessence which even though can be resolved in particular cases (tracker quintessence), can not answer the above questions in a satisfactory way.

The parameter $w(z)$ determines not only the gravitational properties of dark energy but also its evolution. This evolution is easily obtained from the energy momentum conservation

$$d(\rho_X a^3) = -p_X d(a^3) \tag{3}$$

which leads to

$$\rho_X = \rho_{0X} e^{-3\int_1^a \frac{da'}{a'}(1+w(a'))} = \rho_{0X} e^{3\int_0^z \frac{dz'}{1+z'}(1+w(z'))} \tag{4}$$

Therefore the determination of $w(z)$ is equivalent to that of $\rho_X(z)$ which in turn is equivalent to the observed $H(z)$ from the Friedman equation (1) expressed as

$$H(z) = H_0[\Omega_{0m}(1+z)^3 + \Omega_{0X} e^{3\int_0^z \frac{dz'}{1+z'}(1+w(z'))}] \quad (5)$$

Thus, knowledge of Ω_{0m} and $H(z)$ suffices to determine $w(z)$ which is obtained from equation (5) as[12]

$$w(z) = \frac{\frac{2}{3}(1+z)\frac{d\ln H}{dz} - 1}{1 - \frac{H_0^2}{H^2}\Omega_{0m}(1+z)^3} \quad (6)$$

In the second class of approaches the Einstein equations get modified and the new equations combined with the assumption of homogeneity and isotropy lead to a generalized Friedman equation of the form

$$f(H^2) = g(\rho_m) \quad (7)$$

where f and g are appropriate functions determined by the modified gravity theory.[13–15] In this class of models, the parameter $w(z)$ can also be defined from equation (6) but it can not be interpreted as $\frac{p_X}{\rho_X}$ of a perfect fluid.

The simplest (but quite general) examples of modified gravity theories are scalar tensor theories[13,16–19] where the Newton's constant G is promoted to a function of a field Φ: $8\pi G \rightarrow \frac{1}{F(\Phi)}$ whose dynamics at the Lagrangian level is determined by a potential $U(\Phi)$. Assuming homogeneity and isotropy, the modified Friedman equation in these theories take the form

$$H^2 = \frac{1}{3F}(\rho_m + \frac{1}{2}\dot{\Phi}^2 + U - 3H\dot{F}) \quad (8)$$

This equation reduces to a regular minimally coupled scalar field dark energy (quintessence) in the general relativity limit of a constant $F = \frac{1}{8\pi G}$. The positive nature of the kinctic term $\dot{\Phi}^2$ however implies that certain types of behaviors of H may not be reproducible without invoking a time-dependent F. These types of behavior of $H(z)$ which include a $w(z)$ crossing the Phantom Divide Line (PDL) $w = -1$ are potential signatures of extended gravity theories and will be discussed in what follows.

2. Observational Probes

In order to identify the trends encoded in the current cosmological data with respect to dark energy evolution and in particular PDL crossing we group the data[20] in three categories:

(1) SnIA Gold sample[1]
(2) SnIA SNLS[2]
(3) Other Dark Energy Probes (ODEP) that include the CMB,[3] BAO,[4] Cluster Baryon Fraction (CBF)[5] and Perturbations Growth Rate (PGR) at $z = 0.15$[21]

We then fit to the data the parameterization[22,23]

$$w(z) = w_0 + w_1 \frac{z}{1+z} \tag{9}$$

In order to identify the dependence of the resulting best fits on the $\Omega_{0\mathrm{m}}$ prior used we have not marginalized over $\Omega_{0\mathrm{m}}$. Instead we have fixed $\Omega_{0\mathrm{m}}$ and considered two cases ($\Omega_{0\mathrm{m}} = 0.2$ and $\Omega_{0\mathrm{m}} = 0.3$). The best fit form of $w(z)$ for each dataset category[20] is shown in Fig. 1 for both $\Omega_{0\mathrm{m}} = 0.2$ and $\Omega_{0\mathrm{m}} = 0.3$. The corresponding χ^2 contours in the $w_0 - w_1$ parameter space is shown in Fig. 2.

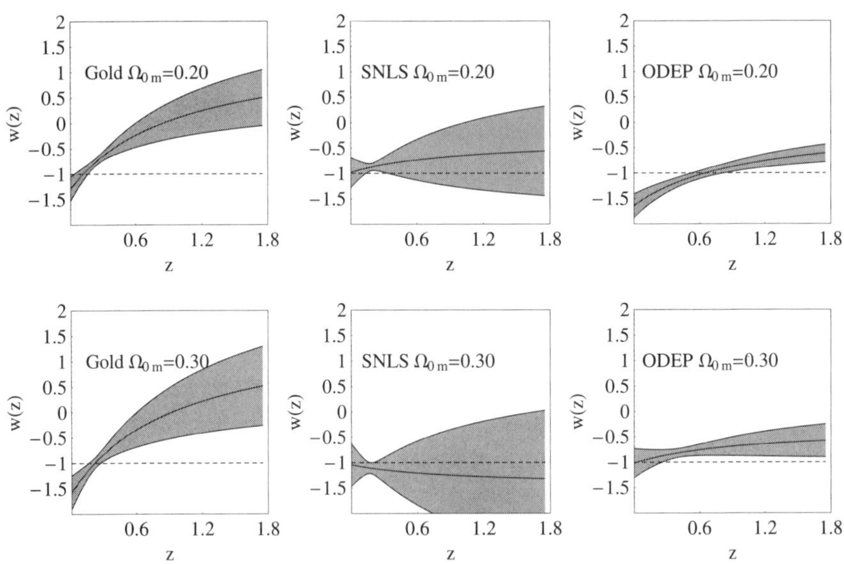

Fig. 1. The best fit form of $w(z)$ for each dataset category for both $\Omega_{0m} = 0.2$ and $\Omega_{0m} = 0.3$ along with the 1σ errors (shaded region).

The following comments can be made based on our results summarized in Figs. 1 and 2:

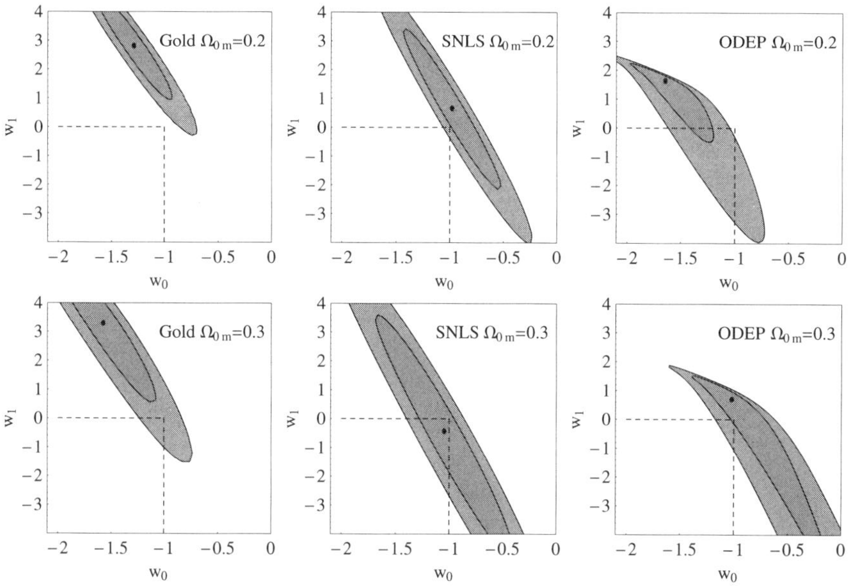

Fig. 2. The 68% and 95% χ^2 confidence contours in the $w_0 - w_1$ parameter space for each dataset category for both $\Omega_{0m} = 0.2$ and $\Omega_{0m} = 0.3$. Notice that for the SNLS dataset ΛCDM is within the 1σ region.

- The Gold dataset mildly favors dynamically evolving dark energy crossing the PDL at $z \simeq 0.2$ over the cosmological constant while the SNLS does not.
- Dark energy probes other than SnIa mildly favor crossing of the PDL for low values of $\Omega_{0\mathrm{m}}$ ($\Omega_{0\mathrm{m}} \lesssim 0.25$) while for $\Omega_{0\mathrm{m}} \simeq 0.3$ this trend is significantly reduced.
- The best probes of dark energy with currently existing data, other than SnIa are the CMB shift parameter and the BAO peak.

The last point emerges by plotting the third column of Figs. 1, 2 utilizing only CMB and BAO data. Such plots are practically unchanged compared to those of Figs. 1, 2.

It is therefore clear that even though currently available dark energy probes agree on the fact that the dark energy equation of state $w(z)$ is close to -1 there is no universal trend with respect to the evolution properties of dark energy and the existing trends appear to depend on the value of $\Omega_{0\mathrm{m}}$ prior considered.

3. Signatures of Extended Gravity Theories

Given the current observational situation described in the previous section which allows for the possibility of crossing of the PDL it becomes clear that the following question arises: *What theory produces crossing of the $w = -1$ PDL?*

The simplest class of physically motivated models generalizing the cosmological constant and producing accelerated universe expansion are based on a simple evolving homogeneous scalar field Φ (quintessence[11]) minimally coupled to gravity whose dynamics is determined by a potential $U(\Phi)$ and has a canonical kinetic term $\frac{1}{2}\dot{\Phi}^2$. It is easy to see however that this class of models can not reproduce a $w(z)$ crossing the PDL $w = -1$ for *any* potential $U(\Phi)$. Indeed, the equation of state w for such models takes the form

$$w(z) = \frac{p_X}{\rho_X} = \frac{\frac{1}{2}\dot{\Phi}^2 - U(\Phi)}{\frac{1}{2}\dot{\Phi}^2 + U(\Phi)} \tag{10}$$

which approaches $w = -1$ in the limit of a small kinetic term but does not cross the PDL as long as $\dot{\Phi}^2 > 0$ for any sign of $U(\Phi)$. This result has been generalized by Vikman[24] who showed that any minimally coupled scalar field with a generalized kinetic term also can not cross the PDL through a stable trajectory.

The simplest but very general (given its simplicity) extension of general relativity is expressed through scalar-tensor theories. In these theories Newton's constant obtains dynamical properties expressed through the potential $F(\Phi)$. The dynamics are determined by the Lagrangian density[13,16]

$$\mathcal{L} = \frac{F(\Phi)}{2}\, R - \frac{1}{2}\, g^{\mu\nu}\partial_\mu\Phi\partial_\nu\Phi - U(\Phi) + \mathcal{L}_m[\psi_m; g_{\mu\nu}] \tag{11}$$

where $\mathcal{L}_m[\psi_m; g_{\mu\nu}]$ represents matter fields approximated by a pressureless perfect fluid. The function $F(\Phi)$ is observationally constrained as follows:

- $F(\Phi) > 0$ so that gravitons carry positive energy.[16]
- $\frac{dF}{d\Phi} < 10^{-5}$ from solar system observations.[25]

Assuming a homogeneous Φ and varying the action corresponding to (11) in a background of a flat FRW metric, we find the coupled system of generalized Friedman equations

$$3FH^2 = \rho + \frac{1}{2}\dot{\Phi}^2 - 3H\dot{F} + U \tag{12}$$

$$-2F\dot{H} = \rho + p + \dot{\Phi}^2 + \ddot{F} - H\dot{F} \tag{13}$$

where we have assumed the presence of a perfect fluid ($\rho = \rho_m, p \simeq 0$) playing the role of matter fields. Expressing in terms of redshift and using a simple ansatz for the kinetic term

$$\Phi'(z)^2 = ae^{-z} \tag{14}$$

its is straightforward to solve the system (12)-(13) cite for $F(z)$, $U(z)$. We may then then use $z(\Phi)$ (obtained from (14) and the best fit form of $H(z) - w(z)$ obtained from the Gold dataset and crossing the PDL (see Fig. 1a) to reconstruct the scalar tensor potentials $F(\Phi)$, $U(\Phi)$ which are consistent with both the solar system constraints and the crossing of the PDL.[19]

It therefore becomes clear that even though the crossing of the PDL $w = -1$ is inconsistent with dark energy in the form of a single minimally coupled scalar field, such a crossing can well be explained by non-minimally coupled scalar field playing the role of dark energy. We therefore conclude that if the crossing of the PDL favored by the Gold dataset is confirmed by future observations, then this may viewed as a signature for extended theories of gravity.

References

1. A. G. Riess *et al.* [Supernova Search Team Collaboration], Astrophys. J. **607**, 665 (2004) [arXiv:astro-ph/0402512].
2. P. Astier *et al.*, Astron. Astrophys. **447**, 31 (2006) [arXiv:astro-ph/0510447].
3. D. N. Spergel *et al.*, arXiv:astro-ph/0603449.
4. D. J. Eisenstein *et al.* [SDSS Collaboration], Astrophys. J. **633**, 560 (2005) [arXiv:astro-ph/0501171].
5. S. W. Allen, R. W. Schmidt, H. Ebeling, A. C. Fabian and L. van Speybroeck, Mon. Not. Roy. Astron. Soc. **353**, 457 (2004) [arXiv:astro-ph/0405340].
6. E. J. Copeland, M. Sami and S. Tsujikawa, arXiv:hep-th/0603057.
7. T. Padmanabhan, arXiv:astro-ph/0603114.
8. L. Perivolaropoulos, arXiv:astro-ph/0601014.
9. V. Sahni and A. A. Starobinsky, Int. J. Mod. Phys. D **9**, 373 (2000) [arXiv:astro-ph/9904398].
10. S. Weinberg, arXiv:astro-ph/0005265.
11. R. R. Caldwell, R. Dave and P. J. Steinhardt, Phys. Rev. Lett. **80**, 1582 (1998) [arXiv:astro-ph/9708069].
12. D. Huterer and M. S. Turner, Phys. Rev. D **64**, 123527 (2001) [arXiv:astro-ph/0012510].
13. B. Boisseau, G. Esposito-Farese, D. Polarski and A. A. Starobinsky, Phys. Rev. Lett. **85**, 2236 (2000) [arXiv:gr-qc/0001066].
14. S. Nojiri, S. D. Odintsov and M. Sami, Phys. Rev. D **74**, 046004 (2006) [arXiv:hep-th/0605039].

15. V. Sahni and Y. Shtanov, JCAP **0311**, 014 (2003) [arXiv:astro-ph/0202346].
16. G. Esposito-Farese and D. Polarski, Phys. Rev. D **63**, 063504 (2001) [arXiv:gr-qc/0009034].
17. D. F. Torres, Phys. Rev. D **66**, 043522 (2002) [arXiv:astro-ph/0204504].
18. R. Gannouji, D. Polarski, A. Ranquet and A. A. Starobinsky, arXiv:astro-ph/0606287.
19. L. Perivolaropoulos, JCAP **0510**, 001 (2005) [arXiv:astro-ph/0504582].
20. S. Nesseris and L. Perivolaropoulos, arXiv:astro-ph/0610092.
21. L. Verde *et al.*, Mon. Not. Roy. Astron. Soc. **335**, 432 (2002) [arXiv:astro-ph/0112161].
22. M. Chevallier and D. Polarski, Int. J. Mod. Phys. D **10**, 213 (2001) [arXiv:gr-qc/0009008].
23. E. V. Linder, Phys. Rev. Lett. **90**, 091301 (2003) [arXiv:astro-ph/0208512].
24. A. Vikman, Phys. Rev. D **71**, 023515 (2005) [arXiv:astro-ph/0407107].
25. P.D. Scharre, C.M. Will, Phys. Rev. D 65 (2002); E. Poisson, C.M. Will, Phys. Rev. D 52, 848 (1995).

AKAIKE MODEL SELECTION CRITERION APPLIED TO SUPERNOVAE DATA

M. BIESIADA

Department of Astrophysics and Cosmology, University of Silesia, Uniwersytecka 4, 40-007 Katowice, Poland

In this contribution we apply model selection approach based on Akaike criterion as an estimator of Kullback-Leibler entropy. In particular, we present the proper way of ranking the competing models based on Akaike weights (in Bayesian language - posterior probabilities of the models). This important ingredient is missing in alternative studies dealing with cosmological applications of Akaike criterion.

Out of many particular models of dark energy we focus on four: quintessence, quintessence with time varying equation of state, brane-world and generalized Chaplygin gas model and test them on Riess' Gold sample.

As a result we obtain that the best model - in terms of Akaike Criterion - is the quintessence model with evolving equation of state. The odds suggest that although there exist differences in the support given to specific scenarios by supernova data all models considered receive similar support. One can also notice that models similar in structure i.e. ΛCDM, quintessence and quintessence with variable equation of state are closer to each other in terms of Kullback-Leibler entropy. Models having different structure i.e. Chaplygin gas or brane-world scenario are more distant (in Kullback-Leibler sense) from the best one.

Keywords: classical tests of cosmology, dark energy theory.

1. Introduction

The problem of "dark energy" in the Universe is one of the most important issues in modern cosmology. It appeared after the discovery of accelerated expansion of the Universe as inferred from the SNIa Hubble diagram.[1] Since then a lot of specific scenarios have been put forward as an explanation of this puzzling phenomenon. They fall into two broad categories: searching an explanation among hypothetical candidates for dark energy (cosmological constant Λ,[1] quintessence - evolving scalar fields,[2] Chaplygin gas[3]) or modification of gravity theory (such like brane world scenarios[4]).

In the problem of statistical inference from empirical data one very often encounters the problem of selecting the best approximating model.[5] This is exactly the problem one has in the context of dark energy where there exist a variety of theoretical ideas of what could be the cause of accelerating Universe and at the same time there is relatively little theoretical guidance of which specific model (or a class thereof) is preferred. Therefore, it is interesting to ask which cosmological model is the most supported by the data which triggered the problem.

In this contribution we approach the above mentioned question from the perspective of Akaike model selection criterion.[6] In cosmology Akaike criterion has first been used by Liddle[7] and then by others.[8] Since AIC is relatively unknown among cosmologists while being being popular, if not standard, in other branches of science (e.g. biostatistics[5]) we start with rudimentary introduction to the ideas underlying AIC closely following.[5]

2. Akaike model selection criterion

Akaike information-theoretical model selection criterion is based on Kullback-Leibler information. Kullback-Leibler information between two distributions $f(x)$ and $g(x)$ is defined as

$$I(f,g) = \int f(x) \ln \frac{f(x)}{g(x)} dx \tag{1}$$

The intuitive meaning of $I(f,g)$ (also called K-L divergence) is the information lost when g is used to approximate f. It can be viewed as an extension of Shannon's entropy and sometimes is thus referred to as relative entropy. In cosmology, a very interesting application of this concept has been made by Hosoya et al.[10] who proposed the Kullback-Leibler Relative Information Entropy as a measure of the distinguishability of the local inhomogeneous mass density field from its spatial average on arbitrary compact domains.

Let us assume that $f(x)$ denotes the true mechanism behind the data and $g(x|\theta)$ its approximating model (parametrized by θ). The problem is that K-L divergence cannot be assessed without prior knowledge of the true model $f(x)$ as well as parameters θ of the approximating model $g(x|\theta)$. However, given $f(x)$ and $g(x|\theta)$ there exists the "best" value of θ for which Kullback-Leibler divergence is minimized. The observation that maximum likelihood estimator $\hat{\theta}$ of θ parameter is exactly this K-L "best" one was the crucial ingredient in Akaike's derivation of his criterion.

The core result of Akaike was in showing that an approximately unbiased estimator of K-L divergence is $ln(\mathcal{L}(\hat{\theta}|data)) - K$ where $\mathcal{L}$ is the likelihood function (more precisely its numerical maximum value - taken at $\hat{\theta}$) and K

is the number of estimable parameters (θs) in approximating model $g(x|\theta)$. For historical reasons Akaike formulated this result in the following form:

$$AIC = -2ln(\mathcal{L}(\hat{\theta}|data)) + 2K \tag{2}$$

which became known as Akaike information criterion. Heuristically one may think of it as of an estimator of K-L divergence between the model at hand $g(x|\theta)$ and an unknown true model $f(x)$ which generated the data. In the expression for AIC one can recognize two terms: the first measuring goodness of model fit (or more precisely the lack thereof) and the second one (competing) measuring model complexity (number of free parameters).

Defined in this way AIC value has no meaning by itself for a single model (simply because the true model $f(x)$ is unknown). What is useful, instead are the differences $\Delta_i := AIC_i - AIC_{min}$ calculated over the whole set of alternative candidate models $i = 1, ..., N$ where by AIC_{min} we denoted $min\{AIC_i; i = 1, ..., N\}$. Comparing several models, the one which minimizes AIC could be considered the best. Then the relative strength of evidence for each model can be calculated as the likelihood of the model given the data $\mathcal{L}(g_i|data) \propto exp(-\frac{1}{2}\Delta_i)$. Relative likelihoods of the models $\mathcal{L}(g_i|data)$ normalized to unity are called Akaike weights w_i. In Bayesian language Akaike weight corresponds to the posterior probability of a model (under assumption of equal prior probabilities). The (relative) evidence for the models can also be judged by the evidence ratios of model pairs $\frac{w_i}{w_j} = \frac{\mathcal{L}(g_i|data)}{\mathcal{L}(g_j|data)}$. If referred to the best model, evidence ratio gives odds against the given model. One can easily see (with these definitions) that AIC differences of 2,4,8,10 correspond to the odds ratios 2.7, 7.4, 54.6 and 148.4 respectively. This justifies the rules of Akaike model selection that Δ_i in the range $0 - 2$ mean that model i has almost the same support from data as the best one, for the range $2 - 4$ this support is considerably less and with $\Delta_i > 10$ model i is practically irrelevant.

3. Cosmological models fitted to supernovae data

Our aim is to find out what is the degree of support (in terms of AIC) given by supernovae data to different cosmological scenarios which might (at least phenomenologically) describe presently accelerating Universe.

The sample we use is the so called "Gold" sample of Riess et al.[9] comprising 157 supernovae compiled from a set of previously observed SNIa with reduced systematic errors from differences in calibrations.

To proceeded with fitting the SNIa data we need the magnitude-redshift relation $m(z, \mathcal{M}, \theta_i) = \mathcal{M} + 5\log_{10} D_L(z, \theta_i)$ where: by θ_i we denoted symbolically cosmological parameters of fitted scenario and D_L is the luminosity distance with H_0 factored out. The fitting is performed according to procedure equivalent to marginalization over the intercept $\mathcal{M}$ (as described in Ref. 12). Further on we will restrict our attention to flat model $k = 0$ because flat Friedman-Robertson-Walker geometry is strongly supported by cosmic microwave background radiation (CMBR) data.[13] In this case the luminosity distance reads:

$$d_L(z) = (1+z)\frac{c}{H_0}\int_0^z \frac{dz'}{H(z')}. \tag{3}$$

The estimation of cosmological model parameters was performed using the maximum likelihood approach. We assumed that supernovae measurements came with uncorrelated Gaussian errors and in this case the likelihood function $\mathcal{L}$ could be determined from chi-square statistic $\mathcal{L} \propto \exp(-\chi^2/2)$.[1] The χ^2 function here is defined as:

$$\chi^2 = \sum_i \frac{(m_i^{th} - m_i^{obs})^2}{\sigma_i^2}$$

where the sum is over the SNIa sample and σ_i denote the (full) statistical error of magnitude determination.

The paragraphs below briefly introduce cosmological models which will then be compared by using the Akaike criterion.

Friedman - Robertson - Walker model with non-vanishing cosmological constant and pressure-less matter including the dark part of it responsible for flat rotation curves of galaxies (the co called ΛCDM model) is a standard reference point in modern cosmology. Sometimes it is referred to as a concordance model since it fits rather well to independent data (such like CMBR data, LSS considerations, supernovae data). It suffers, however from the fine tuning problem.

The most popular explanation of the accelerating Universe is to assume the existence of a negative pressure component called dark energy. One can heuristically assume that it is described by hydrodynamical energy-momentum tensor with $p = w\rho$ where $-1 < w < -1/3$.[14] In such case this component is called "quintessence".

If the origins of the quintessence are in an evolving scalar field, it would be natural to expect that w coefficient should vary in time, i.e. $w = w(z)$. An arbitrary function $w(z)$ can be Taylor expanded. Then, bearing in mind that both SNIa surveys or strong gravitational lensing systems are able to probe the range of small and moderate redshifts it is sufficient to explore first the linear order of this expansion. Such possibility, i.e. $w(z) = w_0 + w_1 z$ has been considered in the literature (e.g. Ref. 15).

In another class of models matter content of the Universe consists of pressure-less gas (representing baryonic and cold dark matter) and of the generalized Chaplygin gas with the equation of state $p_{Ch} = -\frac{A}{\rho_{Ch}{}^{\alpha}}$ with $0 \leq \alpha \leq 1$, representing dark energy responsible for acceleration of the Universe. Generalized Chaplygin gas models have been intensively studied in the literature[16] and in particular they have been tested against supernovae data (e.g. Ref. 17 and references therein).

According to brane-world scenarios,[4] our 4-dimensional Universe is a surface (a brane) embedded into a higher dimensional bulk space-time in which gravity propagates. As a consequence there exists a certain cross-over scale r_c above which an observer will detect higher dimensional effects. Cosmological models in brane-world scenarios have been widely discussed in the literature.[18] It has been shown in Ref. 18 that flat brane-world Universe with $\Omega_m = 0.3$ and $r_c = 1.4\ H_0^{-1}$ is consistent with current SNIa and CMBR data.

Table 1. Expansion rates $H(z)$ in four models tested.

Model	Cosmological expansion rate $H(z)$ (the Hubble function).
ΛCDM	$H^2(z) = H_0^2(\Omega_m\ (1+z)^3 + \Omega_\Lambda)$
Quintessence	$H^2(z) = H_0^2(\Omega_m\ (1+z)^3 + \Omega_Q\ (1+z)^{3(1+w)})$
Var Quintessence	$H^2(z) = H_0^2(\Omega_m\ (1+z)^3 + \Omega_Q\ (1+z)^{3(1+w_0-w_1)}\ \exp(3w_1 z))$
Chaplygin Gas	$H(z)^2 = H_0^2\left[\Omega_m(1+z)^3 + \Omega_{Ch}\left(A_0 + (1-A_0)(1+z)^{3(1+\alpha)}\right)^{\frac{1}{1+\alpha}}\right]$
Braneworld	$H(z)^2 = H_0^2\left[(\sqrt{\Omega_m(1+z)^3 + \Omega_{r_c}} + \sqrt{\Omega_{r_c}})^2\right]$

Note: The quantities Ω_i represent fractions of critical density currently contained in energy densities of respective components (like clumped pressure-less matter, Λ, quintessence, Chaplygin gas or brane effects).

4. Results and conclusions

Table 2 displays the results of fitting the above mentioned models to the "Gold" sample of SNIa. As already mentioned the flat prior $k = 0$ was assumed. The reason for taking prior assumptions was that although Ω_k could have been included as a free parameter in statistical analysis , the information (even more precise than achievable this way) about cosmological parameters like k, H_0, etc. comes from other types of experiments (such like CMBR, LSS power spectrum, BBN, gravitational lensing etc.).

In observational cosmology our goal is in building a consistent picture of the Universe rather than expanding parameter space for statistical analysis. With this in mind we have also took a prior of $\Omega_m = 0.3$ while fitting the quintessence and Chaplygin Gas models, since this value is strongly supported by independent evidence.

Table 2. Values of best fitted parameters of four models tested. While fitting quintessential and Chaplygin gas models $\Omega_m = 0.3$ prior was used.

Model	Best fit model parameters (with 1σ ranges)
ΛCDM	$\Omega_m = 0.31 \pm 0.04$
Quintessence	$w = -1.02 \pm 0.11$
Var Quintessence	$w_0 = 1.40 \pm 0.25$ $w_1 = 1.67 \pm 0.80$
Chaplygin Gas	$A_0 = 0.99 \pm 0.03$ $\alpha = 1.0 \pm 0.59$
Braneworld	$\Omega_m = 0.21 \pm 0.03$

From Table 3 one can see that the best model - in terms of Akaike Criterion - is the quintessence model with evolving equation of state. Therefore this model should be identified as a reference for calculating Akaike differences, weights and odds against alternative models considered. The odds suggest that although there exist differences in the support given to specific scenarios by supernova data all models considered receive similar support by the data. One can also notice that models similar in structure i.e. ΛCDM, quintessence and quintessence with variable equation of state are closer to each other in terms of Kullback-Leibler entropy. In fact, from purely statistical point of view (apart from different physical motivations behind each one of them), they can be considered as a family of nested models. Models having different structure i.e. Chaplygin gas or brane-world scenario are more distant (in K-L sense) from the best one. Consequently one loses more information while fitting these models to supernovae Gold sample. It should be noticed that in other papers[8] referring to Akaike criterion as a tool for cosmological model selection authors ignored the issue of odds against competing model (with respect to the best fitted one) - treating AIC merely as a tool for ranking based just on numerical value of AIC. This paper fills this gap in showing how the model selection procedure should be implemented.

One can hope that the future will shed more light on the nature of dark energy in the Universe. Special surveys (e.g. SNAP) are designed for this purpose. One should realize however, that the emphasis of the ongoing and forthcoming research is shifting from estimating specific parameters of

Table 3. Values of AIC, Akaike differences, Akaike weights w_i (in Bayesian language equivalent to posterior model probabilities) and odds against the model (with respect to the best fitted one).

Model	AIC	Δ_i	w_i	Odds against
ΛCDM	179.072	0.707	0.209	1.42
Quintessence	179.091	0.726	0.207	1.44
Var Quintessence	178.365	0.	0.297	1.
Chaplygin Gas	179.600	1.235	0.160	1.86
Braneworld	180.075	1.710	0.127	2.35

the cosmological model (like the Hubble constant or deceleration parameter or any other physical parameter of the theory) to the model selection. Along with Bayesian techniques[19] information-theoretic model selection approaches are the most promising for this purpose.

References

1. S. Perlmutter, G. Aldering, G. Goldhaber, *et al.*, *Astrophys. J* **517**, 565 (1999)
   A. Riess, A.V. Filipenko, P. Challis, *et al.*, *Astron. J* **116**, 1009 (1998)
2. R. Caldwell, R. Dave and P.J. Steinhardt, *Phys. Rev. Lett.* **80**, 1582 (1998)
3. A. Kamenshchik, V. Moschella and V. Pasquier, *Phys.Lett. B* **511**, 256 (2000)
4. G.Dvali and G.Gabadadze, *Phys.Rev.D* **63**, 065007 (2001)
5. K.P. Burnham, D.R. Anderson, *"Model Selection and Multimodel Inference"*, (Springer - New York, 2002)
6. H. Akaike, *IEEE Transactions on Automatic Control AC* **19**, 716 (1974)
7. A.R. Liddle, *MNRAS* **351**, L49 (2004)
8. W. Godłowski and M. Szydłowski, *Phys.Lett.B* **623**, 10 (2005)
   M. Szydłowski and W. Godłowski , *Phys.Lett.B* **633**, 427 (2006)
9. A.G. Riess *et al.* [Supernova Search Team Collaboration], *Astrophys. J* **607**, 665 (2004)
10. A. Hosoya, T. Buchert and M. Morita, *Phys.Rev.Lett.* **92**, 141302-1-4 (2004)
11. M. Szydłowski, A. Kurek, A. Krawiec, *Phys.Lett. B* **642**, 171 (2006)
12. M. Goliath, R. Amanullah, P. Astier, A. Goobar and R. Pain, *Astron.Astrophys.* **380**, 6 (2001)
13. A. Benoit, *et al.*, *Astron.Astrophys.* **399**, L25 (2003)
14. T. Chiba, N. Sugiyama and T. Nakamura, *MNRAS* **301**, 72 (1998)
15. J. Weller and A. Albrecht, *Phys.Rev.Lett.* **86**, 1939 (2001)
    I. Maor, R. Brustein and P.J. Steinhardt, *Phys.Rev.Lett.* **86**, 6 (2001)
16. M. Makler, S.Q. de Oliveira and I. Waga, *Phys.Lett. B* **555**, 1 (2003)
17. M. Biesiada, W. Godłowski and M. Szydłowski, *Astrophys. J* **622**, 28 (2005)
18. D. Jain, A. Dev and J.S. Alcaniz, *Phys.Rev.D* **66**, 083511 (2002)
19. M. Kunz, R. Trotta, D. Parkinson, *Phys.Rev. D* **74**, 023503 (2006)

DM HALOS INTERACTING WITH DE. IMPLICATIONS TO DIRECT DM DETECTION

NIKOLAOS BROUZAKIS

Department of Physics, University of Athens,
Athens, Greece
**E-mail: ab_nbruzak@phys.uoa.gr*
www.uoa.gr

We study static configurations of dark matter coupled to a scalar field responsible for the dark energy of the Universe. The dark matter is modelled as a Fermi gas within the Thomas- Fermi approximation. The mass of the dark matter particles is a function of the scalar field. We analyze the profile of the dark matter halos in galaxies. In this case our framework is equivalent to the model of the isothermal sphere. In the presence of a scalar field, the velocity of a massive object orbiting the galaxy is not of the order of the typical velocity of the dark matter particles, as in the conventional picture. Instead, it is reduced by a factor that quantifies the dependence of the dark matter mass on the scalar field. This has implications for dark matter searches.

Keywords: dark matter; dark energy theory; Dark matter halos; Dark matter detection.

1. Introduction

Our present understanding of the history of the Universe assumes the presence of various contributions to its energy content. The total energy density is believed to take the critical value that results in a spatially flat Universe. An energy component that has not been directly observed is dark matter: a pressureless fluid that is responsible for the growth of cosmological perturbations through gravitational instability. Its contribution to the total energy density is estimated at $\sim 25\%$. The dark matter is expected to become more numerous in extensive halos, that stretch up to 100–200 kpc from the center of galaxies. Ongoing experiments are trying to detect the dark matter particles in the halo of the Milky Way. The component with the biggest contribution to the energy density has an equation of state similar to that of a cosmological constant. The ratio $w = p/\rho$ is negative and close to −1. This component is responsible for $\sim 70\%$ of the total energy density and induces the observed acceleration of the Universe.[4]

It is conceivable that there is a coupling between the dark matter and the field responsible for the dark energy.[7] In such a scenario it may be possible to resolve the coincidence problem, i.e. the reason behind the comparable present contributions from the dark matter and the dark energy to the total energy density. The required flatness of the potential of the scalar field could lead to phenomenological difficulties. In particular, the coupling between the field and the dark matter could lift the flatness of the potential through radiative corrections. We assume that at the level of the effective potential the required flatness is maintained, possibly guaranteed by some symmetry.[8]

The presence of an interaction between the dark matter and the scalar field responsible for the dark energy has consequences that are potentially observable. The cosmological implications depend on the form of the coupling between the dark matter and the scalar field, as well as on the potential of the latter.[9] The observations constrain a combination of the coupling and the parameters appearing in the potential. If the scale for the field mass is set by the present value of the Hubble parameter, then the field is effectively massless at length scales of the order of the galactic scale. Its coupling to the dark matter particles results in a long range force that can affect the details of structure formation.[10,11] In the following section we develop the necessary formalism for our study. We assume that the interaction between dark matter and dark energy takes the form of an effective field-dependent mass term for the dark matter particles. We derive the equation of motion of the scalar field. In order to derive an equation of state for the dark matter, we model it as a Fermi gas in the Thomas-Fermi approximation. Then we discuss a class of solutions that gives an approximate description of galaxy halos in the presence of the scalar field.

2. Formalism

2.1. *The basic expressions*

We assume that the dark matter consists of a weakly interacting fermionic gas. The mass m of the dark matter particles depends on the value of a slowly varying classical scalar field ϕ.[7] The action of the system can be written as[10]

$$\mathcal{S} = \int d^4x\sqrt{-g}\left(M^2R - \frac{1}{2}\frac{\partial\phi}{\partial x^\mu}\frac{\partial\phi}{\partial x^\nu}g^{\mu\nu} - U(\phi)\right) - \sum_i \int m(\phi(x_i))d\tau_i, \tag{1}$$

with $d\tau_i = \sqrt{-g_{\mu\nu}(x_i)dx_i^\mu dx_i^\nu}$ and the second integral taken over particle trajectories. Variation of the action with respect to ϕ results in the equation of motion[10]

$$\frac{1}{\sqrt{-g}}\frac{\partial}{\partial x^\mu}\left(\sqrt{-g}\; g^{\mu\nu}\frac{\partial\phi}{\partial x^\nu}\right) = \frac{dU}{d\phi} + W, \tag{2}$$

where

$$W = \frac{1}{\sqrt{-g}}\sum_i \int d\tau_i\; \frac{dm(\phi(x_i))}{d\phi}\,\delta^{(4)}(x - x_i) = -\frac{d\ln m(\phi(x))}{d\phi}\; T^\mu_{\ \mu} \tag{3}$$

The energy-momentum tensor associated with the gas of particles is

$$T^{\mu\nu} = \frac{1}{\sqrt{-g}}\sum_i \int d\tau_i\; m(\phi(x_i))\;\frac{dx_i^\mu}{d\tau_i}\frac{dx_i^\nu}{d\tau_i}\delta^{(4)}(x - x_i). \tag{4}$$

In the following we shall look for solutions of eqs. (2), (3), employing an approximation for the form of the energy-momentum tensor $T^{\mu\nu}$. We shall assume that it takes the diagonal form $T^\mu_{\ \nu} = \mathrm{diag}(-\rho, p, p, p)$.

We are interested in stationary, spherically symmetric configurations of the system. We consider a metric of the form

$$ds^2 = -B(r)dt^2 + r^2(d\theta^2 + \sin^2\theta\, d\phi^2) + A(r)dr^2. \tag{5}$$

From now on a prime denotes a derivative with respect to r and the scale M is defined as $M = (16\pi G)^{-1/2}$, where G is Newton's constant. Combining the conservation equation $\nabla_\mu T^{\mu\nu} = 0$ with the equation of motion for the scalar field gives

$$\frac{dp}{dr} = (3p - \rho)\,\frac{d\ln m(\phi)}{d\phi}\,\frac{d\phi}{dr} - (\rho + p)\frac{1}{2B}\frac{dB}{dr} \tag{6}$$

which is a generalized hydrostatic equilibrium equation.

Since we discuss about galaxy halos we will use the Newtonian approximation of the Einstein equations derived from the metric (5). For $B = 1 + 2\Phi, A = 1 - 2\Phi$, with $|\Phi| \ll 1$ we find

$$\Phi'' + \frac{2}{r}\Phi' = \frac{1}{4M^2}\left[\rho + 3p - 2U\right] \tag{7}$$

for the Newtonian potential. In the same limit the equations of motion for the scalar field becomes independent of the gravitational field and reads

$$\phi'' + \frac{2}{r}\phi' = \frac{\partial(U-p)}{\partial\phi} = \frac{dU}{d\phi} + \frac{d\ln m(\phi)}{d\phi}(\rho - 3p). \tag{8}$$

2.2. *The Thomas-Fermi approximation*

We describe the gas of dark matter particles within the Thomas-Fermi approximation. We assume that in a local frame at every point in space the particles are described through the Fermi-Dirac distribution

$$f(p) = \left[\exp\left(\frac{\sqrt{p^2 + m^2(\phi(r))} - \mu(r)}{T(r)}\right) + 1\right]^{-1}. \tag{9}$$

The chemical potential μ and temperature T, measured by a local observer, are functions of the radial coordinate r, while the mass of the particles depends on the local value of the field ϕ. From (9) we can calculate the pressure p, the number density n and the energy density ρ as a function of $T(r), \mu(r)$. Inserting those quantities in the conservation equation (6) results in

$$T(r) = T_0/\sqrt{B(r)}, \qquad \mu(r) = \mu_0/\sqrt{B(r)}. \tag{10}$$

These are the standard expressions that describe the behavior of a Thomas-Fermi fluid in a gravitational field.

3. Dark Matter Haloes

The most prominent feature of the distribution of dark matter particles in galaxy halos can be understood using the formalism of section 2. The dark matter distribution results in approximately flat rotation curves for objects orbiting the galaxies at distances $r > 10$ kpc. An analytical understanding of this behavior is possible within simple models of the dark matter gas, such as the isothermal sphere. If these simple models are extended through the addition of a scalar field to the theory, an analytical treatment is still feasible. For our study we model the dark matter gas as a Thomas-Fermi fluid, described by the formalism of section 2. At the scale of galaxies we can ignore the effect of the potential of the quintessence field.[1] We assume that the gas is non-relativistic and non-degenerate, with $p << \rho$. We also assume that the scalar field is displaced from its asymptotic($r \to \infty$) only by a small amount so that the approximations $m(\phi) = m_0 + m_0'\delta\phi, m'/m \simeq m_0'/m_0, \phi = \phi_0 + \delta\phi$ can be employed.

The relativistic Thomas-Fermi model in the Newtonian limit and when $\mu << T$ is equivalent to the isothermal sphere model as it results to an equation of state of the form $p(r) = \rho(r)\,\langle v_d^2\rangle$, with $\langle v_d^2\rangle$ the velocity dispersion of dark matter in the galaxy. The equations of motion of the scalar field and the Newtonian potential now become

$$\Phi'' + \frac{2}{r}\Phi' = \frac{1}{4M^2}\,\rho_0 \exp\left(-\alpha\Phi - \tilde{\alpha}\delta\phi\right) \tag{11}$$

$$(\delta\phi)'' + \frac{2}{r}(\delta\phi)' = \frac{m_0'}{m_0}\,\rho_0 \exp\left(-\alpha\Phi - \tilde{\alpha}\delta\phi\right), \tag{12}$$

where $T_0 = m_0\langle v_d^2\rangle$ is an effective temperature and $\alpha = m_0/T_0$, $\tilde{\alpha} = m_0'/T_0$ are the effective couplings of dark matter with the gravitational field and the scalar field respectively. A simple analysis[1] of (11,12) indicates that the approximately flat rotation curves outside the galaxy cores are a persistent feature even if the dark matter is coupled to a scalar field through its mass. However, numerical simulations are probably necessary in order to reproduce the detailed form of the curves.

The velocity v of a massive baryonic object in orbit around the galaxy, at a distance r from its center, can be expressed as[1]

$$v^2 = \left(\frac{2}{1+\kappa^2}\right)\langle v_d^2\rangle. \tag{13}$$

The effect of the scalar field is encoded in the factor $\kappa^2 = 4M^2(m_0'/m_0)^2$. When this is small, the velocity of an object orbiting the galaxy is of the order of the square root of the dispersion of the velocity of the dark matter particles. If κ^2 is large the orbital velocity can become much smaller than the typical dark matter velocity. The allowed range of κ is limited by the observable implications of the model that describes the dark sector. The dependence of the mass of dark matter particles on an evolving scalar field during the cosmological evolution since the decoupling is reflected in the microwave background. The magnitude of the effect is strongly model dependent. In the model of ref. [9] the observations result in the constraint $\kappa^2 \leq 0.01$ In the model of refs. [10, 11] the scalar interaction among dark matter particles is screened by an additional relativistic dark matter species. As a result, the model is viable even for couplings $\kappa^2 \simeq 1$.

4. Implications to Direct Dark Matter Searches

The scenario we considered may be interesting for dark matter searches. For large values of κ^2 the velocity of dark matter particles exceeds significantly the observed rotation velocity ($\sim$ 220 km/s for the Milky Way). So is the escape velocity by a factor $\sqrt{1+\kappa^2}$. The kinetic energy of dark matter now becomes

$$T = (1+\kappa^2) 40KeV \frac{M_{DM}}{100GeV}. \tag{14}$$

The effect of this modification to the detection rates in direct dark matter detection experiments is studied in ref.[2]. Their conclusion is that the event rate of recoil experiments will increase for light targets. An other interesting feature of this model is that the increased average energy of dark matter particles offers the possibility of nuclear targets being excited by the WIMP-nucleus.

References

1. N. Brouzakis and N. Tetradis, JCAP **0601**, 004 (2006) [arXiv:astro-ph/0509755].
2. J. D. Vergados and A. Faessler, arXiv:hep-ph/0611230.
3. P. J. E. Peebles and B. Ratra, Rev. Mod. Phys. **75** (2003) 559 [arXiv:astro-ph/0207347].
4. A. G. Riess *et al.* [Supernova Search Team Collaboration], Astron. J. **116** (1998) 1009 [arXiv:astro-ph/9805201];
S. Perlmutter *et al.* [Supernova Cosmology Project Collaboration], Astrophys. J. **517** (1999) 565 [arXiv:astro-ph/9812133];
W. J. Percival *et al.* [The 2dFGRS Collaboration], Mon. Not. Roy. Astron. Soc. **327** (2001) 1297 [arXiv:astro-ph/0105252].
5. C. Wetterich, Nucl. Phys. B **302** (1988) 668.
6. P. J. E. Peebles and B. Ratra, Astrophys. J. **325** (1988) L17.
7. C. Wetterich, Astron. Astrophys. **301** (1995) 321 [arXiv:hep-th/9408025].
8. S. M. Carroll, Phys. Rev. Lett. **81** (1998) 3067 [arXiv:astro-ph/9806099];
M. Doran and J. Jaeckel, Phys. Rev. D **66** (2002) 043519 [arXiv:astro-ph/0203018].
9. L. Amendola, Phys. Rev. D **62** (2000) 043511 [arXiv:astro-ph/9908023];
L. Amendola and D. Tocchini-Valentini, Phys. Rev. D **66** (2002) 043528 [arXiv:astro-ph/0111535].
10. G. R. Farrar and P. J. E. Peebles, Astrophys. J. **604** (2004) 1 [arXiv:astro-ph/0307316].
11. A. Nusser, S. S. Gubser and P. J. E. Peebles, Phys. Rev. D **71** (2005) 083505 [arXiv:astro-ph/0412586].
12. N. Bilic and R. D. Viollier, Eur. Phys. J. C **11** (1999) 173 [arXiv:hep-ph/9809563]; Gen. Rel. Grav. **31** (1999) 1105 [arXiv:gr-qc/9903034];
N. Bilic, R. J. Lindebaum, G. B. Tupper and R. D. Viollier, Phys. Lett. B **515** (2001) 105 [arXiv:astro-ph/0106209].
13. A. K. Drukier, K. Freese and D. N. Spergel, Phys. Rev. D **33** (1986) 3495.
14. M. Pietroni, Phys. Rev. D **72**, 043535 (2005) [arXiv:astro-ph/0505615].

EURECA – THE EUROPEAN UNDERGROUND RARE EVENT CALORIMETER ARRAY

H KRAUS[1], M BAUER[4], I BAVYKINA[2], A BENOIT[13], J BLÜMER[5,6], A BRONIATOWSKI[10], V BRUDANIN[14], G BURGHART[9], P CAMUS[13], A CHANTELAUZE[6], M CHAPELLIER[8], G CHARDIN[7], C CIEMNIAK[3], C COPPI[3], M DE JÉSUS[11], A DE LESQUEN[7], H DESCHAMPS[7], P DI STEFANO[11], L DUMOULIN[10], K EITEL[6], F VON FEILITZSCH[3], M FESQUET[7], J GASCON[11], G GERBIER[7], C GOLDBACH[12], M GROS[7], D HAUFF[2], S HENRY[1], M HORN[6], J IMBER[1], S INGLEBY[1], C ISAILA[3], J JOCHUM[4], A JUILLARD[10], M KIMMERLE[4], J-C LANFRANCHI[3], R LEMRANI[7], A LUBASHEVSKY[14], M LUCA[11], M MALEK[1], S MARNIEROS[10], R MCGOWAN[1], V MIKHAILIK[1], X-F NAVICK[7], T NIINIKOSKI[9], G NOLLEZ[12], E PANTIC[2], P PARI[8], L PEREVOSHCHIKOV[14], F PETRICCA[2], S PFISTER[3], W POTZEL[3], F PRÖBST[2], F RITTER[4], K ROTTLER[4], V SANGLARD[11], S SCHOLL[4], F SCHWAMM[7], W SEIDEL[2], A SMOLNIKOV[14], M STERN[11], M TESHIMA[2], B TOLHURST[1], W WESTPHAL[3], P WIKUS[9], J WOLF[5] and E YAKUSHEV[14]

[1]*University of Oxford, Department of Physics, Keble Road, Oxford OX1 3RH, UK;*

[2]*Max-Planck-Insitut für Physik, Föhringer Ring 6, 80805 Munich, Germany;*

[3]*Technische Universität München, Physik Dept. E15, 85748 Garching, Germany;*

[4]*Eberhard Karls Universität Tübingen, Auf der Morgenstelle 14, 72076 Tübingen, Germany;*

[5]*Institut für Experimentelle Kernphysik, Universität Karlsruhe (TH), Gaedestrasse 1, 76128 Karlsruhe, Germany;*

[6]*Forschungszentrum Karlsruhe, Institut für Kernphysik, Postfach 3640, 76021 Karlsruhe, Germany;*

[7]*CEA, Centre d'Etudes Saclay, DSM DAPNIA, 91191 Gif-Sur-Yvette Cedex, France;*

[8]*CEA, Centre d'Etudes Saclay, DSM DRECAM, 91191Gif-Sur-Yvette Cedex, France;*

[9]*CERN, 1211 Geneva 23, Switzerland;*

[10]*Univ Paris 11 CNRS, IN2P3, Centre Spec Nucl & Spect Masse, 91405 Orsay, France;*

[11]*Institut de Physique Nucléaire de Lyon, IN2P3-CNRS et Université Claude Bernard Lyon 1, 4 rue Enrico Fermi, 69622 Villeurbanne Cedex, France;*

[12]*Institut d'Astrophysique de Paris, 98bis Brd Arago, 75014 Paris, France;*

[13]*CNRS-CRTBT, 25 Avenue des Martyrs, 38042 Grenoble cédex 9, France;*

[14]*Joint Institute for Nuclear Research, DLNP, 141980 Dubna, Moscow Region, Russia*

EURECA (European Underground Rare Event Calorimeter Array) is a new project, searching for dark matter. The collaboration is comprised largely of the present groups of the CRESST and EDELWEISS experiments and several new groups. The aim is to explore scalar cross sections in the $10^{-9} - 10^{-10}$ picobarn region with a target mass of up to one tonne. A major advantage of EURECA is the planned use of more that just one target material (multi target experiment for WIMP identification). In preparation for this large-scale experiment, R&D for EURECA is provided through the current phases of CRESST and EDELWEISS.

1. Motivation

Recent experimental data on the cosmic microwave background, combined with other astronomical and astrophysical data, have significantly improved the precision of the fundamental parameters in our cosmological model [1]. Much of the matter density of the Universe seems to comprise non-luminous, non-baryonic particles [2]. Supersymmetry provides weakly interacting massive particles (WIMPs) as appealing and well-motivated candidates for this dark matter [3]. The WIMP-nucleon cross section is at or below the electroweak scale and the expected event rates are accordingly low. The identification of a WIMP interaction in a detector is therefore challenging, owing to the rate of WIMP interactions being very small compared with the event rates expected from background radioactivity of present detectors with highest purity and from cosmic radiation. In addition, the recoil energies produced by elastic WIMP-nucleus scattering are very small, in the range of a few keV to a few tens of keV.

To address these experimental challenges, a new generation of cryogenic detectors has been developed, exhibiting powerful background discrimination in combination with unprecedented energy threshold and resolution [4, 5]. The detectors allow highly efficient identification of nuclear recoils (caused by WIMP and also neutron interaction) by eliminating electron recoils due to radioactivity. The success of cryogenic detectors is clearly demonstrated as, even with relatively modest exposure, results from cryogenic detectors now exceed by an order of magnitude the performances of present competing experiments [6]. CRESST and EDELWEISS are currently in the process of completing the construction of the second phases of their experiments, and will subsequently enter the exploitation phase, involving dark matter data taking for several years. Construction and running of EDELWEISS II and CRESST II is providing valuable R&D for EURECA. The aim of EURECA is to design, build and operate at a European underground laboratory a cryogenic detector array with a target mass of several hundred kg to search for dark matter particles.

2. Choice of Detector Technology and Scalability

In the early 1980s, a number of groups began researching cryogenic detectors, operating in the milli-kelvin temperature range, for applications in neutrino physics and dark matter searches. The steady progress in this field is well documented by a series of conference proceedings (LTD-1 – LTD-11). After over two decades of development, the technique has matured and there are numerous science results that have been obtained with cryogenic detectors. These results cover a wide range of topics: contributions to x-ray astronomy, the spectrometry of heavy bio-molecules, the detection of extremely rare events (*e.g.*, neutrinoless double beta decay), and several dark matter results.

The main reason for beginning an intense technology development programme more than twenty years ago was the clearly identified need for lower energy threshold and better energy resolution in massive detectors for rare event searches. Cryogenic detectors were, and are, considered to be a most promising technique, requiring only milli-eV for producing a countable information carrier, compared with ~3 eV for semiconductor detectors and in the region of ~100 eV for scintillators. Consequently, at an energy of 5.89 keV (^{55}Fe source), the energy resolutions obtained are 3.2 eV (Si-thermistors [7]) and 3.9 eV (Ti/Au TES [8]) for small cryodetectors. This can be compared to resolutions of ~115 eV for Ge detectors [9] and ~2350 eV for NaI-Tl scintillators [10, 11]. These are, of course, the best resolutions achieved to date. For large absorbers of a few hundred gram, as used in the cryogenic experiments CRESST or EDELWEISS, energy resolutions of a few ~100 eV with thresholds around 1 keV have been demonstrated. Although good energy resolution may not appear to be a top priority for measuring the largely featureless energy spectra of WIMP-induced nuclear recoil, it is rather important for identifying backgrounds. As there will always be some form of background, especially when measurement periods have to be at the scale of many months or years, its identification is of paramount importance for a dark matter experiment which aims to discover WIMP interactions.

A further important issue is scalability. Cryogenic detectors can be scaled to large masses relatively easily. Individual modules have already been developed and optimised; scaling up merely requires the production of more modules and a larger cold space in a dilution refrigerator. Increasing the production rate of modules does not require additional detector development, although further improvement is always possible and likely to occur as a by-product of optimising the manufacturing process. Use of a larger cryostat does not affect the operation of individual detectors. Large dilution refrigerators are already in use, for example for gravitational wave antennae, but cooling large objects to ~10 millikelvin could be a challenge. Large-scale cryogenic equipment, operating at a temperature of 1.8 kelvin, is also used on the 27 km circumference of the Large Hadron Collider. System-engineering will of course be necessary to bring together technologies which so far have been demonstrated separately.

A modular approach is vital to achieving large detector masses. There are likely to be limitations in large-scale detectors due to radioactive backgrounds present in the target materials. With individual sub-kilogram solid targets, modules with abnormally high backgrounds can be isolated and replaced. Fitting individual readout channels to each target mass also helps keep the data stream free from pile-up and makes calibration and monitoring easier.

3. CRESST and EDELWEISS Technology

EURECA is based on the expertise acquired by the European CRESST and EDELWEISS experiments. Both use low-temperature calorimeters with complementary techniques as far as the discrimination of nuclear and electron recoil events are concerned.

EDELWEISS, operating in the Modane Underground Laboratory (LSM, France), is based on charge-phonon detectors [4]. The thermal signal induced by energy deposition in a germanium detector crystal is measured with a high-impedance thermistor attached to its surface. Simultaneously, the ionization signal is read out via electrodes on the crystal surface. The ratio between the measured ionisation and heat signals provides an efficient tool for the identification of the event type.

CRESST, operating in the Gran Sasso Laboratory (LNGS, Italy), is based on scintillation-phonon detectors [5]. The target is a scintillator ($CaWO_4$ crystal) and the thermal signal is measured with a superconducting transition edge sensor (TES) on the crystal surface. Simultaneously, scintillation light from the $CaWO_4$ is detected with thin calorimeters also using a TES sensor, but optimized for detection of scintillation. The scintillation-phonon technique exhibits benefits similar to those of the charge-phonon approach. Both experiments have achieved comparable sensitivity so far and are expected to improve their present sensitivity for direct dark matter detection by one or two orders of magnitude during the exploitation phases of CRESST II and EDELWEISS II.

The aim of the EURECA R&D is to explore concepts and designs appropriate to a large-scale experiment. Moving from ~10 kg dark matter targets to several 100 kg of target mass is a challenge as large as that faced when the experiments were upgraded to their second phases; i.e. from individual detectors to arrays of 30 to 40 targets. The timescale for the R&D and design of EURECA is shorter but similar to the exploitation phases of EDELWEISS II and CRESST II. The collaboration is studying the key topics required for operating efficiently a tonne-scale cryogenic dark matter experiment.

4. Features of EURECA

EURECA, in short, is a further scaling up and a combination of the CRESST and EDELWEISS experiments. The feasibility of cryogenic dark matter searches has been demonstrated by CDMS, CRESST and EDELWEISS. An increase of target mass to several hundred kg requires improvements in a number of aspects of the experiment, such as a reduction of the internal radioactivity of absorbers, improvement of scintillation yield, density of readout channels, etc. The mass of the cryogenic dark matter target is envisaged to be in the region of several hundred kg up to 1 tonne. An experiment of that scale would cover an extensive range of theoretically favoured supersymmetric particle dark matter candidates.

EURECA aims to have a target sensitivity a factor >100 better than projected by current phase II experiments. Although it is not unlikely that a discovery will be made at WIMP-nucleon cross sections above 10^{-8} picobarn, the range between 10^{-8} and 10^{-10} picobarn is currently most favoured [12] and will be covered by EURECA. At the lower end of cross section values, this translates to only a few events per ton per year in typical targets, requiring ultra-low background environments and excellent event type discrimination, neutron moderators and muon vetos.

Crucial to the success of EURECA will be the use of cryogenic detectors with a very low energy threshold to cover a range as wide as possible of recoil energies and excellent energy resolution to identify residual background using discrete-line spectroscopy. Discrimination is necessary to identify nuclear recoil events in the presence of a background from electron recoils. Improving further the robustness of event-by-event discrimination will be crucial and a detailed understanding of the detector response to various types of events will gain in importance. Reducing radioactive background levels further, especially in the keV range, is an important topic in light of substantial target mass increases. In contrast to the MeV energy range, relevant for ββ-decay, the rate of undesirable contributions could be considerably higher for the low energy range.

An important feature of EURECA will be its multi-material target. Having several targets is highly desirable for testing the correct A-scaling of WIMP-nucleon interactions and to determine residual neutron backgrounds, if present.

Further strong motivation for equipping EURECA with a range of target materials is provided by kinematic considerations, as the mass of the WIMP is unknown. A natural initial choice for EURECA is to use germanium and $CaWO_4$ targets, given the expertise of the collaboration. Additional absorbers are being researched; $ZnWO_4$, for example, has been identified as a further target, while other scintillators, such as molybdates, are being optimized [13, 14]. Just after the IDM 2006 workshop, two institutes of the ROSEBUD collaboration joined EURECA and their expertise on scintillating bolometers (see these proceedings) will add new opportunities and possibilities to our choice of absorber materials.

Arranging the detectors in a large array of smaller absorbers has the further advantage of allowing testing for uniform rates within the target, and for providing an additional dark matter signature by requiring single interactions only for a dark matter candidate event. This should allow identification of residual neutron background through coincidences.

5. EURECA – Next Steps

EURECA plans to carry out a design study to address aspects of scaling up from the absorber masses currently installed in CRESST II and EDELWEISS II to the tonne-scale. Such a substantial increase in target mass is likely to involve several smaller steps. The design study will aim to prepare the necessary technical concepts needed to construct a small demonstrator in 2009/10, initially carrying several 10 kg of target material, but with a clear upgrade path to allow increasing the target mass step by step.

5.1. *Cryogenics*

The design of the cryogenic infrastructure requires detailed studies and careful planning. A major challenge in cryostat design will be the desired ability to replace certain detectors without interrupting the operation of the remaining detectors for long periods. Without doubt, improvements to detectors will continue as long as EURECA is running and the ability to exchange detectors quickly is considered a major advantage. Dilution refrigeration will be used to cool the ton-scale cold mass of EURECA. Studies are under way to determine the cooling power required of the dilution refrigerator of EURECA. First results indicate a considerable heat load, which is in good agreement with experimental values obtained for large cold masses of similar size in other experiments. The proximity cryogenics must satisfy the stringent requirements resulting from remote operation in underground laboratories. In addition, much has been learned from the rather different design concepts of CRESST and

EDELWEISS. The lessons learned from the operation of these experiments will have their impact on the design of the EURECA cryogenic system.

5.2. *Electronics*

Electronic readout and DAQ will have to scale to several thousand channels. Some of the scalability issues have already been addressed in providing the readout hardware for the present experiments, but a further ~10-fold increase in density, reduction of heat loads through readout wires, and a reduction in the cost per channel will be necessary for the realization of EURECA.

In order to reduce the cost per readout channel to an acceptable level we will need to develop multi-channel front-end electronics, to replace the present single-channel systems; and redesign detector bias and control systems to take advantage of the economy of scale. Some hardware components can be avoided by providing their function through software.

Multiplexing systems for cryogenic detectors have been developed for astronomical arrays to reduce the number of readout channels. It is not yet clear if these can be adapted for a dark matter search, but it is likely there will be some sort of multiplexing at some level.

Volume of data flow: although the rate of particle interaction is low in a dark matter experiment, the data that need to be acquired and potentially stored can be rather significant. This is in part due to the recording of calibration pulses that allow monitoring of the detector response at all times, but also high rate of data taking is necessary to stabilize the operating point of the detectors with high precision. The most likely implementation of a large-scale data acquisition system will see digitization of signals early on and implementation of as many functions possible in firmware and software. The rate of digitization of phonon-detector signals is 25 kHz and 100 kHz for CRESST and EDELWEISS, respectively, resulting in 4.32 GB/day and 17.32 GB/day (plus house keeping information) of data per readout channel. With ~6000 channels (1 tonne detector mass), this would amount to 25 – 100 TB/day. It is obvious that a staged system of triggers has to be implemented, as is common in high-energy physics experiments, which also will provide separate data paths for signal data and data needed for the slow control of the detectors. Research and development has started already, with "intelligent" electronics, capable of identifying potentially interesting signals above noise. A trigger system with several levels allows balancing the available computing power between the levels of sophistication of the trigger condition, successively reducing the volume of the data stream as it propagates through the data pipeline.

With the number of readout channels increasing, automated control and monitoring becomes more important. We will need to further develop the capability of operating the experiment from outside the underground laboratory, and to keep track of an increasing number of parameters in order to certify the origin of any event in our detector. We will build on the experience from EDELWEISS II and CRESST II to develop suitable remote control and monitoring systems.

A larger experiment means longer cables and a greater number of passages through thermal shields and heat sinks. This is less a problem for low-impedance detector readout channels, but might cause problems for high-impedance read-out due to stray and parasitic capacitance. To somewhat alleviate these potential drawbacks, research will be necessary into specific grounding schemes, possibly involving actively driven shields and a comparison of various possible implementation, such as coax cables or strip lines. Avoiding cross talk between channels despite the high density of cables is also an important issue.

The readout system and DAQ will need to be able to accommodate a variable number of detectors with very different design, operation and speed. We aim to design it as modular as possible so that a large number of sections / components will be the same across the range of readout configurations. Other aspects to include are linking the output from the muon veto into the data acquisition system and ensuring synchronisation of time stamp recording with precision better than 10 mirco-seconds.

Many of the above ideas have already passed their first stage of research and development and are currently implemented in EDELWEISS II and CRESST II. The combination of expertise and experience gained in these two experiments is now being harnessed for the next level of scaling up.

5.3. *Detectors*

The major task faced by EURECA will be the mass production of the detector modules. In moving from prototyping, which by its nature gives only a low rate of detector production, to mass production, the crucial issue of radiopurity has to be addressed. This is a challenging aspect, as raw materials will have to be screened before they enter the production process, and assessment at sensitivity levels required for measuring only a few events per year require long measurement times.

A key feature of EURECA is its operation of different target materials and also slightly complementary readout and discrimination technologies within a single common volume. This approach should combine the advantages of the detector concepts, namely scintillation and heat (as in CRESST and ROSEBUD), and ionization and heat (as in EDELWEISS). It is likely, at least in the early exploratory stages, that a number of different detector concepts will be used. The heat signal, for example, could be measured using high-impedance sensors (EDELWEISS), or superconducting thermometers made from tungsten (CRESST), or another material such as Ir/Au.

Regarding the supply in large quantities of scintillating absorbers, we are working with a number of suppliers to reduce the levels of radioactive impurities and at the same time aim to improve the light yield and size of the scintillators. In this systematic approach we put great emphasis on reproducibility, reliability and quality control. An investigation into the possibility of growing our own crystals is also under way.

5.4. *Low Background Issues*

EURECA will require significant improvements in the radio purity of the immediate environment to achieve the very low background rate required. This concerns gamma ray background and neutron-induced background. Obviously, the depth of the underground site is a key factor for achieving lowest muon-induced backgrounds. Further, dedicated integrated shielding (in case of a new purpose-built underground excavation), an environment with de-radonized air, providing clean room conditions, appears mandatory for the creation of a 'radio-pure' environment into which one would install the EURECA setup.

Further, we will have to improve the radio purity of the bulk detector material but also of any material potentially close to the actual detectors, such as cabling or shielding materials.

We will require thousands of samples to be tested with sensitivities two to three orders of magnitude better than the current ones. For EDELWEISS II, this is currently done using a dedicated low-background n-type coaxial High Purity Ge diode of 210 cm^3 with archaeological lead shielding. Typically, the sensitivity obtained for copper samples is at the mBq/kg level for U/Th and ^{60}Co. This sensitivity limit could be improved by several orders of magnitude. The main limitation of Ge detectors is their poor efficiency, especially at low energies. Increasing masses and exploring different geometries for the crystals would improve this substantially.

Other instruments and methods for measuring very low levels of contamination will have to be investigated, such as: advanced radiochemistry, Inductively Coupled Plasma – Mass Spectroscopy (ICP-MS), and Neutron Activation Analysis (NAA).

The storage of materials, manufacturing of devices and growth of crystals underground would prevent cosmogenic activation. An adequate underground environment should consist of large storage areas and dedicated clean rooms, with sensitive radon monitoring and trapping facilities (~ mBq/m^3 of air). While underground storage is not a huge problem for the small amount of material in the current experiments, it will need a dedicated effort for the increased space requirements of EURECA.

Surface contamination is an important concern for cryogenic detectors, and therefore, the capability to measure α s and β s from surfaces must be developed.

6. Summary

EDELWEISS and CRESST are progressing well and, with new readout systems and detectors, will soon start data taking for dark matter analyses . Once the commissioning and fine-tuning of the new systems are complete, physics data taking will begin and manpower effort will become available to carry out a design study on EURECA. This will build on the experience gained in the construction, commissioning and operation of EDELWEISS and CRESST, and will focus on the scaling up of the otherwise mature technologies.

References

1. Spergel D N et al., *Astrophys. J. Suppl.* **148** (2003) 175.
2. Bergström L, *Rep. Prog. Phys.* **63** (2000) 793.
3. Lee B W L and Weinberg S, *Phys. Rev. Lett.* **39** (1977) 165.
4. Sanglard D et al., *Phys. Rev.* D **71** (2005) 122002.
5. Angloher G et al., *Astropart. Phys.* **23** (2005) 325.
6. Akerib D S et al., *Phys. Rev. Lett.* **96** (2006) 011302.
7. Porter F S et al., *Nucl. Instrum. Meth.* A **559** (2006) 436.
8. Bruijn M P et al., *Nucl. Instrum. Meth.* A **513** (2003) 143.
9. http://www.pgt.com/PDF/Detect_Brochure.pdf
10. Aitken D W et al., *IEEE Trans Nucl Sci* **14** (1967) 468.
11. Mengesha W and Valentine J D, *IEEE Trans Nucl Sci*, **49** (2002) 2420.
12. Ellis J et al., *Phys. Rev.* D **71** (2005) 095007.
13. Kraus H et al., *Phys. Lett.* B **610** (2005) 37.
14. Mikhailik V B et al., *J. Appl. Phys.* **97** (2005) 083523.

LIST OF PARTICIPANTS

Last Name	First Name	Affiliation
ANISIMOV	Alexey	EPFL
ARNOWITT	Richard	Texas A&M University
AXENIDES	Minos	NCSR Demokritos
BAUER	Michael	University of Tuebingen
BELLI	Pierluigi	INFN Roma Tor Vergata
BERTONE	Gianfranco	INFN
BIESIADA	Marek	University of Silesia
BOURQUIN	Maurice	University of Geneva
BROUZAKIS	Nikolaos	University of Athens
BUCCI	Carlo	INFN
BUENO	Antonio	University of Granada
BURIGANA	Carlo	INAF
CANTATORE	Giovanni	INFN - Universite di Trieste
CAROLL	Sean	Caltech
CARR	Bernard	Queen Mary, University of London
CERULLI	Riccardo	INFN - LNGS
CHAGANI	Hassan	University of Sheffield
CHOI	Ki Young	University of Sheffield
CIMINO	Roberto	INFN
CLINE	David	University of California
COAKLEY	Kevin Jerome	NIST
COLAFRANCESCO	Sergio	INAF
CRITTENDEN	Robert	University of Portsmouth
DAW	Edward	University of Sheffield
DE BOER	Wim	University of Karlsruhe
DE FELICE	Fernando	INFN
DELLA VALLE	Federico	INFN - Universite di Trieste
DIAKONIDIS	Theodoros	Aristotle University of Thessaloniki
DIMOPOULOS	Konstantinos	Lancaster University
DOBLER	Gregory	Harvard University
DOKUCHAEV	Vyacheslav	Institute of Nuclear Research
DORO	Michele	INFN
DROBYSHEVSKI	Edward	Ioffe Physico-Technical Institute

EJIRI	Hiroyasu	Osaka University
FANOURAKIS	George	NCSR Demokritos
FINGER	Michael	Charls University Prague
FINKBEINER	Douglas	Harvard University
FRENK	Carlos Silvestre	Durham University
FUSHIMI	Ken - Ichi	Ken - Ichi - Fushimi
GERBIER	Jill	CEA - Saclay
GIOMATARIS	Yannis	Saclay
GODFRIN	Henri	CNRS
GOMEZ MALUENDA	Hector	University of Zaragosa
GONZALEZ	Anthony	University of Florida
GOUNARIS	Georgios J.	Aristotle University of Thessaloniki
GREEN	Anne	University of Nottingham
HAWKINS	Michael	Royal Observatory
HAYASHI	Eric	Max Planck Institute for Astrophysics
HITACHI	Akira	Kochi Medical School
HOOPER	Dan	Fermilab
IRASTORZA	Igor G.	University of Zaragosa
JETZER	Philippe	University of Zurich
JIMENEZ	Raul	University of Pennsylvania
KATSANEVAS	Stavros	CNRS
KATZ	Urlich	University of Erlangen
KAUFMANN	Lilian Dominique	ETH Zurich
KIM	Jihn	Seoul National University
KIM	Sun Kee	Seoul National University
KINION	Darin	LLNL
KLAPDOR-KLEINGROTHAUS	Hans Volker	Max Planck Institut fuer Kernphysik
KOKORELIS	Christos	NCSR Demokritos
KRAUS	Hans	University of Oxford
KRAUSS	Carsten	Queen's University
KRIVOSHEINA	Irina	Max Planck Institut fuer Kernphysik
KUDRYAVTSEV	Vitaly	University of Sheffield
KUSTER	Markus	Tu Darmstadt
LACHENMAIER	Tobias	University of Tuebingen
LANDSMAN	Hagar	University of Wisconsin
LEMRANI – ALAOUI	Rachid	CEA - Saclay
LUSCHER	Roland	Rutherford Appleton Laboratory

LUZON	Gloria	University of Zaragosa
MAJOROVITS	Bela	MPI for Physics
MANDOLESI	Nazzareno	INAF
MARGIOTTA	Annarita	Universita di Bologna e INFN Sezione di Bologna
MARTINEZ	Maria	University of Zaragosa
MAVROKORIDIS	Konstantinos	University of Sheffield
MAVROMATOS	Nikolaos	King's College London
MITROPOULOS	Pantelis	NCSR Demokritos
MIUCHI	Kentaro	Kyoto University
MORSELLI	Aldo	INFN Roma Tor Vergata
MOULIN	Emmanuel	LPTA IN2P3
MUNA	Demitri	University of Sheffield
MUNOZ	Carlos	Universidad Autonoma de Madrid
NATH	Pran	Northeastern University
ODOM	Brian	University of Chicago
PAIN	Reynald	IN2P3/CNRS and Université Paris VI
PALLIS	Constantinos	University of Manchester
PELLICCIA	Diego-Nicola	INFN
PERIVOLAPOULOS	Leandros	University of Ioannina
PETER	Annika	Princeton University
PIERI	Lidia	INAF
PLIONIS	Manolis	National Observatory of Athens
POBES	Carlos	University of Zaragosa
REGENFUS	Christian	University of Zurich
RICH	James	CEA - Saclay
RINGWALD	Andreas	DESY
RODRIGUEZ ARRUGA	Asuncion	University of Zaragosa
ROOS	Matts	University of Helsinki
ROTTLER	Klemens	University of Tuebingen
RUOSO	Giuseppe	Instituto Nazionale di Fisica Nucleare - LNL
SALATI	Pierre	Laboratoire d'Annecy-le-Vieux de Physique Theorique
SANTOS	Daniel	PLSC-CNRS-IN2P3
SARSA	Marisa Luisa	University of Zaragosa
SATO	Joe	Saitama University
SCHOLL	Stephan	University of Tuebingen
SCHWAMM	Frank	CEA - Saclay

SCHWARZ	Dominik	University of Bielefeld
SETO	Osamu	University of Sussex
SHAPOSHNIKOV	Mikhail	EPFL
SIOTIS	Ion	NCSR Demokritos
SKORDIS	Constantinos	Perimeter Institute
SORENSEN	Peter	Brown University
SPADA	Francesca	INFN Roma Tor Vergata
SPOONER	Neil	University of Sheffield
SUMNER	Tim	Imperial College London
TAKAYAMA	Fumihiro	Cornell University
TAKEDA	Atsushi	Institute for Cosmic Ray Research, Univ. of Tokyo
TASITSIOMI	Argyro	Princeton University
TOMASELLO	Vito	University of Sheffield
TROTTA	Roberto	Oxford University
TURRELL	Brian	University of British Columbia
TZAMARIAS	Spyros	Hellenic Open University
TZANAKOS	George	University of Athens
TZIAFERI	Eirni	University of Physics
VAN BIBBER	Karl	Lawrence Livermore National Laboratory
VERGADOS	Ioannis	University of Ioannina
VERMISOGLOU	Grigoris	NCSR Demokritos
YAMASHITA	Masaki	Columbia University
ZIOUTAS	Konstantin	University of Patras